Insect Pests of Potato

Global Perspectives on
Biology and Management

Edited by

**Philippe Giordanengo, Charles Vincent, and
Andrei Alyokhin**

AMSTERDAM • BOSTON • HEIDELBERG • LONDON
NEW YORK • OXFORD • PARIS • SAN DIEGO
SAN FRANCISCO • SINGAPORE • SYDNEY • TOKYO

Academic Press is an imprint of Elsevier

Academic Press is an imprint of Elsevier
The Boulevard, Langford Lane, Kidlington, Oxford, OX5 1GB, UK
225 Wyman Street, Waltham, MA 02451, USA

First published 2013

Notices
Knowledge and best practice in this field are constantly changing. As new research and experience
broaden our understanding, changes in research methods, professional practices, or medical
treatment may become necessary.

Practitioners and researchers must always rely on their own experience and knowledge in evaluating and using any information, methods, compounds, or experiments described herein. In using
such information or methods they should be mindful of their own safety and the safety of others,
including parties for whom they have a professional responsibility.

To the fullest extent of the law, neither the Publisher nor the authors, contributors, or editors, assume
any liability for any injury and/or damage to persons or property as a matter of products liability,
negligence or otherwise, or from any use or operation of any methods, products, instructions, or
ideas contained in the material herein.

British Library Cataloguing in Publication Data
A catalogue record for this book is available from the British Library

Library of Congress Control Number: 2012940399
ISBN: 978-0-12-386895-4

For information on all Academic Press publications
visit our website at **store.elsevier.com**

Printed and bound by CPI Group (UK) Ltd, Croydon, CR0 4YY

Working together to grow
libraries in developing countries

www.elsevier.com | www.bookaid.org | www.sabre.org

ELSEVIER   BOOK AID   Sabre Foundation
            International

Contents

Part III
The Potato Field as a Managed Ecosystem

12 Successional and Invasive Colonization of the Potato Crop by the Colorado Potato Beetle: Managing Spread 339

Gilles Boiteau and Jaakko Heikkilä

Part IV
Management Approaches

13 Chemical Control of Potato Pests 375

Thomas P. Kuhar, Katherine Kamminga, Christopher Philips, Anna Wallingford, and Adam Wimer

Part V
Current Challenges and Future Directions

Cultivated in all continents except Antarctica, the potato (*Solanum tuberosum* L.) is one of the five most important agricultural crops in the world. Nutritionally superior to most other staple crops, rugged, and relatively easy to grow, potato has been instrumental in improving the quality of life in a variety of geographic areas throughout human history. Its cultivation is still of great importance in both industrialized and developing parts of the world.

For over 8,000 years since its domestication in the Central Andes, the potato has been plagued by a number of serious insect pests. These include some of the most prolific and adaptable species known to man. If left uncontrolled, they can completely destroy the affected crops. Currently, insect management in commercial potato production is heavily reliant on synthetic insecticides. This results in well-known undesirable side effects of ecological backlash and environmental pollution.

Not surprisingly, considerable scientific and management efforts have been always invested in potato protection from insect damage. Our readers might be familiar with "Advances in Potato Pest Biology and Management" edited by Zehnder, Jansson, Powelson, and Raman and published by the APS Press in 1994, as well as with "Advances in Potato Pest Management" edited by Lashomb and Casagrande and published by the Hutchinson Ross Publishing Co. in 1981. Both books provided excellent reviews of potato entomology, and are widely quoted in this volume. However, a considerable research effort has been dedicated to studying biology and management of insect pests of potatoes during the last 15 years. Until now, the results of that effort remained dispersed among numerous scientific journals.

This book is made of contributions written by an international team of experts working in major potato-growing areas of the world. Among other things, the book includes a lot of valuable, but often little known, information published over the years in non-English language literature. In Part I, we start by introducing potato as a crop that is essential for meeting the nutritional demands of the humankind, and discuss the challenges of its sustainable production. After that, we proceed to covering the biology of potato pests in Part II of our book. In addition to well-known key pests such as the Colorado potato beetle and potato tuberworms, we also discuss more sporadic and/or local pests such as wireworms and hadda beetles. Part III is dedicated to ecological interactions among the living and non-living components of potato ecosystems, a good understanding of which lays a foundation for developing scientifically sound integrated pests management plans. In Part IV, we talk about practical approaches to managing insect pests of potatoes. Particular emphases are placed on techniques allowing pest suppression in an environmentally friendly manner, and on using evolutionary principles to ensure their sustainability. Part V concludes the book with the discussion of current challenges and future prospects in managing potato pests.

We are deeply grateful to our editors, Pat Gonzalez and Kristi Gomez. Without their patience and understanding, this project would have unraveled long time ago. We also thank Lindsey Miller and Amanda Bailey for their help with proofreading and formatting the manuscript.

It is our hope that this book will be of use and interest to a variety of people involved in potato production. We also always welcome our readers' feedback, including (but not limited to) constructive criticism.

Sincerely,

Andrei Alyokhin
Charles Vincent
Philippe Giordanengo

Contributors

Juan M. Alvarez (11), DuPont Crop Protection, Stine Haskell Research Center, Newark, DE, USA

Andrei Alyokhin (1, 2, 10, 19, 20), School of Biology and Ecology, University of Maine, Orono, ME, USA

Galina Benkovskaya (2, 19), Institute of Biochemistry and Genetics, Russian Academy of Science, Ufa, Russian Federation

Gilles Boiteau (12), Agriculture and Agri-Food Canada, Potato Research Centre, Fredericton, NB, Canada

Felix A. Cervantes (11), Entomology and Nematology Department, University of Florida, Gainesville, FL, USA

R.S. Chandel (8), Department of Entomology, Himachal Pradesh Agriculture University, Palampur, Himachal Pradesh, India

V.K. Chandla (8), Division of Plant Protection, Central Potato Research Institute, Shimla, Himachal Pradesh, India

Yolanda H. Chen (19), Plant and Soil Science Department, University of Vermont, Burlington, VT, USA

David W. Crowder (9), Department of Entomology, Washington State University, Pullman, WA, USA

Beata Gabryś (18), Department of Botany and Ecology, University of Zielona Góra, Zielona Góra, Poland

Philippe Giordanengo (1, 3, 20), Université de Picardie Jules Verne, Amiens, France; CNRS, and INRA, Institut Sophia Agrobiotech, Sophia Antipolis, France; Université de Nice Sophia Antipolis, Sophia Antipolis, France

Serena Gross (10), Department of Entomology, Purdue University, West Lafayette, IN, USA

Jaakko Heikkilä (12), MTT Agrifood Research Finland, Economic Research, Helsinki, Finland

Donald C. Henne (4), Texas AgriLife Research, Subtropical Pest Management Laboratory, Weslaco, TX, USA

Finbarr G. Horgan (15), Crop and Environmental Sciences Division, International Rice Research Institute, Metro Manila, The Philippines

Randa Jabbour (9), Department of Entomology, Washington State University, Pullman, WA, USA

Katherine Kamminga (13), Department of Entomology, Virginia Tech, Blacksburg, VA, USA

Bożena Kordan (18), Department of Phytopathology and Entomology, University of Warmia and Mazury in Olsztyn, Olsztyn, Poland

Jürgen Kroschel (6), Global Program Integrated Crop and System Research, Agroecology/IPM, International Potato Center, Lima Peru

Thomas P. Kuhar (13), Department of Entomology, Virginia Tech, Blacksburg, VA, USA

Lawrence A. Lacey (16), PO Box 8338, Yakima, WA 98908, USA

Leena Lindström (19), Department of Biological and Environmental Science, University of Jyväskylä, Finland

Ning Liu (7), Institute of Zoology, Chinese Academy of Sciences, Chaoyang, Beijing, China

Christine A. Lynch (9), Department of Entomology, Washington State University, Pullman, WA, USA

Joseph E. Munyaneza (4), USDA-ARS, Yakima Agricultural Research Laboratory, Wapato, WA, USA

Mandeep Pathania (8), Department of Entomology, Himachal Pradesh Agriculture University, Palampur, Himachal Pradesh, India

Yvan Pelletier (15), Potato Research Centre, Agriculture and Agri-Food Canada, Fredericton, NB, Canada

Christopher Philips (13), Department of Entomology, Virginia Tech, Blacksburg, VA, USA

Julien Pompon (15), Potato Research Centre, Agriculture and Agri-Food Canada, Fredericton, NB, Canada, and Department of Biology, University of New Brunswick, Fredericton, NB, Canada

Julien Saguez (3), Agriculture and Agri-Food Canada, Saint-Jean-sur-Richelieu, Québec, Canada

Birgit Schaub (6), Global Program Integrated Crop and System Research, Agroecology/IPM, International Potato Center, Lima Peru

William E. Snyder (9), Department of Entomology, Washington State University, Pullman, WA, USA

Marc Sporleder (16), International Potato Center (CIP), ICIMOD Building, Khumaltar, Lalitpur, Nepal

Rajagopalbabu Srinivasan (11), Department of Entomology, University of Georgia, Tifton, GA, USA

Maxim Udalov (2, 19), Institute of Biochemistry and Genetics, Russian Academy of Science, Ufa, Russian Federation

Willem van Herk (5), Pacific Agri-Food Research Center, Agriculture and Agri-Food Canada, Agassiz, British Columbia, Canada

K.S. Verma (8), Department of Entomology, Himachal Pradesh Agriculture University, Palampur, Himachal Pradesh, India

Robert S. Vernon (5), Pacific Agri-Food Research Center, Agriculture and Agri-Food Canada, Agassiz, British Columbia, Canada

Charles Vincent (1, 3, 20), Agriculture and Agri-Food Canada, Saint-Jean-sur-Richelieu, Québec, Canada

Anna Wallingford (13), Department of Entomology, Virginia Tech, Blacksburg, VA, USA

Donald C. Weber (14), Invasive Insect Biocontrol and Behavior Laboratory, USDA Agricultural Research Service, Beltsville, MD, USA

Phyllis G. Weintraub (17), Agricultural Research Organization, Gilat Research Center, D.N. Negev, Israel

Adam Wimer (13), Department of Entomology, Virginia Tech, Blacksburg, VA, USA

Jing Xu (7), Institute of Zoology, Chinese Academy of Sciences, Chaoyang, Beijing, China

Runzhi Zhang (7), Institute of Zoology, Chinese Academy of Sciences, Chaoyang, Beijing, China

Potato as an Important Staple Crop

Chapter 1

Potatoes and their Pests – Setting the Stage

Charles Vincent[1], Andrei Alyokhin[2], and Philippe Giordanengo[3]

[1]Agriculture and Agri-Food Canada, Saint-Jean-sur-Richelieu, Québec, Canada, [2]School of Biology and Ecology, University of Maine, Orono, USA, [3]Université de Picardie Jules Verne, Amiens, France; CNRS, and INRA, Institut Sophia Agrobiotech, Sophia Antipolis, France; Université de Nice Sophia Antipolis, Sophia Antipolis, France

PROLOGUE: A SHORT HISTORY OF THE POTATO

Potatoes are one of the most important staple crops in human history. The cultivated potato, *Solanum tuberosum* (Solanaceae), is a tuber crop that was domesticated about 8000 years ago in the Central Andes region near Lake Titicaca (Peru-Bolivia).The potato (*papa* in Quechua) became instrumental in the rise of the great Inca Empire that extended from present-day Columbia to Argentina (Zuckerman 1999, McEwan 2006, Reader 2009). The Spanish conquistadors introduced the potato to their home country around 1570, where a few farmers started growing potatoes on a small scale. From Spain, potatoes were introduced, mostly as botanical curiosities, to Italy in 1586, to Austria in 1588, to England in 1596, and to Germany in 1601. They arrived in North America in the 1620s, when the British Governor of the Bahamas presented several tubers to the Governor of the colony of Virginia (Brown 1993, Zuckerman 1999). Acceptance of the potato has been an uneasy road. For instance, the French Parliament forbade cultivation of the potato in 1748 as it was thought to cause leprosy, among other things. Thanks mostly to the work of A. Parmentier, the Paris Faculty of Medicine declared potatoes edible in 1772. However, it was not until the second part of the 19th century that potatoes became widely adopted as a food source outside of South America (Zuckerman 1999, Reader 2009).

POTATOES AND PEOPLE

Incorporating potatoes into their daily diets was of tremendous benefit to people around the world. Crop diversification insured against catastrophic losses due to pest outbreaks and unfavorable environmental conditions. Potatoes were also

Insect Pests of Potato. http://dx.doi.org/10.1016/B978-0-12-386895-4.00001-6

nutritionally superior to many other staple crops (Woolfe 1987, Kolasa 1993). As a result, better balanced diets improved human health in general and resistance to infectious diseases in particular. Not surprisingly, wide-scale adoption of potatoes commonly coincided with periods of rapid population growth in a variety of nations (Zuckerman 1999, Reader 2009). Potato cultivation was especially important for improving food security for the economically disadvantaged classes of the society, such as landless peasants, or factory workers flocking to the rapidly growing cities fueled by the Industrial Revolution.

By providing inexpensive, nutritious, and easy-to-cook meals to people with little spare money or time at their disposal (Kolasa 1993, Zuckerman 1999), potatoes were a strong driving force behind the runaway economic growth of the 19th century. Unfortunately, around the same time, this crop also provided a strong and cruel reminder of the importance of sustainable development.

The introduction of the potato to Ireland initially provided a great relief to the local population, which was severely oppressed by their British overlords. While the landless peasants were still forced to divert considerable resources into producing meat and grains for their landlords to pay for tenancy, growing potatoes for personal consumption provided enough dietary calories to remove them from the edge of starvation. Consequently, the Irish population increased from 1.5 million in 1790 to 9 million in 1845, despite little change in the overall system of land ownership or colonial exploitation (Hobhouse 1986, Brown 1993).

Unfortunately, the late blight epidemic of 1845–1849 caused widespread failures of potato crops in Ireland. The farmers were faced with the dilemma of either paying their rent with the usual meat and grain, or consuming them and being evicted from the land for non-payment. Essentially, it was a choice between starving now and starving later. About 40% of the population experienced severe hardship, with 1 million dying of hunger and related diseases, and another 1.3 million emigrating in search of a better life (Zuckerman 1999).

Following up an idea put forward by the Government of Peru, the FAO declared 2008 as the International Year of the Potato (IYP) as a gesture to recognize the impact of potato on mankind. Several events were organized worldwide to celebrate the IYP. For example, two of us edited a Special Issue on Potato in a French multidisciplinary journal (Giordanengo et al. 2008).

BIOLOGY AND AGRONOMIC ISSUES

There are over 4000 edible varieties of potato, mostly found in the Andes of South America. An unusual characteristic of potato is that large quantities (typically 1.8–2.2 tons/ha) of seed are required for planting, mainly because of the large amount of water (ca. 78%) in tuber content.

The potato presents unique challenges and advantages to plant breeders. Because it is propagated vegetatively by tuber cuttings, potato cultivars do not need to be bred to produce homogenous plants from true seed. A major disadvantage of potatoes for breeders is that S. tuberosum is tetraploid, making it difficult to transfer desirable traits between cultivars and have them expressed in progeny. Unfortunately, many wild *Solanum* relatives are diploid, greatly complicating the breeding process.

Recent advances in molecular genetics and understanding of potato physiology is facilitating and speeding up genetic transformation of the potato. This technology has improved our understanding of the molecular basis of plant-pathogen interactions, and has also opened new opportunities for using the potato in a variety of non-food biotechnological applications (see review in Vreugdenhil *et al.* 2007). New potato genomic resources are currently being established to facilitate gene discovery and molecular breeding across several international projects. The chief aim of these projects is to improve resistance traits, thus reducing the environmental impact of potato production and protection. Potato cultivars expressing the *Bacillus thuringiensis* var. *tenebrionis* Cry 3A toxin for resistance to the Colorado potato beetle, *Leptinotarsa decemlineata* Say (Coleoptera: Chrysomelidae) (cv. NewLeaf, Monsanto Corp.) were the first genetically modified food crop approved for human consumption and commercially produced in the USA (1995). Because of consumer concerns, *Bt* potato cultivars were taken off the market in 2000.

Potato is a nearly perfect crop to grow in places where land is limited and labor is abundant, a situation typical of much of the developing world. The poorest and most undernourished rural households often depend on potatoes for their survival, as it is capable of meeting their dietary requirements in a relatively reliable fashion under conditions in which other crops may fail (Lutaladio and Castaldi 2009).

The website (http://www.cipotato.org/) of the International Potato Center (Centro Internacional de la Papa), located north of Lima, Peru, is a valuable resource of information about subjects relevant to the potato. Numerous documents can be downloaded freely from this website.

POTATO MARKETS

Although cultivated potatoes belong to a unique botanical species, thousands of varieties, presenting different biological traits linked with different agricultural characteristics are available. Because of its nutritional qualities and its physiological attributes, the potato can be a basic food for the poor as an untransformed product, and for the rich as transformed products. In several so called "advanced countries", transformed potato is fast food par excellence (e.g., French fries and the famous potato chips (Burhans 2006)). In addition to its role in human diets, the potato is used as livestock feed, and for various industrial purposes, including the production of starch-based products and alcohol.

Because it is heavy and relatively difficult to transport compared to cereals such as rice, corn, or wheat, the potato is less subject to international trade (only *ca*. 5% of the production is traded). As a result, its prices are largely determined by local production costs, not by developments on extremely speculative commodity exchanges located thousands of kilometers away. Thus, it is not surprising that potato prices remain stable compared to those of other staple foods (Lutaladio and Castaldi 2009).

Before the 1990s, most potatoes were grown and consumed in Europe, North America, and countries of the former Soviet Union. Since then, there has been a dramatic increase in potato production and demand in Asia, Africa, and Latin America, where output rose from less than 30 million tonnes in the early 1960s to more than 324 million tonnes in 2010. The top 10 potato producers worldwide are currently China, Russia, India, the United States, Ukraine, Poland, Germany, Belarus, The Netherlands, and France (FAO 2012). China is now the leading potato-producing country, and, as of 2009, *ca.* 33% of the world's potato production was harvested in China and India.

Currently, the potato remains the fourth most important agricultural crop on our planet. As was often the case throughout the history of its cultivation, it continues to greatly benefit people at risk of inadequate nutrition.

INSECTS OF THE POTATO

A wide variety of insects can damage potato crops, either directly, through feeding on tubers and spoiling the harvest, or indirectly, by feeding on leaves or stems, or transmitting pathogens (Radcliffe 1982). If severe enough, indirect damage may reduce harvestable yield and quality.

The literature concerning potato entomology is vast. For example, a search done in April 2012 in the database Scopus with the keywords "potato insects" yielded 1738 entries. The potato is plagued by a number of serious insect pests that can completely destroy the crop if left uncontrolled. These pests can conveniently be assigned to two classes, namely above-ground (indirect) and below-ground (direct) pests. At the present time, their management relies predominantly on synthetic insecticides and poses a serious threat to the environment in several parts of the world. Considerable research effort in the past 15 years has been dedicated to studying the biology and management of insect pests of potatoes. The results of that effort, however, remain dispersed among numerous scientific journals. Their comprehensive review will be of great benefit to a variety of people involved in potato research and production, as well as to people facing similar issues in other crop systems.

Although all concepts, strategies, and tactics discussed in major IPM textbooks (e.g., Radcliffe *et al.* 2009) are relevant to potato protection, it is important to stress here that, depending on the market targeted, protection issues can be managed differently. Also, depending on the geographical area considered, the challenges posed by potato insects differ. For example, in Picardie-Nord Pas de Calais (northern France), where *ca.* 20% of European potatoes are produced, aphids are of major concern and the Colorado potato beetle is of secondary importance. The reverse is true in the Northeastern United States and Maritime Provinces of Canada.

Numerous field manuals for identifying potato pests have been published either on paper (e.g., Zehnder *et al.* 1994, Strand and Rude 2006, Johnson 2007) or online (e.g., International Potato Center 1996, University of Idaho 1999, University of California 2011, Ontario Ministry of Agriculture, Food and Rural Affairs 2011, University of Maine Cooperative Extension 2012) formats. Perusal of these documents will reveal that potato diseases are major drivers of protection programs.

Currently, chemical control is the most popular form of insect pest management in the potato industry. Since the mid-20th century, intense use of insecticides has led to the selection of resistant insect pest populations. Today, as with other major crops, potato culture has to deal with increasing environmental and public concerns that lead to the reduction of new chemical discoveries and development, while also supporting a rapidly rising world demand. Such a challenge requires a major contribution by the whole potato industry, in particular researchers and agricultural engineers, to allow the development of successful management strategies. Depending on the variety used, targeting the different market segments (e.g., seed, fresh market, processing), insect management strategies vary due to crop value and specific quality needs.

Potato has been infamously nicknamed "one of the most chemically-dependent crops in the world." Yet this does not necessarily have to be the case. We hereafter present several chapters written by world experts on potato pests. It is intended that they provide valuable information such that our book will allow advances towards sustainable production of this amazing crop.

REFERENCES

Brown, C.R., 1993. Origin and history of the potato. Am. Potato J. 70, 363–373.

Burhans, D., 2008. Crunch! A history of the great American potato chip. Terrace Books. University of Wisconsin Press, Madison, WI.

FAO (Food and Agriculture Organization of the United Nations). 2012. FAOSTAT. FAO Statistical Databases. http://faostat.fao.org/.

Giordanengo, P., Pelletier, Y., Vincent, C. (Eds.), 2008. La Pomme de terre, enjeux et opportunités, Cahiers Agric, 17, pp. 329–420.

Hobhouse, H., 1986. Seeds of Change. Harper Row, New York, NY.

International Potato Center, 1996. Major Potato Diseases, Insects, and Nematodes. Peru, Limahttp://www.cipotato.org/publications/pdf/002408.pdf.

Johnson, D.A. (Ed.), 2007. Potato Health Management. The American Phytopathological Society Press, St Paul, MN.

Kolasa, K.M., 1993. The potato and human nutrition. Am. Potato J. 70, 375–384.

Lutaladio, N., Castaldi, L., 2009. Potato: the hidden treasure. J. Food Compos. Anal. 22, 491–493.

McEwan, G.F., 2006. The Incas, New Perspectives. ABC-CLIO, Inc., Santa Barbara, CA.

Ontario Ministry of Agriculture, Food and Rural Affairs, 2011. Common Potato Insects Scouting Guidelines. , http://www.omafra.gov.on.ca/english/crops/facts/potato_insects.htm.

Radcliffe, E.B., 1982. Insect Pests of Potato. Annu. Rev. Entomol. 27, 173–204.

Radcliffe, E.B., Hutchison, W.D., Cancelado, R.E., 2009. Integrated Pest Management, Concepts, Tactics Strategies and Case Studies. Cambridge University Press, Cambridge, UK.

Reader, J., 2009. Potato: A History of the Propitious Esculent. Yale University Press, New Haven, CT.

Strand, L.L., Rude, P.A., 2006. Integrated Pest Management for Potatoes in the Western United States, Second ed. University of California Agriculture and Natural Resources Publications, Davis, CA.

University of California, 2011. How to Manage Pests – Potatoes. http://www.ipm.ucdavis.edu/PMG/selectnewpest.potatoes.html.

University of Idaho, 1999. Identification keys for insects in Pacific Northwest Field Crops –Potatoes. http://www.cals.uidaho.edu/edComm/keys/potatoes/potatoes01.htm.

University of Maine Cooperative Extension, 2012. Colorado Potato Beetle, *Leptinotarsa decemlineata* Say – University of Maine Cooperative Extension Factsheet No. 201. University of Maine Cooperative Extension, Orono, ME.

Vreugdenhil, D., Bradshaw, J., Gebhardt, C., Govers, F., Taylor, M.A., MacKerron, D.K.L., Ross, H.A., 2007. Potato Biology and Biotechnology: Advances and Perspectives. Elsevier Science, San Diego, CA.

Woolfe, J.A., 1987. The Potato in the Human Diet. Cambridge University Press, Cambridge, UK.

Zehnder, G.W., Powelson, M.L., Jansson, R.K., Raman, K.V., 1994. Advances in Potato Pest Biology and Management. The American Phytopathological Society Press, St Paul, MN.

Zuckerman, L., 1999. The Potato: How the Humble Spud Rescued the Western World. North Point Press, New York, NY.

Biology of Major Pests

Chapter 2

The Colorado Potato Beetle

Andrei Alyokhin[1], Maxim Udalov[2], and Galina Benkovskaya[2]

[1]*School of Biology and Ecology, University of Maine, Orono, ME, USA,* [2]*Institute of Biochemistry and Genetics, Russian Academy of Science, Ufa, Russian Federation*

INTRODUCTION

The Colorado potato beetle, *Leptinotarsa decemlineata* (Coleoptera: Chrysomelidae), is one of the most notorious insect pests of potatoes. Since becoming a problem in the mid-19th century, this insect has received enormous attention from the scientific community. A comprehensive Colorado potato beetle bibliography (Dill and Storch 1992) contains 3537 references. A more recent on-line bibliography limited to peer-reviewed journal articles written in English (Alyokhin 2011) has over 700 entries. Despite this, the beetle remains a formidable threat to the potato industry in already-colonized potato-growing areas, and it continues to expand its geographic range into new regions of the world. A diverse and flexible life history, combined with a remarkable adaptability to a variety of stressors, makes the Colorado potato beetle a very challenging pest to control.

TAXONOMIC POSITION AND MORPHOLOGICAL DESCRIPTION

The Colorado potato beetle belongs to the family Chrysomelidae, or leaf beetles. With 35,000 species described worldwide, it is the third largest family in the order Coleoptera. Members of this family feed on plants, both as larvae and as adults, with both life stages consuming the same or related plant species. Many species are host-specific (Arnett 2000).

The Colorado potato beetle was first described by Thomas Say in 1824 as a member of the genus *Chrysomela* (Say 1824). Based on morphological characteristics, it was then moved to the genus *Doryphora* (Suffrian 1858). Finally, Stål (1865) included this species in a newly described genus, *Leptinotarsa*, where it remains at the present moment. Jacques (1988) listed a total of 41 species in this genus, of which 9 occur in the United States, 9 in Central and South America, and 27 in Mexico. However, Bechyne (1952) argued that *L. porosa* Baly and *L. paraguensis* Jacoby belong to the genus *Cryptostetha*.

Insect Pests of Potato. http://dx.doi.org/10.1016/B978-0-12-386895-4.00002-8

11

Leptinotarsa is considered to be an evolutionarily recent genus that is still in the process of active speciation, with Southern Mexico most likely being its center of origin (Tower 1906, Medvedev 1981). Host plants are known for 20 *Leptinotarsa* species (Hsiao and Hsiao 1983). All of them are specialized feeders, with 10 species (including the Colorado potato beetle) feeding on plants in the family Solanaceae, 9 in the family Compositae, and 1 in the family Zygophyllaceae.

The Colorado potato beetle is the most notorious member of *Leptinotarsa*. The adult beetles are oval in shape and are approximately 10 mm long by 7 mm wide. They are pale yellow in color, with five black stripes along the entire length of each elytron and black spots on the head and pronotum. The eggs are about 1.5 mm long, and their color changes from yellow right after oviposition to orange for mature eggs that are ready to hatch. The larvae are eruciform, red to orange in color, with black head and legs and two rows of black dots on each side. Four instars are completed before pupation.

Based on morphological characteristics, in particular on spot patterns and coloration of the head, pronotum, and elytra, Tower (1906) originally subdivided what is currently known as *L. decemlineata* into four species and nine varieties. However, later experiments showed that all of those were fully capable of interbreeding (Tower 1918, Hsiao 1985). Furthermore, analysis of male genitalia did not reveal any noticeable differences. As a result, Jacques (1972) merged them into a single species – an approach currently followed by most scientists. However, Jacobson and Hsiao (1983) found distinct differences in isozyme frequencies in the Colorado potato beetle population from southern Mexico and populations from the United States, Canada, and Europe. The difference was large enough to regard the two as separate subspecies. Morphological analyses of spot patterns on the adult pronotae (Fasulati 1985, 1993, 2002, 2007) (Fig. 2.1) also supported the existence of several (American, European-Siberian, and Central Asian) subspecies of the Colorado potato beetle. Interestingly, there were considerable changes in spot patterns within the

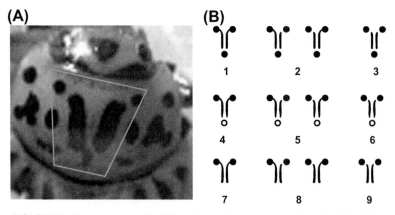

FIGURE 2.1   Spot patterns on the Colorado potato beetle pronotum (see also Plate 2.1).

same populations over several decades (Zeleev 2002, Benkovskaya *et al.* 2004, Kalinina and Nikolaeva 2007). These may indicate active microevolutionary processes within the species. Alternatively, they could be attributed to genetic bottlenecks due to insecticide applications (see Chapter 19 for more details).

ORIGINS AND HISTORY OF SPREAD

The Colorado potato beetle is native to the central highlands of Mexico. Wild populations feed mostly on buffalobur, *Solanum rostratum*, which is considered to be its original ancestral host (Tower 1906, Hsiao 1981, Casagrande 1987). Both buffalobur and the beetles might have been brought into the southern and central plains of the United States by early Spanish settlers moving northwards (Gauthier *et al.* 1981, Casagrande 1987, Hare 1990). The Colorado potato beetle was first collected in the United States in 1811 by Thomas Nuttall. Subsequently, additional collections were made in 1819–1820 near the Iowa-Nebraska border by Thomas Say, who later described it for science (Casagrande 1985, Jacques 1988).

The first major Colorado potato beetle outbreak in cultivated potatoes was reported in 1859, when severe damage was observed on fields about 100 miles west of Omaha, Nebraska (Jacques 1988). Feeding on potatoes represented a host range expansion for this species, which is described in detail in Chapter 19. Following the initial outbreak, eastward expansion of the beetles' geographic range was very rapid, with beetles reaching the Atlantic coast of the US and Canada in 15 years (Casagrande 1987). The beetles crossed the Mississippi river in 1865, reached Ohio in 1869, and arrived at Maine in 1872 (Jacques 1988). They then proceeded to the southern provinces of Canada, which were colonized by 1901 (Ivanschik and Izhevsky 1981).

Westward expansion was somewhat slower, limited in part by scarcity of potatoes (Riley 1877). The first serious damage to potatoes in Colorado was reported in 1874 (Riley 1875). However, 10 years earlier, Walsh (1865) saw a considerable beetle population feeding on *S. rostratum* in Colorado. That observation eventually resulted in the name of that state being incorporated into the generally accepted common name of this species (Jacques 1988). All in all, the beetle's range between 1860 and 1880 expanded by more than 4 million square kilometers (Trouvelot 1936). Colonization of North America was completed in 1919, when the beetles were found in British Columbia (Ivanschik and Izhevsky 1981).

The first European population of Colorado potato beetles was discovered in England in 1875; the beetle then invaded continental Europe via Germany in 1877. Another infestation was discovered 1 year later in Poland. All those populations were successfully eradicated soon after being discovered (Feytaud 1950, Wegorek 1955, Jacques 1988). Quarantine measures and eradication campaigns were largely successful in keeping the pest out of Europe until 1922, when self-propagating populations were finally established in France (Feytaud 1950). After that, the beetle steadily spread throughout Western and Central Europe, reaching the border of Poland by the mid-1940s (Ivanschik and

Izhevsky 1981). Beetle dispersal was greatly facilitated by relaxed quarantine regulations and large-scale movements of military cargo during World War II. In 1949, Colorado potato beetles crossed the border of the Soviet Union but were quickly eradicated. Strict quarantine, combined with field monitoring and eradication programs, kept beetles away for the following 9 years. However, in 1958, warm spring temperatures and strong western winds resulted in massive invasions from the Carpathian Mountains to the Baltic Sea. This led to the establishment of reproducing populations, which have continued their eastward spread ever since (Ivanschik and Izhevsky 1981).

Presently, the Colorado potato beetle damages potato crops all over Europe, Asia Minor, Iran, Central Asia, and western China (Jolivet 1991, Weber 2003). Its current range covers about 16 million square kilometers in North America, Europe, and Asia, and continues to expand (Weber 2003). Potentially, the beetle could spread to temperate areas of East Asia, the Indian subcontinent, South America, Africa, New Zealand, and Australia (Vlasova 1978, Worner 1988, Jolivet 1991, Weber 2003).

Although the Colorado potato beetle is a highly mobile species that is capable of flying over long distances, especially with prevailing winds (Boiteau et al. 2003), its rapid dispersal had been greatly facilitated by human movement. Potato is a common and ubiquitous crop, which is often moved over considerable distances after harvest. Furthermore, there is often a considerable amount of traffic through potato-growing areas. Because of their small size, the beetles can easily hitch a ride with a variety of different cargoes.

The rapid spread of the Colorado potato beetle during World War II and at the onset of the Cold War has been sometimes attributed to its use as a biological weapon. In particular, East German authorities initiated an aggressive propaganda campaign that accused the United States in dropping the beetles on their crops. Similar rumors circulated in the Soviet Union, although the propaganda pitch was much more subdued. There was indeed some research into the possibilities of weaponizing this species, conducted in France, Germany, and possibly Great Britain, right before or during World War II (Garrett 1996, Lockwood 2009). However, there is no evidence that the Colorado potato beetle has ever been released for the purpose of sabotaging enemy crops. On the contrary, the timing and geography of the spread indicate that the spread was attributed to natural range expansion from previously colonized areas. Interestingly, during the current war between NATO and the Taliban in Afghanistan, local farmers have blamed the recent arrival of beetles on United States aid workers bringing the pest in with contaminated shipments of seed potatoes (Arnoldy 2010). Again, in this case, the beetles most likely arrived on their own, from neighboring Tajikistan.

GEOGRAPHIC VARIABILITY

Populations invading new areas are usually subject to the founder effect during the colonization event (Sakai et al. 2001). As a result, they are often genetically depauperate compared to their geographic centers of origin. Grapputo *et al.* (2005) used the analysis of mitochondrial DNA (mtDNA) and amplified fragment length polymorphism (AFLPs) markers to examine the genetic diversity of Colorado potato beetle populations in North America and Europe. They found high levels of both mitochondrial and nuclear variability in North American beetle populations, with the highest genetic diversity detected in populations from the central United States. There was also a strong genetic differentiation between populations on the two continents. European populations showed a significant reduction at nuclear markers (AFLPs) and were fixed for one mitochondrial haplotype. That finding suggested the possibility of a single successful founder event. However, European populations have maintained genetic variability at the nuclear level. When the populations from the two continents were analyzed separately, the level of population differentiation was similar among North American populations and among European populations. Thus, it is probably more likely that the Colorado potato beetle invasion of Europe resulted from multiple introductions of the same haplotype.

The extent of the gene flow between the geographically distinct Colorado potato beetle populations remains somewhat unclear. The beetles are capable of long-distance flights, particularly when assisted by wind (Boiteau *et al.* 2003). However, the frequency of such dispersal, and the resulting gene flow, may be relatively low (Grafius 1995). Zehnder *et al.* (1992) found no evidence of significant separation within North American populations, based on the mtDNA data, which they attributed to the rapid range expansion of this species across the continent. However, other studies on North American beetles using mtDNA markers (Azeredo-Espin *et al.* 1991, 1996, Grapputo *et al.* 2005) and AFLPs (Grapputo *et al.* 2005) detected a strong separation among the studied populations. Furthermore, genetic data were consistent with differences in host-plant affinity, photoperiodic response (Jacobson and Hsiao 1983), and insecticide resistance (Hare 1990). In addition, chromosomal studies suggested the existence of three different races within the species (Hsiao and Hsiao 1983).

Analysis of phenotypic variations in different geographic populations seems to confirm the populations' relative isolation from each other. Based on nine variations of the spot patterns on the pronotae of adults, at least five distinct population complexes of the Colorado potato beetle were identified in Eastern Europe. An additional two complexes were described in North Kazakhstan and Central Asia (Fasulati 1993). The taxonomic statuses of these seven population complexes have not yet been determined. According to Fasulati (1993), they are "probably close to geographic races". Similarly, Eremina and Denisova (1987) reported differences in frequencies of pronotal patterns between different populations collected in the Saratov region of the Russian Federation. In the Lipetsk region, western and eastern groups of populations were identified on the basis of color pattern studies; the boundary between the ranges of these groups coincided with that between the agro-climatic regions (Ovchinnikova and Markelov 1982). Two large population complexes were isolated on the territory of Bashkortostan in the Russian Federation based on the spot patterns and coloration of heads, pronotae, and elytra. The first complex included populations from the central part of the area, and the second complex included populations from the peripheral parts of the range (Udalov *et al.* 2010).

Considerable differences in the levels and mechanisms of insecticide resistance in adjacent Colorado potato beetle populations (Boiteau *et al.* 1987, Heim *et al.* 1990, Ioannides *et al.* 1991, Grafius 1995) also imply low gene flow at the time of resistance development. In addition, de Wilde and Hsiao (1981) and Hsiao (1981) observed significant differentiation in photoperiodic responses and host-plant adaptation among geographically isolated Colorado potato beetle populations. Subsequent crosses confirmed the genetic nature of those differences.

PEST STATUS

Wherever present, the Colorado potato beetle is considered to be the most important insect defoliator of potatoes. Indeed, *ca.* 40 cm^2 of potato leaves are consumed by a single beetle during the larval stage (Ferro *et al.* 1985, Logan *et al.* 1985), and close to 10 cm^2 of foliage are consumed per day during the adult stage (Ferro *et al.* 1985). After removing all foliage from colonized plants, beetles can feed on stems and exposed tubers. However, these constitute a suboptimal diet compared to leaves, and lead to poor larval growth and cessation of oviposition by adults (Alyokhin, unpubl. data). The Colorado potato beetle is very prolific, with one female laying 300–800 eggs (Harcourt 1971).

It is not unusual for the beetles to completely destroy potato crops in the absence of control measures. Nevertheless, potato plants can withstand a considerable amount of defoliation without any reduction in tuber yield, particularly when damage is done before or after the tuber bulking period (Dripps and Smilowitz 1989). For example, Hare (1980) found little effect of beetle feeding except during the middle 4–6 weeks of the season, when 70% defoliation resulted in an approximately 20% reduction in yield, while complete defoliation resulted in an approximately 64% reduction in yield. Similarly, in the study by Cranshaw and Radcliffe (1980), plants completely recovered from up to 33% defoliation inflicted early in the season and suffered only minor yield reduction from 67% defoliation. Wellik *et al.* (1981) also found that losing 29% of the leaf area did not affect potato yield. Zehnder *et al.* (1995) developed the action thresholds of 20% defoliation from plant emergence to early bloom, 30% from early bloom to late bloom, and 60% from late bloom to harvest. Ferro *et al.* (1983) and Zehnder and Evanylo (1989) did not find any effects of up to 100% defoliation during the 2 weeks immediately preceding vine kill.

Unfortunately, yield-loss data are commonly highly variable, and their analysis is often a challenge (Nault and Kennedy 1998). Furthermore, commercial farmers are very risk averse and are generally not willing to tolerate beetle infestations in their crops. Therefore, even non-damaging levels of Colorado potato beetle infestations trigger control measures, usually in the form of insecticidal sprays.

SEASONAL LIFE CYCLE AND DIAPAUSE

Diapause plays a very important role in Colorado potato beetle adaptation to the surrounding environment, and greatly contributes to its success as a pest of cultivated potatoes. In particular, it allows for the colonization of territories with much colder climates compared to that of the beetle's center of origin (Biever and Chauvin 1990). Furthermore, diapausing individuals escape certain catastrophic events, such as insecticide applications, and so can restore population sizes once conditions become favorable.

Colorado potato beetles have a facultative overwintering diapause that takes place during the adult stage. It is induced by a short-day photoperiod, and is modulated by temperature and food condition and availability (de Wilde and Hsiao 1981, de Kort 1990). The exact ratio of light-to-dark hours differs among the populations. For example, beetles from Utah (41°44′ N) entered diapause in response to a 15-hour photophase, while beetles from Arizona (31° 58′ N) responded to a 12-hour photophase. Beetles from Texas (26° 24′ N) did not enter diapause, regardless of photoperiod (Hsiao 1981). A temperature of 31°C has been shown to shorten the critical photophase by three to five hours (de Wilde and Hsiao 1981). Being a complex phenomenon, diapause is regulated by a number of different genes, including those coding for the juvenile hormone esterase (Vermunt *et al.* 1999), vitellogenin (de Kort *et al.* 1997), and a number of specialized diapause proteins (de Kort *et al.* 1997, Yocum 2003).

Diapause phenotypes may also vary within beetle populations. Yocum *et al.* (2011) developed a multiplex PCR protocol using five diapause-regulated genes to monitor diapause development of the Colorado potato beetle under field conditions. They found that some beetles were already in the diapause initiation phase in June when the day length was greater than 17h. There was also noticeable inter-seasonal variation in the timing of diapause development, with the greatest differences being observed before the day length decreased to less than 15h.

After diapause initiation, some beetles burrow into the soil in the field. Others move towards field edges by flight and by walking, presumably navigating towards tall silhouettes of trees and shrubs commonly found in hedgerows (Voss and Ferro 1990a, Weber and Ferro 1993, French *et al.* 1993). Upon arrival at overwintering sites, the beetles seek concealment by burrowing into the soil (Voss and Ferro 1990b). The majority of the beetles dig down to between 10 and 25cm, and overwintering survivorship increases with increasing depth (Lashomb *et al.* 1984, Weber and Ferro 1993, Hunt and Tan 2000). Lashomb *et al.* (1984) calculated that a 10cm in-soil depth decreased winter mortality in loam soils in New Jersey, USA by *ca.* 32%. However, additional digging requires the expenditure of extra energy, which can be in short supply during diapause. To channel additional resources used to survive unfavorable winter conditions, the flight muscles of diapausing beetles undergo significant degeneration (Stegwee *et al.* 1963).

Diapause is terminated by temperatures > 10°C (de Kort 1990). However, there is usually a refractory phase of approximately 3 months during which the beetles do not react to changes in environmental conditions. The post-diapause beetles usually accumulate 50–250 degree-days (DD, > 10°C) before they appear on the soil surface (Ferro *et al.* 1999). Males and females exit their diapause simultaneously and start mating before re-colonizing the host plants (Ferro *et al.* 1999). There is considerable intra-population variation in the times of beetle emergence from the soil. Ferro *et al.* (1999) reported that beetles over-wintering within a woody hedgerow adjacent to a potato field in western Massachusetts, USA, emerged over a 3-month period.

A certain number of overwintering beetles may remain in an extended dia-pause for 2 or more years (Isely 1935, Trouvelot 1936, Wegorek 1957a, 1957b, Ushatinskaya 1962, 1966). Their exact proportion varies among beetle popula-tions and may depend in part on environmental conditions. Ushatinskaya (1962, 1966) reported that, in western Ukraine, 0.4–6.5% of beetles overwintering in sandy soils remained dormant for 2 years, but all beetles overwintering in clay soils emerged after the first winter. In Washington State, 16–21% of overwinter-ing adults emerged after two winters, and up to 2% emerged after three winters (Biever and Chauvin 1990). Tauber and Tauber (2002) found extended diapause in 0–7.2% of the beetles overwintering in upstate New York. Up to 25% of the overwintered population may enter a second diapause (Isely 1935, Jermy and Saringer 1955, de Wilde 1962, Minder and Petrova 1966). However, they usu-ally suffer very high mortality rates (Isely 1935, Minder and Petrova 1966), and probably do not play a significant role in the overall dynamics of beetle populations.

In addition to overwintering diapause, Colorado potato beetles may also enter summer diapause, or aestivation. It is particularly common in arid areas (Tower 1906, Faber 1949), but has also been reported in other locations (Grison 1939, Jermi 1953, Minder and Petrova 1966). Physiologically, summer dia-pause is similar to overwintering diapause, although its duration is usually shorter (Ushatinskaya 1961, 1966, Minder and Petrova 1966). From an ecologi-cal standpoint, it serves as an adaptation to excessive heat and desiccation.

INTERACTIONS WITH HOST PLANTS

The Colorado potato beetle is an oligophagous herbivore that infests about 10 species of solanaceous plants. Of those, the potato (*Solanum tuberosum*), tomato (*S. lycopersicum*), and eggplant (*S. melongena*) are important cultivated hosts. Feeding on potatoes represents a host range expansion for the Colorado potato beetle. There is also some host specialization among geographically-isolated Colorado potato beetle populations. Host-plant associations and the evolution of feeding preferences are discussed in detail in Chapter 19.

To find and colonize potatoes and other host plants, Colorado potato beetles use both visual (de Wilde *et al.* 1976, Zehnder and Speese 1987) and olfactory

(Visser and Nielsen 1977, Visser and Ave 1978, Thiery and Visser 1986, Landolt *et al.* 1999, Dickens 2000) cues. Beetle attraction to potato odor in a laboratory olfactometer was first observed by McIndoo (1926) and was then repeatedly confirmed by a number of other authors (e.g., de Wilde *et al.* 1969, Visser 1976, Thiery and Visser 1986, Landolt *et al.* 1999). Not surprisingly, the attraction is stronger in hungry beetles (Thiery and Visser 1995). At the same time, prior feeding experience enhances the beetle's responses to host-plant odor, probably due to associative or non-associative learning (Visser and Thiery 1986). The beetles are capable of distinguishing between the odors of different host-plant species. In an olfactometer study by Hitchner *et al.* (2008), they chose potato over eggplant or tomato, and eggplant over tomato. Also, damaged plants were more attractive to adult beetles compared to intact plants (Bolter *et al.* 1997, Landolt *et al.* 1999), but no such difference was detected for larvae (Dickens 2002).

Potato odor is comprised of a number of general leaf volatiles, all of which are emitted by most species of flowering plants (Visser *et al.* 1979). When tested individually or in combination, they did not elicit any response in the exposed Colorado potato beetles (Visser and Ave 1978, Visser *et al.* 1979). Therefore, it is thought that beetles distinguish host plants based on the ratios of individual green leaf volatiles in the odor blend (Visser and Ave 1978). Indeed, Dickens (2000, 2002) identified a three-component mixture consisting of common green leaf volatiles (z)-3-hexenyl acetate, (+/−) linalool, and methyl salicilate that was attractive to Colorado potato beetle adults.

Despite strong olfactory responses displayed by the Colorado potato beetles under laboratory conditions, their ability to actively search for and find hosts over long distances in the field is somewhat uncertain (Boiteau *et al.* 2003). It is likely that chance encounters play a significant role in the colonization of new habitats (Jermy *et al.* 1988), although beetle dispersal is definitely not a completely random process.

Once Colorado potato beetles arrive on host plants, they use their sense of taste for final host acceptance. Hsiao and Fraenkel (1968) and Hsiao (1969) identified several carbohydrates, amino acids, phospholipids, and chlorogenic acid that act as phagostimulants. Szafranek *et al.* (2006) found two additional phytochemicals: the alcohols present in the leaf surfaces of potatoes. Other secondary plant compounds are toxic to the beetles and/or serve as feeding deterrents. They might be responsible for plant resistance to beetle herbivory and therefore would be of interest to plant breeders, as discussed in Chapter 15 of this book.

REPRODUCTION AND INDIVIDUAL DEVELOPMENT

Reproductive behavior in the Colorado potato beetle is strongly directed toward maximizing the genetic diversity of its offspring, and might be largely responsible for the evolutionary plasticity and adaptability of this species (see Chapter 19 for more details). The Colorado potato beetle is a highly promiscuous species, with both males and females performing multiple copulations with different partners (Szentesi 1985). Although males guard females following copulation and display aggressive behavior towards other males (Szentesi 1985), duration of such guarding is not sufficient to prevent subsequent mating (Alyokhin, unpubl. data). Instead of trying to protect their parental investment in a single female, mated summer-generation males increase their flight activity, probably to maximize the number of copulations with different mates (Alyokhin and Ferro 1999a).

Sexually mature females produce an airborne sex pheromone which acts as a long-range attractant for males (Edwards and Seabrook 1997). In addition, there is a difference between the sexes in the composition of cuticular hydrocarbons (Dubis *et al.* 1987), which might be perceived by contact chemoreception and may play an important role in sex recognition (Jermy and Butt 1991).

Boiteau (1988) determined that at least three copulations are required to completely fill the female's spermatheca. Moreover, between 5% and 20% of all copulations do not result in sperm transfer (Thibout 1982). Therefore, repeated copulations appear to be necessary for the females to realize their full reproductive potential. However, Orsetti and Rutowski (2003) did not find any correlation between the number of matings and the number of transferred sperm or female fecundity. On the contrary, there was a significant decrease in hatch rate with an increase in the number of copulations. When a summer-generation female mates with two different males, their sperm mixes, and the first male still fertilizes 28–48% of the eggs (Boiteau 1988, Alyokhin and Ferro 1999b, Roderick *et al.* 2003).

Gravid Colorado potato beetle females engage in a considerable amount of flight activity (Ferro *et al.* 1999, Alyokhin and Ferro 1999a), allowing them to distribute eggs within and between host habitats. However, they fly significantly less than unmated females (Alyokhin and Ferro 1999a), probably because of the weight and energy demands of maturing eggs. Alternatively, a higher flight propensity of unmated females may be related to their attempts to find a mate.

Post-diapause females can lay eggs by utilizing sperm from pre-diapause mating from the previous fall, but at a significant fertility cost compared to spring-mated females (Ferro *et al.* 1991, Baker *et al.* 2005). Therefore, beetles usually mate after diapause termination in the spring (Ferro *et al.* 1999). Experiments using radiation-sterilized Colorado potato beetle males revealed that sperm from spring mating takes complete precedence over overwintered sperm from the previous year's mating (Baker *et al.* 2005). Unlike the flight of summer-generation beetles, the flight of post-diapause beetles is not affected by their mating status (Ferro *et al.* 1999).

Summer-generation females do not usually start ovipositing until they accumulate at least 51 DD since emergence from pupae (Alyokhin and Ferro 1999a). The effective developmental threshold for the Colorado potato beetle is 10°C. Development from egg to adult takes between 14 and 56 days (de Wilde 1948, Ferro *et al.* 1985, Logan *et al.* 1985). The fastest development occurs between 25° and 32°C, and appears to differ among populations of different geographic origins. Growth rates follow a curve typical of poikilothermic organisms, including insects (Logan *et al.* 1976). This curve initially ascends from threshold temperature to optimum temperature, and then rapidly descends from optimum to lethally high temperature.

Colorado potato beetle larvae frequently cannibalize each other, especially soon after eclosion from eggs. Harcourt (1971) found that cannibalism accounted for over 10% of total beetle mortality.

Pupation takes place in soil near plants where the larvae have completed their development. Average pupation depth is 5–12 cm (Feytaud 1938).

Both Colorado potato beetle larvae and adults are capable of behavioral thermoregulation, which allows them to maintain body temperatures that are more optimal for physiological development than ambient temperatures (May 1981). The beetles usually rest and feed on the upper surface of leaves when air temperatures are low, thus increasing their exposure to solar radiation (May 1981). As a result, their body temperatures are often elevated to several degrees above air temperature (May 1981). When ambient temperatures increase, larvae tend to move under leaves (Lactin and Holliday 1994) or to the inner part of the potato canopy (May 1981).

MOVEMENT AND DISPERSAL

Dispersal and migration are important adaptive strategies in the Colorado potato beetle. Similar to diapause, migration in this species is facultative. When environmental conditions are benign, the beetles often spend their entire lives in the general vicinity of the place of their larval development (Grafius 1995). However, when the need arises, they are also capable of traveling over considerable distances. For example, the beetles have repeatedly invaded the island of Jersey, located 20 km off the French coast, arriving both by flight and by being carried by sea currents (Small 1948, Small and Thomas 1954). Incursions across the Baltic Sea to Scandinavia imply that, given favorable wind speed and direction, Colorado potato beetles can fly more than 100 km (Wiktelius 1981). Also, a group of beetles was recorded landing on the deck of a ship 110 km away from the nearest shore (van Poeteren 1939).

The Colorado potato beetle has three distinct types of flight that play different ecological roles in its life history (Voss and Ferro 1990a). A low-altitude flight with frequent turning within the host habitat serves to distribute eggs within a field, to find mates, or to move onto less defoliated host plants. A straight, often downwind flight over a distance of several hundred meters or more is used for colonization of new areas. It is a true migratory flight that is not interrupted by the presence of suitable habitats in the vicinity of the beetle's place of origin (Caprio and Grafius 1990). Diapause flight is a low-altitude, directed flight towards tall vegetation bordering potato fields. The flying beetles arrive in wooded hedgerows, where they immediately burrow into the soil to diapause (Voss and Ferro 1990a).

Both male and female beetles engage in flights of all three types (Weber and Ferro 1994a). As discussed above, however, gender and reproductive status may affect their propensity to fly. Voss and Ferro (1990b) showed that significantly more males than females engaged in local flight activity, possibly in search of mates. Hough-Goldstein and Whalen (1996) reported that almost twice as many overwintered males immigrated into fields by flight compared to overwintered females, although a portion of their data probably reflected local flight activity, especially later in the season (Voss and Ferro 1990a, Hough-Goldstein and Whalen 1996). Also, Weber and Ferro (1994b) found that overwintered males departed from a non-host habitat more readily than females, but were more likely to remain in a potato field than females. On the contrary, Zehnder and Speese (1987) reported a 50:50 sex ratio of beetles caught in windowpane traps throughout the growing season.

Movement by flight is instrumental for the Colorado potato beetle to be able to colonize new habitats and escape from hostile environments. It also ensures gene flow between isolated populations. Walking is relatively less important because beetles are able to walk only several hundred meters at a relatively low speed (Ng and Lashomb 1983). However, it plays a role in beetle dispersal within already-colonized host habitats, and for movement between host habitats and overwintering sites. More details on the theoretical and applied aspects of the movement and spread of the Colorado potato beetle are provided by Boiteau in Chapter 12.

MANAGEMENT IMPLICATIONS

Different approaches to Colorado potato beetle control are discussed in Chapters 14–19. However, it is extremely important to realize that no single technique will ever provide a lasting solution for managing this insect. A complex and diverse life history makes the Colorado potato beetle a challenging pest to suppress. The beetles integrate diapause, dispersal, feeding, and reproduction into an ecological "bet-hedging" strategy, distributing their offspring in both space (within and between host habitats) and time (within and between seasons). They are also extremely adaptable to adverse conditions, including human attempts of their control (see Chapter 19). Yet, humans historically rely on a very rigid and simplistic set of techniques, which is largely limited to spraying insecticides.

Twenty-five years ago, Casagrande (1987) described the long history of Colorado potato beetle control as "135 years of mismanagement". Unfortunately, the situation is no different at present. Even replacing insecticides with non-chemical methods, as described later in this book, will never provide a sustainable means of controlling this pest. In order to succeed, we need to become as flexible and adaptable as the Colorado potato beetle itself. The only option for the economically sound and environmentally friendly protection of potato crops is the science-based integration of multiple control techniques into a comprehensive and dynamic pest-management approach.

REFERENCES

Alyokhin, A., 2011. Colorado potato beetle bibliography. Internet at www.potatobeetle.org/Bibliography/index.php accessed November 18, 2011.

Alyokhin, A.V., Ferro, D.N., 1999a. Reproduction and dispersal of summer-generation Colorado potato beetle (Coleoptera: Chrysomelidae). Environ. Entomol. 28, 425–430.

Alyokhin, A.V., Ferro, D.N., 1999b. Electrophoretic confirmation of sperm mixing in mated Colorado potato beetles (Coleoptera: Chrysomelidae). Ann. Entomol. Soc. Am. 92, 230–235.

Arnett, R.H., Jr. "American Insects: A Handbook of the Insects of America North of Mexico." Second ed. CRC Press, Boca Raton, FL.

Arnoldy, B., 2010. Afghanistan war: How USAID loses hearts and minds. Christian Science Monitor July 26, 2010.

Azeredo-Espin, A.M.L., Schroder, R.F.W., Huettell, M.D., Sheppard, M.S., 1991. Mitochondrial DNA variation in geographic population of Colorado potato beetle, *Leptinotarsa decemlineata* (Coleoptera: Chrysomelidae). Experientia 47, 483–485.

Azeredo-Espin, A.M.L., Schroder, R.F.W., Roderickl, G.K., Sheppard, M.S., 1996. Intraspecific mitochondrial DNA variation in the Colorado potato beetle, *Leptinotarsa decemlineata* (Coleoptera: Chrysomelidae). Biochem. Genet. 34, 253–268.

Baker, M.B., Alyokhin, A., Dastur, S.R., Porter, A.H., Ferro, D.N., 2005. Sperm precedence in the overwintered Colorado potato beetles (Coleoptera: Chrysomelidae) and its implications for insecticide resistance management. Ann. Entomol. Soc. Am. 98, 989–995.

Bechyne, I., 1952. Achter Bertrag zur Kenntnisder Gattung Crysolina Motsch. Entomol. Arb. Mus. 3, 2.

Benkovskaya, G.V., Udalov, M.B., Poscryakov, A.V., Nikolenko, A.G., 2004. The phenogenetical polymorphism of Colorado potato beetle *Leptinotarsa decemlineata* Say and susceptibility to insecticides in Bashkortostan. Agrochemistry 12, 23–28.

Biever, K.D., Chauvin, R.L., 1990. Prolonged dormancy in a Pacific Northwest population of the Colorado potato beetle, *Leptinotarsa decemlineata* (Say) (Coleoptera: Chrysomelidae). Can. Entomol. 1/2, 175–177.

Boiteau, G., 1988. Sperm utilization and post-copulatory female-guarding in the Colorado potato beetle, *Leptinotarsa decemlineata*. Entomol. Exp. Appl. 47, 183–187.

Boiteau, G., Parry, R.H., Harris, C.R., 1987. Insecticide resistance in New Brunswick populations of the Colorado potato beetle (Coleoptera: Chrysomelidae). Can. Entomol. 119, 459–463.

Boiteau, G., Alyokhin, A., Ferro, D.N., 2003. The Colorado potato beetle in movement. Can. Entomol. 135, 1–22.

Bolter, C.J., Dicke, M., van Loon, J.J.A., Visser, J.H., Posthumus, M.A., 1997. Attraction of Colorado potato beetle to herbivore-damaged plants during herbivory and after its termination. J. Chem. Ecol. 23, 1003–1023.

Caprio, M., Grafius, E., 1990. Effects of light, temperature and feeding status on flight initiation in postdiapause Colorado potato beetle. Environ. Entomol. 19, 281–285.

Casagrande, R.A., 1985. The "Iowa" potato beetle, *Leptinotarsa decemlineata*. Bull. Entomol. Soc. Am. 31, 27–29.

Casagrande, R.A., 1987. The Colorado potato beetle: 125 years of mismanagement. Bull. Entomol. Soc. Am. 33, 142–150.

Cranshaw, W.S., Radcliffe, E.B., 1980. Effect of defoliation on yield of potatoes. J. Econom. Entomol. 73, 131–134.

de Kort, C.A.D., 1990. Thirty-five years of diapause research with the Colorado potato beetle. Entomol. Exp. Appl. 56, 1–13.

de Kort, C.A.D., Koopmanschap, A.B., Vermunt, A.M.W., 1997. Influence of pyriproxifen on the expression of haemolymph protein genes in Colorado potato beetle. Leptinotarsa decemlineata. J. Insect. Physiol. 43, 363–371.

de Wilde, J., 1948. Developpement embryonnaire et postembryonnaire dy doryphore (*Leptinotarsa decemlineata* Say) en function de la temperature. In: Proceedings of the 8th International Congress of Entomology, Sweden, Stockholm, pp. 320–321.

de Wilde, J., 1962. The relation between diapause research and control of the Colorado potato beetle *Leptinotarsa decemlineata* Say. Ann. Appl. Biol. 50, 605–608.

de Wilde, J., Hsiao, T.H., 1981. Geographic diversity of the Colorado potato beetle and its infestation in Eurasia. In: Lashomb, J., Casagrande, R. (Eds.), Advances in Potato Pest Management, Dowden, Hutchinson & Ross, New York, NY, pp. 47–68.

de Wilde, J., Hille Ris Lambers-Suverkropp, K., Van Tool, A., 1969. Responses to air flow and airborne plant odour in the Colorado beetle. Netherlands J. Plant Path 75, 53–57.

de Wilde, J., de Wilde, J., Jermy, T., 1976. The olfactory component in host-plant selection in the adult Colorado beetle (*Leptinotarsa decemlineata* Say). In: The Host Plant in Relation to Insect Behaviour and Reproduction, Plenum Publishing Co., New York, NY, pp. 291–300.

Dickens, J.C., 2000. Orientation of Colorado potato beetle to natural and synthetic blends of volatiles emitted by potato plants. Agric. For. Entomol. 2, 167–172.

Dickens, J.C., 2002. Behavioral responses of larvae of Colorado potato beetle, *Leptinotarsa decemlineata* (Coleoptera: Chrysomelidae), to host plant volatile blends attractive to adults. Agric. Forest Entomol. 4, 309–314.

Dill, J.F., Storch, R.H., 1992. Bibliography of Literature of the Colorado Potato Beetle, *Leptinotarsa decemlineata* Say. Maine Agricultural Experiment Station Miscellaneous Publication 686, Orono, ME.

Dripps, J.E., Smilowitz, Z., 1989. Growth analysis of potato plants damaged by Colorado potato beetle (Coleoptera: Chrysomelidae) at different plant growth stages. Environ. Entomol. 18, 854–867.

Dubis, E., Malinski, E., Dubis, A., Szafranek, J., Nawrot, J., Poplawski, J., Wrobel, J.T., 1987. Sex-dependent composition of cuticular hydrocarbons of the Colorado beetle, *Leptinotarsa decemlineata* Say. Comp. Biochem. Physiol. Part A: Mol. Integr. Physiol. 87, 839–843.

Edwards, M.A., Seabrook, W.D., 1997. Evidence for an airborne sex pheromone in the Colorado potato beetle, *Leptinotarsa decemlineata*. Can. Entomol. 129, 667–672.

Eremina, I.V., Denisova, I.A., 1987. Variability of several signs in Colorado potato beetle in Saratov province. VINITI 3, 59–70 [in Russian].

Faber, W., 1949. Biologische Untersuchunen zur Diapause des Kartoffelkäfers (*Leptinotarsa decemlineata* Say). Pflanzenschutzberichte 3, 65–94.

Fasulati, S.R., 1985. Polymorphism and population structure of Colorado potato beetle *Leptinotarsa decemlineata* Say (Coleoptera, Chrysomelidae). Russian J. Ecology. 6, 50–56.

Fasulati, S.R., 1993. Polymorphism, environmental groups, and microevolution of Colorado potato beetle *Leptinotarsa decemlineata* Say (Coleoptera, Chrysomelidae). In: Species and Its Productivity in the Range, Gidrometeoizdat, St Petersburg, Russia, pp. 260–262 [in Russian].

Fasulati, S.R., 2002. Territorial radiation of the Colorado Potato Beetle in northern potato growing regions. In: Intl Sci-Pract Conf., Proc. Environmental Aspects of Intensification of Agricultural Production, Penza, Russia, pp. 205–207 [in Russian].

Fasulati, S.R., 2007. The study of adaptive variations in pests for ecologization of plant protection systems: A case study of the Colorado potato beetle. Inform. Byull. VPRS MOBB, 246–250 [in Russian].

Ferro, D.N., Morzuch, B.J., Margolies, D., 1983. Crop loss assessment of the Colorado potato beetle (Coleoptera: Chrysomelidae) on potatoes in western Massachusetts. J. Econ. Entomol. 76, 349–356.

Ferro, D.N., Logan, J.A., Voss, R.H., Elkinton, J.S., 1985. Colorado potato beetle (Coleoptera: Chrysomelidae) temperature-dependent growth and feeding rates. Environ. Entomol. 14, 343–348.

Ferro, D.N., Tuttle, A.F., Weber, D.C., 1991. Ovipositional and flight behavior of overwintered Colorado potato beetle (Coleoptera: Chrysomelidae). Environ. Entomol. 20, 1309–1314.

Ferro, D.N., Alyokhin, A.V., Tobin, D.B., 1999. Reproductive status and flight activity of the overwintered Colorado potato beetle. Entomol. Exp. Appl. 91, 443–448.

Feytaud, J., 1938. Sur l'ecologie du Doryphore. Landwirtsch. Jahrb. Schweiz 22, 698.

Feytaud, J., 1950. Le Doryphore à la conquète de l'Europe. In: Proceedings of the VIII International Congress of Entomology, Sweden, Stockholm, pp. 643–646.

French III, N.M., Follett, P., Nault, B.A., Kennedy, G.G., 1993. Colonization of potato fields in eastern North Carolina by Colorado potato beetle. Entomol. Exp. Appl. 68, 247–256.

Garrett, B.C., 1996. The Colorado potato beetle goes to war. Chemical Weapons Control Bulletin 33, 2–3.

Gauthier, N.L., Hofmaster, R.N., Semel, M., 1981. History of Colorado potato beetle control. In: Lashomb, J.H., Casagrande, R. (Eds.), Advances in Potato Pest Management, Hutchinson Ross Publishing Co., Stroudsburg, PA, pp. 13–33.

Grafius, E., 1995. Is local selection followed by dispersal a mechanism for rapid development of multiple insecticide resistance in the Colorado potato beetle? Am. Entomol. 41, 104–109.

Grapputo, A., Boman, S., Lindström, L., Lyytinen, A., Mappes, J., 2005. The voyage of an invasive species across continents: Genetic diversity of North American and European Colorado potato beetle populations. Mol. Ecol. 14, 4207–4219.

Grison, P., 1939. Notes écologiques sur le Doryphore et éléments pour les prognostics d'invasion qu'elles permettent. In: Verh. VII Intern. Kongr, Entomol., Berlin, pp. 2663–2668.

Harcourt, D.G., 1971. Population dynamics of Leptinotarsa decemlineata (Say) in eastern Ontario. III. Major population processes. Can. Entomol. 103, 1049–1061.

Hare, J.D., 1980. Impact of defoliation by the Colorado potato beetle Leptinotarsa decemlineata on potato yields. J. Econ. Entomol. 73, 369–373.

Hare, J.D., 1990. Ecology and management of the Colorado potato beetle. Annu. Rev. Entomol. 35, 81–100.

Heim, D.C., Kennedy, G.G., Van Duyn, J.W., 1990. Survey of insecticide resistance among North Carolina Colorado potato beetle (Coleoptera: Chrysomelidae) populations. J. Econ. Entomol. 83, 1229–1235.

Hitchner, E.M., Kuhar, T.P., Dickens, J.C., Youngman, R.R., Schultz, P.B., Pfeiffer, D.G., 2008. Host plant choice experiments of Colorado potato beetle (Coleoptera: Chrysomelidae) in Virginia. J. Econ. Entomol. 101, 859–865.

Hough-Goldstein, J.A., Whalen, J.M., 1996. Relationship between crop rotation distance from previous potatoes and colonization and population density of Colorado potato beetle. J. Agric. Entomol. 13, 293–300.

Hsiao, T.H., 1969. Chemical basis of host selection and plant resistance in oligophagous insects. Entomol. Exp. Appl. 12, 777–788.

Hsiao, T.H., 1981. Ecophysiological adaptations among geographic populations of the Colorado potato beetle in North America. In: Lashomb, J., Casagrande, R. (Eds.), Advances in Potato Pest Management, Dowden, Hutchinson & Ross, New York, NY, pp. 69–85.

Hsiao, T.H., 1985. Ecophysiological and genetic aspects of geographic variations of the Colorado potato beetle. In: Ferro, D.N., Voss, R.H. (Eds.), Proceedings, Symposium on the Colorado Potato Beetle. XVIIth International Congress of Entomology, Research Bulletin 704, pp. 63–78.

Hsiao, T.H., Fraenkel, G., 1968. The influence of nutrient chemicals on the feeding behavior of the Colorado potato beetle, *Leptinotarsa decemlineata* (Coleoptera: Chrysomelidae). Ann. Entomol. Soc. Am. 61, 44–54.

Hsiao, T.N., Hsiao, C., 1983. Chromosomal analysis of *Leptinotarsa* and *Labidomera* Species (Coleoptera, Chrysomelidae). Genetics 60, 139–150.

Hunt, D.W.A., Tan, C.S., 2000. Overwintering densities and survival of the Colorado potato beetle (Coleoptera: Chrysomelidae) in and around tomato (Solanaceae) fields. Can. Entomol. 132, 103–105.

Ioannidis, P.M., Grafius, E., Whalon, M.E., 1991. Patterns of insecticide resistance to azinphosmethyl, carbofuran, and permethrin in the Colorado potato beetle (Coleoptera: Chrysomelidae). J. Econ. Entomol. 84, 1417–1423.

Isely, D., 1935. Variations in the seasonal history of the Colorado potato beetle. Journal. Kans. Entomol. Soc. 8, 142–145.

Ivanschik, E.P., Izhevsky, S.S., 1981. The history of Colorado potato beetle, *Leptinotarsa decemlineata* Say dispersal and its current range. In: Ushatinskaya, R.S. (Ed.), The Colorado Potato Beetle, *Leptinotarsa decemlineata* Say, Nauka Publishers, Moscow, Russia, pp. 11–26 [in Russian].

Jacobson, J.W., Hsiao, T.H., 1983. Isozyme variation between geographic populations of the Colorado potato beetle, *Leptinotarsa decemlineata* (Coleoptera: Chrysomelidae). Ann. Entomol. Soc. Am. 76, 162–166.

Jacques, R.L., 1972. Taxonomic revision of the genus *Leptinotarsa* (Coleoptera: Chrysomelidae) of North America. PhD dissertation. Purdue University, West Lafayette, IN.

Jacques, R.L., 1988. The Potato Beetles. E. J. Brill, Leiden, The Netherlands.

Jermy, T., 1953. Egyszerü módszer a burgonyabogár lárvák fejlödése fikozatainak megkülönböztetésére. Növényvéd. Idösz. Kérd. 2, 17–18.

Jermy, T., Butt, B.A., 1991. Method for screening female sex pheromone extracts of the Colorado potato beetle. Entomol. Exp. Appl. 59, 75–78.

Jermy, T., Saringer, G., 1955. A burgonyabogár (*Leptinotarsa decemlineata* Say). Hungary, Mezögazd Kiadó, Budapest.

Jermy, T., Szentesi, A., Horvath, J., 1988. Host plant finding in phytophagous insects – the case of the Colorado potato beetle. Entomol. Exp. Appl. 49, 83–98.

Jolivet, P., 1991. The Colorado beetle menaces Asia (*Leptinotarsa decemlineata* Say 1824)(Col. Chrysomelidae). L'Entomologiste 47, 29–48.

Kalinina, K.V., Nikolaeva, Z.V., 2007. Evaluation of potato varieties for resistance to Colorado potato beetle in North-West Russia. Kartofel and Ovoshchi 8, 16–18 [in Russian].

Lactin, D.J., Holliday, N.J., 1994. Behavioral responses of Colorado potato beetle larvae to combinations of temperature and insolation, under field conditions. Entomol. Exp. Appl. 72, 255–263.

Landolt, P.J., Tumlinson, J.H., Alborn, D.H., 1999. Attraction of Colorado potato beetle (Coleoptera: Chrysomelidae) to damaged and chemically induced potato plants. Environ. Entomol. 28, 973–978.

Lashomb, J.H., Ng, Y.S., 1984. Colonization by the Colorado potato beetle, *Leptinotarsa decemlineata* (Coleoptera: Chrysomelidae) in rotated and non-rotated potato fields. Environ. Entomol 13, 1352–1356.

Lockwood, J.A., 2009. Six-Legged Soldiers. Oxford University Press, New York, NY.

Logan, J.A., Wollkind, D.T., Hoyt, J.C., Tanigoshi, L.K., 1976. An analytic model for description of temperature dependent rate phenomena in arthropods. Environ. Entomol. 5, 1130–1140.

Logan, P.A., Casagrande, R.A., Faubert, H.H., Drummond, F.A., 1985. Temperature-dependent development and feeding of immature Colorado potato beetles, *Leptinotarsa decemlineata* Say (Coleoptera: Chrysomelidae). Environ. Entomol. 14, 275–283.

May, M.L., 1981. Role of body temperature and thermoregulation in the biology of the Colorado potato beetle. Advances in Potato Pest Management. In: Lashomb, J.H., Casagrande, R. (Eds.), Hutchinson Ross Publishing Co., Stroudsburg, PA, pp. 86–104.

McIndoo, N.E., 1926. An insect olfactometer. J. Econ. Entomol. 19, 545–571.

Medvedev, L.N., 1981. Systematic status of *Leptinotarsa decemlineata* Say within the family Crysomelidae, phylogeny, evolution of the species. In: Ushatinskaya, R.S. (Ed.), The Colorado Potato Beetle, *Leptinotarsa decemlineata* Say, Nauka Publishers Russia, Moscow, pp. 27–34 [in Russian].

Minder, I.F., Petrova, D.V., 1966. Ecological and physiological characteristics of the summer rest of the Colorado beetle. In: Arnoldi, K.V. (Ed.), Ecology and Physiology of Diapause in the Colorado Beetle, Nauka, Moscow, Russia, pp. 257–279 [in Russian].

Nault, B.A., Kennedy, G.G., 1998. Limitations of using regression and mean separation analyses for describing the response of crop yield to defoliation: A case study of the Colorado potato beetle (Coleoptera: Chrysomelidae). J. Econ. Entomol. 91, 7–20.

Ng, Y.S., Lashomb, J.H., 1983. Orientation by the Colorado potato beetle (*Leptinotarsa decemlineata* Say). Anim. Behav. 31, 617–618.

Orsetti, D.M., Rutowski, R.L., 2003. No material benefits, and a fertilization cost, for multiple mating by female leaf beetles. Animal Behaviour 66, 477–484.

Ovchinnikova, N.A., Markelov, G.V., 1982. Intraspecific variation of the Colorado potato beetle in the Lipetsk Oblast. Nauch. Dokl. Vyssh. Shkoly. Biol. Nauki 7, 63–67 [in Russian].

Riley, C.V., 1875. Seventh Annual Report on the Noxious, Beneficial, and Other Insects of the State of Missouri. Regan and Carter, Jefferson City, MO.

Riley, C.V., 1877. Ninth Annual Report on the Noxious, Beneficial, and Other Insects of the State of Missouri. Regan and Carter, Jefferson City, MO.

Roderick, G.K., De Mendoza, L.G., Dively, G.P., Follett, P.A., 2003. Sperm precedence in Colorado potato beetle, *Leptinotarsa decemlineata* (Coleoptera: Chrysomelidae): Temporal variation assessed by neutral markers. Ann. Entomol. Soc. Am. 96, 631–636.

Sakai, A.K., Allendorf, F.W., Holt, J.S., Lodge, D.M., Molofsky, J., With, K.A., et al., 2001. The population biology of invasive species. Annu. Rev. Ecol. Systematics 32, 305–332.

Say, T., 1824. Description of *Doryphora 10-lineata* n. sp. J. Philadelphia Acad. Nat. Sci. 3, 453.

Small, T., 1948. Colorado beetle in Jersey 1947. Agriculture 4, 569–574.

Small, T., Thomas, G.E., 1954. Colorado beetle in Jersey. A study of the problem of seaborne invasions. Agriculture 11, 118–122.

Stål, C., 1865. Till kannedomen om Amerikas chrysomeliner. Ofv. Svenska Vet.-Akad. Forh. 15, 469–478.

Stegwee, D., Kimmel, E.C., de Boer, J.A., Henstra, S., 1963. Hormonal control of reversible degeneration of flight muscle in the Colorado potato beetle. *Leptinotarsa decemlineata* Say (Coleoptera). J. Cell Biol. 19, 519–527.

Suffrian, E., 1858. Ubersicht der in den Verein. Staaten von Nord-Amerika einheimischen Chrysomelen. Stettiner Ent. Zeitung 19, 237–278.

Szafranek, B., Chrapkowska, K., Waligora, D., Palavinskas, R., Banach, A., Szafranek, J., 2006. Leaf surface sesquiterpene alcohols of the potato (*Solanum tuberosum*) and their influence on Colorado beetle (*Leptinotarsa decemlineata* Say) feeding. J. Agric. Food Chem. 54, 7729–7734.

Szentesi, A., 1985. Behavioral aspects of female guarding and inter-male conflict in the Colorado potato beetle. In: Ferro, D.N., Voss, R.H. (Eds.), Proceedings, Symposium on the Colorado Potato Beetle. XVIIth International Congress of Entomology, Research Bulletin 704, pp. 127–137.

Tauber, M.J., Tauber, C.A., 2002. Prolonged dormancy in *Leptinotarsa decemlineata* (Coleoptera: Chrysomelidae): A ten-year field study with implications for crop rotation. Environ. Entomol. 31, 499–504.

Thibout, E., 1982. Le comportement sexuel du doryphore, *Leptinotarsa decemlineata* Say et son possible controle par l'hormone juvenile et les corps allates. Behaviour 80, 199–217.

Thiery, D., Visser, J.H., 1986. Masking of host plant odour in the olfactory orientation of the Colorado potato beetle. Entomol. Exp. Appl. 41, 165–172.

Thiery, D., Visser, J.H., 1995. Satiation effects on olfactory orientations patterns of Colorado potato beetle females. Comptes Rendus Academ. Sci. Ser. Life Sciences 318, 105–111.

Tower, W.L., 1906. An investigation in evolution in Chrysomelid beetle of the genus *Leptinotarsa*. Publ. Carnegie Inst. Wash. 48, 1–320.

Trouvelot, B., 1936. Remarques sur l'ecologie du doryphore en 1935 dans le massif central et le centre de la France. Rev. Zool. Agr. Appl. 25, 33–37.

Udalov, M.B., Benkovskaya, G.V., Khusnutdinova, E.K., 2010. Population structure of the Colorado potato beetle in the Southern Urals. Russian J. Ecol. 2, 126–133.

Ushatinskaya, R.S., 1961. Summer diapause and second winter diapause in the Colorado potato beetle in Transcarpathia. Dokl. AN USSR 5, 1189–1191 [in Russian].

Ushatinskaya, R.S., 1962. Colorado potato beetle diapause and development of its multi-year infestations. Zashch. Rast. 6, 53–54 [in Russian].

Ushatinskaya, R.S., 1966. Prolonged diapause in the Colorado beetle and conditions of its development. In: Arnoldi, K.V. (Ed.), Ecology and Physiology of Diapause in the Colorado Beetle, Nauka, Moscow, Russia [in Russian], pp. 120–143.

van Poeteren, N., 1939. Die Entwicklung der Kartoffelkafer. Frage in Niederlanden. In Verh. VII Intern. Kongr. Entomol., Berlin , 1938, 2701–2703.

Vermunt, A.M., Koopmanschap, A.B., Vlak, J.M., de Kort, C.A., 1999. Expression of the juvenile hormone esterase gene in the Colorado potato beetle, *Leptinotarsa decemlineata*: Photoperiodic and juvenile hormone analog response. J. Insect Physiol. 45, 135–142.

Visser, J.H., 1976. The design of a low-speed wind tunnel as an instrument for the study of olfactory orientation in the Colorado beetle (*Leptinotarsa decemlineata*). Entomol. Exp. Appl. 20, 275–288.

Visser, J.H., Ave, D.A., 1978. General green leaf volatiles in the olfactory orientation of the Colorado beetle, *Leptinotarsa decemlineata*. Entomol. Exp. Appl. 24, 738–749.

Visser, J.H., Nielsen, J.K., 1977. Specificity in the olfactory orientation of the Colorado beetle, *Leptinotarsa decemlineata*. Entomol. Exp. Appl. 21, 14–22.

Visser, J.H., Thiery, D., 1986. Effects of feeding experience on the odor-conditioned anemotaxes of Colorado potato beetles. Entomol. Exp. Appl. 42, 198–200.

Visser, J.H., Straten, S.V., Maarse, H., 1979. Isolation and identification of volatiles in the foliage of potato, *Solanum tuberosum*, a host plant of the Colorado beetle, *Leptinotarsa decemlineata*. J. Chem. Ecol. 5, 13–25.

Vlasova, V.A., 1978. A prediction of the distribution of Colorado beetle in the Asiatic territory of the USSR. Zashch. Rast. 6, 44–45 [in Russian].

Voss, R.H., Ferro, D.N., 1990a. Phenology of flight and walking by Colorado potato beetle (Coleoptera: Chrysomelidae) adults in western Massachusetts. Environ. Entomol. 19, 117–122.

Voss, R.H., Ferro, D.N., 1990b. Ecology of migrating Colorado potato beetles (Coleoptera: Chrysomelidae) in western Massachusetts. Environ. Entomol. 19, 123–129.

Walsh, B.D., 1865. The new potato bug and its natural history. Pract. Entomol. 1, 1–4.

Weber, D., 2003. Colorado beetle: Pest on the move. Pestic. Outlook 14, 256–259.

Weber, D.C., Ferro, D.N., 1993. Distribution of overwintering Colorado potato beetle in and near Massachusetts potato fields. Entomol. Exp. Appl. 66, 191–196.

Weber, D.C., Ferro, D.N., 1994a. Colorado potato beetle: Diverse life history poses challenge to management. In: Zehnder, G.W., Jansson, R.K., Powelson, M.L., Raman, K.V. (Eds.), Advances in Potato Pest Biology and Management, APS Press, St Paul, MN, pp. 54–70.

Weber, D.C., Ferro, D.N., 1994b. Movement of overwintered Colorado potato beetles in the field. J. Agric. Entomol. 11, 17–27.

Wegorek, W., 1955. Investigation on spring migration of the Colorado beetle (*Leptinotarsa decemlineata* Say) and possibilities of combating the insect. Ecol. Pol. Ser. A 3, 217–271.

Wegorek, W., 1957a. Badania nad biologia i ekologia stonki ziemniaczanej (*Leptinotarsa decemlineata* Say). Rocz. Nauk Roln. 74, 135–185.

Wegorek, W., 1957b. Badania nad zimovaniem stonki ziemniaczanej (*Leptinotarsa decemlineata* Say). Rocz. Nauk Roln. 74, 316–338.

Wellik, M.J., Slosser, J.E., Kirby, R.D., 1981. Effects of simulated insect defoliation on potatoes. Am. Potato J. 58, 627–632.

Wiktelius, S., 1981. Wind dispersal of insects. Grana 20, 205–207.

Worner, S.P., 1988. Ecoclimactic assessment of potential establishment of exotic pests. J. Econ. Entomol. 81, 973–983.

Yocum, G.D., 2003. Isolation and characterization of three diapause-associated transcripts from the Colorado potato beetle *Leptinotarsa decemlineata*. J. Insect Physiol. 49, 161–169.

Yocum, G.D., Rinehart, J.P., Larson, M.L., 2011. Monitoring diapause development in the Colorado potato beetle, *Leptinotarsa decemlineata*, under field conditions using molecular biomarkers. J. Insect Physiol. 57, 645–652.

Zehnder, G.W., Evanylo, G.K., 1989. Influence of Colorado potato beetle sample counts and plant defoliation on potato tuber production. Am. Potato J. 65, 725–736.

Zehnder, G., Speese III, J., 1987. Assessment of color response and flight activity of *Leptinotarsa decemlineata* (Say) (Coleoptera: Chrysomelidae) using window flight traps. Environ. Entomol. 16, 1199–1202.

Zehnder, G.W., Sandall, L., Tisler, A.M., Powers, T.O., 1992. Mitochondrial DNA diversity among 17 geographic populations of *Leptinotarsa decemlineata* (Coleoptera: Chrysomelidae). Ann. Entomol. Soc. Am. 85, 234–240.

Zehnder, G., Vencill, A.M., Speese III, J., 1995. Action thresholds based on plant defoliation for management of Colorado potato beetle (Coleoptera: Chrysomelidae) in potato. J. Econ. Entomol. 88, 155–161.

Zeleev, R.M., 2002. Evaluation of polymorphism of pronotum and elytra patterns of the Colorado potato beetle, *Leptinotarsa decemlineata*, in the Vicinity of Kazan. Russian Zool. J. 3, 316–322.

Chapter 3

Aphids as Major Potato Pests

Julien Saguez[1], Philippe Giordanengo[2], and Charles Vincent[1]

[1]Agriculture and Agri-Food Canada, Saint-Jean-sur-Richelieu, Québec, Canada, [2]Université de Picardie Jules Verne, Amiens, France; CNRS, and INRA, Institut Sophia Agrobiotech, Sophia Antipolis, France; Université de Nice Sophia Antipolis, Sophia Antipolis, France

INTRODUCTION

The potato is one of the most important crops worldwide and one of the main food sources in many countries. Aphids cause serious losses in potato plants and tubers, mostly because they vector viruses.

Due to their remarkable adaptations and colonization of several ecological systems, including crops, aphids are interesting models of study at different levels. Their biology (viviparity, oviparity, and parthenogenesis), physiology (osmoregulation, regulation of the water balance), and behavior (feeding, virus transmission, and plant manipulation) reveal unique adaptations and fascinating relationships with their host plants.

The main aphid species associated with potatoes worldwide are non-specific to this crop. Most of them are cosmopolitan and polyphagous. Numerous articles have been published on several aphid/crop systems. In this chapter, we first present attributes common to most aphid species and, whenever appropriate, we use potato-colonizing aphids to illustrate our points. Secondly, we discuss characteristics that are specific to the most important aphid species of potato. Finally, we address issues relevant to the management of potato aphids.

LIFE CYCLES, REPRODUCTION AND DISPERSION

Aphids have complex life cycles, and their classification depends on host alternation and on their mode of reproduction (Moran 1992, Blackman and Eastop 2000, Williams and Dixon 2007). Different morphs are associated with these life cycles.

Heteroecy and Monoecy

Heteroecy and monoecy refer to the status of aphids regarding their host plant. Aphids that practice host alternation are heteroecious. They live on a primary host during the winter and colonize secondary hosts during the rest of the year

Insect Pests of Potato. http://dx.doi.org/10.1016/B978-0-12-386895-4.00003-X

before coming back to their primary host. Heteroecy occurs in only 10% of aphid species that generally colonize herbaceous plants, including economically important crop species such as potatoes (Williams and Dixon 2007).

In contrast, a majority of aphid species live on the same plant throughout the year, do not have host plant alternation, and are classified as monoecious species. Some of these species are monophagous. Other species are oligo- or polyphagous and may migrate between plant species. However, they do not have regular alteration of primary and secondary hosts in their life cycles (Williams and Dixon 2007).

Holocycly and Anholocycly

Holocycly and anholocycly refer to the ability of aphids to reproduce using parthenogenesis alone or in combination with sexual reproduction (Blackman and Eastop 2000). Most aphid species alternate parthenogenesis and sexual reproduction and are holocyclic. In this case, parthenogenesis occurs from the first generation in spring to the appearance of sexual morphs in autumn. The appearance of sexual individuals is triggered by seasonal changes in temperature and photoperiod. In contrast, some species are anholocyclic; they do not produce sexual morphs or eggs, and only reproduce by parthenogenesis (Williams and Dixon 2007). Anholocyclic life cycles may occur when climatic conditions are favorable for aphids to maintain populations on various plants during winter. Depending on the region, certain populations in some holocyclic aphid species can lose their sexual phase and become anholocyclic or generate only male populations (androcycly) during winter (Blackman 1971, Fenton *et al.* 1998, Williams and Dixon 2007).

Aphids that colonize potato are mainly heteroecious and holocyclic. Their life cycles include an overwintering phase, during which fertilized eggs constitute the resistant form during periods of cold temperatures. Because potato is an annual plant, its colonizing aphids are heteroecious.

Parthenogenesis and Developmental Rates

Aphids mainly reproduce by cyclical parthenogenesis (Moran 1992) that enables asexual reproduction without males. Females give birth to nymphs that are immediately able to exploit plants. This viviparous mode of reproduction confers a rapid reproduction rate with short developmental times, resulting in population growth that is atypically high, even for insects. For instance, Dixon (1971) estimated that aphid populations in potato fields can reach densities of 2×10^9 individuals per hectare. Douglas (2003) suggested that such rates of population increase reflect nutrient allocation to the reproductive system. Energy is preferentially invested in embryo biomass and larval development rather than in maternal tissues. Aphids have telescoping generations – i.e., ovarian development and embryo formation start at the same time in embryonic mothers (Powell *et al.* 2006).

Parthenogenetic reproduction results in clonal aphid colonies that have the same genotype. With this reproduction mode, an atypical characteristic can be

amplified and become predominant in a given population after several generations. This can explain why aphids are able to adapt quickly to disturbances
in their environment. Aphid populations may crash depending on the weather
(Barlow and Dixon 1980), deteriorating resources, or pesticide treatments.
However, parthenogenesis rapidly generates new populations that are adapted
to their environment and, in some cases, resistant to pesticides. Parthenogenesis
generally occurs during the warmer months of the year and maximizes offspring
production. In fall, it is interrupted and followed by sexual reproduction that
produces overwintering eggs.

Dispersal and Colonization

Aphids produce both apterous (wingless) and alate (winged) morphs. Production of alate morphs is energetically costly (Dixon *et al.* 1993). Alates appear
at different times during the year. They are considered to be colonizers, and use
winds to disperse and locate new hosts.

Wingless fundatrices emerge from eggs laid on the primary host. Their alate
progeny are the spring migrants. Alate production is completed within a 2-week
period (Radcliffe 1982). These individuals fly to secondary hosts (e.g., potato)
and, when conditions are favorable, generate apterous and parthenogenetic
populations. During summer, overpopulation of aphids, degradation of host-
plant nutritional suitability, or variations in light intensity, temperature, and precipitation induce the decline in aphid populations and the appearance of winged
morphs that move to more suitable host habitats.

In autumn, as day length and temperature decrease, the quality of secondary
host plants is altered. These factors generate the appearance of a new generation of virginoparous alates that migrate to the primary host. After the second
generation on the primary host, oviparous females appear and are fertilized by
winged males (Radcliffe 1982). After reproduction, oviparous females lay their
eggs on the secondary host for overwintering (Powell *et al.* 2006). Timing of
flight and the number of migrants are important for colonization, clonal fitness,
and overwintering success.

BIODIVERSITY AND THE MOST IMPORTANT ECONOMIC SPECIES ON POTATO

Aphids belong to the Stenorrhyncha (Hemiptera) (Blackman and Eastop 1984).
The most important genera found on potato crops, i.e. *Aulacorthum* spp., *Aphis*
spp., *Macrosiphum* spp., and *Myzus* spp., belong to the family Aphididae.

Aphids are characterized by high polymorphism. Within a given species,
different morphs may occur in the same population, including apterous and
alate individuals. The color of the individuals can also be highly variable within
a given population and can be influenced by the symbiotic bacteria they host
(Tsuchida *et al.* 2010). Morphology is influenced by several factors, such as

environmental, climatic, and seasonal conditions; quality of the host plants and population densities.

Blackman and Eastop (1984, 2000) listed 15 species of aphids that commonly infest potato plants (Table 3.1), and provided an identification key using

TABLE 3.1 Biodiversity and Characteristics of Aphids Associated with Potato Crop

Species	Subfamilies	Life cycles	Lengths (mm)	
			Apterous	Alate
Aphis fabae	Aphidinae	Heteroecious holocyclic	1.5–3.1	1.5–3.1
Aphis frangulae	Aphidinae	Holocyclic		
Aphis gossypii	Aphidinae	Anholocyclic/holocyclic	0.9–1.8	1.1–1.8
Aphis nasturtii	Aphidinae	Heteroecious holocyclic	1.3–2.0	1.3–2.0
Aphis spiraecola	Aphidinae	Holocyclic/anholocyclic	1.2–2.2	1.2–2.2
Aulacorthum circumflexum	Aphidinae	Anholocyclic	1.2–2.6	1.2–2.6
Aulacorthum solani	Aphidinae	Anholocyclic/holocyclic	1.8–3.0	1.8–3.0
Macrosiphum euphorbiae	Aphidinae	Anholocyclic (Europe + elsewhere) Heteroecious holocyclic (USA) Holocyclic (occasionally)	1.7–3.6	1.7–3.6
Myzus ascalonicus	Aphidinae	Anholocyclic	1.1–2.2	1.1–2.2
Myzus ornatus	Aphidinae	Anholocyclic	1.0–1.7	1.0–1.7
Myzus persicae	Aphidinae	Heteroecious holocyclic Anholocyclic	1.2–2.1	1.2–2.1
Pemphigus sp.	Eriosomatinae			
Rhopalosiphoninus latysiphon	Aphidinae	Anholocyclic	1.4–2.5	1.4–2.5
Rhopalosiphum rufiabdominalis	Aphidinae	Anholocyclic Heteroecious holocyclic (Japan)	1.2–2.2	1.2–2.2
Smynthurodes betae	Eriosomatinae	Heteroecious holocyclic Anholocyclic (possible)	1.6–2.7	1.6–2.7

morpho-anatomic criteria such as body color, length, shape, and segmentation; antennal tubercles, head, siphunculi, legs and femurs, cauda and anal plate; and hairs on these structures. Hereafter, we will briefly introduce the main species of aphids feeding on the potato around the world.

Myzus persicae (Sulzer)

Presumed to originate from China, the green peach aphid, or peach-potato aphid, is the most important potato aphid worldwide. It is a small to medium-sized aphid that is highly polymorphic, with colors varying from green to red. Alates have a shiny black dorsal abdominal patch, and immature alates are generally red or pink (Blackman and Eastop 2007) (Fig. 3.1). While its primary hosts are usually found in the *Prunus* genus, other tree species may also host eggs in some regions. *M. persicae* is a highly polyphagous species that successfully colonizes hundreds of plant species belonging to 40 different families (Flanders *et al.* 1992, Blackman and Eastop 2000). It is generally heteroecious and heterocyclic but, under some conditions (temperature, food availability, and geographical location), it can be anholocyclic (Blackman 1971, Blackman and Eastop 2000).

FIGURE 3.1   Alate (left) and apterous (right) *Myzus persicae. Photo by Sébastien Boquel.* See also Plate 3.1

Macrosiphum euphorbiae (Thomas)

The potato aphid is a medium-sized to large aphid whose color is generally green, but can be pink or magenta with reddish eyes in some morphs (Blackman and Eastop 2007) (Fig. 3.2). It is a polyphagous species of North American origin that feeds on up to 200 plant species belonging to 20 different families, including several *Solanum* species (Flanders *et al.* 1992, Le Roux *et al.* 2010).

M. euphorbiae is often anholocyclic, but may be heteroecious holocyclic in some cases. Its primary hosts are *Rosa* spp., but eggs can also be deposited on herbaceous plants (Blackman and Eastop 2000).

FIGURE 3.2   Alate and nymph (on leaf vein) *Macrosiphum euphorbiae.Photo by Sébastien Boquel.* See also Plate 3.2

Aphis spp.

Several species from the genus *Aphis* have been reported on potatoes, including the most common *Aphis fabae* Scopoli (the blackbean aphid), *A. frangulae* Kaltenbach, and *A. nasturtii* Kaltenbach (the buckthorn-potato aphid). *A. fabae* is dull black and is highly variable in size. White wax markings appear on old aphid colonies. This species is polyphagous and heteroecious holocyclic and lays eggs on a spindle (*Euonymus europaeus*) to overwinter. *A. frangulae* is often confused with *A. gossypii* Glover. It is a small species whose color varies from dark green to black. This species is holocyclic and occurs on potato and other herbaceous plants (Blackman and Eastop 2000). *A. nasturtii* is also a small aphid, but the color of apterous individuals is bright yellowish-green. It is a polyphagous and heteroecious holocyclic aphid that can be anholocyclic depending on the region (Blackman and Eastop 2000).

Aulacorthum solani (Kaltenbach)

The glasshouse-potato aphid, or foxglove aphid, is a medium-sized to large aphid with variable color in apterous morphs, although it is generally green or yellow. This species has a probable European origin and is highly polyphagous and common on potato. Anholocyclic and holocyclic individuals occur on many different crops (Blackman and Eastop 2000).

Other Species

Blackman and Eastop (2000) also described other aphid species that have less economic impact on potato, such as *Aphis spiraecola*, *Aulacorthum circumflexum*, *Myzus ascalonicus*, *Myzus ornatus*, *Pemphigus* sp., *Rhopalosiphoninus latysiphon*, *Rhopalosiphum rufiabdominalis*, and *Smynthurodes betae* (Table 3.1).

FEEDING ON PHLOEM: A REAL CHALLENGE FOR APHIDS

A suitable host plant provides all of the nutrients needed for optimal growth and reproduction. Aphid mouthparts allow them to access chemicals of the plant tissues by puncturing superficial cells and ingesting plant fluids such as phloem and xylem saps. Feeding behavior on potato varies across aphids and morphs (Boquel *et al.* 2011a). Good indicators of plant acceptance by aphids are the colonization of the plant and the initiation of reproduction (Powell *et al.* 2006), both of these parameters being dependent on host-plant quality. Several studies have demonstrated that aphids are mainly phloemophagous, and have developed a range of strategies to optimize their utilization of this resource.

Host-Plant Selection and Colonization

Alate aphids are able to found new colonies and efficiently exploit their hosts. Several sensory mechanisms, including vision, olfaction, and gustation, are involved in host-plant selection before acceptance (Powell *et al.* 2006). During flight, aphids can visually identify a potential host plant and detect volatiles or pheromones that act as repellents or attractants (Gibson and Pickett 1983, Vandermoten *et al.* 2012). After landing on a plant, its surface texture is explored by tarsal contacts, and odors are detected by antennae. Stylet penetration in the epidermis allows aphids to evaluate the phytochemistry of the plant and to detect antifeedant compounds, providing aphids with the information to decide whether to accept or reject the plant. Sustained feeding, characterized by a sequence of ingestions and brief salivations in the sieve-tube elements (phloem and xylem vessels), indicates that the plant is suitable for food ingestion and colony settlement (Powell *et al.* 2006). Stylet penetration enables aphids to exploit plant tissues without macroscopic injuries. This specific interaction between the aphids and their host plants is a crucial feature which explains how aphids are able to develop on phloem sap and transmit viruses on plants.

Phloem Sap as Food Source

Phloem sap is used as a food source by only some hemipterans (Douglas 2006). Phloem sieve tubes transport the sap, which is composed of high amounts of soluble carbohydrates, free amino acids (Douglas 2003), and some proteins.

Sucrose is usually the main sugar in phloem sap, with variable amounts of other carbohydrates, such as galactose, raffinose, and sugar alcohols (polyols), also present (Ziegler 1975, Zimmermann and Ziegler 1975, Douglas 2003).

In some plants, such as rice, sucrose is the only sugar present in phloem sap (Kawabe *et al.* 1980). Phloem sap also contains relatively high levels of free amino acids of variable composition (Ziegler 1975, Rahbé *et al.* 1990, Girousse *et al.* 1991, Sandström and Moran 1999, Sandström 2000). However, it usually contains <20% essential amino acids (Sandström and Moran 1999).

Several studies have shown that sucrose concentration, amino acid concentration and composition, and sucrose:amino acid ratio play an important role in aphid growth and reproduction (Auclair 1963, Dadd 1985, Douglas 1993, 1998a, Karley *et al.* 2002). Feeding on phloem sap has forced these insects to develop adaptive strategies to regulate osmotic pressure, water balance, and essential nutriments required for growth and reproduction.

Osmoregulation

Because phloem sap is their main nutritional source, aphids continuously ingest high concentrations of sucrose. Although sucrose is the main carbon source and respiratory substrate in aphids (Febvay *et al.* 1995, Rhodes *et al.* 1996), its concentration in sap could reach up to 1 M, creating an osmotic pressure that is up to three times higher than that of the insect's body fluids. Consequently, regulation of osmotic pressure represents a challenge for aphids.

In aphids, the digestive tract is simple and composed of an esophagus, a midgut, and a proximal hindgut. Consequently, the regulation of osmotic and ionic pressures between the ingested sap and the hemolymph fluids occurs essentially in the digestive tract, with many exchanges (Downing 1980, Douglas 2003, Shakesby *et al.* 2009). The aphid midgut is comprised of a stomach and a looped intestine, which are closely apposed and consist of a single epithelial cell layer bound by a lamina and muscles. This anatomic organization may favor the digestion of sugars and water exchange in aphids (Rhodes *et al.* 1997, Douglas 2003). Few digestive enzymes are secreted in the gut, and these participate in nutriment assimilation.

Ingested sugars are digested in the posterior midgut of aphids. Sucrose is metabolized by α-glucosidases localized in the proximal intestine. These enzymes possess a sucrase activity (Rhodes *et al.* 1997, Cristofoletti *et al.* 2003, Karley *et al.* 2005, Price *et al.* 2007) that liberates two monosaccharides – i.e., fructose and glucose. Subunits of fructose quickly cross the digestive epithelium and are assimilated to be preferentially used as a respiratory substrate (Ashford *et al.* 2000), while glucose is transformed into oligosaccharides by transglucosidases (Walters and Mullin 1988, Rhodes *et al.* 1997, Ashford *et al.* 2000). Oligosaccharides synthesized by *M. persicae* have been shown to be important molecules for honeydew (i.e., sugar-rich secretion resulting from aphid digestion) osmoregulation (Fisher *et al.* 1984). These oligosaccharides possess a higher molecular weight than the ingested sugars (Douglas 2003, Karley *et al.* 2005), inducing a reduction of the osmotic pressure in the gut lumen. They are excreted in the honeydew fluids, essentially composed of oligosaccharides derived from the glucose moiety (Fisher *et al.* 1984, Walters and Mullin 1988, Rhodes *et al.* 1997, Ashford *et al.* 2000). Honeydew is iso-osmotic with the hemolymph (Downing 1978, Wilkinson *et al.* 1997).

Trehalose is one of the oligosaccharides that can be produced to regulate sugar pressure in hemolymph by redirecting it in the aphid metabolism. After sucrose hydrolysis, two glucose units can be combined to form trehalose in the hemolymph. This disaccharide is used as a source of energy (Wyatt 1967, Rhodes *et al.* 1997) by the fat body (Wyatt 1967, Kono *et al.* 1998). For instance, *M. persicae* and *A. gossypii* hemolymph contains high concentrations of trehalose (Wyatt 1967, Rhodes *et al.* 1997). Trehalose concentrations in aphid hemolymph can reflect the concentration of sucrose in phloem sap (Moriwaki *et al.* 2003).

Water Balance

A consequence of feeding on phloem sap along with a thin cuticle is dehydration. Deprived of the Malpighian tubules that normally recycle water within insects, aphids are very vulnerable to desiccation. Nonetheless, the anatomy of their digestive tract allows for water cycling (Downing 1980, Douglas 2003, Shakesby *et al.* 2009) which prevents dehydration and seems to be highly involved in osmoregulation.

The ingestion of phloem exerts an osmotic stress for aphids that induces water transfer between the hemocoel and the gut lumen (Shakesby *et al.* 2009). This loss of water generates dehydration that can affect aphid metabolism, growth, and reproduction. Proximity of the stomach and hindgut in the pea aphid suggests that oligosaccharide synthesis and sugar assimilation in the midgut induce a low osmotic pressure in the hindgut and can generate a water flux from the hindgut to the stomach, inducing a dilution of the midgut fluids. This mechanism possibly reduces the loss of water from hemolymph to the midgut, and from the hindgut to the honeydew (Rhodes *et al.* 1997, Shakesby *et al.* 2009).

Although aphids are essentially phloem feeders, it has been shown that they occasionally ingest xylem sap to compensate for dehydration and to regulate water balance. Xylem is rich in water and electrolytes. It is therefore suggested that feeding on it allows aphids to compensate for a deficit of water, as demonstrated by Pompon *et al.* (2011) for different stages of development and morphs in *M. euphorbiae*. However, on potato plants, ingestion of xylem varies across morphs and species (Boquel *et al.* 2011a). Xylem consumption increases with the duration of the starvation period. Pompon *et al.* (2010a) compared apterous and alate aphids, and suggested that xylem consumption and water cycling contribute to hemolymph osmoregulation in aphids, and that fecundity is negatively correlated with time spent ingesting the water in xylem sap.

In the midgut, water fluxes may be promoted by water channels in sites where the stomach and intestine are anatomically close (Shakesby *et al.* 2009).

Water fluxes are mediated by aquaporins localized on the midgut membrane, such as the aquaporin ApAQP1 identified in *Acyrthosiphon pisum*. ApAQP1 is supposed to participate in the water flux between the intestine and the stomach to reduce the osmotic gradient through the digestive tract.

Digestive Proteases

Phloem sap contains several proteins (e.g., peptides, enzymes, proteinase inhibitors, lectins) in concentrations that vary from 0.3 to 60 mg/mL, depending on the plant species (Rahbé and Febvay 1993, Kehr 2006, Pyati *et al.* 2011). These proteins, such as chitinases, protease inhibitors, and lectins, are often produced as plant defense compounds. Although providing only small amounts of amino acids, they can still play a role in nutrition. To take advantage of this nitrogenous source, and to protect themselves against their deleterious effects as plant defense compounds, aphids have to degrade them. Aphids lack a peritrophic membrane, a chitinous structure that limits the lumen of the midgut from the mesenteric epithelial cells. In many insects, the peritrophic membrane is involved in the sequestration and cycling of the endogenous digestive proteases. The lack of the peritrophic membrane in aphids is probably linked to the negligible concentrations of proteins in phloem sap and the low digestive proteolytic activities in the aphid midgut (Klingauf 1987, Srivastava 1987, Terra 1990, Tellam 1996).

However, several studies point out some putative protease activities in the guts of *M. persicae* (Rahbé *et al.* 2003), *A. pisum* (Cristofoletti *et al.* 2003), and *A. gossypii* (Deraison *et al.* 2004). Although aphids are insensitive to many protease inhibitors, they apparently use cysteine proteinases for protein digestion. A gut-specific cathepsin L-like cysteine proteinase was characterized in *A. pisum* and *A. gossypii* (Cristofoletti *et al.* 2003, Deraison *et al.* 2004). Several copies of genes coding for cathepsin B-like cysteine proteinases are expressed at high levels in the *A. pisum* gut (Rispe *et al.* 2008). Cysteine proteinases and proteolytic activities have also been demonstrated in the cereal aphid, *Sitobion avenae*. They hydrolyze ingested proteins to supplement nutrition and partially compensate for a lack of free amino acids (Pyati *et al.* 2011). Protein digestion in the aphid gut may allow for a response to variations of phloem sap composition (Gattolin *et al.* 2008). Proteolytic activities can also be a strategy to degrade toxic proteins produced by host plants.

Primary Endosymbionts and Essential Amino Acids

All 20 amino acids are found in aphid hemolymph and are required for protein synthesis. However, aphids are unable to produce or acquire some of them in their diet (Shigenobu *et al.* 2000, Tamas *et al.* 2002, van Ham *et al.* 2003). Essential amino acids required by aphids vary depending on host-plant quality and the aphid species and clones. Furthermore, the quality of plants varies widely between young and old plants and affects the amino acid composition and the physiological parameters of aphid development (Karley *et al.* 2002).

Many insects have established a symbiotic relationship with primary (obligate) and secondary (facultative) bacteria (Buchner 1965, Douglas 1998a,1998b, Gil *et al.* 2004). Endosymbionts are unculturable microorganisms. Primary endosymbionts are hosted in specialized cells called mycetocytes (Munson *et al.* 1991) or bacteriocytes and participate in the nutritional enrichment of the insect diet.

A typical example is the symbiosis between aphids and their primary endosymbionts, the γ-proteobacterium *Buchnera aphidicola* (Munson *et al.* 1991). The mutualistic association is obligatory for both partners because the bacterium cannot live without its host, and itself is essential for the normal growth and reproduction of the aphids (Baumann *et al.* 1995, Douglas 1998a). Endosymbionts are vertically transmitted via a transovariole transfer from the mother to the developing eggs or embryos (Buchner 1965, Houk and Griffiths 1980).

B. aphidicola complements the aphid diet, which lacks essential amino acids (Liadouze *et al.* 1996, Febvay *et al.* 1999). Several genes that code for the biosynthesis of essential amino acids have been described in *B. aphidicola*. In contrast, endosymbionts are defective for many genes and have lost many of the transcription regulation mechanisms involved in metabolic intermediates. Consequently, these are provided by their hosts to ensure the production of essential amino acids. For instance, aphids provide free amino acids, ammonia, other enzymes, and metabolites to their endosymbionts.

B. aphidicola has a versatile metabolism that allows its adaption to aphid requirements. Molecular studies have shown a total of 13 genes directly involved in essential amino acid biosynthesis in *B. aphidicola* (Douglas 1998b). Dadd and Krieger (1968) demonstrated that: (1) *M. persicae* only requires three essential amino acids (i.e. methionine, histidine, and isoleucine) to ensure normal growth on artificial diets, (2) cysteine is a source of sulfur, and (3) a lack of cysteine can be offset by inorganic sulfate or methionine. Using an antibiotic treatment of *M. persicae*, Mittler (1971) showed that at least 10 amino acids (arginine, histidine, isoleucine, leucine, lysine, methionine, phenylalanine, threonine, tryptophane, and valine) are essential for this species.

Several studies on the pea aphid, *A. pisum*, demonstrated bioconversion of asparagine in aspartic acid and glutamate used as nitrogen source by *B. aphidicola* to produce ammonia and glutamic acid (Douglas 1993, Sasaki and Ishikawa 1995, Whitehead and Douglas 2003). It was also shown that glutamic acid is used as a precursor for the synthesis of isoleucine, leucine, phenylalanine and valine (Sasaki and Ishikawa 1995). Douglas (1988) demonstrated that *B. aphidicola* uses sulfate as a sulfur source for providing *M. persicae* in methionine. Tryptophan is an amino acid that can also be provided by *B. aphidicola*, even though aphids possess tryptophan synthase (Douglas and Prosser 1992, Lai *et al.* 1994).

Antibiotics such as rifampicin can inhibit *B. aphidicola* activities, unbalance the profile of free amino acids and proteins in aphids (Prosser and Douglas 1991,

Wilkinson and Douglas 1995, 1996), or kill them (Miao *et al.* 2003). When deprived of their endosymbionts, aphids show delayed growth and become sterile or die (Houk and Griffiths 1980, Ishikawa and Yamaji 1985, Ohtaka and Ishikawa 1991, Douglas and Prosser 1992). Under some conditions (e.g., poor host-plant quality), *B. aphidicola* has the capacity to synthesize essential amino acids to compensate for the deficit of amino acids for optimal aphid growth (Gündüz and Douglas 2009). Bermingham and Wilkinson (2010) demonstrated that tryptophan is also an important amino acid supplied by *B. aphidicola* for *Aphis fabae* embryo growth, and participates in a rapid rate of reproduction.

Secondary Endosymbionts

Aphids also interact with bacterial secondary endosymbionts that are facultative (Oliver *et al.* 2010). Secondary endosymbionts may be transmitted horizontally, and are found free in the hemolymph, as well as within various cell types (Oliver *et al.* 2010). Secondary symbionts can impact important fitness-related traits, such as body pigmentation (Tsuchida *et al.* 2010), offspring production (Simon *et al.* 2011), and parasitoid or pathogen resistance (Montllor *et al.* 2002, Oliver *et al.* 2003, Guay *et al.* 2009). Interestingly, a close relationship has been shown between the pattern of hosted secondary symbionts and the ability of different aphid genotypes to exploit various host plants (Simon *et al.* 2003, Tsuchida *et al.* 2004, Ferrari *et al.* 2007). For example, studies have revealed a complex association between infection by *Regiella insecticola*, aphid genotype, and host-plant use (Ferrari *et al.* 2007). Interactions between aphids, their mutualistic symbionts, and their host plant have been explored in the context of evolution, ecology, behavior, and population genetics (Oliver *et al.* 2010), but the molecular processes underlying these interactions remain largely unknown.

VIRUS VECTORS AND DISEASES

Research on potatoes often involves studying viruses because of their considerable impact on crop quality and yields. No antiviral treatment is available to control virus spread among cultivated plants. Consequently, research effort goes into the management of their aphid vectors. Brunt and Loebenstein (2001) mentioned that at least 37 viruses have been reported to infest potato, several of which were identified in the 1990s. *M. persicae* and *M. euphorbiae* are often reported as the most important virus vectors that can, respectively, transmit over 100 (Kennedy *et al.* 1962) and 45 plant viruses (Fuentes *et al.* 1996).

A very low tolerance for virus infection is allowed for potato seed certification. This forces seed producers to increase their management efforts at different stages of production. Viruses seriously affect tuber size and quality (Hane and Hamm 1999) and often generate necrosis, making tubers unmarketable.

Potato Viruses: Their Symptoms and Transmission by Aphids

Potato can be infected by more than 30 RNA viruses (Salazar 1996), among which 13 are transmitted by aphids (Brunt and Loebenstein 2001). The two most important potato viruses are the potato leafroll virus (PLRV) and the potato virus Y (PVY; Fig. 3.3). The latter includes different strains; the ordinary or common strain (PVY^O) and the potato virus Y^{NTN} (PVY^{NTN}), which causes potato tuber necrotic ringspot disease. In North America other viruses are also described, such as the potato virus A (PVA), potato virus M (PVM), potato virus S (PVS), potato latent virus (PLV), alfalfa mosaic virus, and cucumber mosaic virus (see Chapter 11).

FIGURE 3.3   Potato plant infected with PVY (white arrow). *Photo by Gary Sewell.* See also Plate 3.3

Four modes of virus transmission by aphids are known: (1) non-persistent (i.e., strictly stylet-borne viruses), (2) semi-persistent (i.e. limited to the foregut), (3) persistent and circulative, and (4) persistent and propagative (Nault 1997). PLRV is a persistent and circulative potato virus mainly limited to phloem and companion cells (van den Heuvel *et al.* 1995), while other potato viruses are non-persistent and stylet-borne viruses.

In non-persistent mode, virus particles are quickly acquired, and there is no latency period before virus inoculation to another plant. These kinds of viruses are stylet-borne, and aphids are viruliferous (i.e., capable of transmitting persistent viruses to plants) for only a few minutes or hours after acquisition (see Chapter 11).

When aphids acquire a persistent virus, there is a period of latency between acquisition and becoming viruliferous. Then, the aphids remain infectious for the rest of their lives and can transmit the virus each time they feed on plants. PLRV is a circulative virus: after its acquisition in the phloem of an infected plant, the viral particles migrate in the gut, cross the digestive epithelium to reach the hemocoel of the body cavity, and then reach the salivary glands, from where they can be injected into another plant during bouts of salivation. The latency period varies from 8 to 72 hours. In such cases, there is no multiplication of circulative viruses in the aphid, compared with propagative viruses that multiply in salivary glands before transmission, thus increasing the latency period.

Aphids that Occasionally Feed on the Potato as Potential PVY Vectors

The most cosmopolitan, efficient, and important potato virus vectors are *M. persicae* and *M. euphorbiae*. However, other aphids should not be neglected as potentially significant virus vectors. While several aphid species were demonstrated to transmit PVY under laboratory conditions, most species found in potato fields have not been studied. Advances in PVY detection on stylets of a single aphid (Zhang *et al.*, unpubl. data), and a method to preserve PVY RNA in yellow pan traps (Nie *et al.* 2011), along with the development of Genetic Bar Coding to identify aphids, provided the necessary tools to tackle this topic in the field. Recent studies conducted in New Brunswick (Canada) allowed the identification of *ca.* 70 aphid species and detected PVY-positive on their stylets (Pelletier *et al.*, unpubl. data). These findings support the hypothesis that most aphid species can be efficient vectors of PVY in fields.

Role of Acrostyle in Virus Transmission

Conditions leading to successful virus acquisition and transmission occur during the feeding process and depend on the specific interactions between aphids and potatoes. Several studies have focused on the mouthparts of aphids to elucidate which mechanisms and stylet structures are involved in virus acquisition and transmission for different types of viruses. Uzest *et al.* (2010) identified a distinct anatomical structure located at the extremity of maxillary stylets of several aphid species, including *M. persicae*, *M. Euphorbiae*, and *A. pisum*. This structure, named the acrostyle, is a dense, cuticular surface composed of protein on which a viral protein from the cauliflower mosaic virus can be retained. It may have functions in plant penetration, fluid dynamics, and protein binding. It is also suggested that noncirculative viruses can be attached to the acrostyle during ingestion of contaminated sap, and released with saliva fluxes during salivation in plant tissues (Uzest *et al.* 2010). More research should be done to fully understand these mechanisms.

INTERACTION WITH PLANTS

Aphid fitness depends not only on their ability to obtain nutrients in plants, but also on the nutritional quality of plants (Karley *et al.* 2002), the aphid physiology, and the physiological role of endosymbionts (Powell *et al.* 2006). Host-plant quality is a major factor for the colonization of plants by aphids and for their development.

Plants have developed defenses to limit the damage caused by insects. To escape plant defenses, aphids protect themselves by producing two types of saliva and manipulating the plant metabolism to improve the nutrient composition of the phloem sap. Virus-infected plants have also been described to modify the performance of aphids.

Plant Defenses

Plants have developed a range of defenses to delay colonization and alter insect feeding, growth, development, and fecundity. For example, foliar pubescence constitutes a physical barrier that disturbs plant colonization by aphids. Healthy plants also express base levels of phytochemicals that represent the constitutive plant defenses. An attack of the plant by a pest or a pathogen generates strong responses by inducing the overexpression of these defenses (Kaplan *et al.* 2008). Upon all types of attack, the first plant responses are common and characterized by protein phosphorylation, cell membrane depolarization, calcium influx, and release of reactive oxygen species (e.g. H_2O_2 or OH^-). Then, specific phytohormone-dependent responses are activated depending on the bioaggressors. Activation of these signaling pathways induces the accumulation of defense proteins and secondary metabolites, allowing plant protection locally and constitutively (Walling 2008, Giordanengo *et al.* 2010).

Unlike chewing herbivores, aphids have been reported to induce the salicylate (SA)-dependent plant defense hormone signaling pathway rather than the jasmonate (JA)-dependent pathway (Moran and Thompson 2001, de Vos *et al.* 2005, Mewis *et al.* 2005). It has been shown *in planta* that overexpression of chitinases, which are SA-dependent enzymes, increases *M. persicae* population growth (Saguez *et al.* 2005). On the contrary, reduced populations of aphids were reported during infestations of JA-induced plants (Stout *et al.* 1998) or mutants constitutively expressing JA response (Ellis *et al.* 2002). The most convincing evidence of the efficiency of the defense mediated by the JA pathway was reported on the potato aphid *M. euphorbiae* on plants treated by methyl-jasmonate. Thaler *et al.* (2001) showed a quick decrease in populations while under laboratory conditions. Brunissen *et al.* (2010) reported a reduced attractiveness of the treated plants. Aphid performance was also reduced, due to the alteration of *M. euphorbiae* feeding behavior (Brunissen *et al.* 2010).

Thus, aphids seem to have developed a lure strategy, inhibiting efficient JA-dependent defenses while inducing an SA signaling pathway which appears to be ineffective (Walling 2008; Giordanengo *et al.* 2010).

Aphid Salivary Secretions

Aphids may bypass constitutive and induced plant defenses. To reach phloem sap, aphid stylets move through the apoplasmic compartment. Aphid stylets are protected by a salivary sheath produced by a gelling saliva injected at the beginning of leaf penetration. This sheath constitutes a mechanical and physical barrier composed of proteins, carbohydrates, and phospholipids (Miles 1999) that counteracts plant defenses and protects aphids from recognition by the plant. This gelling saliva also seals the cells punctured during stylet transit towards the sieve tubes (Tjallingii and Hogen Esch 1993, Tjallingii 2006).

Watery saliva, which is different in composition from the gelling saliva described above, is injected in plant tissues during intracellular punctures and while feeding on phloem (Miles 1999, Giordanengo *et al.* 2010). Its chemical composition is highly complex, and depends on the aphid species and their diet. It is mainly composed of enzymes that facilitate stylet penetration in plant tissues by cell alteration, or by repressing plant defenses. However, some small proteins of watery saliva have been shown to elicit a plant defense response (Giordanengo *et al.* 2010).

Modification of Plant Metabolism by Aphids

Phloem sap contains several proteins and metabolites that allow plants to defend themselves against aphid infestations. A few minutes after a puncture, callose accumulates in apoplasmic spaces surrounding phloem vessels and reduces or stops sap fluxes. Interestingly, it has been shown that callose deposition occurs only after the withdrawal of aphid mouthparts from their feeding site (Giordanengo *et al.*, unpubl. data). Once stylets penetrate a phloem cell, and before ingestion, aphids inject watery saliva to counteract plant defense responses. Proteins in salivary secretions suppress plant defenses by inhibiting coagulation of phloem proteins, callose deposit, and sieve-tube occlusion (Will *et al.* 2007, Harmel *et al.* 2008). Aphids also affect plant metabolism to improve phloem sap composition, notably by modifying the nitrogen metabolism (Giordanengo *et al.* 2010). Consequently, aphids are able to increase phloem sap fluxes and modify nutrient allocation in plants, redirecting plant source-sink fluxes to their own advantage (Girousse *et al.* 2003).

Watery saliva also contains enzymes with peroxidase or polyphenoloxidase activities (Cherqui and Tjallingii 2000). These enzymes may convert phenolic compounds produced by plants into derivative products that are less toxic for the aphids (Miles 1969).

In addition, aphids can also interfere with different aspects of plant metabolism involved in cell-wall modelling, photosynthetic activities, plant growth, and the production of secondary metabolites (Giordanengo *et al.* 2010). For instance, *M. persicae* may locally and systematically increase glutamine synthase and glutamate dehydrogenase activities for a remobilization of nitrates and sugars (Divol *et al.* 2005, Giordanengo *et al.* 2010).

Preference for Virus-Infected Potato Plants?

The relationships between the virus, the plant, and the aphid are key issues to understanding vector performance (Hodgson 1981). Plants infected by viruses can positively or negatively affect host-plant selection by aphids, affecting plant colonization behavior, developmental growth, feeding behavior, and aphid performance in general.

Eigenbrode *et al.* (2002) demonstrated that infection with PLRV enhances attractiveness of potato plants to *M. persicae*. In contrast, although colonizing (*M. persicae*) and non-colonizing (*A. fabae, Brevicoryne brassicae,* and *S. avenae*) aphids do not discriminate between healthy and PVY-infected plants, their colonizing behavior is modified. *M. persicae* seems to be a sedentary species, whereas non-colonizing aphids appear to be nomads (Boquel *et al.* 2012). Consequently, their feeding behavior (e.g. probing, phloem ingestion) and their ability to establish colonies on potato plants differ.

The performance of aphids on infested plants generally depends on the aphid species and the virus strain. PLRV-infected plants modify the feeding behavior of *M. persicae* (Castle and Berger 1993, Alvarez *et al.* 2007, Srinivasan and Alvarez 2007), improving its fitness, and PVY-infested potatoes increase the growth rate of *M. persicae* (Srinivasan and Alvarez 2007).

Furthermore, aphids can respond to PVY-infected potato plants, and PVY infection modifies the feeding behavior and the duration of phloem and xylem ingestion (Boquel *et al.* 2011b, 2012). The feeding behavior of *M. persicae* is positively affected. The opposite effects were observed on *M. euphorbiae*, which had a reduced duration of feeding on phloem sap of PVY-infected potatoes. Boquel *et al.* (2011b) also observed that virus transmission rates on healthy plants differ between *M. persicae* (80%) and *M. euphorbiae* (20%).

In contrast, several studies have shown that plant viruses do not impact or negatively impact aphids. As an example, PVX-infested plants have no effects on aphids (Eigenbrode *et al.* 2002, Srinivasan and Alvarez 2007). When reared on PVY-infested plants, *M. euphorbiae* had unchanged population growth parameters compared with potato plants infected by several viruses (Srinivasan and Alvarez 2007). However, *M. euphorbiae* feeding activity was reduced, as evidenced by delayed stylet insertion in potato tissues, an increase of the non-probing phases, reduced salivation bouts, and reduced ingestion of phloem sap (Boquel *et al.* 2011a).

The effects of plants infected with viruses on the biology and physiology of aphids can be the consequences of a modification of the chemical and/or physical properties of potato plants. Several studies have reported that the composition of phloem sap is modified in virus-infected plants, with an increase in carbohydrates and amino acids. Better food quality may favor host-plant acceptance and improve aphid performance (Castle and Berger 1993, Srinivasan and Alvarez 2007). In contrast, negative effects may reflect an increase in plant resistance following the synthesis of plant defense compounds.

CRITICAL ISSUES IN RESEARCH AND MANAGEMENT

Effective management of viruses in potato fields cannot be achieved without taking into consideration the relationship between the viruses and their aphid vectors. Because there are no means of controlling viruses once they penetrate plant tissues, research and management programs focus on the development of methods to reduce aphid populations and limit virus spread. This involves interfering with aphid physiology, reducing the colonization of host plants, and disrupting aphid feeding.

Management of Aphid Populations

Insecticides can be used to manage aphids associated with potatoes and other crops. However, many aphids became resistant to insecticides (Radcliffe 1982, Devonshire *et al.* 1998, Foster *et al.* 2000). For instance, *M. persicae* is resistant to a wide range of compounds. Resistance reflects adaptations of aphids through detoxification of insecticides, or through a modification of their target sites that has repercussions at genotypic, molecular, and biochemical levels (Foster *et al.* 2000). Aphids produce esterases, such as the carboxylesterases E4 and FE4, that degrade insecticide esters. It was also demonstrated that they modify acetylcholinesterases and sodium channels to make them insecticide insensitive (Devonshire *et al.* 1998, Foster *et al.* 2000). These mechanisms have been shown to confer resistance to organophosphorus, carbamates and pyrethroid compounds (Radcliffe 1982, Wheelock *et al.* 2005). For further details, see Chapter 19 . However, developing resistance has fitness costs (Foster *et al.* 2000). For example, insecticide-resistant *M. persicae* has altered behavior and reduced growth rates. Also, resistant aphid populations suffer higher overwintering mortality (Devonshire *et al.* 1998, Foster *et al.* 2000). In addition, most insecticides are toxic to beneficial insects, such as parasitoids and predators (Fernandes *et al.* 2010).

Aphids can be parasitized by parasitoid micro-wasps belonging to Aphelinidae or Aphidiidae (Hymenoptera). These parasitoids lay their eggs in aphids. Parasitoid larvae feed on aphid tissues and eventually kill the aphid. Upon completion of their development, adult parasitoids emerge from the emptied aphid exoskeleton, which is commonly referred to as a mummy. However, the biological control of potato aphids using these natural enemies in the field has limited impact on aphid populations. In contrast, it has been shown that aphid predators, including lady beetles (Coleoptera: Coccinellidae), hover flies (Diptera: Syrphidae), lacewings (Neuroptera: Chrysopidae), and ground beetles (Coleoptera: Carabidae) can sometimes significantly reduce aphid populations (Symondson *et al.* 2002). To favor and maintain the presence of these beneficial insects in or near potato fields, crop rotation, diversified vegetation along field borders, and landscape management (e.g., conservation or establishment of minimally managed habitats) are recommended. Floral or weedy strips may constitute sources of food for natural enemies. However, these hosts may also attract aphids and serve as reservoirs for viruses. In some cases, and depending on the year, strips did not significantly impact the number of aphids in potato fields (Giordanengo, unpubl. data).

To reduce virus spread, it is important to eliminate infected material from potato fields. In particular, volunteer potato plants sprouting from unharvested, potentially infected tubers may constitute reservoirs of viruses. It is also important to carefully clean tools that have been in contact with infected plants, to reduce the risk of contamination by contact.

Mineral Oils for Potato Protection Against Stylet-Borne Virus Transmission

One of the most important challenges for breeders and seed potato growers is to produce certified seed tubers of high quality. To be marketable, foundation tubers used in seed propagation programs may contain only few virus-infected potatoes (1–5%, depending on the country). As previously shown, ineffectiveness of insecticide treatments and cultural methods favors the discovery of alternative methods to prevent aphid build-up and virus spread. For several decades, mineral oil application has been considered the most effective tactic to control and reduce the efficiency of non-persistent virus transmission by aphids in the laboratory and in the field (Bradley *et al.* 1962, 1966, Boiteau and Wood 1982, Powell *et al.* 1998, Hooks and Fereres 2006). In Northern France, weekly treatments with a 1–3% mineral oil emulsion successfully protected potato plants against non-persistent viruses (Ameline *et al.* 2010).

Although mineral oils are effective in protecting potatoes from stylet-borne virus infections, their mode of action is not yet understood. Several studies have shown that mineral oil treatments not only affect the biology and the physiology of the aphids, but also their feeding behavior. As a result, virus transmission is delayed and infections are limited. These effects depend on the mode of aphid exposure (inhalation, contact, or ingestion) and the oil concentrations.

Mineral oils can affect the finding and the chemical perception of the plant by aphids, disrupting the first step of plant colonization. For instance, it has been shown that potato plants treated with mineral oils have a reduced attractiveness to aphids for at least 24 hours (Ameline *et al.* 2009). A repulsive effect on *M. euphorbiae* has been demonstrated 30 minutes after spraying, with plants regaining attractiveness 7 days post-treatment (Ameline *et al.* 2010). Observed changes in the orientation behavior of the aphids can be the direct consequence of repellency exerted by mineral oil, or an indirect effect of a modification of the semiochemicals emitted by the plant (Ameline *et al.* 2009, 2010).

Mineral oils induce an increase in nymphal mortality after topical contact at concentrations >3%. The toxic effects of mineral oil on aphids may be due to death by anoxia or suffocation (Taverner 2002, Najar-Rodríguez *et al.* 2008), or by direct oral intoxication after ingestion (Najar-Rodríguez *et al.* 2007a, 2007b, Martoub *et al.* 2011). Repeated treatments of leaves may induce mineral oil accumulation within plant tissues (Tan *et al.* 2005) and the production of derivative antifeedant compounds (Powell *et al.* 1998), both of which may produce some phytotoxic effects and aphid intoxication. Unexpectedly, mineral oils have probiotic effects on the demographic parameters of aphids. It was shown that a shorter pre-reproductive period, enhanced longevity, and enhanced fecundity led to an increase in the reproductive rate of *M. euphorbiae* (Ameline *et al.* 2010). It was also demonstrated that, at high concentrations (>30% emulsion), mineral oil vapors can increase aphid fecundity (Martoub *et al.* 2011).

The alteration of the feeding behavior of *M. persicae* and *M. euphorbiae* reared on treated potato plants depends on the time after treatment (Ameline *et al.* 2009, 2010). The time to the first probe, the number of probes, and the time spent to reach xylem and phloem elements, salivate, and ingest sap, are good indicators of the preventive effects of oil application on plants. Modification of the stylet activities of *M. persicae* was observed at the surface of the plant, with a delayed first probe revealing that the surface of the plant was not favorable for feeding (Simons *et al.* 1977, Powell *et al.* 1998). Mineral oil present on the surface of the leaves and in the cells of the epidermis can interfere with the binding of the virus in the aphid stylets, leading to a reduced virus transmission (Wang and Pirone 1996). In contrast, oil may dissolve the wax cuticle of the leaves and facilitate stylet penetration in plant tissues (Ameline *et al.* 2009). In several cases, the ingestion of xylem sap increases (Ameline *et al.* 2009, 2010). The modified salivation and ingestion phases observed after treatment suggest that plant physiology and sap composition are modified (Will *et al.* 2007).

Finally, synergistic effects have been observed when mineral oils are used in combination with other methods, including insecticides, such as pyrethroids (Gibson and Rice 1986, Weidemann 1988, Collar *et al.* 1997), or crop borders (Boiteau *et al.* 2009). In contrast, the combination of mineral oils with fungicides may be phytotoxic (Boiteau and Singh 1982).

Transgenic Plants

Genetic transformation of crops with genes conferring resistance to aphids is an attractive option for aphid control. Although different projects have been developed on potato plants expressing different kinds of genes, most of them have failed, and no transgenic plant has been commercialized to manage aphids.

The "Newleaf" potato cultivars that expressed a *Bacillus thuringiensis* δ-endotoxin were introduced in the North American market in 1996 for their resistance against the Colorado potato beetle, *Leptinotarsa decemlineata*, and against potato viruses PVY and PLRV. Ashouri *et al.* (2001) showed that the transgenic lines negatively affected *M. euphorbiae* growth and fecundity, but increased inter-plant flights of this aphid, thus promoting the spread of non-persistent viruses. *Bt* potato also affected parasitoid fitness (reduced immature survival and reduced adult size). Transgenic lines of *Bt* potato were withdrawn from the market in 2001 (Romeis *et al.* 2006).

Protease inhibitors (PIs) have been shown to be efficient against Coleopteran and Lepidopteran pests. Because of the negligible amount of proteolytic activities in the aphid midgut, the effects of PIs on aphids are too weak to consider protease inhibitor-based strategies as a promising management method for aphids. However, Tran *et al.* (1997) demonstrated that the natural potato inhibitors PI-I and PI-II can cause mortality and low fecundity in cereal aphids.

When expressed in potato plants, the cysteine protease inhibitor oryzacystatin I (OC-I) had shown deleterious effects on the Colorado potato beetle (Lecardonnel *et al.* 1999, Cloutier *et al.* 2000), but improved the performance of *M. euphorbiae* (Ashouri *et al.* 2001). In contrast, OC-I significantly reduced nymphal survival and prevented this aphid from reproducing on artificial diet (Azzouz *et al.* 2005). When delivered via artificial diets, OC-I induced a moderate growth inhibition of several aphid species, including *A. gossypii* and *M. persicae* (Rahbé *et al.* 2003). It was suggested that these effects reflect the inhibition of extra-digestive proteolytic activities associated with reproduction instead of the inhibition of digestive proteases. Serine PI-based strategies also failed in many cases. For instance, on artificial diet, Soybean Bowman-Birk inhibitor (SbBBI) did not affect *M. euphorbiae* vitality but altered its demographic parameters, such as fecundity, the intrinsic rate of natural increase, and the doubling time of populations (Azzouz *et al.* 2005). Saguez *et al.* (2010) have shown that potato plants transformed with the mustard trypsin inhibitor gene (*mti-2*) had variable effects on *M. persicae* demographic parameters that could not be attributed to a serine protease inhibitor target.

Lectins were also assessed for their toxic effects on aphids, both via artificial diet delivery (Rahbé *et al.* 1995, Down *et al.* 1996, Sauvion *et al.* 1996) and in transgenic plants (Down *et al.* 1996, Gatehouse *et al.* 1996). On an artificial diet, the snowdrop lectin (*Galanthus nivalis* agglutinin, GNA) affected the development (body length and width) and increased the mortality of the aphid *Aulacorthum solani*. On potato plants expressing a recombinant GNA, fecundity was also decreased in *M. persicae* (Gatehouse *et al.* 1996) and *A. solani* (Down *et al.* 1996). However, no potato plants expressing lectins are commercially available for producers.

Chitinases are plant-defensive compounds that are largely involved in insect molting processes. They were considered to be potential enzymes that could affect insect growth. However, the chitinase-based strategy failed on aphids. Gatehouse *et al.* (1996) have shown that a bean chitinase expressed in potato plants had only weak effects on *M. persicae*, and Saguez *et al.* (2005) reported unexpected probiotic effects on *M. persicae* fed on potato plants expressing insect chitinases.

Natural Resistance Against Aphids

Natural resistance to sap-sucking insects is often mediated by single genes with a gene-for-gene action. Since the late 1990s, several studies have been conducted to find plant genes that confer resistance against aphids or viruses, with the objective of introducing resistance to plants of interest by genetic engineering and/or classical breeding programs. Two of these genes are the *Mi* and *Vat* genes, respectively identified in the tomato and melon. The *Mi* gene confers resistance against nematodes and aphids such as *M. euphorbiae*. It reduces phloem ingestion, increases aphid mortality, and reduces fecundity on resistant plants (Rossi *et al.* 1998).

In contrast, the *Vat* gene confers a resistance by inhibiting the transmission of a non-persistent virus through the induction of an early and strong antixenosis response. However, this resistance seems to be highly specific, because transmission is blocked when *A. gossypii* feeds on melon, while the *Vat* gene does not affect virus inoculation by *M. persicae* (Chen *et al.* 1997, Martin *et al.* 2003). Although very little is known about the resistance mechanisms conferred by the *Vat* gene (Boissot *et al.* 2008, Dogimont *et al.* 2008), Martin *et al.* (2003) demonstrated that it does not directly affect virus infectivity. Instead, they suggested that the mechanism of its action may involve preventing the release of viral particles by the temporary blockage of the aphid stylet tip, or an interaction between the *Vat* product and saliva.

The resistance in several accessions of wild *Solanum* species and their effects on *M. euphorbiae* and *M. persicae* life traits were investigated as an integrated pest management strategy in breeding programs. For instance, some wild *Solanum* species whose leaf surfaces are covered with glandular trichomes present a resistance against aphids (Gibson 1971, Bonierbale *et al.* 1994), but this morphologic character is difficult to select for in breeding programs (see Chapter 15).

Several wild *Solanum* species are resistant to aphids (Gibson 1971, Radcliffe and Lauer 1971, Radcliffe *et al.* 1974, 1988), but the resistant traits were lost in *Solanum tuberosum* because of artificial selection in favor of productive traits. However, the reintegration of these resistant characteristics may provide alternative means of controlling aphids in cultivated potatoes. Several accessions of wild *Solanum* species were screened on *M. persicae* and *M. euphorbiae*, and the results revealed differences in life-history parameters and behavior depending on the accession, the plant age, and the aphid species. Several accessions have been demonstrated to be highly resistant to *M. persicae* and *M. euphorbiae*, inducing >90% mortality of these aphids (Le Roux *et al.* 2007). The demographic parameters, including fecundity, population growth, and doubling time of *M. persicae* and *M. euphorbiae* populations, are also affected by wild *Solanum* species (Le Roux *et al.* 2007, Fréchette *et al.* 2010, 2012, Pelletier *et al.* 2010, Pompon *et al.* 2010b, 2010c).

Behavioral studies using olfactometry and electropenetrography (EPG) experiments have been conducted to identify the origin, nature, and plant tissue localization of the resistance conferred by wild potato accessions. The results demonstrated that wild *Solanum* volatiles may repel aphids (Gibson and Pickett 1983) and contribute to resistance. These experiments revealed that the nature of the resistance of wild *Solanum* is a phloem-based antixenosis resistance (Le Roux *et al.* 2010), but antibiosis was also reported in few cases (Le Roux *et al.* 2008).

CONCLUSION

The potato crop harbors at least 15 aphid species that have a remarkable variety of life cycles (Table 3.1) and have developed many strategies to colonize and exploit plants, including the potato. Several papers reviewed here have researched key attributes of aphids. In spite of considerable advancements in the knowledge of physiological mechanisms and behavior, very few tactics can be used to manage viruses vectored by aphids on potato. This situation is also encountered in other aphid/crop systems.

Other piercing-sucking insects that feed on potato leaves, for example leafhoppers and psyllids, vector pathogens (see Chapter 4). However, their feeding behavior and pathogen transmission differ markedly from those of aphids. Consequently, approaches used to manage them differ.

Because of the specific interactions between the aphids and their host plants, different strategies targeting various aphid metabolisms could be developed. Considerable efforts are currently being invested in the identification of elicitors or suppressors of the plant immune response in aphid salivary secretomes (e.g., Will *et al.*, 2007, Harmel *et al.* 2008, Bos *et al.* 2010, Pitino *et al.* 2011). However, few compounds have been functionally investigated so far (Bos *et al.* 2010, Pitino *et al.* 2011), and neither an elicitor nor a suppressor of plant immune response has been identified yet.

The role of secondary symbionts in aphid performance, including adaptation to the host plant, has been studied from ecological and evolutionary standpoints (Oliver *et al.* 2010), but not at the functional level.

At the present time, spraying mineral oils and, to some extent, insecticides is a tactic that can be used to prevent the establishment and spread of viruses in potato fields. The successful use of mineral oil without incurring phytotoxicity to potato plants actually provides a sustainable tactic. However, much more work has to be done to develop non-pesticidal methods, such as biopesticides (see Chapter 16) or cultural control (see Chapter 18). Resistant cultivars would also allow sustainable management of aphids in potato (see Chapter 15).

REFERENCES

Alvarez, A.E., Garzo, E., Verbeek, M., Vosman, B., Dicke, M., Tjallingii, W.F., 2007. Infection of potato plants with potato leafroll virus changes attraction and feeding behaviour of *Myzus persicae*. Entomol. Exp. Appl. 125, 135–144.

Ameline, A., Couty, A., Martoub, M., Giordanengo, P., 2009. Effects of plant mineral oil treatment on the orientation and feeding behaviour of *Macrosiphum euphorbiae* (Homoptera: Aphidae). Acta. Entomol. Sin. 52, 617–623.

Ameline, A., Couty, A., Martoub, M., Sourice, S., Giordanengo, P., 2010. Modification of *Macrosiphum euphorbiae* colonisation behaviour and reproduction on potato plants treated by mineral oil. Entomol. Exp. Appl. 135, 77–84.

Ashford, D.A., Smith, W.A., Douglas, A.E., 2000. Living on a high sugar diet: the fate of sucrose ingested by a phloem-feeding insect, the pea aphid *Acyrthosiphon pisum*. J. Insect Physiol. 46, 335–341.

Ashouri, A., Michaud, D., Cloutier, C., 2001. Unexpected effects of different potato resistance factors to the Colorado potato beetle (Coleoptera: Chrysomelidae) on the potato aphid (Homoptera: Aphididae). Env. Entomol. 30, 524–532.

Auclair, J.L., 1963. Aphid feeding and nutrition. Ann. Rev. Entomol. 8, 439–490.

Azzouz, H., Cherqui, A., Campan, E.D.M., Rahbé, Y., Duport, G., Jouanin, L., Kaiser, L., Giordanengo, P., 2005. Effects of plant protease inhibitors, oryzacystatin I and soybean Bowman-Birk Inhibitor, on the aphid *Macrosiphum euphorbiae* (Homoptera, Aphididae) and its parasitoid *Aphelinus abdominalis* (Hymenoptera, Aphelinidae). J. Insect Physiol. 51, 75–86.

Barlow, N.D., Dixon, A.F.G., 1980. Simulation of lime aphid population dynamics. Pudoc, Wageningen, The Netherlands.

Baumann, P., Baumann, L., Lai, C., Rouhbakhsh, D., Moran, N., Clark, M., 1995. Genetics, physiology, and evolutionary relationships of the genus *Buchnera*: Intracellular symbionts of aphids. Ann. Rev. Microbiol. 49, 55–94.

Bermingham, J., Wilkinson, T.L., 2010. The role of intracellular symbiotic bacteria in the amino acid nutrition of embryos from the black bean aphid, *Aphis fabae*. Entomol. Exp. Appl. 134, 272–279.

Blackman, R.L., 1971. Variation in the photoperiodic response within natural populations of *Myzus persicae* (Sulz.). Bull. Entomol. Res. 60, 533–546.

Blackman, R.L., Eastop, V.F., 1984. Aphids on the world's crops: An identification guide. Wiley, Chichester, UK.

Blackman, R.L., Eastop, V.F., 2000. Aphids on the world's crops: An identification and information guide, Second edition. Wiley, Chichester, UK.

Blackman, R.L., Eastop, V.F., 2007. Taxonomic Issues. In: van Emden, H.F., Harrington, R. (Eds.), Aphids as Crop Pests, CABI, Wallingford, UK., pp. 1–29.

Boissot, N., Mistral, P., Chareyron, V., Dogimont, C., 2008. A new view on aphid resistance in melon: The role of A. gossypii variability. In: Pitrat, M. (Ed.), Cucurbitaceae 2008, Proceedings of the IXth EUCARPIA meeting on genetics and breeding of Cucurbitaceae, 2008, INRA, Avignon, pp. 163–171.

Boiteau, G., Wood, F., 1982. Persistence of mineral oil spray deposits on potato leaves. Am. J. Potato Res. 59, 55–63.

Boiteau, G., Singh, R., 1982. Evaluation of mineral oil sprays for reduction of virus Y spread in potatoes. Am. J. Potato Res. 59, 253–262.

Boiteau, G., Singh, M., Lavoie, J., 2009. Crop border and mineral oil sprays used in combination as physical control methods of the aphid-transmitted potato virus Y in potato. Pest Manag. Sci. 65, 255–259.

Bonierbale, M.W., Plaisted, R.L., Pineda, O., Tanksley, S.D., 1994. QTL analysis of trichome-mediated insect resistance in potato. Theor. Appl. Genet. 87, 973–987.

Boquel, S., Giordanengo, P., Ameline, A., 2011a. Probing behavior of apterous and alate morphs of two potato-colonizing aphids. J. Insect Sci. 11: Article 164. www.insectscience.org/11.164/i1536-2442-11-164.pdf .

Boquel, S., Giordanengo, P., Ameline, A., 2011b. Divergent effects of PVY-infected potato plant on aphids. Eur. J. Plant Pathol. 129, 507–510.

Boquel, S., Delayen, C., Couty, A., Giordanengo, P., Ameline, A., 2012. Modulation of aphid vector activity by potato virus Y on *in vitro* potato plants. Plant Dis. 96, 82–86.

Bos, J.I.B., Prince, D., Pitino, M., Maffei, E.M., Win, J., Hogenhout, S.A., 2010. A functional genomics approach identifies candidate effectors from the aphid species *Myzus persicae* (Green peach aphid). PLoS Genetics 6, e1001216.

Bradley, R.H.E., Wade, C.V., Wood, F.A., 1962. Aphid transmission of potato virus Y inhibited by oils. Virology 18, 327–329.

Bradley, R.H.E., Moore, C.A., Pond, D.D., 1966. Spread of potato virus Y curtailed by oil. Nature 209, 1370–1371.

Brunissen, L., Vincent, C., Le Roux, V., Giordanengo, P., 2010. Effects of systemic potato response to wounding and jasmonate on the aphid *Macrosiphum euphorbiae* (Sternorryncha: Aphididae). J. Appl. Entomol. 134, 562–571.

Brunt, A.A., Loebenstein, G., 2001. The main viruses infecting potato crops. In: Loebenstein, G., Berger, P.H., Brunt, A.A., Lawson, R.H. (Eds.), Virus and virus-like diseases of potatoes and production of seed potatoes, Kluwer Academic Publishers, Dordrecht, The Netherlands, pp. 65–134.

Buchner, P., 1965. Endosymbiosis of animals with plant microorganisms. Interscience, New York, NY.

Castle, S.J., Berger, P.H., 1993. Rates of growth and increase of *Myzus persicae* on virus-infected potatoes according to type of virus–vector relationship. Entomol. Exp. Appl. 69, 51–60.

Chen, J.-Q., Martin, B., Rahbé, Y., Fereres, A., 1997. Early intracellular punctures by two aphid species on near-isogenic melon lines with and without the virus aphid transmission (*Vat*) resistance gene. Eur. J. Plant Pathol. 103, 521–536.

Cherqui, A., Tjallingii, W.F., 2000. Salivary proteins of aphids, a pilot study on identification, separation and immunolocalisation. J. Insect Physiol. 46, 1177–1186.

Cloutier, C., Jean, C., Fournier, M., Yelle, S., Michaud, D., 2000. Adult Colorado potato beetles, *Leptinotarsa decemlineata* compensate for nutritional stress on oryzacystatin I-transgenic potato plants by hypertrophic behavior and over-production of insensitive proteases. Arch. Insect Biochem. Physiol. 44, 69–81.

Collar, J.L., Avilla, C., Duque, M., Fereres, A., 1997. Behavioral response and virus vector ability of *Myzus persicae* (Homoptera: Aphididae) probing on pepper plants treated with aphicides. J. Econ. Entomol. 90, 1628–1634.

Cristofoletti, P.T., Ribeiro, A.F., Deraison, C., Rahbé, Y., Terra, W.R., 2003. Midgut adaptation and digestive enzyme distribution in a phloem feeding insect, the pea aphid *Acyrthosiphon pisum*. J. Insect Physiol. 49, 11–24.

Dadd, R.H., 1985. Nutrition: organisms. In: Kerkut, G.A., Gilbert, L.I. (Eds.), Comprehensive insect physiology, biochemistry and pharmacology, Vol. 4. Pergamon Press, Oxford, UK, pp. 313–391.

Dadd, R.H., Krieger, D.L., 1968. Dietary amino acid requirements of the aphid *Myzus persicae*. J. Insect Physiol. 14, 741–764.

Deraison, C., Darboux, I., Duportets, L., Gorojankina, T., Rahbé, Y., Jouanin, L., 2004. Cloning and characterization of a gut-specific cathepsin L from the aphid *Aphis gossypii*. Insect Mol. Biol. 13, 165–177.

Devonshire, A.L., Field, L.M., Foster, S.P., Moores, G.D., Williamson, M.S., Blackman, R.L., 1998. The evolution of insecticide resistance in the peach-potato aphid, *Myzus persicae*. Philos. Trans. R. Soc. Lond. B Biol. Sci. 353, 1677–1684.

de Vos, M., van Oosten, V.R., van Poecke, R.M., van Pelt, J.A., Pozo, M.J., Mueller, M.J., Buchala, A.J., Metraux, J.P., van Loon, J.J.A., Dicke, M., Pieterse, C.M.J., 2005. Signal signature and transcriptome changes of *Arabidopsis* during pathogen and insect attack. Mol. Plant–Microbe Interact 18, 923–937.

Divol, F., Vilaine, F., Thibivilliers, S., Amselem, J., Palauqui, J.C., Kusiak, C., Dinant, S., 2005. Systemic response to aphid infestation by *Myzus persicae* in the phloem of *Apium graveolens*. Plant Mol. Biol. 57, 517–540.

Dixon, A.F.G., 1971. The role of intra-specific mechanisms and predation in regulating the numbers of the lime aphid *Eucallipterus tiliae* L. Oecologia 8, 179–193.

Dixon, A.F.G., Horth, S., Kindlmann, P., 1993. Migration in Insects: cost and strategies. J. Animal Ecol. 62, 182–190.

Dogimont, C., Chovelon, V., Tual, S., Boissot, N., Rittener, V., Giovinazzo, N., Bendahmane, A., 2008. Molecular diversity at the *Vat/Pm-W* resistance locus in melon. In: Pitrat, M. (Ed.), Cucurbitaceae 2008, Proceedings of the IXth EUCARPIA meeting on genetics and breeding of Cucurbitaceae, 2008, INRA., Avignon, France, pp. 219–227.

Douglas, A.E., 1988. Sulphate utilization in an aphid symbiosis. Insect Biochem. 18, 599–605.

Douglas, A.E., 1993. The nutritional quality of phloem sap utilized by natural aphid populations. Ecol. Entomol. 18, 31–38.

Douglas, A.E., 1998a. Host benefit and the evolution of specialization in symbiosis. Heredity 80, 599–603.

Douglas, A.E., 1998b. Nutritional interactions in insect–microbial symbioses. Ann. Rev. Entomol. 43, 17–37.

Douglas, A.E., 2003. Nutritional physiology of aphids. Adv. Insect Physiol. 31, 73–140.

Douglas, A.E., 2006. Phloem sap feeding by animals: problems and solutions. J. Exp. Bot. 57, 747–754.

Douglas, A.E., Prosser, W.A., 1992. Synthesis of the essential amino acid tryptophan in the pea aphid (*Acyrthosiphon pisum*) symbiosis. J. Insect Physiol. 38, 565–568.

Down, R.E., Gatehouse, A.M.R., Hamilton, W.D.O., Gatehouse, J.A., 1996. Snowdrop lectin inhibits development and decreases fecundity of the glasshouse potato aphid (*Aulacorthum solani*) when administered *in vitro* and via transgenic plants both in laboratory and glasshouse trials. J. Insect Physiol. 42, 1035–1045.

Downing, N., 1978. Measurements of the osmotic concentrations of stylet sap, haemolymph and honeydew from an aphid under osmotic stress. J. Exp. Biol. 77, 247–250.

Downing, N., 1980. The regulation of sodium, potassium and chloride in an aphid subjected to ionic stress. J. Exp. Biol. 87, 343–350.

Eigenbrode, S.D., Ding, H., Shiel, P., Berger, P.H., 2002. Volatiles from potato plants infected with potato leafroll virus attract and arrest the virus vector, *Myzus persicae* (Homoptera: Aphididae). Proc. R. Soc. Lond. B Biol. Sci. 269, 455–460.

Ellis, C., Karafyllidis, I., Turner, J.G., 2002. Constitutive activation of jasmonate signaling in an Arabidopsis mutant correlates with enhanced resistance to *Erysiphe cichoracearum, Pseudomonas syringae*, and *Myzus persicae*. Mol. Plant–Microbe Interact 15, 1025–1030.

Febvay, G., Liadouze, I., Guillaud, J., Bonnot, G., 1995. Analysis of energetic amino acid metabolism in *Acyrthosiphon pisum*: a multidimensional approach to amino acid metabolism studies in aphids. Arch. Insect Biochem. Physiol. 29, 45–69.

Febvay, G., Rahbe, Y., Rynkiewicz, M., Guillaud, J., Bonnot, G., 1999. Fate of dietary sucrose and neosynthesis of amino acids in the pea aphid, *Acyrthosiphon pisum*, reared on different diets. J. Exp. Biol. 202, 2639–2652.

Fenton, B., Woodford, J.A.T., Malloch, G., 1998. Analysis of clonal diversity of the peach–potato aphid, *Myzus persicae* (Sulzer), in Scotland, UK and evidence for the existence of a predominant clone. Mol. Ecol. 7, 1475–1487.

Fernandes, F.L., Bacci, L., Fernandes, M.S., 2010. Impact and selectivity of insecticides to predators and parasitoids. EntomoBrasilis 3, 1–10.

Ferrari, J., Scarborough, C.L., Godfray, H.C.J., 2007. Genetic variation in the effect of a facultative symbiont on host-plant use by pea aphids. Oecologia 153, 323–329.

Fisher, D.B., Wright, J.P., Mittler, T.E., 1984. Osmoregulation by the aphid *Myzus persicae*: A physiological role for honeydew oligosaccharides. J. Insect Physiol. 30, 387–393.

Flanders, K.L., Hawkes, J.G., Radcliffe, E.B., Lauer, F.I., 1992. Insect resistance in potatoes: sources, evolutionary relationships, morphological and chemical defenses, and ecogeographical associations. Euphytica 61, 83–111.

Foster, S.P., Denholm, I., Devonshire, A.L., 2000. The ups and downs of insecticide resistance in peach–potato aphids (*Myzus persicae*) in the UK. Crop Prot. 19, 873–879.

Fréchette, B., Bejan, M., Lucas, É., Giordanengo, P., Vincent, C., 2010. Resistance of wild *Solanum* accessions to aphids and other potato pests in Quebec field conditions. J. Insect Sci. 10: Article 161. http://www.insectscience.org/10.161/i1536-2442-10-161.pdf.

Fréchette, B., Vincent, C., Giordanengo, P., Pelletier, Y., Lucas, E., 2012. Do resistant plants provide an ennemy-free space to aphids? Eur. J. Entomol. 109, 135–137.

Fuentes, S., Mayo, M.A., Jolly, C.A., Nakano, M., Querci, M., Salazar, L.F., 1996. A novel luteovirus from sweet potato, sweet potato leaf speckling virus. Ann. Appl. Biol. 128, 491–504.

Gatehouse, A.M.R., Down, R.E., Powell, K.S., Sauvion, N., Rahbé, Y., Newell, C.A., Weather, A.M., Hamilton, W.D.O., Gatehouse, J.A., 1996. Transgenic potato plants with enhanced resistance to the peach-potato aphid *Myzus persicae*. Entomol. Exp. Appl. 79, 295–307.

Gattolin, S., Newbury, H.J., Bale, J.S., Tseng, H.-M., Barrett, D.A., Pritchard, J., 2008. A diurnal component to the variation in sieve tube amino acid content in wheat. Plant Physiol. 147, 912–921.

Gibson, R.W., 1971. Glandular hairs providing resistance to aphids in certain wild potato species. Ann. Appl. Biol. 68, 113–119.

Gibson, R.W., Pickett, J.A., 1983. Wild potato repels aphids by release of aphid alarm pheromone. Nature 302, 608–609.

Gibson, R.W., Rice, A.D., 1986. The combined use of mineral oils and pyrethroids to control plant viruses transmitted non- and semi-persistently by *Myzus persicae*. Ann. Appl. Biol. 109, 465–472.

Gil, R., Latorre, A., Moya, A., 2004. Bacterial endosymbionts of insects: insights from comparative genomics. Env. Microbiol. 6, 1109–1122.

Giordanengo, P., Brunissen, L., Rusterucci, C., Vincent, C., van Bel, A., Dinant, S., Girousse, C., Faucher, M., Bonnemain, J.L., 2010. Compatible plant–aphid interactions: How do aphids manipulate plant responses? C.R. Biologies 333, 516–523.

Girousse, C., Bonnemain, J.L., Delrot, S., Bournoville, R., 1991. Sugar and amino acid composition of phloem sap of *Medicago sativa*: a comparative study of two collecting methods. Plant Physiol. Biochem. 29, 41–48.

Girousse, C., Faucher, M., Kleinpeter, C., Bonnemain, J.-L., 2003. Dissection of the effects of the aphid *Acyrthosiphon pisum* feeding on assimilate partitioning in *Medicago sativa*. New Phytol. 157, 83–92.

Guay, J.F., Boudreault, S., Michaud, D., Cloutier, C., 2009. Impact of environmental stress on aphid clonal resistance to parasitoids: role of *Hamiltonella defensa* bacterial symbiosis in association with a new facultative symbiont of the pea aphid. J. Insect Physiol. 55, 919–926.

Gündüz, E.A., Douglas, A.E., 2009. Symbiotic bacteria enable insect to utilise a nutritionally-inadequate diet. Proc. R. Soc. Lond. B Biol. Sci. 276, 987–991.

Hane, D.C., Hamm, P.B., 1999. Effects of seedborne potato virus Y infection in two potato cultivars expressing mild disease symptoms. Plant Dis. 83, 43–45.

Harmel, N., Létocart, E., Cherqui, A., Giordanengo, P., Mazzucchelli, G., Guillonneau, F., De Pauw, E., Haubruge, E., Francis, F., 2008. Identification of aphid salivary proteins: a proteomic investigation of *Myzus persicae*. Insect Mol. Biol. 17, 165–174.

Hodgson, C.J., 1981. Effects of infection with the cabbage black ringspot strain of turnip mosaic virus on turnip as a host to *Myzus persicae* and *Brevicoryne brassicae*. Ann. Appl. Biol. 98, 1–14.

Hooks, C.R.R., Fereres, A., 2006. Protecting crops from non-persistently aphid-transmitted viruses: A review on the use of barrier plants as a management tool. Virus Res. 120, 1–16.

Houk, E.J., Griffiths, G.W., 1980. Intracellular symbiotes of the Homoptera. Ann. Rev. Entomol. 25, 161–187.

Ishikawa, H., Yamaji, M., 1985. A species-specific protein of an aphid is produced by its endosymbiont. Zool. Sci. 2, 285–287.

Kaplan, I., Halitschke, R., Kessler, A., Sardanelli, S., Denno, R.F., 2008. Constitutive and induced defenses of herbivory in above- and belowground plant tissues. Ecology 89, 392–406.

Karley, A.J., Douglas, A.E., Parker, W.E., 2002. Amino acid composition and nutritional quality of potato leaf phloem sap for aphids. J. Exp. Biol. 205, 3009–3018.

Karley, A.J., Ashford, D.A., Minto, L.M., Pritchard, J., Douglas, A.E., 2005. The significance of gut sucrase activity for osmoregulation in the pea aphid, *Acyrthosiphon pisum*. J. Insect Physiol. 51, 1313–1319.

Kawabe, S., Fukumorita, T., Chino, M., 1980. Collection of rice phloem sap from stylets of homopterous insects severed by YAG laser. Plant Cell Physiol. 21, 1319–1327.

Kehr, J., 2006. Phloem sap proteins: their identities and potential roles in the interaction between plants and phloem-feeding insects. J. Exp. Bot. 57, 767–774.

Kennedy, J.S., Day, M.F., Eastop, V.F., 1962. A conspectus of aphids as vectors of plant viruses. Commonwealth Institute of Entomology, London.

Klingauf, F.A., 1987. Host plant finding and acceptance. In: Minks, A.K., Harrewijn, P. (Eds.), Aphids: their biology, natural enemies and control, Elsevier, Amsterdam, pp. 209–223.

Kono, Y., Takahashi, M., Matsushita, K., Nishina, M., Kameda, Y., 1998. *In vitro* trehalose synthesis in American cockroach fat body measured by NMR spectroscopy using ^{13}C–sugars and a trehalase inhibitor, validoxylamine A. Med. Entomol. Zool. 49, 321–330.

Lai, C.Y., Baumann, L., Baumann, P., 1994. Amplification of trpEG: adaptation of *Buchnera aphidicola* to an endosymbiotic association with aphids. Proc. Natl. Acad. Sci. USA 91, 3819–3823.

Le Roux, V., Campan, E.D.M., Dubois, F., Vincent, C., Giordanengo, P., 2007. Screening for resistance against *Myzus persicae* and *Macrosiphum euphorbiae* among wild *Solanum*. Ann. Appl. Biol. 151, 83–88.

Le Roux, V., Dugravot, S., Campan, E., Dubois, F., Vincent, C., Giordanengo, P., 2008. Wild *Solanum* resistance to aphids: antixenosis or antibiosis? J. Econ. Entomol. 101, 584–591.

Le Roux, V., Dugravot, S., Dubois, F., Vincent, C., Pelletier, Y., Giordanengo, P., 2010. Antixenosis phloem-based resistance to aphids: is it the rule? Ecol. Entomol. 35, 407–416.

Lecardonnel, A., Chauvin, L., Jouanin, L., Beaujean, A., Prévost, G., Sangwan-Norreel, B., 1999. Effects of rice cystatin I expression in transgenic potato on Colorado potato beetle larvae. Plant Sci. 140, 71–79.

Liadouze, I., Febvay, G., Guillaud, J., Bonnot, G., 1996. Metabolic fate of energetic amino acids in the aposymbiotic pea aphid *Acyrthosiphon pisum* (Harris) (Homoptera: Aphididae). Symbiosis 21, 115–127.

Martin, B., Rahbé, Y., Fereres, A., 2003. Blockage of stylet tips as the mechanism of resistance to virus transmission by *Aphis gossypii* in melon lines bearing the *Vat* gene. Ann. Appl. Biol. 142, 245–250.

Martoub, M., Couty, A., Giordanengo, P., Ameline, A., 2011. Opposite effects of different mineral oil treatments on *Macrosiphum euphorbiae* survival and fecundity. J. Pest Sci. 84, 229–233.

Mewis, I., Appel, H.M., Hom, A., Raina, R., Schultz, J.C., 2005. Major signaling pathways modulate *Arabidopsis* glucosinolate accumulation and response to both phloem-feeding and chewing insects. Plant Physiology 138, 1149–1162.

Miao, X.-x., Gan, M., Ding, D.-C., 2003. The role of bacterial symbionts in amino acid composition of black bean aphids. Insect Sci. 10, 167–171.

Miles, P.W., 1969. Interaction of plant phenols and salivary phenolases in the relationship between plants and Hemiptera. Entomol. Exp. Appl. 12, 736–744.

Miles, P.W., 1999. Aphid saliva. Biological Reviews 74, 41–85.

Mittler, T.E., 1971. Dietary amino acid requirements of the aphid *Myzus persicae* affected by antibiotic uptake. J. Nutr. 101, 1023–1028.

Montllor, C.B., Maxmen, A., Purcell, A.H., 2002. Facultative bacterial endosymbionts benefit pea aphids *Acyrthosiphon pisum* under heat stress. Ecol. Entomol. 27, 189–195.

Moran, N.A., 1992. The evolution of aphid life cycles. Ann. Rev. Entomol. 37, 321–348.

Moran, P.J., Thompson, G.A., 2001. Molecular responses to aphid feeding in *Arabidopsis* in relation to plant defense pathways. Plant Physiology 125, 1074–1085.

Moriwaki, N., Matsushita, K., Nishina, M., Kono, Y., 2003. High concentrations of trehalose in aphid hemolymph. Appl. Entomol. Zool. 38, 241–248.

Munson, M.A., Baumann, P., Kinsey, M.G., 1991. *Buchnera* gen. nov. and *Buchnera aphidicola* sp. nov., a taxon consisting of the mycetocyte-associated, primary endosymbionts of aphids. Int. J. Syst. Bacteriol. 41, 566–568.

Najar-Rodríguez, A.J., Walter, G.H., Mensah, R.K., 2007a. The efficacy of a petroleum spray oil against *Aphis gossypii* Glover on cotton. Part 2: Indirect effects of oil deposits. Pest Manag. Sci. 63, 596–607.

Najar-Rodríguez, A.J., Walter, G.H., Mensah, R.K., 2007b. The efficacy of a petroleum spray oil against *Aphis gossypii* Glover on cotton. Part 1: Mortality rates and sources of variation. Pest Manag. Sci. 63, 586–595.

Najar-Rodríguez, A.J., Lavidis, N.A., Mensah, R.K., Choy, P.T., Walter, G.H., 2008. The toxicological effects of petroleum spray oils on insects – Evidence for an alternative mode of action and possible new control options. Food Chem. Toxicol. 46, 3003–3014.

Nault, L.R., 1997. Arthropod transmission of plant viruses: a new synthesis. Ann. Entomol. Soc. Am. 90, 521–541.

Nie, X., Pelletier, Y., Mason, N., Dilworth, A., Giguère, M.A., 2011. Aphids preserved in propylene glycol can be used for reverse transcription-polymerase chain reaction detection of potato virus Y. J. Virol. Methods 175, 224–227.

Ohtaka, C., Ishikawa, H., 1991. Effects of heat-treatment on the symbiotic system of an aphid mycetocyte. Symbiosis 11, 19–30.

Oliver, K.M., Russell, J.A., Moran, N.A., Hunter, M.S., 2003. Facultative bacterial symbionts in aphids confer resistance to parasitic wasps. Proc. Natl. Acad. Sci. USA 100, 1803–1807.

Oliver, K.M., Degnan, P.H., Burke, G.R., Moran, N.A., 2010. Facultative symbionts in aphids and the horizontal transfer of ecologically important traits. Ann. Rev. Entomol. 55, 247–266.

Pelletier, Y., Pompon, J., Dexter, P., Quiring, D., 2010. Biological performance of *Myzus persicae* and *Macrosiphum euphorbiae* (Homoptera: Aphididae) on seven wild *Solanum* species. Ann. Appl. Biol. 156, 329–336.

Pitino, M., Coleman, A.D., Maffei, M.E., Ridout, C.J., Hogenhout, S.A., 2011. Silencing of aphid genes by dsRNA feeding from plants. PLoS ONE 6, e25709.

Pompon, J., Quiring, D., Giordanengo, P., Pelletier, Y., 2010a. Role of xylem consumption on osmoregulation in *Macrosiphum euphorbiae* (Thomas). J. Insect Physiol. 56, 610–615.

Pompon, J., Quiring, D., Giordanengo, P., Pelletier, Y., 2010b. Characterization of *Solanum cho-matophilum* resistance to 2 aphid potato pests, *Macrosiphum euphorbiae* (Thomas) and *Myzus persicae* (Sulzer). Crop Prot. 29, 891–897.

Pompon, J., Quiring, D., Giordanengo, P., Pelletier, Y., 2010c. Role of host plant selection in resistance of wild *Solanum* species to *Macrosiphum euphorbiae* (Thomas) and *Myzus persicae* (Sulzer) (Hemiptera: Aphididae). Entomol. Exp. Appl. 137, 73–85.

Pompon, J., Quiring, D., Goyer, C., Giordanengo, P., Pelletier, Y., 2011. A phloem-sap feeder mixes phloem and xylem sap to regulate osmotic pressure. J. Insect Physiol. 57, 1317–1322.

Powell, G., Hardie, J., Pickett, J.A., 1998. The effects of antifeedant compounds and mineral oil on stylet penetration and transmission of potato virus Y by *Myzus persicae* (Sulz.) (Hom., Aphididae). J. Appl. Entomol. 122, 331–333.

Powell, G., Tosh, C.R., Hardie, J., 2006. Host plant selection by aphids: behavioral, evolutionary, and applied perspectives. Ann. Rev. Entomol. 51, 309–330.

Price, D.R.G., Karley, A.J., Ashford, D.A., Isaacs, H.V., Pownall, M.E., Wilkinson, H.S., Gatehouse, J.A., Douglas, A.E., 2007. Molecular characterisation of a candidate gut sucrase in the pea aphid, *Acyrthosiphon pisum*. Insect Biochem. Mol. Biol. 37, 307–317.

Prosser, W.A., Douglas, A.E., 1991. The aposymbiotic aphid: An analysis of chlortetracycline-treated pea aphid, *Acyrthosiphon pisum*. J. Insect Physiol. 37, 713–719.

Pyati, P., Bandani, A.R., Fitches, E., Gatehouse, J.A., 2011. Protein digestion in cereal aphids (*Sitobion avenae*) as a target for plant defence by endogenous proteinase inhibitors. J. Insect Physiol. 57, 881–891.

Radcliffe, E.B., 1982. Insect pests of potato. Ann. Rev. Entomol. 127, 173–204.

Radcliffe, E.B., Lauer, F.I., 1971. Resistance to green peach aphid and potato aphid in introductions of wild tuber-bearing *Solanum* species. J. Econ. Entomol. 64, 1260–1266.

Radcliffe, E.B., Lauer, F.I., Stucker, R.E., 1974. Stability of green peach aphid resistance in tuber-bearing *Solanum* introductions and its effect on screening procedures. Env. Entomol. 3, 1222–1226.

Radcliffe, E.B., Tingey, W.M., Gibson, R.W., Valencia, L., Raman, K.V., 1988. Stability of green peach aphid (Homoptera: Aphididae) resistance in wild potato species. J. Econ. Entomol. 81, 361–367.

Rahbé, Y., Febvay, G., 1993. Protein toxicity to aphids: an *in vitro* test on *Acyrthosiphon pisum*. Entomol. Exp. Appl. 67, 149–160.

Rahbé, Y., Delobel, B., Calatayud, P.A., Febvay, G., 1990. Phloem sap composition of lupine analyzed by aphid stylectomy: methodolgy, variations in major constituents and detection of minor solutes. In: Peters, D.C., Webster, J.A., Chlouber, C.S. (Eds.), Aphid–plant interactions: from populations to molecules, 1990, Oklahoma State University, Stillwater, p. 307.

Rahbé, Y., Sauvion, N., Febvay, G., Peumans, W.J., Gatehouse, A.M.R., 1995. Toxicity of lectins and processing of ingested proteins in the pea aphid *Acyrthosiphon pisum*. Entomol. Exp. Appl. 76, 143–155.

Rahbé, Y., Deraison, C., Bonadé-Bottino, M., Girard, C., Nardon, C., Jouanin, L., 2003. Effects of the cysteine protease inhibitor oryzacystatin (OC-I) on different aphids and reduced performance of *Myzus persicae* on OC-I expressing transgenic oilseed rape. Plant Sci. 164, 441–450.

Rhodes, J.D., Croghan, P.C., Dixon, A.F.G., 1996. Uptake, excretion and respiration of sucrose and amino acids in the pea aphid *Acyrthosiphon pisum*. J. Exp. Biol. 199, 1269–1276.

Rhodes, J.D., Croghan, P.C., Dixon, A.F.G., 1997. Dietary sucrose and oligosaccharide synthesis in relation to osmoregulation in the pea aphid, *Acyrthoslphon pisum*. Physiol. Entomol. 22, 373–379.

Rispe, C., Kutsukake, M., Doublet, V., Hudaverdian, S., Legeai, F., Simon, J.-C., Tagu, D., Fukatsu, T., 2008. Large gene family expansion and variable selective pressures for cathepsin B in aphids. Mol. Biol. Evol. 25, 5–17.

Romeis, J., Meissle, M., Bigler, F., 2006. Transgenic crops expressing *Bacillus thuringiensis* toxins and biological control. Nat. Biotech. 24, 63–71.

Rossi, M., Goggin, F.L., Milligan, S.B., Kaloshian, I., Ullman, D.E., Williamson, V.M., 1998. The nematode resistance gene *Mi* of tomato confers resistance against the potato aphid. Proc. Natl. Acad. Sci. USA 95, 9750–9754.

Saguez, J., Hainez, R., Cherqui, A., Van Wuytswinkel, O., Jeanpierre, H., Lebon, G., Noiraud, N., Beaujean, A., Jouanin, L., Laberche, J.C., Vincent, C., Giordanengo, P., 2005. Unexpected effects of chitinases on the peach-potato aphid (*Myzus persicae* Sulzer) when delivered via transgenic potato plants (*Solanum tuberosum* Linné) and *in vitro*. Transgenic Res. 14, 57–67.

Saguez, J., Cherqui, A., Lehraiki, S., Beaujean, A., Jouanin, L., Laberche, J.C., Giordanengo, P., 2010. Effects of *mti-2* expressing potato plants *Solanum tuberosum* on the aphid *Myzus persicae* (Sternorrhyncha: Aphididae). Intl. J. Agronomy (Art. ID 653431), p. 7.

Salazar, L.F., 1996. Potato viruses and their control, International Potato Center. Peru, Lima.

Sandström, J., 2000. Nutritional quality of phloem sap in relation to host plant-alternation in the bird cherry-oat aphid. Chemoecology 10, 17–24.

Sandström, J., Moran, N., 1999. How nutritionally imbalanced is phloem sap for aphids? Entomol. Exp. Appl. 91, 203–210.

Sasaki, T., Ishikawa, H., 1995. Production of essential amino acids from glutamate by mycetocyte symbionts of the pea aphid, *Acyrthosiphon pisum*. J. Insect Physiol. 41, 41–46.

Sauvion, N., Rahbé, Y., Peumans, W.J., Van Damme, E.J.M., Gatehouse, J.A., Gatehouse, A.M.R., 1996. Effects of GNA and other mannose binding lectins on development and fecundity of the peach-potato aphid *Myzus persicae*. Entomol. Exp. Appl. 79, 285–293.

Shakesby, A.J., Wallace, I.S., Isaacs, H.V., Pritchard, J., Roberts, D.M., Douglas, A.E., 2009. A water-specific aquaporin involved in aphid osmoregulation. Insect Biochem. Mol. Biol. 39, 1–10.

Shigenobu, S., Watanabe, H., Hattori, M., Sakaki, Y., Ishikawa, H., 2000. Genome sequence of the endocellular bacterial symbiont of aphids *Buchnera* sp. APS. Nature 407, 81–86.

Simon, J.C., Carré, S., Boutin, M., Prunier-Leterme, N., Sabater-Munoz, B., Latorre, A., Bournoville, R., 2003. Host-based divergence in populations of the pea aphid: insights from nuclear markers and the prevalence of facultative symbionts. Proc. R. Soc. Lond. B Biol. Sci. 270, 1703–1712.

Simon, J.C., Boutin, S., Tsuchida, T., Koga, R., Le Gallic, J.-F., Frantz, A., Outreman, Y., Fukatsu, T., 2011. Facultative symbiont infections affect aphid reproduction. PLoS ONE 6, e21831.

Simons, J.N., McLean, D.L., Kinsey, M.G., 1977. Effects of mineral oil on probing behavior and transmission of stylet-borne viruses by *Myzus persicae*. J. Econ. Entomol. 70, 309–315.

Srinivasan, R., Alvarez, J.M., 2007. Effect of mixed viral infections (potato virus Y-potato leafroll virus) on biology and preference of vectors *Myzus persicae* and *Macrosiphum euphorbiae* (Hemiptera: Aphididae). J. Econ. Entomol. 100, 646–655.

Srivastava, P.N., 1987. Nutritional physiology. In: Minks, A.K., Harrewijn, P. (Eds.), Aphids: their biology, natural enemies and control, Elsevier, Amsterdam, The Netherlands, pp. 99–121.

Stout, M.J., Workman, K.V., Bostock, R.M., Duffey, S.S., 1998. Specificity of induced resistance in the tomato, *Lycopersicon esculentum*. Oecologia 113, 74–81.

Symondson, W.O.C., Sunderland, K.D., Greenstone, M.H., 2002. Can generalist predators be effective biocontrol agents? Ann. Rev. Entomol. 47, 561–594.

Tamas, I., Klasson, L., Canbäck, B., Näslund, A.K., Eriksson, A.-S., Wernegren, J.J., Sandström, J.P., Moran, N.A., Andersson, S.G.E., 2002. 50 million years of genomic stasis in endosymbiotic bacteria. Science 296, 2376–2379.

Tan, B.L., Sarafis, V., Beattie, G.A.C., White, R., Darley, E.M., Spooner-Hart, R., 2005. Localization and movement of mineral oil in plants by fluorescence and confocal microscopy. J. Exp. Bot. 56, 2755–2763.

Taverner, P.D., 2002. Drowning or just waving? A perspective on the ways petroleum-based oils kill arthropod pests of plants. In: Beattie, G.A.C., Watson, D.M., Stevens, M.L., Rae, D.J., Spooner-Hart, R.N. (Eds.), Spray oils beyond 2000, University of Western Sydney, Hawkesbury, Australia, pp. 78–87.

Tellam, R.L., 1996. The peritrophic matrix. In: Billingsley, P.F., Lehane, M.J. (Eds.), The Biology of the Insect Midgut, Chapman and Hall, London, UK, pp. 86–114.

Terra, W.R., 1990. Evolution of Digestive Systems of Insects. Ann. Rev. Entomol. 35, 181–200.

Thaler, J.S., Stout, M.J., Karban, R., Duffey, S.S., 2001. Jasmonate-mediated induced plant resistance affects a community of herbivores. Ecol. Entomol. 26, 312–324.

Tjallingii, W.F., 2006. Salivary secretions by aphids interacting with proteins of phloem wound responses. J. Exp. Bot. 57, 739–745.

Tjallingii, W.F., Hogen Esch, T., 1993. Fine structure of aphid stylet routes in plant tissues in correlation with EPG signals. Physiol. Entomol. 18, 317–328.

Tran, P., Cheesbrough, T.M., Keickhefer, R.W., 1997. Plant proteinase inhibitors are potential anticereal aphid compounds. J. Econ. Entomol. 90, 1672–1677.

Tsuchida, T., Koga, R., Fukatsu, T., 2004. Host plant specialization governed by facultative symbiont. Science 303, 1989.

Tsuchida, T., Koga, R., Horikawa, M., Tsunoda, T., Maoka, T., Matsumoto, S., Simon, J.C., Fukatsu, T., 2010. Symbiotic bacterium modifies aphid body color. Science 330, 1102–1104.

Uzest, M., Gargani, D., Dombrovsky, A., Cazevieille, C., Cot, D., Blanc, S., 2010. The "acrostyle": A newly described anatomical structure in aphid stylets. Arthropod Struct. Dev. 39, 221–229.

van den Heuvel, J.F.J.M., de Blank, C.M., Peters, D., van Lent, J.W.M., 1995. Localization of potato leafroll virus in leaves of secondarily-infected potato plants. Eur. J. Plant Pathol. 101, 567–571.

van Ham, R.C.H.J., . Kamerbeek, J., Palacios, C., Rausell, C., Abascal, F., Bastolla, U., Fernández, J.M., Jiménez, L., Postigo, M., Silva, F.J., Tamames, J., Viguera, E., Latorre, A., Valencia, A., Morán, F., Moya, A., 2003. Reductive genome evolution in *Buchnera aphidicola*. Proc. Natl. Acad. Sci. USA 100, 581–586.

Vandermoten, S., Mescher, M.C., Francis, F., Haubruge, E., Verheggen, F., 2012. Aphid alarm pheromone: An overview of current knowledge on biosynthesis and functions. Insect Biochem. Mol. Biol. 42, 155–163.

Walling, L.L., 2008. Avoiding effective defenses: strategies employed by phloem-feeding insects. Plant Physiol. 146, 859–866.

Walters, F.S., Mullin, C.A., 1988. Sucrose-dependent increase in oligosaccharide production and associated glycosidase activities in the potato aphid *Macrosiphum euphorbiae* (Thomas). Arch. Insect Biochem. Physiol. 9, 35–46.

Wang, R.Y., Pirone, T.P., 1996. Oil interferes with the retention of tobacco etch potyvirus in the stylets of *Myzus persicae*. Phytopathology 86, 820–823.

Weidemann, H.L., 1988. Importance and control of potato virus YN (PVYN) in seed potato production. Potato Res. 31, 85–94.

Wheelock, C.E., Shan, G., Ottea, J., 2005. Overview of carboxylesterases and their role in the metabolism of insecticides. J. Pestic. Sci. 30, 75–83.

Whitehead, L.F., Douglas, A.E., 2003. Metabolite comparisons and the identity of nutrients translocated from symbiotic algae to an animal host. J. Exp. Biol. 206, 3149–3157.

Wilkinson, T.L., Douglas, A.E., 1995. Why pea aphids (*Acyrthosiphon pisum*) lacking symbiotic bacteria have elevated levels of the amino acid glutamine. J. Insect Physiol. 41, 921–927.

Wilkinson, T.L., Douglas, A.E., 1996. The impact of aposymbiosis on amino acid metabolism of pea aphids. Entomol. Exp. Appl. 80, 279–282.

Wilkinson, T.L., Ashford, D.A., Pritchard, J., Douglas, A.E., 1997. Honeydew sugars and osmo-regulation in the pea aphid *Acyrthosiphon pisum*. J. Exp. Biol. 200, 2137–2143.

Will, T., Tjallingii, W.F., Thönnessen, A., van Bel, A.J.E., 2007. Molecular sabotage of plant defense by aphid saliva. Proc. Natl. Acad. Sci. USA 104, 10536–10541.

Williams, I.S., Dixon, A.F.G., 2007. Life cycles and polymorphism. In: van Emden, H., Harrington, R. (Eds.), Aphids as crop pests, CAB International, Wallingford, UK, pp. 69–86.

Wyatt, G.R., 1967. The biochemistry of sugars and polysaccharides in insects. Adv. Insect Physiol. 4, 287–360.

Ziegler, H., 1975. Nature of transported substances. In: Zimmermann, M.H., Milburn, J.A. (Eds.), Encyclopedia of plant physiology. New Series, Vol. 1. Transport in plants, Springer-Verlag, Berlin, pp. 59–100.

Zimmermann, M.H., Ziegler, H., 1975. List of sugars and sugar alcohols in sieve-tube exudates. In: Zimmermann, M.H., Milburn, J.A. (Eds.), Encyclopedia of plant physiology. New series, Vol. 1. Transport in plants, Springer-Verlag, Berlin, pp. 480–503.

Chapter 4

Leafhopper and Psyllid Pests of Potato

Joseph E. Munyaneza[1], and Donald C. Henne[2]

[1]*United States Department of Agriculture - Agricultural Research Service, Yakima Agricultural Research Laboratory, Wapato, WA, USA,* [2]*Texas A&M University, Texas AgriLife Research, Subtropical Pest Management Laboratory, Weslaco, TX, USA*

INTRODUCTION

Several species of leafhoppers and psyllids are important pests of potato. These insects are not only capable of causing serious damage to potato by direct feeding, but may also transmit potato pathogens, including phytoplasmas, bacteria, and viruses.

Leafhoppers (Hemiptera: Cicadellidae) are small to medium-sized insects rarely exceeding 12 mm in length and narrow-bodied in shape. They generally have a sharply or bluntly pointed head. The wings are normally fully formed, extending over the length of the abdomen, but occasionally leafhoppers have wings that are shorter than their abdomen. The front wings are slightly thickened. The antennae are thread-like, originating from between or beneath the eyes.

Psyllids (Hemiptera: Psylloidea) are small insects, measuring 2–5 mm. These insects resemble aphids superficially, and tiny cicadas upon close examination. In adults, the antennae are long and consist of 9–11 segments. The hind legs are stout and capable of producing long jumps, hence the common name of these insects: "jumping plant lice". The nymphs have bodies that are dorsoventrally flattened, with a short fringe of filaments along the lateral edge. While psyllid adults are very active, the nymphs are sedentary and generally move only when disturbed. The nymphs also excrete large quantities of honey dew and white fecal pellets during feeding.

LEAFHOPPERS

Several species of leafhoppers are common and serious pests of potato. These insects cause damage to the potato by direct feeding or by acting as vectors of

Insect Pests of Potato. http://dx.doi.org/10.1016/B978-0-12-386895-4.00004-1

potato diseases. Leafhopper-transmitted potato phytoplasmas in particular are of great importance because diseases caused by these plant pathogens are on the rise worldwide.

Diseases Caused By Phytoplasmas In Potatoes

Phytoplasmas, previously called mycoplasma-like organisms (MLO), are unculturable, phloem-limited insect-transmitted plant pathogens. These small prokaryotes are related to bacteria and belong to the class Mollicutes (Seemüller *et al.* 1998). In contrast to bacteria, phytoplasmas do not have a rigid cell wall. Phytoplasmas have been associated with diseases affecting hundreds of plant species, including many economically important food crops, ornamentals, and trees (Seemüller *et al.* 1998). In recent years, emerging phytoplasma diseases of potato have become increasingly important in many potato-producing areas around the world. Epidemics of purple top disease, caused by phytoplasmas, have recently occurred in North America (Leyva-Lopez *et al.* 2002, Khadhair *et al.* 2003; Lee *et al.* 2004, Munyaneza 2005, 2010a; Munyaneza *et al.* 2006a, 2006b, 2008a, 2009a, 2010a, 2010b, Rubio-Covarrubias *et al.* 2006; Secor *et al.* 2006, Olivier *et al.* 2009, Santos-Cervantes *et al.* 2010), Central and South America (Secor and Rivera-Varas 2004, Jones *et al.* 2004), Central and Eastern Europe (Linhartova *et al.* 2006, Paltrinieri and Bertaccini 2007, Bogoutdinov *et al.* 2008, Girsova *et al.* 2008, Fialova *et al.* 2009, Ember *et al.* 2011), India (Khurana *et al.* 1988), and New Zealand (Liefting *et al.* 2009a). Countries severely affected by phytoplasma diseases of potato include the United States, Canada, Mexico, Guatemala, India, Romania and Russia (Leyva-Lopez *et al.* 2002, Munyaneza 2005, 2010a, Rubio-Covarrubias *et al.* 2006, Munyaneza *et al.* 2007a, 2009a, 2010a, Girsova *et al.* 2008, Olivier *et al.* 2009, Santos-Cervantes *et al.* 2010, Ember *et al.* 2011). These emerging potato diseases have caused significant yield losses and a reduction in tuber processing and seed quality (Munyaneza 2005, 2010a, Munyaneza *et al.* 2007a, 2010a, Paltrinieri and Bertaccini 2007, Ember *et al.* 2011).

Based on visual symptoms, the diseases caused by phytoplasmas in potatoes can be classified in two general groups: aster yellows-related phytoplasmas and potato witches'-broom (Salazar and Javasinghe 2001, Slack 2001). The potato disease related to the aster yellows group has several different names, including purple top wilt, haywire, apical leafroll, bunch top, purple dwarf, yellow top, potato hair sprouts, stolbur, potato phyllody, and potato marginal flavescence (e.g., Rich 1983, Banttari *et al.* 1993, Salazar and Javasinghe 2001, Slack 2001). Potato phytoplasmas in the aster yellows group occur worldwide, and include stolbur phytoplasma in Europe (Paltrinieri and Bertaccini 2007, Bogoutdinov *et al.* 2008, Girsova et al. 2008, Fialova *et al.* 2009, Ember *et al.* 2011). The potato witches'-broom disease occurs in Europe, Asia, and North America, and is usually of minor economic importance (Brčák *et al.* 1969, Harrison and Roberts 1969, Maramorosch *et al.* 1970, Hodgson *et al.* 1974, Rich 1983, Khadhair *et al.* 1997, 2003, Slack 2001).

Symptoms in potato plants infected with phytoplasmas in the aster yellows group usually include upward rolling of the apical leaves often associated with reddish or purplish discoloration, secondary bud proliferation, shortened internodes, swollen nodes, aerial tubers, and early senescence. In the case of stolbur, symptoms are often more severe and infected plants may wilt and die soon after they exhibit initial infection symptoms. In addition, stolbur-infected tubers often produce chips with a discoloration defect, rendering them unmarketable (Ember *et al.* 2011). Potato plants affected by witches'-broom disease are dwarfed and have numerous axillary buds at the base of the plant. Instead of the compound leaves typical of healthy potato plants, leaves from infected plants are simple, rounded, and later develop chlorosis. Primary witches'-broom infection may result in an upright (erect) stand growth, rolling of leaflets, and some apical leaves turning purple or red. If the infected plants flower, inflorescences become green (virescence) and adopt the shape of leaves (phyllody). Unlike aster yellows, potato witches'-broom phytoplasmas are tuber-perpetuated (Rich 1983, Slack 2001). First-year infected potato plants usually produce tubers that appear normal but give rise to infected plants with witches'-broom symptoms the subsequent year. Tubers from the second year of infection are small and frequently produce elongated hair sprouts; these miniscule tubers are often borne in chains along the stolons that grow out of the eyes, and usually lack the normal dormancy period (Rich 1983, Slack 2001).

Potato phytoplasma diseases were for a long time diagnosed only on the basis of visual symptoms, presence of insect vectors, and/or with the help of electron microscopy of infected phloem tissues. However, as different microorganisms can produce almost identical symptoms in different potato cultivars and in different plant species, visual symptomatology of phytoplasma infection is no longer considered a very reliable characteristic, and the use of modern molecular techniques such as polymerase chain reaction (PCR) is essential to determine accurately the etiology of phytoplasma diseases. Based on modern classification of phytoplasmas, which uses sequence comparisons within the 16S-23S rRNA region (Davis and Sinclair 1998, Lee *et al.* 1998, 2000), at least eight groups of phytoplasmas have so far been identified on potatoes around the world: aster yellows (16SrI), peanut witches'-broom (16SrII), X-disease (16SrIII), clover proliferation (16SrVI), apple proliferation (16SrX), stolbur (16SrXII), Mexican periwinkle virescence (16SrXIII), and American potato purple top wilt (16SrXVIII) (Lee *et al.* 1998, 2000, 2006a, Leyva-Lopez *et al.* 2002, Paltrinieri and Bertaccini 2007, Santos-Cervantes *et al.* 2010, Ember *et al.* 2011).

The epidemiology of phytoplasmas in potatoes is poorly understood, and the insect vectors, primarily leafhoppers and planthoppers, have been identified for only a relatively few phytoplasmas (Sinha and Chiykowski 1967, McCoy 1979, Purcell 1982, Weintraub and Beanland 2006). Phytoplasmas are transmitted by their insect vectors in a persistent manner. They reproduce within their insect vectors and are found in the alimentary canal, hemolymph, salivary glands, and intracellularly in various body organs (Purcell 1982, McCoy 1983, Weintraub and Beanland 2006). Phloem-feeding insects acquire phytoplasmas passively from infected plants during feeding. The acquisition access period can be as short as a few minutes, but it is generally measured in hours; the longer the acquisition access period, the greater the chance of phytoplasma acquisition. Also, acquisition success may depend on the titer of phytoplasmas in the plants (Purcell 1982, Weintraub and Beanland 2006).

The latent, or incubation, period of phytoplasmas in their insect vectors is temperature dependent, and ranges from a few days to about 3 months (Nagaich *et al.* 1974, Murral *et al.* 1996, Weintraub and Beanland 2006). During the latent period, the phytoplasmas move through and reproduce inside the insect vector. To be successfully transmitted to plants, phytoplasmas must penetrate specific cells of the salivary glands, and high levels of these pathogens must accumulate in the posterior acinar cells of the salivary gland before they can be transmitted (Kirkpatrick 1992). At each point in this process, should the phytoplasmas fail to enter or exit a tissue, the insect can become a dead-end host and would be unable to transmit the phytoplasmas (Wayadande *et al.* 1997). Thus, leafhoppers can be infected with a phytoplasma and yet may be unable to transmit it to healthy plants (Lefol *et al.* 1993, 1994, Vega *et al.* 1993, 1994), perhaps because of the salivary gland barriers (Weintraub and Beanland 2006). Once an efficient leafhopper vector acquires a phytoplasma, it can transmit the pathogen for life.

Management of phytoplasmas in potatoes is primarily accomplished by controlling the vectors. Thus, accurate identification of phytoplasma and insect vectors, coupled with a better understanding of disease epidemiology and vector population dynamics, is essential to effective management of phytoplasma diseases in potatoes (Munyaneza 2010a).

Leafhopper Pests of Potato

To minimize taxonomic confusion, the term "leafhoppers" (normally applied to members of Cicadellidae) will be used herein to also include planthoppers (e.g., Cixiidae and Delphacidae). The focus of this section will be on the well-known and important leafhopper species, including the potato leafhopper (*Empoasca fabae*), aster leafhopper (*Macrosteles fascifrons*), and beet leafhopper (*Circulifer tenellus*). However, other important but less-studied leafhopper species affecting potato will also be discussed.

Potato Leafhopper (Empoasca fabae Harris)

Identification

A brief description of this leafhopper is provided by Capinera (2001). Potato leafhopper adults are pale green, marked with a row of white spots on the anterior margin of the pronotum, and measure an average of 3.5 mm long. Eggs are transparent to pale yellow and measure about 1 mm long. Total egg production is about 200–300 per female. Eggs are inserted into the veins and petioles of leaves and hatch in about 10 days, but hatching occurs over a range of 7–20 days. Nymphal development requires 8–25 days, depending on temperature. The lower temperature threshold for development is estimated to be 8.4°C and the upper threshold to be 29°C. The average development time for the five instars is typically 15 days. Wing pads develop in instars three to five and extend over the first, second, and fourth abdominal segments in instars three, four, and five, respectively.

Geographic Distribution

The potato leafhopper is thought to be native to North America (DeLong 1931, Capinera 2001), and is found throughout the humid, low-altitude regions of the eastern United States, occurring as far west as eastern Colorado (DeLong 1931, 1971). This insect occurs in eastern Canada, including the Prairie provinces, but it is most damaging in southern Ontario. The potato leafhopper is known to overwinter in Gulf Coast states from Louisiana to Florida, and disperses northward annually. It typically arrives with warm fronts in midwestern states during April to mid-May, and in northern states and the Canadian provinces during June (Medler 1957, Pienkowski and Medler 1964). In late summer and fall months, the leafhoppers are carried southward again by cold fronts (Taylor and Reling 1986).

Biology

The potato leafhopper feeds on over 200 wild and cultivated plants, though fewer species are suitable for nymphs than adults, and males have a wider host range than females (Lamp *et al.* 1994). Both vegetable and field crops are attacked by this insect. The most suitable crop hosts for potato leafhopper are alfalfa, bean, cowpea, and potato. Adults normally mate within 48 hours after emergence, and the pre-oviposition period is 3–8 days. The potato is reported as the most preferred oviposition host plant for this leafhopper (Poos and Smith 1931). Depending on the region, two to six generations may be produced, beginning in the spring. Potato leafhopper adult longevity is typically 30–60 days. This insect overwinters as an adult.

Pest Significance

Although the potato leafhopper is not known to transmit plant pathogens (Radcliffe 1982, Capinera 2001), this insect is nevertheless considered to be one of the most destructive potato pests in North America, particularly in the northeastern and midwestern United States (DeLong 1931, Radcliffe 1982). Potato leafhoppers feed on phloem or mesophyll tissue (Backus and Hunter 1989) and secrete a toxin into plants. Plant respiration is increased and photosynthesis is decreased by leafhopper feeding, thereby depleting reserves available for growth and potato tuber development, in addition to occlusion of vascular elements (Ladd and Rawlins 1965). Potato leafhopper feeding results in the curling, stunting, and yellowing (chlorosis) of potato foliage. The chlorotic foliar tissue eventually becomes necrotic, beginning initially in the leaf margins. The damage is known as "hopperburn" because the plant appears to have been singed by fire (Ball 1918). The toxin is not systemic, and the level of damage is directly proportional to the number of leafhoppers feeding on the plant and is exacerbated by drought. Both nymphs and adults are toxicogenic; however, late nymphal instars are the most damaging (Radcliffe 1982). Reduction in crop yield is often significant. Although the normal number of tubers may be produced, they are often very small in size.

Management

The potato leafhopper has few natural enemies, and no biological controls effectively reduce populations of this leafhopper. Thus, routine insecticide applications provide the only effective means of controlling this insect (Cancelado and Radcliffe 1979, Radcliffe 1982). The standard practice in the United States is to spray insecticides targeted against this pest on a regular schedule (Radcliffe 1982). However, timing of insecticide applications is crucial to minimize their impact on beneficial insects and avoid outbreaks of aphids, psyllids, and mites. In Minnesota, it was shown that as few as two well-timed foliar insecticide applications aimed at peak nymphal populations can be adequate to prevent injury by potato leafhoppers (Radcliffe 1982). Systemic insecticides applied at planting can also provide excellent control through the early season. In the Midwest, Cancelado and Radcliffe (1979) recommended an action threshold of one leafhopper per potato leaf.

The wide host range and highly dispersive nature of the potato leafhopper limits the use of crop rotation and many other cultural practices to manage this insect. However, because leafhoppers are affected by hairiness of foliage and petioles, plant resistance offers potential for potato leafhopper management. Glycoalkaloids also have been implicated in resistance to leafhoppers, and glandular trichomes (hooked leaf hairs) associated with wild *Solanum* impede leafhopper mobility and feeding, resulting in death of the insect (Radcliffe 1982). Nymphs are especially vulnerable to mortality from glandular exudates (Tingey and Gibson 1978, Tingey and Laubengayer 1981). Some potato varieties display considerable tolerance to potato leafhopper, but none is immune to damage caused by this insect pest (DeLong 1971).

Aster Leafhopper (Macrosteles fascifrons Stal)

Identification

There are several species of *Macrosteles* worldwide, but the aster leafhopper, *M. fascifrons* (also known as *M. quadrilineatus*), is a well-known serious pest of potatoes and several other cultivated crops in the United States and Canada. The aster leafhopper is a morphologically inseparable species complex of considerable biological, ecological, and physiological variability (DeLong 1971).

Adult aster leafhoppers are small, measuring about 3.2–3.8 mm long, and are light green, with the front wings tending toward grayish-green and the abdomen yellowish-green. There are six pairs of black spots, including some that are elongated almost into bands, starting at the vertex of the head and extending along the frons of the head almost to the base of the mouthparts. The six pairs of spots on the head are the basis for the other common name of this insect, "six-spotted leafhopper". The eggs are deposited singly in leaf, petiole, or stem tissue. These eggs measure an average of 0.80 mm long and 0.23 mm wide and are translucent when first laid, but soon turn white. The egg incubation period is about 7–8 days. Newly hatched aster leafhopper nymphs are nearly white (teneral), but soon become yellow and gain brownish markings, including dark markings on the head. There are five instars that develop into adults in about 19–26 days. As the nymphs mature, they gain spines on the hind tibiae and the tip of the abdomen. The wing pads become apparent in the fourth instar and overlap the abdominal segments in the fifth instar. Total generation time from egg to adult stage requires about 27–34 days (Capinera 2001).

Geographic Distribution

The aster leafhopper is native to North America, where it is found in almost every one of the United States and the Canadian provinces. However, it is most common in the central states and provinces. Most areas with aster leafhopper problems are invaded annually by leafhoppers originating in the southern Great Plains. In eastern North America, wind-borne migrants are carried northward from overwintering sites in Arkansas and adjoining states (Chiykowski and Chapman 1965, Drake and Chapman 1965, Hoy *et al.* 1992). In the mild-climate states of the Pacific Northwest such as Washington, however, aster leafhoppers are able to overwinter successfully in the egg stage and long-distance dispersal is not an important factor (Hagel *et al.* 1973). *M. fascifrons* was found to be the most abundant leafhopper in the main potato-producing areas of Alaska, accounting for about 34% of all collected leafhoppers during a survey conducted from 2004 to 2006 (Pantoja *et al.* 2009).

Biology

The aster leafhopper is polyphagous, but not all plants that are suitable for adult maintenance are suitable for adult reproduction and nymphal development. Aster leafhoppers have been reported as having three to five generations per year (Westdal *et al.* 1961, Capinera 2001). Adults tend to overwinter on grains such as wheat and barley, as well as grasses, clover, and several weeds, and will later disperse to vegetables in the summer months. The vegetable crops damaged by aster leafhopper include carrot, celery, lettuce, parsley, potato, and radish. Among these vegetables, only lettuce is consistently suitable for leafhopper reproduction. Other crops that are fed upon by the aster leafhopper are barley, clover, dill, field corn, flax, oat, rice, rye, sugar beet, and wheat. Low, sparse, and young vegetation provide the ideal habitat for aster leafhopper. Adults over-winter poorly in cold areas; the aster leafhopper generally overwinters in the egg stage in cold northern locations and in the adult stage in warmer climates (Wallis 1962, Hagel and Landis 1967, Hagel et al. 1973). Strong winds moving north in the spring transport leafhopper adults into midwestern and northern crop production areas annually (Chiykowski and Chapman 1965, Drake and Chapman 1965, Hoy et al. 1992). Adults usually arrive in advance of the hatching of overwintering eggs in northern regions, and populations of long-distance dispersants greatly exceed resident leafhoppers. The arrival time in the north varies, but usually occurs sometime in May (Wallis 1962).

Pest Significance

Although seldom abundant on potatoes, a host on which it cannot reproduce, the aster leafhopper can nevertheless seriously damage this crop. Similar to several other leafhopper species, aster leafhoppers pierce the leaf tissue of plants and remove the sap. The feeding punctures cause death and discoloration of individual plant cells, resulting in a yellow, speckled appearance in affected plants. However, the most economically important damage is due to the transmission of phytoplasmas to numerous host plants that causes "aster yellows disease" in cultivated crops such as carrot, celery, cucumber, lettuce, potato, pumpkin, and squash. Losses of 50–100% in some of these crops have been reported due to this disease. Phytoplasma-infected plants are discolored, stunted, and deformed.

In potato, the aster yellows disease is known as "purple top", the symptoms of which normally do not appear until flowering. The bases of young leaflets turn purplish, reddish, or yellowish. The petiole stands erect, internodes shorten, and the whole plant grows vertically straight and exhibits stunting. Chlorosis may be generalized to the entire plant and the leaves usually turn upward. Proliferation of axillary buds is common on infected potato plants. Aerial tubers may be formed in the axillary buds due to phloem tissue damage, preventing carbohydrates from moving to the developing underground tubers. The vascular bundles lose color, in some cases very severely, and the root neck rots, causing the plant to wilt. The stolons show browning, which can spread to the attached tubers. Phytoplasmas are infrequently passed to daughter tubers, although some aster yellows strains prevent infected tubers from sprouting or else produce elongated hair sprouts that, in turn, produce weak plants that do not survive (Conners 1967, Salazar and Javasinghe 2001, Leyva-López et al. 2002, Jones et al. 2004) and which produce few marketable tubers.

Severely infected potato fields may have substantially reduced yields. Tubers from initially infected potato plants at harvest usually appear normal and, depending on the time of infection, only some of the tubers on individual plants may be infected (Banttari et al. 1993). Infections can alter the sugar balance in stored tubers, leading to undesirable color development upon processing that results from high concentrations of sucrose and reducing sugars, namely glucose and fructose (Banttari *et al.* 1990, 1993, Munyaneza 2006, Munyaneza *et al.* 2006b). Tubers from infected plants may also develop stem-end necrosis or "sugar ends" (Rich 1983).

Aster leafhoppers acquire the phytoplasma via horizontal transmission by feeding on infected perennial and biennial weeds and/or crop plants other than potato. Pathogen acquisition requires a prolonged period of feeding, usually at least 2 hours, before the leafhopper is infected. Normally less than 2% of migrating leafhoppers become infected. There is evidence that the phytoplasma multiplies in the body of the leafhopper, and there is an incubation period of about 2 weeks in nymphs and 6–10 days in adults before the insects are capable of transmitting the aster yellows phytoplasma. Leafhoppers remain infective for the duration of their life, but the phytoplasma is not vertically transmitted between generations through the egg stage (Capinera 2001).

Management

Aster leafhoppers are easily sampled with sweep nets, especially from grasses and grain fields. Yellow sticky traps are also effective and easy to use (O'Rourke *et al.* 1998). Light traps equipped with fans for suction also have been used effectively to collect aster leafhoppers (Hagel *et al.* 1973). Cool and wet weather and wind limit leafhopper activity and decrease the ability to effectively sample these insects (Durant 1973). In addition to sampling for leafhopper abundance, it is also desirable to determine the proportion of leafhoppers that harbor the phytoplasma. Formulas based on both insect number and disease incidence, referred to as "the aster yellows index", have been developed in the Midwestern states to allow the initiation of control measures before the pathogen is widely transmitted to susceptible crops. A phytoplasma infection rate of 2% in leafhoppers was recommended as a standard to trigger insecticide treatments for the aster leafhopper in the Midwest (Mahr *et al.* 1993, Foster and Flood 1995). In the past, to estimate phytoplasma infection rates, leafhoppers were collected before they entered an area and were fed on aster plants. The plants were then scored for the disease. The use of contemporary molecular techniques, such as PCR, has offered a means to easily and quickly detect and identify phytoplasmas in both the plant hosts and insect vectors. This is unlike earlier approaches, which mainly relied on actual transmission and were lengthy and laborious (Crosslin *et al.* 2006, Munyaneza *et al.* 2010b). The use of the aster-yellow index works effectively to alert large areas, such as entire states, but is not useful for local prediction.

Insecticides are commonly used to control aster leafhoppers, thereby minimizing disease transmission. As there are protracted acquisition and incubation periods associated with this disease, chemical-based disease suppression is feasible (Eckenrode 1973, Koinzan and Pruess 1975). Insecticides are especially effective in the absence of long-distance dispersal by leafhoppers. Systemic insecticides are often favored due to their persistence, but contact insecticides can also be effective (Thompson 1967, Henne 1970). Insecticides are often applied at 5- to 7-day intervals. As it takes 10–15 days for infected plants to show symptoms of infection, it is not necessary to treat plants just before harvest.

Other approaches to aster leafhopper management include biological and cultural control methods. Natural enemies of the aster leafhopper are not well known, nor do they seem to be very important in the population ecology of this insect. However, up to 37% parasitism of the aster leafhopper by the parasitoid *Pachygonatopus minimus* (Hymenoptera: Dryinidae) was reported by Barrett *et al.* (1965) in Canada. Crop varieties differ in their susceptibilities to infection with aster yellows disease. Therefore, cultural manipulation can also enhance resistance. For example, straw mulch and row covers have been shown to provide good protection against aster leafhoppers and reduced disease transmission in a number of vegetable crops (Lee and Robinson 1958, Setiawan and Ragsdale 1987). Destruction of weed species known to harbor aster yellows phytoplasmas is also effective.

Beet Leafhopper (Circulifer tenellus Baker)

Identification

The beet leafhopper adult is a small and variably-colored insect that measures about 3.4–3.7 mm long. Beet leafhopper adults are usually a uniform whitish or greenish color during the summer months but acquire some dark spots dorsally during the fall, particularly on the forewings. During the winter, the adults become mostly dark. Eggs are whitish to yellowish in color, elongate, and slightly curved, with the posterior end tapering almost to a point. Each egg measures 0.06–0.07 mm long and 0.02 mm wide. The eggs are deposited individually within a slit in the tissue of the leaves and stems. The petiole and leaf midrib are the preferred oviposition sites for the beet leafhopper, but leaf margins are sometimes selected. Oviposition commences 5–10 days after mating, around the time when winter host plants begin their spring growth. Each female may deposit 300–400 eggs when conditions are favorable. Eggs hatch in about 5–7 days under optimal temperature conditions (30°C), but egg incubation can last as long as 26 days under temperatures of 18°C or lower (Harries and Douglass 1948). Nymphs are transparent-white upon egg eclosion but acquire a greenish color within a few hours. There are five nymphal instars, and the later instars are typically spotted with black, red, and brown on the thorax and abdomen. Head width ranges from 0.33 to 0.84 mm, depending on the nymphal stage. Nymphal body length is between 1.13 and 3.2 mm. Development time from egg to adult normally ranges from 2 to 6 weeks, depending on temperature (Capinera 2001).

Geographic Distribution

The beet leafhopper is believed to be native to the Mediterranean region and was apparently introduced to the western hemisphere by Spanish explorers, eventually spreading throughout Central and South America, the Caribbean, Hawaii, and Australia. In North America, the beet leafhopper is a common and damaging insect pest throughout the western United States, from southwest Texas to Washington (Cook 1967, Capinera 2001). It also occurs in low numbers in the eastern United States, where it is not considered a pest.

Biology

The beet leafhopper has been widely studied in California, Washington, Oregon, Idaho, New Mexico, Utah, and Arizona, particularly because of its unique association with the curly top virus disease (Hills 1937, Cook 1941, 1967, Lawson *et al.* 1951, DeLong 1971, Thomas and Martin 1971, Capinera 2001). It is well adapted for life in the desert, where it feeds on many plant species, develops rapidly, and disperses readily to find new food sources. Normally, this leafhopper breeds on desert weeds in early spring, migrating into culti-vated areas in late spring or early summer. On summer hosts it produces one or more broods, the last of which usually migrates back to the desert in the fall. In the western United States, major breeding areas are the San Joaquin Valley of California; the lower Colorado River area of southern California, southwestern Arizona, southern Nevada, and southern Utah; the Rio Grande River area of New Mexico and Texas; the lower Snake River plains of Idaho and Oregon; the Columbia River area of Oregon and Washington; and some small, scattered areas in western Colorado, northern Utah, and northern Nevada (Hills 1937, Cook 1941, Douglass and Hallock 1957, Cook 1967, Capinera 2001). These areas differ in climate, host-plant complex, and leafhopper seasonal history. Permanent breeding grounds for the beet leafhopper are areas that have low annual precipitation (<24 cm), low relative humidity, and desert-type vegetation (Hills 1937, Cook 1967). This leafhopper often has complex host-plant require-ments that vary regionally. The main winter and summer hosts of this leafhopper in the Pacific Northwest of the United States are filaree (*Erodium cicutarium* L.), tumble mustard (*Sisymbrium altissimum* L.), flixweed (*Descurainia sophia* L.), and Russian thistle (*Salsola iberica* Sennen) (Hills 1937, Cook 1941, Douglass and Hallock 1957, Cook 1967).

The beet leafhopper is known to overwinter primarily on winter annuals of the mustard family. As winter annuals mature and die in the spring, the beet leafhopper moves to host plants in other habitats, primarily other mustards (*Brassica* spp.), kochia (*Kochia scoparia* L.), hoary cress (*Cardaria draba* L.), pigweed (*Amaranthus* spp.), halogeton (*Halogeton glomeratus* (Bieb.)), Russian thistle, and several cultivated crops. Beet leafhoppers prefer sparse veg-etation that allows maximum sunlight and heat to penetrate through the plant canopy. A complete generation occurs over a span of 1–2 months, with several generations developing annually. In northern areas, such as Washington, Oregon, Idaho, and Utah, three generations are generally produced, but in warmer areas, such as California and Arizona, five generations are typical (Cook 1967, Capinera 2001). In the Pacific Northwest area, beet leafhoppers overwinter only as fertilized females that are inactive during cold weather and become active again when the weather is favorable; males apparently perish during the winter months (Hills 1937, Cook 1941). The beet leafhopper is highly dispersive, moving north to British Columbia and east to the Great Plains area. Even within the generally infested area west of the Rocky Mountains, there is considerable annual movement from Arizona to Utah and Colorado. However, it can also remain fairly resident, feeding throughout the year on crops and weeds in localized areas, especially areas where irrigation is practiced and where a succession of crops and weeds allows adequate survival, precluding the need for dispersal (Lawson *et al.* 1951, Cook 1967, DeLong 1971).

Pest Significance

For a long time the beet leafhopper has been known as a serious pest in the western United States, principally because it transmits curly top virus to several crops, including beans, beet, cantaloupe, cucumber, pepper, spinach, sugar beet, squash, tomato, watermelon, and several ornamental plants (e. g., Hills 1937, Cook 1941, 1967, Lawson *et al.* 1951, Thomas and Martin 1971, Capinera 2001). Because of this insect pest, production of sugar has been abandoned in several western sugar beet-producing areas, and commercial vegetable production is infrequent in some southwestern areas, owing to high incidence of curly top virus disease. The beet leafhopper also transmits *Spiroplasma citri* that causes ailments known as stubborn disease in citrus and brittle root in horseradish (O'Hayer *et al.* 1984).

Recently, the beet leafhopper has become a major concern in the Pacific Northwest, as it was identified as the primary vector of the potato purple top and other vegetable diseases in this important production region of the United States (Lee *et al.* 2004, Crosslin *et al.* 2005, Munyaneza *et al.* 2006a, Lee *et al.* 2006a, Munyaneza 2010a). Since 2002, serious epidemics of purple top disease of potato have occurred in the Columbia Basin of Washington and Oregon, causing significant yield losses to potato fields and a reduction in tuber quality (Munyaneza 2005, 2010a, Munyaneza and Upton 2005, Munyaneza *et al.* 2006a, 2006b). The disease has also been observed in Idaho (Munyaneza 2005). With symptoms similar to aster yellows, apical leaves of affected potato plants roll upward with yellowish, reddish or purplish discoloration, proliferation of buds, shortened internodes, swollen nodes, aerial tubers, leaf scorching, and early plant decline (Lee *et al.* 2004, Munyaneza 2005, Crosslin *et al.* 2005, Munyaneza *et al.* 2006a, 2006b). In the Pacific Northwest it has been determined that the potato purple top disease is caused by the beet leafhopper-transmitted virescence agent (BLTVA) phytoplasma (Fig. 4.1), also known as "Columbia Basin potato purple top phytoplasma", and is vectored by the beet leafhopper (Lee *et al.* 2004, Crosslin *et al.* 2005, 2006, Munyaneza *et al.* 2006a). contrasts with the phytoplasma associated with potato purple top disease in the north-central United States (Banttari *et al.* 1993) and Mexico (Leyva-López *et al.* 2002) that is related to aster yellows phytoplasmas and

FIGURE 4.1 Potato plants infected with BLTVA phytoplasma and showing purple top disease symptoms. See also Plate 4.1. *(Photo: J. Munyaneza)*

whose major vector is the aster leafhopper (*Macrosteles* spp.). BLTVA phytoplasma has also been associated with the dry bean phyllody and carrot purple leaf disease (Lee *et al.* 2004, 2006b) in Washington; the latter disease is caused by *Spiroplasma citri* (Lee *et al.* 2006b, Mello *et al.* 2009), but had long been considered to be associated with the aster yellows phytoplasma. This phytoplasma has also been reported to infect tomatoes, radish, beets, and several other vegetable crops (Golino *et al.* 1989, Shaw *et al.* 1990, 1993, Schultz and Shaw 1991, Munyaneza *et al.* 2006a).

Little is known about phytoplasma transmission through potato tubers and its impact on potato seed quality. However, a recent study by Crosslin *et al.* (2011) reported that BLTVA phytoplasma was transmitted to potato tubers and daughter tubers at a relatively high rate. The 3-year study evaluated eight different potato varieties commonly grown in the Pacific Northwest of the United States, and the results showed that the frequency of BLTVA tuber transmission ranged from 4% to 96% among eight cultivars. Up to 50% transmission of the phytoplasma from infected tubers to daughter plants was observed among the cultivars, with Russet Burbank showing the lowest rate of tuber transmission (less than 5% over the 3 years).

Management

Sampling of the beet leafhopper is accomplished by sweeping plants with insect nets and deploying yellow sticky traps (Munyaneza *et al.* 2008a). Foliar insecticides currently provide the only effective means of controlling beet leafhoppers on potatoes. However, similar to aster leafhopper control, the timing of insecticide applications targeted against the beet leafhopper is crucial to minimize their impact on beneficial insects, while still preventing aphids, psyllids, and mites from flaring up in potatoes. In the Pacific Northwest, the BLTVA infection rate in leafhoppers in and adjacent to potato fields is often high (5–30%) and varies from year to year. This phytoplasma infection rate in beet leafhoppers is far higher relative to the aster yellows index in the Midwestern states. Because beet leafhoppers are in potato fields throughout the growing season, conditions appear conducive to disease transmission all season long (Munyaneza *et al.* 2010c). In response to this threat, growers in the Pacific Northwest generally treat beet leafhoppers with repeated foliar sprays of broad-spectrum insecticides throughout the growing season. Ultimately, nearly all insecticide application decisions on Pacific Northwest potatoes are driven by the need to control the beet leafhopper.

Insecticide sprays are made largely on a zero-tolerance basis once this insect is detected in a field or region. Systemic insecticides appear ineffective against the beet leafhopper. At planting, insecticide sprays are followed by frequent and continuing foliar applications of broad-spectrum organophosphate, neonicotinoid, carbamate or pyrethroid insecticides. However, Munyaneza *et al.* (2010b) have shown that spraying in May and early June could be enough to provide season-long BLTVA control, because the potato appears to be predominantly susceptible to the phytoplasma during early development. Moreover, Munyaneza *et al.* (2009a) have demonstrated that there were significant differences in susceptibility of different potato cultivars to BLTVA, with Russet Burbank being relatively resistant to the phytoplasma. Although no action thresholds have been established for the beet leafhopper, the plethora of information concerning the susceptibility of different plant growth stages and cultivars of potato (Munyaneza *et al.* 2009a, 2010b) could be applied to significantly reduce the amount of insecticide used to control the beet leafhopper in the Pacific Northwest.

Little information is available on the impact of natural enemies of the beet leafhopper. However, some parasitoids have been reported to parasitize all life stages of the beet leafhopper, including big-headed flies (Diptera: Pipunculidae) that attack leafhopper adults, wasp egg parasitoids (Hymenoptera: Mymaridae and Trichogrammatidae), wasp nymphal parasitoids (Hymenoptera: Dryinidae), and twisted-wing parasites (Strepsiptera: Halictophagidae). In the United States, although egg parasitoids sometimes cause up to 90% parasitism of eggs during the summer, and overwintering populations of beet leafhopper may experience up to 25% parasitism, parasitoids are often not reliable for suppressing beet leafhopper populations, mostly due to the high dispersal ability of this leafhopper. Several parasitoids have been introduced to the western United States from the Mediterranean area, particularly northern Africa; however, they have failed to establish (Clausen 1978, Capinera 2001). Furthermore, where economically feasible, management could be targeted against weed hosts around potato fields that could potentially lead to the reduction of beet leafhoppers and purple top disease in potatoes.

Other Important Leafhoppers

The potato leafhopper *(E. fabae)* was originally thought to be the only empoascan attacking the potato in North America (DeLong 1931, Radcliffe 1982). However, at least three closely related species have been reported on potatoes in the United States, including *E. filamenta* in arid, high-altitude western intermountain regions, and *E. abrupta* and *E. arida* in arid, low-altitude Pacific Coast regions (DeLong 1931, Radcliffe 1982). Unlike *E. fabae*, these three economically important species do not cause hopperburn damage in potato, but rather cause a physiological condition that results in a speckled or white stippled appearance on the leaves (DeLong 1931).

Several leafhopper species are common pests of potato in India. *Amrasca devastans*, *A. biguttula biguttula*, and *Empoasca devastans* are important pests throughout potato-growing regions of India, where they damage the potato crops by direct feeding. In contrast, *Seriana equate*, *Alebroides nigriscutulatus*, and *Orosius albicinctus* are known vectors of potato phytoplasma diseases, including potato marginal flavescence and purple top roll (Nagaich *et al.* 1974, Saxena *et al.* 1974, Khurana *et al.* 1988, Slack 2001). Nagaich and Giri (1973) reported that purple top roll of the potato in India was tuber-transmitted to 6–33% of plants grown from tubers produced on infected plants.

Potato purple top phytoplasma from Japan is reported to be transmitted by *Scleroracus flavopictus* and not *Macrosteles* spp. (Shiumi and Sugiura 1984). Recently, purple top disease associated with "*Candidatus* Phytoplasma australiense" was reported on potatoes in New Zealand (Liefting *et al.* 2009a); however, the insect transmitting this phytoplasma to potato in this region has not yet been determined.

Potato purple top has become a limiting factor for potato production in several areas of Mexico, Guatemala, and Panama, where it has become the second most important disease of potato after late blight (Cadena-Hinojosa 1996, 1999, Leyva-López 2002). In Mexico, this disease is also associated with potato hair sprouts disease, which is observed during germination of potato tubers (Martinez-Soriano *et al.* 1999, Leyva-Lopez 2002); both diseases have caused major yield losses to the potato industry in this country. Phytoplasmas from at least four different groups have been reported in Mexico (Leyva-Lopez 2002, Santos-Cervantes *et al.* 2010). Severe damage to potato production due to phytoplasma infections has also been reported in Peru, Bolivia, and Argentina (Salazar and Javasinghe 2001, Jones *et al.* 2004). Insects vectoring purple top phytoplasmas in Mexico and Central and South America have not yet been identified (Leyva-López *et al.* 2002, Santos-Cervantes *et al.* 2010). In Mexico, purple top (also referred to as "punta morada" in Spanish) damage has historically been confounded with damage caused to potatoes by the psyllid *Bactericera (=Paratrioza) cockerelli*, which has recently been associated with zebra chip potato disease and the bacterium "*Candidatus* Liberibacter" (Rubio-Covarrubias *et al.* 2006, 2011); more details are provided in the section below on the psyllids.

Leafhopper vectors of the potato stolbur phytoplasma in Europe include *Hyalesthes obsoletus*, *H. phytoplasmakosiewiczi*, *Euscelis plebejus*, *Aphrodes bicinctus*, and *Macrosteles quadripunctulatus* (Brčák 1979, Salazar and Javasinghe 2001, Slack 2001). A recent study by Kolber *et al.* (2010) showed that several species of the planthopper genus *Reptalus*, including *R. panzeri* and *R. quinquecostatus*, are also important vectors of stolbur phytoplasma in potatoes in Romania and southern Russia.

Known leafhoppers vectoring the potato witches'-broom phytoplasmas include *Scleroracus flavopictus*, *S. dasidus*, and *S. balli* (Raine 1967, Slack 2001); however, in many parts of the world the vectors of this disease are still unknown. Leafhoppers appear unable to acquire witches'-broom phytoplasmas from potato since it is not their preferred host, and instead acquire the phytoplasmas from other infected host plants, including perennial legumes.

PSYLLIDS

Only a few psyllid species have so far been reported to cause damage to solanaceous crops; these include the potato psyllid (*Bactericera* (=*Paratrioza*) *cockerelli* (Šulc)), *B. nigricornis* (Förster), *Russelliana solanicola* Tuthill, and *Acizzia solanicola* (Kent & Taylor). By far, *B. cockerelli* has received the most study among these psyllids because it is known as a serious and economically important pest of potatoes, tomatoes, and other solanaceous crops in the western United States, southern Canada, Mexico, Central America and New Zealand. The potato psyllid has historically been linked to psyllid yellows disease (Fig. 4.2), but more recently has been implicated as a vector of a liberibacter pathogen that causes zebra chip disease of potato and also affects other solanaceous species. Zebra chip is an emerging and economically important disease that has

FIGURE 4.2   Psyllid yellows disease symptoms. See also Plate 4.2. *(Photo: J. Munyaneza)*

devastated some potato growers within the psyllid's range, often leading to the abandonment of entire fields. Therefore, much of the discussion in this section will focus on *B. cockerelli*. However, a brief description of each of the three other psyllid species that attack the potato and other solanaceous crops is provided.

POTATO/TOMATO PSYLLID (*BACTERICERA COCKERELLI*)

Identification

Bactericera cockerelli adults are quite small, measuring about 2.5–2.75 mm long. In general, the adults resemble tiny cicadas, largely because they hold their wings angled and roof-like over their body (Wallis 1955). *B. cockerelli* adults possess two pairs of clear wings; the front wings bear conspicuous veins and are considerably larger than the hind wings. The antennae are moderately long, extending almost half the length of the body. The overall body color ranges from pale green at emergence to dark green or brown within 2–3 days, and eventually becomes gray or black thereafter. Prominent white or yellow lines are found on the head and thorax, and dorsal whitish bands are located on the first and terminal abdominal segments. These white markings are distinguishing characteristics of the psyllid, particularly the broad, transverse white band on the first abdominal segment and the inverted V-shaped white mark on the last abdominal segment (Pletsch 1947, Wallis 1955). Adults are active, in contrast to the largely sedentary nymphal stages. These insects are exceptional fliers and readily jump when disturbed. The pre-oviposition period is normally about 10 days, with oviposition lasting up to 53 days. Total adult longevity ranges from 20 to 62 days, and females usually live two to three times longer than males, depending on the host plants they are reared on (Pletsch 1947, Abernathy 1991, Abdullah 2008, Yang and Liu 2009). Females lay an average of 300–500 eggs over their lifetime (Knowlton and Janes 1931, Pletsch 1947, Abdullah 2008, Yang and Liu 2009). A sex ratio of 1:1 has been reported (Abernathy 1991, Yang and Liu 2009).

The eggs of *B. cockerelli* are deposited singly, principally on the lower surface of leaves and usually near the leaf edge, but some eggs can be found throughout suitable host plants. Often, females will lay numerous eggs on a single leaf. The eggs are initially light yellow, and become dark yellow or orange with time. The eggs measure about 0.32–0.34 mm long, 0.13–0.15 mm wide, and are mounted on a stalk of about 0.48–0.51 mm. Eggs hatch 3–7 days after oviposition (Pletsch 1947, Wallis 1955, Capinera 2001, Abdullah 2008). Following eclosion, the young nymph crawls down the egg stalk to search for a place to feed. Because nymphs prefer sheltered and shaded locations, they are mostly found on the lower surfaces of leaves and usually remain sedentary during their entire development. Nymphs and adults produce large quantities of whitish excrement particles, which may adhere to foliage and fruit. Nymphs are elliptical when viewed from above, but are very flattened in profile, appearing almost scale-like. Potato psyllid nymphs may be confused with the nymphs of whiteflies, although the former move when disturbed. There are five nymphal instars, with each instar possessing very similar morphological features besides size. Nymphal body widths are variable, ranging from 0.23 to 1.60 mm, depending on instar (Rowe and Knowlton 1935, Pletsch 1947, Wallis 1955). Initially the nymphs are orange, but they become yellowish-green and then green as they mature. The compound eyes are reddish and quite prominent. During the third instar the wing pads, light in color, are evident, and become more pronounced with each subsequent molt. A short fringe of wax filaments is present along the lateral margins of the body. Total nymphal development time depends on temperature and host plant, and has been reported to have a range of 12–24 days (Knowlton and Janes 1931, Abdullah 2008, Yang and Liu 2009).

Geographic Distribution

The potato psyllid is native to North America and occurs mainly in the Rocky Mountain region of the United States and Canada, from Colorado, New Mexico, Arizona, and Nevada, north to Utah, Wyoming, Idaho, Montana, Alberta, and Saskatchewan (Pletsch 1947, Wallis 1955, Cranshaw 1993). This insect pest is common in southern and western Texas, and has also been documented in Oklahoma, Kansas, Nebraska, South Dakota, North Dakota, Minnesota, and as far west as California and British Columbia; interestingly, this insect has not been documented east of the Mississippi River (Richards and Blood 1933, Pletsch 1947, Wallis 1955, Cranshaw 1993, Capinera 2001). The potato psyllid also occurs in Mexico and Central America, including Guatemala and Honduras (Pletsch 1947, Wallis 1955, Rubio-Covarrubias *et al.* 2006, 2011, Crosslin *et al.* 2010, Munyaneza 2010b). Contrary to previous reports, *B. cockerelli* has recently been documented to occur in Washington and Oregon (Munyaneza *et al.* 2009b, Munyaneza 2010b). The psyllid was accidentally introduced into New Zealand, apparently sometime in the early 2000s (Gill 2006, Thomas *et al.* 2011), and is now established on both the North and South Islands, where it is causing extensive damage to the potato, tomato, pepper, and tamarillo (*Solanum betaceum*) (Teulon *et al.* 2009). It is not clear how the insect arrived in New Zealand; however, it has been suggested that it was introduced from the western United States, probably through smuggled primary host-plant material (Thomas *et al.* 2011).

Biology

Bactericera cockerelli is found primarily on plants within the family Solanaceae, but also attacks, and reproduces and develops on, a variety of cultivated and weedy plant species (Essig 1917, Knowlton and Thomas 1934, Pletsch 1947, Jensen 1954, Wallis 1955), including crop plants such as the potato (*Solanum tuberosum*), tomato (*Solanum lycopersicon*), pepper (*Capsicum annuum*), and eggplant (*Solanum melongena*), and non-crop species such as nightshade (*Solanum* spp.), groundcherry (*Physalis* spp.), and matrimony vine (*Lycium* spp.).

Adults have been collected from plants in numerous plant families, including the Pinaceae, Salicaceae, Polygonaceae, Chenopodiaceae, Brassicaceae, Asteraceae, Fabaceae, Malvaceae, Amaranthaceae, Lamiaceae, Poaceae, Menthaceae, and Convolvulaceae, but this is not an indication of the true host range of this psyllid (Pletsch 1947, Wallis 1955, Cranshaw 1993). Besides solanaceous species, *B. cockerelli* has been shown to reproduce and develop on some *Convolvulus* species, including field bindweed and sweet potato (Knowlton and Thomas 1934, List 1939, Wallis 1955, Puketapu and Roskruge 2011, Munyaneza, unpubl. data).

Weather is an important element governing the biology of *B. cockerelli* and its damage potential. The potato psyllid seems to be adapted for warm (but not hot) temperatures. Cool weather during migrations, or at least the absence of elevated temperatures, has been associated with several outbreaks of this insect (Pletsch 1947, Wallis 1955, Capinera 2001, Cranshaw 2001). Optimum psyllid development occurs at approximately 27°C, whereas oviposition, hatching, and survival are reduced at 32°C and cease at 35°C (List 1939, Pletsch 1947, Wallis 1955, Cranshaw 2001, Abdullah 2008). A single generation may be completed in 3–5 weeks, depending on temperature. The number of generations varies considerably among regions, usually ranging from three to seven. However, once psyllids invade an area, prolonged oviposition by adults causes the generations to overlap, making it difficult to distinguish between generations (Pletsch 1947, Wallis 1955). Both adults and nymphs are very cold tolerant, with nymphs surviving exposure to temperatures of −15°C and 50% of adults surviving exposure to −10°C for 24 hours (Henne *et al.* 2010).

In North America, driven primarily by wind and hot temperatures in late spring, *B. cockerelli* annually migrates from its overwintering and breeding areas in southern and western Texas, southern New Mexico, Arizona, California, and northern Mexico. It moves into northerly regions of the United States and southern Canada. This migration occurs especially through the Midwestern states and Canadian provinces along the Rocky Mountains (Romney 1939, Pletsch 1947, Jensen 1954, Wallis 1955). In these regions, damaging outbreaks of potato psyllid in potatoes and tomatoes occurred at regular intervals, beginning in the late1800s and extending into the 1940s (List 1939, Wallis 1946, Pletsch 1947). In more recent years, unprecedented outbreaks have also occurred in regions outside of the midwestern United States, including in southern California, Baja California, Washington State, Oregon, and Central America (Trumble 2008, 2009, Munyaneza *et al.* 2009b, Wen *et al.* 2009; Crosslin *et al.* 2010, Munyaneza 2010b). Outbreaks in Baja California and coastal California led to the discovery that potato psyllid in those regions is genetically distinct from psyllids that overwinter in southern Texas and eastern Mexico, suggesting the existence of different potato psyllid biotypes (Liu *et al.* 2006). These western psyllids differ from the southern Texas populations in several life-history traits (Liu and Trumble 2007), and possibly overwinter in geographic areas that differ from regions used by psyllids of the midwestern United States (Trumble 2008). It is not yet known whether populations in southern California are a source of insects in more northern latitudes (Jensen 1954). Information on *B. cockerelli* migration movements within Mexico and Central America is lacking. In the southwestern United States, potato psyllids reappear in overwintering areas between October and November, presumably dispersing southward from northern locations (Capinera 2001); however, their origin has not been determined. In countries and regions where there are no significant seasonal changes during the winter, temperatures are relatively cool, and suitable host plants are available (e.g., Mexico, Central America), the potato psyllid is able to reproduce and develop all year round.

Pest Significance

The potato psyllid is one of the most destructive potato pests in the western hemisphere. It was recognized early on (Šulc 1909, Compere 1915, 1916, Essig 1917) that *B. cockerelli* had the potential to be an explosive and injurious insect. Crawford (1914) described *B. cockerelli* as occurring throughout the southwestern United States, often in "great numbers", and occasionally reaching pest status on cultivated plants. By 1917 it had already been considered a minor pest in Colorado for a number of years, but it was not yet considered a pest in California at that time, despite being widely distributed there (Compare 1916). By the 1920s and 1930s *B. cockerelli* had become a serious and destructive pest of potatoes in many of the southwestern United States, giving rise to the description of a new disease that became known as "psyllid yellows" (Richards 1928, 1931, 1933, Binkley 1929, Richards and Blood 1933, List and Daniels 1934). Historically, the extensive damage to solanaceous crops observed during the potato psyllid outbreak in the early 1900s is thought to have been due to *B. cockerelli*'s association with the psyllid yellows condition. This condition induces a physiological disorder in plants, presumably caused by a toxin that is transmitted during the insect's feeding activities, especially in nymphs (Eyer and Crawford 1933, Eyer 1937). However, the nature of this toxin has not yet been demonstrated.

Above-ground plant symptoms of psyllid yellows include arrested growth, erectness of new foliage, chlorosis and purpling of new foliage with basal cupping of leaves, upward rolling of leaves throughout the plant, shortened and thickened terminal internodes resulting in rosetting, enlarged nodes, axillary branches, and aerial potato tubers (List 1939, Pletsch 1947, Daniels 1954, Wallis 1955). The below-ground symptoms include setting of excessive numbers of tiny misshaped potato tubers, production of chain tubers, and early breaking of dormancy of tubers (Richards and Blood 1933, Eyer 1937, List 1939, Pletsch 1947, Wallis 1955). In areas where outbreaks of psyllid yellows have occurred, the disorder was often present in 100% of plants in affected fields, with yield losses exceeding 50% in some areas (Pletsch 1947). Many of the reported outbreaks in the early 1900s occurred well north of the insect's overwintering range, such as the states of Montana and Wyoming (Pletsch 1947), which is a testimony to the migratory reach of this psyllid.

In recent years, potato, tomato, and pepper growers in a number of geographic areas have suffered extensive economic losses associated with potato psyllid outbreaks (Trumble 2008, 2009, Munyaneza *et al.* 2007b, 2007c, 2008b, 2009a, 2009c, 2009d, 2009e, Liefting *et al.* 2008, 2009b, Secor *et al.* 2009, Crosslin *et al.* 2010, Munyaneza 2010b, Rehman *et al.* 2010, Crosslin *et al.* 2012a, 2012b). Damage is due to a previously undescribed species of the bacterium liberibacter, tentatively named "*Candidatus* Liberibacter solanacearum" (syn. *Ca.* L. psyllaurous) (Hansen *et al.* 2008, Liefting *et al.* 2008, 2009b), now known to be vectored by the potato psyllid (Munyaneza *et al.* 2007b, 2007c). The pathogen is closely related to the "*Ca.* Liberibacter" species that causes huanglongbing ("citrus greening") in citrus crops (Hansen *et al.* 2008, Liefting *et al.* 2009b, Lin *et al.* 2009). Potato psyllids acquire and spread the pathogen by feeding on infected plants (Munyaneza *et al.* 2007b, 2007c). The bacterium is also transmitted transovarially in the psyllid (Hansen *et al.* 2008), which contributes to the spread of the disease between geographic regions by dispersing psyllids, and also helps maintain the bacterium in geographic regions during the insect's overwintering period (Crosslin *et al.* 2010).

Recent studies have shown that liberibacter within the adult psyllid is sensitive to high and low temperatures, with the highest bacterial titers occurring at approximately 28°C (Henne, unpubl. data). This temperature is very similar to the preferred optimum temperature (27°C) for potato psyllid reproduction and liberibacter development (Munyaneza 2010b, Munyaneza *et al.* 2012a). It has also been determined that liberibacter titer does not differ between males and females, exists at very low titer in fifth-instar nymphs, and increases with age of the adult psyllid (Henne, unpubl. data). Very recently, Munyaneza *et al.* (2010d, 2010e, 2011a, 2012b, 2012c) detected "*Ca.* L. solanacearum" in carrots attacked by the psyllid *Trioza apicalis* in Finland, Sweden, and Norway. This constitutes the first report of liberibacter in Europe and "*Ca.* L. solanacearum" in a non-solanaceous species. Subsequently, the same liberibacter species was discovered in carrot in Spain and the Canary Islands, where it is suspected of having been vectored to carrot by the psyllid *Bactericera trigonica* (Alfaro-Fernandez *et al.* 2012a, 2012b). Damage to carrots by liberibacter-infected carrot psyllids can cause up to 100% crop loss if the psyllid is not controlled (Munyaneza *et al.* 2010d, 2010e, 2012b, 2012c). Mixed infections of "*Ca.* L. solanacearum" and phytoplasmas have been reported in potatoes (Liefting *et al.* 2009a, Munyaneza, unpubl. data) and carrots (Munyaneza *et al.* 2011a).

Foliar symptoms associated with liberibacter in potato closely resemble those caused by psyllid yellows and purple top diseases (Munyaneza *et al.* 2007b, 2007c, Sengoda *et al.* 2010) (Fig. 4.3). However, tubers from liberibacter-infected plants develop a defect referred to as "zebra chip", which is not induced by the putative toxin causing psyllid yellows (Munyaneza *et al.* 2007b, 2007c, 2008b, Sengoda *et al.* 2010). Characteristic symptoms of zebra chip in potato tubers consist of collapsed stolons, browning of vascular tissue concomitant with necrotic flecking of internal tissues, and streaking of the medullary

FIGURE 4.3   Foliar symptoms of the zebra chip disease. See also Plate 4.3. *(Photo: J. Munyaneza)*

FIGURE 4.4   Tubers with zebra chip symptoms. See also Plate 4.4. *(Photo: J. Munyaneza)*

ray tissues, all of which can affect the entire tuber (Fig. 4.4). Upon frying, these symptoms become more pronounced and chips or fries processed from affected tubers show very dark blotches, stripes, or streaks, rendering them commercially unacceptable (Munyaneza *et al.* 2007b, 2007c, 2008b, Secor *et al.* 2009, Crosslin *et al.* 2010, Miles *et al.* 2010, Buchman *et al.* 2011a, 2001b, 2012) (Figs. 4.5, 4.6).

Zebra chip was first documented in 1994 in potatoes growing near Saltillo, Mexico (Secor *et al.* 2009). Initial records of the disease in the United States are from 2000, for potato fields in southern Texas (Secor *et al.* 2009). Infected fields of potatoes have since been documented in several other states, including Nebraska,

FIGURE 4.5   Chips processed from tubers infected with zebra chip. See also Plate 4.5. *(Photo: J. Munyaneza)*

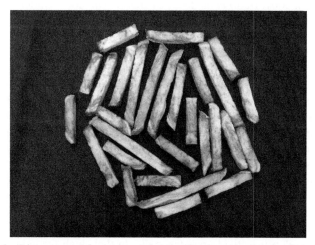

FIGURE 4.6   Fries processed from tubers infected with zebra chip. See also Plate 4.6. *(Photo: J. Munyaneza)*

Colorado, Kansas, Wyoming, New Mexico, Arizona, Nevada, California, Oregon, Washington, and Idaho (Munyaneza *et al.* 2007a, 2007c, Secor *et al.* 2009, Munyaneza 2010b, Crosslin *et al.* 2012a, 2012b). The defect was of sporadic importance until 2004, when it began to cause millions of dollars in losses to potato growers in the United States, Mexico, and Central America (Rubio-Covarrubias *et al.* 2006, Munyaneza *et al.* 2007a, 2007c, 2009c, Crosslin *et al.* 2010, Rehman *et al.* 2010). In some regions, entire fields have been abandoned because of zebra chip (Secor and Rivera-Varas 2004, Munyaneza *et al.* 2007a, 2007c, Crosslin *et al.*

2010). The potato industry in Texas estimates that zebra chip could affect over 35% of the potato acreage there, with potential losses to growers annually exceeding US$25 million (CNAS 2006). Finally, quarantine issues have begun to emerge in potato psyllid-affected regions because some countries now require that shipments of solanaceous crops from certain growing regions be tested for the pathogen before the shipments are allowed entry (Crosslin *et al.* 2010).

Management

Monitoring *B. cockerelli* is crucial to effective management of this insect pest. The adult populations are commonly sampled using sweep nets or vacuum devices, but egg and nymphal sampling requires visual examination of foliage. The adults can also be sampled with yellow sticky traps and yellow water-pan traps, but other colors have shown improved detection (Henne *et al.* 2010). Typically, psyllid populations are highest initially along field edges (Workneh *et al.* 2012), but if not controlled, the insects will eventually spread throughout the crop.

B. cockerelli control is currently dominated by insecticide applications (e.g., Goolsby *et al.* 2007, Berry *et al.* 2009, Gharalari *et al.* 2009, Butler *et al.* 2011), but even with conventional insecticides *B. cockerelli* tends to be difficult to manage. Good coverage is important because all psyllid life stages are commonly found on the undersides of leaves (Nansen *et al.* 2010). Also, the different life stages require the use of specific insecticides, as it has been shown that chemicals that control adults do not necessary control nymphs or eggs. Therefore, caution is necessary when selecting and applying insecticides targeting the potato psyllid by considering which life stages are present in the crop and timing insecticide applications accordingly. Until the mid-2000s, potato growers in Texas relied on pyrethroids and organophosphates to control psyllids, but these insecticides did not provide adequate control, and many growers still incurred severe damage and losses due to zebra chip disease. Many potato growers in Texas now routinely apply a neonicotinoid insecticide at planting, followed by applications of abamectin, spirotetramat ,and pymetrozine, but novaluron, flonicamid, and spiromesifen are also used (Zens *et al.* 2010). Psyllids have been shown to develop insecticide resistance due to their high fecundity and short generation times (McMullen and Jong 1971). Therefore, alternative strategies should be considered to limit the impact of the potato psyllid and its associated diseases.

It has been determined that liberibacter is transmitted to potato very rapidly by the potato psyllid. Research has shown that groups of 20 psyllids per plant successfully transmitted liberibacter to potato, ultimately causing zebra chip after an inoculation access period of 1 hour, whereas a 6 hour-inoculation access period was required for a single psyllid per plant to do the same (Buchman *et al.* 2011a, 2011b, 2012). This observed low psyllid density, coupled with a short inoculation access period, represents a substantial challenge for growers in controlling the potato psyllid and preventing zebra chip transmission. Just a few infective psyllids feeding on potato for a short period could result in substantial spread of the disease within a potato field or region (Henne *et al.* 2012). Most importantly, conventional pesticides may have a limited direct disease control, as they may not quickly kill the potato psyllid to prevent liberibacter and zebra chip transmission, although they may be useful for reducing the overall population of psyllids. The most valuable and effective strategies to manage zebra chip would likely be those that discourage vector feeding, such as the use of plants that are resistant to psyllid feeding or less preferred by the psyllid. Unfortunately, no potato variety has so far been shown to exhibit sufficient resistance or tolerance to zebra chip or potato psyllid (Munyaneza *et al.* 2011b). However, some conventional and biorational pesticides, including plant and mineral oils and kaolin, have shown some substantial deterrence and repellency to potato psyllid feeding and oviposition (Gharalari *et al.* 2009, Yang *et al.* 2010, Butler *et al.* 2011, Peng *et al.* 2011) and could be useful tools in integrated pest management programs to manage zebra chip and its psyllid vector.

Several predators and parasites of *B. cockerelli* are known, and include chrysopid larvae, coccinellids, geocorids, anthocorids, mirids, nabids, syrphid larvae, and the parasitoid *Tamarixia triozae* (Hymenoptera: Eulophidae), but little is known regarding their effectiveness against the psyllid (Pletsch 1947, Wallis 1955, Cranshaw 1993, Al-Jabar 1999). In addition, several entomopathogenic fungi have been determined to be effective natural enemies of *B. cockerelli* (Lacey *et al.* 2009, 2011). Furthermore, in some areas, such as southern Texas, early planted crops are more susceptible to psyllid injury than crops planted mid- or late season (Munyaneza *et al.* 2010a); however, the reasons behind this differential are not well understood. They may possibly reflect differences in liberibacter infection rate in psyllids colonizing potato fields.

Other psyllids

Bactericera nigricornis

This Old World potato psyllid has a distribution that ranges from Europe into Central Asia, south to the Middle East and North Africa. The host range of this psyllid is uncertain because of its complicated taxonomy (Ossiannilsson 1992, Burckhardt and Lauterer 1993), but includes potatoes, carrots, various cole crops, and possibly onions (Hodkinson 1981, Burckhardt and Freuler 2000).

B. nigricornis is an important pest of potato in Iran (Fathi 2011), where it causes serious damage to potato crops. In potato fields infested with the psyllid, yield is decreased and symptoms in potato tubers suggestive of zebra chip disease have been reported (Fathi 2011). However, no study has investigated the presence of the zebra chip pathogen (liberibacter) in either psyllids or affected potato plants. Eggs of *B. nigricornis* are similar to those of *B. cockerelli* and are also oviposited singly, usually on the lower surface of the leaves. Similarly to *B. cockerelli*, nymphs excrete honeydew droplets that crystallize into white granules (Fathi 2011). Insecticides are the main means to control this psyllid in Iran. However, some potato cultivars have been shown to significantly reduce reproduction and development of *B. nigricornis*, and could be useful in integrated pest management programs for this insect (Fathi 2011). Further studies are needed to clarify whether *B. nigricornis* may also be associated with liberibacter and zebra chip.

Russelliana solanicola

Little information on the biology of this South American potato psyllid is available. The psyllid is not known to occur outside of South America and has been reported in Argentina, Chile, Peru, and Brazil (Tuthill 1959, Brown and Hodkinson 1988). *R. solanicola* was first described from *Datura* sp. and has been reported as a pest of potato (Tuthill 1959, Burckhardt 1987, Tenorio *et al.* 2000, Chávez *et al.* 2003, Salazar 2006). The psyllid is also possibly a pest of other solanaceous crops, including tomato and pepper, and several plants in the Compositae have been reported as hosts to the insect (Burckhardt 1987, Chávez *et al.* 2003).

R. solanicola has been confirmed to readily transmit a new virus to potato coded SB26/29 in Peru; it is the first psyllid that has been found to vector a plant virus (Tenorio *et al.* 2000, Salazar 2006). The virus has not yet been purified and characterized, but viral isometric particles have been observed or isolated from infected potato plants (Salazar 2006). Symptoms in infected potato plants initially consist of a mosaic in leaves that later develops into severe foliar deformation. Severe plant dwarfing and arrested growth have been observed in some varieties (Salazar 2006). Depending on potato cultivars, up to 85% yield reduction due to the virus has been reported in Peru (Salazar 2006). To date, no resistance to the virus has been found in potatoes (Salazar 2006). The virus has successfully been inoculated mechanically to a number of solanaceous species, including *Nicotiana occidentalis*, *N. benthamiana*, and *Physalis floridana* (Salazar 2006). However, there is no report on whether the diseased potato plants or the psyllids have been tested for fastidious prokaryotes, including liberibacter. Further research is needed to elucidate the identity of the virus and other potential pathogens that may be associated with *R. solanicola*.

Acizzia solanicola

Little information on the biology of this psyllid is available. It has recently been discovered as a pest of the eggplant (*Solanum melongena*) and is known only from coastal New South Wales and Adelaide in Australia, but its origin is still unrecognized (Kent and Taylor 2010). The insect's host range includes eggplant and wild tobacco bush (*Solanum mauritianum*), but the psyllid does not appear to feed on tomato and pepper. No information concerning this psyllid's ability to feed on the potato is available. Nymphs of *A. solanicola* are covered with long setae and excrete honey dew in bags of various sizes. Feeding by nymphs deforms eggplant leaves and causes wilting and premature senescence of new leaves and flowers. Damage to flowers causes crop loss, which can be severe. There is no information on pathogen transmission by this psyllid. However, the potential for *A. solanicola* to become a serious pest of commercial solanaceous crops, including potato, in Australia and other countries, cannot afford to be overlooked (Kent and Taylor 2010). Therefore, more studies are needed to elucidate and assess the risk that this new psyllid species may pose to potato production.

REFERENCES

Abdullah, N.M.M., 2008. Life history of the potato psyllid *Bactericera cockerelli* (Homoptera: Psyllidae) in controlled environment agriculture in Arizona. Afr. J. Agric. Res. 3, 60–67.

Abernathy, R.L., 1991. Investigation into the nature of the potato psyllid toxin. MS Thesis. Colorado State University, Fort Collins, CO.

Alfaro-Fernández, A., Cebrián, M.C., Villaescusa, F.J., Hermoso de Mendoza, A., Ferrándiz, J.C., Sanjuán, S., Font, M.I., 2012a. First report of "*Candidatus* Liberibacter solanacearum" in carrot in mainland Spain. Plant Dis. 96, 582.

Alfaro-Fernández, A., Siverio, F., Cebrián, M.C., Villaescusa, F.J., Font, M.I., 2012b. "*Candidatus* Liberibacter solanacearum" associated with *Bactericera trigonica*-affected carrots in the Canary Islands. Plant Dis. 96, 581.

Al-Jabar, A.M., 1999. Integrated pest management of tomato/potato psyllid, *Paratrioza cockerelli* (Sulc) (Homoptera: Psyllidae) with emphasis on its importance in greenhouse grown tomatoes. PhD Dissertation. Colorado State University, Fort Collins, CO.

Backus, E.A., Hunter, W.B., 1989. Comparison of feeding behavior of the potato leafhopper, *Empoasca fabae* (Homoptera: Cicadellidae) on alfalfa and broad bean leaves. Environ. Entomol. 18, 473–480.

Ball, E.D., 1918. Leaf burn of the potato and its relation to the potato leafhopper. Science 48, 194.

Banttari, E.E., Orr, P.H., Preston, D.A., 1990. Purple top as a cause of potato chip discoloration. Trans. ASAE 33, 221–226.

Banttari, E.E., Ellis, P.J., Khurana, S.M.P., 1993. Management of diseases caused by viruses and virus-like pathogens. In: Rowe, R.C. (Ed.), Potato Health Management, APS Press, St Paul, MN, pp. 127–133.

Barrett, C.F., Westdal, P.H., Richardson, H.P., 1965. Biology of *Pachygonatopus minimus* Fenton (Hymenoptera: Dryinidae), a parasite of the six-spotted leafhopper, *Macrosteles fascifrons* (Stal), in Manitoba. Can. Entomol. 97, 216–221.

Berry, N.A., Walker, M.K., Butler, R.C., 2009. Laboratory studies to determine the efficacy of selected insecticides on tomato/potato psyllid. N. Z. Plant Protect 62, 145–151.

Binkley, A.M., 1929. Transmission studies with the new psyllid yellows disease of solanaceous plants. Science 70, 615.

Bogoutdinov, D.Z., Valyunas, D., Navalinskene, M., Samuitene, M., 2008. About specific identification of phytoplasmas in Solanaceae crops. Agric. Biol. 1, 77–80.

Brčak, J., 1979. Leafhopper and planthopper vectors of plant disease agents in central and southern Europe. In: Maramorosch, K., Harris, K.F. (Eds.), Leafhopper Vectors and Plant Disease Agents, Academic Press, London, UK, pp. 97–146.

Brčak, J., Králik, O., Limberk, J., Ulrychová, M., 1969. Mycoplasma-like bodies in plants infected with potato witches'-broom disease and the response of plants to tetracycline treatment. Biol. Plantarum 11, 470–476.

Brown, R.G., Hodkinson, I.D., 1988. Taxonomy and ecology of the jumping plant-lice of Panama (Homoptera: Psylloidea). Entomonograph 9. E.J. Brill/Scandinavian Science Press Ltd, New York, NY.

Buchman, J.L., Heilman, B.E., Munyaneza, J.E., 2011a. Effects of *Bactericera cockerelli* (Hemiptera: Triozidae) density on zebra chip potato disease incidence, potato yield, and tuber processing quality. J. Econ. Entomol. 104, 1783–1792.

Buchman, J.L., Sengoda, V.G., Munyaneza, J.E., 2011b. Vector transmission efficiency of liberibacter by *Bactericera cockerelli* (Hemiptera: Triozidae) in zebra chip potato disease: effects of psyllid life stage and inoculation access period. J. Econ. Entomol. 104, 1486–1495.

Buchman, J.L., Fisher, T.W., Sengoda, V.G., Munyaneza, J.E., 2012. Zebra chip progression: from inoculation of potato plants with liberibacter to development of disease symptoms in tubers. Am. J. Pot. Res. 89, 159–168.

Burckhardt, D., 1987. Jumping plant lice (Homoptera: Psylloidea) of the temperate neo-tropical region Part 1: Psyllidae (subfamilies Aphalarinae, Rhinocolinae, and Aphalaroidinae). Zool. J. Linn. Soc-Lond. 89, 299–392.

Burckhardt, D., Freuler, J., 2000. Jumping plant-lice (Hemiptera: Psylloidea) from sticky traps in carrot field in Valais. Switzerland. Mit. DSEG 73, 191–209.

Burckhardt, D., Lauterer, P., 1993. The jumping plant-lice of Iran (Homoptera, Psylloidea). Revue Suisse Zool. 100, 829–898.

Butler, C.D., Byrne, F.R., Karemane, M.L., Lee, R.F., Trumble, J.T., 2011. Effects of insecticides on behavior of adult *Bactericera cockerelli* (Hemiptera: Triozidae) and transmission of *Candidatus* Liberibacter psyllaurous. J. Econ. Entomol. 104, 586–594.

Cadena-Hinojosa, M.A., 1996. La punta morada de la papa en México: Efecto de cubiertas flotantes, genotypos y productos químicos. Rev. Mex. Fitopatol. 14, 20–24.

Cadena-Hinojosa, M.A., 1999. Potato purple top in Mexico: Effects of plant spacing and insecticide application. Rev. Mex. Fitopatol. 17, 91–96.

Cancelado, R.E., Radcliffe, E.B., 1979. Action thresholds for potato leafhopper on potatoes in Minnesota. J. Econ. Entomol. 72, 566–569.

Capinera, J.L., 2001. Handbook of Vegetable Pests. Academic Press, San Diego, CA.

Chavez, R., Salazar, L., Upadhya, M., Chujoy, E., Cabello, R., Garcia, A., Linares, J., 2003. The occurrence of genetic resistance and susceptibility to the new potato virus SB-29 among tetraploid potato populations (*Solanum tuberosum* L., 2n = 4× = 48 AAAA) in an arid agroecosystem. IDESIA (Chile) 21, 9–22.

Chiykowski, L.N., Chapman, R.K., 1965. Migration of the six-spotted leafhopper *Macrosteles fascifrons* (Stal): migration of the six-spotted leafhopper in central North America. Univ. Wis. Res. Bull. 261, 21–45.

Clausen, C.P., 1978. Introduced parasites and predators of arthropod pests and weeds: A world review. USDA Agric. Handbook 480, p. 545.

CNAS, 2006. Economic impacts of zebra chip on the Texas potato industry. Center for North American Studies http://cnas.tamu.edu/zebra%20chip%20impacts%20final.pdf.

Compere, H., 1915. *Paratrioza cockerelli* (Sulc). Monthly Bull. Calif. State Commiss. Hort 4, 574.

Compere, H., 1916. Notes on the tomato psylla. Monthly Bull. Calif. State Commiss. Hort 5, 189–191.

Conners, J.L., 1967. An annotated index of plant diseases in Canada. Can. Dep. Agric. Pub. No. 381.

Cook, W.C., 1941. The beet leafhopper. USDA Farmers' Bull. 1886.

Cook, W.C., 1967. Life history, host plants, and migrations of the beet leafhopper in the western United States. USDA Tech. Bull. 1365.

Cranshaw, W.S., 1993. An annotated bibliography of potato/tomato psyllid, Paratrioza cockerelli (Sulc) (Homoptera; Psyllidae). Colo. State Univ. Agric. Exp. Stn. Bull., TB93–5.

Cranshaw, W.S., 2001. Diseases caused by insect toxin: psyllid yellows. In: Stevenson, W.R., Loria, R., Franc, G.D., Weingartner, D.P. (Eds.), Compendium of potato diseases, 2nd edn. APS Press, St Paul, MN, pp. 73–74.

Crawford, D.L., 1914. A monograph of the jumping plant-lice or Psyllidae of the new world. US Natl Mus. Bull. 85, p. 186.

Crosslin, J.M., Munyaneza, J.E., Jensen, A.S., Hamm, P.B., 2005. Association of the beet leafhopper (Hemiptera: Cicadellidae) with a clover proliferation group phytoplasma in the Columbia Basin of Washington and Oregon. J. Econ. Entomol. 98, 279–283.

Crosslin, J.M., Vandemark, G.J., Munyaneza, J.E., 2006. Development of real-time, quantitative PCR for detection of the Columbia Basin potato purple top phytoplasma in plants and beet leafhoppers. Plant Dis. 90, 663–667.

Crosslin, J.M., Munyaneza, J.E., Brown, J.K., Liefting, L.W., 2010. Potato zebra chip disease: A phytopathological tale. Plant Health Progress. DOI 10.1094/PHP-2010-0317-01-RV.

Crosslin, J.M., Hamlin, L.L., Buchman, J.L., Munyaneza, J.E., 2011. Transmission of potato purple top phytoplasma to potato tubers and daughter plants. Am. J. Potato Res. 88, 339–345.

Crosslin, J.M., Hamm, P.B., Eggers, J.E., Rondon, S.I., Sengoda, V.G., Munyaneza, J.E., 2012a. First report of zebra chip disease and "*Candidatus* Liberibacter solanacearum" on potatoes in Oregon and Washington State. Plant Dis. 96, 452.

Crosslin, J.M., Olsen, N., Nolte, P., 2012b. First report of zebra chip disease and "*Candidatus* Liberibacter solanacearum" on potatoes in Idaho. Plant Dis. 96, 453.

Daniels, L.B., 1954. The nature of the toxicogenic condition resulting from the feeding of the tomato psyllid *Paratrioza cockerelli* (Sulc). PhD dissertation. University of Minnesota, Minneapolis and St Paul.

Davis, R.E., Sinclair, W.A., 1998. Phytoplasma identity and disease etiology. Phytopathology 88, 1372–1376.

DeLong, D.M., 1931. A revision of the American species of *Empoasca* known to occur north of Mexico. USDA Tech. Bull. 231.

DeLong, D.M., 1971. The bionomics of leafhoppers. Annu. Rev. Entomol. 16, 179–210.

Douglass, J.R., Hallock, H.C., 1957. Relative importance of various host plants of the beet leafhopper in southern Idaho. US Dept Agr. Tech. Bull. 1155, p. 11.

Drake, D.C., Chapman, R.K., 1965. Migration of the six-spotted leafhopper *Macrosteles fascifrons* (Stal): evidence for long distance migration of the six-spotted leafhopper into Wisconsin. Univ. Wis. Res. Bull. 261, 3–20.

Durant, J.A., 1973. Notes on factors influencing observed leafhopper (Homoptera: Cicadellidae) population densities on corn. J. Georgia Entomol. Soc. 8, 1–5.

Eckenrode, C.J., 1973. Foliar sprays for control of the aster leafhopper on carrots. J. Econ. Entomol. 66, 265–266.

Ember, I., Acs, Z., Munyaneza, J.E., Crosslin, J.M., Kolber, M., 2011. Survey and molecular detection of phytoplasmas associated with potato in Romania and Southern Russia. Eur. J. Plant Pathol. 130, 367–377.

Essig, E.O., 1917. The tomato and laurel psyllids. J. Econ. Entomol. 10, 433–444.

Eyer, J.R., 1937. Physiology of psyllid yellows of potatoes. J. Econ. Entomol. 30, 891–898.

Eyer, J.R., Crawford, R.F., 1933. Observations on the feeding habits of the potato psyllid (*Paratrioza cockerelli* Sulc.) and the pathological history of the "psyllid yellows" which it produces. J. Econ. Entomol. 26, 846–850.

Fathi, S.A.A., 2011. Population density and life history parameters of the psyllid *Bactericera nigricornis* (Forster) on four commercial cultivars of potato. Crop Prot. 30, 844–848.

Fialova, R., Valova, P., Balakishiyeva, G., Danet, J.L., Safarova, D., Foissac, X., Navratil, M., 2009. Genetic variability of stolbur phytoplasma in annual crop and wild plant species in south Moravia. J. Plant Pathol. 91, 411–416.

Foster, R., Flood, B., 1995. Vegetable insect management with emphasis on the Midwest. Meister Publ. Co., Willoughby, OH, p. 265.

Gharalari, A.H., Nansen, C., Lawson, D.S., Gilley, J., Munyaneza, J.E., Vaughn, K., 2009. Knockdown mortality, repellency, and residual effects of insecticides for control of adult *Bactericera cockerelli* (Hemiptera: Psyllidae). J. Econ. Entomol. 102, 1032–1038.

Gill, G., 2006. Tomato psyllid detected in New Zealand. Biosecur. NZ 69, 10–11.

Girsova, N., Bottner, K.D., Mozhaeva, K.A., Kastalyeva, T.B., Owens, R.A., Lee, I.-M., 2008. Molecular detection and identification of Group 16SrI and 16SrXII phytoplasmas associated with diseased potatoes in Russia. Plant Dis. 92, 654.

Golino, D.A., Oldfield, G.N., Gumpf, D.J., 1989. Experimental hosts of the beet leafhopper-transmitted virescence agent. Plant Dis. 73, 850–854.

Goolsby, J.A., Adamczyk, J., Bextine, B., Lin, D., Munyaneza, J.E., Bester, G., 2007. Development of an IPM program for management of the potato psyllid to reduce incidence of zebra chip disorder in potatoes. Subtrop. Pl. Sci. 59, 85–94.

Hagel, G.T., Landis, B.J., 1967. Biology of the aster leafhopper, *Macrosteles fascifrons* (Homoptera: Cicadellidae) in eastern Washington and some overwintering sources of aster yellows. Ann. Entomol. Soc. Am. 60, 591–595.

Hagel, G.T., Landis, B.J., Ahrens, M.C., 1973. Aster leafhopper: source of infestation, host plant preference, and dispersal. J. Econ. Entomol. 66, 877–881.

Hansen, A.K., Trumble, J.T., Stouthamer, R., Paine, T.D., 2008. A new huanglongbing species, "*Candidatus* Liberibacter psyllaurous" found to infect tomato and potato, is vectored by the psyllid *Bactericera cockerelli* (Sulc). Appl. Environ. Microbiol. 74, 5862–5865.

Harries, F.H., Douglass, J.R., 1948. Bionomic studies on the beet leafhopper. Ecol. Mono. 18, 45–79.

Harrison, B.D., Roberts, I.M., 1969. Association of mycoplasma-like bodies with potato witch's broom disease from Scotland. Ann. Appl. Biol. 63, 347–349.

Henne, D.C., Paetzold, L., Workneh, F., Rush, C.M., 2010. Evaluation of potato psyllid cold tolerance, overwintering survival, sticky trap sampling, and effects of liberibacter on potato psyllid alternate host plants. In: Workneh, F., Rush, C.M. (Eds.), Proceedings of the 10th Annual Zebra Chip Reporting Session, pp. 149–153. Dallas, TX (November 7–10, 2010).

Henne, D.C., Workneh, F., Rush, C.M., 2012. Spatial patterns and spread of potato zebra chip disease in the Texas Panhandle. Plant Dis. 96, 948–956.

Henne, R.C., 1970. Effect of five insecticides on populations of the six-spotted leafhopper and the incidence of aster yellows in carrots. Can. J. Plant Sci. 50, 169–174.

Hills, O.A., 1937. The beet leafhopper in the central Columbia River breeding area. J. Agr. Res. 55, 21–31.

Hodgson, W.A., Pond, D.D., Munro, J., 1974. Diseases and pests of potatoes. Can. Dep. Agric. Pub. No. 1492.

Hodkinson, I.D., 1981. Status and taxonomy of the *Trioza (Bactericera) nigricornis* Förster complex (Hemiptera: Triozidae). Bull. Entomol. Res. 71, 671–679.

Hoy, C.W., Heady, S.E., Koch, T.A., 1992. Species composition, phenology, and possible origins of leafhoppers (Cicadellidae) in Ohio vegetable crops. J. Econ. Entomol. 85, 2336–2343.

Jensen, D.D., 1954. Notes on the potato psyllid, *Paratrioza cockerelli* (Sulc) (Hemiptera: Psyllidae). Pan-Pac. Entomol. 30, 161–165.

Jones, P., Arocha, Y., Antezana, O., Montellano, E., Franco, P., 2004. Brotes grandes (big bud) of potato: A new disease associated with a 16SrI-B subgroup phytoplasma in Bolivia. New Dis. Rep. 10, 18.

Kent, D., Taylor, G., 2010. Two new species of Acizzia Crawford (Hemiptera: Psyllidae) from the Solanaceae with a potential new economic pest of eggplant, *Solanum melongena*. Aust. J. Entomol. 49, 73–81.

Khadhair, A.-H., Hiruki, C., Hwang, S.F., .Wang, K., 1997. Molecular identification and relatedness of potato witches'-broom phytoplasma isolates from four potato cultivars. Microbiol. Res. 152, 281–286.

Khadhair, A.-H., Duplessis, H., McAlister, P., Ampong-Nyarko, K., Bains, P., 2003. Transmission and characterization of phytoplasma diseases associated with infected potato cultivars in Alberta. Acta Hort. 619, 167–176.

Khurana, S.M.P., Singh, R.A., Kalay, D.M., 1988. Mycoplasma-associated potato diseases and their control in India. In: Maramorosch, K., Raychandhuri, J.P. (Eds.), Mycoplasma Diseases of Crops: Basic and Applied Aspects, Springer-Verlag, Berlin, Germany, pp. 285–316.

Kirkpatrick, B.C., 1992. Mycoplasma-like organisms – plant and invertebrate pathogens. In: Balows, A., Truper, H.G., Dworkin, M., Harder, W., Schleifer, K.H. (Eds.), The Prokaryotes, Vol. 4Springer, New York, NY, pp. 4050–4067.

Knowlton, G.F., Janes, M.J., 1931. Studies on the biology of *Paratrioza cockerelli* (Sulc). Ann. Entomol. Soc. Am. 24, 283–291.

Knowlton, G.F., Thomas, W.L., 1934. Host plants of the potato psyllid. J. Econ. Entomol. 27, 547.

Koinzan, S.D., Pruess, K.P., 1975. Effects of a wide-area application of ULV malathion on leafhoppers in alfalfa. J. Econ. Entomol. 68, 267–268.

Kolber, M., Z. Acs, I. Ember, Z. Nagy, C. Talaber, J. Horrocks, I., Hope-Johnstone, S. Marchenko, V. Gonchar, I. Kiselev, A. Lupascu, M. Munteanu, and N. Filip. 2010. Phytoplasma infection of chips potatoes in Romania and south Russia. In Proceedings of the 10th Annual Zebra Chip Reporting Session, pp. 50–54, F. Workneh and C.M. Rush (eds). Dallas, TX (November 7–10, 2010).

Lacey, L.A., de la Roza, F., Horton, D.R., 2009. Insecticidal activity of entomopathogenic fungi (Hypocreales) for potato psyllid, *Bactericera cockerelli* (Hemiptera: Triozidae): development of bioassay techniques, effect of fungal species and stage of the psyllid. Biocontrol Sci. Techn. 19, 957–970.

Lacey, L.A., Liu, T.-X., Buchman, J.L., Munyaneza, J.E., Goolsby, J.A., Horton, D.R., 2011. Entomopathogenic fungi (Hypocreales) for control of potato psyllid, *Bactericera cockerelli* (Šulc) (Hemiptera: Triozidae) in an area endemic for zebra chip disease of potato. Biol. Control 56, 271–278.

Ladd Jr., T.L., Rawlins, W.A., 1965. The effects of the feeding of the potato leafhopper on photosynthesis and respiration in the potato plant. J. Econ. Entomol. 58, 623–628.

Lamp, W.O., Nielsen, G.R., Danielson, S., 1994. Patterns among host plants of potato leafhopper, *Empoasca fabae* (Homoptera: Cicadellidae). J. Kans. Entomol. Soc. 67, 354–368.

Lawson, F.R., Chamberlin, J.C., York, G.T., 1951. Dissemination of the beet leafhopper in California. U. S. Dept. Agr. Tech. Bull. 1030.

Lee, P.E., Robinson, A.G., 1958. Studies on the six-spotted leafhopper, *Macrosteles fascifrons* (Stal), and aster yellows in Manitoba. Can. J. Plant Sci. 38, 320–327.

Lee, I.-M., Gundersen-Rindal, D.E., Davis, R.E., Bartoszyk, I.M., 1998. Revised classification scheme of phytoplasmas based on RFLP analyses of 16S rRNA and ribosomal protein gene sequences. Intl. J. Syst. Bacteriol. 48, 1153–1169.

Lee, I.-M., Davis, R.E., Gundersen-Rindal, D.E., 2000. Phytoplasma: phytopathogenic mollicutes. Annu Rev. Microbiol. 54, 221–255.

Lee, I.-M., Bottner, K.D., Munyaneza, J.E., Secor, G.A., Gudmestad, N.C., 2004. Clover proliferation group (16SrVI) Subgroup A (16SrVI-A) phytoplasma is a probable causal agent of potato purple top disease in Washington and Oregon. Plant Dis. 88, 429.

Lee, I.-M., Bottner, K.D., Secor, G., Rivera-Varas, V., 2006a. "*Candidatus* phytoplasma americanum", a phytoplasma associated with a potato purple top disease complex. Intl. J. Syst. Evol. Micr. 56, 1593–1597.

Lee, I.-M., Bottner, K.D., Munyaneza, J.E., Davis, R.E., Crosslin, J.M., du Toit, L.J., Crosby, T., 2006b. Carrot purple leaf: a new spiroplasmal disease associated with carrots in Washington State. Plant Dis. 90, 989–993.

Lefol, C., Caudwell, A., Lherminier, J., Larrue, J., 1993. Attachment of the flavescence dorée pathogen (MLO) to leafhopper vectors and other insects. Ann. Appl. Biol. 123, 611–622.

Lefol, C., Lherminier, J., Boudon-Padieu, E., Larrue, J., Louis, C., 1994. Propagation of flavescence dorée MLO (mycoplasma-like organisms) in the leafhopper vector *Euscelidius variegates* Kbm. J. Invertebr. Pathol. 63, 285–293.

Leyva-López, N.E., Ochoa-Sánchez, J.C., Leal-Klevezas, D.S., Martínez-Soriano, J.P., 2002. Multiple phytoplasmas associated with potato diseases in Mexico. Can. J. Microbiol. 48, 1062–1068.

Liefting, L.W., Rez-Egusquiza, Z.C., Clover, G.R.G., Anderson, J.A.D., 2008. A New "*Candidatus* Liberibacter" Species in *Solanum tuberosum* in New Zealand. Plant Dis. 92, 1474.

Liefting, L.W., Veerakone, S., Ward, L.I., Clover, G.R.G., 2009a. First report of "*Candidatus* Phytoplasma australiense" in potato. Plant Dis. 93, 969.

Liefting, L.W., Sutherland, P.W., Ward, L.I., Paice, K.L., Weir, B.S., Clover, G.R.G., 2009b. A new "*Candidatus* Liberibacter" species associated with diseases of solanaceous crops. Plant Dis. 93, 208–214.

Lin, H., Doddapaneni, H., Munyaneza, J.E., Civerolo, E., Sengoda, V.G., Buchman, J.L., Stenger, D.C., 2009. Molecular characterization and phylogenetic analysis of 16S rRNA from a new species of "*Candidatus* Liberibacter" associated with zebra chip disease of potato (*Solanum tuberosum* L.) and the potato psyllid (*Bactericera cockerelli* Sulc). J. Plant Pathol. 91, 215–219.

Linhartova, S., Cervana, G., Rodova, J., 2006. The occurrence of potato stolbur phytoplasma on different hosts in the Czech Republic. Mitt. Bundesforsch 400, 456.

List, G.M., 1939. The effect of temperature upon egg deposition, egg hatch and nymphal development of *Paratrioza cockerelli* (Sulc). J. Econ. Entomol. 32, 30–36.

List, G.M., Daniels, L.B., 1934. A promising control for psyllid yellows of potatoes. Science 79, 79.

Liu, D., Trumble, J.T., 2007. Comparative fitness of invasive and native populations of the potato psyllid (*Bactericera cockerelli*). Entomol. Exp. Appl. 123, 35–42.

Liu, D., Trumble, J.T., Stouthamer, R., 2006. Genetic differentiation between eastern populations and recent introductions of potato psyllid (*Bactericera cockerelli*) into western North America. Entomol. Exp. Appl. 118, 177–183.

Mahr, S.E.R., Wyman, J.A., Chapman, R.K., 1993. Variability in aster yellows infectivity of local populations of the aster leafhopper (Homoptera: Cicadellidae) in Wisconsin. J. Econ. Entomol. 86, 1522–1526.

Maramorosch, K., Granados, R.R., Hirumi, H., 1970. Mycoplasma diseases of plants and insects. Adv. Virus Res. 16, 135–193.

Martínez-Soriano, J.P., Leyva-López, N.E., Zavala-Soto, M.E., Bères, M., Leal-Klevezas, D.S., 1999. Detección molecular del agente causal del síndrome "bola de hilo" de la papa en semillas infectadas y asintomáticas. Biotecnología Aplicada 16, 93–96.

McCoy, R.E., 1979. Mycoplasmas and yellows disease. In: Whitcomb, R.F., Tully, J.G. (Eds.), Academic Press, New York, NY, pp. 229–264.

McCoy, R.E., 1983. Wall-free prokaryotes of plants and invertebrates. In: Starr, M.P., Stolp, H, Truper, H.G., Truper, H.G., Bolows, A., Schlegel, H.G. (Eds.), The Prokaryotes, Vol. 2. Springer-Verlag, New York, NY, pp. 2238–2246.

McMullen, R.D., Jong, C., 1971. Dithiocarbamate fungicides for control of pear psylla. J. Econ. Entomol. 64, 1266–1270.

Medler, J.T., 1957. Migration of the potato leafhopper – a report on a cooperative study. J. Econ. Entomol. 50, 493–497.

Mello, A.F., Wayadande, A.C., Yokomi, R.K., Fletcher, J., 2009. Transmission of different isolates of *Spiroplasma citri* to carrot and citrus by *Circulifer tenellus* (Hemiptera: Cicadellidae). J. Econ. Entomol. 102, 1417–1422.

Miles, G.P., Samuel, M.A., Chen, J., Civerolo, E.L., Munyaneza, J.E., 2010. Evidence that cell death is associated with zebra chip disease in potato tubers. Am. J. Potato Res. 87, 337–349.

Munyaneza, J.E., 2005. Purple top disease and beet leafhopper-transmitted virescence agent (BLTVA) phytoplasma in potatoes of the Pacific Northwest of the United States. In: Haverkort, A.J, Struik, P.C., Struik, P.C. (Eds.), Potato in Progress: Science Meets Practice, Wageningen Academic Publishers, Wageningen, The Netherlands, pp. 211–220.

Munyaneza, J.E., 2006. Research update: Potato purple top disease and beet leafhoppers in the Columbia Basin. Potato Country 22, 28–29.

Munyaneza, J.E., 2010a. Emerging leafhopper-transmitted phytoplasma diseases of potato. Southwest. Entomol. 35, 451–455.

Munyaneza, J.E., 2010b. Psyllids as vectors of emerging bacterial diseases of annual crops. Southwest. Entomol. 35, 471–477.

Munyaneza, J.E., Upton, J.E., 2005. Beet leafhopper (Hemiptera: Cicadellidae) settling behavior, survival, and reproduction on selected host plants. J. Econ. Entomol. 98, 1824–1830.

Munyaneza, J.E., Crosslin, J.M., Upton, J.E., 2006a. The beet leafhopper (Hemiptera: Cicadellidae) transmits the Columbia Basin potato purple top phytoplasma to potatoes, beets, and weeds. J. Econ. Entomol. 99, 268–272.

Munyaneza, J.E., Crosslin, J.M., Jensen, A.S., Hamm, P.B., Schreiber, A., 2006b. Beet leafhopper and potato purple top disease: 2005 recap and new research directions. In: Jensen, A., (Eds.), Proceedings of the 45th Annual Washington State Potato Conference, Moses Lake, WA (February 7–9), pp. 107–118.

Munyaneza, J.E., Crosslin, J.M., Upton, J.E., 2007a. Association of *Bactericera cockerelli* (Homoptera: Psyllidae) with "zebra chip", a new potato disease in southwestern United States and Mexico. J. Econ. Entomol. 100, 656–663.

Munyaneza, J.E., Crosslin, J.M., Lee, I.-M., 2007b. Phytoplasma diseases and insect vectors in potatoes of the Pacific Northwest of the United States. Bull. Insectol. 60, 181–182.

Munyaneza, J.E., Goolsby, J.A., Crosslin, J.M., Upton, J.E., 2007c. Further evidence that zebra chip potato disease in the lower Rio Grande Valley of Texas is associated with *Bactericera cockerelli*. Subtrop. Pl. Sci. 59, 30–37.

Munyaneza, J.E., Jensen, A.S., Hamm, P.B., Upton, J.E., 2008a. Seasonal occurrence and abundance of beet leafhopper in the potato growing region of Washington and Oregon Columbia Basin and Yakima Valley. Am. J. Potato Res. 85, 77–84.

Munyaneza, J.E., Buchman, J.L., Upton, J.E., Goolsby, J.A., Crosslin, J.M., Bester, G., Miles, G.P., Sengoda, V.G., 2008b. Impact of different potato psyllid populations on zebra chip disease incidence, severity, and potato yield. Subtrop. Pl. Sci. 60, 27–37.

Munyaneza, J.E., Buchman, J.L., Crosslin, J.M., 2009a. Seasonal occurrence and abundance of the potato psyllid, Bactericera cockerelli, in South Central Washington. Am. J. Potato Res. 86, 513–518.

Munyaneza, J.E., Crosslin, J.M., Buchman, J.L., 2009b. Susceptibility of different potato cultivars to purple top disease. Am. J. Potato Res. 86, 499–503.

Munyaneza, J.E., Sengoda, V.G., Crosslin, J.M., De la Rosa-Lozano, G., Sanchez, A., 2009c. First report of "*Candidatus* Liberibacter psyllaurous" in potato tubers with zebra chip disease in Mexico. Plant Dis. 93, 552.

Munyaneza, J.E., Sengoda, V.G., Crosslin, J.M., Garzon-Tiznado, J., Cardenas-Valenzuela, O., 2009d. First report of "*Candidatus* Liberibacter solanacearum" in tomato plants in Mexico. Plant Dis. 93, 1076.

Munyaneza, J.E., Sengoda, V.G., Crosslin, J.M., Garzon-Tiznado, J., Cardenas-Valenzuela, O., 2009e. First report of "*Candidatus* Liberibacter solanacearum" in pepper in Mexico. Plant Dis. 93, 1076.

Munyaneza, J.E., Buchman, J.L., Goolsby, J.A., Ochoa, A.P., Schuster, G., 2010a. Impact of potato planting timing on zebra chip incidence in Texas. In: Workneh, F., Rush, C.M. (Eds.), Proceedings of the 10th Annual Zebra Chip Reporting Session, Dallas, TX, pp. 106–109.

Munyaneza, J.E., Crosslin, J.M., Buchman, J.L., Sengoda, V.G., 2010b. Susceptibility of different potato plant growth stages to purple top disease. Am. J. Potato Res. 87, 60–66.

Munyaneza, J.E., Crosslin, J.M., Upton, J.E., Buchman, J.L., 2010c. Incidence of beet leafhopper-transmitted virescence agent phytoplasma in local populations of the beet leafhopper, *Circulifer tenellus,* in Washington State. J. Insect Sci. 10, 18 available online: insectscience.org/10.18.

Munyaneza, J.E., Fisher, T.W., Sengoda, V.G., Garczynski, S.F., Nissinen, A., Lemmetty, A., 2010d. First report of "*Candidatus* Liberibacter solanacearum" in carrots in Europe. Plant Dis. 94, 639.

Munyaneza, J.E., Fisher, T.W., Sengoda, V.G., Garczynski, S.F., Nissinen, A., Lemmetty, A., 2010e. Association of "*Candidatus* Liberibacter solanacearum" with the psyllid *Trioza apicalis* (Hemiptera: Triozidae) in Europe. J. Econ. Entomol. 103, 1060–1070.

Munyaneza, J.E., Lemmetty, A., Nissinen, A.I., Sengoda, V.G., Fisher, T.W., 2011a. Molecular detection of aster yellows phytoplasma and "*Candidatus* Liberibacter solanacearum" in carrots affected by the psyllid *Trioza apicalis* (Hemiptera: Triozidae) in Finland. J. Plant Pathol. 93, 697–700.

Munyaneza, J.E., Buchman, J.L., Sengoda, V.G., Fisher, T.W., Pearson, C.C., 2011b. Susceptibility of selected potato varieties to zebra chip potato disease. Am. J. Potato Res. 88, 435–440.

Munyaneza, J.E., Sengoda, V.G., Buchman, J.L., Fisher, T.W., 2012a. Effects of temperature on "*Candidatus* Liberibacter solanacearum" and zebra chip potato disease symptom development. Plant Dis. 96, 18–23.

Munyaneza, J.E., Sengoda, V.G., Stegmark, R., Arvidsson, A.K., Anderbrant, O., Yuvaraj, J.K., Ramert, B., Nissinen, A., 2012b. First report of "*Candidatus* Liberibacter solanacearum" associated with psyllid-affected carrots in Sweden. Plant Dis. 96, 453.

Munyaneza, J.E., Sengoda, V.G., Sundheim, L., Meadow, R., 2012c. First report of "*Candidatus* Liberibacter solanacearum" associated with psyllid-affected carrots in Norway. Plant Dis. 96, 454.

Murral, D.J., Nault, L.R., Hoy, C.W., Madden, L.V., Miller, S.A., 1996. Effects of temperature and vector age on transmission of two Ohio strains of aster yellows phytoplasma by the aster leafhopper (Homoptera: Cicadellidae). J. Econ. Entomol. 89, 1223–1232.

Nagaich, B.B., Giri, B.K., 1973. Purple top roll disease of potato. Am. Potato. J. 50, 79–85.

Nagaich, B.B., Puri, B.K., Sinha, R.C., Dhingra, M.K., Bhardwaj, V.P., 1974. Mycoplasma-like organisms in plants affected with purple top-roll, marginal flavescence and witches'-broom diseases in potatoes. Phytopathol. Z. 81, 273–379.

Nansen, C., K. Vaughn, Y. Xue, C. M. Rush, F. Workneh, J. A. Goolsby, N. Troclair, J. Anciso, and X. Martini. 2010. Spray coverage and insecticide performance, pp. 78–82. In F. Workneh and C. M. Rush (Eds.), Proceedings of the 10th Annual Zebra Chip Reporting Session, 7–10 November, 2010, Dallas, TX.

O'Hayer, K.W., Schultz, G.A., Eastman, C.E., Fletcher, J., 1984. Newly discovered plant hosts of *Spiroplasma citri*. Plant Dis. 68, 336–338.

Olivier, C.Y., Lowery, D.T., Stobbs, L.W., 2009. Phytoplasma diseases and their relationship with insect and plant hosts in Canadian horticultural and field crops. Can. Entomol. 141, 425–462.

O'Rourke, P.K., Burckness, E.C., Hutchison, W.D., 1998. Development and validation of a fixed-precision sequential sampling plan for aster leafhopper (Homoptera: Cicadellidae) in carrot. Environ. Entomol. 27, 1463–1468.

Ossiannilsson, F., 1992. The Psylloidea (Homoptera) of Fennoscandia and Denmark. Fauna Entomologica Scandinavica. Vol. 26. E.J. Brill, Leiden, The Netherlands.

Paltrinieri, S., Bertaccini, A., 2007. Detection of phytoplasmas in plantlets grown from different batches of seed-potatoes. Bull. Insectol. 60, 379–380.

Pantoja, A., Hagerty, A.M., Emmert, S.Y., Munyaneza, J.E., 2009. Leafhoppers (Homoptera: Cicadellidae) associated with potatoes in Alaska: Species composition, seasonal abundance, and potential phytoplasma vectors. Am. J. Potato Res. 86, 68–75.

Peng, L., Trumble, J.T., Munyaneza, J.E., Liu, T.-X., 2011. Repellency of a kaolin particle film to potato psyllid, *Bactericera cockerelli* (Hemiptera: Psyllidae), on tomato under laboratory and field conditions. Pest Manage. Sci. 67, 815–824.

Pienkowski, R.L., Medler, J.T., 1964. Synoptic weather conditions associated with long-range movement of the potato leafhopper, *Empoasca fabae*, into Wisconsin. Ann. Entomol. Soc. Am. 57, 588–591.

Pletsch, D.J., 1947. The potato *psyllid Paratrioza cockerelli* (Sulc) its biology and control. Mont. Ag. Exp. Stn. Bull. 446.

Poos, F.W., Smith, F.S., 1931. A comparison of oviposition and nymphal development of *Empoasca fabae* (Harris) on different host plants. J. Econ. Entomol. 24, 361–371.

Puketapu, A., Roskruge, N., 2011. The tomato-potato psyllid lifecycle on three traditional Maori food sources. P. Ag. Soc. NZ 41, 167–173.

Purcell, A.H., 1982. Insect vector relationship with prokaryotic plant pathogens. Ann. Rev. Phytopathol. 20, 397–417.

Radcliffe, E.B., 1982. Insect pests of potato. Annu. Rev. Entomol. 27, 173–204.

Raine, J., 1967. Leafhopper transmission of witches'-broom and clover phyllody viruses from British Columbia to clover, alfalfa, and potato. Can. J. Bot. 45, 441–445.

Rehman, M., Melgar, J., Rivera, C., Urbina, N., Idris, A.M., Brown, J.K., 2010. First report of "*Candidatus* Liberibacter psyllaurous" or "*Ca.* Liberibacter solanacearum" associated with severe foliar chlorosis, curling, and necrosis and tuber discoloration of potato plants in Honduras. Plant Dis. 94, 376.

Rich, A.E., 1983. Potato Diseases. Academic Press, New York, NY.

Richards, B.L., 1928. A new and destructive disease of the potato in Utah and its relation to the potato psylla. Phytopathology 18, 140–141.

Richards, B.L., 1931. Further studies with psyllid yellows of the potato. Phytopathology 21, 103.

Richards, B.L., 1933. Psyllid yellows of the potato. J. Agric. Res. 46, 189–216.

Richards, B.L., Blood, H.L., 1933. Psyllid yellows of the potato. J. Agric. Res. 46, 189–216.

Romney, V.E., 1939. Breeding areas of the tomato psyllid, *Paratrioza cockerelli* (Sulc). J. Econ. Entomol. 32, 150–151.

Rowe, J.A., Knowlton, G.F., 1935. Studies upon the morphology of *Paratrioza cockerelli* (Sulc). J. Utah Acad. Sci. 12, 233–237.

Rubio-Covarrubias, O.A., Almeyda-Leon, I.H., Moreno, J.I., Sanchez-Salas, J.A., Sosa, R.F., Borbon-Soto, J.T., Hernandez, C.D., Garzon-Tiznado, J.A., Rodriguez, R.R., Cadena-Hinojosa, M.A., 2006. Distribution of potato purple top and *Bactericera cockerelli* Sulc. in the main potato production zones in Mexico. Agr. Tec. Mex. 32, 201–211.

Rubio-Covarrubias, O.A., Almeyda-Leon, I.H., Cadena-Hinojosa, M.A., Lobato-Sanchez, R., 2011. Relation between *Bactericera cockerelli* and presence of *Candidatus* Liberibacter psyllaurous in commercial fields of potato. Rev. Mex. Cienc. Ag. 2, 17–28.

Salazar, L.F., 2006. Emerging and re-emerging potato diseases in the Andes. Potato Research 49, 43–47.

Salazar, L., Javasinghe, U., 2001. Diseases caused by phytoplasmas in potato. In: CIP (Ed.), Techniques in Plant Virology, Lima, Peru, International Potato Center (CIP).

Santos-Cervantes, M.E., Chávez-Medina, J.A., Acosta-Pardini, J., Flores-Zamora, G.L., Méndez-Lozano, J., Leyva-López, N.E., 2010. Genetic diversity and geographical distribution of phytoplasmas associated with potato purple top disease in Mexico. Plant Dis. 94, 388–395.

Saxena, K.N., Gandhi, J.R., Saxena, R.C., 1974. Patterns of relationship between certain leafhoppers and plants: responses to plants. Entomol. Exp. Appl. 17, 303–318.

Schultz, T.R., Shaw, M.E., 1991. Occurrence of the beet leafhopper-transmitted virescence agent in red and daikon radish seed plants in Washington State. Plant Dis. 75, 751.

Secor, G.A., Rivera-Varas, V.V., 2004. Emerging diseases of cultivated potato and their impact on Latin America. Rev. Latinoam. Papa (Suppl.) 1, 1–8.

Secor, G.A., Lee, I.-M., Bottner, K.D., Rivera-Varas, V., Gudmestad, N.C., 2006. First report of a defect of processing potatoes in Texas and Nebraska associated with a new phytoplasma. Plant Dis. 90, 377.

Secor, G.A., Rivera, V.V., Abad, J.A., Lee, I.-M., Clover, G.R.G., Liefting, L.W., Li, X., De Boer, S.H., 2009. Association of "*Candidatus* Liberibacter solanaceraum" with zebra chip disease of potato established by graft and psyllid transmission, electron microscopy, and PCR. Plant Dis. 93, 574–583.

Seemüller, E., Marcone, C., Lauer, U., Ragozzino, A., Göschl, M., 1998. Current status of molecular classification of the phytoplasmas. J. Plant Pathol. 80, 3–26.

Sengoda, V.G., Munyaneza, J.E., Crosslin, J.M., Buchman, J.L., Pappu, H.R., 2010. Phenotypic and etiological differences between psyllid yellows and zebra chip diseases of potato. Am. J. Potato Res. 87, 41–49.

Setiawan, D.P., Ragsdale, D.W., 1987. Use of aluminum-foil and oat-straw mulches for controlling aster leafhopper, *Macrosteles fascifrons* (Homoptera: Cicadellidae), and aster yellows in carrots. Great Lakes Entomol. 20, 103–109.

Shaw, M.E., Golino, D.A., Kirkpatrick, B.C., 1990. Infection of radish in Idaho by beet leafhopper transmitted virescence agent. Plant Dis. 74, 252.

Shaw, M.E., Kirkpatrick, B.C., Golino, D.A., 1993. The beet leafhopper-transmitted virescence agent causes tomato big bud disease in California. Plant Dis. 77, 290–295.

Shiumi, T., Sugiura, M., 1984. Differences among *Macrosteles orientalis*-transmitted MLO, potato purple-top wilt MLO in Japan and aster yellows MLO from USA. Ann. Phytopathol. Soc. Jpn.

Sinha, R.C., Chiykowski, L.N., 1967. Initial and subsequent sites of aster yellows virus infection in a leafhopper vector. Virology 33, 702–708.

Slack, S.A., 2001. Diseases caused by phytoplasmas. In: Stevenson, W.R., Loria, R., Franc, G.D., Weingartner, D.P. (Eds.), Compendium of Potato Diseases, APS Press, St Paul, MN, pp. 56–57.

Sulc, K., 1909. *Trioza cockerelli* n.sp., a novelty from North America, being also of economic importance. Acta Soc. Entomol. Bohemiae 6, 102–108.

Taylor, R.A., Reling, D., 1986. Preferred wind direction of long distance leafhopper (*Empoasca fabae*) migrants and its relevance to the return migration of small insects. J. Anim. Ecol. 55, 1103–1114.

Tenorio, J., Chuquillanqui, C., Garcia, A., Guillen, M., Chavez, R., Salazar, L.F., 2000. Symptomology and effect on potato yield of the new virus transmitted through the psyllid *Russelliana solanicola* [in Spanish]. Fitopatologia 38, 32–36.

Teulon, D.A.J., Workman, P.J., Thomas, K.L., Nielsen, M.C., 2009. *Bactericera cockerelli*: incursion, dispersal and current distribution on vegetable crops in New Zealand. NZ Plant Protect 62, 136–144.

Thomas, K.L., Jones, D.C., Kumarasinghe, L.B., Richmond, J.E., Gill, G.S.C., Bullians, M.S., 2011. Investigation into the entry pathway for the tomato potato psyllid *Bactericera cockerelli*. N. Z. Plant Prot. 64, 259–268.

Thomas, P.E., Martin, M.W., 1971. Vector preference, a factor of resistance to curly top virus in certain tomato cultivars. Phytopathology 61, 1257–1260.

Thompson, L.S., 1967. Reduction of lettuce yellows with systemic insecticides. J. Econ. Entomol. 60, 716–718.

Tingey, W.M., Gibson, R.W., 1978. Feeding and mobility of the potato leafhopper impaired by glandular trichomes of *Solanum berthaultii* and *S. Polyadenium*. J. Econ. Entomol. 71, 856–858.

Tingey, W.M., Laubengayer, J.E., 1981. Defense against the green peach aphid and potato leafhopper by glandular trichomes of *Solanum berthaultii*. J. Econ. Entomol. 74, 721–725.

Trumble, J., 2008. The tomato psyllid: a new problem on fresh market tomatoes in California and Baja Mexico. Univ. Calif. Coop. Ext. http://ceventura.ucdavis.edu/Vegetable_Crops/Tomato_Psyllid.htm.

Trumble, J., 2009. Potato psyllid. Center for Invasive Species Research, University of California Riverside. http://cisr.ucr.edu/potato_psyllid.html.

Tuthill, L.D., 1959. Los Psillidae del Perú Central (Insecta: Homoptera). Rev. Peru. Entomol. Ag. 2, 1–27.

Vega, F.E., Davis, R.E., Barbosa, P., Dally, E.L., Purcell, A.H., Lee, I.-M., 1993. Detection of a plant pathogen in a nonvector insect species by the polymerase chain reaction. Phytopathology 83, 621–624.

Vega, F.E., Davis, R.E., Dally, E.L., Barbosa, P., Purcell, A.H., Lee, I.-M., 1994. Use of a biotinylated DNA probe for detection of the aster yellows mycoplasma-like organism in *Dalbulus maidis* and *Macrosteles fascifrons* (Homoptera: Cicadellidae). Fla. Entomol. 77, 330–334.

Wallis, R.L., 1946. Seasonal occurrence of the potato psyllid in the North Platte Valley. J. Econ. Entomol. 39, 689–694.

Wallis, R.L., 1955. Ecological studies on the potato psyllid as a pest of potatoes. USDA Tech. Bull. 1107.

Wallis, R.L., 1962. Spring migration of the six-spotted leafhopper in the western Great Plains. J. Econ. Entomol. 55, 871–874.

Wayadande, A.C., Baker, G.R., Fletcher, J., 1997. Comparative ultrastructure of the salivary glands of two phytopathogen vectors, the beet leafhopper, *Circulifer tenellus* (Baker), and the corn leafhopper, *Dalbulus maidis* Delong and Wolcott (Homoptera: Cicadellidae). Int. J. Insect Morphol. Embryol. 26, 113–120.

Weintraub, P.G., Beanland, L., 2006. Insect vectors of phytoplasmas. Annu. Rev. Entomol. 51, 91–111.

Wen, A., Mallik, I., Alvarado, V.Y., Pasche, J.S., Wang, X., Li, W., Levy, L., Lin, H., B Scholthof, H., Mirkov, T.E., Rush, C.R., Gudmestad, N.C., 2009. Detection, distribution, and genetic variability of "*Candidatus* Liberibacter" species associated with zebra complex disease of potato in North America. Plant Dis. 93, 1102–1115.

Westdal, P.H., Barrett, C.F., Richardson, H.P., 1961. The six-spotted leafhopper, *Macrosteles fascifrons* (Stål.) and aster yellows in Manitoba.. Can. J. Plant Sci. 41, 320–331.

Workneh, F., Henne, D.C., Childers, A.C., Paetzold, L., Rush, C.M., 2012. Assessments of the edge effect in intensity of potato zebra chip disease. Plant Dis. 96, 943–947.

Yang, X.-B., Liu, T.-X., 2009. Life history and life tables of *Bactericera cockerelli* (Homoptera: Psyllidae) on eggplant and bell pepper. J. Environ. Entomol. 38, 1661–1667.

Yang, X.-B., Zhang, Y.-M., Hau, L., Peng, L.-N., Munyaneza, J.E., Liu, T.-X., 2010. Repellency of selected biorational insecticides to potato psyllid, *Bactericera cockerelli* (Hemiptera: Psyllidae). Crop Prot. 29, 1324–1329.

Zens, B., C. M. Rush, D. C. Henne, F. Workneh, E. Bynum, C. Nansen, and N. C. Gudmestad. 2010. Efficacy of seven chemical programs to control potato psyllids in the Texas Panhandle. In: Workneh, F., Rush, C. M. (Eds.), Proceedings of the 10th Annual Zebra Chip Reporting Session, 7–10 November 2010. Dallas, TX, pp. 83–87.

Chapter 5

Wireworms as Pests of Potato

Robert S. Vernon, and Willem G. van Herk
Pacific Agri-Food Research Center, Agriculture and Agri-Food Canada, Agassiz, British Columbia, Canada

INTRODUCTION

Wireworms, the name commonly given to the larval stage of click beetles (Coleoptera: Elateridae), are among the most important and challenging agricultural pests worldwide. Mostly generalist herbivores in nature, wireworms have caused serious economic damage to a multitude of key agricultural crops, including: grain (i.e. wheat, barley, oats), forage (corn, maize), tobacco, most vegetables, sugar beets, sugar cane, sweet potatoes, small fruits (e.g., strawberries), and potatoes (Thomas 1940, Chalfant and Seal 1991, Jansson and Seal 1994, Vernon *et al.* 2000, 2003, Parker and Howard 2001). Wireworms are somewhat unique as a group of pests in many respects. The number of pest species is quite numerous, with an estimated 39 species of wireworms from 21 genera known to attack potatoes worldwide (Jansson and Seal 1994). In addition, wireworms vary considerably in species occurrence and abundance across the worldwide agricultural landscape, with field populations often reaching millions per ha (Miles and Cohen 1941), consisting of one or more dominant or co-dominant key species (MacLeod and Rawlins 1935, Ward and Keaster 1977, Parker and Howard 2001). Typical elaterid life cycles present additional challenges, in that wireworms generally require several years to complete their larval stage (e.g., 4–5 years for certain *Agriotes* species; Miles 1942), during which time economic damage to susceptible crops may occur. In contrast to foliage-feeding insects, wireworms are subterranean in nature and are therefore very difficult to monitor with the precision required of threshold-based integrated pest management (IPM) programs. Also, different wireworm species have differing life cycles and behaviors, differences in availability of monitoring tools (e.g., pheromone traps) and differences in the effectiveness of controls (e.g., biological, natural, and synthetic insecticides). Finally, the general neglect of wireworms as research subjects following the widespread and highly effective use of organochlorine insecticides post-World War II puts our general biological/ecological knowledge of this diverse group of insects far behind that of other common agricultural pests.

Insect Pests of Potato. http://dx.doi.org/10.1016/B978-0-12-386895-4.00005-3

Interest in wireworms has ebbed and flowed over the past century in concert with a number of important agricultural events. Wireworms first gained conspicuous, wide-ranging prominence as crop pests during World Wars I and II (Roberts 1921, Miles and Cohen 1941). The increased necessity for edible crops at that time resulted in the rapid conversion of millions of hectares of grassland (1,000,000 ha in the UK alone) into mostly cereal and potato production (Miles and Cohen 1941). Since wireworms are commonly associated with grassland habitats, damage to subsequent crops by these polyphagous, long-lived pests became an immediate and urgent problem. Without the availability of effective insecticides at that time, considerable research in North America and Europe was devoted to studying the biology and management of various wireworm species. Much of the research conducted during this period is summarized nicely in a review article by Thomas (1940), and it is from these early studies that the bulk of our knowledge of the biology, behavior, and control of various key wireworm species by alternative and integrated control methods comes.

After the advent of modern-day insecticides, beginning with the organochlorines (OCs) in the late 1940s, wireworm researchers became somewhat preoccupied with insecticide efficacy testing, with a declining amount of effort being devoted to the general biology of wireworms and alternative methods of control. The impressive effectiveness of the OCs in controlling wireworms of all species for many years (discussed below), effectively reduced wireworms to sporadic, field-specific importance, which continued with the registration of various carbamate and organophosphate (OP) insecticides after the abandonment of the more persistent OCs in the 1970s and 1980s. The decline in non-pesticide related research on wireworms between 1950 and 2000 essentially paralleled the decline in wireworms as key agricultural pests during that time. In articles by Radcliffe *et al.* (1991) and Jansson and Seal (1994), centering on surveys of potato researchers and their knowledge of potato wireworm problems in North America, it was apparent that wireworm problems had become quite sporadic or absent, and general knowledge of wireworms was often lacking in extension professionals. To most researchers worldwide, wireworms had become minor pests, and the impetus to study them relative to other more important economic insect pests was limited.

The status of wireworms as agricultural pests began to shift from minor to major in various regions of the world in the 1990s, and in some potato and cereal production areas has now, as of this writing, reached a level of urgency comparable to that observed in the early 1900s. A number of reasons have been proposed to account for the resurgence of wireworms as pests, including the loss of all residual OCs and declining residues of these in the soil, the attrition or complete loss of most other effective wireworm insecticides (primarily carbamates and OPs) in many countries, an increase in rotational practices favoring wireworm build-up, adoption of minimum tillage practices, grass set-aside requirements prior to organic production, erosion control initiatives (e.g., the Conservation Reserve Program in the USA), and the growing use of modern-day insecticides that do not reduce wireworm populations (Jansson and Seal 1994, Parker and

Howard 2001, Horton and Landolt 2002, Vernon *et al.* 2009). An awareness of the increasing problems occurring with wireworms, especially in potato crops, prompted an excellent review article covering pertinent biological information and control options for wireworms, particularly with reference to the historical and existing wireworm situation in the UK (Parker and Howard 2001).

In writing this chapter we have drawn on the contributions of many researchers over the past century, but the reader is also directed to the key review articles by Thomas (1940), Miles (1942), Keaster *et al.* (1988), Chalfant and Seal (1991), Jansson and Seal (1994), and Parker and Howard (2001). Some of the topics covered in these reviews, especially those relating to wireworm damage in potatoes, sampling, and management options, will also be covered in the current chapter, albeit with additional up-to-date research. Where this writing differs from other reviews is in the discussion of pest elaterid biology/ecology on a more global scale, which we feel is necessary to illustrate the complexities and differences among this pest complex that must be considered in the development of contemporary management approaches. Where wireworms of all species were generically controlled by OCs in the past, this cannot necessarily be said of modern-day insecticides (van Herk *et al.* 2007, Vernon *et al.* 2008) or other tactics (e.g., biological controls), and many integrated management approaches developed or under development have become more species-specific and regional in scope. It is our intent, therefore, to emphasize the fundamental need to first understand wireworm diversity, biology and ecology, and their potential economic impact(s) in an agricultural landscape, prior to developing regional management strategies.

Since the target audience of this writing is meant to be global, whereas research related to pest management in potatoes has been more regional in scope, we have chosen to present information related to management in a more generalized and conceptual format. In some cases, certain generalities represent the opinions of the authors, and are derived from a synthesis of literature evidence, discussions with other experts and personal experience.

ELATERID BIOLOGY

Wireworm Diversity

Knowing the species of wireworm(s) present in a field is of fundamental importance in developing effective management strategies. While there may be as many as 100 different economic species in the Holarctic region (many listed below), larvae found in a field, even if in abundance, may not cause crop damage. MacLeod and Rawlins (1935) found large numbers of *Cryptohypnus* (later: *Hypolithus*; now: *Hypnoides*) *abbreviatus* (Say) larvae in close proximity to potato plants, but no damage was done to the tubers, suggesting to them that this species was not a pest of the potato. In the Czech Republic, Jedlička and Frouz (2007) found *Agriotes obscurus* L., one of the most important wireworm pests in Europe and Canada, in the same field as species that are saprophagous (e.g.,

Athous niger (L.), *A. subfuscus* (Müller), *A. vittatus* (Fab.)) or carnivorous (e.g., *Dalopius marginatus* (L.), *Agrypnus murinus* (L.)). Wireworms may be a pest of one crop but not another, or in one region but not another. *Glyphonyx bimarginatus* Schaeffer is the second most important wireworm pest of sugarcane in Florida, but is not considered to be of economic importance to potato (Deen and Cuthbert 1955); larvae of *A. sputator* L. and *Athous niger*, both serious pests of vegetables in Europe, entered the USA with nursery stock in the 1920s (Sasscer 1924) but have not been reported as pests there. A wireworm species may also be a pest in one area but beneficial in another. Stirret (1936) lists *Aeolus mellillus* Say as a pest in Ontario, but Doane (1977) reports that it might reduce populations of *Ctenicera destructor* (Brown) and *Hypnoides bicolor* Esch., the two most important pest species on the Canadian prairies, by feeding on their eggs. One of the main wireworm pests in California, *Limonius californicus* (Mann), is an effective predator of root maggots (Stone 1953). *Pyrophorus luminosus* Ill., native to Puerto Rico, was introduced to Mauritius as a biological control agent of white grubs (Bartlett 1939).

Wireworm species vary considerably in size and life history. In the southern USA, *Glyphonyx recticollis* (Say) grow to only 12–14 mm, while *Melanotus communis* (Gyl.), with which it is often found, grow to 25–30 mm (Kulash and Monroe 1955). On the Canadian prairies, *H. bicolor* and *C. destructor* are often found together and reach similar sizes to *G. recticollis* and *M. communis*, respectively (Fig. 5.1). Presumably, both a wireworm's size and duration as larvae will affect the amount of damage it can do. The larval period varies considerably among pest species: *Conoderus bellus* Say completes its larval stage in approximately 30 days (Jewett 1945), *M. communis*, often found in the same field, takes up to 6 years (Fenton 1926). The life histories of some species (e.g., *A. obscurus*) have been studied in detail (Langenbuch 1932, Subklew 1934), but unfortunately very little work has been done in recent decades to determine the life histories of economic species, a notable exception being the excellent work of Furlan on *Agriotes sordidus* (Ill.) and *A. ustulatus* Schäller (Furlan, 2004, 1998).

What is not surprising, but unfortunately often overlooked, is that different species respond to insecticides differently. In both lab and field studies in California, Lange *et al.* (1949), found *Aeolus* spp. was more susceptible to lindane-treated wheat seed than *Limonius canus* LeC., *L. californicus*, and an *Anchastus* species, in that order. Similarly, in China, Chung and Wei (1956) found that two different species of wireworms differed in their susceptibility to lindane-treated wheat seed in both field and lab studies. More recently, van Herk *et al.* (2007) topically exposed larvae of five economic species to two insecticides and found that *Agriotes sputator* L. and *A. obscurus* were more susceptible to clothianidin and chlorpyrifos than *L. canus*, and that the time required for both *Agriotes* species to die after exposure to these chemicals was more than twice as long as required for *L. canus*, *Ctenicera pruinina* (Horn), and *C. destructor*. Carpenter and Scott (1972), who reported that Dasanit® may control *Ctenicera* spp. but not *L. californicus* in Idaho, aptly caution that insecticides which are species-specific "should not be recommended for wireworm control, since most farmers

FIGURE 5.1   Wireworms are the larval stage of click beetles. (A) Late instar larva of *Aeolus mellillus* Say. (B) Adult click beetle of *Agriotes obscurus* L. (C) Late instar larvae of *Selatosomus destructor* (top) and *Hypnoides bicolor*, the two most important pest species on the Canadian Prairies, and often found together.

cannot determine the species of wireworm which they are trying to control, and there usually is not time to get an expert opinion."

Identification: Traditional Methods

Determining the species present in a field is often difficult. Determining what species are present in a field is often difficult. Wireworms, even if present at economic levels, often cannot be found when looked for, due to their moulting cycles and their vertical migration in the soil to escape drought or heat conditions. When wireworms are found, it can be difficult to determine what species they are, as the larval form of most wireworm species remains unknown.

In Europe, some 20 species of *Agriotes* are known to occur in arable land, of which only 8 are described in Klausnitzer's identification key (1994). In some countries, most of the important larvae have been described (e.g., Japan: Ôhira, 1962); in Canada only about 10% of larvae have been described, nearly all of them prior to 1950, and in the USA this percentage is even smaller. As a result, there are a limited number of keys available in North America, and available keys must be used with caution. Most keys for larvae either key to genus (e.g., that of J.R. Dogger, in Becker 1991) or separate known species within a genus (e.g., *Limonius:* Lanchester 1946; *Melanotus:* Riley and Keaster 1979). Unfortunately it is not always possible to use one key to identify down to genus and then switch to another key, as there has been considerable revision of genera and renaming of species since most keys were published. For example, North American larvae of the genus *Ctenicera* can be identified using Glen (1950), but this genus has recently been separated into more than 10 other genera. Unfortunately, new names are often slow to be accepted; Becker renamed *Athous niger* to *Hemicrepidius niger*, but *A. niger* is still used today, as are *Ctenicera destructor* and *C. pruinina*, despite these species now belonging to *Selatosomus*. Other keys attempt to separate economic wireworms of a particular region (e.g., Wilkinson 1963 for B.C., Glen *et al.* 1943 for Canada), but these keys are limited to a very small number of larvae and often end up being used outside of their intended geographic region (e.g., Toba and Campbell (1992) used them to identify Oregon species). To complicate things further, larvae of closely related species (e.g., *A. obscurus* and *A. lineatus* L., *M. communis* and *M. dietrichi* Quate) are virtually impossible to distinguish visually, and one can often only conclude that a particular larva fits the description of a known species rather than reaching a definitive conclusion.

Identification: Non-Traditional Methods

Molecular Approaches

The difficulty in identifying larvae using traditional methods has given impetus to the use of molecular approaches, some of which have demonstrated the difficulties and uncertainties inherent in traditional identification methods. Staudacher *et al.* (2011a) sequenced the mitochondrial COI gene from adults of 17 *Agriotes* species occurring in central Europe and developed specific multiplex PCR assays to identify the 9 most abundant species. This revealed a total of 22 haplotypes, though some species could not be distinguished from each other molecularly (e.g., *A. lineatus* from *A. proximus* Schwarz; *A. brevis* Cand. from *A sputator*). Testing primers developed for these species on larvae thought to be *Agriotes* revealed some actually belonged to the genus *Adrastus*. In subsequent work, Staudacher *et al.* (2011b) have used these primers to identify *Agriotes* larvae from 85 sites in Austria to species, revealing that some species (e.g., *A. brevis*, *A. ustulatus*) prefer warmer, drier climates and alkaline soils, while other species (e.g., *A. obscurus*, *A. lineatus*, *A. proximus*) prefer higher altitudes with lower temperatures, more precipitation, and more acidic soils. Similarly, Lindroth and Clark (2009) have sequenced the COI gene of

11 species of wireworms of economic importance in the Midwestern US, and used these in phylogenetic analyses. Although preliminary, the analyses allowed them to separate 10 species of *Melanotus*, including morphologically indistinguishable species such as *M. communis* from *M. dietrichi*, and *M. opacicollis* LeC. from *M. lanei* Quate. Interestingly, this analysis revealed seven distinct haplotypes within one species (*M. depressus* (Mels.)) alone.

In a joint Canada/UK project, the mitochondrial 16S rRNA of economic wireworms collected from over 100 locations in Canada has been sequenced (Benefer *et al.*, 2013). This analysis has revealed considerable genetic variability related to particular geographic location for species such as *L. californicus* and *H. bicolor*, within-species variability at a particular location for some species (e.g., *Aeolus mellillus*, and *M. similis*) but not others (e.g., *Agriotes criddlei* Van Dyk, *A. stabilis* (LeC.)), cryptic species within *H. bicolor*, and confirmation of the genetic similarity of *C. destructor* (Brown) and *C. aeripennis* (Kby.), often considered subspecies of each other. This analysis also found that some larvae keyed to *C. pruinina* and *A. obscurus/lineatus* were actually other species. The 16S rRNA sequence was also used by Ellis *et al.* (2009) to distinguish between *A. sputator*, *A. lineatus*, and *A. obscurus* in the UK.

Pheromones

Pheromone traps for adult click beetles have also been used to infer what species may be present in an area. This approach has been more successful in Europe, where lures exist for all known economic species (Tóth *et al.* 2003), than in North America, where pheromones of many important species (e.g., *C. destructor*, *H. bicolor*, or most *Conoderus* spp.), if they exist, have not yet been isolated. Certain factors should be kept in mind when considering the potential of this approach. Pheromone trapping will not be effective for species that are (in some geographic regions) parthenogenic, as is the case for *H. bicolor* and *Aeolus mellillus* in Canada (Zacharuk 1958, Stirrett 1936). Some pest species (e.g., *A. lineatus* and *A. proximus* Schwarz) respond very similarly to the same pheromone blend (Tóth *et al.* 2008), potentially leading to misidentification. Some species (e.g., *A. sputator*, *A. obscurus*, *A. lineatus*, and *A. ustulatus*) respond differently to particular pheromone blends based on their geographic location (i.e., they have pheromone dialects; Yatsynin *et al.* 1996), possibly due to genetic differences within their populations. Also, different species will have different periods of flight activity. In Russia the mating flight of *A. sputator*, a northern species, occurs at the end of April and early May, that of *A. obscurus* and *A. lineatus* in mid-May, and that of more southerly species (e.g., *A. gurgistanus* (Fald.), *A. tauricus* Heyden, and *A. ustulatus*) at the end of June and in early July (Kudryavtsev *et al.* 1993).

SPECIES OF ECONOMIC IMPORTANCE IN THE HOLARCTIC

Above we argue the importance of identifying the wireworm species responsible for causing damage in a field. This importance is owed largely to the considerable spatial and temporal variability in wireworm species composition; the species most important can vary from one field to another and from one decade to another. Fields often have more than one economic species, and this species composition can change over time depending on changes in cropping, tillage, or irrigation practices. Hence, knowing what species used to be the predominant pest in an area does not mean it will be the predominant pest today. These points will be illustrated while listing the most important pest species known to occur in various parts of the Holarctic.

Pacific Northwest, Montana, California

Toba and Campbell (1992) state that most of the wireworms of economic importance in the Pacific Northwest (PNW) region of North America belong to six genera (*Aeolus*, *Agriotes*, *Ctenicera*, *Dalopius*, *Limonius*, and *Melanotus*), and report *C. pruinina* to be the predominant dryland species in northern Oregon, followed by *L. californicus*, *M. longulus oregonensis* (LeC.), and *L. infuscatus* (Mots.). Andrews *et al.* (2008) considers the five most important species in the PNW to be *L. canus, L californicus, L infuscatus, L subauratus* LeC., all on irrigated land, and *C. pruinina* on dry land. *C. pruinina* can become a serious pest of vegetables in this region when dryland is first brought into agricultural production (van Herk, personal observation). In central Montana, Hastings and Cowen (1954) list *Athous* spp. as the most important pests, with *L. canus* being the main pest in irrigated areas (Mail 1932). In California, *L. californicus* and *L. canus* are listed as the main pests on irrigated land, though the dominant pest species varied from field to field (Stone and Campbell 1933, Campbell 1942). These species are also listed as serious pests in southwest Idaho (Shirck 1945) and the Yakima Valley (Gibson 1939). Interestingly, Lane (1925) describes the species of wireworms of principal economic importance in Washington, Oregon, and Idaho as *Ludius [Ctenicera] inflatus* Say (the name mistakenly given to *L. glaucus* (Germar)), *L. [C.] noxius* Hyslop, and *Pheletes [Limonius] occidentalis* Cand., but these species were no longer considered economically important by 1950 (Glen 1950). If the predominant species in the PNW changed over the last century, it will likely undergo more change in the near future with the spread of *Agriotes obscurus* and *A. lineatus*. These species were first reported in the northernmost area of Washington state in the 1990s (Vernon and Päts 1997), but by 2005 were found in southern Washington (LaGasa *et al.* 2006) and in potato fields in Oregon (Andrews *et al.* 2008).

Midwestern USA

Lindroth and Clark (2009) suggest 15 species belonging to 6 genera are economic pests of cereal crops in the midwestern USA, including *A. mancus* (Say), *L. dubitans* (LeC.), *C. lividus* (De Geer), *Hemicrepidius memnonius* (Herbst), *Aeolus mellillus*, and 10 species of *Melanotus* – i.e., *M. depressus, M. verberans*

(LeC.), *M. lanei*, *M. opacicollis*, *M. similis* (Kirby), *M. cribulosus* (LeC.), *M. pilosus* Blatchley, *M. communis*, *M. dietrichi*, and *M. indistincus* Quate. To this list, Lefko *et al.* (1998) adds *Conoderus vespertinus* (Fab.), *Ctenicera inflata* (Say), *Hemicrepidius hemipodus* (Say), and *Hypnoides abbreviatus* for Iowa. Many of these species were also listed as being economically important in the past. Ward and Keaster (1977) listed *M. depressus*, *M. verberans* (LeC.), *M. opacicollis*, *L. dibutans* (LeC.) (now considered the same species as *L. agonus* (Say), Al Dhafer 2009) as the most common wireworm species of Missouri, Illinois, Indiana, and Iowa; Fenton (1926) describes *M. communis* and *M. pilosus* as the most important species in Iowa; and McColloc *et al* (1927) lists the most destructive wireworms belonging to the *Melanotus*, *Monocrepidius* (*Conoderus*), *Lacon*, *Agriotes*, and *Ludius* (*Ctenicera*) genera in Kansas.

As elsewhere, the predominant species in this region varies from field to field. In baiting studies in different fields in Missouri, Ward and Keaster (1977) found a 3 : 1 ratio of *M. depressus* and *Aeolus mellillus* in one field, and a nearly pure population of *M. verberans* in another. Similarly Lefko *et al.* (1998) found wireworm species composition to differ considerably among 89 fields sampled in Iowa, and concluded that species composition changes with latitude, soil moisture, and other soil characteristics. The predominant species appears to change when moving north and east within this region. In his survey of potato farmers in North Dakota, Munro *et al.* (1938) lists 58% of potato farmers suffering crop loss due to wireworms and identifies the most important pest as *Ctenicera destructor*, followed by *Melanotus* spp., *Hemicrepidius* spp., and *Limonius* spp., but in Wisconsin Dogger and Lilly (1949) found *M. communis* and *Dalopius pallidus* Brown to be the predominant pest species, followed by *Agriotes mancus* and *Aeolus mellillus*.

Some of the species identifications mentioned in the literature above must be treated with caution, however, as some species are not distinguishable as larvae (e.g., *M. dietrichi* and *M. communis*; *M. lanei* and *M. opacicollis*; Riley and Keaster 1979), and as the larval form of most species remains unknown. To date only two (of 204) species of *Aeolus* have been described as larvae, one of which does not occur in North America (Casari 2006).

Mid-Atlantic, Central Eastern

The economic species in the mid-Atlantic and Central Eastern states are mostly *M. communis* and various *Conoderus* spp., though the predominant species differ considerably between and within states. In Georgia, *Conoderus rudis* (Brown) and *C. scissus* (Schaeffer) are the predominant pest species in sweet potatoes (Seal 1990), while *M. communis* is listed as the most important species in Virginia (Kuhar *et al.* 2008), though Herbert *et al.* (1992) lists the predominant species in Virginia and North Carolina as *C. vespertinus* (F.), followed by *C. lividus*, *Glyphonyx* sp., and *M. communis*. Willis (2010) also reports *C. vespertinus* as the predominant species (in the sweet potato) in North Carolina, but follows it with *C. amplicollis* (Gyl.), *C. bellus*, *C. falli* (Lane), *C. lividus*,

C. scissus, M. communis, and *Glyphonyx bimarginatus*. Interestingly, Kulash and Monroe (1955) found *M. communis* to be the most abundant and damaging pest in sections of North Carolina, followed by *C. lividus* and *Glyphonyx recticollis*, and, earlier, Kulash (1947) listed *C. bellus* and *C. auritus* (Herbst) to be the main pest (on corn) in North Carolina. These discrepancies may indicate that the species composition in these states is changing over time, or reflect bias in how the surveys were conducted (e.g., if only on a certain crop the survey may be biased to wireworms with a particular soil preference), or may simply reflect species variability occurring on small scale. Evidence for the latter explanation comes from Seal *et al.* (1991): in a study in Tifton, Georgia, on sweet potato in two fields with different soil types at one research farm, *C. scissus* was dominant in one field and *C. rudis* in the other, and there were differences in the varieties that caused the most damage. Similarly, Willis *et al.* (2010) report *C. amplicollis* was by far the predominant species in one North Carolina field, and very scarce in all other fields sampled in this survey. Some caution must again be used as to the accuracy of some of these identifications. The importance of *C. rudis* as a pest of sweet potato was not appreciated until the larvae was described in recent years (Seal and Chalfant 1994), after which it was considered one of the most important pest species in this area (Jansson and Seal 1994).

Southeastern USA

As in the mid-Atlantic and Central Eastern states, the most important species in southeastern USA appears to be *M. communis* and various *Conoderus* species. Jansson and Lecrone (1991) list *C. rudis, C. amplicollis, C. falli, C. vespertinus*, and *C. scissus* as the most important pests of potato. Of these, Griffin *et al.* (1953) list *C. amplicollis* as the primary pest on sweet potato and Day *et al.* (1964) list *C. falli* as the most important wireworm pest. Deen and Cuthbert (1955) name *C. falli* (listed by them as *C. vagus* Cand. and appearing in literature as *Monocrepidius difformis* Fall., *M. vagus* Cand., *M. falli* Lane, *C. difformis* Fall., and *Heteroderes vagus* Cand., *cf.* Dobrovsky 1953, Lane 1956, Stone 1975) as the most common pest species of potato in southeastern USA, and do not consider *C. vespertinus, C. lividus, C. bellus*, and *C. rudis* as serious pests of potato. *Horistonotus uhlerii* Horn is listed as the most important wireworm pest of Louisiana (Floyd 1949) and economic in Missouri (Keaster and Fairchild 1960), while Bynum *et al.* (1949) consider *Aeolus* sp. as one of the main pest species in Louisiana.

Part of these conflicting claims are owing to change over time due to the introduction of exotic species. *C. amplicollis* was listed as the chief pest of potato in the 1920s and 1930s in southern Alabama (Cockerham and Deen 1936), but became less important than *C. falli* (a species that, like *C. rudis*, was introduced from South America) in the 1950s (Deen and Cuthbert 1955). The relative importance of these two pest species varied with location, with *C. falli* predominating in coastal areas of South Carolina, Georgia, Florida, Alabama, but not elsewhere in these states (Deen and Cuthbert 1955). The relative importance ascribed by an author to a pest species also depends on the crop in which wireworm damage is assessed. *Glyphonyx bimarginatus* is considered to be the second most important pest species in Florida (after *M. communis*) due to its effect on sugarcane (Cherry 1988), and while it can damage potato in the southeastern USA is not considered to be of economic importance (Deen and Cuthbert 1955). Another consideration in assessing the importance of a pest species should be the geographical range over which it is found. For example, *C. falli* can be a pest species from North Carolina to Alabama (Lane 1953), and *C. amplicollis*, formerly known as *Heteroderes laurentii* (Guer.), is also a pest in Texas and as far west as California (Deen and Cuthbert 1955).

Northeastern USA

Although little has been published from the Northeastern USA in recent years on wireworm species of economic importance, the most important species appear to be *Agriotes mancus* and *Limonius agonus* (listed as *L. ectypus* in older literature, but not the same as *L. infuscatus*, which is now called *L. ectypus*; Al Dhafer 2009) in Maine (Hawkins 1930, Hawkins *et al.* 1958), New York State (MacLeod and Rawlins, 1935), Pennsylvania (Horsfall and Thomas 1926), and New Jersey (Pepper *et al.* 1947). *A. mancus* is also listed as a key potato pest in Michigan, along with *Melanotus* spp. (Merrill 1952), while *L. agonus* and *Ctenicera* spp. are listed as pests in cultivated land, and *Melanotus* spp. in sod in the Connecticut River Valley of Massachusetts (Kulash 1943). Other species of importance include *Hemicrepidius decoloratus* Say and *Hypnoides abbreviatus* in Maine (Hawkins 1936). It is somewhat misleading to state what species is most important on a per state basis, as *Melanotus* spp. were the most damaging species in some Maine fields (Hawkins 1930) and in Pennsylvania *A. mancus* is the main pest in the north and *L. agonus* in the south (Rawlins 1934), at least historically.

Canada, Alaska

The main pest species in Alberta, Saskatchewan, and Manitoba are *Hypnoides bicolor* and *Ctenicera destructor* on non-irrigated land and *Limonius californicus* on irrigated land (Brooks 1960), the latter two being capable of considerable damage to potato (Lilly 1973). *C. destructor* is also found in North and South Dakota, Minnesota, Washington, Oregon, Wyoming, and Alaska (Zacharuk 1962a). Other Prairie species of more local economic importance include *Oestodes puncticollis* Horn in saline fields; *Agriotes criddlei, C. sexualis* (Brown), and *C. glauca* (Germar) in mixed prairie regions of southern Alberta and Saskatchewan; *Limonius pectoralis* (LeConte), *C. kendalli* (Kirby), and *A. limosus* (LeC.) in newly cleared forest land; and *A. mancus, Melanotus castanipes* (Paykull), *M. fissilis* (Say), *Aeolus mellillus*, and *H. abbreviatus* in parkland areas of Manitoba (Brooks 1960). The main native species in the rest of Canada include *C. triundulata* (Randall) and *C. nitidulus* (LeC.) in New Brunswick (Morris 1951); *Dalopius pallidus*, the *M. communis-fissilis* complex, *A. mancus, Ctenicera cylindriformis* Herbst, *D. pallidus, Hemicrepidius memnonius,* and *Hypnoides abbreviatus* in Nova Scotia (Fox 1961); *A. mancus* and *Limonius aeger* LeC. in Quebec (Lafrance 1963); and *Ctenicera aeripennis, C. lobata* (Esch.), *L. canus, L. infuscatus,* and *A. sparsus* LeC. in British Columbia (BC).

We have identified wireworms causing economic damage in Canada since 2004, and have found the most important pest species to have remained the same on the Prairie provinces since 1960. Samples taken from over 300 locations indicate that *C. destructor* and *H. bicolor* are often found in the same fields, and that the latter is a serious pest of wheat (W. van Herk, unpubl. data). Introduction of three European species has changed the picture in the rest of Canada, however. *Agriotes obscurus, A. lineatus,* and *A. sputator* have become the dominant wireworm pest species in Canada where they have become established (Vernon and Tóth, 2007): *A. obscurus* and *A. lineatus* in British Columbia, Nova Scotia, and Newfoundland (Vernon *et al.* 2001, Vernon 2005), and *A. sputator* in Nova Scotia (Fox 1961) and Prince Edward Island, where it has become a serious pest of potato (R. S. Vernon, unpubl. data). *A. obscurus* was probably introduced to Agassiz, BC, with soil on nursery stock (e.g., hops) between 1895 and 1900 (MacNay 1954) and was causing serious crop damage in that area by the 1950s (King *et al.* 1952). In Atlantic Canada, *A. obscurus, A. lineatus,* and *A. sputator* may have arrived as early as the 1850s, likely with ballast unloaded from sailing ships (Eidt 1953). *A. lineatus* has also established itself in Brazil, Haiti, and New Zealand (Afonin *et al.* 2008).

Wireworms also cause economic damage to potatoes in Alaska, although the species responsible varies with region. Pantoja *et al.* (2010) list *Hypnoides bicolor* as the predominant wireworm species in Delta Junction and *Limonius pectoralis* as predominating in Fairbanks.

Russia and Eastern Europe

As in North America, economically important wireworm species in Russia vary with latitude. In a landmark study, Kudryavtsev *et al.* (1993) placed 300,000 pheromone traps across Russia, monitoring some 3 million hectares over 7 years. This research outlined the distribution of six key *Agriotes* species in western Russia, revealing that *A. obscurus* and *A. lineatus* are distributed further north than *A. sputator*, and that two forms of *A. lineatus* exist in the European part of Russia. In general, *A. sputator, A. obscurus,* and *A. lineatus* are found in central and northwestern Russia (though *A. obscurus* and *A. lineatus* are also found in Mongolia and Siberia, and *A. sputator* in North Africa and Asia Minor), and *A. gurgistanus, A. ustulatus,* and *A. tauricus* are the predominant pest species in Southern Russia, the Crimea, eastern Europe, the Caucasus, Asia Minor, and the Balkans (Kudryavtsev *et al.* 1993). Considering the large distribution of some species, it is not surprising that the pheromone composition of *A. sputator,*

A. obscurus, *A. lineatus*, and *A. ustulatus* differs considerably between beetles collected near Moscow, in Estonia, or in the Caucasus mountains (near Krashnodar) (Yatsynin *et al.* 1996). Excellent information about wireworm distribution in Russia can be found at www.agroatlas.ru. According to Afonin *et al.* (2008) the economic threshold in Russia for *A. obscurus* in potato is 3–15 larvae per m^2, though the population density may reach 180 larvae per m^2 in some fields.

Other species of economic importance belong to genera of the so-called "soft-bodied" wireworms – e.g., *Selatosomus*, *Ctenicera*, *Athous*, *Pleonomus*. In a survey of injurious soil insects of arable land in Ukraine and Russia, Ghilarov (1937) considers *S. aeneus* (L.) more injurious to crops than *A. sputator*. Near Moscow, *A. lineatus* and *A. obscurus* were found in established farm fields, while *Athous niger* predominated in uncultivated fields and were more damaging when these fields were brought into agricultural production. Ghilarov (1937) lists *Pleonomus* spp. as the most important species on non-irrigated crops grown in Kazakhstan and Uzbekistan, while the predominant species of irrigated land was *Agriotes meticulosus* Cand., which he did not consider to be of economic importance. Other important pest species include *S. latus* F., which is found also in Asia Minor, the Caucasus, Kazakhstan, Northern Mongolia, and Northern China, though not in southern Siberia; *S. aeneus* is also found in the Caucuses, Northern Kazakhstan and Northern Mongolia, though most damaging in eastern Europe; and *S. spretus* (Mann.) most important in southeastern Russia and Northern Mongolia and generally occuring on south-facing mountainsides up to 1000m above sea level (Afonin *et al.* 2008). Of these, *S. aeneus* is said to prefer light mineral podzolic soils, *S. latus* heavier soils, and *S. spretus* chernozem-like meadow soils.

Western and Central Europe

Of the 672 elaterid species known in Europe (Cate 2004), most of the economic species belong to the genus *Agriotes*. Tóth *et al.* (2003) lists the important species as *A. brevis*, *A. lineatus*, *A. litigiosus*, *A. obscurus*, *A. rufipalpis*, *A. sordidus*, *A. sputator*, and *A. ustulatus*, to which Staudacher *et al.* (2011a) adds *A. proximus*. The distribution of these species varies considerably, as shown on the Fauna Europaea website (www.faunaeur.org). For example, while *A. obscurus*, *A. lineatus*, and *A. sputator* are more northern species, *A sordidus* and *A. ustulatus*, two serious soil pests in Italy, are widespread throughout central and southern Europe, Turkestan, Armenia, and North Africa, though they are also present in the UK, Belgium, Switzerland, and Germany (Furlan 1996, 2004). Of these, *A. ustulatus* has recently been recorded as new to Denmark, indicating that the species distribution changes over time. Of the 60 species known in the UK, only *A. obscurus, A. lineatus*, and *A. sputator* are considered pests (Parker and Howard 2001). Only the larvae of these three, and *A. brevis* and *A. ustulatus*, are pest species described in Klausnitzer's key (1994). Pest species of lesser importance include *Athous haemorrhoidalis* (Fab.) and other *Athous* species in the UK, France, and Germany (Blot 1999, Parker and Howard 2001, Hemerik *et al.* 2003).

Asia

In recent years, wireworm research in Japan appears to have focused on *Melanotus okinawensis* Ohira and *M. sakishimensis* Ohira, two serious pests of sugarcane with a somewhat different distribution in Okinawa and Kagoshima Prefectures (Ôhira 1988). A third species, *M. tamusyensis* Bates, is mentioned in older literature. Earlier research indicates that in Hokkaido the most important pest species of potato is *Agriotes fuscicollis* Miwa (Hayakawa *et al.* 1985), followed by *Ctenicera puncticollis* (Mots.) and *Melanotus caudex* Lewis (Kuwayama *et al.* 1960). Of these, *A. fuscicollis* is found in ill-drained peat soil and appears to be restricted to the Ishikari basin, *M. caudex* is most common in light volcanic ash soil and is widely distributed in Japan, and *C. puncticollis*, thought to be the most damaging species where present, is limited to clay soil in the northern part of the Sorachi Subprefecture (Kuwayama *et al.* 1960).

Wireworms are emerging as serious pests in China on a variety of crops (e.g., bamboo, Jinping Shu, pers. comm.). In a recent survey, Zhao and Yu (2010) list some 50 economic species, of which 5 are of particular importance, and Wu and Li (2005) mention that 600–700 species are known in China, of which 20–30 occur in arable land and 4 are of economic importance. These species are *Melanotus cribricollis* Horn (also ascribed to Candéze and Faldermann in recent literature), a major pest of bamboo in Zhejiang province; *A. fuscicollis*, a pest in northern China, including the Gansu province (Liu *et al.* 1989); *M. caudex*; *Selatosomus latus*; and *Pleonomus canaliculatus* (Fald.), a pest in 13 provinces in northern China, often in dry and poor soil (Wu and Li 2005), and in loamy soil along riverbanks in Honan province (Wu 1966). Other known pest species in China include *Agriotes subvittatus* (Mots.) and *Limonius minutus* (L.), though their importance is less well known.

Wireworms are listed as pests of potato in Korea in the 1960s, though no species names are mentioned (Choi 1972). In a recent review of the most important species in Korea, Park *et al.* (1989) identified 14 species as pests of potato in Korea, listing *Selatosomus (Ctenicera) puncticollis* as the most important. In their evaluation of resistance of potato cultivars to wireworm damage, Kwon *et al.* (1999) lists *S. puncticollis*, *Melanotus legatus* Cand., *Agrypnus argillaceus* (Solsky), and *Agrypnus binodulus coreanus* Kishii as the main species (Kwon *et al.* 1999). Pemberton (1962) lists *Lacon musculus* Cand., *Melanotus tamsuyensis*, and *Sephilus formosanus* Schwarz as the most serious wireworm pests of sugarcane in Taiwan, but their damage to potato is not clear.

Little is known to the western world about the wireworm situation in the Middle East and Turkey, an exception being Iran. Bagheri and Nematollahi (2007) and Bagheri and Ardebili (2005) evaluated the resistance of 10 commercial

potato cultivars to wireworms in the Shahr-e-Kord region of Iran, and used pheromone traps to determine that the pest species at their research centre were a
mixture of *Agriotes baghrii* Platia, Furlan & Gudenzi, *A. iranicus* Platia, Furlan
& Gudenzi, and *A. proximoides* Platia, Furlan & Gudenzi (Platia *et al.* 2002). In
Hormozgan province, Ranjbaraqdam *et al.* (2001) surveyed the main economic
species attacking potato and found *Melanotus* spp. the most damaging, also
observing their life cycle to be more than 1 year.

DIFFERENCES WITHIN ECONOMIC SPECIES

Mating, Oviposition, and Larval Development

Describing a typical elaterid life cycle is misleading, since both beetle and larval behavior differ widely among species and are dependent on food availability
and environmental factors such as soil temperature and moisture. While little is
known about the life history of most economic species, the little that is known
indicates considerable variability both within and among genera.

In northern Italy, larval *Agriotes sordidus* pupate in late summer, overwinter
as adults, and emerge from their soil cocoons in late March and early April, their
emergence peaking in May and females emerging 1–2 weeks after males (Furlan 2004). Mating occurs immediately after emergence, and oviposition takes
place several days later. An average of 150 eggs (range: 120–200) are laid in
the field, in clusters of 3–30, apparently at random. In comparison, *A. ustulatus*
lays approximately 80 eggs (range: 50–140), in similar clusters, 2–4 days after
mating (Furlan 1996). In *A. ustulatus* mating can be repeated several times, and
after mating both sexes live for another 10–15 days (Furlan 1996). Similarly,
on the Canadian prairies, larval *Ctenicera destructor* pupate in the fall after
10 instars and also overwinter as adults which emerge in spring (Strickland
1933, 1939, Zacharuk 1962a). After emergence, adults mate for 10–25 minutes, and females begin to lay eggs 7–16 days thereafter and continue to do so
for 1–22 days (Zacharuk 1962a). Neither sex mates more than once (Zacharuk
1962a). According to Zacharuk (1962a), *C. destructor* lays an average of 180
eggs (range: up to 480), while Doane (1963) reports an average of 160–950 eggs
(range: up to >1400). Both report that egg laying typically follows an interrupted pattern of several days of laying followed by several days of not laying,
repeated several times. In comparison, the average fecundity of *A. mancus* in
New York was calculated to be 100 eggs (Rawlins 1940), that of *A. litigiosus* in
Russia 200 eggs (range: up to 370, Kosmatshevsky 1960), and that of *M. caudex*
in Japan only 17 (Yoshida 1961). Not surprisingly, there are also differences in
where females lay their eggs. Gibson *et al.* (1958) reports that *L. canus* prefers
to lay its eggs in bare soil while *L. californicus* prefers soil shaded by vegetation.

The number of larval instars and duration of development also differ considerably among and within species. *Lacon variabilis* Cand., an Australian species, develops through 6–8 instars (Zacharuk 1962b), while in the southern USA

Conoderus rudis develops through 3–4 instars (Seal and Chalfant 1994) and *C. scissus* through 7–10 (Chalfant and Seal 1991). *A. sputator* develops through 10–12 instars in Russia and 7–9 in England (Zacharuk 1962b). Larvae of *Glyphonyx recticollis* complete development in 1 year, while larvae of *M. communis* and *C. lividus* in the same field require 4 and 2 years, respectively (Kulash and Monroe 1955). In China, *Melanotus cribricollis* and *S. latus* have a life cycle of 4–5 years, *M. caudex* and *P. canaliculatus* 3 years, and *A. fuscicollis* 2 years (Wu and Li 2005, Zhou *et al.* 2008). In Japan, *M. okinawensis* has a 2- to 3-year life cycle (Arakaki 2010).

The variation of life cycles is perhaps most notable in the genus *Conoderus*. Of the pest species often found together in southern USA, *C. scissus* has a 2-year life cycle (Seal *et al.* 1992); *C. lividus* and *C. amplicollis* complete their larval development in 1 or 2 years (Cockerham and Deen 1936, Jewett 1946, Seal *et al.* 1992); *C. vespertinus* and *C. bellus* have a 1-year life cycle, although *C. vespertinus* is a larva for 300–350 days (Eagerton 1914, Rabb 1963) and *C. bellus* for approx. 30 days (Jewett 1945); and *C. rudis* and *C. falli* complete larval development in 2–3 months (Seal *et al.* 1992) with at least two generations per year (Norris 1957, Chalfant *et al.* 1979).

Unfortunately, the life cycle of many pest species remains unknown, and that of others can vary considerably based on latitude and diet. The larval stage of *Ctenicera destructor* can vary from 4 to 11 or more years during which the larvae have been observed to reduce in size despite availability of food (Strickland 1942). This delayed development and "regressive moulting" has also been observed for *L. canus* and *A. obscurus* larvae (W. van Herk, unpubl. data).

Larval Activity

Pest wireworm species do not always cause damage when present, as the larvae appear to spend a relatively small amount of time of each instar actively feeding. In his studies of *A. ustulatus* and *A. sordidus* development, Furlan (1998, 2004) reports the amount of time these larvae spend per instar on (consecutively) mandible hardening (10 d, 8%, respectively), feeding (19 d, 24%), and pre-moulting (71 d, 68%). Similarly, Kosmatshevsky (1960) describes *A. litigiosus* as undergoing four phases of activity in each instar: moulting, intensive feeding, intensive burrowing with little feeding, and quiescence in earthen cells. Zacharuk (1962a) observed these phases in *C. destructor*, though not the intensive burrowing phase.

As conspecific larvae in a field undergo these phases in synchrony (e.g., moult at the same time; W. van Herk, unpubl. data), there are periods of peak wireworm activity. The number of such peak activity periods will depend on numerous factors (e.g., species, latitude, weather), and the time of activity periods may vary from year to year for a particular species in a particular field (R. S. Vernon, unpubl. data). Doane (1981) reports that on the Canadian prairies *C. destructor* and *H. bicolor* have a period of peak activity in June, followed by a sharp decline in activity, and a second, less intense period of activity in August–September; similar periods of activity have been observed for

A. obscurus in Agassiz, BC (R. S. Vernon, unpubl. data). For these species the periods of reduced activity generally coincide with the periods of warmest and driest soil temperatures, which are avoided by vertical migration downward in the soil (Falconer 1945). As periods of activity vary with species and location, it is critical for effective management that they be assessed on a local level, as has been done for *M. communis* in Florida (Cherry 2007) and various *Limonius* species in the Pacific Northwest (Jones and Shirck 1942, Stone and Foley 1955).

Preferences in Soil Type and Soil Moisture Content

Economic wireworms differ considerably in soil type and soil moisture preferences. Some genera (e.g., *Limonius*) prefer moist soils, others are obligate dryland species that disappear when it is brought under irrigation (e.g., *Ctenicera*, with the notable exception of *C. cylindriformis*; Fox 1961). For a review of wireworm soil moisture preferences we refer to Thomas (1940) and Chalfant and Seal (1991). Little research has been done on this in recent years, notable exceptions being Zacharuk (1962b), Lees (1943), and Schaerffenberg (1942). Wireworm soil preferences can be quite specific also. In a comparison of the distribution of *Ctenicera aeripennis* and *C. destructor* in Canada, Zacharuk (1962a) found *C. destructor* in brown and black soil zones and *C. aeripennis* in the more northern grey soils, with virtually no overlap of the two. The degree of specificity is species-dependent: MacLeod and Rawlins (1935) found both *A. mancus* and *L. agonus* prefer low-lying parts of New York fields, but this was considerably more pronounced in *A. mancus*. Such preferences help explain why the species composition can vary between nearby fields.

Feeding Preferences

Many wireworm species do not feed on crops (Zacharuk 1963, Turnock 1968), but there is considerable variability in the feeding preferences of conspecific larvae of pest species. In his study of the feeding ecology of *A. obscurus* in Austria, Germany, and Italy, Traugott *et al.* (2008) measured their carbon ($^{12}C/^{13}C$) and nitrogen ($^{14}N/^{15}N$) isotope ratios to determine their trophic level and the importance of soil organic matter and weeds within their diet, and observed that 10% of the *A. obscurus* larvae fed primarily on animal prey. Interestingly, phytophagous *A. obscurus* show a clear preference for the roots of some types of grass over others (Hemerik *et al.* 2003), but remain with a particular food source as long as its supply remains sufficient (Schallhart *et al.* 2011). Not surprisingly, a wireworm's food source will affect its growth, as has been demonstrated for *L. dibutans* and *M. depressus* (Keaster *et al.* 1975). Wireworm feeding preferences have been reported for various species (e.g., *A. sputator*; Fox 1973), and help explain the differences in tuber damage when different potato cultivars are planted in the same fields, as reported from studies conducted in Sweden, Iran, Korea, and the UK (Jonasson and Olsson 1994, Kwon *et al.* 1999, Bagheri and Nematollahi 2007, Johnson *et al.*

2008). It is interesting to note that different species may also cause different types of damage to a crop. Willis *et al.* (2010) report that *C. amplicollis* appears to cause more extensive surface scarring of sweet potato than other *Conoderus* species.

WIREWORMS AND THE POTATO CROP

Typically, common above-ground insect pests of potato arrive as immigrants to the already established potato crop. The opposite is true with wireworms, where potatoes are planted into fields that have already been occupied by one or more species for one to several years, depending on the species and cropping history of the field. There is absolutely no question that wireworm populations can build up to enormous numbers in grassland or pasture, especially where these fields have been in this state for several years (Thomas 1940, Parker and Howard 2001), and this seems to be a generality worldwide. A mature field of pasture may contain cohorts of more than one species of wireworm from adult oviposition events occurring over a number of consecutive years, and populations can be distributed in various spatial patterns throughout a field depending on species and the various habitat variables (e.g., soil moisture) mentioned above. When pasture is removed, generally by ploughing, wireworm populations remain in place in the soil and, depending on the length of their life cycle, can feed on subsequent crops for years until all cohorts have left the soil as adults. It is generally when potatoes are planted following the removal of pasture that damage from wireworms can be severe, and this damage can be as or even more severe if potatoes are grown in the second year (Miles and Cohen 1938). The severity of wireworm damage to a potato crop in the first or second year is likely related to how the preceding pasture is removed from the field. Pasture that is ploughed green just before planting potatoes will delay and reduce damage to them in the first year, since wireworm populations will occupy and feed on the slowly decomposing green manure that typically lies about 20–30 cm below ground (Miles and Cohen 1938). When this green manure has fully decayed, usually by late summer, wireworms will then gradually move to other food sources such as daughter potato tubers, and cause damage commensurate with population size and the amount of time they spend feeding. If potatoes are planted the following year, the only food source throughout the growing season is generally the potato crop, and a higher amount of daughter tuber damage can occur. In fields where the pasture is killed with herbicide in fall or early spring, and subsequently disked rather than ploughed prior to planting potatoes, wireworm populations will feed primarily on the potato crop throughout the initial growing season, and damage will be highest in the initial year of planting (R. S. Vernon, unpubl. data). This principle also has a direct bearing on the success or failure of various wireworm monitoring and management strategies, which will be discussed later.

In addition to pasture, significant oviposition will also occur in cereal crops, and high wireworm populations can arise in these fields in a single year (Thomas 1940, Salt and Hollick 1944, Andrews *et al.* 2008). This is of particular relevance to potatoes, especially in Canada, in that the majority of potatoes grown in Canada have 1 or more years of a cereal crop in their rotations (Noronha 2011, Vernon *et al.* 2011). On the Canadian prairies, where an increasing amount of wireworm damage is occurring to wheat each year (R. S. Vernon, unpubl. data), growers routinely apply the organophosphate phorate (Thimet 15G) in rotated potato fields to prevent cosmetic damage to tubers destined mostly for the processing industry. In Prince Edward Island (PEI), growers typically plant barley or wheat undersown with clover in 3-year rotations with potatoes (Noronha 2011). This practice, which is known to give rise to damaging wireworm populations in fields (Landis and Onsager 1966), has also resulted in the requirement for prophylactic phorate use on an increasing scale, and, despite this, severe damage is occurring on many farms.

The potato crop itself has not been reported to be a favored site for oviposition by click beetles, although our work has shown that some oviposition and production of a small cohort of wireworms (*A. obscurus*) will occur in well-weeded potato fields (R. S. Vernon and W. van Herk, unpubl. data).

During a typical potato growing season in the northern hemisphere, many species of wireworms manifest two distinct periods of feeding activity in potatoes (e.g., Doane 1981, Gratwick 1989, Parker and Howard 2001), but this can vary between species and between species between years (Doane 1977, 1981) as discussed above. Typically, the first activity period occurs in spring (between April and June), which usually coincides with the planting of mother tubers and the development of roots and stems. As will be discussed later, it is believed that wireworms in the soil respond to carbon dioxide (CO_2) (Doane *et al.* 1975) produced by the sprouting mother tubers, and follow these cues to the planted rows. Damage to the tuber or roots appears as holes of about the same diameter as the wireworms that made them, and they can often be seen partially or wholly inside tubers at that time. In cases of extremely high populations some tuber mortality may occur, but the crop generally establishes normally even in the presence of high populations. This period of feeding is followed in the hot, dry summer months of July and August by a quiescent period where little feeding or damage occurs. The second period of feeding generally occurs from late August through to the end of potato harvest; during this time daughter tubers are developing and two types of characteristic damage may occur. Wireworms feed on daughter tubers in much the same way as on mother tubers, making wireworm-sized holes into the tuber flesh. However, feeding on smaller daughter tubers produces holes that expand and suberize as the tuber matures, often giving rise to misshapen and unmarketable tubers (Fig. 5.2). Entry holes made to larger tubers are the most common form of damage, however, and these can be quite numerous under high wireworm pressure (Fig. 5.2). As would be expected, the higher the population of potato-feeding wireworms, the higher the number of holes generally made in tubers (Menusan and Butcher 1936, Thomas 1940), although the number of holes per tuber is not necessarily in direct proportion to the population, and damage to tubers may be clumped (Gui 1935). Potato varieties

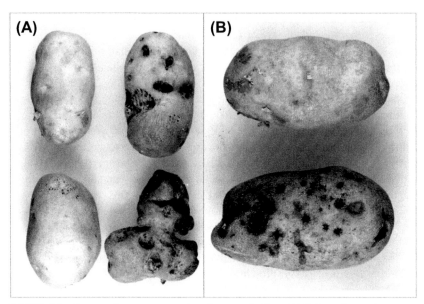

FIGURE 5.2 Damage done to potatoes by wireworm feeding. (A) Left, undamaged potatoes; Right, potatoes damaged early in the growing season, causing misshaping. (B) Extensive (bottom) and slight (top) damage to tubers late in the growing season.

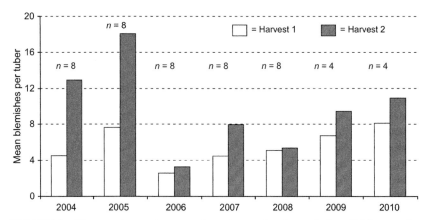

FIGURE 5.3 Wireworm damage increases when tubers are kept longer in the soil. Mean number of wireworm blemishes per market-sized tuber in samples taken approximately 20 days apart from control treatments of insecticide efficacy studies conducted in by the authors 2004–2010.

with a lower number of tubers per hill have also been shown to have more damage per tuber than varieties with larger numbers per hill (MacLeod 1936). In addition, the longer the tubers are left in the ground before harvest, the higher the amount of damage that will occur, as we have found in several years of potato trials (Fig. 5.3) and has also been reported elsewhere (Anonymous 1948,

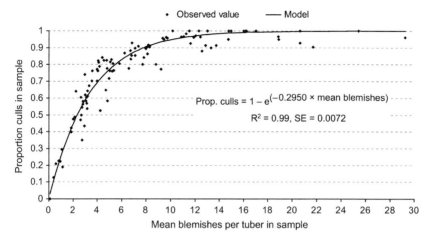

FIGURE 5.4   The proportion of tubers culled per sample is significantly correlated to the number of mean blemishes per tuber. Shown are data from control treatments of insecticide efficacy studies conducted in 2003–2010; $n = 104$ independent samples.

Parker and Howard 2001, Kuhar and Alvarez 2008). This increase in holes with time is likely due to continued feeding by individuals, by increasing numbers of wireworms feeding with time, or both, although this has not been thoroughly researched (Kuhar and Alvarez 2008, Parker and Howard 2001).

The impact of wireworm feeding on the marketability of a potato crop is a function of the amount of visible wireworm damage and all other defects on tubers, and tolerance for tuber blemishes varies according to various industry standards and the final destination of the crop. Organic and seed potatoes, for example, may have a higher tolerance for blemishes than ware or processing potatoes, but even these rules of thumb may vary depending on the collective type(s) of blemishes involved, the client, product abundance, competition, and so on. In our work to develop management approaches for wireworms, we have established a toler-ance of only one obvious wireworm hole per tuber, with two or more holes or wireworm-caused deformities constituting a "cull". In addition, we aim for less than 5% overall cullage caused by wireworm feeding as being a threshold that would generally meet most industry standards. The relationship of the percentage of culls due to wireworm damage (≥ 2 blemishes/tuber) versus the mean number of holes per tuber is shown in Fig. 5.4 for feeding by *A. obscurus* on Chieftain pota-toes in British Columbia, Canada, and a virtually identical relationship has been found for wireworm damage (*Melanotus* spp. and *Limonius agonus*) on potatoes in Ontario, Canada (R. S. Vernon, unpubl. data). Such a relationship, however, would be expected to change depending on whether feeding by other species is more or less aggregated, or whether various soil amendments (e.g., repellent insecticides) steer wireworm feeding activities to more limited regions of a hill.

Although wireworm damage to potatoes is fairly obvious once seen, there are other types of damage that might cause some confusion. In Europe and the UK, injury by small slugs, for example, resembles entrance holes made by wireworms, although slugs will hollow out cavities within tubers, which is not a characteristic of wireworm damage (Gratwick 1989, Parker and Howard 2001). Subterranean larvae of the tuber flea beetle (*Epitrix tuberis* Gent.), a pest of potatoes in the Pacific Northwest region of North America, also make holes in the flesh of potatoes, but these holes are generally smaller in diameter than those of wireworms. Millipedes are often found in association with wireworms and damage (Parker and Howard 2001), especially in former fields of sod or pasture, but direct damage by millipedes to tubers, although suspected, has not been confirmed.

SAMPLING

Wireworm Sampling and Risk Assessment

Threshold-based monitoring programs have been developed and implemented for many potato insect pests, including Colorado potato beetles (*Leptinotarsa decemlineata*), aphids (e.g., *Myzus persicae* (Sulzer)) and flea beetles (e.g., *E. cucumeris*) (e.g., Anonymous 2010). The development of efficient and accurate monitoring programs and risk assessment methods for use in management strategies for wireworms, however, has proven to be somewhat elusive. The fact that wireworms occupy a subterranean habitat has greatly impeded the development of accurate sampling methods. Not only are wireworms underground, they may also be anywhere from near the soil surface to 1.5 m deep (Andrews *et al.* 2008), depending on the species, geography, temperature and moisture, cycles of feeding and moulting, presence or absence of ground cover (e.g., grass vs fallow), and location of food (e.g., Zacharuk 1962b, Doane *et al.* 1975, Toba and Turner 1983, Parker 1996). Thus, the timing and protocols selected for sampling are absolutely critical to the success of any sampling objective, be it simple detection, species census/survey, population estimation, or threshold-based monitoring. Since the results of various sampling approaches will vary quite radically both spatially and temporally (Parker 1996, Simmons *et al.* 1998, Horton 2006), the realities and limitations of wireworm sampling need to be considered at the very outset of developing such programs. This is particularly true if the sampling approach is to provide information to growers regarding the need to "apply" or, more importantly, "not apply" pre-emptive wireworm controls (i.e., field avoidance or insecticide use).

Sampling for wireworms can be segregated into "absolute" versus "relative" population sampling methods (Southwood 1978). Absolute sampling methods, such as soil coring to extract wireworms *in situ* from their soil habitat, typically involve a higher level of expertise, time, infrastructure, and associated expense to conduct, and as such are generally relegated to more research than extension programs. Relative sampling methods, including the use of attractive baits or traps to draw wireworms from variable distances in the soil, are also used for research purposes, but have mostly been developed with the intention of providing indications of wireworm presence and relative abundance for use in management programs.

Absolute Sampling Methods

The main objective of absolute sampling for wireworms has historically been to estimate the population size, spatial distribution (horizontal and vertical), and temporal activities of wireworms in a field. This has typically involved removing field soil in layers (e.g., Gibson 1939, Salt and Hollick 1944) or by soil coring (e.g., Yates and Finney 1942, Anonymous 1948) at various depths and at enough locations in a field over time that wireworm population estimates can be made with reasonable precision. Probably the best example of an absolute sampling program is from the UK during World War II in the early 1940s, where core samples were taken in pasture to estimate wireworm populations (*Agriotes* spp.) prior to conversion to cultivated crops, including potato. Sample sizes of 20 cores (10 cm diameter × 15 cm deep) per field (4–10 ha) were used, and hundreds of fields were surveyed annually (Yates and Finney 1942). Although providing acceptable population estimates where wireworm levels were high, sampling error increased at lower population levels, with the lower limit of detection being about 62,500 wireworms/ha (Yates and Finney 1942, Parker and Howard 2001). In determining the suitability of a field (i.e., under cultivation or in pasture) for eventual potato production, however, higher levels of accuracy are required at lower population levels, since significant cosmetic damage can occur at levels even below 62,500 wireworms/ha (French and White 1965, Parker and Howard 2001). Up to 5% of tubers have been observed to have some wireworm damage at populations of 25,000 wireworms/ha, and the number of tubers with 9 or more blemishes was well above tolerance at 75,000 wireworms/ha (Hawkins 1936). In order to detect such low economic levels in a field with acceptable precision, a larger number of core samples would be required, along with a proportionately higher amount of labor and associated sampling costs. It is interesting to note that increasing the number of soil samples from 20 to 40 cores per field did not reduce wireworm sampling variability in cultivated and grass pasture fields in Iowa (Simmons *et al.* 1998), suggesting that a much higher number of cores per field would be required to reduce both the variability and the detection level of 62,500 wireworms/ha.

With soil sampling methods, a major drawback has always been the difficulty in extracting wireworms of all instars from soil samples in a timely, cost-effective manner. Although methods have been designed to remove wireworms from soil or turf by flotation (Salt and Hollick 1944), mechanical methods (Smith *et al.* 1981, Lafrance and Tremblay 1964), or with heat (Vernon *et al.* 2009), these approaches still require considerable labor and laboratory processing, and by modern standards would be expensive and/or inconvenient. Where abundant labor and extraction infrastructure are available, however, soil coring with a sample size sufficient to detect low population levels of wireworms would provide an accurate means of sampling in fields destined for potatoes. As has been amply covered in the previous section on wireworm biology, soil sampling programs to guide management decisions would have to be customized for specific agricultural regions, and would have to consider the species complex involved, their field distribution(s), the optimal times and conditions for sampling, and the development of conservative action thresholds (e.g., Robinson 1976). The level of expertise required to establish and interpret these sampling programs would be commensurate with the complexity of the wireworm complex involved, and the risk of making economic errors.

Relative Sampling Methods

The time required to examine soil samples, and the large number of samples required to accurately estimate wireworm populations in fields, has led to the development of several attractant-based sampling methods in North America and Europe, and a number of these are cited in Chalfant and Seal (1991) and Parker and Howard (2001). Essentially, wireworms are attracted in soil to sources of CO_2 production (Doane *et al.* 1975), which in the field would include germinating seeds, respiring plants, decomposing plant material, and so on. Of the large number of CO_2 -producing baits tested, including fruits and vegetables (i.e., melons, carrot, potato) and processed cereals (i.e., bran, rolled oats, flour), baits containing germinating cereal seed (e.g., wheat, barley) and/or other seeds (e.g., corn, sorghum) have been found most effective and are now most commonly used (Bynum and Archer 1987, Jansson and Lecrone 1989, Parker 1994, 1996, Simmons *et al.* 1998, Parker and Howard 2001, Horton and Landolt 2002, Vernon *et al.* 2003, Furlan *et al.* 2011).

Among the more important characteristics of wireworm trapping systems for use in research and especially in management programs include: consistency among traps; some control over CO_2 production; ease of trap assembly and deployment; and rapid, accurate methods of wireworm extraction. Of the methods developed, we have found that traps similar to those described by Chabert and Blot (1992), consisting of 450-ml plastic pots filled with medium-grade vermiculite, and with 100 ml each of untreated corn and hard red spring wheat spread in layers in the middle of the pots, adequately meet these criteria (Vernon *et al.* 2009). Traps are soaked with warm water to runoff twice, placed the same day in 15-cm deep holes, covered with soil on all sides, and a 20-cm diameter inverted tray positioned 5 cm above the trap and level with the ground. Traps are generally left for 12–14 d, are removed without surrounding soil, and the trap contents (vermiculite and germinated wheat and corn seed) are sorted by hand to find larger wireworms (> 1 cm), and/or are placed in Tullgren funnels which effectively extracts all instars (Vernon *et al.* 2009). Traps such as this have been shown to be as or more efficient than other "relative" methods in terms of ease of deployment (i.e., placement, retrieval and sorting), consistency, and quantity of wireworm catch and cost (e.g., Simmons *et al.* 1998). Other bait trapping systems, including the various food baits mentioned, are likely to be inconsistent in CO_2 production among baits, and wireworm counts are prone to greater variability since variable amounts of soil surrounding the baits often has to be sampled as well.

Various methods of improving the efficacy of bait traps have been developed, such as covering the baits at the soil surface with black plastic (Ward and Keaster 1977, Chabert and Blot 1992, Simmons *et al.* 1998) or charcoal dust (Ward and Keaster 1977, Bynum and Archer 1987). Essentially these approaches raise soil temperature, which provides better conditions for the germination of living baits (e.g., wheat and corn seed) or microbial respiration in non-living baits (e.g., bran, rolled oats, flour), all of which increase CO_2 production. These methods have facilitated earlier trapping and higher catches for some species (Ward and Keaster 1977, Bynum and Archer 1987), but they also increase the time and cost of sampling, which may not be necessary for all monitoring strategies.

Baits or baited traps draw wireworms in the soil from distances as far away as wireworms can detect distinct CO_2 gradients. Doane *et al.* (1975) found that wireworms (e.g., *Ctenicera destructor*) can detect and orient to CO_2 sources from as far away as 20 cm (the limit of their soil bioassay arena), and Vernon *et al.* (2000) showed that the majority (83%) of wireworms (*A. obscurus*) in field plots will orient to trap crops of wheat spaced 1 m apart. It has also been found that bait stations with untreated wheat (11 cultivars), barley, oats, or fall rye seed will increase in attractiveness to wireworms (*A. obscurus*) as the density of seed increases, but will reach plateaus of catch at seeding densities specific to each variety (Vernon *et al.* 2003). The draw of any CO_2-producing bait or trap to wireworms, therefore, will reach well beyond the physical boundaries of the trap, but this distance will vary with the type of trap, the trap bait, the temperature and moisture of soil, competing CO_2 sources (e.g., ploughed pasture), the soil texture, and the species, instar, and feeding status of wireworms in the field (Parker 1996, Vernon *et al.* 2003, Horton 2006). In our experience, we have found that bait traps are much less effective at soil temperatures below 10°C, in freshly ploughed fields of pasture with high levels of green manure (Parker 1996), and at various times of the growing season when wireworms are not feeding (Furlan 1998, 2004). Where we have found bait trapping to be relatively successful and consistent is when the area surrounding a bait trap (1 m radius) has been cleared of all living plant material or other potential sources of competing CO_2 production. Trapping is also done only when soil reaches a fairly stable temperature of greater than 10°C, which is typically when wireworms become more active, germination of seed in traps is high, and seedling growth in baits is about 1 cm/d, which is ideal for a 12- to 14-day trap placement. Lower temperatures reduce wireworm activity, seed germination, and growth, and higher temperatures result in rapid seedling growth and increased biomass for sorting, all of which alter the effectiveness and consistency of trapping. In the development of any bait trap sampling program for wireworms, therefore, especially where consistency in trapping is important, the field conditions most suitable for sampling all economic species present should be identified concurrent with determining the most appropriate baits and sampling protocols.

Absolute versus Relative Sampling

Bait trapping, which draws in wireworms from areas outside of the trap, would be expected to collect higher numbers of wireworms than direct soil sampling techniques, and this has been verified in a number of studies (e.g., Parker 1994, 1996, Simmons *et al.* 1998). Of these studies, that of Simmons *et al.* (1998) is of particular note in that they compared the "absolute" soil sampling method developed in the UK (Yates and Finney 1942, Salt and Hollick 1944, Anonymous 1948) with a number of candidate "relative" sampling methods to develop farmer- or consultant-run sampling programs in the USA. Of the relative methods tested, wheat/corn baits (placed in holes in the ground) had consistently higher wireworm counts (collective counts of *Melanotus pilosus* Blatchley, *Conoderus auritus* (Herbst), *Ctenicera inflata* (Say), *Hypnoides abbreviatus* (Say), and *M. similis* (Kirby)) and a lower variability index (Buntin 1994) than core samples. They also determined that a high positive correlation existed between the wheat/corn baits and core samples in fields with higher population pressure. Their data further indicated that wheat/corn bait catch could be calibrated to estimate field populations, and that sample sizes of 50, 25, or 10 baits/ha would be needed to achieve 10, 15, or 25% levels of precision, respectively. Although only a single study, the methodologies used by Simmons *et al.* (1998) to determine sample size and efficiency (trap effectiveness and cost) for various "relative" wireworm sampling methods versus "absolute" sampling methods can be considered for other regions and species.

Attempts at using relative sampling methods to consistently predict wireworm damage to potato have met with discouraging results. Using cereal-baited traps in the UK, Parker (1996) found that high levels of damage occurred (> 60% tubers with blemishes) in plots where no wireworms (*Agriotes* spp.) were detected. This was also observed in trials conducted in the USA, where plots baited with rolled oats had between 3.3% and 6.8% damaged tubers despite no wireworms (*L. canus*) being caught over a 7-week period (Horton 2006). Horton's work also found that damage predictions based on wireworm counts could vary dramatically from week to week, indicating the variability in efficacy over time that can be expected with relative sampling methods. Both authors concluded that using baits to predict damage to tubers would be difficult to implement with a great deal of confidence (Parker 1996, Horton 2006). This is not to say that the use of baits or bait traps has no value in wireworm management programs. Parker (1996) further concluded that bait traps are more effective than core samples at indicating the presence or absence of wireworm infestations, and would be of value in fields where wireworm populations are below the limit of detection using soil cores (62,500 wireworms/ha), and bait trapping and/or soil core sampling is now recommended alongside pheromone trapping programs (see below) in the UK for improving risk assessments in fields (Anonymous 2011).

Timing of Sampling in Potato Fields

What should be obvious from the above discussion is that the optimum conditions for sampling wireworms to determine the risk of damage to potato crops often occur when growers intend to plant potatoes, and this is a major impediment to the implementation of pre-planting wireworm monitoring (Horton 2006). Optimal conditions for deploying absolute sampling methods such as soil coring also occur at this time, since populations are predominantly in the upper regions of the soil and within coring depth (Simmons *et al.* 1998, Horton 2006). The selection of any sampling system, therefore, must consider the time it takes to complete the sampling procedure and provide timely input to the grower prior to planting. Because of these current-season time constraints, wireworm sampling can also be conducted the year prior to planting potatoes. Fields at greatest risk of damage by wireworms will often have been in pasture the year previously, and sampling of pasture using absolute or relative sampling techniques can be timed to coincide with wireworm feeding activity periods nearest the soil surface. Although bait trapping techniques are less effective in pasture due to competition with grassy hosts (Parker and Howard 2001), the strategic removal of sections or strips of pasture with herbicide and/or shallow cultivation well in advance of trapping would enhance the competitiveness of baits. Such considerations also apply to the use of bait traps deployed in cultivated fields the year prior to potato planting.

Habitat and Risk to Potatoes

Attempts to correlate field-specific characteristics, including soil type, moisture, topography, cropping history, grower activities, etc., on the distribution, abundance, and risk of economic injury from a species complex of wireworms has met with limited success (Thomas 1940, Parker and Howard 2001). Nevertheless, abiotic factors such as soil moisture content have been used to model the likelihood of wireworms occurring in "Conservation Reserve Program" grass fields in Iowa (Lefko *et al.* 1998), and soil sampling has been stratified within fields according to soil moisture characteristics (MacLeod and Rawlins 1935). Biotic factors, such as the presence and duration of grassland in a field, are obvious but not guaranteed indicators of wireworm presence or absence (Parker and Seeney 1997). As a general rule, the authors consider any field that has had a history of grassland, cereals, or grassy cover crops (present during adult oviposition periods) within the past 4–5 years to be at higher risk for wireworm damage. This is not to say that fields without a history of these crops will not be at potential risk. Parker and Howard (2001) have observed that wireworm damage to potato crops in the UK has become an increasing problem in all arable rotations, including fields with no history of long-term grass. Although there are uncertainties associated with using field variables and cropping history to predict the risk of wireworm damage in fields, improved knowledge of the species complex present in an area, their habitat, oviposition and feeding preferences, and the cropping history of individual fields would improve the accuracy of site risk assessments.

Click Beetle Sampling

Pheromone Traps

Pheromones have been discovered for a number of pest elaterids, including *L. californicus*, *L. canus* (Lehman 1932, Jacobson *et al.* 1968) and possibly *Melanotus depressus* in North America (Weires 1976), several *Agriotes* species in Europe and Russia (e.g., Yatsinin and Lebedeva 1984, Tóth *et al.* 2003), and *M. okinawensis* and *M. sakishimensis* in Japan (Tamaki *et al.* 1986, 1990). A number of pheromone trap designs have also been described and are commercially available for monitoring *Agriotes* spp. (e.g., Tóth *et al.* 2003, Ester and Rozen 2005, Vernon and Tóth 2007) and *Melanotus* spp. (Kawamura *et al.* 2002). Elaterid pheromone traps, assuming species specificity, are particularly useful in rapidly determining the presence or absence of various pest species in agricultural areas. This is particularly true of the *Agriotes* genera, where aggressive surveys of eight species (*A. ustulatus*, *A. litigiosus*, *A. sputator*, *A. obscurus*, *A. lineatus*, *A. rufipalpis*, *A. sordidus*, and *A. brevis*) have been carried out in Europe since 1999 (Furlan *et al.* 2001a, Furlan and Tóth 2007), and delimitation surveys of *A. sputator*, *A. obscurus*, and *A. lineatus*, introduced from Europe to the western and eastern coasts of North America, have been conducted since 2000 (Vernon *et al.* 2001, LaGasa *et al.* 2006). Since the starting point in developing pest management strategies for wireworms in any agricultural region is to know the key species present, pheromone trapping, if available for the species involved (as is the case in Europe) is the most efficient and inexpensive method available, and does not require a high level of expertise to conduct and interpret. In addition, pheromone traps are effective at detecting species at very low population levels, which cannot generally be achieved with soil sampling or bait trapping (Furlan and Tóth 2007).

Pheromone traps are currently being used by growers to estimate the risk of wireworm damage to potatoes in the UK. Traps for the key species *Agriotes lineatus*, *A. obscurus*, and *A. sputator* (www.syngenta-crop.co.uk) are placed in fields in the year prior to planting potatoes (Anonymous 2011). The total number of beetles *Agriotes* caught over the course of the season is then used by growers to determine appropriate risk-aversion measures for the following growing season, including avoidance of planting, growing an early potato crop and/or harvesting earlier to reduce damage, selecting a more damage-tolerant variety, or using an insecticide at planting. According to this system, collective beetle captures of 0, 1–49, 50–150, and >150 correspond to wireworm population estimates and risk of damage of, respectively, 0 wireworms (little or no damage); 25,000–150,000 wireworms (some damage); 150,000–250,000 wireworms (significant damage), and >250,000 wireworms (severe damage) (Anonymous 2011). These categories, however, are somewhat subjective, and are meant to be used in conjunction with soil sampling or bait trapping, generally in the year prior to planting potatoes (Anonymous 2011).

The interpretation of click beetle pheromone trap data is in its infancy, and there are a number of important problems to be considered when developing these traps as monitoring tools for use in management programs (e.g., Blackshaw *et al.* 2009). Fundamentally, the number of male click beetles captured is the result of an oviposition event that occurred in the field or surrounding areas (wireworm reservoirs) several years previously (e.g., 4–5 years for various *Agriotes* spp.). In the case of permanent grassland, where oviposition occurs annually and populations are relatively stable, it is likely that wireworm populations of various yearly cohorts will be present in the soil during the year of pheromone trapping in pasture. On the other hand, in fields that have been out of pasture for 1 or more years, and where favorable crops for oviposition have been planted on one or more occasion during the past 4–5 years, wireworm populations may be much more variable in abundance and age structure. For example, a field that was converted from pasture to arable crops 4–5 years ago may have high numbers of adults in pheromone traps 4–5 years later (from oviposition occurring during the last year of pasture), but have correspondingly low numbers of wireworms in the field due to unfavorable crops for oviposition after the pasture is removed. Alternatively, levels of adults in pheromone traps might be low due to an unfavorable crop for oviposition 4–5 years previous, but one or more large wireworm cohorts could be present in the field due to the planting of favorable oviposition crops thereafter (e.g., cereal crops). In addition, if pheromone traps are placed too close to headland areas or adjacent fields harboring populations of wireworms/click beetles, the trap catch can simply reflect immigrants arriving at traps from outside of the field. The presence and stability of reservoir populations of wireworms and adults in grassy areas surrounding arable fields has been reported (Blackshaw and Vernon 2006, 2008), and we have demonstrated the movement of large populations of *A. obscurus* and *A. lineatus* into fields from grassy verges (R. S. Vernon, unpubl. data). The means of immigration (flight and/or walking) and the relative mobility of various species are other considerations that present additional problems of data interpretation where several species are involved. For example, Hicks and Blackshaw (2008) reported the relative speed of the three primary UK species to be of the order *A. lineatus > A. obscurus > A. sputator*, which suggests that the rate of pheromone trap catch will vary between species, and therefore risk cannot be accurately estimated by simply adding catches of these traps together.

The confounding effect of immigrant adults on pheromone trap interpretation can be mitigated somewhat by trapping for shorter periods of time. In the UK, the possibility of placing pheromone traps in the field in the morning and removing them in the afternoon is being explored (Anonymous 2011), and has been used to collect adults over smaller, more localized areas in various field studies (R. Blackshaw, unpubl. data). If this is done early enough during adult emergence, it would more accurately reflect populations arising from within as opposed to outside of a field (R. Blackshaw, pers. comm.). In addition, Hicks and Blackshaw (2008) estimated the effective sampling areas for *A. obscurus*, *A. lineatus*, and *A. sputator* pheromone traps after a 15-day deployment period to be 2580 m^2, 2588 m^2, and 1698 m^2, respectively. To reduce immigrants from outside areas entering pheromone traps inside fields, therefore, the outside areas should lie well outside of the effective trapping areas of the traps.

Pheromone traps also provide a tool to study the occurrence of click beetle populations spatially and temporally in fields or in larger agricultural landscapes during single or multiple seasons, which would otherwise have required considerable resources, and which promises to enrich our understanding and even management of this previously ignored, but critical elaterid life stage. For example, Blackshaw and Vernon (2006) were able to study the spatial and temporal occurrence of male *A. lineatus* and *A. obscurus* beetles using pheromone traps relative to wireworm populations with bait traps within large fields of strawberry, and later studied the spatial stability of *A. lineatus* and *A. obscurus* populations in non-farmed wireworm habitats over a 3-year period (Blackshaw and Vernon 2008). A better understanding of click beetle distribution in space and time using pheromone traps has provoked investigations into various methods for controlling wireworms by pre-emptively controlling the click beetle stage, which will be discussed below.

Other Click Beetle Trapping Systems

Pitfall traps have been used to collect elaterid adults, primarily for species survey purposes (e.g., *A. lineatus* and *A. obscurus*, Vernon and Päts 1997) or to assess adult activity and dispersal (e.g., *Melanotus depressus* (Melsheimer) and *M. verberans* (LeConte), Brown and Keaster 1986; and *A. obscurus* and *A. lineatus*, Blackshaw *et al.* unpubl. data). Pitfall traps provide a relative measure of click beetle abundance, since they typically capture click beetles during their trivial movements within a habitat, and have the advantage over pheromone traps in that they collect both sexes. As is the case with absolute soil sampling methods for wireworms, however, pitfall traps catch far fewer click beetles per sample than attractant-based methods, and pheromone traps for *A. obscurus* captured about 200 times more *A. obscurus* males than pitfall traps in one study (Vernon and Tóth 2007). Light traps and colored sticky traps are not commonly used to monitor adult elaterids, other than when they show up as incidental specimens in general surveys, or when surveys of elaterid flight activity have been undertaken (e.g., Boiteau *et al.* 2000) .

Another useful adult elaterid sampling method is the use of what has been generally termed "forage traps" (L. Furlan, pers. comm.). These traps include a clear sheet of plastic (about 60 cm × 90 cm), secured flat to the ground (usually bare ground) with soil along the edges, and with a good handful (5 cm diameter) of wild grasses about 40–60 cm long placed down the middle of the sheet. The grass can be held in place by a wire hoop through the sheet into the ground. These traps appear to act as harborage or as food sites for both sexes of click beetles, and hundreds can be collected in this manner overnight. We have used this method for collecting both sexes of *A. obscurus* and *A. lineatus* in non-pasture fallowed fields for various mark–release–recapture studies. Since these traps collect more beetles than pitfall traps (R. S. Vernon, unpubl. data), it appears that there may be some level of attraction occurring, possibly to the grass component. The use of these traps for monitoring click beetles for management approaches has not been explored.

WIREWORM CONTROL

The availability and efficacy of management tools and approaches for the prevention of economic injury to potato crops from wireworms has varied considerably over the past century, and the reader is directed to review articles by Thomas (1940) and Parker and Howard (2001) for listings of some of the early and more contemporary control options, respectively. The review of Thomas (1940) essentially pre-dates the inception of the synthetic insecticide era (e.g., organochlorines, organophosphates, carbamates, etc.), and the management options at that time included cultural methods such as cultivation, fallowing, crop rotation, fertilization, mulching; time of potato harvest; physical methods such as attractants and repellents; and insecticidal methods such as fumigants (e.g., carbon disulfide, chloropicrin, cyanides, paradichlorobenzene), seed treatments, and soil amendments (e.g., arsenicals, mercury compounds, pyrethrum, rotenone). It is interesting to note that although the management options available for wireworms pre-1940 would likely have been more expensive and much less effective than the synthetic insecticides which came later, the tolerance of consumers for wireworm damage to potatoes was also higher, and the availability of potato products was quite limited in contrast to todays' higher quality standards and diverse product selection. Whereas primarily ware potatoes were produced pre-1940, modern-day production markets for potatoes include regular table grade; pre-packed specialty varieties; numerous processed products (such as French fries, potato chips, hash browns, instant potatoes); seed potatoes; and versions of these products as conventional or organically produced. In many of these modern products there is essentially a zero tolerance for wireworm blemishes (e.g., Anonymous 2011), and the evolution in diversity and quality of the potato industry over the past century has somewhat paralleled our ability to effectively control wireworm damage. However, since our effective arsenal of effective wireworm insecticides is dwindling in many countries, and reports of growing populations and damage are occurring in many key crops, our ability to sustain the quality and abundance of our current potato industries will be challenged in the future.

The availability of various contemporary methods for wireworm management in potatoes was reviewed by Parker and Howard (2001), and covered research relating to cultural, biological, physical, and insecticidal controls. In recent years, however, additional scientific research has expanded our knowledge into these and other management approaches that warrant discussion in this chapter. Since our ability to generically and persistently manage wireworms has virtually vanished with the loss of the persistent organochlorines, our discussion of management options will also focus on the need for more region- and species-specific strategies, and the opportunities and considerations involved in researching and developing these strategies for the future.

Cultural Methods

The potential economic importance of wireworms in a field is a function of the quantity of eggs deposited over several years, and the survivorship of these eggs and wireworms on the hosts available in the field throughout their life history. As such, wireworm populations can be managed by a number of cultural methods that either prevent or reduce oviposition, or decrease the survivorship of wireworms at all stages in the field. These techniques, as they apply to the production of potatoes, include field avoidance, crop rotation, cultivation, and other modifications to field conditions that reduce the economic impact of wireworms.

Crop Avoidance and Rotation

Since wireworm populations generally build up to economic levels in fields with a recent history (e.g., within the past 4–5 years) of pasture, cereals or grass seed (Thomas 1940, Andrews *et al.* 2008, Huiting and Ester 2009), an obvious cultural control method is to avoid planting potatoes immediately into these fields (Parker and Howard 2001). After grassland has been removed, and crops favoring oviposition by click beetles (such as cereal crops, certain forages, grassy cover crops, etc.) are not included in subsequent crop rotations, populations of wireworms will typically decline to sub-economic levels as resident wireworms gradually complete their life cycles (Fox 1961). In general, the presence of any preferred wireworm crop coincident with click beetle oviposition can give rise to economic wireworm populations in a single year (Landis and Onsager 1966, Jansson and Lecrone 1991). It is important, therefore, to know the economic species involved, the oviposition activity periods of the adults, and their preferred oviposition hosts to avoid recurring wireworm problems in a field. This strategy, especially in the absence of effective insecticidal control options (e.g., organic production systems), requires long-term planning and patience in planting potatoes until wireworm populations have dropped to sub-economic levels. Conventional potato production in many countries, however, often involves land that is leased during the year of planting and/or is often pasture, and prophylactic insecticidal control measures are generally required to avoid economic damage (Ester and Huiting 2007).

Considerable work has been done to determine the positive or negative effects of various crop rotations on wireworm populations, and this topic is extensively covered in the reviews of Thomas (1940), Miles (1942), and Parker and Howard (2001). Studies have focused on crops tolerant to or damaged by wireworms, or that favor or disfavor wireworm population survivorship. The findings often vary considerably between and even within areas worldwide, however, and generalities outside of the general preference of wireworms for pasture, cereals or other grassy crops are tenuous. For example, leguminous plants such as alfalfa and clover (several varieties) were found by many researchers in the 1930s to be rotational crops unfavorable for increase of wireworm populations (19 publications cited in Thomas 1940); however, increases in wireworm populations have been found in these same crops in other research (12 publications cited in Thomas 1940). These contradictions may be due to differences in species, geographic regions, soil types, soil moisture, methods of evaluation used, etc., but underpin the need for contemporary, region- and species-specific research into this subject that follow harmonized assessment methodologies.

In recent work, brown mustard (*Brassica juncea*) or buckwheat planted in 3-year rotations with potatoes was found to reduce wireworm (*Agriotes sputator*) damage to daughter tubers relative to high levels of damage observed in rotations of barley (undersown to clover) or alfalfa in Prince Edward Island, Canada (Noronha 2011). This work was significant in that it demonstrated to growers in PEI (who were incurring increasing wireworm damage to potatoes) that wireworm populations and damage to potatoes were exacerbated by the common practice of rotating with a cereal crop (barley or wheat) undersown to clover (Landis and Onsager 1966), and further demonstrated that damage could be reduced by planting rotational crops with allelopathic properties (discussed below).

Cultivation

Mechanical methods of disturbing the soil, such as plowing, harrowing, disking, and rotovating are known to reduce various stages of wireworms (Thomas 1940, Parker and Howard 2001), and although not a primary control method can sometimes be considered part of an IPM program. The objective of cultivation is to directly destroy eggs, larvae, pupae, and adults in the soil, or indirectly kill them by bringing them to the surface and exposing them to heat or to natural enemies such as birds and arthropod predators (Thomas 1940, Seal *et al.* 1992). Pupae of many wireworm species, which are very soft-bodied and generally found in the upper 38 cm of soil during July and August, are particularly vulnerable to shallow plowing, and up to 90% mortality has been reported (Andrews *et al.* 2008). Cultivation at that time might also expose eggs and small larvae to desiccation and mechanical injury (Thomas 1940). Some reductions in larger wireworms by cultivation have been reported, and a 91% drop in wireworms caught in traps occurred when soil was plowed three times during the summer (Seal *et al.* 1992). The aforementioned studies, however, were conducted in fallowed fields during summer months, which are not typical field conditions available to most growers. In the UK, where summer cultivation is not possible, cultivation practices are thought to be most effective in reducing wireworm populations if done in the autumn when wireworms are active near the soil surface (Gratwick 1989).

Soil Amendments

Plant tissues or tissue extracts from a variety of cruciferous plants have been shown to have insecticidal properties in soil (Lichtenstein *et al.* 1964, Kirkegaard *et al.* 1993). This allelochemical activity has been linked to the presence of high levels of glucosinolates found in certain crops (e.g., *Brassica oleracea* L., *B. juncea* L., *B. carinata* L., and *B. nigra* (L.) Koch and *B. napus* L.) (Williams *et al.* 1993). Although the biological activity of glucosinolates is limited, they break down to more toxic molecules when tissues are damaged, especially upon incorporation into soil (Williams *et al.* 1993, Borek *et al.* 1994). Among the more notable of these breakdown products, allyl isothiocyanate has been shown to have toxic as well as antifeedant effects on wireworms (*L. californicus*) in the lab (Williams *et al.* 1993). Rapeseed meal (from *B. napus*) incorporated into soil was found to be repulsive (Brown *et al.* 1991) as well as toxic (Elberson *et al.* 1996) to *L. californicus* in the lab, but it took almost a 1 : 1 mixture of meal/soil to produce a high level of mortality (90%). It was concluded from these studies that plant material with higher levels of glucosinolates would be required to effectively and more economically control wireworms in the field (Elberson *et al.* 1996).

The need for a plant variety with higher levels of glucosinolates appears to have been met with defatted seed meal (DSM) from a particular variety of Ethiopian mustard (*B. carinata* sel. ISCI 7) (Patane and Tringali 2010, Furlan 2007, Furlan *et al.* 2010). In a number of laboratory studies, *B. carinata* DSM was shown to protect maize and lettuce seedlings from wireworm damage (one or more of *A. sordidus* Illiger, *A. ustulatus* Schaller, and *A. brevis* Candeze), and killed 100% of larvae at some dosages (Furlan *et al.* 2010). In field trials conducted in Italy, *B. carinata* DSM applied to the soil surface and worked homogeneously into the top 20 cm of soil provided stand protection in maize equivalent to and sometimes better than the insecticide standard (fipronil). In a potato trial, *B. carinata* DSM provided significant protection of young daughter tubers, but inconclusive protection of mature tubers at harvest (Furlan *et al.* 2010). It was noted in these trials that wireworms were sometimes found in a moribund or dead state at the soil surface in *B. carinata* DSM plots, which has also been observed in laboratory and field trials with neonicotinoid (e.g., Vernon *et al.* 2008) and pyrethroid insecticides (e.g., van Herk *et al.* 2011).

It was concluded in the work of Furlan *et al.* (2010) that for *B. carinata* DSM to be practical in the field certain conditions had to be fulfilled concurrently, including a suitable dosage of glucosinolates in the DSM (at least 160 µmol of glucosinolates per liter of soil), a homogeneous broadcast application of DSM, effective and prompt soil incorporation to 20 cm, suitable soil and humidity conditions, and presence of wireworms predominantly in the upper 20 cm of soil. Given that the persistence of the toxic metabolites of glucosinolates is short (less than 48 h), it is vital that all of these conditions are met within the first 2 days of application for this approach to be fully effective. Once again, a thorough knowledge of all species present and their temporal and spatial behavior in the soil is required to determine the practicality and potential effectiveness of this technique or modified versions thereof in other areas. Nevertheless, the preliminary success of Furlan *et al.* (2010) suggests that further work on biofumigants such as *B. carinata* DSM is warranted, especially for organic production.

Potato Varietal Tolerance

The tolerance of certain potato cultivars to wireworm damage can be an important component of management strategies, particularly in organic production systems, and most of the relevant scientific research has been reviewed by Parker and Howard (2001) and Andrews *et al.* (2008). Among the more notable research in recent years is that of Kwon *et al.* (1999) in Korea, and Johnson *et al.* (2008) in Scotland, who found significant differences between various cultivars in the field (data summarized in Andrews *et al.* 2008). The relative susceptibility of various potato cultivars to feeding by wireworms has been shown to be related to the total glycoalkaloid (TGA) content present in daughter tubers (Jonasson and Olsson 1994, Olsson and Jonasson 1995). Unfortunately, glycoalkaloids are also toxic to humans, and there are regulatory limits to the amounts of these TGAs allowed in new potato varieties. It has not been until recently that breeding of potatoes for resistance specifically to wireworms has taken place (Novy *et al.* 2006). Germplasm from wild relatives of potato from South America, *Solanum berthaultii* and *S. etuberosum*, crossed with a cultivated potato variety has produced a number of resistant clones with wireworm damage reductions as good as observed with insecticide-treated crops (Suszkiw 2011). Some of these resistant clones contain levels of TGAs suitable for human consumption, which opens the door to their use in the development of wireworm-resistant commercial varieties for the future.

Early Harvest

As has been mentioned, damage to daughter tubers typically increases as the growing season progresses, and the longer potatoes are left in the ground, the greater the amount of damage that will occur (Fig. 5.4) (Anonymous 1948, Parker and Howard 2001, Kuhar and Alvarez 2008). Therefore, if potatoes must be planted in infested fields, varieties should be grown that can be lifted before wireworms begin to actively feed on tubers, or later season varieties should be harvested as soon as possible (Parker and Howard 2001, Andrews *et al.* 2008). In Germany, Schepl and Paffrath (2005) found less wireworm damage on tubers harvested in late July to early August (8–50% tubers damaged) than in September (72–77% damage), and an increase in damage with later harvest dates was observed by Neuhoff *et al.* (2007).

Chemical Methods

Wireworm Controls

Chemical controls to manage wireworm damage to potatoes have historically involved prophylactic treatments applied before, at, or after planting to control the wireworm stage in soil. In addition to soil fumigation, pre-planting treatments have included insecticides either broadcasted on the soil surface and worked into the ground, or as insecticide-laced fertilizers (Parker and Howard 2001), with the intent of intercepting and killing wireworms in their movements near the soil surface in spring. At the time of planting, insecticides have also been applied in-furrow either as granular or spray formulations, or as seed treatments applied to mother tubers just prior to planting. Post-planting applications include side-band applications intended to kill wireworms near the surface of potato plants at the start of daughter tuber formation. The effectiveness of these various application methods has varied considerably (Parker and Howard 2001, Kuhar and Alvarez 2008), but in general, at-planting in-furrow applications of insecticides appear to be the most widely used method worldwide.

Organochlorines

The fact that wireworms became a pest of low worldwide importance following World War II can be attributed almost certainly to the introduction and widespread use of the organochlorine (OC) insecticides in the 1950s. At that time, insecticides such as DDT and aldrin applied as pre-planting broadcast sprays or as "aldrinated" fertilizers to the soil became standard treatments for wireworms in many parts of the world (Merrill, 1952, Strickland *et al.* 1962, Parker and Howard 2001, Kuhar and Alvarez 2008). One characteristic of these and other OCs (e.g., heptachlor, dieldrin, chlordane, etc.) that made them particularly effective was their persistence in soil for years following application. In the case of aldrin and heptachlor, one application to soil was reported to kill wireworms (*A. obscurus*) for 13 years (Wilkinson *et al.* 1964, 1976). Multiple applications to fields during the tenure of these insecticides would have been very effective in preventing population build-ups even well beyond their global de-registrations (due primarily to their long persistence) in the 1970s and 1980s. It is thought by some that the present upsurge of wireworm populations is due in part to the gradual decline of these persistent OC residues in fields to levels non-toxic to neonate wireworms (Jansson and Seal 1994, Parker and Howard 2001, Horton and Landolt 2002).

In addition to the persistent soil-applied OCs, it is relevant to this discussion that lindane, an OC with shorter persistence, became a standard seed treatment for control of wireworms in cereal crops and corn in many countries as early as the 1940s. These seed treatments reportedly reduced field populations of wireworms (e.g., *Ctenicera destructor* (Brown) and *Hypolithus nocturnus* Esch.) by about 70% (Arnason and Fox 1948), and in Canada wireworm damage in the prairies declined gradually from 1954 to 1961, coincident with the increasing use of lindane seed treatments (Burrage 1964). With the eventual de-registration of lindane in most countries by 2004, however, it was expected that wireworm populations would increase, since cereal crops are preferred oviposition hosts for many species, and no contemporary cereal seed treatments significantly kill wireworms (Vernon *et al.* 2009, 2011). A dramatic increase in wireworm populations has in fact been observed over the past decade in many major cereal production areas of Canada (Vernon *et al.* 2011), and this has had direct implications for wireworm control in potato crops now typically grown in rotation with cereals.

Organophosphates and Carbamates

With the gradual demise of the persistent organochlorines, a number of organo-phosphate (OP) and carbamate insecticides were registered for wireworm control in potatoes on a global scale, and some of these insecticides remain the first line of defense in most countries (Edwards and Thompson 1971, Parker and Howard 2001, Kuhar and Alvarez 2008). Research in Europe and North America, however, generally found that the OPs and carbamates were not as effective as organochlorines such as aldrin (Caldicott and Isherwood 1967, Hancock *et al.* 1986, Parker *et al.* 1990, Parker and Howard 2001), and OPs have generally proved more effective than carbamates (Arnoux *et al.* 1974, Finlayson *et al.* 1979, Toba 1987, Parker *et al.* 1990). Nevertheless, control of wireworm damage with these insecticides in various countries (i.e., the OPs phorate, chlorpyrifos, ethoprop, and the carbamate carbofuran) has generally been acceptable, although the efficacy of these products can be somewhat inconsistent or even fail (Hancock *et al.* 1986, Parker and Howard 2001, Kuhar and Alvarez 2008).

The reasons for inconsistency in efficacy of OPs and carbamates for wireworm control in potatoes and other crops has not been adequately explained, although an understanding of these factors would likely lead to methods for improving the consistency and efficacy of controls. A common belief is that since most wireworm damage to potatoes occurs late in the growing season, an insecticide applied at or before planting must be residual at levels adequate to kill wireworms up to the time of harvest (Parker and Howard 2001). Although this was likely the case with the persistent organochlorines, there is currently no evidence that OP or carbamate insecticides will provide significant residual control late in the potato-growing season. In fact, the degradation curves of certain commonly used wireworm insecticides in soil (e.g., fonofos and carbofuran) indicted that less than 10% of the parent compounds remained after 70 days (Onsager and Rusk 1969). Late season control with OPs and carbamates also assumes that wireworms feeding near harvest would not have contacted or fed upon the crop within the treated area up to that time. It is the understanding of the authors, however, that most if not all wireworms in a field will feed at some time in the spring, which generally coincides with the planting of potatoes. If the primary source of food at that time is the planted potato crop (i.e., there is no green-ploughed pasture or weeds, etc., in the field), then most wireworms will orient to and feed on mother tubers located within the insecticide-treated furrow. Also, maturing daughter tubers near harvest often lie outside the initial treatment area, which would reduce exposure of wireworms to any residual insecticides left at that time. It is likely, therefore, that the effective control of wireworms by OP and carbamate insecticides in potatoes is primarily through high early-season mortality, rather than mortality occurring later in the season. Following this line of reasoning further, if potatoes are planted in fields with no competing sources of CO_2 in soil at a time when wireworms are fully active in spring, this would ensure that the majority of wireworms would be attracted to the mother tubers (primary source of CO_2), and encounter in-furrow insecticides at a time when insecticide titer was highest. In situations where these conditions are not met – for example, if planting occurs outside of the spring wireworm feeding period, or if alternative sources of CO_2 are present in soil (e.g., green ploughed pasture or sod) – it is likely that part of the population would not encounter the in-furrow treatment in spring, and therefore survive to feed on maturing tubers later on. In efficacy work by the authors over a 12-year period, trials were always conducted under the optimum field conditions described above, and efficacy of various standard OPs (i.e., phorate and chlorpyrifos) has always been consistently high (Vernon *et al.* 2007). When failures of phorate to control wireworm damage have been reported to the authors, they are generally associated with situations where the field has been in pasture and was ploughed green just prior to planting potatoes. Research to validate this line of reasoning is currently underway by the authors and others, but might be taken into account when conducting or interpreting efficacy trials, or in developing IPM programs for growers.

Just as the organochlorines were banned globally due largely to their persistence in soil, the relatively high toxicity of the OPs and carbamates to humans and the environment has resulted in the loss of many of these wireworm insecticides over the past decade, and this attrition is expected to continue. In Canada, for example, the OP phorate Thimet® 15G is the primary insecticide used for wireworm control in potatoes. This highly toxic insecticide has already been withdrawn from use in British Columbia due to raptor poisonings coincident with increased usage in potatoes for wireworm control in the 1990s (Elliott *et al.* 1997, Wilson *et al.* 2002), and is scheduled to be withdrawn from the rest of Canada in 2015. At present there are no replacements for phorate for wireworm control on potatoes, and, as has been recognized by others (Parker and Howard 2001, Kuhar and Alvarez 2008), research into new control options is desperately needed.

Most of the focus in developing new insecticides for control of wireworms over the past decade or so has involved three classes of chemicals: the pyrethroids, the neonicotinoids, and the phenyl pyrazoles. All of these groups have demonstrated effectiveness in reducing damage by wireworms to field crops such as corn and wheat (Wilde *et al.* 2004, DeVries and Wright 2005, Vernon *et al.* 2009, 2011), and registrations on these and other crops have been granted in many countries worldwide. In addition, some of these insecticides have recently been registered for the control of wireworms in potatoes, particularly in the USA (e.g., the pyrethroid, bifenthrin, and the phenyl pyrazole, fipronil), and for wireworm "suppression" in Canada (e.g., the neonicotinoid, clothianidin). However, in contrast to the OCs, OPs, and carbamates, which have been proven to kill wireworms as well as protecting crops from damage (Lange *et al.* 1949, Lane 1954, Edwards and Thompson 1971, van Herk *et al.* 2008, Vernon *et al.* 2009), there is evidence that some of the new chemical classes (e.g., pyrethroids and neonicotinoids), while providing crop protection, may not provide significant mortality of wireworms in the field (Vernon *et al.* 2007, 2009, 2011). Since these new classes of insecticides will likely play an important role in wireworm management in many crops over the next decades, a discussion of their actual effects on damage protection in potatoes and, just as importantly, on wireworm health and behavior is warranted. In addition, these discussions will also cover the effects of these novel insecticides on wireworm populations in crops such as cereals, which are commonly planted in rotation with potatoes.

Neonicotinoids

Neonicotinoids (e.g., imidacloprid, clothianidin, and thiamethoxam) applied as seed treatments to cereal crops have been shown to provide good stand and yield protection from wireworm feeding (e.g., Vernon *et al.* 2009), and registrations exist for cereal crops in a number of countries. Initially, at least in Canada, it was hoped that these insecticides might replace the de-registered OC lindane in providing both yield protection and wireworm reduction in cereal crops, and thus reduce populations leading up to potato rotations. However, several years of field data by the authors have shown that wireworm populations are not significantly reduced by any of these neonicotinoids at the field rates registered, and damage protection in cereals is likely due to wireworms becoming reversibly intoxicated or moribund, rather than dying during the crop establishment phase (Vernon *et al.* 2007, 2009, 2011). Laboratory studies have also shown that contact exposure of several economic species of wireworms (*Agriotes obscurus*, *A. sputator*, *L. canus*, *Ctenicera destructor*, and *C. pruinina* (Horn)) to chloronicotinoid (imidacloprid, acetamiprid) and thianicotinoid (clothianidin, thiamethoxam) insecticides causes rapid and prolonged periods of morbidity (>150 days), during which feeding ceases and after which wireworms make a full recovery (van Herk *et al.* 2007, 2008, Vernon *et al.* 2008). It was also found that toxicities of these neonicotinoids differed among species (van Herk *et al.* 2007, 2008). The conclusion from these laboratory and field studies is that although neonicotinoid-treated cereal crops are protected from early season feeding through wireworm intoxication, populations eventually recover to full health and can thereafter continue their life cycle in subsequent crops (i.e., potato). In addition, neonicotinoid seed treatments have no effect on neonate wireworms arising in the field later in the summer, and this clutch of wireworms will also carry over to subsequent rotational crops (Vernon *et al.* 2009, 2011).

In potatoes, a number of neonicotinoids have been registered as seed piece treatments, or as in-furrow sprays for the systemic control of various aboveground pests (e.g., Colorado potato beetles and leafhoppers; Kuhar *et al.* 2007), and are listed on some labels as providing wireworm damage suppression (e.g., clothianidin in Canada). The reduction of wireworm blemishes by these treatments, however, has been very inconsistent in the field, with reports of acceptable and consistent levels of control in British Columbia, Canada (*A. obscurus*; Vernon *et al.* 2007) and Virginia, USA (*Melanotus communis*; Kuhar and Alvarez 2008), and reports of unacceptable control in Ontario (*Melanotus* spp.; Tolman *et al.* 2005) and Prince Edward Island, Canada (*Agriotes* spp.; Noronha *et al.* 2007). It is interesting to note that potato seed piece treatments with clothianidin and thiamethoxam provided excellent blemish control in one year in a 2006 PEI trial (Noronha *et al.* 2006), but no control in the years follow-ing (e.g., Noronha *et al.* 2007). It is thought that wireworms become intoxicated initially upon contact with neonicotinoids on potato seed or in treated furrows, and that blemish control is dependent on whether or not wireworms remain intoxicated throughout tuber maturation. The duration of intoxication, and thus damage to tubers, is likely to vary according to soil type, climate, or species (Vernon *et al.* 2007). As was observed with neonicotinoid-treated cereal crops, there was no significant mortality of wireworms in plots of potatoes treated with neonicotinoids when plots were sampled the following spring, in several years of study by the authors (Vernon *et al.* 2007, R.S.Vernon, unpubl. data).

Synthetic Pyrethroids

Pyrethroids, although generally formulated as above-ground foliage sprays, have also shown varying degrees of promise as in-furrow applications for wire-worm control. Tefluthrin, for example, although providing acceptable protec-tion from wireworm damage in corn as an at-planting granular application, has not provided acceptable blemish control in potatoes (R.S.Vernon, unpubl. data). As an experimental seed treatment on wheat, tefluthrin provided good crop stand and yield protection under heavy wireworm pressure (*A. obscurus*), but, similar to neonicotinoid treatments, did not reduce resident or neonate wire-worm populations (Vernon *et al.* 2009). In laboratory studies, tefluthrin applied to wheat seed was found to be repulsive and non-lethal to wireworms, and it is hypothesized that repulsion, rather than mortality of wireworms, accounts for the stand protection observed in wheat and corn (van Herk and Vernon 2007).

Bifenthrin, a pyrethroid with long persistence in the soil (half-life of 122–345 days; Fecko 1999), has recently been registered (2006–2007) as an in-furrow spray for wireworm control in the USA, and has been shown to reduce wire-worm damage similar to the commonly used OP, phorate (Kuhar and Alvarez 2008; R. S.Vernon, unpubl. data). It is interesting to note, however, that bifen-thrin at field application rates was also found to be repulsive but not lethal to wireworms (*A. obscurus*) in the laboratory, and that soil from bifenthrin-treated potato trials was repulsive to wireworms 1 year after application (van Herk and Vernon 2011, van Herk *et al.* 2011). This suggests that an in-furrow applica-tion of bifenthrin at planting will establish a repellent zone along seeded potato rows that prevents wireworms from approaching the mother and daughter tubers through the harvest period. Although further research is needed to determine the ultimate fate of various economic wireworm species in bifenthrin-treated fields, the available data suggest that blemish protection can be significant, and offers a much needed alternative to existing OP insecticides.

Phenyl Pyrazols

The phenyl pyrazole, fipronil, is the most effective wireworm insecticide avail-able since the organochlorines, and has been registered for wireworm control in corn and potatoes in the USA. What makes fipronil unique relative to the neonicotinoids and pyrethroids is that it rapidly kills wireworms of all species upon contact, and at much lower rates will kill wireworms several months after exposure (van Herk *et al.* 2007, Vernon *et al.* 2008). This latent toxicity has presented a number of options for effective, lower-risk management of wireworms that will be discussed below. In potatoes, fipronil applied as an in-furrow spray at planting has been shown to provide blemish control comparable to the OP phorate (Sewell and Alyokin 2004, Vernon *et al.* 2007, Kuhar and Alvarez 2008), and high mortality of resident and neonate wireworms has also been observed (Vernon *et al.* 2007, R. S.Vernon, unpubl. data). In cereal crops, fipronil applied to seed at rates 10 times lower than the formerly used OC lindane was shown to provide excellent wheat stand protection, and significantly reduce resident and neonate wireworm populations in lab and field trials (Vernon *et al.* 2007, 2009, 2011). Stand protection and wireworm reduction with fipronil at these low rates were actually superior to lindane (Vernon *et al.* 2011), suggesting that the general reductions in wireworm economic importance coincide with increasing lindane use in the past (Burrage 1964) could again be achieved with fipronil seed treatments.

Insecticide Combinations

Combinations of various insecticides have been shown to enhance the scope and efficacy of management of wireworms and other pest species of potatoes beyond the effects of the individual insecticides alone (Kuhar and Alvarez 2008, Tolman *et al.* 2008, Tolman and Vernon 2009, R. S. Vernon, unpubl. data). For example, combining the non-systemic OP chlorpyrifos applied as an in-furrow spray, with a systemic neonicotinoid (e.g., clothianidin, thiamethoxam) applied either as seed piece treatments or in-furrow sprays, provided excellent blemish control as well as enhanced reductions in wireworm populations (Tolman and Vernon 2009, R.S.Vernon, unpubl. data). Applied alone, chlorpyrifos (not systemic) will provide acceptable control of wireworms but will not control above-ground insect pests (e.g., Colorado potato beetles), and clothianidin or thiamethoxam will provide some wireworm control as well as systemic control of above-ground pests. Together, these insecticides match the broad spectrum efficacy of the commonly used systemic OP phorate, but with reduced environmental risk. Similarly, neonicotinoids applied as potato seed piece treatments or as in-furrow sprays (e.g., clothianidin, thiamethoxam) in combination with in-furrow sprays of a pyrethroid (non-systemic) such as bifenthrin have also been shown to provide slight improvements to wireworm blemish control and will also control above-ground pests (Kuhar and Alvarez 2008, Tolman *et al.* 2008, 2009). The actual effects of these neonicotinoid+pyrethroid combinations on the mortality of wireworm populations are currently being studied by the authors (van Herk *et al.* 2011).

Combinations of insecticides are also being developed to control wireworm populations in rotational crops such as wheat. The OC lindane, formerly registered in Canada and elsewhere as cereal and corn seed treatments, reduced resident wireworm populations in fields of wheat by 65–70% (Arnason and Fox 1948, Vernon *et al.* 2009). It was also found that the residual action of lindane was sufficient to kill about 85% of newly formed neonate wireworms (*A. obscurus*) (Vernon *et al.* 2009, 2011). This clean-up of existing and neonate populations of wireworms meant that the field would have low numbers of wireworms for at least 2 years, during which time potatoes could be planted with low economic risk of wireworm damage (Vernon *et al.* 2009, 2011). As discussed above, neonicotinoids (e.g., imidacloprid, clothianidin, and thiamethoxam) applied as seed treatments to cereals, although providing excellent stand protection and yield, did not significantly reduce wireworm populations (Vernon *et al.* 2009, 2011). The phenyl pyrazole fipronil, on the other hand, provided excellent reduction of both resident and neonate wireworms, but had reduced stand protection when applied at low rates to cereal seed (Vernon *et al.* 2011). To circumvent these problems, neonicotinoids (e.g., thiamethoxam) at lower registered rates (about 10 g a.i./100 kg wheat seed), blended with low rates of fipronil (1–5 g a.i./100 kg seed) have been shown to preserve crop stand and yield, as well as reduce resident and neonate populations even more effectively than lindane applied at 60 g a.i./100 kg seed (Vernon *et al.* 2011). Such blends could be used on wheat and other cereals in future to reduce wireworm populations to sub-economic levels in fields destined for potatoes, and field trials are currently underway in Canada by the authors to confirm this. In addition, neonicotinoid + fipronil blend-treated wheat seed has been incorporated into potato seed furrows for control of wireworms in experimental trials (R. S.Vernon, unpubl. data). The principle of this method is to attract the majority of wireworms to the lethal wheat seed, which germinates before the mother tubers. Such treatments have been shown to reduce wireworm damage to daughter tubers as effectively as phorate (Tolman and Vernon 2009, R. S. Vernon, unpubl. data), and significantly reduce wireworm populations with low amounts of insecticide per ha.

Click Beetle Controls

Click beetles set the stage for wireworm problems by ovipositing in fields of their preferred crops (i.e., pasture, cereals). By preventing oviposition in these fields, wireworm populations will not reach economic levels, and this approach is currently being explored in The Netherlands (Ester *et al.* 2004, van Rozen *et al.* 2007). With the use of pheromone traps for the three primary species (*A. obscurus*, *A. lineatus*, and *A. sputator*), peak activity of male click beetles is determined, and fields are sprayed once or twice with foliar applications of a pyrethroid (i.e., deltamethrin or lambda cyhalothrin). These sprays are very effective in killing both male and female click beetles, and research is currently underway to determine whether wireworm populations in fields have been reduced through reduced oviposition (A. Ester, pers. comm.). This strategy would be applied each time a preferred oviposition crop was grown in the field, and reduced oviposition would be required for a number of years equivalent to the life cycle of the wireworms involved (i.e., 4–5 years for *Agriotes* spp.).

Such a strategy is somewhat more limited in scope than soil-applied insecticides for wireworm control, in that management activities required are heavier and protracted over several years (i.e., routine pheromone trapping for multiple species, trap interpretation, field spraying); simple adult monitoring tools such as pheromone traps must be available for all key species; the fields must be under the control of one grower, and not leased land (as is common in many potato-growing areas worldwide); and insecticides need to be registered specifically for click beetle control in each country using this technique. It is expected, however, that this approach will grow in popularity as effective soil-applied wireworm control options dwindle, and as pheromones become available for more species worldwide.

Biological Controls

Predators

A number of arthropod predators of wireworms have been recorded, including several genera of carabids, staphylinids, and therevids (Thomas 1940, Fox and MacLellan 1956), but there have been no records of significant population reductions occurring due to these species. Birds, especially crows, are commonly cited as feeding on wireworms concurrent with cultivation activities (Thomas 1940, Gratwick 1989), and they have been observed uprooting cabbage (Thomas 1940) and transplanted strawberry seedlings (R.S.Vernon, unpubl. data) to feed on wireworms assembled at the roots. Although crows can eat numerous wireworms (e.g., 72 wireworms; Kalmbach 1920), their impact on field populations is not considered significant because populations can be in the millions per ha (Miles and Cohen 1939), and only a small portion of a population is exposed during cultivation. A more comprehensive listing of the known parasites and predators of wireworms can be found in Thomas (1940).

Microbial Pathogens

Microbial control agents attacking wireworms and click beetles have been commonly observed in nature or in outbreaks occurring in laboratory colonies, and the reader is referred to reviews by Thomas (1940), Parker and Howard (2001), and Wraight et al. (2009) that list the early and more contemporary literature on this subject. Historically, most of the attention has focused on the fungal pathogen *Metarhizium anisopliae* (Metschn.) Sorokin, which has been observed to infect, for example, *Melanotus* spp. (Hyslop 1915); *Agriotes mancus* (Gorham 1923); *Limonius californicus* (Rockwood 1950); and *A. obscurus, A. lineatus, A. sputator,* and *L. canus* (Fox and Jaques 1958, Kabaluk et al. 2005). Early attempts at controlling wireworm populations with inundative releases of endemic strains of *M. anisopliae* in soil were unsuccessful (Hyslop 1915, Fox and Jaques 1958); however, with the development of improved methods of producing and formulating *M. anisopliae* there has been a renewed interest in evaluating various isolates for wireworm biocontrol (Wraight et al. 2009).

The results thus far suggest that although wireworms of a number of species can be infected and killed under defined laboratory conditions, (e.g., Ericsson *et al.* 2007, Kabaluk *et al.* 2007a), attempts to control wireworms with inundative releases in the field are typically variable (e.g., Kabaluk *et al.* 2005, Tharp *et al.* 2007, Kuhar and Doughty 2008). Among the more optimistic of these field trials, Kabaluk *et al.* (2005, 2007a) reported a 33.3% reduction in wireworm blemishes to daughter tubers with a pre-plant broadcast application of *M. anisopliae*, and, in another trial, infected *A. obscurus* cadavers were retrieved from treated field soil, confirming some in-field mortality was achievable with this approach (Kabaluk *et al.* 2007b). Opportunities for enhancing the efficacy of *M. anisopliae* have also been explored – for example, Ericsson *et al.* (2007) found that the natural insecticide spinosyn synergized efficacy against wireworms (*A. lineatus*) in the lab, and novel application techniques to draw wireworms to living baits inoculated with *M. anisopliae* have been suggested (Wraight *et al.* 2009) and are under evaluation (T. Kabaluk, pers. comm.). Although the wireworm stage is generally the main target, Kabaluk *et al.* (2005) reported that adult click beetles (*A. obscurus*) were as susceptible as wireworms to *M. anisopliae* infections in the laboratory, and this has been observed in other wireworm species in nature (Parker and Howard 2001).

The fungal pathogen *Beauveria bassiana* has also been evaluated with some success as a biocontrol agent for wireworms attacking potatoes (e.g., Ester and Huiting 2007, Ladurner 2007), and a liquid formulation of *B. bassiana* conidia has been approved for wireworm control in potatoes in Italy (Naturalis-L®, *B. bassiana* strain ATCC 74040; www.intrachem.com), primarily against several species of *Agriotes*. It should be noted, however, that the effectiveness of fungal pathogens observed against one or more wireworm species cannot be assumed to apply to all species (Wraight *et al.* 2009), and differential susceptibility of various wireworm species to *M. anisopliae* or *B. bassiana* has been observed in the laboratory (Tinline and Zacharuk 1960, Zacharuk and Tinline 1968, Kabaluk *et al.* 2007a). Present and future areas of research identified for microbial control of wireworms include the search for superior species and isolates (in virulence, productivity and persistence; genetic modification; blended entomopathogens or entomopathogens with other agents (e.g., insecticides); and optimization of delivery at the field level (Wraight *et al.* 2009).

Nematodes

Entomopathogenic (EPN) nematodes have shown limited success in controlling wireworms at the commercial level. Toba *et al.* (1983), although documenting reductions in *Limonius californicus* populations of 29% with *Steinernema feltiae* (Filipjev) in the field, concluded that the lethal dose required to achieve higher levels of control would be cost-prohibitive. In addition, residual control of wireworms was not observed within months of *S. feltiae* application (Toba *et al.* 1983). In more recent work, Ester and Huiting (2007) did not find *S. feltiae* to be effective in the field against *Agriotes* spp. Between 24% and 39% mortality has been reported under laboratory conditions with *S. carpocapsae*, *H. bacteriophora*, and *S. riobrave* against small to medium-sized *L. canus* wireworms (Wraight *et al.* 2009). When used in combination with resistant cultivars and/ or insecticide, *S. carpocapsae* reduced damage by wireworms (*Conoderus* spp.) to sweet potatoes by up to 25% (Schalk *et al.* 1993). With any living biological control agent targeted for use in soil, however, careful consideration must be given to determine the conditions required for optimal efficacy to occur (e.g., soil variables including texture, temperature, and moisture). Under abnormally wet conditions, for example, *S. carpocapsae* was leached from the rhizosphere of sweet potatoes and provided no wireworm control (Schalk *et al.* 1993).

Semiochemical Controls

Where pheromones have been developed for key wireworm species, their potential for reducing click beetle populations or disrupting mating in order to reduce oviposition and the ensuing build-up of wireworm populations has been investigated. In Russia, pheromones applied to fields at the rate of 120 g pheromone/ ha caused the "disorientation" (confusion) of male *Agriotes* (species not disclosed), resulting in over 70% of females remaining unmated (Ivashchenko and Chernova 1995). This abstract also alluded to mass trapping, but this technique was not discussed except to state that it was less effective than disorientation. Balkov and Ismailov (1991) found that effective direct control of *A. sputator* and *A. gurgistanus* was achieved with the intensive use of pheromone traps over 3–4 years. In other work, Balkov (1991) found that 30 *A. sputator* pheromone traps/ha reduced larvae by 86% after 4 years of mass trapping in a field with medium wireworm infestation (up to 5 individuals/m^2). At high levels of infestation (> 10 individuals/m^2), 120 traps/ha were required.

In Japan, Kishita *et al.* (2003) used mark–release–recapture studies with pheromone traps to estimate the population density of *Melanotus okinawensis* on Ikei Island, and Yamamura *et al.* (2003) used the same method to estimate its average dispersal distance (144 m in 4 days). From these studies, Arakaki *et al.* (2008a) was able to reduce the population of *M. okinawensis* on Ikei Island (158.3 ha) by about 90% over 6 years of mass trapping (10 pheromone traps/ha), but observed no reduction in population in a similar study with *M. sakishimensis* on Kurima Island (Arakaki *et al.* 2008b). In a long-term mating disruption study on Minami-Daito Island (3057 ha), using one, 80-m long pheromone dispenser roll/ha, numbers of adult *M. okinawensis* captured by hand had decreased by 89.3% after 7 years, and mating rates were significantly lower (range 14.3– 71.4%) than in untreated areas (96.9–100%) (Arakaki *et al.* 2008c). In these studies, however, there were no surveys taken to determine if the number of wireworms in soil were proportionately reduced.

The large-scale removal of adults will not necessarily reduce larval populations, as demonstrated by Campbell and Stone (1939) in California with *L. californicus*. This is particularly a concern if the adults can fly or repopulate a field quickly from refuge areas, in which case mass trapping may be a more effective strategy when combined with topical applications of insecticides in areas where the beetles are known to be concentrated, provided this is done prior to mating and oviposition (Ester and Rozen 2005).

The limited interest in mass trapping or disorientation of click beetles as an indirect wireworm control method can be attributed to several factors. A major obstacle is the cost and inconvenience of deploying and maintaining from 30 to 120 pheromone traps/ha in large fields during a mixed 4-year crop rotation (using Balkov's 1991 study as an example). Also, since multiple species of wireworms (e.g., *Agriotes* spp. and *Melanotus* spp.) are often present in fields, the cost of trapping increases proportionately with the number of economic species present.

The use of semiochemicals for disruption or mass trapping to pre-emptively reduce click beetle oviposition has historically targeted cultivated or soon-to-be cultivated agricultural land. These fields can be considered non-permanent wireworm population reservoirs, however, in that they are subjected to a wide variety of wireworm population-disrupting or -enhancing activities, including favorable or unfavorable crop rotations, field cultivation practices, and various field/crop amendments (e.g., irrigation, insecticide treatments, etc.). Another approach to reduce click beetle oviposition would be to target the more permanent wireworm population reservoirs that often surround cultivated fields in the general agricultural landscape. These permanent reservoirs may include grassy headlands, ditch banks, dykes, etc., which generally contain most stages of various wireworm species, and which often produce the adult beetles that chronically invade adjacent fields (i.e., cereals, pasture) to oviposit. Also, these permanent reservoirs tend to occupy only a very small fraction of land in intensively farmed areas, making higher cost-control methods (such as mass trapping, mating disruption, etc.) potentially more affordable. In addition, once wireworm populations are removed from permanent reservoirs, control efforts in those areas could be abandoned for several years until new click beetle population build-ups warrant renewed control efforts. Management of click beetle populations in reservoir areas is currently being explored in Canada by the authors.

CONCLUSIONS

In their review of the biology and management of wireworms on potato, Jansson and Seal (1994) found that wireworms were generally considered a minor pest of potato in most regions of the world at that time. Since then, wireworms have become increasingly problematic in potato crops in Europe and North America (Parker and Howard, 2001, Vernon *et al.* 2001, Horton 2006, Kuhar *et al.* 2008, Noronha 2011), and scientific interest in this complex group of insect pests has experienced a resurgence over the past decade. During the writing of the present review, a number of generalizations were revealed that have relevance to the present and future direction of wireworm research relating to potatoes. Due to the growing severity of the problem in some areas, and since the number of researchers is somewhat limited, it is hoped that the suggestions presented will help to identify the more relevant research paths.

The first generalization, arising from our review of elaterid species associated with agriculture globally, is that we are dealing with an extremely complex and diverse group of insect pests and non-pests from the worldwide agricultural landscape right down to the field level. Further to this, it was shown that different species may have vastly different life histories and other biological/behavioral traits, and may have differing responses to monitoring approaches and controls; moreover, several of these species may concurrently occupy the same field or fluctuate over time. A fundamental prerequisite for any contemporary management program, therefore, is to know the wireworms involved to the species level, and to preferentially have taxonomic methods available that can be used by researchers at the regional or even local level. Due to the difficulties and errors inherent with identification using morphological characteristics (e.g., *Melanotus* spp.), there has been growing activity and success in the development of molecular diagnostics for a number of wireworm genera. Such diagnostic tools, as they are developed for wireworms, will likely become the taxonomic methods of choice, and ultimately facilitate the rapid and accurate identification of single or mixed species in the field. Following the identification of specific wireworm species in a region, subsequent studies should be directed at unknown but relevant aspects of their general biology, including life history, larval food preferences, spatial and temporal movements in soil, mortality factors, adult oviposition hosts, and so on. The knowledge and tools gained from these research activities are requisite for the development of accurate and effective monitoring and management approaches likely to occur at the regional/field level in the future.

The second generalization relates to fact that although absolute and relative sampling methods have been developed for wireworms in many countries and for many species, none of these methods is entirely reliable. This is the view not only of the authors but also of other researchers intimately involved with the development and implementation of wireworm sampling programs in potatoes (e.g., Parker 1996, Parker and Howard, 2001, Horton 2006). Although absolute or relative sampling reliability is not necessarily crucial for general survey purposes, it becomes profoundly important if the intention of sampling is to provide timely, threshold-based wireworm control recommendations to growers. Typically, the intention of most monitoring programs is to determine whether or not a control action is required to prevent economic pest damage from occurring to a crop. In the case of wireworms and potatoes, the ultimate goal of monitoring would be to indicate whether or not one or more prophylactic controls (field avoidance, planting and harvesting date, insecticide, soil amendment, etc.) is required. Unfortunately, many of the absolute and relative sampling methods developed have been shown to underestimate (or even fail to detect) economic populations of wireworms (Parker and Howard 2001), which would ensure economic damage on occasion in commercial fields. As discussed in this chapter, much of the variability associated with relative wireworm sampling approaches lies in the consistency of the bait or bait traps used, as well as in the biotic and abiotic factors surrounding deployment in the field. The general principles, requirements, and sources of variability of the relative sampling methods themselves (such as various baits and baited traps) have been well covered in the literature, and efforts should continue towards the development of even more consistent, convenient, and cost-effective sampling tools in the future. Where there are gaps in our knowledge of sampling, however, is in the identification of those biotic (e.g., competing sources of CO_2 in the field) and abiotic (soil moisture, temperature) factors in soil over time that impact positively or negatively on the efficacy and accuracy of a sampling approach. If wireworm sampling for management purposes in potatoes is ever to be implemented with confidence, the physical, environmental, and temporal field conditions contributing to a consistent level of sampling efficacy must be determined.

Probably one of the more exciting and applicable research directions of the past decade has been in the development of species-specific pheromone trapping systems for certain genera of wireworms in Europe (*Agriotes*; Tóth *et al.* 2003) and Japan (*Melanotus*; Tamaki *et al.* 1986, 1990). Such systems have facilitated: (1) large-scale surveys of various species in North America, Europe and Russia; (2) spatial and temporal distribution studies of important species at landscape and field levels; (3) monitoring of adult populations to predict wireworm risk in potato fields; (4) monitoring the time to spray adults to reduce oviposition; and (5) investigations into semiochemical-based control programs. Although further research is underway or required to fully develop and interpret these tools and methods, pheromones have already become an important tool in our wireworm research and management arsenal, and should be expanded to other wireworm genera and key species worldwide.

The third and final generalization relates to our belief that to maintain our present standards of quality and abundance in a diverse potato industry, we will require more regional, integrated approaches using the more effective management tools described in this review. The need for development of future management strategies at the regional level is obvious. The conditions favoring wireworm outbreaks will vary from region to region according to many factors, including species complex, agronomic practices (irrigation, tillage), rotations favoring wireworm build-up (pasture, cereal crops), and availability of effective controls (insecticides, entomopathogens, soil amendments, etc.). Due to the attrition of many of the more persistent and effective synthetic insecticides used for wireworm control in potatoes and other crops (e.g., organochlorines, organophosphates, and carbamates), our ability to actually reduce wireworm populations in fields has also diminished in some countries. This is particularly true in crops often rotated with potatoes, such as cereals and forages, where contemporary insecticides now used for wireworm control (e.g., neonicotinoids or pyrethroids), although preventing stand and yield damage, do not actually kill wireworms (Vernon *et al.* 2009). In addition, various existing (organophosphate) and novel insecticides (neonicotinoids or pyrethroids), as well as alternative methods of control (entomopathogens, soil amendments), have been shown to vary in efficacy between certain species, and are more suited or even restricted to some regions over others. Therefore, in designing management approaches for the future, researchers will need to know the key species present at the regional level, and the effectiveness of insecticides and alternative approaches to collectively manage these species. As was discussed above for the development of monitoring approaches, research into or selection of management approaches must also consider the biotic and abiotic effects in fields that favor or disfavor the efficacy of various controls. Knowledge of these factors is currently a major gap in our knowledge of wireworm control efficacy, and will be especially important if management of wireworms with entomopathogens, nematodes, soil amendments, and even novel insecticides is to be optimally realized.

REFERENCES

Afonin, A.N., Greene, S.L., Dzyubenko, N.I., Frolov, A.N., 2008. Interactive agricultural ecological atlas of Russia and neighboring countries: Economic plants and their diseases, pests and weeds. www.agroatlas.ru (online resource).

Al Dhafer, H.M., 2009. Revision of the North American species of Limonius (Coleoptera: Elateridae). Trans. Am. Entomol. Soc. 135, 209–352.

Andrews, N., Ambrosino, M., Fisher, G., Rondon, S.I., 2008. Wireworm biology and nonchemical management in potatoes in the Pacific Northwest. Publication PNW 607, Oregon State University, Corvallis, OR.

Anonymous, 1948. Wireworms and Food Production. A Wireworm Survey of England and Wales (1939–42). Bull. Minist. Agric. Fish. and Food, HMSO, Lond. No. 128.

Anonymous, 2010. Common Potato Insects: Scouting guidelines. Ontario Ministry of Agriculture, Food and Rural Affairs. www.omafra.gov.on.ca (online resource).

Anonymous, 2011. Potato review – Potato Crop Knowledge Centre. www.potatocrop.com/knowledge-june2011/knowledge-JAN11.htm (online resource).

Arakaki, N., Nagayama, A., Kobayashi, A., Kishata, M., Sadoyama, Y., Mougi, N., Kawamura, F., Wakamura, S., Yamamura, K., 2008a. Control of the sugarcane click beetle *Melanotus okinawensis* (Coleoptera: Elateridae) by mass trapping using synthetic sex pheromone on Ikei Island, Okinawa, Japan. Appl. Entomol. Zool. 43, 37–47.

Arakaki, N., Nagayama, A., Kobayashi, A., Tarora, K., Kishata, M., Sadoyama, Y., Mougi, N., Kijima, K., Suzuki, Y., Akino, T., Yasui, H., Fukaya, M., Yasuda, T., Wakamura, S., Yamamura, K., 2008b. Estimation of abundance and dispersal distance of the sugarcane click beetle *Melanotus sakishimensis* Ohira (Coleoptera: Elateridae) on Kurima Island, Okinawa, by mark–recapture experiments. Appl. Entomol. Zool. 43, 409–419.

Arakaki, N., Nagayama, A., Kobayashi, A., Hokama, Y., Sadoyama, Y., Mougi, N., Kishata, M., Adaniya, K., Ueda, K., Higa, M., Shinzato, T., Kawamitsu, H., Nakama, S., Wakamura, S., Yamamura, K., 2008c. Mating disruption for control of *Melanotus okinawensis* (Coleoptera: Elateridae) with synthetic sex pheromone. J. Econ. Entomol. 101, 1568–1574.

Arakaki, N., Hokama, Y., Yamamura, K., 2010. Estimation of the dispersal ability of *Melanotus okinawensis* (Coleoptera: Elateridae) larvae in soil. Appl. Entomol. Zool. 45, 297–302.

Arnason, A.P., Fox, W.B., 1948. Wireworm control in the Prairie Provinces. Can. Dept. Agr. Publ. 111.

Arnoux, J., Brunel, E., Missonnier, J., 1974. Essais de protection des cultures de pomme de terre contre les taupins (*Agriotes* sp.): méthodes d'estimation des résultats et efficacité de différents insecticides, organophosphores notamment. Phytiatrie-Phytopharmacie 23, 135–152.

Bagheri, M.R., Ardebili, Z., 2005. Collection and identification of wireworm fauna of Iran, determination of dominant and harmful species in potato fields and effective factors on the pest management, through country. Published by Agricultural and Natural Resource Research Center, Esfahan, Iran 38pp [in Farsi, abstract in English].

Bagheri, M.R., Nematollahi, M.R., 2007. Reaction of ten commercial potato cultivars to wireworms and influence of type and amount of tuber's sugars on infestation rate. Seed and Plant Improvement Journal 22, 503–512 [in Farsi, abstract in English].

Balkov, V.I., 1991. Attractant traps for control of wireworms. Zashchita Rastenii (Moskova) 5, 24–25.

Balkov, V.I., Ismailov, V.Y., 1991. Attractant traps for elaterids. Zashchita-Rastenii (Moskova) 10, 21.

Bartlett, K.A. 1939. The results of shipments of the predatory Elaterid beetle *Pyrophorus luminosus* from Puerto Rico to England. Bull. Ent. Res. 30, 209–210 .

Becker, E.C., 1979. Notes on some New World and Palearctic species formerly in *Athous* Eschscholtz and *Harminius* Fairmaire with new synonymies (Coleoptera: Elateridae). Can. Entomol. 111, 401–415 .

Becker, E.C., 1991. Elateridae (Elateroidae). In: Stehr, F.W. (Ed.), Immature Insects, Kendal/Hunt, Dubuque, IA, pp. 409–410.

Benefer, C., van Herk, W., Ellis, J., Blackshaw, R., Vernon, R., Knight, M., 2012. Genetic diversity and phylogenetic relationships of economically important Elaterid species. J. Pest Sci. in press.

Blackshaw, R.P., Hicks, H., Vernon, R.S., 2009. Sex pheromone traps for predicting wireworm populations: limitations to interpretation. IOBC/wprs Bulletin Working Group "Integrated Protection of Field Vegetables", Proceedings of the Meeting at Porto (Portugal), 23-29 September, 2007. Edited by: Rosemary Collier. ISBN 92-9067-225-8 51, 17–21.

Blackshaw, R.P., Vernon, R.S., 2006. Spatio-temporal stability of two beetle populations in non-farmed habitats in an agricultural landscape. J. Appl. Ecol. 43, 680–689.

Blackshaw, R.P., Vernon, R.S., 2008. Spatial relationships between two *Agriotes* click-beetle species and wireworms in agricultural fields. Agric. Forest Entomol. 10, 1–11.

Blot, Y., Brunel, E., Courbon, R., 1999. Survey on the infection of wheat and maize by larvae of wireworms of *Agriotes* and *Athous* genera (Coleoptera: Elateridae) in some areas of West France. Annales Soc. Entomol. Fr. 35, 453–457.

Boiteau, G., Bousquet, Y., Osborn, W., 2000. Vertical and temporal distribution of carabidae and elateridae in flight above an agricultural landscape. Environ. Entomol. 29, 1157–1163.

Borek, V., Morra, M.J., Brown, P.D., McCaffrey, J.P., 1994. Allelochemicals produced during sinigrin decomposition in soil. J. Agric. Food Chem. 42, 1030–1034.

Brooks, A.R., 1960. Adult Elateridae of southern Alberta, Saskatchewan and Manitoba (Coleoptera). Can, Entomol. (Suppl. 20).

Brown, E.A., Keaster, A.J., 1986. Activity and dispersal of adult *Melanotus depressus* and *Melanotus verberans* (Coleoptera: Elateridae) in a Missouri cornfield. J. Kansas Entomol. Soc. 59, 127–132.

Brown, P.D., Morra, M.J., McCaffrey, J.P., Auld, D.L., Williams III, L., 1991. Allelochemicals produced during glucosinolate degradation in soil. J. Chem. Ecol. 17, 2021–2034.

Buntin, G.D., 1994. Developing a primary sampling program. In: Pedigo, L.P., Buntin, G.D. (Eds.), Handbook of sampling methods for arthropods in agriculture, CRC Press, Boca Raton, FL, pp. 99–115.

Burrage, R.H., 1964. Trends in damage by wireworms (Coleoptera: Elateridae) in grain crops in Saskatchewan, 1954–1961. Can. J. Plant Sci. 44, 515–519.

Bynum Jr., E.D., Archer, T.L., 1987. Wireworm (Coleoptera: Elateridae) sampling for semiarid cropping systems. J. Econ. Entomol. 80, 164–168.

Bynum, E.K., Ingram, J.W., Charpentier, L.J., 1949. Control of wireworms attacking sugarcane in Louisiana. J. Econ. Entomol. 42, 556–557.

Caldicott, J.J.B., Isherwood, R.J., 1967. The use of phorate for the control of wireworm in potatoes. Proceedings of the 4th British Insecticide and Fungicide Conference, Brighton, pp. 314–318.

Campbell, R.E., 1942. Dichloroethyl ether for protecting melon plants from wireworms. J. Econ. Entomol. 35, 26–30.

Campbell, R.E., Stone, M.W., 1939. Trapping Elaterid beetles as a control measure against wireworms. J. Econ. Entomol. 32, 47–53.

Carpenter, G.P., Scott, D.R., 1972. Sugarbeet wireworm control in potatoes in Idaho. J. Econ. Entomol. 65, 773–775.

Casari, S.A., 2006. Larva, pupa and adult of *Aeolus cinctus* Candèze. (Coleoptera, Elateridae, Agrypninae) Revista Brasileira de Entomologia 50, 347–351.

Cate, P., 2004. Fauna Europaea: Elateridae, Coleoptera. Fauna Europaea, Version 1.1, www.faunaeur.org (online resource).

Chabert, A., Blot, Y., 1992. Estimation des populations larvaires de taupins par un piège attractif. Phytoma 436, 26–30.

Chalfant, R.B., Seal, D.R., 1991. Biology and management of wireworms on sweet potato. In: Jansson, R.K., Raman, K.V. (Eds.), Sweet potato pest management: A global perspective, Westview Press, San Francisco, CA, pp. 304–326.

Chalfant, R.B., Harmon, S.A., Stacey, L., 1979. Chemical control of the sweet potato flea beetle and southern potato wireworm on sweet potatoes in Georgia. J. Ga. Entomol. Soc. 14, 354–358.

Cherry, R., 2007. Seasonal population dynamics of wireworms (Coleoptera: Elateridae) in Florida sugarcane fields. Florida Entomol. 90, 426–430.

Cherry, R.H., 1988. Correlation of crop age with populations of soil insect pests in Florida sugarcane. J. Agric. Entomol. 5, 241–245.

Choi, C.I., 1972. Potato growing and its problems in Korea. In: French, E.R. (Ed.), Prospects for the potato in the developing world: An international symposium on key problems and potentials for greater use of the potato in the developing world, Centro Internacional de la Papa, La Molina, Lima, pp. 40–45.

Chung, C., Wei, H., 1956. Protection against wireworm injury with seed treatment. Acta Entomologica Sinica 4 [abstract only].

Cockerham, K.L., Deen, O.T., 1936. Notes on the life history, habits and distribution of *Heteroderes laurentii* Guér. J. Econ. Entomol. 29, 288–296.

Day, A., Cuthbert, F.P., Reid, W.J., 1964. Control of the southern potato wireworm, *Conoderus falli*, on early-crop potatoes. J. Econ. Entomol. 57, 468–470.

Deen, O.T., Cuthbert, F.P., 1955. The distribution and relative abundance of wireworms in potato-growing areas of the southeastern states. J. Econ. Entomol. 48, 191–193.

DeVries, T., Wright, R.J., 2005. Larval wireworm control 2004. Arthropod Manag. Tests 29, F17.

Doane, J.F., 1963. Studies on oviposition and fecundity of *Ctenicera destructor* (Brown) (Coleoptera: Elateridae). Can. Entomol. 95, 1145–1153.

Doane, J.F., 1977. The flat wireworm, *Aeolus mellillus*: Studies on seasonal occurrence of adults and incidence of the larvae in the wireworm complex attacking wheat in Saskatchewan. Env. Entomol. 6, 818–822.

Doane, J.F., 1981. Evaluation of a larval trap and baits for monitoring the seasonal activity of wire-worms in Saskatchewan. Env. Entomol. 10, 335–342.

Doane, J.F., Lee, Y.W., Klinger, J., Westcott, N.D., 1975. The orientation response of *Ctenicera destructor* and other wireworms (Coleoptera: Elateridae) to germinating grain and carbon dioxide. Can. Entomol. 107, 1233–1252.

Dobrovsky, T.M., 1953. Another wireworm of Irish potatoes. J. Econ. Entomol. 46, 1115.

Dogger, J.R., Lilly, J.H., 1949. Seed treatment as a means of reducing wireworm damage to corn. J. Econ. Entomol. 42, 663–665.

Eagerton, H.C., 1914. The spotted click beetle (*Monocrepidius vespertinus*, Fab.). South Carolina Agric. Exp. Sta Bull 179.

Edwards, C.A., Thompson, A.R., 1971. Control of wireworms with organophosphorus and carbamate insecticides. Pesticide Science 2, 185–189.

Eidt, D.C., 1953. European wireworms in Canada with particular reference to Nova Scotian infestations. Can. Entomol. 85, 408–414.

Elberson, L.R., Borek, V., McCaffrey, J.P., Morra, M.J., 1996. Toxicity of rapeseed meal-amended soil to wireworms *Limonius californicus* (Coleoptera: Elateridae). J. Agric. Entomol. 13, 323–330.

Elliott, J.E., Langelier, K.M., Mineau, P., Wilson, L.K., 1996. Poisoning of bald eagles and red-tailed hawks by carbofuran and fensulfothion in the Fraser delta of British Columbia, Canada. J. Wildlife Dis. 32, 486–491.

Ellis, J.S., Blackshaw, R., Parker, W., Hicks, H., Knight, M.E., 2009. Genetic identification of morphologically cryptic agricultural pests. Agric. Forest Entomol. 11, 115–121.

Ericsson, J.D., Kabaluk, J.T., Goettel, M.S., Myers, J.H., 2007. Spinosyns interact synergistically with the insect pathogen *Metarhizium anisopliae* (Deuteromycete: Hyphomycete) against lined click beetle larvae, *Agriotes lineatus* (Coleoptera: Elateridae). J. Econ. Entomol. 100, 31–39.

Ester, A., Huiting, H., 2007. Controlling wireworms (*Agriotes* spp.) in a potato crop with biologicals. IOBC/WPRS Bull. 30, 189–196.

Ester, A., van Rozen, K., 2005. Monitoring and control of *Agriotes lineatus* and *A. obscurus* in arable crops in the Netherlands. IOBC/WPRS Bull. 28, 81–85.

Ester, A., van Rozen, K., Griepink, F.C., 2004. Tackling wireworms in a new way. International Pest Control, March–April, 80–81.

Falconer, D.S., 1945. On the behaviour of wireworms of the genus *Agriotes* Esch. (Coleoptera: Elateridae) in relation to temperature. J. Exp. Biol. 21, 17–32.

Fecko, A., 1999. Environmental fate of bifenthrin. Environmental Hazards Assessment Program, Dept. Pest. Reg., Sacaramento, CA. www.cdpr.ca.gov/docs/emon/pubs/fatememo/bifentn.pdf (online resource).

Fenton, F.A., 1926. Observations on the biology of *Melanotus communis* and *Melanotus pilosus*. J. Econ. Entomol. 19, 502–504.

Finlayson, D.G., Wilkinson, A.T.S., MacKenzie, J.R., 1979. Efficacy of insecticides against tuber flea beetles, wireworms and aphids in potatoes. J. Entomol. Soc. Br. Columbia 76, 6–9.

Floyd, E.H., 1949. Control of the Sand Wireworm in Louisiana. J. Econ. Entomol. 42, 900–903.

Fox, C.J.S., 1961. The distribution and abundance of wireworms in the Annapolis Valley of Nova Scotia. Can. Entomol. 93, 276–279.

Fox, C.J.S., 1973. Some feeding responses of a wireworm, *Agriotes sputator* (L.) (Coleoptera: Elateridae). Phytoprotection 54, 43–45.

Fox, C.J.S., Jaques, R.P., 1958. Note on the green-muscardine fungus, *Metarhizium anisopliae* (Metch) Sor., as a control for wireworms. Can. Entomol. 90, 314–315.

Fox, C.J.S., MacLellan, C.R., 1956. Some Carabidae and Staphylinidae shown to feed on a wireworm, *Agriotes sputator* (L.), by the precipitin test. Can. Entomol. 88, 228–231.

French, N., White, J.H., 1965. Observations on wireworm populations causing damage to ware potatoes. Plant Pathology 14, 1–3.

Furlan, L., 1996. The biology of *Agriotes ustulatus* Schäller (Col., Elateridae). I. Adults and oviposition. J. Appl. Entomol. 120, 269–274.

Furlan, L., 1998. The biology of *Agriotes ustulatus* Schäller (Col., Elateridae) II. Larval development, pupation, whole cycle description and practical implications. J. Appl. Entomol. 122, 71–78.

Furlan, L., 2004. The biology of *Agriotes sordidus* Illinger (Col., Elateridae). J. Appl. Entomol. 128, 696–706.

Furlan, L., 2007. Management and biological control of wireworm populations in Europe: current possibilities and future perspectives. IOBC/WPRS Bull. 30, 11–16.

Furlan, L., Tóth, M., 2007. Occurrence of click beetle pest spp. (Coleoptera: Elateridae) in Europe as detected by pheromone traps: survey results of 1998–2006. IOBC/WPRS Bull. 30, 19–25.

Furlan, L., Tóth, M., Yatsinin, V., Ujvary, I., 2001. The project to implement IPM strategies against *Agriotes* species in Europe: what has been done and what is still to be done. Proc. XXI IWGO Conf., Legnaro Italia, 253–262.

Furlan, L., Bonetto, C., Finotto, A., Lazzeri, L., Malaguti, L., Patalano, G., Parker, W., 2010. The efficacy of biofumigant meals and plants to control wireworm populations. Industrial Crops Products 31, 245–254.

Ghilarov, M.S., 1937. The fauna of injurious soil insects of arable land. Bull. Entomol. Res. 28, 633–637.

Gibson, K.E., 1939. Wireworm damage to potatoes in the Yakima Valley of Washington. J. Econ. Entomol. 32, 122–124.

Gibson, K.E., Lane, M.C., Cook, W.C., Jones, E.W., 1958. Effect of some crop rotations on wireworm populations in irrigated lands. USDA Tech. Bull. 1172.

Glen, R., 1950. Larvae of the elaterid beetles of the tribe Leptuoidini (Coleoptera: Elateridae). Smithsonian Misc. Coll 111 (11).

Glen, R., King, K.M., Arnason, A.P., 1943. The identification of wireworms of economic importance in Canada. Can. J. Res. 21, 358–387.

Gorham, R.P., 1923. Notes on *Agriotes mancus* Say at Dartmouth, N.S. Proc. Acadian Entomol. Soc. 9, 69–72.

Gratwick, M., 1989. Potato Pests. *MAFF Reference Book* 187. HMSO, London.

Griffin, J.A., Eden, W.G., 1953. Control of the Gulf Wireworm in sweet potatoes in Alabama. J. Econ. Entomol. 46, 948–951.

Gui, H.L., 1935. Soil types as factors in wireworm distribution. Am. Potato J. 12, 107–113.

Hancock, M., Green, D., Lane, A., Mathias, P.L., Port, C.M., Tones, S.J., 1986. Evaluation of insecticides to replace aldrin for the control of wireworms on potatoes. Tests Agrichemicals Cultivars No. 7, Ann. Appl. Biol. 108 (Suppl.), 28–29.

Hastings, E., Cowan, T.E., 1954. Seed treatment of fall planted wheat for wireworm control. J. Econ. Entomol. 47, 597–599.

Hawkins, J.H., 1930. Wireworm control in Maine. J. Econ. Entomol. 23, 349–352.

Hawkins, J.H., 1936. Relation of soil utilization to wireworm injury. J. Econ. Entomol. 29, 728–731.

Hawkins, J.H., McDaniel, I.N., Murphy, E., 1958. Wireworms affecting the agricultural crops of Maine. Maine Agric. Exp. Sta. Bull. 578.

Hayakawa, H., Tsutsui, H., Goto, C., 1985. Survey of wireworms *Agriotes fuscicollis*, mass-infested on carrots in the Tokachi district, Hokkaido, Japan. Ann. Rep. Soc. Plant Protect. North Japan 36, 81–82.

Hemerik, L., Gort, G., Brussaard, L., 2003. Food preference of wireworms analyzed with multinomial logit models. J. Insect. Behavior 16, 647–665.

Herbert, D.A., Brandenburg, R.L., Day, E.R., 1992. Survey of wireworms (Coleoptera: Elateridae) in Virginia and North Carolina peanut fields. Peanut Sci. 19, 98–100.

Hicks, H., Blackshaw, R.P., 2008. Differential responses of three *Agriotes* click beetle species to pheromone traps. Agric. Forest Entomol. 10, 443–448.

Horsfall, J.L., Thomas, C.A., 1926. A preliminary report on the control of wireworms on truck crops. J. Econ. Entomol. 19, 181–185.

Horton, D., 2006. Quantitative relationship between potato tuber damage and counts of Pacific coast wireworm (Coleoptera: Elateridae) in baits: Seasonal effects. J. Entomol. Soc. Br. Columbia 103, 37–48.

Horton, D., Landolt, P., 2002. Orientation response of Pacific coast wireworm (Coleoptera: Elateridae) to food baits in laboratory and effectiveness of baits in field. Can. Entomol. 134, 357–367.

Huiting, H.F., Ester, A., 2009. Neonicotinoids as seed–potato treatments to control *Agriotes* spp. Comm. Agric. Appl. Biol. Sci. 74, 207–216.

Hyslop, J.A., 1915. Wireworms attacking cereal and forage crops. USDA Bull. 156.

Ivanshchenko, I.I., Chernova, S.V., 1995. Biologically active substances against click beetles. Zashchita-Rastenii (Moskva) 9, 16–17.

Jacobson, M., Lilly, C.E., Harding, C., 1968. Sex attractant of sugar beet wireworm: identification and biological activity. Science 159, 208–210.

Jansson, R.K., Lecrone, S.H., 1989. Evaluation of food baits for pre-plant sampling of wireworms (Coleoptera: Elateridae) in potato fields in southern Florida. Florida Entomol. 72, 503–510.

Jansson, R.K., Lecrone, S.H., 1991. Effects of summer cover crop management on wireworm (Coleoptera: Elateridae) abundance and damage to potato. J. Econ. Entomol. 84, 581–586.

Jansson, R.K., Seal, D.R., 1994. Biology and management of wireworm on potato. In: Zehnder, G.W., Powelson, M.L., Jansson, R.K., Raman, K.V. (Eds.), Advances in Potato Pest Biology and Management, American Phytopathological Society Press, St Paul, MN, pp. 31–53.

Jedlička, P., Frouz, J., 2007. Population dynamics of wireworms (Coleoptera, Elateridae) in arable land after abandonment. Biologia 62, 103–111.

Jewett, H.H., 1945. Life history of the wireworm *Conoderus bellus* (Say). Ky. Agric. Exp. Sta. Bull. 472.

Jewett, H. H, 1946. Identification of some larval Elateridae found in Kentucky. Ky. Agric. Exp. Sta. Bull. 489.

Johnson, S.N., Anderson, E.A., Dawson, G., Griffiths, D.W., 2008. Varietal susceptibility of potatoes to wireworm herbivory. Agric. Forest Entomol. 10, 167–174.

Jonasson, T., Olsson, K., 1994. The influence of glycoalkaloids, chlorogenic acid and sugars on the susceptibility of potato tubers to wireworm damage. Potato Res. 37, 205–216.

Jones, E.W., Shirck, F.H., 1942. The seasonal vertical distribution of wireworm in the soil in relation to their control in the Pacific Northwest. J. Agric. Res. 65, 123–124.

Kabaluk, J.T., Goettel, M.S., Erlandson, M.A., Ericsson, J.D., Duke, G.M., Vernon, R.S., 2005. *Metarhizium anisopliae* as a biological control for wireworms and a report of some other naturally-occurring parasites. IOBC/WPRS Bull. 28, 109–115.

Kabaluk, J.T., Goettel, M.S., Ericsson, J.D., Erlandson, M.A., Vernon, R.S., Jaronski, S.T., Mackenzie, K., Cosgrove, L., 2007a. Promise versus performance: Working toward the use of *Metarhizium anisopliae* as a biological control for wireworms. IOBC/WPRS Bull. 30, 69–76.

Kabaluk, J.T., Vernon, R.S., Goettel, M.S., 2007b. Field infection of wireworms with inundative applications of. Metarhizium anisopliae. Phytoprotection 88, 51–56.

Kalmbach, E.R., 1920. The crow in its relation to agriculture. USDA Farmers Bull. 1102.

Kawamura, F., Arakaki, N., Kishita, M., 2002. Trapping efficacy of funnel-vane trap baited with synthetic sex pherom one of the sugarcane wireworms, *Melanotus* sakishimensis Ohira and M. *okinawensis* Ohira (Coleoptera: Elateridae) and influence of trap height on capture. Appl. Entomol. Zool. 37, 373–377.

Keaster, A.J., Fairchild, M.L., 1960. Occurrence and control of sand wireworm in Missouri. J. Econ. Entomol. 53, 963–964.

Keaster, A.J., Chippendale, G.M., Pill, B.A., 1975. Feeding behaviour and growth of the wireworms *Melanotus depressus* and *Limonius dibutans*: Effect of host plants, temperature, photoperiod, and artificial diets. Env. Entomol. 4, 591–595.

Keaster, A.J., Jackson, M.A., Ward, S.S., Krause, G.F., 1988. A worldwide Bibliography of Elateridae. Missouri Agric. Expt. Stn. University of Missouri-Columbia, Columbia.

King, K.M., Glendenning, R., Wilkinson, A.T.S., 1952. A wireworm (*Agriotes obscurus* (L.)). Can. Insect Pest Rev. 30, 269–270.

Kirkegaard, J.A., Gardner, P.A., Desmarchelier, J.M., Angus, J.F., 1993. Biofumigation using *Brassica* species to control pests and diseases in horticulture and agriculture. In: N. Wratten, R.J. Mailer (Eds.), "Proceedings 9th Australian Research Assembly on Brassicas." Agricultural Research Institute, Wagga Wagga, NSW, Australia, pp. 77–82.

Kishata, M., Arakaki, N., Kawamura, F., Sadoyama, Y., Yamamura, K., 2003. Estimation of population density and dispersal parameters of the adult sugarcane wireworm *Melanotus okinawensis* Ohira (Coleoptera: Elateridae), on Ikei Island, Okinawa, by mark–recapture experiments. Applied Entomol. Zool. 38, 233–240.

Klausnitzer, B., 1994. Familie Elateridae. In: Klausnitzer, B. (Ed.), Die Larven der Käfer Mitteleuropas: Myxophaga/ Polyphaga, Vol. 2. Gustav Fischer Verlag, Jena, Germany, pp. 118–189.

Kosmatshevsky, A.S., 1960. Biology of *Agiotes litigiousus* var. *tauricus* Heyd. and *Agriotes sputator L.* (Coleoptera: Elateridae). Entomol. Rev. 38, 663–672.

Kudryavtsev, I., Siirde, K., Lääts, K., Ismailov., V., Pristavko, V., 1993. Determination of distribution of harmful click beetle species (Coleoptera: Elateridae) by synthetic sex pheromones. J. Chem. Ecol. 19, 1607–1611.

Kuhar, T.P., Alvarez, J.M., 2008. Timing of injury and efficacy of soil-applied insecticides against wireworms on potato in Virginia. Crop Protection 27, 792–798.

Kuhar, T.P., Doughty, H., 2008. Evaluation of *Metarhizium anisopliae* treatments for the control of insect pests in potatoes. Final Report submitted to IR-04 Biopesticides Research Program, 7–10.

Kuhar, T.P., Doughty, H., Hitchner, E., Chapman, A., Cassell, M., Barlow, V., 2007. Evaluation of seed-applied insecticide treatment on potatoes, 2006. Arthropod Manag. Tests 32, E38.

Kuhar, T.P., Doughty, H.B., Speese, J., Reiter, S., 2008. Wireworm pest management in potatoes. Virginia Coop. Ext. Factsheet 2812–1026.

Kulash, W.M., 1943. The ecology and control of wireworms in the Connecticut River Valley. J. Econ. Entomol. 36, 689–693.

Kulash, W.M., 1947. Soil treatment for wireworms and cutworms. J. Econ. Entomol. 40, 851–854.

Kulash, W.M., Monroe, R.J., 1955. Field tests for control of wireworms attacking corn. J. Econ. Entomol. 48, 11–19.

Kuwayama, S., Sakurai, K., Endo, K., 1960. Soil insects on Hokkaido, Japan, with special reference to the effects of some chlorinated hydrocarbons. J. Econ. Entomol. 53, 1015–1018.

Kwon, M., Hahm, Y.I., Shin, K.Y., Ahn, Y.J., 1999. Evaluation of various potato cultivars for resistance to wireworms (Coleoptera: Elateridae). Am. J. Potato Res. 76, 317–319.

Ladurner, E., 2007. Naturalis-L bioinsecticide based on *Beauvaria bassiana* strain ATCC 74040. www.abim-lucerne.ch/archive/conference.html (online resource).

LaFrance, J., 1963. Emergence and flight of click beetles (Coleoptera: Elateridae) in organic soils of southwestern Quebec. Can. Entomol. 95, 873–878.

LaFrance, J., Tremblay, R., 1964. An apparatus for separating grass and soil from turf for collecting wireworm larvae (Coleoptera: Elateridae) in organic soils. Can. J. Plant Sci. 44, 212–213.

LaGasa, E.H., Welch, S., Murray, T., Wraspir, J., 2006. 2005 Western Washington delimiting survey for *Agriotes obscurus*and *A. lineatus* (Coleoptera: Elateridae), exotic wireworm pests new to the United States. Wash. State Dept Agric. Pub. 805–144.

Lanchester, H.P., 1946. Larval determination of six economic species of *Limonius* (Coleoptera: Elateridae). Ann. Entomol. Soc. Am. 39, 619–626.

Landis, B.J., Onsager, J.A., 1966. Wireworms on irrigated lands in the west: How to control them. USDA Farmers Bull. 2220.

Lane, M.C., 1925. The economic wireworms of the Pacific Northwest (Elateridae). J. Econ. Entomol. 18, 90–95.

Lane, M.C., 1953. Distribution of the wireworm *Conoderus vagus* Cand. USDA Coop. Econ. Insect Rep. 3, 536.

Lane, M.C., 1954. Wireworms and their control on irrigated lands. USDA Farmers Bull. 1866.

Lane, M.C., 1956. New name for an economic wireworm. J. Kansas Entomol. Soc. 29, 35–36.

Lange, W.H., Carlson, E.C., Leach, L.D., 1949. Seed treatments for wireworm control with particular reference to the use of lindane. J. Econ. Entomol. 42, 942–955.

Langenbuch, R., 1932. Beiträge zur kenntnis der biologie von *Agriotes lineatus* L. und *Agriotes obscurus* L. Z.Ang. Entomol. 19, 278–300.

Lees, A.D., 1943. On the behaviour of wireworm of the genus *Agriotes* Esch. (Coleoptera: Elateridae). I. Reactions to humidity. J. Exp. Biol. 20, 43–53.

Lefko, S.A., Pedigo, L.P., Batchelor, W.D., Rice, M.E., 1998. Wireworm (Coleoptera: Elateridae) incidence and diversity in Iowa Conservation Reserve environments. Env. Entomol. 27, 312–317.

Lehman, R.S., 1932. Experiments to determine the attractiveness of various aromatic compounds to adults of the wireworms. *Limonius (Pheletes) canus* Lec. and *Limonius (Pheletes) californicus* (Mann.). J. Econ. Entomol. 25, 949–958.

Lichtenstein, E.P., Morgan, D.G., Mueller, C.H., 1964. Naturally occurring insecticides in cruciferous crops. J. Agric. Food Chem. 12, 158–161.

Lilly, C.E., 1973. Wireworms: Efficacy of various insecticides for protection of potatoes in Southern Alberta. J. Econ. Entomol. 66, 1205–1207.

Lindroth, E., Clark, T.L., 2009. Phylogenetic analysis of an economically important species complex of wireworms (Coleoptera: Elateridae) in the Midwest. J. Econ. Entomol. 102, 743–749.

Liu, C., Zhang, X., Feng, Y., Yan, J., 1989. Study on the damage and occurrence of the barley wireworm, *Agriotes fuscicollis Miwa*. Hexi Corridor of Gansu Province Acta Phytophylacica Sinica 16, 13–19 [in Chinese, abstract in English].

MacLeod, G.F., 1936. Notes on wireworms. Annual Report 48 Cornell Agric. Exper. Sta., Ithaca, NY, p.99.

MacLeod, G.F., Rawlins, W.A., 1935. A comparative study of wireworms in relation to potato tuber injury. J. Econ. Entomol. 28, 192–195.

MacNay, C.G., 1954. New records of insects in Canada in 1952: A review. Can. Entomol. 86, 55–60.

Mail, A.G., 1932. pH and wireworm incidence. J. Econ. Entomol. 25, 836–840.

McColloc, J.W., Hayes, W.P., Bryson, H.R., 1927. Preliminary notes on the depth of hibernation of wireworms (Elateridae, Coleoptera). J. Econ. Entomol. 20, 561–564.

Menusan, H., Butcher, F.G., 1936. A review of some current research in entomological potato problems. Am. J. Potato Res. 13, 64–70.

Merrill, L., 1952. Reduction of wireworm damage to potatoes. J. Econ. Entomol. 45, 548–549.

Miles, H.W., 1942. Wireworms and Agriculture. J. Royal Agric. Soc. England 102, 1–13.

Miles, H.W., Cohen., M., 1938. Investigations on wireworms and their control. Ann. Rep. 1937 Entomol. Field. Sta. Warburton, UK.

Miles, H.W., Cohen., M., 1939. Investigations on wireworms and their control. Ann. Rep. 1938 Entomol. Field. Sta. Warburton, UK.

Miles, H.W., Cohen., M., 1941. Investigations on wireworms and their control. Ann. Rep. 1939–40 Entomol. Field. Sta. Warburton, UK.

Morris, R.F., 1951. The larval Elateridae of eastern spruce forests and their role in the natural control of *Gilpinia hercyniae* (Htg.) (Hymenoptera: Diprionidae). Can. Entomol. 83, 133–147.

Munro, J.A., Schifino, L.A., 1938. Preliminary studies of wireworms affecting potato tubers in North Dakota. J. Econ. Entomol. 31, 487–488.

Neuhoff, D., Christen, C., Paffrath, A., Schepl, U., 2007. Approaches to wireworm control in organic potato production. Bulletin-OILB/SROP 30, 65–68.

Noronha, C, 2011. (2011). Crop rotation as a management tool for wireworm in potatoes. IOBC/wprs Bulletin, 66, 467–471.

Noronha, C., Smith, M., Vernon, R.S., 2006. Efficacy of seed-piece or in-furrow insecticide treatments against wireworm in potatoes, 2006. Pest Management Research Report 45, 71–72, www.cps-scp.ca/publications.htm (online resource).

Noronha, C., Smith, M., Vernon, R.S., 2007. Efficacy of seed-piece or in-furrow insecticide treatments against wireworm in potatoes, 2007. Pest Management Research Report 46, 71–72, www.cps-scp.ca/publications.htm (online resource).

Norris, D.M., 1957. Bionomics of the southern potato wireworm, *Conoderus falli* Lane. 1. Life history in Florida. Fla. State Hortic. Soc. 70, 109–111.

Novy, R.G., Alvarez, J.M., Sterrett, S.B., Kuhar, T.P., Horton, D., 2006. Progeny of a tri-species potato somatic hybrid express resistance to wireworm in eastern and western potato production regions of the US. Am. J. Potato Res. 83, 126.

Ôhira, H., 1962. Morphological and taxonomic study on the larvae of Elateridae in Japan (Coleoptera). 179 pp. Published by H. Ôhira, Entomol. Lab., Aichi Gakugei University, Okazaki, Japan.

Ôhira, H., 1988. Notes on *Melanotus okinawensis* Ôhira, 1982 and its allied species from the Ryukyu Archipelago (Coleoptera: Elateridae). Edaphologia 38, 27–38.

Olsson, K., Jonasson, T., 1995. Genotypic differences in susceptibility to wireworm attack in potato: mechanisms and implications for plant breeding. Plant Breeding 114, 66–69.

Onsager, J.A., Rusk, H.W., 1969. Potency of the residues of some non-persistent insecticides in soil against wireworms. J. Econ. Entomol. 62, 1060–1064.

Pantoja, A., Hagerty, A., Emmert, S., 2010. A seasonal survey of click beetles in two potato production areas of interior Alaska. Am. J. Potato Res. 87, 531–536.

Park, C.S., Hahm, Y.I., Cheong, S.R., Lee, S.H., 1989. On the kinds, occurrence and chemical control of wireworms collected in Daekwallyong, Korea. J. Agric. Sci. (Crop Protect.) 31, 34–37.

Parker, W.E., 1994. Evaluation of the use of food baits for detecting wireworms (*Agriotes* spp., Coleoptera: Elateridae) in fields intended for arable crop production. Crop Protection 13, 271–276.

Parker, W.E., 1996. The development of baiting techniques to detect wireworms (*Agriotes* spp., Coleoptera: Elateridae) in the field, and the relationship between trap catches and wireworm damage to potatoes. Crop Protection 15, 521–527.

Parker, W.E., Howard, J.J., 2001. The biology and management of wireworms (*Agriotes* spp.) on potato with particular reference to the UK. Agric. Forest Entomol. 3, 85–98.

Parker, W.E., Seeney, F.M., 1997. An investigation into the use of multiple site characteristics to predict the presence and infestation levels of wireworms (*Agriotes* spp., Coleoptera: Elateridae) in individual grass fields. Ann. Appl. Biol. 130, 409–425.

Parker, W.E., Clarke, A., Ellis, S.A., Oakley, J.N., 1990. Evaluation of insecticides for control of wireworms (*Agriotes* spp.) on potato. Tests of Agrochemicals and Cultivars No. 11, Ann. Appl. Biol. 116 (Suppl.), 28–29.

Patane, C., Tringali, S., 2010. Hydrotime analysis of Ethiopian Mustard (*Brassica carinata* A. Braun) seed germination under different temperatures. J. Agronomy Crop Sci. 197, 94–102.

Pemberton, E.C., 1962. Important Pacific insect pests of sugarcane. Pacific Science 17, 251–252.

Pepper, B.B., Wilson, C.A., Campbell, J.C., 1947. Benzene hexachloride and other compounds for control of wireworms infesting potatoes. J. Econ. Entomol. 40, 727–734.

Platia, G., Furlan, L., Gudenzi, I., 2002. Descrizione di sei nuove specie di Elateridi dell' Iran. (Insecta, Coleoptera, Elateridae). IC Naturalista Valtelinese-Atti del Museo Civicio di Storia Naturale di Morbegna 13, 67–77.

Rabb, R.L., 1963. Biology of *Conoderus vespertinus* in the Piedmont section of North Carolina (Coleoptera: Elateridae). Ann. Entomol. Soc. Am. 56, 669–676.

Radcliffe, E.B., Flanders, K.L., Ragsdale, D.W., Noetzel, D.M., 1991. Pest management systems for potato insects. In: Pimentel, D. (Ed.), Second ed. CRC Handbook of Pest Management in Agriculture, Vol. 3. CRC Press, Boca Raton, FL, pp. 587–621.

Ranjbaraqdam, H., Ardabili, A., Fathi, A., Sarihi, S., Habibi, J., 2001. Identification of the wireworms and primary investigation on biology of dominant species in Ardabil potato fields. Published by Ardebil Province Agricultural and Natural Resource Research Centre, Moqan, Iran. 15 pp. [in Farsi, abstract in English].

Rawlins, W.A., 1934. Experimental studies on the wheat wireworm, *Agriotes mancus* Say. J. Econ. Entomol. 27, 308–314.

Rawlins, W.A., 1940. Biology and control of the wheat wireworm *Agriotes mancus* Say. Cornell Univ. Agr. Expt Sta. Bull 783.

Riley, T.J., Keaster, A.J., 1979. Wireworms associated with corn: identification of nine species of *Melanotus* from the north central states. Ann. Entomol. Soc. Am. 72, 408–414.

Roberts, A.W.R., 1921. On the life history of "wireworms" of the genus *Agriotes*, Esch., with some notes on that of *Athous haemorrhoidalis*, F. Ann. Appl. Biol. 8, 193–215.

Robinson, R.R., 1976. Controlling wireworms in potatoes. Publication EC 861, Oregon State University Extension Service.

Rockwood, L.P., 1950. Entomogenous fungi of the family Entomophthoraceae in the Pacific northwest. J. Econ. Entomol. 43, 704–797.

Salt, G., Hollick, F.S.J., 1944. Studies of wireworm populations. I. A census of wireworms in pasture. Ann. Appl. Biol. 31, 52–64.

Sasscer, E.R., 1924. Important foreign insect pests collected on imported nursery stock in 1923. J. Econ. Entomol. 17, 443–444.

Schaerffenberg, B., 1942. Der einfluss von humusgehalt und feuchtigkeit des bodens auf die fraßtätigkeit der Elateridenlarven. Anzeiger für Schädlingskunde 18, 133–136.

Schalk, J.M., Bohac, J.R., Dukes, P.D., Martin, W.R., 1993. Potential of nonchemical control strategies for reduction of soil insect damage in sweet potato. J. Am. Soc. Hort. Sci. 118, 605–608.

Schallhart, N., Tusch, M.J., Staudacher, K., Wallinger, C., Traugott, M., 2011. Stable isotope analysis reveals whether soil-living elaterid larvae move between agricultural crops. Soil Biol. Biochem. 43, 1612–1614.

Schepl, U., Paffrath, A., 2005. Regulierungsmaßnahmen des Drahtwurmbefalls im ökologischen Kartoffelanbau: Ergebnisse einer Status-quo-Analyse. In: Hess, J., Rahmann, G. (Eds.), Ende der Nische, Beiträge zur 8. Wissenschaftstagung Ökologischer Landbau, Kassel University Press, GmbH, Kassel, Germany.

Seal, D.R., 1990. The biology of wireworms affecting sweet potatoes in Georgia. Unpublished PhD dissertation. University of Georgia, Athens, GA.

Seal, D.R., Chalfant, R.B., 1994. Bionomics of *Conoderus rudis* (Coleoptera: Elateridae): Newly reported pest of sweet potato. J. Econ. Entomol. 87, 802–809.

Seal, D.R., Chalfant, R.B., Hall, M.R., 1991. Distribution and density of wireworms and their damage in relation to different cultivars of sweetpotato. Proc. Fla State. Hort. Soc. 104, 284–286.

Seal, D.R., Chalfant, R.B., Hall, M.R., 1992. Effects of cultural practices and rotational crops on abundance of wireworms (Coleoptera: Elateridae) affecting sweetpotato in Georgia. Env. Entomol. 21, 969–974.

Sewell, G.H., Alyokin, A., 2004. Control of wireworm on potato, 2003. Arthropod Management Tests 29, 69.

Shirck, F.H., 1945. Crop rotations and cultural practices as related to wireworm control in Idaho. J. Econ. Entomol. 38, 627–633.

Simmons, C.L., Pedigo, L.P., Rice, M.E., 1998. Evaluation of seven sampling techniques for wireworms (Coleoptera: Elateridae). Environ. Entomol. 27, 1062–1068.

Smith, D.B., Keaster, A.J., Cheshire, J.M., Wards, R.H., 1981. Self-propelled soil sampler: evaluation for efficiency in obtaining estimates of wireworm populations in Missouri cornfields. J. Econ. Entomol. 74, 625–629.

Southwood, T.R.E., 1978. Ecological Methods. Chapman and Hall, New York, NY.

Staudacher, K., Pitterl, P., Furlan, L., Cate, P.C., Traugott, M., 2011a. PCR-based species identification of *Agriotes* larvae. Bull. Entomol. Res. 101, 201–210.

Staudacher, S., Schallhart, N., Pitterl, P., Wallinger, C., Brunner, N., Landl, M., Kromp, B., Glauninger, J., Traugott, M., 2011b. Occurrence of *Agriotes* wireworms in Austrian agricultural land. J. Pest Sci. DOI: 10.1007/s10340-011-0393-y.

Stirret, G.M., 1936. Notes of the "flat wireworm," *Aeolus mellillus* Say. Can. Entomol. 68, 117–118.

Stone, M.W., 1953. Sugar-beet wireworm predaceous on seed-corn maggot. J. Econ. Entomol. 46 1100–1100.

Stone, M.W., 1975. Distribution of four introduced *Conoderus* species in California (Coleoptera: Elateridae). Coleopt. Bull. 29, 163–166.

Stone, M.W., Campbell, R.E., 1933. Chloropicrin as a soil insecticide for wireworms. J. Econ. Entomol. 26, 237–243.

Stone, M.W., Foley, F.B., 1955. Effect of season, temperature, and food on the movement of the sugarbeet wireworm. Ann. Entomol. Soc. Am. 48, 308–312.

Strickland, A.H., Bardner, H.M., Waines, R.A., 1962. Wireworm damage and insecticidal treatment of the ware potato crop in England and Wales. Plant Path. 11, 93–105.

Strickland, E.H., 1933. The biology of prairie-inhabiting wireworms. Proc. World Grain Exhib. Conf., Canada 2, 520–529.

Strickland, E.H., 1939. Life cycle and food requirements of the Northern grain wireworm, *Ludius aeripennis destructor* (Brown). J. Econ. Entomol. 32, 322–329.

Strickland, E.H., 1942. Variations in the length of the lifecycle of wireworms. J. Econ. Entomol. 35, 109–110.

Subklew, W., 1934. *Agriotes lineatus* L. und *A. obscurus* L. (Ein beitrag zu ihrer morphologie und biologie.). Z. Ang. Entomol. 21, 96–122.

Sufyan, M., Neuhoff, D., Furlan, L., 2011. Assessment of the range of attraction of pheromone traps to *Agriotes lineatus* and. *Agriotes obscurus*. Agric. Forest Entomol. 13, 313–319.

Suszkiw, J., 2011. New potatoes withstand destructive wireworms. Agricultural Research 59, 22.

Tamaki, Y., Sugie, H., Nagamine, M., Kinjo, M., 1986. Female sex pheromone of the sugarcane wireworm *Melanotus okinawensis* Ohira (Coleoptera: Elateridae). Jpn. Kokai Tokkyo Koho JP, 61–12601 [in Japanese].

Tamaki, Y., Sugie, H., Nagamine, M., Kinjo, M., 1990. 9,11-Dodecadienyl-butyrato and 9,11-dodecadienyl-hexanoate female sex pheromone of the sugarcane wireworm *Melanotus sakishimensis* Ohira (Coleoptera: Elateridae). Jpn. Kokai Tokkyo Koho JP 2-53753 [in Japanese].

Tharp, C.I., Blodgett, S.L., Jaronski, S.T., 2007. Control of wireworms (Elateridae) in potatoes with microbial *Metarhizium*, 2006. Arthropod Management Tests 32, E43.

Thomas, C.A., 1940. The biology and control of wireworms: A review of the literature. Penn. State Coll. School Agric. Exp. Sta Bull. 392.

Tinline, R.D., Zacharuk, R.Y., 1960. Pathogenicity of *Metarrhizium anisopliae* (Metch.) Sor., and *Beauveria bassiana* (Bals.) Vuill. to two species of Elateridae. Nature 187, 794–795.

Toba, H.H., 1987. Treatment regimes for insecticidal control of wireworms on potato. J. Agric. Entomol. 4, 207–212.

Toba, H.H., Campbell, J.D., 1992. Wireworm (Coleoptera: Elateridae) survey in wheat-growing areas of northcentral and northeastern Oregon. J. Entomol. Soc. Br. Columbia 89, 25–30.

Toba, H.H., Turner, J.E., 1983. Evaluation of baiting techniques for sampling wireworms (Coleoptera: Elateridae) infesting wheat in Washington. J. Econ. Entomol. 76, 850–855.

Toba, H.H., Lindegren, J.E., Turner, J.E., Vail, P.V., 1983. Susceptibility of the Colorado Potato Beetle and the Sugarbeet wireworm to *Steinernema feltiae* and S. *glaseri*. J. Nematol. 15, 597–601.

Tolman, J.H., Vernon, R.S., 2009. Planting treatments for control of damage to potato tubers by field wireworms, 2009. Pest Management Research Report 48, 25–28, www.cps-scp.ca/publications. htm (online resource).

Tolman, J.H., Sawinski, T.A., Vernon, R.S., Clodius, M., 2005. Planting treatments for control of damage to potato tubers by field wireworms, 2005. Pest Management Research Report 44, 72–73, www.cps-scp.ca/publications.htm (online resource).

Tolman, J.H., Alhemzawi, A., Vernon, R.S., 2008. Planting treatments for control of damage to potato tubers by field wireworms, 2008. Pest Management Research Report 47, 55–57. www. cps-scp.ca/publications.htm (online resource).

Tóth, M., Furlan, L., Yatsynin, V.G., Ujváry, I., Szarukán, I., Imrei, Z., Tolasch, T., Francke, W., Jossi, W., 2003. Identification of pheromones and optimization of bait composition for click beetle pests (Coleoptera: Elateridae) in Central and Western Europe. Pest Manag. Sci. 59, 417–425.

Tóth, M., Furlan, L., Xavier, A., Vuts, J., Toshova, T., Subchev, M., Szarukán, I., Yatsynin, V., 2008. New sex attractant composition for the click beetle *Agriotes promixus*: Similarity to the pheromone of *Agriotes lineatus*. J. Chem. Ecol. 34, 107–111.

Traugott, M., Schallhart, N., Kaufmann, R., Juen, A., 2008. The feeding ecology of elaterid larvae in central European arable land: New perspectives based on naturally occurring stable isotopes. Soil Biol. Biochem. 40, 342–349.

Turnock, W.J., 1968. Comparative food preferences of wireworms (Coleoptera: Elateridae). Manitoba Entomol. 2, 76–80.

van Herk, W.G., Vernon, R.S., 2007. Soil bioassay for studying behavioral responses of wireworms (Coleoptera: Elateridae) to insecticide-treated wheat seed. Environ. Entomol. 36, 1441–1449.

van Herk, W.G., Vernon, R.S., 2011. Categorization and numerical assessment of wireworm mobility over time following exposure to bifenthrin. J. Pest Science. DOI: 10.1007/s10340-011-0381-2.

van Herk, W.G., Vernon, R.S., Clodius, M., Harding, C., Tolman, J.H., 2007. Mortality of five wireworm species (Coleoptera: Elateridae), following topical application of clothianidin and chlorpyrifos. J. Entomol. Soc. Br. Columbia 104, 55–63.

van Herk, W.G., Vernon, R.S., Tolman, J.H., Ortiz Saavedra, H., 2008. Mortality of a wireworm, *Agriotes obscurus* (Coleoptera: Elateridae), following topical application of various insecticides. J. Econ. Entomol. 101, 375–383.

van Herk, W.G., Vernon, R.S., McGinnis, S., 2011. Response of the dusky wireworm, *Agriotes obscurus* (Coleoptera: Elateridae), to residual levels of bifenthrin in field soil. J. Pest. Sci. DOI: 10.1007/s10340-011-0386-x.

van Rozen, K., Ester, A., Hendrickx, T., 2007. Practical Dutch experience introducing a monitoring system of click beetles by pheromone traps. IOBC/WPRS Bull. 30 (2007), 53–58.

Vernon, B., Päts, P., 1997. Distribution of two European wireworms, *Agriotes lineatus* and *A. obscurus* in British Columbia. J. Entomol. Soc. Br. Columbia 94, 59–61.

Vernon, B., LaGasa, E., Philip, H., 2001. Geographic and temporal distribution of *Agriotes obscurus* and *A. lineatus* (Coleoptera: Elateridae) in British Columbia and Washington as determined by pheromone trap surveys. J. Entomol. Soc. Br. Columbia 98, 257–265.

Vernon, R.S., 2005. Aggregation and mortality of *Agriotes obscurus* (Coleoptera: Elateridae) at insecticide-treated trap crops of wheat. J. Econ. Entomol. 98, 1999–2005.

Vernon, R.S., Tóth, M., 2007. Evaluation of pheromones and a new trap for monitoring *Agriotes lineatus* and *Agriotes obscurus* in the Fraser Valley of British Columbia. J. Chem. Ecol. 33, 345–351.

Vernon, R.S., Kabaluk, T., Behringer, A., 2000. Movement of *Agriotes obscurus* (Coleoptera: Elateridae) in strawberry (Rosaceae) plantings with wheat (Gramineae) as a trap crop. Can. Entomol. 132, 231–241.

Vernon, R.S., Kabaluk, T., Behringer, A., 2003. Aggregation of *Agriotes obscurus* (Coleoptera: Elateridae) at cereal bait stations in the field. Can. Entomol. 135, 379–389.

Vernon, R.S., van Herk, W., Moffat, C., Harding, C., 2007. European wireworms (*Agriotes* spp.) in North America: toxicity and repellency of novel insecticides in the laboratory and field. IOBC/WPRS Bull. 30, 35–41.

Vernon, R.S., van Herk, W.G., Tolman, J.H., Ortiz Saavedra, H., Clodius, M., Gage, B., 2008. Transitional sublethal and lethal effects of insecticides after dermal exposures to five economic species of wireworms (Coleoptera: Elateridae). J. Econ. Entomol. 101, 365–374.

Vernon, R.S., van Herk, W.G., Clodius, M., Harding, C., 2009. Wireworm management I: stand protection versus wireworm mortality with wheat seed treatments. J. Econ. Entomol. 102, 2126–2136.

Vernon, R.S., van Herk, W.G., Clodius, M., Harding, C., 2011. Crop protection and mortality of *Agriotes obscurus* wireworms with blended insecticidal wheat seed treatments. J. Pest. Sci. DOI 10.1007/s10340-011-0392-z.

Ward, R.H., Keaster, A.J., 1977. Wireworm baiting: use of solar energy to enhance early detection of *Melanotus depressus, M. verberans* and *Aeolus mellilus* in midwest cornfields. J. Econ. Entomol. 70, 403–406.

Weires, R.W., 1976. Evidence of an attractant for the click beetle, *Melanotus depressus* in the pheromones of two *Platynota* spp. (Lepidoptera: Tortricidae). Environ. Entomol. 5, 920–921.

Wilde, G., Roozeboom, K., Claasen, M., Janssen, K., Witt, M., 2004. Seed treatment for control of early-season pests of corn and its effect on yield. J. Agric. Urban Entomol. 21, 75–85.

Wilkinson, A.T.S., 1963. Wireworms of cultivated land in British Columbia. Proc. Entomol. Soc. Br. Columbia 60, 3–17.

Wilkinson, A.T.S., Finlayson, D.G., Morley, H.V., 1964. Toxic Residues in Soil 9 Years after Treatment with Aldrin and Heptachlor. Science 143, 681–683.

Wilkinson, A.T.S., Finlayson, D.G., Campbell, C.J., 1976. Controlling the European wireworm, *Agriotes obscurus* L., in corn in British Columbia. J. Entomol. Soc. Br. Columbia 73, 3–5.

Williams III, L., Morra, M.J., Brown, P.D., McCaffrey, J.P., 1993. Toxicity of allyl isothiocyanate-amended soil to *Limonius californicus* (Mann.) (Coleoptera: Elateridae) wireworms. J. Chem. Ecol. 19, 1033–1046.

Willis, R.B., Abney, M.R., Kennedy, G.G., 2010. Survey of wireworms (Coleoptera: Elateridae) in North Carolina sweetpotato fields and seasonal abundance of *Conoderus vespertinus*. J. Econ. Entomol. 103, 1268–1276.

Wilson, L.K., Elliott, J.E., Vernon, R.S., Smith, B.D., Szeto, S.Y., 2002. Persistence and retention of active ingredients in four granular cholinesterase-inihibiting insecticides in agricultural soils of the lower Fraser River valley, British Columbia, Canada, with implications for wildlife poisoning. Env. Tox. Chem. 121, 260–268.

Wraight, S.P., Lacey, L.A., Kabaluk, J.T., Goettel, M.S., 2009. Potential for microbial biological control of Coleopteran and Hemipteran pests of potato. In: Tennant, P., Benkeblia, N. (Eds.), Fruit, Vegetable and Cereal Science and Biotechnology 3 (Special Issue 1), pp. 25–38.

Wu, I., 1966. Crop rotation and the wheat wireworm control. Acta Entomol. Sinica 2 [abstract only].

Wu, J., Li, Y., 2005. Ground pest insects. http://210.27.80.89/2005/nongyekongchun/page/Agricultural%20Entomplogy%20%28PDF%29/Chapter%203%20Ground%20Pest%20Insects.pdf (online resource).

Yamamura, K., Kishata, M., Arakaki, N., Kawamura, F., Sadoyama, Y., 2003. Estimation of dispersal distance by mark-recapture experiments using traps: Correction of bias caused by the artificial removal by traps. Popul. Ecol. 45, 149–155.

Yates, F., Finney, D.J., 1942. Statistical problems in field sampling for wireworms. Ann. Appl. Biol. 29, 156–167.

Yatsynin, V.G., Lebedeva, K.V., 1984. Identification of multicomponent pheromones in click beetles *Agriotes lineatus* L and *A. ustulatus* L. Khemoretseptsiya Nasekomykh, Vilnius 8, 52–57 [in Russian].

Yatsynin, V.G., Rubanova, E.V., Okhrimenko, N.V., 1996. Identification of female-produced sex pheromones and their geographical differences in pheromone gland extract composition from click beetles (Col., Elateridae). J. Appl. Entomol. 120, 463–466.

Yoshida, M., 1961. Ecological and physiological researches on the wireworm *Melanotus caudex* Lewis. Fac. Agr. Shizuoka Univ. Spec. Rept. 1.

Zacharuk, R.Y., 1958. Note on two forms of *Hypolithus bicolor* Esch. (Coleoptera: Elateridae). Can. Entomol. 90, 567–568.

Zacharuk, R.Y., 1962a. Distribution, habits, and development of *Ctenicera destructor* (Brown) in Western Canada, with notes on the related species *C. aeripennis* (Kby) (Coleoptera: Elateridae). Can. J. Zool. 40, 539–552.

Zacharuk, R.Y., 1962b. Seasonal behaviour of larvae of *Ctenicera* spp. and other wireworms (Coleoptera: Elateridae), in relation to temperature, moisture, food, and gravity. Can. J. Zool. 40, 697–718.

Zacharuk, R.Y., 1963. Comparative food preferences of soil-, sand-, and wood-inhabiting wireworms (Coleoptera, Elateridae). Bull. Entomol. Res. 54, 161–165.

Zacharuk, R.Y., Tinline, R.D., 1968. Pathogenicity of *Metarrhizium anisopliae*, and other fungi to five elaterid (Coleoptera) in Saskatchewan. J. Invert. Pathol. 12, 294–309.

Zhao, J., Yu, Y., 2010. Overview of researches of wireworm in China. J. Agric. Sci. 3 [abstract only].

Zhou, Y., Bai, H., Shu, J., 2008. Study on biological characteristics of *Melanotus cribricollis*. J. Zhejiang Forest. Sci. Tech. 4 [abstract only].

Chapter 6

Biology and Ecology of Potato Tuber Moths as Major Pests of Potato

Jürgen Kroschel and Birgit Schaub

Agroecolgy/IPM, Integrated Crop and Systems Research Global Program, International Potato Center, Lima, Peru

INTRODUCTION

Insect pests account for 16% of the crop losses of potato (*Solanum tuberosum* L.) worldwide (Oerke *et al.* 1994), and reductions in tuber yield and quality can be between 30% and 70% for various insect pests (Raman and Radcliffe 1992). Among those pests are three species of the Gelechiidae (Lepidoptera) family, and these make up the so-called potato tuber moth complex, which are the common potato tuber moth, *Phthorimaea operculella* (Zeller), the Andean potato tuber moth, *Symmetrischema tangolias* (Gyen), and the Guatemalan potato tuber moth, *Tecia solanivora* Povolny. Although the three species have quite similar appearance and biology at first glance, there are differences among the species' worldwide and regional distribution. This chapter describes the biology and ecology of the three species, and analyzes factors that favor or limit the different species' distribution and damage potential.

ORIGIN, DISTRIBUTION, AND HOST RANGE

The potato tuber moth *P. operculella*, also referred to as the potato tuberworm, probably originated in the tropical mountainous regions of South America (Graf 1917) at the center of the potato's origin. Today it has become a cosmopolitan pest. Its distribution is reported in more than 90 countries worldwide (Fig. 6.1). The moth occurs in almost all tropical and subtropical potato production systems in Africa and Asia, as well as in North, Central, and South America. While it can still be of economic significance in subtropical regions of southern Europe (e.g., Italy), the long cold winters in temperate regions generally restrict its development and reduce its pest status. *P. operculella* is a pest of potato and other solanaceous crops such as tomato (*Lycopersicon esculentum* Mill.),

Insect Pests of Potato. http://dx.doi.org/10.1016/B978-0-12-386895-4.00006-5

tobacco (*Nicotana tabacum* L.), and eggplant (*Solanum melongena* L.). In addition, wild species of the Solanaceae family, including important weeds in potato (e.g., black nightshade, *Solanum nigrum* L.), serve as host plants. However, of all host plants, potato, followed by eggplant, is preferred for the deposition of eggs by females (Meisner *et al.* 1974). Further, nutrition has a profound effect both on the length of development and the weight of larvae and pupae, and on the emergence of adults, which are all optimal when larvae are fed with potato tubers. Adults also lived longer, females laid more eggs, and the egg-laying period was prolonged when compared to larvae fed with potato or tomato leaves (Gomaa *et al.* 1978).

S. tangolias was formerly synonymously referred to as *Symmetrischema plaesiosema* (Turner) (Sánchez *et al.* 1986). The common name is the spotted or Andean potato tuber moth; in Australia it is called the tomato stem borer (Osmelak 1987). It is assumed that *S. tangolias* also originated in the mountainous regions of Peru and Bolivia (Povolny 1967) (Fig. 6.1). *S. tangolias* is widely distributed at mid-elevation in the Andes in Colombia, Ecuador, Peru, and Bolivia, and is also present in Australia (Osmelak 1987), Tasmania (Terauds *et al.* 1984), and New Zealand (Martin 1999); it has also been reported in Indonesia (Keller 2003). The host plants of *S. tangolias* include a number of species of the Solanaceae family. Besides its main host plant, potato, the moth also attacks tomato, pepino (*Solanum muricatum* L.) (Osmelak 1987), and poroporo (*Solanum aviculare* L. and *Solanum laciniatum* L.) (Martin 1999).

The Guatemalan potato tuber moth, *T. solanivora*, is a pest in Central and South America. Guatemala is supposed to be the country of origin because of historical reports (Povolny 1973), and in view of the high genetic diversity

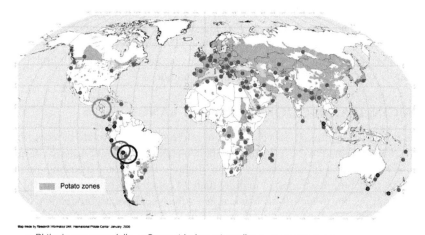

● *Phthorimaea operculella*   ● *Symmetrischema tangolias*   ● *Tecia solanivora*

FIGURE 6.1   The worldwide distribution of the three potato tuber moth species, *Phthorimaea operculella*, *Symmetrischema tangolias*, and *Tecia solanivora*, indicating their center of origin. *Reproduced courtesy of International Potato Center (CIP), Lima, Peru.* See also Plate 6.1

(Puillandre *et al.* 2008, Torres-Leguizamón *et al.* 2011) (Figure 6.1). In 1970, it was accidentally introduced, via infested seed, into Costa Rica (Povolny 1973). In 1983, infested seeds further distributed the pest into several potato-growing regions of Venezuela (Salasar and Escalante 1984) and, in 1985, into potato-growing regions of Colombia (Rincón and López-Ávila 2004). In 2010, *T. solanivora* was reported for the first time in South Mexico (Cruz *et al.* 2011). In the absence of natural enemies, it established itself rapidly in mountainous regions of the Andes between 1350 and 3000 m a.s.l., where in many cases half of the potato harvest was lost and potato stocks were infested and destroyed (Torres 1989). In 1996 the moth was detected in Ecuador (Gallegos and Suquillo 1997), where it spread quickly through trade movements into the country's interior (Barragán *et al.* 2002). In 1999, *T. solanivora* appeared on Tenerife in the Canary Islands (Trujillo *et al.* 2004). Since then, the pest has been considered to be a major threat to potato crops throughout southern Europe, and is listed as a quarantine pest by the European and Mediterranean Plant Protection Organization (EPPO 2005). Its quick dissemination through South America, the sudden occurrence of high damage, and the failure of chemical pesticides to provide reliable control have caused serious problems in potato cultivation. Alerted by these pest outbreaks, measures against its invasion into Peru were taken (Oyala 2002). No presence of the moth has been reported from Peru so far. However, uncontrolled trade of potatoes between Ecuador and Peru is frequent, posing a potential risk for a further expansion of the pest. Peru and other South American countries are therefore taking all possible measures in their quarantine program to keep the pest out of the country. Potato is reported to be the only host plant of *T. solanivora.*

INFESTATION, DAMAGE DEVELOPMENT, AND YIELD LOSS

P. operculella attacks all vegetative plant parts of potato. Females prefer the green foliage over potato tubers or soil for oviposition (Rondon 2010). Typical symptoms of leaf damage are mines caused by larvae feeding in the mesophyll, without damaging the upper and lower epidermis (Fig. 6.2A). Mining in the main vein can cause the loss of a pinnate leaf. Other entry points are leaf axils and the growing points of young plants. The foliage can be completely destroyed (Fig. 6.2B), which can result in substantial yield loss (Broodryk 1979). Field

(A)   (B)   (C)   (D)

FIGURE 6.2   Leaf (A, B) and tuber (C, D) damage caused by the potato tuber moth, *P. operculella.* See also Plate 6.2

experiments to evaluate the effect of *P. operculella* leaf- and stem-mining on tuber yield showed that only high infestations early in the season directly affect tuber yield. In the Republic of Yemen, with 35 mines/plant at growth stage 50 (bud formation stage), tuber yields were reduced by 25% (Kroschel 1995). In comparison, 9 mines/plant at growth stage 70 (formation of berries stage) did not significantly affect tuber yield in Egypt (Keller 2003). However, the strong correlation established between leaf and consequent tuber infestation suggest that reducing the *P. operculella* population density during the potato-growing period is key to reducing tuber infestation at harvest (Figs. 6.2C, 6.2D). Hence, the most devastating yield losses are largely a result of tuber infestation, generally where moths have laid eggs through soil cracks on the developing tubers, or when harvest is delayed. Keller (2003) observed infestation rates of tubers at harvest of between 30% and 40% after the rainy season (June–July) compared to infestations of 3–15% in the months March–April, in the Mantaro Valley, Peru. Tuber infestation in the field before harvest is not only due to eggs directly laid on tubers. As the nutritive value of the potato foliage declines, the larvae leave the aerial plant parts in search of tubers. As the latter expand, they produce soil cracks, thereby creating access for larvae to the tubers (Broodryk 1979). The period from foliar senescence to harvest is therefore considered crucial for control of *P. operculella* (Rondon 2010).

S. tangolias feeds in the stems of potato as well as on tubers, like *P. operculella*. The neonate larvae penetrate the stem by the axils between the stem and lateral petioles, and mine within the stem. Excrement is pushed out through the holes that larvae make at initial penetration. The mining activity of the larvae may result in wilting of the upper part of the stem. Up to 12 larvae have been found in a single stem (Rodriguez 1990).

Larvae of *T. solanivora* feed exclusively on tubers during potato cultivation, and after harvest in potato stores (Fig. 6.3). Damage is caused by the larvae, which bore galleries into the tubers, making them unsuitable for consumption (EPPO 2005). In potato fields, *T. solanivora* attack occurs from tuberization until harvest. Eggs are laid in groups on the soil close to the potato plants;

FIGURE 6.3   Damage on potato tubers caused by *T. solanivora* larvae. See also Plate 6.3

few or no eggs were observed on the plant itself (Barreto *et al.* 2003, Karlsson *et al.* 2009). Larvae hatch, and search for potato tubers in soil. Field infestation depends greatly on the weather conditions. During the rainy season, no damage in the field is observed; a sampling conducted during 1 year from 85 fields in Central Ecuador revealed an average of 7% of damaged potato tubers (Carpio 2008). Occasionally, high damage of 38% of tubers was reached in Ecuador (Barragán *et al.* 2002). These outbreaks are also favored by overproduction and consequently low potato prices during the dry season that lead farmers to delay or to give up harvest, which allows undisturbed multiplication of *T. solanivora* (Barragán *et al.* 2004).

Tuber infestation caused by first-instar larvae of either potato tuber moth species can be hardly noticed, so that even with precautionary measures infested tubers are transferred to potato stores, where further propagation of the pest and infestation of the whole stock can take place. Characteristic piles of feces indicate infestation; inside tubers, larvae bore irregular galleries which may run into the interior of the tubers or remain directly under the skin (Figs. 6.2C, 6.2D). The mining produces weight losses of the tubers, which is exacerbated by increased transpiration through the wounds, causing them to shrink. The wounds also provide entry points for microorganisms and can cause secondary infestation with pathogens. Infestation of stored tubers can result in total destruction; however, this largely depends on the length of the storage period and the prevalent storage temperature, which determines how fast the different potato tuber moth species can reproduce and build up several generations. Hence, in the absence of cold stores, the damage to potatoes in rustic stores can be total within a few months if tubers are left untreated. In these cases, losses up to 45% have been reported in stores in the Republic of Yemen (Kroschel 1995), 50% losses in the Andean region (Palacios and Cisneros 1997), and 90% losses in Kenya (Raman 1987). Infested tubers at harvest in the Mantaro Valley showed a clear dominance of larvae of *P. operculella* compared to larvae of *S. tangolias* (ratio 15:1); in potato stores, though, *S. tangolias* larvae were present in greater numbers, at a ratio of 1:0.22 (Keller 2003), which indicates the adaptation of the two species to different environments. Levels of damage caused by *T. solanivora* in Ecuador reach up to 100% when no preventive measures are taken (Carpio 2008). Infested tubers are unsuitable not only for human consumption but also for use as seed. Infested tubers produce fewer yields and initiate a rapid development of a new field population (Kroschel 1995, 1994, Palacios *et al.* 1999).

Infested tubers at harvest are not the only source of tuber infestation in storage; different developmental stages of the potato tuber moths might persist in reused potato bags and on leftover infested tubers; and often rustic potato stores have no physical protection to stop flying moths entering the stores. In addition to seed, sources for field infestation are leftover tubers from the previous harvest or infested potato tubers discarded close to the field.

MORPHOLOGY

The morphology of *P. operculella* has been described by Al-Ali and Talhouk (1970), Broodryk (1979) and Winning (1941). The female and male moths are brownish-gray, with fraying on the posterior edge of the forewings and on both posterior and inner edges of the hindwings (Fig. 6.4). The wings are folded to form a roof-like shape. The length of the resting moth is 7–9 mm and the wingspan 12–16 mm. The tip of the female's abdomen is cone-shaped, whereas the male possesses two claspers at the hairy tip of its abdomen. The male's sexual organs are situated in the middle of the ninth abdominal segments, the female's in the middle of the eighth. Freshly laid eggs are whitish, and are deposited singly or in small clutches resembling strings of beads. As they develop they take on a yellowish tinge, and before hatching the black head capsule of the tiny larva can be seen through the thin eggshell. The eggs are too small (0.5×0.35 mm) to be visible to the naked eye on potato leaves or tubers. The first-instar larva is about 1 mm long, while the fourth-instar larva reaches 9–13 mm in length before pupation. The color of the larvae depends on their diet; in tubers they are whitish purple, but those on potato leaves are purple to green (Fig. 6.4). A fully developed larva has six ocelli on each side of the head, mouthparts with a silk gland, a prothoracic and anal plate, nine pairs of spiracles, and five pairs of prologs on abdominal segments III–VI and X. The pupa is 7–8 mm long. At first it is brownish in color, turning to dark brown, and almost black 1 day before emergence.

S. tangolias has been described by Vera (1999), Tenorio (1996) and Sánchez *et al.* (1986). The male and female moths are brownish-gray, with a characteristic black triangular spot at the lateral edges of the forewings (Fig. 6.4). Fine hairs cover the edges of the forewings while the hindwings are covered with pale-ocher scales. The length of the moth is 9–12 mm, and the wingspan is 18–19 mm. Males are smaller than females with a cone-shaped abdomen, while

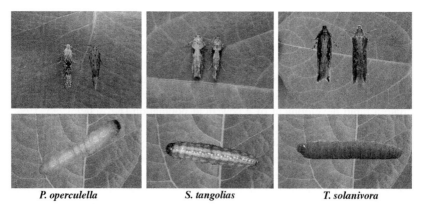

| *P. operculella* | *S. tangolias* | *T. solanivora* |

FIGURE 6.4 Morphology of adults (left: ♀; right: ♂) and fourth-instar larvae of the three potato tuber moth species *Phthorimaea operculella, Symmetrischema tangolias*, and *Tecia solanivora*. See also Plate 6.4

the females' abdomen is wider with a blunt ending. *S. tangolias* is bigger than *P. operculella*. The freshly laid eggs are whitish, oval (0.7×0.4 mm), and are deposited singly or in small clutches. During embryogenesis the eggs become orange-yellow and turn dark-gray shortly before larval hatch. The first-instar larva is barely 1 mm long, while the fourth-instar larva reaches 13 mm in length before pupation. As for *P. operculella*, the color of the larvae depends on the diet. From the third instar on, three characteristic reddish longitudinal stripes on the upper part of the thorax and abdomen can be seen, which can be used to differentiate between *S. tangolias* and *P. operculella* larvae (Fig. 6.4). The full-grown larva usually leaves tubers to pupate, and to spin a soft white cocoon, but pupation also occurs in tubers and stems of potato. The pupa is 7–8 mm long, and is the same color as that of *P. operculella*.

The morphology of *T. solanivora* was first described by Povolny (1973). Further descriptions can be found by Torres *et al.* (1997) and Palacios *et al.* (1997). Adults present sexual dimorphism; females are larger in size than males (12.7 versus 11.8 mm) and their brown coloration is brighter than in the case of males. Both sexes have two dots on their wings and a dark line over the whole length of the forewing (Fig. 6.4). Wing size and shape varies with altitude; at high altitudes wings are larger in size but narrower than at low altitudes (Hernández *et al.* 2010). Eggs are laid in groups during the first days of oviposition and separately when oviposition decreases. They are oval shaped with a size of 0.5×0.4 mm and their coloration turns from pale white just after oviposition to yellow after some days of incubation. One day before hatching the eggs appear black because of the dark head capsule of the larva. First-instar larvae have a length of 1.2 to 1.4 mm; the body is pale white and transparent, and the head capsule and pronotum have a dark brown coloration. Second-instar larvae are not transparent anymore and have some small, dark brown spots on the whole length of their body. Third-instar larvae look similar but are larger in size, and fourth-instar larvae have purple coloration on their dorsal part and are green on their ventral part (Fig. 6.4). Three pairs of thoracic podia and five pairs of pseudopodia are now clearly visible. Larvae reach a length of 12–7 mm and a diameter of 2–3 mm. Last-instar larvae leave the potato tubers in search of a protected pupation site. Before pupation, movements are reduced and larvae shrink in size. A cocoon is formed, integrating particles of soil if available. Female pupae are on average larger than male pupae (8.5×3 mm versus 7.6×2.6 mm) and have a higher weight. They can be distinguished by the position of their genital aperture, which is located in the eighth abdominal segment in females and in the ninth abdominal segment of males (Rincón and López-Ávila 2004).

BIOLOGY AND ECOLOGY

The biology and ecology have been best studied and described for *P. operculella*. As for all poikilotherm organisms, *P. operculella* development depends largely on temperature, but among all three potato tuber moth species it can adapt to the most diverse conditions. Development is possible within a temperature range of between 10°C and 35°C, and is unaffected by air humidity. The lower developmental threshold determined for immature stages by different authors ranges between 4.25°C and 13.5°C, and suggests a high adaptability of the species to different environmental conditions (Rondon 2010). The moth can survive low temperatures around the freezing point for short periods of time at all developmental stages. For development to continue, however, short spells of higher daily temperatures are necessary (Lal 1987). According to the studies of Beukema and Zaag (1990), *P. operculella* becomes inactive at constant temperatures of 10°C and dies at below 4°C. The species does not respond to unfavorable conditions by entering diapause (Mitchell 1978); instead, development ceases for short periods at low temperatures, to be resumed when conditions improve.

After cold or longer no-cropping periods, moths surviving in the open, mainly on leftover potato, are the initial population at the start of a potato season. In smallholder production systems where own instead of improved certified seed is used, infested seed contributes substantially to a rapid population build-up, as demonstrated in the Republic of Yemen for *P. operculella* (Kroschel 1994, Kroschel and Koch 1994). Adults can also recolonize potato fields from nearby potato storages (Haines 1977), and certain weeds (see Chapter 2) can serve as breeding sites, however, they are also sensitive to frost and become less significant where winters are severe. Adults disperse in short "hopping" flights near ground level, with the aid of prevailing winds (Broodryk 1979). Adults become visible in potato fields at mean temperatures of 16°C and first symptoms of leaf infestation can be observed. Population growth rates peak between 20°C and 25°C (Haines 1977). Mean temperatures of more than 21°C lead to development times of less than 30 days to complete one-generation cycle (Kroschel and Koch 1994), and reproduction rates of females increase. The optimal temperature range for egg-laying is 20–30°C, the lower limit being 11°C and the upper limit 39°C (Broodryk 1971). Haines (1977) reported that females lay 100–300 eggs, depending on temperature and food availability, at a female to male ratio of 1:1. Deposition of eggs is inhibited by high light intensity. The best conditions are a low intensity of 5–8 lux, or complete darkness (Broodryk 1971). Different photoperiods have, in contrast to the larval and pupal stages, a significant effect on egg development, which is fastest at a day-to-night ratio of 12:12 hours. Above and below this ratio, development takes longer. Egg-laying capacity and longevity of adults also reach a maximum at this ratio (Gomaa *et al.* 1979). At a constant temperature of 26°C, the egg development takes 4–5 days, the larval stage is completed within 12 days, and the pupal stage is 6–7 days. Adults pair about 16 hours after emergence and begin egg-laying soon after. At 32°C they live only up to 6 days, at 25°C about 13 days, and at 20°C 15 days (Broodryk 1970, 1971). Females live between 2 and 3 days longer than males (Al-Ali and Thalouk 1970). Hence, according to temperature, the number of generations that *P. operculella* produces in its range of geographical distribution varies between 3–4 in the Andean highlands of Peru (Keller 2003), 8 in the Republic of Yemen (Kroschel 1995), and 10 in Egypt (Keller 2003).

From the diverse studies on the adaptation to extreme temperatures it can be assumed that various ecotypes of *P. operculella* have been developed. Attia and Mattar (1939) determined 13.7°C as the lower temperature limit for egg-laying, whereas Broodryk (1971) fixed it at 9.5°C. Winning (1941) was unable to demonstrate any further development at constant temperatures under 14°C. Furthermore, molecular studies proved the existence of at least two different *P. operculella* genotypes in the USA (Medina *et al.* 2010).

In addition to temperature, precipitation also influences *P. operculella* development and abundance. Leaf infestation becomes highest when potato is cultivated during warm, dry seasons, especially under furrow irrigation, where leaf infestation can reach up to 35 mines/plant (e.g., Republic of Yemen, Kroschel 1995). In contrast, heavy rains or regular sprinkler irrigation influences the flight activity of adults and limits leaf or tuber infestation. In rain-fed potato in the Andean highlands (e.g., the Mantaro Valley, Peru), with an annual precipitation of more than 700 mm during the vegetation period, leaf infestation is low; however, delaying harvest during the dry season increased tuber infestation (Keller 2003). Although temperature is favorable for *P. operculella* development in Egypt during the summer, from June to October the population declines significantly because potato is not cultivated (Keller 2003). In addition to abiotic factors such as temperature and precipitation, which mainly influence the *P. operculella* distribution and abundance, cultural practices such as the use of infested seed, or harvesting and storage practices, as well as biotic factors such as the occurrence of natural enemies (for parasitoids see below, and for entomopathogens see Chapter 15), are also important ecological drivers for the *P. operculella* problem in different agroecological zones (Fig. 6.5).

Development of *S. tangolias* life stages has been studied at 13°C and 19.4°C, respectively. Tenorio (1996) reported that the average number of eggs laid by females was 179.6–190.7 at a constant temperature of 13°C. This amount was less in the experiments conducted by Sánchez and Aquino (1986) and Tineon (1993), with 125.6 and 107.6 eggs per female at 15°C and 19°C, respectively. Rodriguez (1990) counted 236 eggs at temperatures between 15°C and 18°C. The sex ratio was stated to be 1:1. At 13°C and 19.4°C, first-instar larvae hatched after 17.4 days and 10.5 days and larvae development lasted 57 and 30 days, respectively. The pupae development was completed after another 32 and 18 days, leading to a total life cycle of 110 and 62 days, respectively (Sánchez and Aquino 1986, Rodriguez 1990, Tenorio 1996, Vera 1999).

Potato-growing zones in the South and Central American countries with *T. solanivora* incidence are located in elevated and mid-elevated mountainous regions between 2400 and 3400 m a.s.l. High temperatures shorten the development time of *T. solanivora*, but moderate temperatures are more conducive to the survival of immature life stages and oviposition (Torres 1989,

Notz 1996, Castillo 2005, Dangles *et al.* 2008). Within potato-growing regions, *T. solanivora* incidence decreased with rising altitude (Povolny 1973, Barreto *et al.* 2004, Dangles *et al.* 2008). *T. solanivora* damage has not yet been reported from high elevated potato regions above 3400 m a.s.l. Yield losses are generally low during cool and rainy periods (Pollet *et al.* 2003) but might rise quickly under warm and dry weather conditions, when damages of 50% and higher were reported in Colombia (Palacios *et al.* 1997), Ecuador (Suquillo 2005), Costa Rica (Hilje 1994) and Tenerife (Trujillo and Perera 2008). At constant temperatures development is possible within a temperature range of about 10°C to 25°C, where 2–10 generations per year might develop, and the optimum temperature for population growth was estimated to be 15°C (Notz 1996). In affected potato fields in Tibaitatá, Colombia, at an altitude of 2547 m a.s.l. and an average temperature of 14°C, development of one generation lasted approximately 94 days (Corpoica 2004); in Villapinzón, Colombia, at 2980 m a.s.l. and 11°C, it lasted 83.3 ± 13.6 days (Corredor and Flórez 2003).

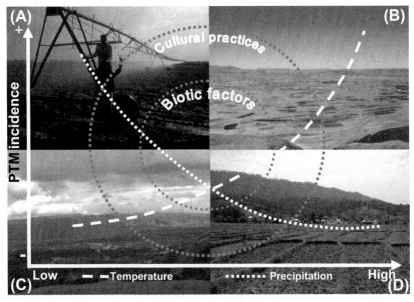

FIGURE 6.5 *Phthorimaea operculella* incidence in different potato agroecosystems affected by temperature and precipitation; further important factors are cultural practices applied in potato production such as the use of uninfested seed and biotic factors such as the occurrence of natural enemies (parasitoids, predators, entomopathogens). Characteristics of different potato agroecosystems where *P. operculella* occurs as major potato pest problem: (A) Egypt: 100 m a.s.l., <20 mm annual rainfall, sprinkler and furrow irrigation, one potato crop/year; (B) Republic of Yemen (Qa Jahran): 2200 m a.s.l., <200 mm rainfall, furrow irrigation, two potato crops/year; (C) Peru (Mantaro Valley): 3250 m a.s.l., 750 mm rainfall, mainly rain-fed potato cultivation, one potato crop/year; (D) India (Kangra Valley): 1400 m a.s.l., 1500 mm rainfall, irrigation and rain-fed potato cultivation, one potato crop/year. *Reproduced courtesy of J. Kroschel.* See also Plate 6.5

Temperature: Key for Ecological Sorting Among Species – Advances in Phenology Modeling

Insect life table data, developed under a wide range of temperatures, give good predictions for the best temperature conditions that insects require for optimal growth and development. Based on such life table data, temperature-based phenology models can be developed to gain an understanding of how temperature affects pest population growth potentials in different agroecologies. We developed temperature-based phenology models for *P. operculella*, which have been validated both in the laboratory and in the field (Keller 2003, Sporleder *et al.* 2004), and for comparison also for *S. tangolias* and *T. solanivora*. The life table studies comprised detailed observations on the development time and mortality of immature life stages, lifespan of male and female adults, oviposition and sex ratio, at temperatures of 5°C, 10°C, 15°C, 20°C, 25°C, 28°C, 30°C, 32°C, and 35°C, depending on the potato tuber species. The models were constructed using best fitting functions in the Insect Life Cycle Modeling (ILCYM) software recently developed by The International Potato Center, Lima, Peru (Sporleder *et al.* 2011, 2012). The development distribution observed for each temperature, species, and life stage was represented by the Logit function, using the same slope at all temperatures for each life stage and species, respectively. The median development time was calculated from these Logit functions, and was inverted to obtain the mean development rate. The effect of temperature on the development rate was described by the Sharpe and DeMichele function (Sharpe and DeMichele 1977), which is based on the impact of the temperature-dependent enzyme activity on the development rate of poikilothermic organisms. Within a medium temperature range, no enzyme inactivation takes place and the development rate rises exponentially; at higher and lower temperatures, enzymes are inactivated and development slows down. Development time decreased with rising

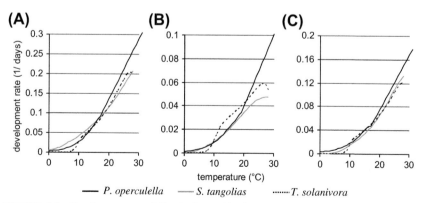

FIGURE 6.6   Development rate (1/days) of eggs (A), larvae (B), and pupae (C) of the species *Phthorimaea operculella*, *Symmetrischema tangolias* and *Tecia solanivora*.

temperatures (Fig. 6.6). Immature development of *P. operculella* was fastest at temperatures above 17–20°C. The adult lifespan across temperature was described by exponential functions and was negatively related with temperature, likewise for both sexes (Fig. 6.7). Lifespans of *S. tangolias* and *T. solanivora* were similar; *P. operculella* adults lived longer. Oviposition peaked at 15°C to 16°C for *S. tangolias* and *T. solanivora* with 160 and 268 eggs laid per female, while the fecundity of *P. operculella* was the highest at 21°C, with 170 eggs (Fig. 6.8A). Female age was normalized by division through the median time of survival, and functions for all temperatures fell on top; one single function was adapted to the accumulated percentage of oviposition for each species (Fig. 6.8B). At least 95% of eggs were laid by all species when half of the females had died. Immature mortality was U-shaped and

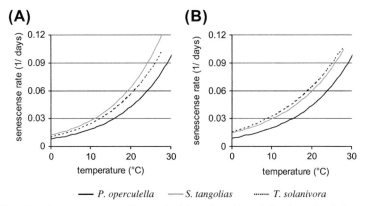

FIGURE 6.7   Senescense rate (1/days) of adult males (A) and females (B) of the species *Phthorimaea operculella, Symmetrischema tangolias* and *Tecia solanivora*.

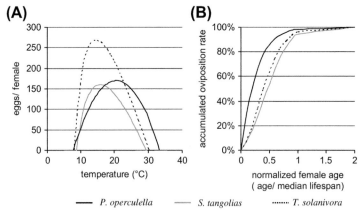

FIGURE 6.8   Total oviposition (A) and age-dependent fecundity (B) of the species *Phthorimaea operculella, Symmetrischema tangolias* and *Tecia solanivora*.

S. tangolias life stages were the most cold-tolerant ones, while *P. operculella* showed higher survival at temperatures above 17–25°C (Fig. 6.9). The sex ratio was independent of temperature and was equal for *P. operculella* and *S. tangolias*, while 53.2% of the *T. solanivora* adults developed were female.

The functions were compiled in an overall temperature-driven (computer-based) phenology model which uses rate summation and a cohort updating algorithm for simulating population growth. Parameters describing the population development, such as doubling time and the intrinsic rate of increase, were calculated as described by Southwood and Henderson (2000) (Fig. 6.10). Prediction of doubling time varied around 1 day from observation in most cases, and the intrinsic rate of increase varied less than 0.5% points from the prediction. The mean generation time of *P. operculella*, *S. tangolias* and *T. solanivora* was shortest at 30°C, 28°C, and 27°C, at 24, 38 and 37 days, respectively (Fig. 6.10A). The net reproduction rate peaked at 16°C in the case of *S. tangolias* and *T. solanivora*, with 26 and 65 female offspring per female, while net reproduction of *P. operculella* was highest at 21°C, with 57 female offspring (Fig. 6.10B). As a result, the population growth potential was highest at 29°C for *P. operculella* and at 22°C for *S. tangolias* and *T. solanivora*; at these temperatures the population doubling time was shortest, at 5.3, 14.4, and 9.6 days, respectively, and the intrinsic rate of increase was highest at 13.0%, 4.8%, and 7.2% of growth per day, respectively (Fig. 6.10C, 6.10D).

The phenology models were used to calculate the number of generations per year and the population growth potential according to the prevailing temperatures for each of the potato tuber moth species for different locations of potato-growing regions where the presence for at least two of the species has been reported (Table 6.1). The number of generations calculated was highest at San Ramon, Peru, where the climate is tropically warm year-round, with

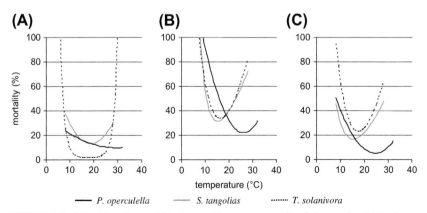

FIGURE 6.9 Mortality of eggs (A), larvae (B), and pupae (C) of the species *Phthorimaea operculella*, *Symmetrischema tangolias* and *Tecia solanivora*.

11.1, 7.0, and 7.4 generations calculated for *P. operculella*, *S. tangolias*, and *T. solanivora*, respectively. With rising altitude or latitude, the number of generations decreases. In very highly elevated potato-growing regions, like Copacabana at Lake Titicaca, Bolivia, where the average daily minimum temperatures fall below 0°C in some months, only 1.6 generations of *P. operculella* or 1.4 generations of *T. solanivora* might develop. Potential population growth of *P. operculella* was positively correlated with the number of generations per year, and was highest at San Ramon, where a 10^{18}-fold population increase per year was estimated. The highest potential population growth of *S. tangolias* and *T. solanivora*, however, was estimated at Arequipa, Peru, where temperatures range between 12°C and 26°C on average; here, a 10^8- and 10^{10}-fold increase per year could be expected, respectively. The predictions for *P. operculella* are well in line with the results of the life table experiments carried out by Keller (2003), who found that 3–4, 7, and 12 generations of *P. operculella* developed at Huancayo, Arequipa, and San Ramon, respectively. For *S. tangolias* and *T. solanivora*, observations on the number of generations per year are not available for comparison.

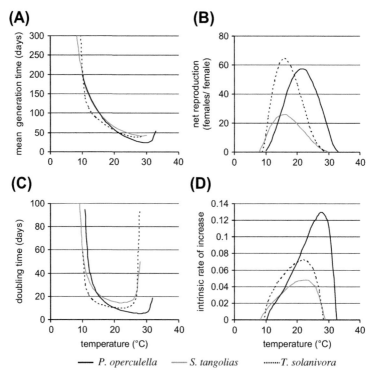

FIGURE 6.10   Life table parameters of the species *Phthorimaea operculella*, *Symmetrischema tangolias* and *Tecia solanivora*.

TABLE 6.1 Potential Population Development of *Phthorimaea operculella* (*P. o.*), *Symmetrischema tangolias* (*S. t.*) and *Tecia solanivora* (*T. s.*) at Different Locations using Phenology Modeling and Application in Insect Life Cycle Modeling (ILCYM) Software (Sporleder et al. 2011, 2012)

Country	Locality	m a.s.l.	Generations/year			Potential Population Growth/year (log x)		
			P. o.	S. t.	T. s.	P. o.	S. t.	T. s.
Spain, Tenerife	Northern Airport	623	4.8	(4.3)	5.2	8.2	(6.2)	8.3
Guatemala	Huehuetenango	1901	6.5	(5.2)	6.1	10.3	(6.8)	8.0
Columbia	Bogota	2628	3.3	3.1	3.5	3.8	4.0	3.9
Ecuador	San Gabriel	2860	2.8	2.7	3.0	2.6	3.3	2.8
	Salcedo	2628	3.7	3.4	4.0	4.9	4.7	5.1
Peru	Huancayo	3259	3.1	2.7	(3.1)	2.7	2.9	(2.6)
	Arequipa	2335	7.0	5.6	(6.7)	12.0	7.6	(9.5)
	San Ramon	770	11.1	7.0	(7.4)	18.2	6.2	(6.5)
Bolivia	Cochabamba	2531	6.9	5.2	(6.1)	10.1	6.1	(6.7)
	Copacabana	3857	1.6	0.0	(1.4)	0.1	0.0	(0.1)

Parentheses indicate that the respective species is not reported from the corresponding location.

PARASITOIDS AS AN IMPORTANT FUNCTIONAL GROUP OF NATURAL ENEMIES

A total number of 20 parasitoid species of the families Braconidae (9 species), Encyrtidae (2 species), and Ichneumonidae (9 species) have been reported parasitizing *P. operculella* (Table 6.2). Only some species, such as *Apanteles subandinus*, *Orgilus lepidus*, *Copidosoma koehleri* – widely used species in classical biological control of *P. operculella* in different parts of the world – or *Copidosoma desantisi* and *Campoplex haywardi*, have originated in South America and have very likely co-evolved with *P. operculella*. Many other species have evolved in North America, Africa, and Asia, where they have been found parasitizing *P. operculella*, but most likely not as their primary host. For cxample, *Diadegma molliplum* has been found parasitizing *P. operculella* with very high parasitism rates (>80%) in the Republic of Yemen (Kroschel 1995), but is also reported to be very effectively controlling the cabbage moth *Plutella xylostella* L. in South Africa (Sarfraz *et al.* 2005).

In comparison, species reported for *S. tangolias* and *T. solanivora* are rather rare. *A. subandinus*, *Copidosoma koehleri*, and *Macrocentrus ancylivora* are the only species reported to parasitize *S. tangolias* in Peru (Tenorio 1996, Vera 1999). *M. ancylivora* has its origin in the USA and is a parasitoid of fruit pests; for mass rearing purposes, *P. operculella* has been used (http://jenny. tfrec. wsu. edu/opm/displaySpecies. php?pn=950).

T. solanivora was reported only from Guatemala up to the year 1970, where the two braconids *Chelonus* sp. and *Orgilus* sp. and *Trichogramma pretiosum* were observed (Leal 1983). However, surveys conducted recently in Guatemala were unable to identify any parasitoid of *T. solanivora*. In surveys conducted in Columbia, the egg parasitoid *Trichogramma* sp. (Hymenoptera: Trichogrammatidae), the larval parasitoid *Apanteles* sp. (Hymenoptera: Braconidae), and flies (Diptera: Tachinidae) supposed to be egg parasites emerged from *T. solanivora* (Osorio *et al.* 2003). In Ecuador, larvae parasitized by Encyrtidae species (Hymenoptera) and *Apanteles* sp. were observed (Barragán *et al.* 2004, own observation). In Tenerife, *Copidosoma koehleri* (Hymenoptera: Encyrtidae) was isolated from *T. solanivora* (Ricón *et al.* 2004). *C. koehleri* was mass reared and released in the field in Tenerife and Venezuela (Ricón *et al.* 2004, Torres 1989), but only 1.15% of *T. solanivora* larvae were parasitized in Venezuela. On the whole, parasitoids infecting *T. solanivora* outside of its region of origin are probably not well adapted to this host, as high parasitism rates have never been observed; however, it must also be considered that the high use of insecticides in potato production systems of the Andes in general (Orozco *et al.* 2009) has a significant impact on parasitoids and natural enemies as described for Peru (Kroschel *et al.* 2012). *T. solanivora* invasion in new areas was facilitated by human transport, and the fact that most of its genetic diversity was lost during its spread (Puillandre *et al.* 2008, Torres-Leguizamón *et al.* 2010) might indicate that a limited number of individuals were introduced. They might have been

TABLE 6.2 Parasitoids of *Phthorimaea operculella, Symmetrischema tangolias* and *Tecia solanivora* Indicating their Origin and Distribution, as well as Countries of Successful Introduction within Classical Biological Control Programs

Parasitoid family, species, origin		P. operculella	S. tangolias	T. solanivora
Braconidae				
Agathis unicolor Schrank	South America	Argentina[1], Brazil[1], and Uruguay[1]		
Apanteles scutellaris Muesebeck	North America	Mexico[2] and USA[2]		Colombia[4,**], Ecuador[5,**]
	Introduced to:	Australia[3], Cyprus[3], Hawaii[3], India[3], Madagascar[3], New Zealand[3], St Helena[3], and Zambia[3,*]		
Apanteles subandinus Blanchard	South America	Argentina[1], Brazil[1], Peru[6], and Uruguay[1]	Peru[12]	
	Introduced to:	Australia[7,8,*], Bermuda[3], Cyprus[3,*], India[3], Madagascar[3,*], Mauritius[3,*], New Zealand[9,10,*], South Africa[3,11,*], St Helena[3], USA[3], Zambia[3,*], and Zimbabwe[3,*]		
Bracon gelechiae Ashmead	North America[13]			
	Introduced to:	Australia[1,3], Bermuda[3,*], Chile[3], Cyprus[3,*], France[3,*], Hawaii[3], India[3,*], Malta[3], New Zealand[3], South Africa[3], St Helena[3], Zambia[3,*], and Zimbabwe[3]		
Chelonus curvimaculatus Cameron; Syn.: *Microchelonus curvimaculatus* Cameron	Africa, Asia	South Africa[3] and India[3]		
	Introduced to:	Bermuda[3], Cyprus[3,*], New Zealand[3], St Helena[3], and USA[3]		

Continued

TABLE 6.2 Parasitoids of *Phthorimaea operculella, Symmetrischema tangolias* and *Tecia solanivora* Indicating their Origin and Distribution, as well as Countries of Successful Introduction within Classical Biological Control Programs—*Cont'd*

Parasitoid family, species, origin		*P. operculella*	*S. tangolias*	*T. solanivora*
Chelonus kellieae Marsh	Central America	Costa Rica[3]		Guatemala[14,**]
	Introduced to:	India[3] and USA[3]		
Chelonus phthorimaeae Gahan; Syn.: *Microchelonus phthorimaeae* Gahan	North America	USA[3]		
	Introduced to:	Australia[1,3], Bermuda[3], Canada[3], Chile[3], Hawaii[3], South Africa[3], and Yemen[15]		
Macrocentrus ancylivora Rohwer	North America	USA	Peru[12,+]	
Orgilus jennieae Marsh.	Central America	Costa Rica[3]		Guatemala[14,**]
	Introduced to:	India[3] and USA[3]		
Orgilus parcus Turner	Africa	South Africa[3]		
	Introduced to:	Cyprus[3], India[3], New Zealand[3], St Helena[3], USA[3], Zambia[3], Zimbabwe[3]		
Orgilus lepidus Muesebeck	South America	Argentina[1] and Uruguay[1]		
	Introduced to:	Australia[7,8,*], Bermuda[3], Cyprus[3,*], India[3,*], Israel[3], New Zealand[3], South Africa[3] St Helena[3], Tanzania[3], USA[3,*], and Zambia[3]		

TABLE 6.2 Parasitoids of *Phthorimaea operculella, Symmetrischema tangolias* and *Tecia solanivora* Indicating their Origin and Distribution, as well as Countries of Successful Introduction within Classical Biological Control Programs—*Cont'd*

Parasitoid family, species, origin		*P. operculella*	*S. tangolias*	*T. solanivora*
Ichneumonidae				
Campoplex haywardi Blanch.	South America	Argentina[1,16] and Uruguay[1]		
	Introduced to:	Australia[3,16], Bermuda[3], Cyprus[3,*], India[3,16,*], Madagascar[3], Mauritius[3,*], New Zealand[3,1,6], South Africa[3], St Helena[3], Tanzania[3], USA[3], Zambia[3], and Zimbabwe[3]		
Campoplex phthorimaea Cushman	North America	USA[3]		
	Introduced to:	Australia[3], Bermuda[3], and Hawaii[3]		
Diadegma stellenboschense Cameron; Syn.: *D. molliplum* Holmgren	Africa	South Africa[3]		
	Introduced to:	Cyprus[3], India[3], Madagascar[3], New Zealand[3], St Helena[3], USA[3], Kenya, and Republic of Yemen[1]		
Diadegma turcator Aubert	Europe	Cyprus[3]		
	Introduced to:	India[3], New Zealand[3], St Helena[3], Tanzania[3], and Zambia[3]		
Eriborus trochanteratus Morley	Asia	India[3]		
	Introduced to:	Cyprus[3], New Zealand[3], St Helena[3,*], and Zambia[3,*]		
Pristomerus sp. nr. *vulnerator* (Panz.)	Asia	India[3]		
	Introduced to:	Cyprus[3]		

Continued

TABLE 6.2 Parasitoids of *Phthorimaea operculella, Symmetrischema tangolias* and *Tecia solanivora* Indicating their Origin and Distribution, as well as Countries of Successful Introduction within Classical Biological Control Programs—*Cont'd*

Parasitoid family, species, origin		*P. operculella*	*S. tangolias*	*T. solanivora*
Pristomerus spinator Fabricius	North America	Mexico[17]		
Temelucha picta Holmgren	Africa	South Africa[11]		
Temelucha sp.	South America	Argentina[1] and Uruguay[1]		
	Introduced to:	Australia[3], Cyprus[3,*], India[3], New Zealand[3], St Helena[3], USA[3], and Zambia[3]		
Encyrtidae				
Copidosoma desantisi Annecke & Mynhardt	South America[13]			
	Introduced to:	Australia[3,*]		
Copidosoma koehleri Blanchard; Syn.: *C. uruguayensis* Tachikawa	South America	Argentina[18], Bolivia[5], Brazil[18], Chile[18], Colombia[19], Ecuador[***], Peru[13], and Uruguay[1]	Bolivia[29], Peru[12]	Ecuador[30,**]
	Introduced to:	Australia[7,20,21,*], Bermuda[3], Cyprus[3,*], India[19,22,23,24,*], Israel[20], Italy[25], Japan[3], Kenya[3], Madagascar[3], Mauritius[6,*], New Zealand[3], Seychelles[3], South Africa[11,20,26,*], St Helena[3,*], Tanzania[5], USA[27*], Yemen[28], Zambia[3,*], and Zimbabwe[3,*]		Spain (Tenerife)[31+] Venezuela[32]

TABLE 6.2 Parasitoids of *Phthorimaea operculella, Symmetrischema tangolias* and *Tecia solanivora* Indicating their Origin and Distribution, as well as Countries of Successful Introduction within Classical Biological Control Programs—*Cont'd*

Parasitoid family, species, origin		*P. operculella*	*S. tangolias*	*T. solanivora*
Trichogrammatidae				
Trichogramma pretosium	South America			Colombia[4,**], Guatemala[14]

*Established, **Species not specified (sp.), ***Personal communication, +no introduction of the species reported. [1]Lloyd 1972; [2]Whitfield 1995; [3]Sankaran and Girling 1980; [4]Osorio et al. 2003; [5]own observation; [6]Ramachandran and Rao 1967; [7]Briese 1981; [8]Horne and Page 2008; [9]Herman 2008; [10]Foot 1979; [11]Neuenschwander et al. 2003; [12]Vera 1999; [13]Raven 1966; [14]Leal 1983; [15]Kroschel 1995; [16]Leong and Oatman 1968; [17]Domínguez et al. 2000; [18]Doutt 1947; [19]López 2006; [20]Keasar and Steinberg 2008; [21]Callan 1974; [22]Dalaya and Patil 1973; [23]Khandge et al. 1979; [24]Pokharkar and Jogi 2003; [25]Pucci et al. 2003; [26]Watmough et al.1973; [27]GBIF 2011; [28]Nasseh and Al-Furassy 1989; [29]Calderon et al. 2002; [30]Barragán et al. 2004; [31]Rincón et al. 2004; [32]Torres 1989.

free from parasitoids, or parasitoids might have failed to establish themselves in the new environment.

CONCLUSIONS

Compared to *P. operculella*, the slow invasion of *S. tangolias* and *T. solanivora* into new potato-growing regions might be due to their more narrow feeding preferences and reduced host range, which might affect establishment and population growth in the absence of potato production. Further, the optimum temperature of development for *S. tangolias* and *T. solanivora*, which is clearly below that of the species *P. operculella*, restricts their development and establishment under warmer temperature conditions. Hence, both species find more suitable conditions under potato field and storage conditions of mid-elevated Andean potato regions; on the other hand, *P. operculella* was able to invade subtropical and tropical potato-growing regions. It is characterized by a fast generation turnover and a high population growth rate on potato foliage at warm temperatures, and attacks potato tubers primarily after foliar senescence. For its control it is crucial to prevent population build-up early in the potato-growing season by controlling larvae on potato foliage, whose feeding behavior offers more options for effective control by biological and chemical means as compared to *S. tangolias* and *T. solanivora*. *S. tangolias* damage at harvest is usually less severe, perhaps due to its longer development time and preference for stems as a food source for larvae. Since *T. solanivora* larvae feed on potato tubers only and adults lay eggs on soil close to potato stems, prevention of tuber infestation at harvest by application of biocontrol agents (e.g., baculovirus) or chemical means is more limited and difficult. For all three potato tuber moth species, cultural practices such as crop rotation, pest-free seed, regular hilling-up, and timely harvest seem to be equally important to prevent tuber damage. Interestingly, *P. operculella* is host of a much higher number of parasitoids of the families Braconidae, Encyrtidae, and Ichneumonidae, any of which species have co-evolved in the center of host origin or have adapted to *P. operculella* in its new regions of invasion and have been successfully used in biological control programs. As the two species *S. tangolias* and *T. solanivora* have a more limited distribution worldwide, they have been obviously less exposed to parasitoids outside their native range of distribution. Generally, though, the low number and abundance of parasitoids indicate that their more cryptic behavior of infestation and development inside stems and tubers might have limited parasitism by larvae parasitoids.

ACKNOWLEDGEMENTS

The authors are grateful for the financial support supplied by the Federal Ministry for Co-operation and Development (BMZ), Germany, the Regional Fund for Agricultural Technology (FONTAGRO), Washington, DC, and the German Academic Exchange service (DAAD) for conducting research on the different potato tuber moth species. Furthermore, we thank our modeling team, Henri Tonnang, Juan Carlos Gonzalo, and Pablo Carhuapoma, for their support in phenology modeling of the pest species, and Jaris Veneros for compiling the parasitoid information.

REFERENCES

Al-Ali, A.S., Talhouk, A.S., 1970. The potato tuber moth (*Gnorimoschema operculella*) (Zeller). 1: Its life history, 2: Its control in storage. American University of Beirut, Beirut, Lebanon, Publication. No. 44.

Attia, R., Mattar, B., 1939. Some notes on the potato tuber moth (*Phthorimaea operculella*, Zell.). Bull. Minist. Agric. Egypt 216 (44), 75–92.

Barragán, A., Pollet, A., Prado, M., Onore, G., Aveiga, I., Ruiz, C., 2002. Avances sobre la distribución y dinámica poblacional de *Tecia solanivora* en Ecuador. Taller Internacional: Prevención y control de la polilla guatemalteca (*Tecia solanivora*) de la papa. SENASA, CIP, Lima, Perú, pp. 43–51.

Barragán, A., Pollet, A., Prado, M., Lagnaoui, A., Onore, G., Aveiga, I., Lery, X., Zeddam, J.L., 2004. La polilla guatemalteca *Tecia solanivora* (Povolny) (Lepidoptera: Gelechiidae) en Ecuador. Diagnóstico y perspectivas de manejo bajo un método de predicción. Memorias – II Taller Internacional de Polilla Guatemalteca Pontificia, Universidad Católica del Ecuador, pp. 5–23.

Barreto, N., Espitia, E., Galindo, R., Sánchez, M., Suárez, A., López-Avila, A., 2003. Determinación de parámetros reproductivos y hábitos de *Tecia solanivora* (Povolny 1973) (Lepidoptera: Gelechiidae) en condiciones de laboratorio y campo. II Taller Nacional: Presente y futuro de la investigación en Colombia sobre polilla guatemalteca Memorias, CEVIPAPA, CNP, Bogotá, Colombia, pp. 31–36.

Barreto, N., Espitia, E., Galindo, R., Gordo, E., Cely, L., Sánchez, G., López-Avila, A., 2004. Fluctuación de la población de *Tecia solanivora* (Povolny) en tres intervalos de altitud en Cundinamarcay Boyacá, Colombia. Memorias – II Taller Internacional de Polilla Guatemalteca Pontificia, Universidad Católica del Ecuador, pp. 25–44.

Beukema, H.P., van der Zaag, D.E., 1990. Introduction to Potato Production. J. Agric. Sci. 116 169–169.

Briese, D., 1981. The incidence of parasitism and disease in field populations of the potato moth *Phthorimaea operculella* (Zeller) in Australia. J. Aust. Ent. 20, 319–326.

Broodryk, S.W., 1970. Dimension and development values for potato tuber moth *Phthorimaea operculella* in South Africa. Phytophylactica 2, 213–214.

Broodryk, S.W., 1971. Ecological investigations on the potato tuber moth, *Phthorimaea operculella* (Zeller) (Lepidoptera: Gelechiidae). Phytophylactica 3, 73–84.

Broodryk, S.W., 1979. *Phthorimaea operculella* (Zell.). In: Kranz, J., Schmutterer, H., Koch, W. (Eds.), Krankheiten, Schädlinge und Unkräuter im tropischen Pflanzenbau. Paul Parey, Berlin, Germany.

Calderon, R., Barea, O., Ramos, J., Crespo, L., Bejarano, C., Herbas, J., Lino, V., 2002. Desarrollo de componentes del manejo integrado de las polillas de la papa (*Phthorimaea operculella* y *Symmmetrischema tangolias*) en Bolivia y el bioinsecticida baculovirus (Matapol). Fundación PROINPA, Cochabamba, Bolivia.

Callan, E., 1974. Changing status of the parasites of potato tuber moth *Phthorimaea operculella* (Lepidoptera: Gelechiidae) in Australia. Entomophaga 19, 97–101.

Carpio, C., 2008. Evaluación de la eficiencia de diferentes formulaciones de bioplaguicidas virales y de un insecticida químico para el control de *Tecia solanivora* (Povolny) y *Symmetrischema plaesiosema* (Turner) en papa almacenada en una localidad del cantón Salcedo de la provincia de Cotopaxi, Ecuador. ESPE, Santo Domingo, Ecuador, MSc Thesis.

Castillo, G., 2005. Determinación del ciclo de vida de las polillas de la papa *Symmetrischema tangolias* (Gyen) y *Tecia solanivora* (Povolny) bajo condiciones controladas de laboratorio. Universidad Central del Ecuador, Quito, Ecuador, MSc Thesis.

Corpoica, 2004. Generación de componentes tecnológicos para el manejo integrado de la polilla guatemalteca de la papa *Tecia solanivora* (Povolny) con base en el conocimiento de la biología, comportamiento y dinámica de población de la plaga. Convenio Pronatta-Corpoica 981251134, Informe Técnico, Bogotá, Colombia.

Corredor, D., Flórez, E., 2003. Estudios básicos de biología y comportamiento de la polilla guatemalteca de la papa en un área de piloto en el municipio de Villapinzón. II Taller Nacional: "Presente y futuro de la investigación en Colombia sobre polilla guatemalteca". Memorias, CEVIPAPA, CNP, Bogotá, Colombia, pp. 37–45.

Cruz, E., Castillo, A., Malo, E., 2011. First report of *Tecia solanivora* (Lepidoptera: Gelechiidae) attacking the potato *Solanum tuberosum* in Mexico. Fla. Entomol. 94, 1055–1056.

Dalaya, V., Patil, S., 1973. Laboratory rearing and field release of *Copidosoma koehleri* Blanchard, an exotic parasite, for the control of *Gnorimoschema operculella* Zeller. Research Journal of Mahatma Phule Agricultural University 4, 99–107.

Dangles, O., Carpio, C., Barragan, A., Zeddam, J., Silvain, J., 2008. Temperature as a key driver of ecological sorting among invasive pest species in the tropical Andes. Ecol. Appl. 18 (7), 1795–1809.

Domínguez, I., Llanderal, C., Nieto, R., 2000. *Pristomerus spinator* Fabricius Hymenoptera: Ichneumonidae, un parasitoide de la palomilla de la papa. Agrociencia 34 (5), 611–617.

Doutt, R., 1947. Polyembryony in *Copidosoma koehleri* Blanchard. Am. Nat. 81 (801), 468–473.

EPPO, 2005. *Tecia solanivora*. EPPO Bulletin 35, 399–401.

Foot, M., 1979. Bionomics of the potato tuber moth, *Phthorimaea operculella* (Lepidoptera: Gelechiidae), at Pukekohe. New Zeal. J. Zool. 6, 623–636.

Gallegos, P., Suquillo, J., 1997. Monitoreo de la polilla de la papa (*Tecia solanivora*). Primer Seminario Taller Internacional sobre Manejo Integrado de *Tecia solanivora*. Ibarra, Ecuador. CIP, INIAP, PNRT-Papa, *FORTIPAPA*, pp. 29–37.

GBIF, 2011. CSIRO Ichthyology data base provided by OZCAM (on line). Consulted Nov 1, 2011. Global Biodiversity Information Facility, DK. Available at http://data.gbif.org/welcome.htm.

Gomaa, A.A., El-Sherif, S., Salem, A.A., Hemeida, I.A., 1978. On the biology of potato tuber worm, *Phthorimaea operculella* Zeller, (Lepidoptera, Geleichiidae). I. Effect of larval diet. Zeitschrift für Angewandte Entomologie 86 (3), 290–294.

Gomaa, A.A., El-Sherif, S., Salem, A.A., Hemeida, I.A., 1979. On the biology of potato tuber worm, *Phthorimaea operculella* (Zeller), (Lepidoptera, Geleichiidae). II. Reaction of photoperiodism. Zeitschrift für Angewandte Entomologie 87 (4), 430–435.

Graf, J.E., 1917. The potato tuber worm. US Department of Agriculture, Washington DC. Bulletin 4217, 56.

Haines, C.P., 1977. The potato tuber moth, *Phthorimaea operculella* (Zeller): a bibliography of recent literature and a review of its biology and control on potatoes in field and store. Report Tropical Products Institute G112, 15.

Herman, T., 2008. Biological control of potato tuber moth by *Apanteles subandinus* Blanchard in New Zealand. In: Kroschel, J., Lacey, L. (Eds.), Integrated pest management for the potato tuber moth, *Phthorimaea operculella* (Zeller) – a potato pest of global importance, Tropical Agriculture 20; Advances in Crop Research 10. Margraf Publishers, Weikersheim; Germany, pp. 73–80.

Hernández, N., Barragán, Á, Dupas, S., Silvain, J., Dangles, O., 2010. Wing shape variations in an invasive moth are related to sexual dimorphism and altitude. Bull. Ent. Res. 100 (5), 529–541.

Hilje, L., 1994. Characterization of the damage by the potato moths *Tecia solanivora* and *Phthorimaea operculella* (Lepidoptera: Gelechiidae) in Cartago, Costa Rica. Manejo Integrado de Plagas 31, 43–46.

Horne, P., Page, J., 2008. Integrated Pest Management dealing with potato tuber moth and all other pests in Australian potato crops. In: Kroschel, J., Lacey, L. (Eds.), Integrated pest management for the potato tuber moth, *Phthorimaea operculella* Zeller – a potato pest of global importance, Margraf Publishers, Weikersheim, Germany, pp. 111–117.

Karlsson, M., Birgersson, G., Cotes, A., Bosa, F., Bengtsson, M., Witzgall, P., 2009. Plant odor analysis of potato: Response of Guatemalan moth to above- and below-ground potato volatiles. J. Agric. Food Chem. 57, 5903–5909.

Keasar, T., Steinberg, S., 2008. Evaluation of the parasitoid *Copidosoma koehleri* for biological control of the potato tuber moth, *Phthorimaea operculella*, in Israeli potato fields. Biocontrol Sci. Techn. 18 (4), 325–336.

Keller, S., 2003. Integrated pest management of the potato tuber moth in cropping systems of different agro-ecological zones. In: Kroschel, J. (Ed.), Advances in Crop Research, Vol. 1, Margraf Verlag, Weikersheim, Germany.

Khandge, S., Parlekar, G., Naik, L., 1979. Inundative releases of *Copidosoma koehleri* Blanchard (Hymenoptera: Encyrtidae) for control of the potato tuber worm *Phthorimaea operculella* Zeller. J. Maharashtra Agric. Universities 4, 165–169.

Kroschel, J., 1994. Population dynamics of the potato tuber moth, *Phthorimaea operculella*, in Yemen and its effects on yield. Vol. I. Proceedings of the Brighton Crop Protection Conference, Pests and Diseases, Brighton, UK, pp. 241–246.

Kroschel, J., 1995. Integrated pest management in potato production in Yemen with special reference to the integrated biological control of the potato tuber moth (*Phthorimaea operculella* Zeller). Tropical Agriculture 8 Margraf Verlag, Weikersheim, Germany.

Kroschel, J., Koch, W., 1994. Studies on the population dynamics of the potato tuber moth (*Phthorimaea operculella* Zell. (Lep., Gelechiidae)) in the Republic of Yemen. J. Appl. Entomol. 118, 327–341.

Kroschel, J., Mujica, N., Alcazar, J., Canedo, V., Zegarra, O., 2012. Developing integrated pest management for potato: Experiences and lessons from two distinct potato production systems of Peru. In: Zhongqi, He, Larkin, R.P., Honeycutt, C.W. (Eds.), Sustainable Potato Production: Global Case Studies, Springer, UK, pp. 419–450.

Lal, L., 1987. Studies on natural repellents against potato tuber moth (*Phthorimaea operculella* Zeller) in country stores. Potato Res. 30 (2), 329–334.

Leal, H., 1983. Avances sobre el control biológico de polilla de la papa en Guatemala. Memoria Seminario Internacional sobre Biología y Control de la Palomilla y Polilla de la Papa y *Scrobipalpopsis solanivora* Povolny, Instituto Nacional de Investigaciones Agrícolas, Programa regional cooperativo de papa, Celaya, Guanajuato, Mexico.

Leong, J., Oatman, E., 1968. The biology of *Campoplex haywardi* (Hymenoptera: Ichneumonidae), a primary parasite of the potato tuberworm. Ann. Entomol. Soc. Am. 61 (1), 26–36.

Lloyd, D., 1972. Some South American parasites of the potato tuber moth *Phthorimaea operculella* (Zeller) and remarks on those in other continents. Technical Bull. Commonwealth Inst. Biol. Control 15, 35–49.

López, E., 2006. Influencia de la temperatura en el ciclo biológico de *Copidosoma koehleri* Blanchard; parasitoide de *Phthorimaea operculella* (Zeller). Universidad Nacional Federico Villarreal, Tesis Bióloga. Lima, Perú.

Martin, N.A., 1999. Arthropods and mollusks associated with poroporo (*Solanum aviculare* and *S. laciniatum*): an annotated species list. J. R. Soc. New Zeal. 29 (1), 65–76.

Medina, R.F., Rondon, S.I., Reyna, S.M., Dickey, A.M., 2010. Population structure of the potato tuberworm *Phthorimaea operculella* (Zeller) (Lepidoptera: Gelechiidae) in the United States. Environ. Entomol. 39 (3), 1037–1042.

Meisner, J., Ascher, K.R.S., Lavie, D., 1974. Factors influencing the attraction to oviposition of the potato tuber moth, *Gnorimoschema operculella* Zell. J. Appl Entomol. 77, 179–189.

Mitchell, B.L., 1978. The biological control of potato tuber moth *Phthorimaea operculella* (Zeller) in Rhodesia. Rhodesia Agric. J. 75 (3), 55–58.

Nasseh, O.M., Al-Furassy, M.A., 1989. Mass production and utilization of *Copidosoma koehleri* (Blanch) (Encyrtidae – Hymenoptera) to control *Phthorimaea operculella* (Zeller) (Gelechiidae: Lepidoptera) in the Yemen Arab Republic. Proceedings Integrated Pest Management in Tropical and Subtropical Cropping Systems, 8–15 February 1989, Bad Durkheim, Germany, Deutsche Landwirtschafts-Gesellschaft, Frankfurt, Germany, pp. 663–668.

Neuenschwander, P., Borgemeister, C., Langewald, J., 2003. Biological control in IPM systems in Africa. CABI Publishing, Wallingford, UK.

Notz, A., 1996. Influencia de la temperatura sobre la biología de *Tecia solanivora* (Povolny) (Lepidoptera: Gelechiidae) criadas en tubérculos de papa *Solanum tuberosum* L. Bol. Entomol. Venez. N. S. 11 (1), 49–54.

Oerke, E.C., Weber, A., Dehne, D.W., Schönbeck, F., 1994. In: Oerke, E.-C., Dehne, H.-W., Schönbeck, F., Weber, A. (Eds.), Crop Production and Crop Protection – Estimated Losses in Major Food and Cash Crops, Elsevier, Amsterdam, The Netherlands, pp. 742.

Orozco, F.A., Cole, D.C., Forbes, G., Kroschel, J., Wanigaratne, S., Arica, D., 2009. Monitoring adherence to the International Code of Conduct – highly hazardous pesticides in central Andean agriculture and farmers' rights to health. Intl J. Occup. Env. Heal 15 (3), 255–268.

Osmelak, J.A., 1987. The tomato stemborer *Symmetrischema tangolias* (Turner), and the potato tuber moth *Phthorimaea operculella* (Zeller), as stemborers of pepino: first Australian record. Plant Protection Quarterly 2, 44.

Osorio, P., Espitia, E., Rincón, D., Barreto, N., Lopez-Avila, A., 2003. Reconocimiento de enemigos naturales de *Tecia solanivora* y cría de depredadores de *Tecia solanivora*, II Taller Nacional: Presente y futuro de la investigación en Colombia sobre polilla guatemalteca. Memorias, CEVIPAPA, CNP, Bogotá, Colombia, pp. 137–147.

Oyala, J., 2002. Riesgo de ingreso de *Tecia solanivora* en Perú. Memorias del I Taller Internacional sobre Prevención y Control de la Polilla Guatemalteca de la Papa. Lima, Perú, pp. 29–37 11–14 de Setiembre 2001.

Palacios, M., Cisneros, F., 1997. Integrated pest management for the potato tuber moth in pilot units in the Andean Region and the Dominican Republic. Program report 1995–1996, International Potato Center (CIP), Lima, Peru, pp. 161–168.

Palacios, M., Sotelo, G., Saenz, E., 1997. La polilla de la papa *Tecia solanivora* (Povolny). Primer seminario taller internacional sobre manejo integrado de *Tecia solanivora*, pp. 1–22. Ibarra, Ecuador. CIP, INIAP, PNRT-Papa, *FORTIPAPA*.

Palacios, M., Tenorio, J., Vera, M., Zevallos, F.E., Lagnaoui, A., 1999. Population dynamics of the Andean potato tuber moth, *Symmetrischema tangolias* (Gyen), in three different agro-ecosystems in Peru. Impact on a Changing World, International Potato Center (CIP), Lima, Peru, pp. 153–160.

Pokharkar, D.S., Jogi, R., 2000. Biological suppression of potato tubermoth, *Phthorimaea operculella* (Zeller) with exotic parasitoids and microbial agents under field and storage conditions. J. Biol. Control 14, 23–28.

Pollet, A., Barragán, A., Zeddam, J.L., Lery, X., 2003. *Tecia solanivora*, a serious biological invasion of potato cultures in South America. Intl. Pest. Control 45 (3), 139–144.

Povolny, D., 1967. Genitalia of some neotropic members of the tribe Gnorimoschemini (Lepidoptera: Gelechiidae). Acta Entomologica Musei Nationalis Pragae 37, 51–127.

Povolny, D., 1973. *Scrobipalpopsis solanivora* sp. n. – a new pest of potato (*Solanum tuberosum*) from Central America. Acta Universitatis Agriculturae, Facultas Agronomica, Brno. 21, 133–146.

Pucci, C., Spanedda, A., Minutoli, E., 2003. Field study of parasitism caused by endemic parasitoids and by the exotic parasitoid *Copidosoma koehleri* on *Phthorimaea operculella* in Central Italy. B. Insectol. 56 (2), 221–224.

Puillandre, N., Dupas, S., Dangles, O., Zeddam, J.L., Capdevielle-Dulac, C., Barbin, K., Torres-Leguizamon, M., Silvain, J.F., 2008. Genetic bottleneck in invasive species: the potato tuber moth adds to the list. Biol. Invasions 10, 319–333.

Ramachandran, K., Rao, V., 1967. Introduction of *Copidosoma koehleri* Blanchard for the control of the potato tuber moth, *Gnorimoschema operculella* (Zeller) in India and its supply to Tanzania and Mauritius. Technical Bull. Commonwealth Inst. Biol. Control 8, 139–147.

Raman, K.V., 1987. Survey of diseases and pests in Africa: pests. Acta Horticulturae 213, 145–150.

Raman, K.V., Radcliffe, E.B., 1992. In: Harris, P.M. (Ed.), The Potato Crop, The scientific basis for improvement, second ed. Chapman and Hall, London.

Raven, K., 1966. Lista de especies de la super-familia Chalcidoidea registradas en el Perú con la inclusión de recientes identificaciones. Revista Peruana de Entomología 8 (1), 145–156.

Ricón, A., Giménez, C., Lorenzo, C., Ríos, D., Cabrera, R., 2004. Control de *Tecia solanivora* (Povolny) (Lepidoptera: Gelechiidae) en Tenerife, Islas Canarias, España. Memorias: II Taller Internacional de polilla guatemalteca, Pontificia Universidad Católica del Ecuador, Quito, Ecuador, pp. 189.

Rincón, D., López-Ávila, A., 2004. Dimorfismo sexual en pupas de *Tecia solanivora* (Povolny) (Lepidoptera: Gelechiidae). Revista Corpoica 5 (1), 41–42.

Rodríguez, A.R., 1990. Ciclo biológico de *Symmetrischema plaesiosema* (Turner 1919) Lep. – Gelechiidae y evaluación de daños en 16 cultivares de papa bajo condiciones de laboratorio. Tesis (Ing. Agr.) Univ. Nac. San Antonio Abad del Cuzco. Fac, Agronomía, Peru.

Rondon, S., 2010. The potato tuberworm: A literature review of its biology, ecology, and control. Am. J. Potato Res. 87, 149–166.

Salasar, J., Escalante, W., 1984. La polilla guatemalteca de la papa, *Scrobipalpopsis solanivora*, nueva plaga del cultivo de la papa en Venezuela. Sociedad Venezolana de Ingenieros Agrónomos 9, 24–28.

Sánchez, G., Aquino, C., 1986. La polilla de la Papa *Symmetrischema tangolias* (Turner, 1919). Boletín Técnico Instituto Nacional de Investigación y Promoción Agropecuaria, Lima, Peru.

Sánchez, G.A., Aquino, V., Aldana, R., 1986. Contribución al conocimiento de *Symmetrischema plaesiosema* (Lep. : Gelechiidae). Revista Peruana de Entomología 29, 89–93.

Sankaran, T., Girling, D., 1980. The current status of biological control of the potato tuber moth. Biocontrol News and Information 1 (3), 207–211.

Sarfraz, M., Keddie, A.B., Dosdall, L.M., 2005. Biological control of the diamondback moth, *Plutella xylostella*: A review. Biocontrol Sci. Techn. 15 (8), 763–789.

Sharpe, J., DeMichele, D., 1977. Reaction kinetics of poikilotherm development. J. Theor. Biol. 64, 649–670.

Southwood, R., Henderson, P.A., 2000. The construction, description, and analysis of age-specific life-tables. Ecological Methods, third ed. Blackwell Scientific, Malden, MA, pp. 404–436.

Sporleder, M., Kroschel, J., Gutierrez Quispe, M.R., Lagnaoui, A., 2004. A temperature-based simulation model for the potato tuberworm, *Phthorimaea operculella* Zeller (Lepidoptera: Gelechiidae). Environ. Entomol. 33, 477–486.

Sporleder, M., Tonnang, H.E.Z., Carhuapoma, P., Gonzales, J., Juarez, H., Simon, R., Kroschel, J., 2011. ILCYM – Insect Life Cycle Modeling. A software package for developing temperature-based insect phenology models with applications for regional and global pest risk assessments and mapping. International Potato Center, Lima, Peru.

Sporleder, M., Tonnang, H.E.Z., Carhuapoma, P., Gonzales, J.C., Juarez, H., Kroschel, J., 2012. Insect Life Cycle Modeling (ILCYM) software – a new tool for Regional and Global Insect Pest Risk Assessments under Current and Future Climate Change Scenarios. CABI Publishing, Wallingford, UK (in press).

Suquillo, J., 2005. Alternativas culturales y biológicas en el control de *Tecia solanivora* en campo y almacenamiento. Escuela politécnica del ejército, Sangolquí, Quito, Ecuador Masters Thesis.

Tenorio, J.E., 1996. Biología, comportamiento y control de las polillas de la papa *Symmetrischema tangolias* (Gyen) y *Phthorimaea operculella* (Zeller) en Cajamarca. Tesis (Ing. Agr.) Universidad Nacional Agraria La Molina, Lima (Perú).

Terauds, A., Rapley, P.E.L., Williams, M.A., Ireson, J.E., Miller, L.A., Brieze–Stegeman, R., McQuillan, P.B., 1984. Insect pest occurrences in Tasmania 1982/83. Insect Pest Survey, Tasmanian Department of Agriculture, Hobart, Australia No. 16.

Tineo, E., 1993. Biología y comportamiento de *Symmetrischema tangolias* (Turner, 1919), polilla de la papa en ambiente controlado. Facultad de ciencias biológicas. Universidad Nacional de San Cristobal de Huamanga, Ayacucho, Peru.

Torres., F., 1989. Algunos aspectos de la biología y comportamiento de la polilla de la papa, *Scrobipalpopsis solanivora* Povolny 1973 (Lepidoptera: Gelechiidae) en el Estado Táchira. Universidad Central de Venezuela, Facultad de Agronomía, Venezuela Tesis de Grado.

Torres, F., Notz, A., Valencia, L., 1997. Ciclo de vida y otros aspectos de biología de la papa *Tecia solanivora* Povolny (Lepidoptera: Gelechiidae) en el estado Táchira, Venezuela. Bol. Entomol. Venez. N. S. 12 (1), 81–94.

Torres-Leguizamón, M., Dupas, S., Dardon, D., Gómez, Y., Niño, L., Carnero, A., Padilla, A., Merlin, I., Fossoud, A., Zeddam, J.L., Lery, X., Capdevielle-Dulac, C., Dangles, O., Silvain, J.F., 2011. Inferring native range and invasion scenarios with mitochondrial DNA: the case of *T. solanivora* successive north–south step-wise introductions across Central and South America. Biol. Invasions 13 (7), 1505–1519.

Trujillo, E., Perera, S., 2008. Polilla guatemalteca de la papa – medidas preventivas. Información Técnica. Agrocabildo, Cabildo de Tenerife, España.

Trujillo, E., Rios, D., Cabrera, R., 2004. Distribución de *Tecia solanivora* (Povolny) (Lepidoptera: Gelechiidae) en Tenerife, Islas Canarias, España. Memorias – II Taller Internacional de Polilla Guatemalteca, Pontificia Universidad Católica del Ecuador.

Vera, M., 1999. Control de la polilla de la papa, *Symmetrischema tangolias* (Gyen) en el Valle del Mantaro. Departamento de Entomología, Facultad de Agronomía, Universidad Nacional del Centro del Peru, Huancayo, Peru.

Watmough, R., Broodryk, S., Annecke, D., 1973. The establishment of two imported parasitoids of potato tuber moth (*Phthorimaea operculella*) in South Africa. Entomophaga 18, 237–249.

Whitfield, J., 1995. Annotated checklist of the microgastrinae of North America north of Mexico (Hymenoptera: Braconidae). J. Kansas Entomol. Soc. 68 (3), 245–262.

Winning, E.V., 1941. Zur Biologie von *Phthorimaea operculella* als Kartoffelschädling. Arbeiten über Physiologische und Angewandte Entomologie aus Berlin Dahlem 8 (2), 112–128.

Chapter 7

Other Pests – China

Jing Xu, Ning Liu, and Runzhi Zhang

Institute of Zoology, Chinese Academy of Sciences, Chaoyang, Beijing, China

AN OVERVIEW OF POTATO CULTIVATION IN CHINA

History of Potato Cultivation

In the mid-16th century, the Dutch introduced potato, which originated in the Andes in Peru, Chile and Bolivia, to Taiwan, China. By the mid-17th century, potato production had spread to mainland China. There was also a record number of people in Dinghai County on the Zhoushan Islands planting potatoes in the thirty-ninth year of the Kangxi Period (1700) (Tang 2002).

At first, the potato species introduced to Yunnan, Sichuan, Guizhou, and other provinces were unable to adapt to the subtropical climate, and thus failed to establish. However, with their gradual spread to high-altitude temperate regions in Eastern Gansu, Southern Shaanxi and Northern Shanxi, the planting area and production of potato in China continually increased (Zhai 2001).

Current Situation

Currently, China is one of the main potato-producing countries. In the past decade, the cultivation scope and the cultivated area have expanded annually (Chen and Qu 2007). According to FAO statistics, the potato planting area in China was 5.083 million hectares in 2009, accounting for 27.3% of the world's cultivated area, or 56.3% of the cultivated area in Asia. The potato-cultivated area of China was much larger than that of the United States, Russia, Canada, Ukraine, and India, ranking first in the world. Due to its large area of potato cultivation, potato production in China is also greater than in other counties; potato production in China reached 73.28 million tonnes, accounting for 22.2% of global production, or 50.2% of total production in Asia. However, the current yield of potato in China is only 14.47 tonnes/ha, which is 3.2 tonnes/ha less than the world's average yield (17.67 tonnes/ha), and is far below the average yield of about 45 tonnes/ha in the United States, The Netherlands, France, Britain,

Insect Pests of Potato. http://dx.doi.org/10.1016/B978-0-12-386895-4.00007-7

and New Zealand. Therefore, increasing this yield is a major challenge facing China's potato industry.

At present, the provinces in China that have more than 300,000 hectares of potatoes planted include Inner Mongolia, Guizhou, Gansu, Sichuan, Yunan, and Chongqing. The cultivated area of the above-mentioned six provinces reaches 3.37 million hectares, accounting for 65% of China's total cultivated area (China Agriculture Yearbook 2009).

MAJOR POTATO PESTS IN CHINA

In China, potatoes are attacked by a relatively diverse complex of insect pests. Above-ground pests include the Colorado potato beetle (*Leptinotarsa decemlineata*), ladybird beetles (*Henosepilachna vigintioctomaculata* and *Henosepilachna vigintioctopunctata*), the green peach aphid (*Myzus persicae*), the potato tuberworm (*Phthorimaea operculella*), whiteflies (*Bemisia tabaci* and *Trialeurodes vaporariorum*), leafhoppers (*Empoasca spp.*), and plant bugs (*Apolygus lucorum* and *Adelphocoris lineolatus*). Pests that damage the underground tubers include grubworms (*Amphimallon solstitialis* and *Holotrichia oblita*), cutworms (*Agrotis spp.*), wireworms (*Pleonomus canaliculatus* and *Agriotes subrittatus*), mole crickets (*Gryllotalpa spp.*), and the pharaoh ant *(Monomorium pharaonis)*. Several other species occasionally cause minor damage. The following sections describe the most important species in detail.

Potato Ladybird *Henosepilachna vigintioctomaculata* (Motschulsky)

Taxonomic Position and Morphological Description

Taxonomic Information

Class: Insecta; Order: Coleoptera; Family: Coccinelidae; Subfamily: Epilachninae.

Due to intraspecific variation, different populations usually have significant variations in external morphology and host plants. Therefore, some scholars propose the use of the "*H. vigintioctomaculata* complex" to address this phenomenon. According to morphological characters, Katakura (1980) divided this complex into two groups. Group A only has one species (*H. vigintioctomaculata*); group B has three "species" (*H. niponica* (Lewis), *H. pustulosa* (Kòno) and *H. yasutomii* Katakura). These species differ not only in morphology but also in biological characteristics and distribution range, and the two groups are distinguished from each other based primarily on a number of morphological features. There is also some degree of reproductive isolation between group A and group B. Limited mating between *H. vigintioctomaculata* and *H. pustulosa* is possible; however, the hatching ratio was found to be distinctly lower in eggs laid by heterogamic pairs than those by homogamic pairs (Katakura and Nakana

1979), because the estimated number of sperm preserved by the heterospecific females was approximately one-fifth to one-tenth of those preserved by the conspecific females, and the majority of sperm were lost during the migration from the bursa copulatrix to the place for sperm storage (Katakura 1986). In addition, incompatibility between the sperm and the female genital tract caused the death of the majority of heterospecific embryos during development (Katakura and Sobu 1986).

Although the three "species" in group B differ in body shape, body color, habits, host-plant preference, distribution range, and other aspects, they readily mate with each other and their offspring are fertile. These three "species" should be regarded as "form". Currently, the complex mainly includes species *H. vigintioctomaculata* (Motschulsky) in group A and several forms in group B.

Main Morphological Characteristics

Adult: 7–8 mm long and about 5.5 mm wide, hemispherical. Dorsal plate and elytra are yellowish-brown to reddish-brown, covered with dense fine brown hair. Head flat and small, usually hidden under the prothorax. Antennae clavate, 11-segmented, terminal 3 segments enlarged. Pronotum concave, angles prominent, with a longitudinal sword-shaped spot in the center and two small spots on each side (sometimes combined into one). Each elytron has 14 spots: 3 spots at the base, 4 spots behind them in a non-straight line, one or two pairs of spots in contact at the conjunction of elytra (Fig. 7.1). Male genitalia with median lobe having 4–7 teeth (Liu 1963).

Egg: about 1.5 mm long, shell-shaped, yellowish at the beginning, gradual transition to orange. Eggs in egg mass distributed haphazardly with 20–30 eggs vertically arranged at the lower surface of the leaf (Liu 1963).

Larva: mature larvae about 9 mm long, yellowish-brown, fusiform, enlarged at the center, tergite convex. Dorsal surface of each segment has setae; the prothorax and the eighth and ninth abdominal segments each have 4 setae, while the remaining segments each have 6 setae, with 6–8 small spines on each seta.

Pupa: about 6 mm long, flat oval, light yellow with black strip. Terminal enclosed with the molt at last instar. Setae visible (Liu 1963).

Current Distribution in China

The potato ladybird is mainly distributed in Northern China and is a common palaearctic species (Katakura 1980). It has been found in Heilongjiang, Shaanxi, Hebei, Henan, Liaoning, Jilin, Shaanxi, Gansu, Jiangsu, Anhui, Zhejiang, Fujian, Guangdong, Taiwan, Guangxi, Yunnan, Guizhou, and Sichuan (Liu 1963, Casagrande 1985, Zhang *et al.* 1993).

Host Plant

The potato ladybird is a typical polyphagous pest that feeds on at least 29 plant species belonging to 13 families, with different host plants utilized throughout the

FIGURE 7.1   Adult *Henosepilachna vigintioctomaculata* on damaged plant. See also Plate 7.1

season (Zhang *et al.* 1993, Wu and Wang 2008). Most damage is done to solanaceous plants, including Chinese wolfberry (*Lycium chinense* Miller), nightshade (*Solanum nigrum* L.), eggplant (*Solanum melongena* L.), tomato (*Lycopersicon esculentum* Miller), and especially potato (*Solanum tuberosum* L.) (Cui *et al.* 2007).

In the past, people used to believe that the potato ladybird bred only after feeding on potato leaves. However, subsequent field investigations and laboratory rearing proved that potato ladybirds can successfully complete their development by feeding on eggplant, tomato, nightshade, and datura (*Datura stramonium* L.) (Zhang *et al.* 1993).

Damage to Potato

In China, the potato ladybird is the most important pest of the potato plant. The adults and larvae cause severe damage to potato crops by chewing the leaves, stems and flowers. They feed on the lower epidermis and mesophyll of potato leaves, leaving only the upper epidermis intact. As a result, the damaged leaves and stems have many transparent concave lines running parallel to each other, leaving the vein and epidermis seriously damaged. The damaged parts gradually turn into rust-colored patches of necrotic tissue, the leaf dries up, and photosynthesis is hampered. Finally, the death of the plant occurs as a result (Liu 1963).

A survey of potato crops showed that, in the case of severe infestation, the number of potato ladybirds on one hundred potato plants can reach 5000 or more (Qi 1997). Potato ladybird damage may promote secondary infections by gray mold (*Botrytis cinerea*) and many other diseases (Yao *et al.* 1992). In recent years, as the potato planting area has expanded, the infestation of potato ladybird has increased year by year, causing a substantial reduction of potato production by 10–15% in normal years and by 20%–30% in bad years (Song *et al.* 2008).

Annual Population Dynamics

The potato ladybird has been reported to have one to two generations in a single year in the Heilongjiang province and in the northern Shanxi province (Lin 2001, Qin *et al.* 2006, Zhuang 2010), two generations in Tianjin (Guo *et al.* 2005), two to three generations in Qingjian County in the Shanxi province (Zhuang 2010), three generations in Heze City in the Shandong province, and four generations in years with high temperatures (Hao *et al.* 2006). The generations overlap almost completely. The adults overwinter in aggregations in cracks on the sunny leeward sides of rocks or other shelters in the vicinity of their summer habitats. Overwintering adults resume activities in mid- or late May, feeding on nightshade, Chinese wolfberry, lucerne (*Medicago sativa* L.), and other wild plants. In early June, adult potato ladybirds switch to feeding on newly emerged seedlings of cucurbits, eggplant, and pepper (*Capsicum annuum* L.), and afterwards move to potato plants for feeding, mating, and ovipositing. The oviposition season is between mid-June and mid-August, with peak activity from late June to early July. Eggs are laid in clutches on the lower surfaces of the leaves. Larvae undergo four life stages of development and pupate on the lower surfaces of leaves from late June to late August. The pupal stage lasts for 5–7 days. The new adults emerge from early July to early September, and some oviposit in mid-July to mid-August. The second-generation larvae appear in late July, and the heaviest damage occurs in mid-August. After pupation, the second-generation adults emerge from late August to early September, and some of them overwinter without mating and ovipositing (Zhuang 2010).

Biological and Ecological Characteristics

Habits

Adult: Adults usually emerge from the pupal stage during the day. The newly eclosed adults are soft and pale yellow, with six spots on the elytra; 1 hour later, the remaining 22 spots appear on the elytra and the adults begin to crawl; 2–3 hours later, adults start feeding. Adults feed during the day and the night, but the largest amount of plant consumption happens on sunny days. Two to three days after emergence, adults begin to mate; the length of mating time differs, ranging from 10 minutes to several hours. Four days after mating, females begin to oviposit. Unmated females can also lay a small number of eggs, but these eggs are unable to hatch. There are multiple matings and multiple ovipositions; a female can lay 10–15 egg masses during her lifetime, and each egg mass contains 9–56 eggs. The longevity of an adult is about 30–80 days, and the longevity of an over-wintered adult is up to 240 days. Adults show phototactic responses and feign death when disturbed; their flying ability is low, as they can only fly for a few meters (Duanmu *et al.* 1995, Ma and Yin 2001, Hao *et al.* 2006).

Egg: Most eggs are laid on the lower surfaces of leaves, with very few laid on upper surfaces. Most eggs of the first generation are laid on the lower leaves of plants, while most eggs of the second, third, and fourth generations are laid on the middle and upper leaves. The newly laid eggs are bright yellow, and later become dark brown before hatching. Eggs from the same egg mass hatch almost simultaneously. Laboratory rearing found that the egg stage for the first-generation eggs lasted for 7–8 days, and the second-generation egg stage lasted for 3–5 days (Hao *et al.* 2006).

Larva: Most larvae hatch between 5 am and 8 am. Newly hatched larvae start feeding in the vicinity of the egg mass after 2–5 hours (Fig. 7.2). Larvae gradually disperse from their places of hatching, but usually do not move further than an adjacent plant. The third and fourth instars consume the most foliage. Towards pupation time, the last instars become pale and stop feeding. Four instars are usually completed in 12–16 days (Hao *et al.* 2006).

FIGURE 7.2   *Henosepilachna vigintioctomaculata* larva. See also Plate 7.2

Pupa: Larvae usually pupate on the lower surfaces of leaves. The new pupa is milky white, while the mature pupa is yellow (Fig. 7.3). The pupal stage lasts for 3–5 days (Hao *et al.* 2006).

Development Threshold

Xiong (1991) determined that the threshold temperature of development and the effective temperature for potato ladybird larvae are 14.78°C and 141.15 degree-days, respectively. Zhang (1997) investigated the potato ladybird population in Luanchuan County in Hebei province and found that the threshold temperature of development of egg, larva, and pupa was 10.43°, 9.48°, and 9.35°C, respectively, and the effective accumulated temperature was 68.6587, 239.2038 and 75.1445 degree-days, respectively.

Effects of Temperature and Humidity on the Development of the Potato Ladybird

Zhang (1997) performed laboratory tests to study the life table of potato ladybird under five conditions of constant temperature with 16L: 8D and RH 65–85%. The results showed that there was a positive linear relationship between humidity and development time of egg, larva, and pupa within the range of 15–30°C. Temperatures above 27°C inhibited adult ovaries and the

FIGURE 7.3   *Henosepilachna vigintioctomaculata* pupa. See also Plate 7.3

larval development rate. Eggs could not hatch when the relative humidity was lower than 50%; suitable humidities for different development stages were 90% RH for egg, 65%–75% RH for the first to third instars, and 65%RH for pupa; the fourth instar was sensitive to high humidity, but its development time showed no significant difference under 35%–75% RH. The supercooling point and body freezing point of the overwintered adults were −7.52 ± 2.80°C and −5.05 ± 2.85°C, respectively.

Effects of Abiotic Factors on Population

Cui *et al.* (2007) investigated the occurrence of potato ladybirds in Luliang County in Shaanxi province and found that temperature and humidity were the major factors affecting their occurrence. Cold and dry conditions in winter and early spring resulted in high mortality of overwintered adults. Frequent rainfalls in June and July, a daily average relative humidity greater than 70%, and a daily average temperature of 20–25°C were the most suitable conditions for the occurrence of potato ladybird adults, eggs, and larvae. With little rainfall, relative humidity below 50%, and a daily average temperature higher than 23°C for several days, the number of withered eggs in the fields and the mortality of newly hatched larvae significantly increased. Heavy rain can wash away the egg masses and young larvae, reducing pest occurrence. Zhuang and Sun (2009) studied the natural population life table of potato ladybirds in Heilongjiang province. The results showed that, under natural conditions, the population of the subsequent generation was 37.28 times that of the previous generation, and the survival rate from egg to adult in one generation was 16%. The comparison results of the control index indicated that predation and precipitation were the most important factors impeding a population's survival.

Control Methods

Most farmers rely on chemical control for suppressing populations of this pest. The optimal time period for spraying is between plant colonization by overwintering adults and the peak hatching of first-generation larvae. Commonly used pesticides include phoxim, phoxim-deltamethrin tank mix, and cyhalothrin. The control efficiency of conventional spraying methods reaches over 90% (Li *et al.* 2001, Chen *et al.* 2003, Dong *et al.* 2007, Sun 2008).

In addition to insecticides, potato ladybirds can be controlled by manual destruction of adults and eggs. This approach is greatly facilitated by the tendency of diapausing adults to aggregate at their overwintering sites, where they can be easily discovered and killed. Timely removal of egg masses during the oviposition season also can reduce subsequent damage by hatching larvae.

Practicing good sanitation also helps in reducing beetle numbers. Clearing away the residual parts of solanaceous plants immediately after harvest can increase beetle mortality through starvation.

Twenty-Eight-Spotted Ladybird *Henosepilachna vigintioctopunctata* (F.)

Taxonomic Position and Morphological Description

Taxonomic Position

Order: Coleoptera; Family: Coccinelidae; Subfamily: Epilachninae.

Main Morphological Characteristics

Adult: 5.2–7.4 mm long and 5–5.6 mm wide, hemispherical. Body color and shape are similar to potato ladybird. Pronotum with a horizontal double diamond spot in the center. Each elytron has 14 spots; 3 spots at the base and 4 spots behind in a straight line, with the spots at the conjunction of elytra not in contact. Median lobe of male genitalia without teeth. (Pang 1979).

Egg: about 1.2 mm long, shell-shaped. Eggs in egg mass closely arranged.

Larva: mature larva about 7 mm long, white. White setae, with dark brown ring at the base (Pang 1979).

Pupa: about 5.5 mm long, light yellow, with black spots on tergite.

The differences from the potato ladybird include the following: 28-spotted ladybird adults are small, about 5.2–7.4 mm long; pronotum has a horizontal double diamond spot in the center. Each elytron has 14 spots, but the 4 spots in the second column at the elytral base are in a straight line; spots at the conjunction of elytra are not in contact; white larval setae (Pang 1979).

Zhang *et al.* (2002) used scanning electron microscopy to further compare the body surface ultrastructures of the potato ladybird and the 28-spotted ladybird. The results revealed several differences between the two species. First, chaetae of the potato ladybird were located in the center of the pits, and those of the 28-spotted ladybird were located on the margin of the pits. Secondly, the pits on the elytra were deeper in the potato ladybird.

Current Distribution in China

The 28-spotted ladybird is a common species widely distributed in Oriental, Palearctic, and Australasian regions. In China, the 28-spotted ladybird is found from Heilongjiang and Inner Mongolia in the north to Taiwan, Hainan, Guangdong, Guangxi, and Yunnan provinces in the south, and from the national boundary in the east to the Shaanxi, Gansu, Sichuan, Yunnan, and Tibet in the west. The 28-spotted ladybird has greater density in areas south of the Yangzi River.

Host Plant

The host plants of the 28-spotted ladybird include potato, eggplant, tomato, and pepper in the Solanaceeae family, and cucumber (*Cucumis sativus* L.), white gourd (*Benincasa hispida* (Thunb.)), and towel gourd (*Luffa cylindrica* Roem.) in the Cucurbitaceae family. The damage on eggplant is the most serious.

The 28-spotted ladybird can also damage Chinese cabbage (*Brassica rapa pekinensis*) and several species of the family Leguminosae.

In addition to vegetables, the 28-spotted ladybird also causes damage to Chinese wolfberry, morelberry (*Physalis alkekengi* L.), datura, and other medicinal herbs.

Damage to Potato

The most important host plant for the 28-spotted ladybird is eggplant. However, it also causes severe damage to potato. The gregarious newly-hatched larvae chew mesophyll on the lower surfaces of leaves, leaving only the epidermis. Similar to the potato ladybird, feeding damage first appears as many translucent fine concave lines running parallel to each other. As larvae mature, they gradually disperse from the place of initial eclosion. Chewing by adults and larvae makes the leaves perforated, eventually leaving only thick veins. The damaged leaves become dry and brown, and this results in the death of the plant. The 28-spotted ladybird also damages stems, petals, sepals, fruits, and other parts of plants. In years of high density, feeding by this pest can completely destroy large areas of spring potato, seriously affecting the quality and yield of tubers (Li *et al.* 2001).

Annual Population Dynamics

The 28-spotted ladybird has three to five generations every year in the Yangtze River valley, and the beetles overwinter as adults. However, in areas with a warm climate, such as the Guangdong, Guangxi, Taiwan, Yunnan, and Hainan provinces, there is no overwintering diapause. Because the oviposition period of the overwintering adults lasts for 2–3 months, the generations overlap. The overwintering adults move out in mid-April, flying to spring potato and eggplant fields near their overwintering sites for feeding and breeding. At the beginning of the potato harvest in mid- to late May, overwintering adults and some larvae of the first generation move to eggplant, tomato, pepper, and other crops, where they also cause damage. The peak periods of larval abundance in each generation are as follows: first generation, late May; second generation, late June to early July; third generation, late July to early August; fourth generation, mid- to late August. Generally, the number of overwintering pests is small, so the damage caused by overwintering adults and the first-generation larvae and adults is not very serious. The occurrence periods of the second, third, and fourth generations are consistent with the peak growth of solanaceous vegetables; rich diets are beneficial for breeding, and the population increases sharply. These three generations are the main generations that cause damage. From late August to early September, solanaceous vegetables are harvested and fields are plowed. At this point, food shortages increase the mortality of larvae and pupae, thus decreasing the field population. After that, some larvae and adults migrate to nightshade, morelberry, and other wild host plants in the same field. A few beetles move to sword bean (*Canavalia gladiata* (Jacq.) DC.), cowpea (*Vigna unguiculata* (L.) Walp.), and cucumber plants, but their numbers are usually insufficient to cause significant damage. After early to mid-October, the adults fly to overwintering sites and enter diapause.

Biological Characteristics

Habits

Adult: Adults typically emerge from pupae during the day. Adults begin feeding and flying 3–5 hours after emergence and begin mating 3–4 days after emergence. Males and females can mate several times during their lifetime, and the pre-oviposition period lasts for 3–13 days. Eggs are laid during the day on the lower surface of leaves. Most of them are laid on middle and upper leaves, and a small number are laid on stems, shoots, and the lower surface of the lower leaves. Fecundity is, on average, about 300 eggs per female (range: 51–511). Oviposition periods of different generations vary greatly; the oviposition period of overwintering adults averages about 2 months, ranging from about 7 days at the least to 3 months at the most, and the oviposition periods of other generations are usually 10–20 days. Adults feed during the day and at night, preferring potato and eggplant leaves, and the fruits and leaves of sweet pepper and tomato. Females can live 10–12 days without food, and males can live 8–9 days. Adults feign death when disturbed and are phototactic, but avoid strong direct light. During food shortages, adults become cannibalistic, especially the males (Chen *et al.* 1989).

Eggs: Most eggs hatch in the morning. Hatching in an egg mass starts from the margin and proceeds towards the center, with all eggs hatching within 1–3 hours. The hatching rate is greatly affected by temperature, and decreases in subsequent generations. At temperatures higher than 32°C, eggs suffer high mortality (Chen *et al.* 1989).

Larva: Newly hatched larvae often gather around the egg masses, and they will start dispersing and feeding 5–6 hours after hatching. The dispersal ability of larvae is weak, as larvae hatched from the same egg mass generally feed on the plant where they hatched and on immediately adjacent plants. Larvae feed during the day and at night. They avoid strong illumination more than adults, and often stay on the lower surfaces of leaves. During food shortages, larvae also become cannibalistic. Larvae go through four instars in 11–34 days, and molt during the day and at night (Chen *et al.* 1989).

Pupa: Most mature larvae pupate on the lower surfaces of middle or lower leaves. Between 24 and 36 hours before pupation, larvae become motionless, with the end of their abdomens close to the host; their central uplift and body length grow shorter. A larva will exuviate the last epidermis, which remains at the tail of the pupa. A newly pupated pupa is white with setae. The pupal period is usually about 3–5 days (Chen *et al.* 1989).

Effects of Temperature on Development

Chen *et al.* (1989) studied the threshold temperature and degree-day accumulation for the 28-spotted ladybird population in Jiangsu province. The threshold temperatures were 10.7°C for eggs, 11.7°C for larvae, 14.3°C for pupae, and 12.0°C for adults. Egg development was completed after accumulating 63.2 degree-days, larval development after 216.7 degree-days, and pupation after 53.1 degree-days. Adults accumulated 476.0 degree-days between eclosion from pupae and death.

Chen *et al.* (1989) performed laboratory tests to study the development of the 28-spotted ladybird under five conditions of constant temperature. The results showed that, under the temperature conditions of 18°C, 20°C, and 22°C, the 28-spotted ladybird could not complete a generation. At temperatures of 18–30°C, the rate of development was positively correlated with temperature. When the temperature was above 30°C, the developing rate decreased as temperature increased.

Effects of Host Plants on the Development of Twenty-Eight Spotted Ladybird

Wang (2002) investigated the effects of four eggplant varieties (including Shanghaiziqie, Huza 34#, Toutaomo 3#, and the Nanchang County local species Yanghongqie) on the development of the 28-spotted ladybird. The results showed that the beetles developed more quickly when feeding on the Shanghaiziqie variety than on other varieties. The duration of development for larvae and pupae was 5.69 days faster compared to beetles feeding on the Toutaomo variety. This indicates that the 28-spotted ladybird prefers varieties with a smooth surface rather than varieties with rough or prickly surfaces.

Control Methods

Chemical control is the most common approach to suppressing populations of this pest in China. The best time for spraying is when overwintering adults are present and when hatching in the first-generation larvae is at its peak; in this period, most larvae are aggregated. Commonly used pesticides include fenvalerate, malathion, deltamethrin, cypermethrin, phoxim, and cyhalothrin.

In addition to chemical control, 28-spotted ladybirds can be controlled by some physical approaches, such as installations of insect-proof screens in the field to prevent oviposition of adults, and the use of light traps to attract and kill adults. These two methods can reduce the egg number in the field and will thereby decrease the damage dealt to plants by enclosing larvae.

Manual destruction of adults and eggs is helpful in controlling this pest. The optimal time for adult destruction is between winter and early spring, when it is easy to capture diapausing adults at their overwintering sites. Adults can also be destroyed during the outbreak season; they feign death, and can then be easily dislodged from the plants. Timely removal of egg masses can reduce subsequent damage by hatching larvae, and this approach is greatly facilitated by the bright color of the egg mass.

Practicing good sanitation also helps in reducing beetle numbers. Clearing away the residual parts of solanaceous plants and plowing fields after harvest can reduce beetle numbers.

Potato Tuberworm *Phthorimaea operculella* (Zeller)

Taxonomic Position

Class: Insecta; Order: Lepidoptera; Family: Gelechiidae.

Current Distribution in China

The potato tuberworm was first recorded in China in 1937, when Chen (1937) reported damaged tobacco (*Nicotiana tabacum* L.) plants in Liuzhou City in Guangxi province. At the present, the potato tuberworm is widely distributed in Sichuan, Yunnan, Guizhou, Guangdong, Guangxi, Hunan, Huber, Jiangxi, Anhui, Gansu, Shaanxi, Henan, Shanxi, Taiwan, and some other provinces in China. The damage is most serious in Yunnan, Guizhou, Sichuan, and Shaanxi (Xu 1985, Hu 2008).

Host Plant

The host plant that the potato tuberworm most prefers is tobacco, followed by potato and eggplant. Potato tuberworm also damages tomato, pepper, datura, Chinese wolfberry, night shade, morelberry, deadly nightshade (*Atropa belladonna* L.), flos daturae (*Datura metel* L.), and other solanaceous plants.

Damage to Potato

The potato tuberworm is the most important storage pest of potato, with a wide distribution in warm, dry potato-planting regions. This pest can also seriously damage potato plants in the field, where it damages stems, leaves, apical meristems, and buds. Damaged apical meristems and buds turn yellow and die. In serious cases, the entire seedling dies. Larvae eat the mesophyll of leaves, leaving only the upper and lower epidermis intact and making the leaves translucent. Yield losses caused by potato tuberworm infestation in a field can reach 20–30% (Anonymous 1990).

Potato tuberworm damage is even more serious in storage. Larvae burrow under the potato epidermis, forming tunnels inside the tubers and piling brown or white frass outside of tunnel openings. The damaged potatoes are usually completely destroyed. In southwest areas of China with single-cropping systems, between 50% and 100% of tubers are damaged due to the long storage time of potato tubers and suitable temperatures for pest breeding (Anonymous 1990).

Annual Population Dynamics

The number of annual generations of potato tuberworm varies with the climate in different regions. In Chongqing, six to nine generations occur each year; six to seven generations occur in Changsha City in Hunan province; five to six generations occur in Guizhou and Yunnan provinces; and three to five generations occur in Shaanxi, Shanxi, and Henan provinces. Multiple generations overlap with each other. The pest does not have an obligatory diapause, developing normally in winter with suitable temperature, humidity, and food conditions. In South China, the potato tuberworm survives the winter in all life stages. In Henan, Shaanxi, Shanxi, and other provinces in North China, the tubermoth overwinters at the pupal stage.

In spring, the overwintering adults fly to potato plants and oviposit on exposed potato tubers, and the larvae eat leaves, stems and tubers after hatching. During the spring potato harvest, the pests enter warehouses and basements in infested tubers; some individuals move to tobacco, eggplant, and tomato, where they continue to cause significant damage. After the tobacco harvest, adult moths lay eggs on autumn potato crops and on tubers in storage. During the autumn potato harvest, larvae enter warehouses with the tubers; some individuals stay and overwinter on residual potato plants in the fields. The damage by potato tuberworm in storage is heavier than in the field, and the damage to tobacco crops is heavier than the damage to potato crops. Damage rates are the worst if the potato-growing season is in May or November, and if the storage period is from July to September (Xu 1993, Qiu 2000, Li *et al.* 2005).

Biological and Ecological Characteristics

The biology and ecology of this pest is covered in Chapter 6 of this book.

Control Methods

In order to control the spread of potato tuberworm several quarantine measures were implemented in China, including inspection and fumigation of transported potatoes and a ban on moving potatoes from areas where tuberworm is established.

Chemical control is the most common approach to suppressing tuberworms in China. Cymperator, cypermethrin, and abamectin are used during periods of adult abundance; chlorophos, malathion, dimethoate, and fenvalerate are usually used when larval damage is detected in the field. Empty warehouses are fumigated with methyl bromide, and the tubers are dipped in chlorophos or phoxin before storage.

In addition to chemical control, practicing sound agricultural measures also helps in reducing potato tuberworm numbers. Avoiding adjacent planting of tobacco and potato plants is important, as well as covering potatoes with soil in the later growth stage to avoid exposure of the tubers for adult oviposition. Timely cultivation is crucial in destroying pupae, and manual killing of larvae on the leaves is also recommended. Exposed tubers should be picked out and treated 1–3 days before harvest, good tubers should not be piled in the field overnight after harvest, and potato piles should be tightly covered with dry bran, rice husk, plant ash, etc., to prevent the oviposition of adults during storage. Timely removal and disposal of residual potato tubers in post-harvest fields reduces the number of overwintering pests.

Cutworms (*Agrotis segetum* (Schiffermueller), *Agrotis ipsilon* Rottemberg and *Agrotis exclamationis* (L.))

Taxonomic Position and Morphological Description

Taxonomic Position

A. segetum (Schiffermueller), *A. ipsilon* Rottemberg, and *A. exclamationis* (L.) all belong to Order Lepidoptera, Family Noctuidae.

Main Morphological Characteristics

A. segetum *Adult*: 14–19 mm long, with 32- to 43-mm wingspan, gray-brown to brown. At the front there is a blunt cone-shaped process with a depression in the center. Forewing brown, with extensively distributed small brown points; transverse line is a double curve but often unobvious; reniform spots, ring spots and sword spots obvious, circled by dark brown fine edge; other parts are yellowish-brown. Hindwing gray, translucent, with more than 40 wavy, curved longitudinal veins, 15 of which reach micropylar area; horizontal veins less than 15, forming mesh pattern (Qu 2011).

Egg: oblate, bottom flat. Newly laid eggs are white, then gradually darken and develop a pink ripple, eventually turning black right before hatching (Qu 2011).

Larva: 33–45 mm long, light brown with a brown head; granules on body surface unobvious and wrinkles on body densely distributed and light in color; pygidium has two large, yellowish-brown spots, disconnected at the center, with dense black specks; abdominal tergum with hairs, with the last two tergi slightly larger than the first two (Qu 2011).

Pupa: 16–19mm long, red-brown; tergum of abdominal segments 5–7 have 9–10 rows of very dense punctures; terminal abdominal segment with a pair of thick spines (Qu 2011).

A. ipsilon *Adult*: 16–23 mm long, with 42- to 54-mm wingspan, dark brown. Forewing divided into three sections by internal transverse line and external transverse line, having obvious reniform spot, ring spot, clavate spot, and two sword spots; hindwing gray and without markings (Xiang and Yang 2008a).

Egg: 0.5 mm long, hemispherical, having vertical and horizontal uplift patterns. Newly laid egg is milky, later developing red markings; mature, ready-to-hatch egg is grayish-black (Xiang and Yang 2008a).

Larva: 37–47 mm long, gray-black; body surface covered with points of variable sizes; pygidium yellowish-brown, having two dark brown vertical bands (Xiang and Yang 2008a).

Pupa: 18–23 mm long, reddish brown, glossy; punctures on the tergi of abdominal segments 5–7 are bigger than punctures on the pleurum of these segments; cremaster (hooked spine at the apex of the last abdominal segment of a pupae of Lepidoptera) has a pair of short spines (Xiang and Yang 2008a).

A. exclamationis *Adult*: about 16–18 mm long, with 36- to 38-mm wingspan; body gray, with head and thorax slightly brown-gray; jugular plate grayish-brown, with a black line; forewing gray to grayish brown, and costal and exterior margins of forewing in some individuals are slightly purple; transverse lines unobvious, and internal transverse line dark brown and wavelike; sword spot black; reniform spot large, brown with black margin; ring spot and rod spot obvious, especially the rod spot, thick and long, with black color and easy to identify; hind wing white and slightly brown, with light brown costal margin (Yang 1966) .

Egg: hemispherical, 0.75 mm long, with vertical and horizontal uplifted patterns on the surface; newly laid eggs white, but darken with maturity until turning grayish-black before hatching (Yang 1966).

Larva: mature larva about 30–40 mm long, narrowing at each end; head yellowish brown, without mesh pattern; body grayish yellow, covered with points of variable sizes, and with little wrinkles; dorsal line and sub-dorsal line brown, spiracular line unobvious; thoracic feet yellowish brown, abdominal feet pale yellow; spiracles black and oval (Yang 1966).

Pupa: 16–18 mm long, brown; maxilla, mesopedes and antenna stretch near the wing ends and expose the ends of metapedes; spiracles protruding; red-brown region near the front margin of fifth abdominal segments have many punctures of variable sizes, and posterior margin of puncture is unclosed; apex abdominis have two cremasters (Yang 1966).

Current Distribution in China

A. segetum is widely distributed in China. It has been reported from all provinces in China except Guangdong, Guangxi, and Hainan (He and Fu 1984).

A. ipsilon is distributed throughout China, but causes the most severe damage in the arid lands and hills of South China. In North China, the heavy damage caused by *A. ipsilon* mainly occurs in farmlands surrounding the coast, lakes and rivers, as well as in swampy areas and on irrigated lands (Xiang and Yang 2008a)

A. exclamationis is distributed in Inner Mongolia, Gansu, Ningxia, Xingjiang, Tibet, and Qinghai provinces in China. The most severe damage occurs in the Hexi Corridor in Gansu province, and in the southern and northern feet of the Tianshan Mountain in Xinjiang province. *A. exclamationis* often co-occurs with *A. segetum*, and it also has sporadic occurrences in northeast China and southern regions of Liangshan Mountain in Sichuan province (He and Fu 1984).

Host Plant

All three species are polyphagous with broad host ranges. The main host plants of *A. segetum* include vegetables belonging to the families Solanaceae, Leguminosae, Cruciferae, and Liliaceae, as well as spinach (*Spinacia oleracea* L.), lettuce (*Lactuca sativa* L.), Shepherd's purse (*Capsella bursa-pastoris* (L.)), fennel (*Foeniculum vulgare* Miller), sesame (*Sesamum indicum* L.), millet (*Setaria italica* (L.)), corn (*Zea mays* L.), sorghum (*Sorghum bicolor* (L.)), cotton (*Gossypium hirsutum* L.), tobacco, and other crops. *A. segetum* also causes damage to seedlings of various fruit trees.

Host plants of *A. ipsilon* include corn, cotton, soybean (*Glycine max* (L.)), cowpea, hyacinth bean (*Lablab purpureus* (L.)), kidney bean (*Phaseolus vulgaris* L.), tobacco, millet, sorghum , wheat (*Triticum aestivum* L.), barley (*Hordeum vulgare* L.), sweet potato (*Ipomoea batatas* (L.)), yam bean (*Pachyrhizus erosus* (L.)), Chinese wild yam (*Dioscorea opposita* Thunb.), ramie (*Boehmeria nivea* (L.)), hemp (*Cannabis sativa* L.), flax (*Linum usitatissimum* L.), lucerne (*Medicago sativa* L.), sugar beet (*Beta vulgaris* L.), oilseed rape (*Brassica campestris* L.), melons, and a variety of other vegetables. Medicinal plants, grasses, and tree seedlings are also often damaged by *A. ipsilon*. Various weeds are also important host plants of *A. ipsilon*.

The host plants of *A. exclamationis* include oilseed rape, radish (*Raphanus sativus* L.), potato, green onion (*Allium fistulosum* L.), sugar beet, lucerne, and flax.

Damage to Potato

The larvae of *A. segetum* and *A. ipsilon* bite stems of potato seedlings near the ground, leading to the death of the whole plant. Damage by these two pests results in poor sprouting in parts of the field, or, in more serious cases, destroys seed tubers.

The damage characteristics of *A. exclamationis* are different from *A. segetum* and *A. ipsilon*. Larvae of *A. exclamationis* often bore into potato tubers in the soil, feeding internally.

Annual Population Dynamics

A. segetum   *A. segetum* are univoltine in Heilongjiang, Liaoning, Inner Mongolia, and northern Xinjiang, have two to three generations in the Hexi district in Gansu province, and three generations in the southern Xinjiang and Shaanxi provinces. They generally overwinter as mature larvae in the soil. When the temperature rises from March to April, the overwintering larvae exit diapause and pupate at about 3 cm below the soil surface. The pupal stage lasts for 20–30 days. April and May are the months for peak occurrence of *A. segetum* in various regions (Xia and Ding 1989).

The larvae have six instars. In Shaanxi province, larvae of the first generation appear in mid-May to early June, larvae of the second generation in mid-July to mid-August, and larvae of the overwintering generation in mid-August.

In the Hexi district in Gansu province, *A. segetum* pupates in early to mid-April and emerges in late April; the larval stage of the first generation lasts for 54–63 days, and the larval stage of the second generation lasts for 51–53 days. Larvae hatched in the late stage of the second generation and in the early stage of the third generation mature in late August and begin overwintering from late September. In the Kuche district in Xinjiang province, *A. segetum* adults appear in late April. Larvae of the first generation hatch in early May and pupate in early June. Every year there are three adult occurrence peaks: late April to early May; early July; and late August (Anonymous 1990).

A. ipsilon   The number of generations of *Agrotis ipsilon* varies from north to south. *A. ipsilon* has two generations per year in Heilongjiang, three to four generations in Beijing, five generations in Jiangsu, and six generations in Fuzhou. In the northern regions, the overwintering populations of *A. ipsilon* are still unknown. It is assumed that adults appear in the spring as they migrate from other regions. In the Yangtze River Valley, *A. ipsilon* overwinter as mature larvae or pupae. In Guangdong, Guangxi, and Yunnan provinces, *A. ipsilon* reproduces and damages crops throughout the year without overwintering. Adults migrate to North China from the south in late February of the following year, with peak migration between mid-March and early April. Duration of the egg stage ranges from 4.5 to 10 days, depending on temperature. Small larvae first feed on weeds after hatching, then move to vegetable fields, causing damage to seedlings from the end of April to early May. Larvae of the first generation stop feeding and begin to pupate in mid-May. Adults of the first generation occur from late May to early June, and adults of the last generation occur in mid-October (Xiang and Yang 2008a).

A. exclamationis   In the Shache district in Xinjiang province and the Wuwei district in Gansu province in Northwest China, *A. exclamationis* has two generations every year, and overwinters as mature larvae in the soil (Wang and Dai 1966, Zhao *et al.* 1982).

In the Shache district in Xinjiang province, overwintering larvae pupate in mid- to late March, and peak pupation appears in late April. Overwintering adults start to emerge in mid-April, and peak occurrence is in early May. The adult stage lasts for about 2 months. Larvae of the first generation occur from early May to early July, and the damage period lasts for 2 months. Adults of the first generation appear from July to September. Larvae of the second generation mature in early to mid-October and then overwinter in the soil (Wang and Dai 1966).

Biological Characteristics
Habits of A. segetum

Adults are nocturnal and are most active on windless nights with high temperatures and high humidity; they are strongly phototactic and chemotactic. Adults need to feed on nectar before oviposition, and have high fertility rates. Adults prefer to oviposit on lower leaves of the plant, near the ground. Female fecundity is 300–600 eggs. Duration of the egg stage varies with temperature, generally lasting for 5–9 days. The first and second instars feed on the leaf buds of seedlings; the third instars burrow into stems near the ground and feed inside. Third instars begin to disperse. They normally hide in the soil near the roots of damaged crops or weeds during the day, and come out to feed at night. Mature larvae pupate about 3 cm below the soil surface (Qu 2011). The pupal stage lasts for 34–48 days at 14–15°C, and for 14–16 days at 23–24°C (Dong 1983).

Biological Characteristics of *A. ipsilon*

Habits Adults of *A. ipsilon* hide in soil cracks, weed gaps, eaves, and other shelters during the day, and come out at night. Adults have strong chemotaxis, and can be trapped using baits consisting of a mixture of sugar, vinegar and wine. They are also attracted to black-light lamps. Adults oviposit on clods, ground cracks, dry plant residue, exposed roots, and the lower surfaces of leaves of weeds and seedlings of many crop species, including potato. Larvae undergo six instars. First and second instars usually stay on the soil surface or on the lower surfaces of leaves. They are active during the day and at night, and do not burrow into the soil. Starting at the third instar, larvae burrow into the soil to about 1.6 cm below the soil surface. They come out to feed at night, biting roots and stems or dragging seedlings into their burrows. The fifth and sixth instars consume the most foliage, with the amount of damage caused in this period accounting for 95% of the total damage done over the entire lifespan. Larvae curl into a ring and play dead when startled. In the case of food shortages, larvae can move to a new host plant. Most mature larvae migrate to the dry soil at the ridge of the field or near weed roots, pupating 6–9 cm below the soil surface (Xiang and Yang 2008a).

Mating Behavior Xiang and Yang (2008b) studied the mating behavior of *A. ipsilon* under laboratory conditions (25 ± 1°C, RH70% ± 7%, L14 : D10). The results indicated that adults were not able to mate until 1 day after emergence, and that all males and females observed could mate several times. Mating activity was influenced by sex ratio. Few matings were observed when one male was confined with one female. The activity increased significantly when one male was confined with more than two females, or one female was confined with more than two males.

Effects of Temperature on the Development and Reproduction of *A. ipsilon* Xiang *et al.* (2009) studied the development and reproduction of *A. ipsilon* at seven different temperatures (16°C, 19°C, 22°C, 25°C, 28°C, 31°C, and 34°C) in the laboratory. The results showed that the growth rate of *A. ipsilon* increased and development duration shortened with the rising temperatures. The growth rate of eggs, larvae, and pupae began to slow down at 31°C. The female-to-male ratio declined with rising temperatures. At either low or high temperature extremes, adult longevity was shortened, and survival and fecundity declined. The optimal temperature for the development and reproduction of *A. ipsilon* was 25°C. The threshold temperature of development and the effective accumulated temperature for an entire generation of *A. ipsilon* were 10.74°C and 620.64 degree-days, respectively.

Effects of Mating on Longevity and Reproduction of *A. ipsilon* Xiang *et al.* (2010) studied the effects of mating on *A. ipsilon* in the laboratory. The unmated adults lived longer than mated ones, and the females lived longer than the males. The mated females had a shorter pre-oviposition period, a longer oviposition period, and a higher fecundity than the unmated females. Eggs laid by unmated females did not hatch. The fecundity of adults and the hatching rate of the larvae significantly increased with an increase in the number of matings.

Habits of *A. exclamationis*

Adults are phototactic. In the study by Wang and Dai (1966), female moths preferred feeding on flowers of sweet (*Iris ensata* Thunb). Females feeding on that species had higher fecundity compared to females feeding on the flowers of black locust (*Robinia pseudoacacia* L.), and Chinese cabbage (Wang and Dai 1966). Eggs are laid individually or in small groups, usually on lower leaves near the ground or directly on soil clods. Larvae of the second generation enter diapause. Non-diapausing larvae can still pupate, but less than half of such pupae successfully develop to adulthood. *A. exclamationis* often co-occurs with *A. segetum*. However, *A. exclamationis* withstands dry weather better than *A. ipsilon*, and causes heavier damage in dry areas.

Control Methods

Chemical control is the most commonly used approach to control cutworms in China. Foliar and soil sprays are used against early instars, and poisonous baits made of fried bran mixed with trichlorfon-treated trap plants (often paulownia, *Paulownia fortunei* (Seem.) or lettuce, *Lactuca sativa* L.) are used to kill larvae after the third instar. Adults can be attracted by a mixture of sugar, vinegar, and wine, and a fermentation liquor of sweet potato or carrot. Lambda-cyhalothrin and chlorfluazuron are commonly used as foliar sprays, and the commonly used pesticide in poisonous bait is trichlorfon. Phoxim or imidacloprid are often used to treat seed tubers before planting.

Good sanitation also helps in reducing cutworm numbers. Intensive plowing in spring, and clearing away weeds when early instars are present, can help in reducing populations of cutworms.

Grubworms (*Amphimallon solstitialis* Reitter and *Holotrichia oblita* (Faldermann))

Taxonomic Position and Morphological Description

Taxonomic Position

A. solstitialis and *H. oblita* belong to Order Coleoptera, Superfamily Scarabaeoidea, Family Melolonthidae.

Main Morphological Characteristics

A. solstitialis *Adult*: 14.2–17.4 mm long, 7.2–9.5 mm wide, with a medium-sized and narrow body. Head and venter dark chestnut-brown; clypeus, mouthparts, antenna, escutcheon, elytra, pygidium and legs light brown; pectoral plate chestnut-brown; pronotal disc and lateral area with three longitudinal yellowish-brown bands, disc dark chestnut-brown. Head densely covered with rough setiferous punctures, and front center often sunk into a short cannelure. Pronotum long and densely covered with setiferous punctures; lateral area of the disc has oblique bands formed by gray hairs, and with rim in each margin. Elytra narrow, densely covered with setiferous punctures, and less dense at the ends. Each abdominal segment covered with white hair, with the end of sternum smooth. Male sternum with obvious longitudinal groove; legs slender, with strong setae behind the femora of mesopedes and metapedes. Outer margin of protibia with three teeth (Zhao 2009).

Egg: The newly laid egg is milky in color and oval, and gradually turns subround and white. The average size is 2.5 ± 0.1 mm long and 1.9 ± 0.1 mm wide (Zhao 2009).

Larva: mature larva are 28–32 mm long, with 4.3–4.8-mm head width. Front vertex has two setae on each side, the postvertex has one seta on each side, and the midvertex has two to three setae on each side (Zhao 2009).

Pupa: light yellow to yellow, exorate, 20–25 mm long and 8–14 mm wide, with ventral bending. Pygal segment slender, with a pair of caudal horns at the terminal. Tuberculi on pygofer venter of male pupae obvious; venter of female pupae concave. The new pupa is white, and then gradually turns yellow (Zhao 2009).

H. oblita *Adult*: oblong oval, 21–23 mm long and 11–12 mm wide, shiny black or dark brown, with yellow long hair on thorax and abdomen. Pronotum twice as wide as it is long; costal angle and posterior angle almost at right angles. Each elytron has three carinae. Outer margin of protibia with three teeth, terminal of mesotibia and metatibia with two calcaria. Paratelum venter concave in the center in male and convex in female (Zhang *et al.* 2007).

Egg: oval and milky (Zhang *et al.* 2007).

Larva: 35–45 mm long, culus triradiate and slit-like, with a cluster of flat hamulus in front (Zhang *et al.* 2007).

Pupa: pupa yellowish white, oval, with a pair of protuberances in periproct (Zhang *et al.* 2007).

Current Distribution in China

The distribution of *A. solstitialis* in China ranges from Heilongjiang, Inner Mongolia, and Xinjiang in the north to Hebei, Shanxi, and Shaanxi in the south, and from Bohai Bay in the east to Qinghai and Tibet in the west. *H. oblita* is mainly found in the Hebei, Henan, Shandong, Shanxi, Shaanxi, Inner Mongolia, and Gansu provinces in China, but it has also been reported in Northeast, Southwest, and East China.

Host Plant

The main host plants of *A. solstitialis* include potatoes, oilseed rape, legumes, and other crops.

H. oblita mainly damages grains, corn, sorghum, potato, peanut (*Arachis hypogaea* L.), sugar beet, cotton, and other field crops, and it also causes damages to a variety of vegetables, fruit trees, and seedlings.

Damage to Potato

A. solstitialis and *H. oblita* feed on potato roots and stems, leading to the die-off of newly emerged plants. These two pests also eat potato tubers; the wounding of tubers makes potato plants more susceptible to fusarium and soft-rot fungi.

Zhang *et al.* (2007) found that, in serious cases, the density of *A. solstitialis* was up to 50.16 larvae per square meter, accounting for more than 30% of the total number of underground pests.

Annual Population Dynamics

A. solstitialis　*A. solstitialis* has one generation every 1–2 years in Bashang district in Hebei province, and overwinters as a larva. Pupation starts in mid-June of the following year, and adults emerge in July. Adults mate immediately after emergence and begin to oviposit 10 days after mating. Eggs hatch in about 13 days. Peak larval hatch is in mid-July, development to the third instar is completed in early September, and the third instars burrow into the soil to the depth of about 1 cm in mid- to late November. Larvae move to the soil surface from late April to early May in the following spring and cause damage to crops. A small number of larvae do not pupate in the following year and continue feeding throughout the next year. They overwinter once more at the third instar, start to pupate in mid-June of the third year, and then complete the entire life cycle (Tian 1958).

In the Ili Mountains in Xinjiang province, *A. solstitialis* has one generation every 2 years; generations overlap and overwinter as egg and larva. The overwintering eggs start to hatch in early April of the following year. The mature larvae begin to pupate in late May, and the pupation peak appears in early June. Adults emerge in mid-June; larvae hatch in mid- to late July, develop to the third instar in early September, and then overwinter in the soil. Larvae move to soil surface in late March of the following spring and cause damage to crops. A small number of larvae do not pupate in the following year. They overwinter once more at the third instar, start to pupate in mid-June of the third year, and then complete the entire life cycle (Zhang *et al.* 2007).

H. oblita　*H. oblita* has one generation every 2 years in Northwest, Northeast, and East China, and one generation each year in Central China and in Jiangsu-Zhejiang area. The pests overwinter as adults or larvae. The overwintering adults come out in approximately mid-April in Hebei province and overwinter in September; their activity period lasts for 5 months. Adults oviposit from late

May to mid-August. Larvae hatch starting in mid-June and continuously cause damage until December. Then they overwinter at the second or third instar. The overwintering larvae continue to develop and damage crops from the following April and start to pupate from early June. The pupation peak happens in late June; adults emerge at the beginning of July and overwinter in the soil successively. The overwintering adults come out in the spring of the third year. The life cycle of *H. oblita* in Northeast China is delayed by about half a month compared to that in Hebei province (Ma 1985).

Biological Characteristics

Habits of *A. solstitialis*

The grub stage of *A. solstitialis* is spent underground, feeding on roots and decaying organic matter. *A. solstitialis* appear to move very little in the soil, and the migration is not obvious. Mature larvae burrow 5–10 cm below the ground for pupation; the pupal period lasts for 15 days. After emergence, adults hide in the soil or plant roots, remaining motionless during the day, and males come out to search for females at sunset on sunny evenings. Adults often fly less than 1 m above the ground. Females release a sex hormone to attract males. Male and female copulate immediately after encounter, and the copulation lasts for 15–30 minutes. After copulation, the female digs into the soil 5–10 cm below the ground and begins ovipositing about 5 days later. Fecundity is 15–40 eggs that are scattered around the field. Adults do not feed and live for 10–25 days (Zhang *et al.* 2007).

H. oblita

Habits Adults hide in the soil during the day and come out at dusk; the activity peak is between 8 and 9 pm. Adults have feign death and are phototactic. After coming out of the soil, adults prefer to feed and copulate in bushes or weedy road-sides, and then oviposit in the soil nearby. As a result, the damage is most serious at field edges. Adults live for approximately 27 days, during which they copulate multiple times and have up to eight bouts of oviposition. Eggs are scattered in the moist soil 6–15 cm below the ground. A female lays 32–193 eggs (average: 102 eggs) that hatch in 19–22 days. Larvae have three instars. They are solitary, and often kill each other during encounters. Larvae often move and cause damage within crop rows, so it is easy to find them under newly damaged plants. In spring, larvae move up from the deeper soil when the temperature is about 10°C at 10 cm below the ground and feed at 5–10 cm below the ground when the temperature is about 20°C; in autumn, larvae move down to deeper soil when the temperature is below 10°C and overwinter at 30–40 cm below the ground.

Larvae stop feeding and move deeper in case of precipitation or irrigation. In flooded soil, larvae hide inside burrows and drown if submerged in water for more than 3 days. Mature larvae make pupation cells at 20 cm below the ground. The pupal stage lasts 15–22 days (Hu and Xu 1986).

Effects of Soil Moisture on Development and Reproduction of H. oblita Liu *et al.* (2008) found that *H. oblit* could oviposit when water content of the soil was 10–25 %, with the optimal water content for oviposition being 15–20%. The optimal water content for egg hatching and survival of newly hatched larvae was 18 –20%.

Dang *et al.* (2009) studied the influence of soil moisture on the growth and development of *H. oblit* in the laboratory. The results showed that soil moisture had different effects on different developmental stages. The optimal water content was 10–20% for eggs, 10–15% for larvae, and 15–20% for pupae. The sex ratio of the adult was close to 1 : 1 when the water content was 10–15%. Interestingly, the sex ratio became 1 female : 1.43 male when the water content was 25%, indicating that high water content is not conducive to female survival. The overall survival rate was the highest when the water content was 15%.

Zhou *et al.* (2009) studied the effect of soil moisture on the development of *H. oblit* in the laboratory. The results suggested that soil moisture had a great effect on egg hatching. The hatching rate was up to 75.6% when the soil moisture was 18%, but dropped to 2.1% at 6% soil humidity. No eggs hatched at 0% soil humidity. Soil moisture between 15%–18% was most favorable to the development of larvae, and the death rate increased at either higher or lower moisture levels.

Effects of Diet on Development and Reproduction of H. oblita Feeding on a wide range of diets, all adults of *H. oblita* have two oviposition peaks. Compared with peanut leaves, potato leaves, corn leaves, and wheat leaves, adults feeding under field conditions on elm (*Ulmus pumila* L.) leaves have the longest oviposition period, which lasts for about 90 days; the adults feeding on elm leaves have the highest fecundity, of 82.4 eggs/female (Dang *et al.* 2007).

Liu *et al.* (2008) conducted a laboratory study investigating the effects of various diets on the development and reproduction of *H. oblita* under laboratory conditions. The results showed that adults feeding on elm leaves had longer pre-oviposition periods and the longest life expectancies; adults feeding on oilseed rape had the shortest life expectancies and were almost infertile. Adults feeding on Japanese hop herb (*Humulus scandens* (Lour.) Merr) had the highest female fecundity, averaging 156.7 eggs, followed by the adults feeding on elm leaves, averaging 145.4 eggs. Eggs produced by adults feeding on peanuts and elm leaves had higher hatching rates of 91.7% and 90.2%, respectively; eggs produced by adults feeding on ryegrass (*Lolium perenne* L.) had the lowest hatching rate, which was 46.7%. The survival rates of the newly hatched larvae produced by adults feeding on different diets showed no significant difference.

Zhou *et al.* (2009) studied the effects of diets on the development and reproduction of *H. oblita* in the laboratory. Their results also suggested that female fecundity of *H. oblita* feeding on elm leaves was the highest (107.1 eggs/female), followed by *H. oblita* feeding on Chinese white poplar (*Populus tomentosa* Carr.). Female fecundity of *H. oblita* feeding on corn and black locust leaves was the lowest. The larvae showed a preference for potato tubers over peanuts and sweet potato tubers. The larvae feeding on potato tubers had the highest survival rate, and their weight was significantly higher than those feeding on peanuts and sweet potato tubers.

Control Methods

Most farmers rely on applications of the insecticides phoxim, quinalphos, rogor, and dipterex for suppressing populations of grubworms.

In addition to insecticides, initial populations of grubworms can be reduced by destruction of adults using insect-killing lamps, manual collection, and planting trap crops of castor-oil.

Cultural control approaches are also helpful in reducing grubworm numbers. These include destroying weeds and crop residues immediately after the fall harvest to reduce the oviposition of adults and the feeding of larvae; applying mature, well-decomposed organic fertilizers to reduce field attraction for adults; and deep plowing and careful cultivation during the larval stage of pests in order to kill the underground worms. Also, irrigation forces grubs to move deeper into the soil, thus reducing damage to more vulnerable small plants.

Oriental Mole Cricket *Gryllotalpa orientalis* Burmeister

Taxonomic Position and Morphological Description

Taxonomic Position

Class: Insecta; Order: Orthoptera; Family: Gryllotalpidae.

Main Morphological Characteristics

Adult: 30–35 mm long, grayish-brown with abdomen of lighter color, whole body covered with dense hairs (Fig. 7.4). Head conical with long filiform antenna. Pronotum oval, with distinct long, dark red, heart-shaped, concaved spots in the middle. Forewings (also known as tegmina) grayish-brown, short, extending to the middle of the abdomen; hind wings membranous, fan-shaped and longer than the terminal of abdomen. Terminal of abdomen with a pair of cerci. Front legs are fossorial (Luh and Hwang 1951).

Egg: 2.8 mm long when newly laid and 4 mm long before hatching, oval; the newly laid egg is milky, then gradually turns to yellowish-brown, and is dark purple before hatching (Luh and Hwang 1951).

Nymph: with 8–9 instars; 25 mm long at the last instar, and similar to the adult morphologically (Luh and Hwang 1951).

Current Distribution in China

G. orientalis is distributed extensively throughout China, causing heavier damage in the south than in the north.

FIGURE 7.4   Adult *Gryllotalpa orientalis*. See also Plate 7.4

Host Plant

The main host plants of oriental mole cricket include sugar beet, eggplant, pepper, potato, sweet potato, legumes, melons, corn, cotton, hemp (*Cannabis sativa* L.), and other crops. It can also damage carnation (*Dianthus caryophyllus* L.), lucky bamboo (*Dracaena sanderiana*), kumquat (*Fortunella margarita* Swingle), and other ornamental plants, especially the annual or biennial herbaceous flowers and cutting seedlings, which may suffer serious damage.

Damage to Potato

The oriental mole cricket damages potato both as a nymph and as an adult. It bites off young roots and tears underground stems and roots into filaments, causing wilting and sometimes death of vines above the ground. It also occasionally bites sprouts germinating on seed tubers, interfering with plant emergence in the spring. In addition, tunneling in the soil causes mechanical damage to plant roots. In the fall, the oriental mole cricket also feeds on potato tubers. Feeding holes lower tuber quality and create entry routes for fungal pathogens.

Annual Population Dynamics

Oriental mole crickets have one generation every year in southern China and one generation every 2 years in northern China. They overwinter as nymphs at different instars below the freezing horizon in the soil. Crickets move up to the ground surface in April–May of the following year and damage spring crops. Early May to mid-June is the most active period for oriental mole crickets, and it is also the first period of peak damage to potato crops. From late June to late August, oriental mole crickets oviposit in the soil, with June to July being the peak oviposition time. Starting in September, oriental mole crickets move to the lower soil layer as the temperature drops. They continue feeding, causing some damage to autumn crops. In October, they enter diapause (Gao 2009).

Biological Characteristics

Habits

The egg stage lasts for 15–18 days; the nymph stage, with 8 to 9 instars, lasts for 400 days; adult longevity is 8–12 months.

The oriental mole cricket is nocturnal, and its most active period is between 9 and 10 pm. In evenings with high temperatures and humidity, a large number of crickets come out of the soil. In early spring or late autumn with cool temperatures, oriental mole crickets are active only in the topsoil and not above the ground. They move down to deeper soil layers in the hotter hours around noon. Oriental mole crickets are phototactic, and show strong chemotaxis for substances with a fragrant and sweet smell, such as half-cooked millet, parched bean cake, wheat bran, horse manure, and other fertilizers.

Adults and larvae of oriental mole cricket prefer soft and humid loam soils or sandy loam soils. The optimal water content of 200-mm deep topsoil is above 20%, and cricket activities decrease when water content is below 15%. The optimal air temperatures and the optimal soil temperatures at a depth of 200 mm are 12.5–19.8°C and 15.2–19.9°C, respectively. Crickets move down to deeper soil when the temperature is too high or too low (Gao 2009).

Control Methods

Baits are often used to trap and kill this pest. Commonly used baits are made of horse manure or wetted fresh weeds mixed with dipterex or half-cooked wheat bran mixed with dimethoate.

Soil treatment with insecticides also helps in controlling this pest. In severely infested plots, phoxim mixed with fine soil is sprinkled on the soil surface before planting and then incorporated into the soil by shallow plowing.

Using the phototactic habit of this pest, black light lamps can be used to trap and kill emerging adults. The optimal weather for trap deployment is hot, cloudless, and windless, and the recommended density is one lamp for every 2 hectares.

Some cultural control approaches are also used to suppress the population of this pest. These include intensive cultivation, deep plowing followed by repeated harrowing, and applications of fully decomposed farmyard manure. Paddy-upland rotation is effective in destroying the habitat and oviposition sites of this pest.

Two-Spotted Leaf Beetle *Monolepta hieroglyphica* Motachulsky

Taxonomic Position and Morphological Description

Taxonomic Position

Class: Insecta; Order: Coleoptera; Superfamily: Chrysomeloidea; Family: Chrysomelidae: Subfamily: Galerucinae.

Main Morphological Characteristics

Adult: 3.6–4.8 mm long, 2.0–2.5 mm wide, oval, brownish yellow with gloss; head and pronotum of darker color, sometimes orange-red; elytra light yellow, with a sub-rotund light-colored spot near the wing base on each elytron; light-colored spot with black margins, its posterolateral part not completely closed; the black band behind the spot stretching to the posterior process, forming a horn; black band ambiguous or even completely absent in some individuals. Epipleura and scutellum usually black. Antennae 11-segmented, and two-thirds as long as its body length; the first to the third segments counting from the antennae base are brownish yellow; the second and the third segments are shorter than other segments, and the sum of length of the second and third segment is equal to the length of any other segment. The terminal parts of tibiae and tarsi are black. Pronesurface of mesothorax and metathorax is black; triangular frontal area of head slightly convex; frontal tubercle wide, with a stria with ultrafine punctures between two tubercles; compound eyes big, oval, and obviously convex. Pronotum wide, with ratio of length to width about 2 : 3; surface convex, covered with fine punctures, with hair in each corner. Scutellum triangular, without punctures. Elytra covered with dense and ultrafine punctures, and the lateral margin slightly convex; elytra with circular terminal when folded. Terminal of metatibia with a long spine; the first segment of metatarsus long, its length exceeding the sum of the lengths of the other three segments. Post margin of paratelum sternum with three lobes in male, but integral in female (Li 2008).

Egg: oval, and color associated with the diets of adults – adults lay brownish red eggs when feeding on cotton leaves, and light yellow or yellow eggs when feeding on Chinese cabbage, corn or elm leaves; about 0.6 mm long and 0.4 mm wide, with roughly equilateral hexagon reticular pattern on the shell surface (Li 2008).

Larva: elongated, light yellow when newly hatched, and gradually turning to yellow with the increase in instar. Larva have three instars; wide head capsule is 0.19–0.23 mm for the first instar, 0.29–0.32 mm for the second instar, and 0.42–0.45 mm for the third instar. Regularly arranged tuberculi and setae on body surface, with deep transverse folds on abdominal segments. Larval segments expand and contract considerably during movement. Newly hatched larvae about 2.0 mm; at the third instar about 10 mm long, and even up to 11.2 mm long and 1.2 mm wide; mature larvae shortened, thickened and slightly curved before pupation. Larvae with a pair of antennae on head, frontal suture and coronal suture distinct; terminal of mandibles narrow, with three denticles. Thorax 3-segmented, with a pair of legs on each segment; pronotum sclerotized and deeper in color; abdomen slightly flat and 9-segmented, obviously swollen from the third segment. Paratelum dark brown, as a spade-shaped sclerotized plate, with long hairs on the terminal. Spiracle 10 pairs, with 2 pairs on thorax and 1 pair on each abdomen segment. The spade-shaped sclerotized plate on paratelum of the two-spotted leaf beetle larvae is the key feature to distinguish it from other larvae (Li 2008).

Pupa: 2.8–3.5 mm long and about 2.0 mm wide, yellow, with setae on body surface. Front end of the body is pronotum, head beneath it; scutellum triangular; forewings and hind wings on the lateral side of the body, with forewings covering hind wings. Most part of metanotum visible. Abdomen nine-segmented, with a pair of particles on each segment from the first to the seventh segments; terminal of the ninth segment with a pair of slightly outward-curving spines. Head, legs, wings and some segments visible on the ventral side. Antennae stretch outward from between the two compound eyes (Li 2008).

Current Distribution in China

The two-spotted leaf beetle is distributed in Heilongjiang, Liaoning, Jilin, Inner Mongolia, Xinjiang, Ningxia, Gansu, Hebei, Shanxi, Shaanxi, Jiangsu, Zhejiang, Jiangxi, Fujian, Hubei, Hunan, Sichuan, Yunnan, Guizhou, Guangxi, Guangdong, and Taiwan provinces in China (Li 2008).

Host Plant

The two-spotted leaf beetle is a polyphagous insect, and the adults feed on cruciferous vegetables, corn, sorghum, sugarcane (*Saccharum officenarum* L.), Chinese jute (*Abutilon theophrasti* Medic), broomcorn millet (*Panicum miliaceum* L.), potato, cotton, millet, legumes, salicaceous plants, and many other plants. Its host plants also include wild plants in the genera *Viburnum, Rubus,* and *Pyrenacantha.*

Damage to Potato

Adults of two-spotted leaf beetle skeletonize potato leaves, seriously affecting photosynthesis.

Annual Population Dynamics

Two-spotted leaf beetles have one generation every year and overwinter as eggs under the topsoil. In Hebei and Shanxi provinces, larvae hatch in mid-May following the overwintering period and stay in the soil, damaging roots of gramineous crops and weeds. After 30 days, they make pupation cells in the soil and pupate inside. The pupal stage lasts for 7–10 days. Newly emerged adults inhabit roadside weeds and then migrate to grain fields. The population increases from early July, and the adult peak occurs from late August to early September. Adults emerge in mid- to late August, mating after feeding on supplemental nutrition. The pre-oviposition period lasts for more than 20 days, and the copulation and oviposition peak is in early September. Adults migrate to vegetable fields, including potatoes, after grain ripening in late September.

The two-spotted leaf beetle has one generation in Xinjiang province and overwinters as an egg. The overwintering eggs hatch in mid- to late April of the following year. Larvae complete their development by feeding on cotton, corn, potato, weeds, and other plants, and the larval stage lasts for about 30–40 days. Adults emerge from late May to early June, damaging crop leaves. The hatching time of overwintering eggs is positively correlated with field humidity (Li 2008).

For a small number of overwintering eggs, hatching is delayed until June or even July. Late June to early July is the peak time for adult occurrence in the field, and also for crop damage. Early to mid-June is the peak oviposition period of adults. Adult life expectancy is fairly long, and the adult population generally decreases in late August. A large number of adults die in late September when the leaves senesce and their nutritional quality deteriorates (Li 2008).

Biological Characteristics

Habits

Adults feed on plant leaves after emergence and copulate after 5–8 days. Eggs are laid individually or in small clutches. Adults lay about 40 or 50 eggs a day in the topsoil about 1–5 cm below the ground (Li 2008).

Females oviposit many times, and both females and males engage in multiple matings. The mean life expectancy of females is longer than that of males. At 25–28°C, the mean longevity of females is about 55–60 days, and the mean longevity of males is about 40–45 days (Li 2008).

Adult activity increases with increasing illumination. Adults are distributed unevenly in the field and usually form aggregations. Adults are weak fliers, as they generally can only fly for a distance of 2–5 m. They prefer hiding under plant roots of dry leaves in the mornings and nights, when temperatures are low. Feeding and mating activities are all performed during the daytime (Li 2008).

Larvae have three instars and live in the topsoil. They are small, photophobic, and mainly feed on roots of crops and weeds to complete their development. Larval feeding is usually insufficient to cause serious crop damage (Li 2008).

Mature larvae make pupation cells in the soil, and the pupation is preceded by a pre-pupal stage. The body segments of mature larva contract before pupation, and the body gradually curls towards the ventral side, eventually forming a shortened and widened pupa (Li 2008).

Effects of Temperature on the Development of Two-Spotted Leaf Beetle

Li *et al.* (2008) studied the temperature-dependent development of two-spotted leaf beetle. The threshold temperatures of egg, larva, pupa, and pre-oviposition adult were estimated to be 9.8, 10.8, 12.6, and 10.1°C, respectively, and the effective accumulated temperatures were 1182.2, 401.2, 111.9, and 269.0 degree-days, respectively. The effective accumulated temperature required to complete the whole generation was 1971.6 degree-days.

Li *et al.* (2007, 2010) studied the effects of temperature on the egg development, adult longevity, and female fecundity of the two-spotted leaf beetle at five constant temperatures (19, 22, 25, 28, and 31°C). They found that temperature had significant effects on egg development and adult longevity. The egg stage and adult longevity were significantly shortened as the temperature increased. Temperature also had a significant effect on the pre-oviposition period, oviposition period, and fecundity of adults. The pre-oviposition period was shortened as the temperature increased, and the oviposition periods under 19, 22, and 25°C were significantly longer than under the other two temperatures. Adult fecundities under 22 and 25°C were significantly higher than under the other tested temperatures.

Control Methods

Most farmers rely on spraying with the insecticides fenvalerate and phoxim. The best time period for insecticide applications is during the peak occurrence of adults before oviposition. In addition to insecticides, light traps are used to attract and kill adults.

Cultural control approaches to reducing pest numbers include deep plowing in the autumn to eliminate pest eggs, practicing good weed control to eliminate alternative plant hosts, and collecting adults by hand or using modified cultivators with attached sweep nets.

REFERENCES

Anonymous, 1990. Agriculture Cyclopedia of China, Volume Insect. Chinese Agricultural Press, Beijing, China.

Casagrande, R.A., 1985. The "Iowa" potato beetle, its discovery and spreads to potatoes. Bull. Entomol. Soc. Am. 31, 27–29.

Chen, Y.L., Qu, D.Y., 2007. Progress of potato cultivation and processing. Harbin Engineering University Press, Harbin, China.

Chen, J.B., 1937. Observation of the tobacco moth and its prevention in Guangxi. Interesting Insects. 2, 15.

Chen, L.F., Lu, Z.Q., Zhu, S.D., 1989. Biology and effective accumulated temperature of *Henosepilachna vigintioctopunctata*. Plant Prot. 15, 7–8.

Chen, B., Li, Z.Y., Gui, F.R., Sun, Y.X., Yan, N.S., 2003. Integrated control of potato pest insects in Yunnan province. Yunnan Agric. Sci. Technol. 1, 136–141.

Cui, N.Z., Bai, X.E., Gao, Y.C., Han, Y.G., 2007. Infection law and the control of 28-star ladybird in potato. J. Shanxi Agric. Sci. 35, 77–79.

Dang, Z.H., Gao, Z.L., Li, Y.F., Pan, W.L., 2007. Effect of diet on adult fecundity of *Holotrichia oblita*. In: Cheng, Z.M. (Ed.), Plant protection and modern agriculture: proceedings of annual conference of China plant protection society, China Agricultural Science and Technology Press, Beijing, China, pp. 394–395.

Dang, Z.H., Li, Y.F., Gao, Z.L., Pan, W.L., 2009. Influence of soil moisture on growth and development of *Holotrichia oblita*. Chinese Bull. Entomol. 46, 135–138.

Dong, J.T., 1983. Studies on the biology of *Agrotis segetum*. Entomol. Knowl. 20, 17–20.

Dong, F.L., Guo, Z.Q., Ma, G.Y., 2007. Occurrence characteristics and integrated control methods of potato pests and diseases in Guyuan City. Chinese Potato J. 21, 238–239.

Duanmu, T.Z., Li, Z.G., Zhang, Z.Y., Li, D.X., Lei, T.S., Ma, X.L., 1995. Life history and developmental biology of potato ladybird. J. Luoyang Agric. Coll. 15, 7–10.

Editorial Committee of China agriculture yearbook, 2009. China agriculture yearbook- 2009. Chinese Agricultural Press, Beijing, China.

Gao, K., 2009. Characteristic, function, mechanics and bionic analysis of oriental mole cricket (*Gryllotalpa orientalis* Burmeister). Ph.D. thesis Jilin University, Jilin, China.

Guo, W.Y., Yu, X.T., Zhou, C.M., 2005. Occurrence and the control of potato ladybird. Sci. Technol. Tianjin Agric. For. 6, 24–25.

Hao, W., Lu, M., Jiang, X.L., Chao, G.D., Cao, X.F., 2006. Study on biological characteristics of potato ladybird. Chinese Plant Prot. 26, 22–23.

He, J.L., Fu, T.Y., 1984. Description of eight cutworm larvae. J. Shanghai Agric. Coll. 2, 41–51.

Hu, H.Y., Xu, C.Y., 1986. Studies on biological characteristics of *Holotrichia oblita*. J. Anhui Agric. Sci. 4, 69–70.

Hu, J., 2008. Occurrence and control methods of potato tuberworm. Plant Doctor 21, 46.

Katakura, H., 1980. Classification and evolution of the phytophagous ladybirds belonging to *Henosepilachna vigintioctomaculata* complex (Coleoptera, Coccinellidae). PhD thesis Hokkaido University, Japan.

Katakura, H., 1986. Evidence for incapacitation of heterospecific sperm in female genital tract in a pair of closely related ladybirds (Insecta, Coleoptera, Coccinellidae). Zool. Sci. 3, 115–121.

Katakura, H., Nakano, S., 1979. Preliminary experiments on the crossing between two puzzling phytophagous ladybirds, *Henosepilachna vigintioctomaculata* and *H. pustulosa* (Coleoptera, Coccinellidae). Jpn J. Entomol. 47, 176–184.

Katakura, H., Sobu, Y., 1986. Cause of low hatchability by the inter specific mating in a pair of sympatric ladybirds (Insecta, Coleoptera, Coccinellidae): Incapacitation of alien sperm and death of hybrid embryos. Zool. Sci. 3, 315–322.

Li, G.W., 2008. Studies on biological and ecological characteristics and integrated control of *Monolepta hieroglyphica* Motschulsky. M.Sc. thesis Shihezi University, Shihezi, Xinjiang.

Li, G.W., Zhang, J.P., Chen, J., Liu, J., 2007. Threshold temperature and effective accumulated temperature to complete development for the eggs of *Monolepta hieroglyphica*. J. Shihezi Univ. 25, 703–705.

Li, G.W., Zhang, J.P., Chen, J., Liu, J., 2008. Threshold temperature and effective accumulated temperature of *Monolepta hieroglyphica*. Chinese Bull. Entomol. 45, 621–624.

Li, G.W., Chen, X.L., Zhang, J.P., Chen, J., 2010. Effect of temperature on adult longevity and fecundity of *Monolepta hieroglyphica*. Chinese Bull. Entomol. 47, 322–325.

Li, T.J., Li, Y., Wang, X.H., 2001. Occurrence and control of *Henosepilachna vigintioctopunctata* (Fabricius) on potato in Youyang County. Southwest Chin. J. Agric. Sci. 14, 90–91.

Li, X.J., Jin, X.P., Li, Z.Y., 2005. The present status and developing tendency in Phthorimaea oper-culella research. J. Qinghai Norm. Univ. 21, 67–70.

Lin, S.J., 2001. Occurrence and the control of potato ladybird. Agric. Technol. 21, 55–57.

Liu, C.L., 1963. Economic insect fauna of China. Fasc. 5. Coleoptera: Coccinellidae. Science Press, Beijing, China.

Liu, C.Q., Li, K.B., Zhang, P., Xi, G.C., Feng, X.J., Li, H.J., Wang, Q.L., Cao, Y.Z., 2008. Effect of diet and soil moisture on adult longevity and fecundity of *Holotrichia oblita*. In: Cheng, Z.M. (Ed.), Innovation and development of plant protection technology: proceedings of annual con-ference of China plant protection society, Beijing: China Agricultural Science and Technology Press, Beijing, China, pp. 682–686.

Luh, P.L., Hwang, K.H., 1951. The External Morphology of the Mole-cricket, *Gryllotalpa afrlcana* de Beauvois (Orthoptera: Gryllotalpidae). 1. The Head and Feeding Mechanism. Acta Entomol. Sin. 1, 308–320.

Ma, L., Yin, Z.H., 2001. Preliminary study on occurring regulation of *Henosepilachna vigintioc-tomaculata* in Yulin area and its economic thresholds. Plant Prot. Technol. Ext. 21, 14–15.

Ma, P.S., 1985. Life history and control method of *Holotrichia oblita*. J. Shanxi Agric. Sci. 5, 24–25.

Ma, Y.F., Li, Z.Y., Ren, M.J., Li, R.R., Xiao, C., 2010a. Ovipositional choices of potato tuber moth, *Phthorimaea operculella* Zeller for host plants. Agrochemicals 49, 380–382.

Ma, Y.F., Xu, Y., Li, N., Li, Z.Y., He, Q.Y., Xiao, C., 2010b. Effect of larval density on growth, development and reproduction of potato tuber moth, *Phthorimaea operculella*. Chin. Bull. Entomol. 47, 694–699.

Qi, H.S., 1997. Damage caused by 28-spotted ladybird is becoming increasingly severe. Plant Prot. Technol. Ext. 17, 43.

Qin, Y.X., Cui, N.Z., Bai, X.E., Gao, Y.C., 2006. Occurrence and integrated control of potato lady-bird in northern Shanxi. Chinese Plant Prot. 26, 27–28.

Qiu, N.X., 2000. Occurring regulation and integrated control methods of potato tuberworm. Plant Doct. 13, 29.

Qu, N.H., 2011. Observation on biological characteristics of *Agrotis segetum* (Schiffermueller) and its control in western Liaoning. Jilin Agric. 9, 74.

Song, G.H., Wu, W.W., Zhao, Q.L., 2008. Infection law and the control of 28-spot ladybird. Jilin Veg. 1, 50.

Sun, Y.P., 2008. Current occurrence status and integrated control methods of potato pests and dis-eases in Yuzhong County. Gansu Sci. Technol. 24, 141–143.

Tang, L., 2002. The introduce of crop varieties and the mergence of national economy. Study Natl. Guangxi 4, 96–105.

Tian, F., 1958. Preliminary observation of the life history of *Amphimallon solstitialis* in Bashang district. Entomol. Knowledge 4, 281–282.

Wang, G.H., 2002. Effects of eggplant varieties on growth and development *of Henosepilachna vigintioctopunctata* and its parasitoid. Entomol. Knowledge 39, 373–376.

Wang, J.R., Dai, S.H., 1966. Studies on the bionomics of *Euxoa exclamationis* (L.) (Lepidoptera, Noctuidae). Acta Entomol. Sin. 15, 120–130.

Wu, D., Wang, H.P., 2008. Preliminary study on the rearing potato ladybird *Henosepilachna vigin-tioctomaculata* with potato tubers. J. Henan Agric. Sci. 4, 75–76.

Xia, Z.X., Ding, F.L., 1989. Occurrence and integrated control measures of *Agrotis segetum*. Chin. Cotton 2, 47–48.

Xiang, Y.Y., Yang, M.F., 2008a. Mating behavior and ability of the black cutworm moth, *Agrotis ipsilon*. Chinese Bull. Entomol. 45, 50–53.

Xiang, Y.Y., Yang, M.F., 2008b. Occurrence and control techniques of the black cutworm *Agrotis ipsilon* in China. J. Anhui Agric. Sci. 36, 14636–14639.

Xiang, Y.Y., Yang, K.L., Liao, Q.R., Yang, M.F., Li, Z.Z., 2009. Effects of temperature on the development and reproduction of the black cutworm. J. Anhui Agric. Univ. 36, 365–368.

Xiang, Y.Y., Yang, M.F., Li, Z.Z., 2010. Effects of mating on longevity and reproduction in the black cutworm moth. Sichuan J. Zool. 29, 85–86.

Xiong, J., 1991. Study on the development threshold temperature and the effective accumulated temperature of 28-spot ladybird. Chinese Potato J. 5, 175–178.

Xu, G.G., 1985. Potato tuberworm. Institute of Plant Protection, Chinese Academy of Agricultural Sciences, Beijing, China.

Xu, S.Y. Control methods of tobacco pests. Henan Science and Technology Press, Zhenzhou, China.

Yang, W., 1966. Brief observation of *Agrotis exclamationis* in Inner Mongolia. Entomol. Knowledge 4, 56–57.

Yao, X.L., Li, C.Q., Gao, Z.L., Lu, H.M., Lan, L., Liu, H., 1992. Preliminary observations on biology of potato ladybird. J. Shaanxi Agric. Sci. 5, 28–29.

Zhai, Q.X. Preliminary exploration of the introduction time of potato in China. Agric. Hist. China 20: 91–92.

Zhang, Q., Wumaer, B.K., Airan, T.J., Lou, S.Q., Li, H., 2007. Preliminary study on biological characteristics of *Amphimallon solstitialis* Reitter in Ili Mountain prairie. Xinjiang Anim. Husb. S1, 59.

Zhang, Y.C., Liu, H.J., Zhen, Z.M., 2002. Ultrastructure of *Henosepilachna vigintioctomaculata* and *H. vigintioctopunctata*. Entomol. Knowledge 39, 132–135.

Zhang, Z.Y., 1997. Effects of temperature and humidity on the development of 28-spot ladybird *Henosepilachna vigintioctomaculata* (Coleoptera, Coccinellidae). Acta Agric. Sin. 6, 30–34.

Zhang, Z.Y., Lei, T.S., Li, D.X., Ma, X.L., 1993. Preliminary study on the host plants of potato ladybird *Henosepilachna vigintioctomaculata* (Coleoptera, Coccinellidae). Chinese Potato J. 7, 96–99.

Zhao, J.S., 2009. The research history of scarab beetles classification system and classified identification of 18 common scarab beetles. Inn. Mong. Agric. Sci. Technol. 5, 68–70.

Zhao, Z.J., He, C.N., Zhang, Y., 1982. *Euxoa exclamationis* (L.) (Lepidoptera, Noctuidae). Plant Prot. 6, 13.

Zhou, H.X., Tan, X.M., Li, C.Y., Zhen, G.L., Li, G.X., 2009. Effect of nourishment and humidity on the development and reproduction of *Holotrichia oblita*. Acta Agric. Boreali Sin. 24, 201–204.

Zhuang, H.D., 2010. Study on biological characteristics and chemical control of *Henosepilachna vigintioctomaculata* (Mots.). MS thesis Heilongjiang Bayi Agricultural University, China.

Zhuang, H.D., Sun, Q., 2009. Life tables of natural population of *Henosepilachna vigintioctomaculata* (Mots.). Jiangxi Plant Prot. 32, 103–106.

Chapter 8

Insect Pests of Potato in India: Biology and Management

R.S. Chandel[1], V.K. Chandla[2], K.S. Verma[1], and Mandeep Pathania[1]

[1]*Department of Entomology, Himachal Pradesh Agriculture University, Palampur, Himachal Pradesh, India,* [2]*Division of Plant Protection, Central Potato Research Institute, Shimla, Himachal Pradesh, India*

INTRODUCTION

The potato is the world's fourth most important food crop after rice, wheat, and maize. During the triennium ending in 2006–2007, India was the third largest potato producer (24.61 million tonnes) after China (71.09 mt) and Russia (37.55 mt), and followed by Ukraine. Scientific advances made by Indian scientists have led to higher average potato productivity in India than in those three countries. India produces 7.72% of the world's potatoes from 7.57% of the total global potato-growing area, with productivity levels higher than the world's average (Rana 2011).

In India, potato is cultivated in almost all states under very diverse agroclimatic conditions. On the basis of geographical variability and climatic differences, the potato-growing areas of India have been divided into six zones (Table 8.1). More than 85% of India's potatoes are grown in the vast Indo-Gangetic plains of north India during short winter days from October to March. The states of Uttar Pradesh, West Bengal, and Bihar account for more than 75% of the potato-growing area in India and for about 80% of total production. Hilly areas, where the crop is grown during the summer from April to September, account for less than 5% of production. In the plateau regions of south-eastern, central, and peninsular India, which constitute about 6% of the potato-growing area, potato is mainly a rain-fed crop or is irrigated as winter crop. In the Nilgiri and Palni hills of Tamil Nadu, the crop is grown year-round under both irrigated and rain-fed conditions. Most potatoes are produced by large-scale commercial farmers (Pandey and Kang 2003).

Insect pests cause variable and complex problems for potato farmers. India has a great diversity of insect pests that attack potato. Some of these insects were transported to new locations with seed tubers; others were already present in locations where potato was introduced and expanded their host ranges to take

Insect Pests of Potato. http://dx.doi.org/10.1016/B978-0-12-386895-4.00008-9

TABLE 8.1 Potato-Growing Zones in India

Zones	States	Crop seasons
North-Western plains	Punjab, Haryana, Rajasthan	September–November October–January/February January–April/May
West-Central plains	Madhya Pradesh, West-Central Uttar Pradesh, North-Western Gujrat	October–January/February
North-Eastern plains	Assam, Bihar, Jharkhand, West Bengal, Orissa, Eastern Uttar Pradesh, North-Eastern Madhya Pradesh, Eastern Chattisgarh	November–February/March
Plateau region	Maharashtra, Karnataka, parts of Gujrat, Madhya Pradesh, and Orissa	June/July–September/October November–January/February
North-Western and Central hills	Jammu & Kashmir, Himachal Pradesh, and Uttrakhand	April–September January–May
North-Eastern hills	Meghalaya, Manipur, Mizorum,Tripura, Nagaland, Arunachal Pradesh	January–May August–December
Southern hills	Nilgiri and Palni hills of Tamil Nadu	April–September August–December January–May

advantage of the new plant. Because potato crops are vegetatively propagated from tubers, which easily carry some pathogens and pests, many pest problems have followed potatoes to areas where they are grown (Chandel *et al.* 2007). These pests can damage potato plants by feeding on leaves, reducing the photosynthetic area and efficiency by attacking stems, weakening plants and inhibiting nutrient transport, and by attacking the potato tubers destined for consumption or for use as seed (Chandel and Chandla 2003). In India, approximately 60 billion rupees (US$1.2 billion) worth of potato tubers are lost annually due to pest damage, which accounts for 10–20% of total production. The annual demand for pesticides in India is approximately 80,000 tonnes and is likely to increase in coming years. However, productivity trends indicate that heavy application of insecticides will not proportionately increase crop productivity (Misra *et al.* 2003).

Potato is one of the most input-intensive crop and is the heaviest user of chemical pesticides of all major food crops. Pesticide consumption in potato can often reach up to 20% of its cost of production (Anonymous 1991). The potato pests are grouped into soil pests, foliage feeders, sap feeders, and storage pests. In seed production, the pests of greatest concern are typically aphid vectors of potato viruses, especially *Myzus persicae* (Sulzer). In ware production, the key pests may be insects which attack tubers, such as tuber moth, white grubs, and cutworms. In some situations, foliage feeders such as noctuid moths and coccinellid beetles are also important.

ROOT AND TUBER-EATING PESTS

Soil insect pests pose one of the most difficult problems for potato growers. To a large degree, the difficulty can be attributed to the very persistent nature of these pests, coupled with the fact that new insecticides are less effective in the soil. Moreover, these organisms often go unnoticed for several years, building up their numbers slowly with each successive potato crop. In India, there are many pests which damage potato roots and tubers inside the soil. Those of major concern are cutworms, white grubs, and potato tuber moth. These pests cause significant economic losses, although they rarely cause substantial damage in one field during a single season. Secondary infection from various diseases can follow, further rendering tubers unfit for marketing. Geographic location, soil characterization, and production practices usually favor specific pests. In this chapter the important soil pests are discussed individually, as well as specific management practices that are effective for each pest.

White Grubs

White grubs (Coleoptera: Scarabaeidae) are most destructive and troublesome soil insects, threatening potato production in hilly states. These soil-dwelling larvae of scarab beetles are present in the soil at a depth of 5–20 cm during the crop season (Chandla 1985). Being polyphagous both in grub and adult stages, they feed on a wide variety of cultivated and uncultivated plants. After hatching, the young grubs orient themselves toward roots and start feeding (Musthak Ali 2001). Almost all field crops grown during rainy season – i.e., potato, vegetables, groundnut, sugarcane, maize, pearl millet, sorghum, cowpea, pigeon pea, green grass, cluster bean, soybean, rajmash, upland rice, ginger, etc. are damaged (Mishra 2001). In potato the damage is only caused by grubs, which feed on rootlets, roots, and tubers. The first-instar grubs can survive on the organic matter present in the soil, but roots are preferred and are fed upon when encountered (Mehta *et al.* 2010). They often remain unnoticed but may suddenly increase in population in places with enough food and with low disturbance of the soil.

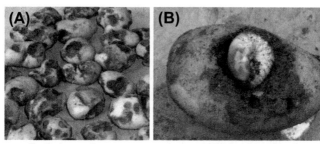

Brahmina coriacea   *Melolontha indica*

FIGURE 8.1   Tuber damage by white grubs; (A) *Brahmina coriacea*; (B) *Melolontha indica*.

The damage to potato is mainly caused by the second and third instars, which make large, shallow, and circular holes in tubers. In cases of heavy infestation, an entire tuber can be transversed by deep tunnels (Figs. 8.1A & B). Tubers infested by the white grubs have poor market value and are sold at a highly reduced price. The white grubs damage the tubers without causing any symptoms on the foliage; thus, farmers remain unaware of the damage done to tubers until harvest. In India, 20 species of white grubs have been reported on potato (Table 8.2). Of these, *Brahmina coriacea* (Hope), *Holotrichia seticollis* Moser, *Holotrichia longipennis* (Blanchard), *Anomala dimidiata* Hope, and *Melolontha indica* Hope are most destructive in the north-western hills. *Holotrichia serrata* (Fab.) damages potato in Karnataka (Misra and Chandel 2003).

Biology of White Grubs

Most of the white grubs are similar in shape and color and have fleshy, curved bodies with brown heads and well-developed legs which are hardly used for locomotion (Mehta *et al.* 2010). Adult beetles usually remain unnoticed throughout the year, and their appearance in large numbers occurs just after break in a monsoon. The beetles' attack on potato persists for a month or two. Most of the beetle's life cycle is spent in the larval stage underground (Chandel and Kashyap 1997). The beetles spend the winter in the soil as larvae in hard earthen cells.

Brahmina coriacea

B. coriacea was first reported in India from the Kullu valley of Himachal Pradesh feeding on pear, apple, plum, fig, and grapevine (Beeson 1941). Sharma *et al.* (1969) observed *B. coriacea* adults defoliating peach, plum, apricot, and pear in the mid-hills of Himachal Pradesh. It has become a serious problem in the north-western Himalaya, which comprise Himachal Pradesh, Uttrakhand, and Jammu & Kashmir (Mehta *et al.* 2010).

In the spring, when apples and other fruits have produced leaves, the adults become active, disperse by flight at night, and feed on the foliage of apples and other plants (Chandel and Kashyap 1997). They leave the soil at dusk and

TABLE 8.2 Different Species of White Grubs Damaging Potato in India

Species	Place of occurrence	Reference(s)
A. Subfamily: Melolonthinae		
1. *Brahmina coriacea* (Hope)	Himachal Pradesh	Butani and Jotwani (1984) Misra and Chandla (1989) Chandel *et al.* (1997)
2. *B. flavoserica* Brenske	Himachal Pradesh	Mehta *et al.* (2008)
3. *Melolontha indica* Blanch.	Himachal Pradesh	Bhalla and Pawar (1977)
4. *Holotrichia longipennis* Blanch.	Himachal Pradesh & Uttranchal	Butani and Jotwani (1984), Haq (1962), Misra and Chandla (1989); Rai and Joshi (1988)
5. *Holotrichia repetita* Sharp	Karnataka	Veeresh *et al.* (1991)
6. *H. rustica* Burmeister	Karnataka	Veeresh *et al.* (1991)
7. *H. serrata* (F.)	Karnataka	Veeresh *et al.* (1991), Butani and Jotwani (1984)
8. *H. conferta* Sharp	South India	Butani and Jotwani (1984)
9. *H. excisa* Moser	Tamil Nadu	Regupathy *et al.* (1997)
10. *H. nototiocollis*	Tamil Nadu	Regupathy *et al.* (1997)
11. *Holotrichia* sp.	North-Eastern India	Anonymous (1989)
12. *H. seticollis* Moser	Himachal Pradesh & Uttranchal	Chandel *et al.* (1997) Sushil *et al.* (2006)
B. Subfamily: Rutelinae		
13. *Anomala dimidiata* Hope	Himachal Pradesh	Misra and Chandla (1989)
14. *Anomala polita* (Blanch.)	Himachal Pradesh	Misra and Chandla (1989)
15. *A. rugosa* Arrow	Himachal Pradesh	Misra and Chandla (1989)
16. *A. rufiventris* Redt.	Uttar Pradesh	Rai and Joshi (1988)
17. *Anomala* sp.	Karnataka	Lingappa and Giraddi (1995)
18. *A. communis* Brenske	Tamil Nadu	Regupathy *et al.* (1997)
19. *A. nathani* Frey	Tamil Nadu	Regupathy *et al.* (1997)
C. Subfamily: Dynastinae		
20. *Phyllognathus dionysius* F.	Himachal Pradesh	Misra and Chandla (1989)

FIGURE 8.2   Fully-fed grubs of *Brahmina coriacea.*

remain on the leaves during night, mating and feeding. Mating normally lasts for 7–11 minutes. The pre-oviposition period ranges from 2–4 days, with an average of 3.14 days. Mated females under laboratory conditions lay 10–31 eggs, with a mean of 20 eggs per female (Chandel *et al.* 1995). At the first streaks of dawn the adults return promptly to the soil, where the females lay their pearly white eggs. The eggs are generally laid in grassland or on patches of grassy weeds in cultivated fields. The eggs hatch in 9–12 days, and the young grubs feed on the roots of grasses until they are about 11 mm long. After hatching, the grubs burrow into the soil.

There are three distinct larval instars. Larvae are about 5–6 mm long following hatching, and attain a length of about 11.6, 20.7, and 29.9 mm by the end of instars 1, 2, and 3, respectively. The head capsule width is about 1.5, 2.4, and 4.3 mm, respectively. In the mid-hills (650–1800 m), development of the first- and second-instar grubs requires about 14.4 and 20 days (Chandel *et al.* 1995). In higher hills (1800–2000 m), the first and second instars are completed in 20.1 and 29.6 days, respectively (Chandla *et al.* 1988). The third-instar grubs (Fig. 8.2) are most active, and have immense capacity for damage. These grubs continue feeding until October and are therefore responsible for heavy damage. Most of the third-instar grubs attain their full size in approximately 50 days before the onset of winter (Misra and Chandel 2003).

During the later part of the rainy season, the white grubs stop feeding, construct earthen cells, remain inside these cells until spring, and then metamorphose into pupae. Pupation takes place in March and April. Ecdysis from pupa to adult occurs in earthen cells in 20 days (Chandel and Kashyap 1997). The pupa is about 17 mm long and 7 mm wide, and is light brown to creamy in color. The pupa generally resembles the adult, although the wings are short and twisted towards the ventral surface. Adults remain confined in these cells for some time. With the onset of rain, the earthen cells are softened, and the beetles emerge from these cells. Adults are black beetles (Fig. 8.3) with an average length and width of 13–15 mm and 7–8 mm, respectively. They are commonly found on apple. Maximum adult emergence takes place in mid-June (Chandel *et al.* 2003). The adult longevity ranges from 32 to 46 days for females, and from 17 to 44 days for males (Chandel *et al.* 1995).

FIGURE 8.3   Adult beetles of *Brahmina coriacea.*

FIGURE 8.4   Adult beetles of *Holotrichia longipennis.*

Holotrichia longipennis

Beetle emergence begins during the first fortnight of May. The beetles (Fig. 8.4) have been reported to feed upon the foliage of a wide variety of fruit/forest trees, but *Rubus ellipticus* Smith, apple, walnut, chestnut, and plum are its preferred hosts (Shah and Shah 1990). Eggs are laid singly inside earthen cells, and the incubation period is 11–18 days. The young grubs soon begin to feed on rotten organic matter in soil, and grow rapidly. The first instar stage lasts for 38–44 days. Two molts occur during the rainy season, and the second instar can be seen damaging tubers by the end of July. The full-grown grubs measure 38.12 mm in length. They cease to feed, move 15–20 cm deep into the soil, and construct earthen cells for overwintering by mid-November. They can go as deep as 2.5 m into the soil to overwinter. The total larval period is 243–282 days (Shah and Shah 1990). The diapausing grubs start their upward movement by the end of March, and pupate at a soil depth of 20–30 cm. The pupal period lasts from 22–28 days. During the first week of May, almost all the pupae are transformed into adults (Mishra and Singh, 1993). Adult beetles live for 28–56 days.

Holotrichia seticollis

H. seticollis is an important species in the hilly tracts of Uttrakhand (Yadava and Sharma 1995). Chandel *et al.* (1994a) also reported this species from Himachal Pradesh. The grubs cause damage to all rainy season crops. Beetle emergence may start in the month of May, after the area has received a good amount of pre-cipitation. Emergence may vary in different localities depending on the amount of rain, and may be observed till the end of August.

Immediately after emergence, adults mate on host trees such as walnut. Cop-ulation lasts for 6–11 minutes. Females may lay 10–20 elongate white eggs. The average length and width of eggs is 2.65 mm and 1.70 mm, respectively, and the incubation period ranges from 9 to 11 days. The newly hatched grubs measure about 8.32 mm in length; the second and third instars average about 17.8 mm and 35.5 mm in length, respectively. Fully-fed third instars transform into pupae in the beginning of October at a depth of 30–50 cm inside an earthen cell. The pupal period ranges from 15–20 days. The adults remain in the soil until emer-gence, which is triggered by pre-monsoon rains during May. The beetles are dark brown in color and medium in size (*ca.* 15–16 mm long). One generation of beetles occurs per year (Yadava and Sharma 1995).

Holotrichia serrata

This species is prevalent in Karnataka, Maharashtra, Andhra Pradesh, Tamil Nadu, Kerala, South Rajasthan, the Tarai belt of Uttrakhand, and South Bihar. The beetles of *H. serrata* may start emerging from soil prior to rain during April–May, and will continue until the onset of a monsoon (Mathur *et al.* 2010). The adults are attracted to neem, palas, babul, guava, grapevine, etc. The grubs cause extensive damage to vegetables, pulses, oilseed, cereals, millets, tobacco, sugarcane, and sorghum (Yadava and Sharma 1995). A gravid female lays up to 40 eggs in her lifetime. Newly laid eggs are spherical and measure, on average, 2 mm in diameter. Eggs hatch in 12–15 days, and 1–2 days before hatching the egg swells to up to 4.0 mm in diameter.

The first instar comes out of the earthen cell if there is sufficient moisture. Under drought, it remains inside the cell until favorable conditions occur. First-, second- and third-instar grubs are 10.8, 20.0, and 47.0 mm long, respectively, and the average durations of each instar are 22.5, 35 and 124.5 days, respectively. The grubs become full grown in October, stop feeding, burrow deeper in the soil, and construct earthen cells for pupation. The pupal stage continues for 15.5 days. Adults are 22.4 mm long and 14.0 mm wide. The color of the pupa is dull brown, with a light brown abdomen and dark brown legs (Yadava and Sharma 1995).

Anomala dimidiata

Anomala dimidiata is a prevalent species in the Himalayan ranges and causes severe crop damage in Himachal Pradesh, Uttrakhand, and Jammu & Kash-mir. The adults are strongly phototactic (Mehta *et al.* 2008). Beetle emergence begins by the end of the May. Adults prefer to feed on leaves of apple, walnut, plum, toon, poplar, and shisham. The beetles typically lay eggs in slightly sandy soil with 30–40% soil moisture and rich decaying vegetative matter. The mated female lays, on an average, 29 eggs. The newly laid eggs are oval in shape and

FIGURE 8.5   Adult beetles of *Anomala dimidiata.*

white in color. Prior to hatching, the eggs turn spherical in shape and dark brown in color. Eggs hatch in 13–15 days.

By completion of the first instar, the larva usually attains a length of 11–13 mm. The second and third instars are similar in appearance but larger in size, with body lengths of about 23 mm and 38 mm attained for the second and third instars, respectively. The duration of instars is about 15, 38 and 255 days for instars 1, 2, and 3, respectively. The grubs cause maximum damage during September. With the onset of winter, the grubs make earthen cells and enter overwintering diapause. At the beginning of March, the grubs again resume their activity and feed. After feeding for about 1–2 weeks, they pupate at the end of March; the duration of the pupal stage is about 2 weeks.

Pupae transform into adults during the month of April. The adults are shiny, metallic green beetles (Fig. 8.5). The length and width of the beetles range from 20–22 mm and 12–15 mm, respectively. Adults normally live for 21–28 days after mating. There is only one generation in a year (Mishra 2001).

Management

As a rule, adults do not deposit eggs in clover and alfalfa unless there is a considerable admixture of grasses or other weeds. Thus, grub populations could be reduced by rotating these crops with potato. One of the best ways to clean grubs out of a field is to pasture the land with pigs, as when pigs are allowed to forage on heavily infested land they will usually root out and eat the grubs. Plowing infested fields when most the grubs are pupating kills many of the pupae and newly formed adults (Misra and Chandel 2003). The beetles are also collected at night and killed in water mixed with kerosene. Fertilization with compost, as opposed to fresh farmyard manure, provides less nutrients to early instars. Preferred hosts (see above) which attract adult beetles could be planted to trap them so they can be killed.

Chemical control options used by Indian farmers include spraying methyl parathion, carbaryl, or monocrotophos (Chandel *et al.* 1994b, Chandla *et al.* 1988, Anonymous, 2000). Damage can also be minimized by the application of phorate at the time of hilling. Chlorpyriphos application at the time of hilling is also equally effective. To obtain the best results, insecticide application should occur soon after adult emergence, and should coincide with egg laying or egg hatching (Chandel *et al.* 2008a).

Cutworms

Cut worms, *Agrotis* spp. (Lepidoptera: Noctuidae), are polyphagous insects of cosmopolitan distribution. In India, *Agrotis ipsilon* (Hufn.), *A. segetum* (Schiff.), *A. flammatra* Schiff., *A. interacta* Wlk., and *A. spinifera* Hb. occur on potato. Of those, *A. segetum* and *A. ipsilon* are the most serious pests; the former is common in the hills and the latter is common in the plains. Peak activity occurs during May–June in the Shimla hills, in August in peninsular India, and in March–April in Bihar and Punjab (Singh 1987). Cutworms are found in the upper layer of soil. They come out during the night and cut plant shoots at the base. In some cases entire rows of the plants are cut, making replanting necessary. The attack is more pronounced during dry periods when potato vines have reduced turgor. Tuber damage does not occur on rainy season crops, but larvae inflict considerable tuber damage on the spring crop.

Smooth, grayish-brown, greasy, and plump-looking caterpillars are found during the daytime hiding in soil close to the stems of plants. Newly hatched larvae feed on potato haulms for the first week after hatching, then drop from the plants and feed underground on stems and tubers. While still on the plant, young caterpillars are susceptible to death by drowning in rain or irrigation water.

The cutworms chew off the plants just above, or at a short distance below, the surface of soil. Most of the plant remains intact, but enough tissue is usually removed from the stem to cause it to fall over. Consequently, these caterpillars have a great capacity for causing damage. Tuber injury is manifested in the form of deep irregular holes in the flesh (Fig. 8.6), which reduce tuber quality and may allow secondary pathogens and pests to invade and cause further damage. The holes can look like those caused by slugs, but slugs are typically only a problem in wet, heavy soils. Cutworm damage is usually only significant in non-irrigated crops in lighter soils during hot and dry summers. Das and Ram (1988) reported 12.7% tuber damage due to cutworms in Bihar. In Himachal Pradesh, 9.0–16.4% tubers were found to be damaged by cutworms (Kishore and Misra 1988).

FIGURE 8.6   Tuber damage by larva of *A. segetum*.

Surface Cutworm, *Agrotis spinifera*

This species occurs in Punjab, Bihar, Andhra Pradesh, and Karnataka. Adult moths appear in August, and their peak population is found in September, followed by a gradual decline during October–November. Damage is usually first noticed in December. Larvae and pupae are found during February and April, respectively (Trivedi and Rajagopal 1999). The moths become active at 7 pm, and a female lays 431–901 eggs over a period of 4–10 days. The mated males and females survive for 4–18 and 6–12 days, respectively. Larvae feed only on leaves and growing shoots (Chandel *et al.* 2008a).

Greasy Cutworm, *Agrotis ipsilon*

This is generally a cool-climate pest. In India, it is a more serious pest in the northern region than in the south. On the plains it is active from October onwards, and it migrates to hilly regions at the onset of summer (Butani and Jotwani 1984). The female starts laying eggs 4–6 days after emergence, and lays 649–1711 eggs over a period of 4–11 days. Eggs are laid during the night, starting at 9 pm, either singly or in batches of 7–42 eggs. Eggs are laid on the ventral surfaces of leaves or on the surface of moist soil. Freshly plowed fields are preferred for oviposition (Srivastava and Butani 1998). Eggs hatch in 3–5 days. The caterpillars are light brown with a reddish tinge, which turns greenish thereafter. Fully grown larvae are 40–50 mm long and feel greasy to the touch, hence the common name "greasy cutworms". The larval period lasts from 22 to 30 days. Occasionally, the caterpillars may also nibble on tubers. Pupation takes place in the soil and lasts for 12–15 days during March. The moth's life cycle is completed in 39–53 days (Singh, 1987). Moths are medium-sized, stout, and dark greenish brown with a reddish tinge, and have grayish-brown wavy lines and spots on their forewings; the hindwings are hyaline with a dark terminal fringe (Fig. 8.7).

Common Cutworm, *Agrotis segetum*

The moths are a pale whitish brown with forewings ochreous-brown, having double-waved sub-basal ante- and post-medial lines and a marginal series of specks; the hindwings are iridescent white with dark marginal lines (Fig. 8.8). The wingspan is 40–48 mm. Adult moths emerge from late May to the end of June and lay eggs in clusters of 18–40 on the leaves and stems of plants. Some of the cutworms feed until late July or August, and then pupate and emerge to

FIGURE 8.7   Adult moth of *A. ipsilon.*

FIGURE 8.8   Adult moth of *A. segetum.*

produce a second generation of moths. Most, however, overwinter as caterpillars in the soil and pupate during the following April or May; the moths emerge in May and June. The eggs hatch in 4–7 days. The eggs are dome-shaped and creamy white in color. Full-grown caterpillars are 35 mm in length. The larvae complete their development in 22–30 days. The pre-pupal and pupal periods range between 2–3 and 12–20 days. Normally, pupation takes place in soil or between the folds of dried potato leaves. Male moths live for 2–4 days and female moths live for 5–8 days. Mean fecundity is 161 eggs per female. Two generations are completed during a potato crop season (Misra *et al.* 1995).

Gram Cutworm, *Agrotis flammatra*

This pest is distributed in Punjab and the sub-Himalayan region. Moths of *A. flammatra* are much bigger in size than other cutworm species, with an average wing span of 56 mm. The forewings have characteristic markings and smoky patches, with two-thirds of the proximal areas being pale. On each wing, there is a semicircular spot below the pale area and a grayish-brown, kidney-shaped spot towards the apical area. The caterpillars are dark gray or dull green, measuring 40–50 mm (Chandel *et al.* 2008a).

The pest is active from October–April in the plains and migrates to the mountains in the summer. In October, the moths lay eggs on the undersides of leaves, on shoots, stems, or in the soil. A female lays up to 980 eggs during her lifespan of 7–13 days. The eggs hatch in 4–7 days during summer and in 10–14 days during winter. Larvae complete their development in 4–7 weeks. The pupal stage lasts 12–15 days. The life cycle is completed in 7–11 weeks, and there are generally two generations in a year (Chandel *et al.* 2007).

Agrotis interacta

A. interacta moths are more or less similar in appearance and size to *A. segetum.* The moth is exclusively subterranean and feeds generally on roots and tubers by chewing inside cavities. Information on its biology is lacking.

Management

Forking the soil exposes the larvae and makes them readily available for feeding by generalist avian predators. Efficient chemical control of cutworms can only be achieved by properly applying sprays when the young caterpillars are still on the haulms and are therefore vulnerable. Once below ground, cutworms are unlikely to be significantly affected by insecticides applied to the soil or to foliage. Older caterpillars are generally less susceptible to insecticides than young caterpillars. Routine treatments are likely to be applied at the wrong time, or when the risk of damage is small. Young caterpillars can be killed by rain or irrigation, and an insecticidal spray would probably be unnecessary in these conditions (Chandel *et al.* 2008a).

Treatments should be applied when the soil is dry and the weather is warm. Good control of cutworms depends on a thorough coverage of the foliage with a high-volume application, preferably using at least 1000L/ha of water. Chlorpyriphos, cypermethrin, and triazophos are used by Indian farmers for the control of cutworms on potatoes. However, applying pesticides to crops suffering from drought may result in phytotoxicity. Severe losses caused in a spring crop can be reduced significantly by using oxydemeton methyl. The economic threshold level is 1 larva per 10 plants (Trivedi and Rajagopa, 1999).

Wireworms

Wireworms are the larvae of various click beetles (Coleoptera: Elateridae), most commonly *Drasterius* spp., *Agronichis* spp., or *Lacon* spp. Occasionally, they can cause a lot of damage to potatoes. Wireworms are often a problem when potatoes follow cereal crops or are planted in fields taken out of sod, pasture, or grass (Chandel and Chandla, 2003). Seed tubers may be attacked by wireworms in the spring to early summer, but such damage rarely affects the establishment of the crop. Major damage occurs from the time of tuber initiation until harvest, and can reduce the marketable quality of the tubers (Chandel *et al.* 2008a). Wireworms bore into the tubers, making cylindrical holes (Fig. 8.9). Secondary infection from various diseases can follow, further reducing the quality of the crop. Because tuber quality is so important, very low levels of wireworm damage can have a large effect on the price of the crop. Sometimes tubers still contain wireworms when they are lifted, but stored potatoes rarely contain them (Chandel

FIGURE 8.9   Tubers showing wireworm damage.

FIGURE 8.10   Fully-fed wireworms.

and Chandla 2003). The economic threshold level is low, and treatment may be initiated if any wireworms are detected in a pre-planting soil sample.

Biology

The wireworms take 4–5 years to complete their development, and spend the entire time feeding in the soil. Eggs are laid in grassland or grassy stubble in May and June. After about 1 month, the eggs hatch and the young wireworms initially feed on organic matter in the soil. When newly hatched, the wireworms are about 1.5 mm long and whitish in color. Fully developed wireworms are about 25 mm long and yellow (Fig. 8.10). Older wireworms feed on the roots of many crops and weeds and bore into stems and other plant organs, including potato tubers. The mature larvae pupate in small cells that they form in the soil and emerge as adults during the following spring. Most damage to potato is caused by the larvae in their second and third years of development.

The presence of wireworms can be monitored by using buried baits. Pieces of carrot can be buried about 7.5 cm deep at 10–20 marked sites throughout the field. In 2–3 days, the carrot pieces are retrieved and checked for wireworms. Another type of bait can be prepared by wrapping 2–3 tablespoons of coarse whole-wheat flour in a small piece of netting or nylon stocking and tying it shut. If more than 1.32 wireworms/m² are found, the field should either be treated before planting potatoes, or not be used for potato production (Chandel and Chandla 2003). However, this action threshold may vary from one region to another.

Management

When the risk of damage is very high and wireworm populations in the soil are large, it may be preferable to avoid growing potatoes and plant a more resistant crop, or one in which quality is less important. Legume crops are good rotational choices in fields prone to wireworm infestation, as long as they are kept weed-free. Insecticides incorporated into ridges immediately before planting can reduce tuber damage, but are unlikely to give complete control where wireworm levels are high. Phorate applied in furrow is approved for the control of wireworms in potatoes in India (Chandel and Chandla 2003). It can also be side-dressed after potato shoots begin to emerge.

Termites and Ants

Several species of termites, such as *Microtermes obesi* (Holmgren), *Odontotermes obesus* (Rambur), and *Eromotermes* spp. (Isoptera: Termitidae), have been reported as damaging potato crops (Butani and Jotwani 1984). Rain-fed crops are more prone to termite damage than frequently irrigated crops. Deep black soils and continuously irrigated areas are free from termite damage. More damage occurs under drought conditions.

Termites are soft-skinned, slender insects varying in color from creamy white to dark brown. They are about 2.5 cm long and dirty white in color with brown heads. They are social in habit, and live in colonies in nests that they build below the ground. Workers are wingless, and can be easily recognized by their vertically-carried head with small, broad jaws.

Termite nests contain fungal combs which are lodged either in the central chamber of the nest or far apart in the soil with interconnected galleries. Fungi are maintained and harvested for food.

Odontotermes obesus builds tall, sub-cylindrical mounds up to 2–4 m in height with series of buttresses on the surface. The inner walls are pitted, without any openings to the outside. A large central cavity contains a number of fungal combs arranged unilocularly. The work of the colony is efficiently organized, and there is division of labor. The winged forms, or reproductives, leave the nest in swarms, generally at the start of the rainy season. Most of the individuals perish, but the few that survive mate, shed their wings, and burrow in the soil to form a new colony of which they become kings and queens. The queen lays the first batch of 10–130 eggs about a week after swarming, but continues to lay in very large numbers (about 70,000–80,000 eggs per day) throughout her life, which may be as long as 5–10 years.

The worker caste of termites is responsible for crop damage by damaging roots and making deep holes in potato tubers. The tubers become hollow and are often filled with soil. The leaves of affected plants start to yellow and wilt, and will ultimately dry up (Srivastava and Butani 1998). When infested plants are pulled out, numerous feeding holes are present on the roots and tubers.

The red ant species *Dorylus orientalis* Westwood and *D. labiatus* Shuck (Hymenoptera: Formicidae) have a termite-like habit of attacking plants underground. Unlike termites, they do not shun light. They live in colonies that each has several specialized castes for performing different duties (Fig. 8.11). In a nest, there may be one or several queens – the reproductive females – and two or three forms of sterile females and the workers. The ants that are commonly seen in the field are workers.

Red ants are reported as a pest of potato, cauliflower, cabbage, groundnut, sugarcane, and coconut seedlings in the North-Eastern states, Bihar, and

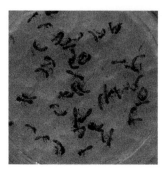

FIGURE 8.11   Damage-causing workers of red ants.

FIGURE 8.12   Tubers showing red ants damage.

Uttar Pradesh (Roonwal 1976). The pest appears during December and remains active until April, causing more than 10% of the damage in irrigated potato crops. High temperatures and dry weather favor population build-up (Kishore *et al.* 1993). The pest damages potato stems and tubers by chewing holes (Fig. 8.12). Severely damaged plants wilt in direct sunlight and will eventually dry up (Trivedi and Rajagopal 1999).

Management

Once a colony is established, it is not easy to eradicate. To avoid damage caused by termites and ants, raising potatoes in sandy soils or areas infested with these pests (especially termites) should be avoided whenever possible. The use of fresh or incompletely decomposed farmyard manure should be avoided. The damage can also be reduced by irrigation (Pandey 2002). Applying a soil spray with chlorpyriphos to the soil as a pre-planting dust formulation or as a post-planting spray has been found effective, particularly in managing termites (Raj and Parihar 1993).

The best cultural method in termite management involves locating and destroying termite nests. This is achieved by breaking open the mounds and removing the queen termite. Clearing and burning of crop residues deprives termites of food and helps to keep their populations at low levels. Mound-building termites are also suppressed by treating the mounds with fumigants, either aluminum phosphide or phorate, as well as by direct applications of liquid formulations of chlorpyriphos or fenvalerate.

Potato Tuber Moth

The potato tuber moth, *Phthorimaea operculella* (Zeller), is a serious pest of stored potato tubers (Fig. 8.13) that also causes considerable damage to potato plants and developing tubers in the field (Fig. 8.14). The potato tuber moth was introduced into India in 1906 through seed potato from Italy (Lefroy 1907). Thereafter, damage to potato tubers under country storage conditions was reported in different parts of the country (Woodhouse 1912, Rahaman 1944, Lal 1945, Kumar and Nirula 1967). Fletcher (1919) reported tuber moth occurring in Pune (Maharashtra), Sitamarihi, Pusa, Purnea (Bihar), Pratapgarh (Uttar Pradesh), and Chhindwara (Madhya Pradesh). Infestation rates ranged between 30% and 70% of the stored tubers (Nirula 1960). Recently, Chandel *et al.* (2001) observed 31.3–60.0% infestation in different villages of the Kangra valley,

FIGURE 8.13   Heavily infested tubers by potato tuberworms.

FIGURE 8.14   Leaf mining by potato tuber worms.

Himachal Pradesh after 3 months of storage. Heavy damage occurs in country stores in Maharashtra, Tamil Nadu, Himachal Pradesh, and the North-Eastern hill states and the plateau region. Trivedi *et al.* (1994) found up to 100% tuber infestation in Karnataka. In Himachal Pradesh, Singh *et al.* (1990) reported 30–60% infestation of tubers from the Kangra valley.

Moths emerge from overwintering larvae in the early spring and lay eggs, chiefly on the undersides of leaves. The eggs hatch in 4–7 days. The larvae are fully fed within 15–20 days. The pupal period ranges from 6–8 days. An entire generation may be completed in 22–25 days in warm weather conditions, and five to six generations are produced per year in the Kangra Valley of Himachal Pradesh (Chandel *et al.* 2001). The damage is most severe in years of low rainfall and high temperatures. In a year, as many as 11 generations have been reported to occur in Assam, 8–9 generations in the northern plains, and 10–13 generations in the plateau region.

Management

Chandel *et al.* (2008b) reported that cultural practices significantly contribute to the reduction of tuber infestation at harvest. It is crucial to reduce initial infestation of stored potatoes and the subsequent population build-up in storage. To control this pest, healthy seed tubers should be planted at 10 cm depth, and potato fields should be kept well cultivated and deeply hilled during their growth. Irrigating the field well before the soil dries, so that there is no formation of cracks in the soil, has been observed to reduce tuber infestation. Tuber infestation is also lower on fields with sprinkler irrigation than on fields with in-furrow irrigation (Chandel *et al.* 2005).

Infestations in growing potatoes are controlled by spraying the foliage with monocrotophos or synthetic pyrethroids after mid-March, when worms begin to infest the leaves (Chandel *et al.* 2000). The infested wilting vines are cut and removed from the field a few days before digging, and should be never piled over dug potatoes. Harvesting should be done before 75% of the foliage dries up. The tubers should not be left exposed to egg-laying moths during late afternoon or overnight (Chandel *et al.* 2008a).

Storage of healthy, uninfested tubers in cold stores is the best way to control tuber moth. In traditional rustic stores, tuber infestation can be prevented by covering stores with a 2-cm layer of dry leaves of *Lantana* or *Eucalyptus* (Anonymous 2000). The application of insecticides to tubers intended for human consumption is highly undesirable, and illegal. Granulosis virus (GV) is extremely effective in reducing PTM damage under storage conditions (Chandel and Chandla 2005). Traditional country stores could be improved by blocking air holes. Openings should have good air circulation to keep the inside temperature cool, but wire-mesh screens should be used to limit access by potato tuber moths. Seed potatoes that are not to be used for food may be treated at the time of storage with malathion dust. A thorough clean-up of potato storage sheds after tuber removal helps to prevent re-infestation (Chandel *et al.* 2000).

Mole Cricket, *Gryllotalpa africana* Palisot

This is a sporadically severe pest that has been reported from Bengal. It is especially damaging to young seedlings in moist soils. The species is widely distributed in warm regions of Eurasia. Konar *et al.* (2005) reported 5–6% plant damage,
along with 10–15% tuber damage, in West Bengal. Eggs are laid during the rainy
season, 100–150 mm deep in the soil in earthen chambers prepared by females. A
female makes 3–4 chambers in her lifetime and deposits 20–30 eggs in each chamber. Eggs are oval in shape, about 1.5 mm long, and are brown in color. Nymphs
live underground in branching burrows and feed on the roots of cultivated and wild
plants. They also tunnel into newly planted seed tubers. Both nymphs and adults
come out of the soil during the night and feed on the leaves of plants. Adults are
22–28 mm long, brown in color, with short wings folded over the abdomen without covering the abdomen completely (Butani and Jotwani 1984).

Management

Mole crickets are usually controlled by applying phorate and chlorpyriphos
(Konar *et al.* 2005).

Minor Pests

Darkling beetles *Gonocephalum hofmannseggi* (Steven) and *Hopatroides
seriatoporus* Fairmaire (Coleoptera: Tenebrionidae) have been reported from
Karnataka (Srivastava and Butani, 1998). These beetles feed on roots and occasionally on tubers. Feeding damage may result in the death of young sprouting
plants, but has little effect on a mature crop.

SAP-FEEDING PESTS

Sap-feeding insects can affect the health of a potato crop, both directly by causing feeding damage as well as indirectly by transmitting viruses and virus-like
pathogens that cause important diseases. As vectors, aphids and whiteflies are
especially critical in the production of seed potatoes because tuber-borne viruses
can severely limit yields in subsequent crops. They have elongated mouthparts
which form a tube composed of paired mandibular and maxillary stylets, with
a central food canal and a separate salivary canal. This tube is inserted into the
plant, typically to reach and withdraw sap from the phloem. Sap removal is, by
itself, damaging to plants, but toxic saliva and infectious virus particles may
also be injected during feeding (Chandel *et al.* 2008c).

Aphids

Aphids (Hemiptera: Aphididae) can injure a potato plant directly by sap feeding, and are capable of transmitting several important potato viruses. High aphid

populations can have substantial direct effects on yield, but such populations are uncommon in commercial potato production. The primary concern with aphids is usually their role as virus vectors. The honeydew excreted by aphids is deposited on the plant, where it provides a good growth medium for sooty molds (Chandel *et al.* 2008c).

Five aphid species are commonly found on potato crops in the India; the green peach aphid (*Myzus persicae* (Sulzer)), cotton aphid (*Aphis gossypii* Glover), potato root aphid (*Rhopalosiphum rufiabdominalis* (Sasaki)), tuber aphid (*Rhopalosiphoninus latysiphon* (Davidson)), and bean aphid (*Aphis fabae Scopoli*). The cosmopolitan aphid, *M. persicae*, is the most important pest on potatoes. (Chandel *et al.* 2008c).

Biology of M. persicae

In *M. persicae*, only eggs are produced by sexual reproduction, whereas all subsequent reproduction is viviparous and parthenogenetic (Verma and Chandla 1999). The aphids have both winged and wingless forms. Wingless forms (Fig. 8.15) are predominant on potato during most of the year. *Myzus persicae* overwinter as eggs on a very restricted number of primary host species, usually woody plants, with peach being the most important host. In spring, wingless aphids called stem mothers hatch from eggs, feed on the primary host, mature, and produce young asexually. The offspring of stem mothers are generally wingless. Aphids molt four times, with the mean number of offspring per wingless aphid ranging from 60 to 75 (Chandla and Verma 2000). Optimum temperature for reproduction is around 21 °C. There may be several generations on a primary host, but eventually winged adults (spring migrants) develop and fly away to colonize "secondary", often herbaceous, host plants (Chandel *et al.* 2007).

FIGURE 8.15   Sucking wingless green peach aphid, *M. persicae*.

Winged spring migrants are not produced until at least the second generation on the primary host, and peak production occurs in the third generation. Spring migrants leave the primary host in search of suitable secondary hosts, usually herbaceous plants, one of which is the potato. These spring migrants are capable of traveling long distances – up to 1600 km – and have been found at altitudes of up to 3048 m (Chandel et al. 2008c). Winged aphids land at random, since they cannot visually distinguish a host from a non-host plant. To find a suitable host, winged aphids feed for short periods on sap from epidermal tissues of plants on which they land. This is called "sap sampling". They move from plant to plant until they locate a suitable secondary host. Sap sampling can result in the transmission of certain viruses, even by aphid species incapable of colonizing potato.

Once an acceptable host is found, the spring migrants settle and reproduce. Their offspring are mostly wingless, but a small proportion of each succeeding generation is winged. Each unfertilized fundatrix produces around 50–60 young ones viviparously. As many as eight apterous generations have been found on primary host plants (Verma and Chandla 1999). As the quality of host plants declines, more winged summer migrants are produced, which then fly to other secondary host plants. As the day length shortens, fall migrants are produced with both males and females. They return to the primary host plant, on which the females give birth asexually to wingless females, which then mate with the male fall migrants and lay fertilized overwintering eggs.

The eggs measure about 0.6 mm long and 0.3 mm wide and are elliptical in shape. They are initially yellow or green, but soon turn black. Nymphs are initially greenish, but soon turn yellowish, greatly resembling viviparous adults. Verma and Chandla (1999) reported four instars in this aphid, with duration of each instar ranging from 1–2, 2–3, 2–3, and 4–5 days, respectively. Alate aphids have a black head and thorax and a yellowish–green abdomen, with a large dark patch dorsally. They measure 1.8–2.1 mm long. The apterous aphids are yellowish or greenish. They measure about 1.7–2.0 mm long. Medial and lateral green stripes may be present. The cornicles are moderately long, unevenly swollen along their length, and match the body in color. The appendages are pale. Each gynopara produces about 5–15 ovipare. A male is capable of fertilizing several ovipare. Each oviparous female lays 4–13 eggs. Each viviparous female produces 30–80 nymphs (Verma and Chandla 1999). The eggs are deposited near buds of the primary host. In mild climates, winged migrants also develop from aphid populations that survive the winter as adults. Occasionally the life cycle is incomplete, in which case asexual breeding takes place throughout the year and overwintering takes place on stored, sprouting potatoes, on hardy herbaceous plants, or on glasshouse crops (Chandel et al. 2008c).

Management of Aphids

Young plants are particularly susceptible to viral infection, and seed plots must therefore be kept free of aphids whenever possible, in particular early in the season. High-altitude areas of the north-western Himalaya (>2000 m above sea level) are the best seed-growing areas in India because they remain practically aphid-free during the summer season when the potato seed crop is grown. The selection and rouging of plants infected with a virus must also be done, even under aphid-free conditions. An action threshold of 20 aphids per 100 compound leaves is strictly followed in Indian seed production (Chandla *et al.* 2004).

The most vulnerable period in the aphid life cycle is passed on the overwintering hosts. The application of chemical defoliant to peach trees in fall, the denial of foliage to fall migrants, and the pruning of trees to remove most overwintering eggs are useful. When peach trees are pruned the twigs that are removed must be destroyed, or else the eggs can still hatch (Chandel *et al.* 2008c).

Several soil-applied systemic insecticides give good early-season control of aphids. However, late-season aphid pressure is often more severe in potatoes treated with a soil systemic insecticide at planting. Late-season outbreaks in aphid populations may be a consequence of early-season control measures that prevent the establishment of the aphids' natural enemies (Chandel *et al.* 2007). Foliar sprays of dimethoate or metasystox and phorate applied in-furrow at planting are widely recommended in seed potato for aphid control (Chandla *et al.* 2004). Imidacloprid and thiamethoxam also provide effective protection for about 2 weeks (Chandel *et al.* 2008c).

Leafhoppers

Leafhoppers (Hemiptera: Cicadellidae) are strong fliers and are much more mobile than aphids. Unlike aphids, leafhoppers are important mainly because of the direct feeding damage that they cause. The potato leaf hopper (*Empoasca devastans* Distant) is the most important species, and has long been recognized as a major pest of potato. Several other species are important on potato in certain regions. These include *Amrasca biguttula biguttula* (Ishida), *Alebroides nigroscutulatus* Distant, *Seriana equata* Singh, *Empoasca solanifolia* Pruthi, *Empoasca kerri motti* Pruthi, *E. fabae* Harris, and *E. punjabensis* Pruthi (Butani and Jotwani 1984, Misra 1995).

Nature of Damage

Prolonged feeding by the potato leafhopper causes a condition known as "hopper burn", manifested in the form of brown triangular lesions at the tips of the leaves. Both adults and nymphs are injurious, but late-instar nymphs can reduce yields more than twice as much as an equal number of adults. Damage results from disruption of phloem (Trivedi and Rajagopal 1999). Toxins in the saliva of potato leaf hopper induce swelling of cells, which eventually crushes the phloem. There is also depletion of plant reserves due to an increase in plant respiration subjected to leafhopper attack. Infestations are most damaging during early tuber bulking (growth stage IV).

Transmission of Diseases

Potato yellow dwarf virus and beet curly top virus are transmitted by leafhoppers (*Empoasca* spp.). Purple top in potato, which is caused by aster yellows mycoplasma- like organisms, is also transmitted by leafhoppers (Misra 1995).

Biology of Leafhoppers

The leafhoppers have a broad host range. On potato, they usually complete two to four generations in a year. The population density is dependent upon the date of aphid arrival on the crop and the temperature.

Amrasca biguttula biguttula

Amrasca biguttula biguttula, commonly known as cotton leafhopper, is a polyphagous pest. Although its main hosts are cotton and okra, it also causes serious damage to brinjal and potato (Shankar *et al.* 2008). Both nymphs and adults suck cell sap, usually from the ventral surfaces of leaves. Adults are wedge-shaped, about 2 mm long, and pale green in color with a black dot on the posterior portion of each forewing. Eggs, which are pear-shaped, elongated, and yellowish-white in color, are deposited individually on leaves. A single female lays 15–30 eggs that hatch within 4–10 days. The nymphal period lasts for 7–21 days, and the nymphs are whitish to pale green in color. Peculiarly, they move diagonally across leaves. The longevity of the adults is about 2 weeks (Srivastava and Butani 1998).

Empoasca kerri motti

This species breeds throughout the year, but it is most active from October to March. Beside potato, this hopper attacks brinjal, chili pepper, cowpea, and tomato. In the absence of these hosts, it migrates to castor, alfalfa, and barseem (Butani and Jotwani 1984).

The adults are 3 mm long and yellowish-green in color; the vertex is flat, greenish-yellow, and smaller than the protonum, which is also smooth and flat; the front wings are long, narrow, semi-transparent, and pale green in color, with green at the costal and gray at the distal regions (Srivastava and Butani 1998). A female lays 25–60 eggs in 25–30 days. The incubation period is 4–11 days, and the nymphal period averages 25 days. Adult longevity is about 2 weeks in males and as many as 13 weeks in females (Butani and Jotwani 1984).

Empoasca punjabensis

E. punjabensis produces symptoms of hopper burn on leaves. These symptoms are manifested in the forms of etiolated spots and patches on leaves, browning, rotting of margins and tips, and drying of leaves. Females lay eggs in leaf veins. Eggs hatch in 4–9 days, nymphal development takes 19–21 days, and adult longevity is 7–15 days (Butani and Jotwani 1984). Adults are, on average, 3.5 mm long and yellowish-green in color. The vertex is flat, smooth, and pointed, and the protonum is transparent and longer than the vertex. The abdomen is yellowish, and the ovipositor is stout and green in color. The front wings are longer than the body and are transparent greenish- yellow in color, with deep-green coastal margins (Srivastava and Butani 1998).

E. solanifolia

This species causes similar damage to E. punjabensis. The adults are bigger in size, being about 4.0 mm long, robust, and pale brown in color; the vertex is flat and slightly raised; the protonum is 1½ times longer than the vertex; the abdomen is tinged with yellow; the ovipositor is stout and the pygopher is covered with a few minute hairs. The hemelytra are transparent and are twice as long as the abdomen, having thin, distinct veins (Butani and Jotwani 1984).

E. fabae

Commonly called potato leafhopper, this species is more common outside India. Similar to other leafhoppers, it produces hopper burn symptoms such as stunted growth and crinkling of the leaves (Srivastava and Butani 1998). The adults are pale green and marked with a row of white spots on the anterior margins of the protonum. They are about 3.5 mm long. Females lay transparent to pale yellow eggs, which are inserted into the veins and petioles of leaves. Total egg production is about 200–300 eggs per female, and the average incubation period is 10 days. The average development time for nymphs is 15 days. Adult longevity is typically 30–60 days.

Control of Leafhoppers

Foliage of early-maturing cultivars is generally more susceptible to leafhopper damage. However, they bulk more rapidly, and their yield may actually be less affected compared to later-maturing cultivars. It is more important to control leafhoppers under drought conditions, when potato is more susceptible to leafhopper injury. In seed crop, soil systemic insecticides applied in furrow at planting or side-dressed at plant emergence give 6–8 weeks of control. On fresh market potatoes, the standard practice is to apply foliar sprays. Dimethoate and methyl demeton applied at the appearance of the pest, and phorate applied at planting, are approved for the control of leafhoppers in India (Pandey 2002).

Thrips

Thrips are the vectors of tospo viruses, which cause stem necrosis in potato. Seven species of thrip are associated with potato. Of these, *Thrips palmi* Karny, *Scirtothrips dorsalis* Hood, *Caliothrips collaris* (Bagnall) (Thysanoptera: Thripidae), and *Haplothrips* sp. (Thysanoptera: Phlaeothripidae) are important. These are tiny, slender, fragile insects, with adults having heavily fringed wings. The females have extremely slender wings with a fringe of long hairs around their margins.

Nature of Damage

Both adults and larvae scrape the epidermal tissues of leaves, usually near the tips, and rasp the oozing sap (Butani and Jatwani 1984). The surface of leaves becomes whitened and somewhat flecked in appearance. The tips of leaves wither, curl up, and die. The undersides of leaves become spotted with small, brownish-blackish specks (Fig. 8.16). When damage is severe, the whole field has a "dry blight" appearance, where most of the infected plants have dry leaves hanging on blighted stems. Eventually, such plants wilt and die (Khurana *et al.* 2001).

Biology

Parthenogenesis is common. The males are wingless and very scarce; the females regularly reproduce without mating. In case of *T. palmi*, eggs are deposited in leaf tissue, in a slit cut by the female. Females produce up to 200 eggs, but average about 50 per female. The bean-shaped egg is colorless to pale white. Duration of the egg stage is about 16 days at 15°C, 7.5 days at 26°C, and 4.3 days at 32°C (Khurana *et al.* 2001).

The larvae (Fig. 8.17) resemble the adults in general body form, though they lack wings and have a smaller body size. There are two instars during the larval period. Larvae require about 14, 5, and 4 days to complete their development at 15°C, 26°C, and 32°C, respectively. At the completion of the larval stage the insect usually descends to the soil or leaf litter, where it constructs a small

FIGURE 8.16   Spots due to thrips feeding.

FIGURE 8.17   Nymphs of thrips.

earthen chamber for a pupation site. There are two instars during the pupal period. The pre-pupal instar is nearly inactive, and the pupal instar is inactive. Both instars are non-feeding stages (Khurana *et al.* 2001).

The combined pre-pupal and pupal development time is about 12, 4, and 3 days at 15°C, 26°C, and 32°C, respectively. After the fourth molt, the adult females return to the plants and soon lay eggs for another generation. The total time required for each generation is about 20 days. The high temperatures (30–35°C) and dry weather during September–October are highly favorable for aphid/thrip activity and, consequently, higher disease incidence (Khurana *et al.* 2001). Adults are pale yellow or whitish, measuring 0.8–1.0 mm in body length, with females being slightly larger than males. Unlike the larval stage, the adults tend to feed on young growth, and thus are found on new leaves. Adult longevity is 10–30 days for females and 7–20 days for males.

Transmission of Tospo Viruses

These viruses are acquired by the first instars shortly after they hatch from the eggs and start feeding on infected plants. The latent period is 3–10 days. Transmission is mainly via the adult thrips. During the latent period, the virus circulates in the vector and replicates throughout the life of viruliferous thrips (Chandel *et al.* 2008c).

Control

In potato, there is a strong positive correlation between early planting and thrip activity. Therefore, early planting (September/October) must be avoided whenever possible. Certain varieties are resistant to injury by this pest. In the plains, the Kufri Sutlej, Kufri Badshah, and Kufri Jawahar strains are comparatively resistant (Singh *et al.* 1997), and the Kufri Chandramukhi and Kufri Bahar strains are susceptible. After the plants start growing they should be watched carefully, and if the thrips or the characteristic injury appears upon them then the plants should be sprayed with imidacloprid. Imidacloprid, applied as a foliar spray or as a side-dressing at the first signs of thrip damage, usually provides good control (Singh *et al.* 2000). Generally, chemical applications are limited to the first and second weeks of crop emergence, when thrip activity is at its maximum and viruliferous thrips land on the germinating crop.

White Flies

Bemisia tabaci (Gennadius) (Hemiptera: Aleyrodidae) is widely distributed throughout the world. This pest is distributed throughout the northern and western regions of the Indian subcontinent, and has recently emerged as a very serious pest in potato seed production, particularly in the autumn crop in the Indo-Gangetic plains (Kumar *et al.* 2003). *B. tabaci* is a highly polyphagous insect and is a serious pest of cotton, tobacco, okra, and various other vegetables and weed plants (Puri *et al.* 1995). The population of *B. tabaci* is highly diverse, and many biotypes have been identified. Infestation is heavier on early potato crops planted in September. The maximum population on potato occurs in November, followed by a sharp decline by December (Chandel *et al.* 2010).

Nature of Damage

Adults and nymphs use their piercing-sucking mouthparts to feed on the phloem of host plants. This results in direct damage, which is manifested in localized spotting, yellowing, or leaf drop (Broad and Puri 1993). Under heavy feeding pressure, wilting and severe growth reduction may occur (Malik *et al.* 2005). Systemic effects may occur, with non-infested leaves and other tissues becoming severely damaged as long as feeding whiteflies are present on the plant (Butter and Kular 1999). The affected plants remain stunted, and their leaves show distinct upward or downward curling. Leaves of affected plants show dark green veins as compared to the normal translucent veins of healthy plants.

Whiteflies excrete honeydew that promotes the growth of sooty molds. Those, in turn, adversely affect plant photosynthesis, leading to a reduction in yield (Reddy and Rao 1989). Once the whiteflies are removed, new plant growth is normal.

In addition to direct damage, *B. tabaci* also causes damage indirectly by transmitting gemini viruses. Some viruses, such as potato apical leaf curl virus (PALCV), cause more damage than insect feeding alone (Lakra 2002). The first report of potato apical leaf curl virus in India was made around the year 2000 (Garg *et al.* 2001). Tuber formation is adversely affected in virus-infected plants. Dhawan and Mandal (2008) reported that potato plants infected with apical leaf curl virus showed stunting, crinkling, vein thickening, curling, waviness of leaf margins, and leaf distortion.

Biology of B. tabaci

B. tabaci can complete a generation in about 20–30 days under favorable weather conditions (Saini, 1998). Whiteflies produce many generations in a year and quickly reach high population densities. At least three generations are completed on potato. Temperatures in the range of 26–32°C, with a RH of 60–70%, are optimal for whitefly development (Chandel *et al.* 2010).

Whiteflies insert their eggs in leaf tissues. The egg is about 0.2 mm long, elongate, and tapers distally; it is attached to the plant by a short stalk (Rao and Reddy 1989). The whitish eggs turn brown before hatching, which occurs in 4–7 days. The female deposits 90–95% of her eggs on the lower surface of young leaves (Arneja 2000). The older stages prevail on older leaves. There are four nymphal instars (Dhawan *et al.* 2007), all of which are greenish and somewhat shiny. The flattened first instar is mobile, and is commonly called the "crawler" stage. The crawler stage measures about 0.27 mm long and 0.15 mm wide. Movement is usually limited to the first few hours after hatching, and only to a distance of 1–2 mm. The remaining instars are scale-like and stationary.

FIGURE 8.18   Adults of *Bemisia tabaci.*

The wax that the ovipositing whiteflies deposit, the spines adorning the nymphs, and the exuviae of early instars retained by the later instars help in protecting the whiteflies against natural enemies (Reddy and Rao 1989).

Whiteflies normally feed on the lower surface of leaves. Duration of the first instar is usually 2–4 days. The second and third instars are each completed in about 2–3 days (Chandel *et al.* 2010). Body length and width are 0.36 and 0.22 mm and 0.49 and 0.29 mm for the second and third instars, respectively. The sessile fourth instar is usually called the "pupa", although it may still participate in some feeding. The fourth instar measures about 0.7 mm in length and 0.4 mm in width. Duration of the fourth instar is about 4–7 days (Arneja 2000). The total nymphal period ranges from 10 to 14 days. Nymphs transform into pupae on the leaves, and in 2–3 days adult whiteflies emerge from the pupae (Sharma and Rishi, 2004). Total pre-adult development time averages 15–18 days (Dhawan *et al.* 2007). The lower and upper developmental thresholds are about 10°C and 30°C (Chandel *et al.* 2010). The adult is white in color and measures 1.0–1.3 mm in length (Fig. 8.18). The antennae are pronounced, and the eyes are red. Oviposition begins 2–5 days after emergence of the adult, often at a rate of about five eggs per day. Adults typically live 10–20 days and may produce 50–150, or even up to 300, eggs (Reddy and Rao 1989).

Transmission of Potato Apical Leaf Curl Virus

Geminiviruses are transmitted in a persistent circulative mode. For efficient transmission, an acquisition-access feeding period of 2–24 hours, followed by an inoculation-access feeding period of 2–3 days, is required. Transmission occurs only after a latent period of 4–10 hours. After acquisition, whiteflies can transmit virus for 5–20 days (Chandel *et al.* 2010).

Control

Whiteflies are very difficult to control with conventional insecticides. Seed treatment with imidacloprid and foliar applications at emergence, with the second application occurring after 15 days, has been found to be effective (Malik *et al.* 2005). However, under conditions of severe whitefly attack none of the control measures are effective to prevent virus spread.

The whitefly population can be successfully managed only by adopting a package of practices specifically targeted for its control. The cultivation of varieties susceptible to leaf curl virus should be avoided. Varieties like Kufri Bahar have shown high tolerance (Lakra 2003), while cultivars like Kufri Sutlej and Kufri Anand are highly susceptible. Elimination of alternative hosts of the virus and the virus vector, as well as of the infected potato plants, helps to reduce virus transmission to potato plants. Excessive use of nitrogenous fertilizers may promote whitefly growth (Puri *et al.* 1995).

Sap-Sucking Bugs

Green Potato Bug, *Nezara viridula* (Linn.)

The green potato bug (Hemiptera: Pentatomidae) is cosmopolitan in distribution, and has been recorded from South Europe and Japan at its northernmost range to Australia and South Africa at its southernmost range. These bugs occasionally cause economic damage to potato. The green potato bug is a polyphagous pest; its main hosts are castor and coriander, but it also breeds on coffee, citrus, cotton, millets, pulses, potato, rice, tomato, wheat, etc. Nymphs and adults suck the cell sap from tender leaves and shoots, devitalizing the same. Adults are medium-sized, 15 mm long, and light green in color, and nymphs are brownish-red with multicolored spots. A female lays up to 300 eggs in clusters of 50–60 eggs on the dorsal surfaces of leaves. The eggs are deposited in regularly shaped, hexagonal clusters, with the individual eggs ordered in uniform rows and glued together. Eggs are barrel-shaped and measure about 1.3 mm long and 0.9 mm wide. They are yellowish-white to pinkish-yellow, and the top, or cap, is clearly indicated by a ring and 28–32 spines. The eggs darken in color during the incubation process, and hatching occurs after about 5 days. Nymphs remain aggregated in the first instar before they disperse and start feeding during the second instar (Butani and Jotwani 1984). There are five instars. Nymphs are brownish-red with multicolored spots, and the body length of the fifth instar is about 10 mm. Total nymphal development time is about 32 days, and egg-to-adult development requires 35–37 days. The optimal temperature for development is 30°C. The adult is uniformly light green, both dorsally and ventrally, though the ventral surface is paler. Adults measure about 13–17 mm long and 8 mm wide (Srivastava and Butani 1998). No separate control measures are generally required for these bugs, as damage is very limited. The chemical control of aphids is generally sufficient in preventing damage.

Creontiades pallidifer (Walker)

C. pallidifer (Hemiptera: Miridae) is another polyphagous pest with a wide range of host plants, including brinjal, crucifers, melons, okra, and potato.

Nymphs and adults suck the cell sap from leaves and cause small, irregular brown spots on young leaves and growing tips; gradually, the affected leaves die. This bug has been found to breed throughout the year in the Delhi area; from January to April it feeds on brinjal, peas, and potato; from April to June on melons, and from July to September on other cucurbits; during October it is found on maize and pulses; finally, during November–December, the bugs migrate and attack cole crops (Butani and Jotwani 1984).

Adult bugs are delicate, 7 mm long, and ochreous-green in color, with transparent light-green wings. A female lays 100–200 eggs, and the incubation period is 4–5 days. The eggs are deposited within tender tissues of growing points such as petioles and axils of branches. Nymphal development takes, on average, 18 days, and the total life cycle is completed in about 22 days. Adult longevity of females is 15 days; in males, the longevity is less than a week (Srivastava and Butani 1998).

Piezodorus hybneri (Gmelin) (Hemiptera: Pentatomidae)

This bug is a minor pest and has been reported feeding on potato leaves. Eggs are laid in clusters of 25–30 eggs each, on the dorsal surfaces of leaves. The incubation period is 3–4 days, and nymphal development takes 22–26 days.

Recaredus sp.

The lace-wing bug (Hemiptera: Tingidae) attacks stored tubers in some parts of India. Nymphs and adults suck sap from cortical tissues of tubers. Eggs are laid on tubers. The total life cycle is completed in about 30 days, and as many as seven generations have been reported in a year. Adult longevity is 7–8 months, and hibernation takes place in the adult stage (Butani and Jotwani 1984).

LEAF-EATING AND DEFOLIATING INSECTS

Leaf defoliators are either coleopteran or lepidopteran insects causing variable damage, depending on the geographic region and environmental conditions (Trivedi and Rajagopal 1999). In addition to defoliation, some lepidopteran larvae also burrow into potato stems.

Defoliating Caterpillars

Cabbage Semilooper, *Plusia orichalcea* (Fab.)

Cabbage semilooper (Lepidoptera: Noctuidae) is a polyphagous pest that has a wide host range including pea, cole crops, radish, turnip, celery, indigo, linseed, etc. It is widely distributed in India and often inflicts damage to potato throughout its range (Misra *et al.* 1995). The leaves are riddled with large holes of irregular shape and size and covered with masses of greenish to brown frass. In case of severe infestation, the entire plant may be defoliated.

FIGURE 8.19   Adult moth of *Plusia orichalcea.*

FIGURE 8.20   Larva of *Plusia orichalcea.*

The adult is light brown, with a large golden patch on each forewing (Fig. 8.19). Wing span is 42 mm. The caterpillars are pale green (Fig. 8.20) and form characteristic half loops when they walk. The moths are very active at dusk on flowers during spring season. Each female lays 350–450 eggs singly on leaves. Eggs hatch in 3–5 days following oviposition. The caterpillars feed individually. Larvae are fully grown in 20–28 days, and they pupate in debris on the ground. The pupal stage lasts 8–15 days (Misra *et al.* 1995).

Oriental Armyworm, *Mythimna separata* (Walker)

The oriental armyworm (Lepidoptera: Noctuidae) is a polyphagous pest that is found all over the Indian subcontinent, South-east China, Japan, South-east Asia, Korea, the Philippines, Indonesia, Australia, and New Zealand. Young larvae scrape the leaf tissues and skeletonize the leaves, while the advanced instars feed gregariously and voraciously on whole leaves and migrate from one leaf to another. Even medium-level infestation may result in complete defoliation of vines.

The moths are pale brown with dark specks, and the hindwings are white. The eggs are laid in rows or in clusters and hatch in 4–5 days. Freshly emerged larvae are very active and are dull white in color, later turning green. Larvae are fully grown in about 2 weeks. Mature larvae are cylindrical in shape, 45 mm in length, and dark green to greenish-brown, with four distinct longitudinal black green stripes on either side of the mid-dorsal line. Pupation takes place in soil, under dry leaves, or among stubbles. The pupal period is completed in 9–13 days. The entire life cycle is completed in 4–5 weeks (Chandel *et al.* 2011).

Bihar Hairy Caterpillar, *Spilosoma obliqua* (Walker)

The bihar hairy caterpillar (Lepidoptera: Arctiidae) is most common during late winter and spring. It is a polyphagous pest and has been reported damaging a number of fruit trees, tobacco, pulses, vegetables, potato, and sweet potato. The newly hatched caterpillars feed gregariously, skeletonizing leaves. More mature caterpillars segregate and feed voraciously on leaves, often completely defoliating vines. The caterpillars move from plant to plant and from field to field; older leaves of older plants are preferred (Chandel *et al.* 2011).

Wings of adult moths are pinkish-buff with numerous black spots (Fig. 8.21), spanning 40–45 mm. The head, thorax, and ventral side of the body are dull yellow. The female lays more than 400 eggs in clusters on the undersides of leaves. The eggs are spherical in shape and light green in color. They hatch in 8–13 days. Larvae pass through seven instars and are ready to pupate in 4–8 weeks. Fully grown caterpillars are stout, about 40 mm long, and have seven orange-colored, broad, transverse bands with tufts of yellow hairs (Fig. 8.22).

FIGURE 8.21   Adult moth of Bihar hairy caterpillar, *Spilosoma obliqua*.

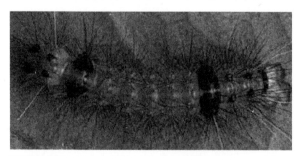

FIGURE 8.22   Larva of Bihar hairy caterpillar, *Spilosoma obliqua*.

Pupation takes place in plant debris or in the soil, and adults emerge in 1–2 weeks. The life cycle is completed in 4–5 weeks, and there may be three to eight generations per year. The caterpillars of a winter brood burrow into soil to diapause (Srivastava and Butani 1998).

Hairy Caterpillar, *Dasychira mendosa* (Hubner)

The hairy caterpillar (Lepidoptera: Lymantriidae) is a polyphagous pest that feeds on potato, coffee, red gram, castor, cauliflower, and many other plant species. The larva is grayish-brown with dark prothoracic and preanal tufts. The prolegs are crimson-red. Fully grown larvae are 38–44 mm long with red stripes on their heads. The adult is smoky brown with hindwings that are pale gray in color. Forewings are uniformly brown, with black specks and a pale patch outside the subbasal line. The wing span of a female moth is 46–54 mm (Chandel *et al.* 2011).

Tobacco cutworm, *Spodoptera litura* (Fab.)

The tobacco cutworm (Lepidoptera: Noctuidae) is a sporadically serious pest. It is highly polyphagous and is widely distributed in many parts of the world. The caterpillars hide during day in crevices and feed at night. Eggs are laid in clusters on the lower sides of leaves and are covered with brown hairs. A single female (Fig. 8.23) lays, on average, 400 eggs in 3–4 clusters of 80–150 eggs. Freshly hatched larvae feed gregariously, scraping the leaves. Later, these larvae disperse. During severe infestation, an entire crop may be defoliated overnight. Fully grown larvae are 40–50 mm long and are pale brown in color with a green to violet tinge (Fig. 8.24). Incubation, larval, and pupal stages last for 3–5, 20–28, and 7–11 days, respectively. The entire life cycle is completed in 30–40 days (Butani and Jotwani 1984).

A related species, *Spodoptera exigua* (Hubner), also occurs on potato as a serious defoliator (Butani and Jotwani 1984). A female lays up to 1300 eggs in batches of 50–200 eggs on the ventral surfaces of leaves. The immature stages

FIGURE 8.23   Female moth of *Spodoptera litura.*

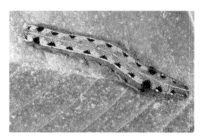

FIGURE 8.24　Fully-fed larva of *Spodoptera litura.*

FIGURE 8.25　Adult moth of *H. armigera.*

are more or less similar in appearance as those of *S. litura*. The larvae grow to a length of about 3.8 cm. Adults have a wingspan of 25–35 mm. Egg, larval, and pupal stages last for 2, 15, and 16–17 days, respectively. Pupation takes place on the soil surface or, rarely, at depths of 5–10 cm.

Gram Pod Borer, *Heliothis armigera* (Hubner)

H. armigera (Lepidoptera: Noctuidae) is a highly polyphagous pest distributed throughout India. The adult is a medium-sized light-brown moth (Fig. 8.25). Eggs are deposited individually on tender leaves and hatch in 3–7 days. The young larvae are leaf scrapers, while older larvae are leaf chewers (Fig. 8.26). The mature larvae are about 35 mm long and greenish-brown in color, with dark yellow stripes. The larvae become fully grown in 17–22 days (Misra *et al.* 1995). They pupate in the soil for 6–12 days and hibernate as pupae. The total life cycle is completed in 4–6 weeks Chandel *et al.* 2011). There are five to eight generations in a year.

Eggplant Borer, *Leucinodes orbonalis* Guenee

Eggplant borer (Lepidoptera: Crambidae) is a minor pest of potato across India (Regupathy *et al.* 1997). Caterpillars bore into the shoots, causing them to wilt and droop. Females deposit eggs individually on leaves and shoots. The eggs hatch in 3–5 days. The pink-colored larva becomes fully grown in 10–15 days,

FIGURE 8.26   Leaf defoliation by larva of *H. armigera.*

measuring about 1.6 cm in length. Pupation takes place in a cocoon on the plant for a period of 6–8 days (Chandel *et al.* 2011). To control this pest, the affected shoots may be clipped off and destroyed by burning or deep burying.

Management of Lepidopterous Defoliators

Every effort should be made to destroy early instars. Alternate weed hosts of these insects, on which the first generation of caterpillars may develop, should be destroyed. The visible egg masses, as well as leaves with gregarious young larvae, should be collected and destroyed mechanically. Plowing could be used to expose and kill pupae in the soil. Flood irrigation may drown the hibernating caterpillars and pupae. In case of severe attack, spraying fields with carbaryl, endosulfan, monocrotophos, quinalphos, chlorpyriphos, or malathion is recommended. Two to three applications are usually required, beginning soon after initial infestation and repeated at 2-week intervals, if needed (Dass 2000, Pandey 2002).

Leaf-Eating Beetles

Hadda Beetles

Epilachna beetles (Coleoptera: Coccinellidae) and their larvae are important pests of potato in India. The two species commonly found all over India are the 12-spotted *(Epilachna ocellata* Redt.) and 28-spotted beetles *(Epilachna vigintioctopunctata* (Fab.)). The former is generally found in higher hills, while the latter is restricted to lower elevations (Misra *et al.* 2003). Plant damage is caused by adults and larvae skeletonizing the leaves. The adults (Fig. 8.27) are active fliers and readily move from plant to plant. The larvae (Fig. 8.28) stay on the leaves and occur in large numbers. In case of heavy infestations, plants can be completely defoliated before tuber maturation.

The yellowish, cigar-shaped eggs are laid in clusters on the lower surfaces of leaves. A female lays 500–750 eggs that hatch in 3–4 days. The larval period

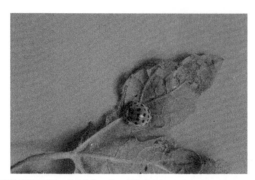

FIGURE 8.27   Adult beetle of *Epilachna vigintioctopunctata.*

FIGURE 8.28   Grubs of *Epilachna vigintioctopunctata.*

lasts 8–10 days on potato, and pupation takes place on leaves and lasts from 3–6 days (Trivedi and Rajagopal 1999). The life cycle is completed in 21–36 days. Adults overwinter under grass and weeds.

Control: Where the beetles are abundant, it is suggested that eggs and larvae are collected and destroyed mechanically. Good control may be obtained by thoroughly spraying the foliage with malathion, dichlorvos, endosulfan, chlorpyriphos, or carbaryl (Dass 2000, Pandey 2002). Application should occur as soon as the beetles or their eggs are found on the plants.

Flea Beetles, *Psyllodes plana* Maulik

Flea beetles (Coleoptera: Chrysomelidae) have enlarged hind legs and jump vigorously when disturbed. The adult flea beetles chew small rounded holes in the leaves (Fig. 8.29), often starting on the lower side of a leaf (Chandla 1985). They attack plants as soon as they emerge from the soil, and the damage continues until the crop is harvested. When flea beetles are abundant, the foliage is so badly damaged (Fig. 8.30) the plant dies.

Flea beetles overwinter as adults under leaves, grass, and in other protected places. Beetles terminate their diapause in April. They feed on weeds before

FIGURE 8.29 Initial symptoms of flea beetle attack.

FIGURE 8.30 Potato leaves severely damaged by flea beetles.

migrating onto potatoes. The inconspicuous eggs are scattered in the soil around the plants and hatch after about a week (Chandel *et al.* 2011). The whitish larvae burrow into the soil and feed for 2–3 weeks on the rootlets within 3–8 cm of the surface. After spending about a week as a whitish pupa in the soil, the new adult emerges. There are generally one or two generations in a year.

Control: Keeping down weeds around the fields is often the most important method of holding these pests in check, since the adults often feed on weeds in early spring and late fall, and the larvae may develop in great numbers on the roots of certain weeds. If the beetles get into the field and attack the plants in large numbers, they can be controlled by spraying malathion, endosulfan, or chlorpyriphos. Application of Bordeaux mixture with endosulfan gives excellent control of flea beetles in potato.

Blister Beetle, *Epicauta hirticornis* Hagg

Epicauta hirticornis (Coleoptera: Meloidae) are slender black beetles, about four times as long as wide, and are soft, with the head distinctly separated from

FIGURE 8.31   Potato leaves damaged by blister beetles, *Epicauta hirticornis*.

the prothorax and the tip of the abdomen exposed beyond the tip of the elytra. Adults feed on the foliage (Fig. 8.31) and may be very damaging. They are very active and are usually found in large groups. The females lay their elongated yellow eggs in clusters of 100–200 in holes they make in the soil (Chandel *et al.* 2011). Newly hatched larvae burrow through the soil until they find a grasshopper egg mass. They then gnaw into the egg pod and feed on the eggs. During the next 3–4 weeks the larvae molt four times, with instars being very morphologically different from one another (a phenomenon known as hyper-metamorphosis).

Control: Where the beetles are very abundant, they may be controlled by spraying with chlorpyriphos or carbaryl. Bordeaux mixture acts as a repellent to the beetles and will give fair protection to potato vines. Blister beetles may be hand-collected in polythene bags and emptied into a pan of kerosene.

Gray Weevil, *Myllocerus subfasciatus* Guerin

Myllocerus subfasciatus (Coleoptera: Curculionidae) has been reported from Tamil Nadu (Nair 1975). Both larvae and adults cause damage. Adult feeding leaves notches on leaf margins, while larvae make small feeding holes in young tubers (Chandla 1985).

The brown weevils lay about 500 eggs in soil. The egg hatches in about a week, and the grubs are fully fed in 2–2½ months. The grubs are small, white, and legless. They feed on roots. Pupation takes place in soil in earthen cocoons. The pupal period lasts from 10 to 12 days.

Adults can be manually collected and mechanically destroyed. Spraying with dichlorvos or endosulfan is also effective.

REFERENCES

Anonymous, 1989. Potato in North Eastern India. Tech. Bull. No. 18 CPRI, Shimla, India.

Anonymous, 1991. Annual Report. International Potato Centre (CIP), Lima, Peru.

Anonymous, 2000. Package of Practices for Rabi Crops. Directorate of Extension Education, HPKV Palampur (Himachal Pradesh).

Arneja, A.K., 2000. Biology of whitefly, *Bemisia tabaci* (Gennadius) on American cotton. MS thesis, Punjab Agricultural University, Ludhiana, India.

Beeson, C.F.C., 1941. The ecology and control of the forest insects of India and neighboring countries. Vasant Press, Dehradun, India.

Bhalla, O.P., Pawar, A.D., 1977. A survey study of insect and non-insect pests of economic importance in Himachal Pradesh. Tikku and Tikku, Kitab Mahal, Bombay, India.

Broad, V.K., Puri, S.N., 1993. Some studies on the behaviour of whitefly. J. Maharashtra Univ. 18, 101–103.

Butani, D.K., Jotwani, M.G., 1984. Insects in Vegetables. Colour Publications, Mumbai, India.

Butter, N.S., Kular, J.S., 1999. Resurgence of whitefly in cotton and its management. Indian J. Entomol. 6, 85–90.

Chandel, R.S., Chandla, V.K., 2003. Managing tuber damaging pests of potato. Indian Hortic. 48, 15–17.

Chandel, R.S., Chandla, V.K., 2005. Integrated control of potato tuber moth (*Phthorimaea operculella*) in Himachal Pradesh. Indian J. Agric. Sci. 75, 837–839.

Chandel, R.S., Kashyap, N.P., 1997. About white grubs and their management. Farmer And Parliament XXXVII, 29–30.

Chandel, R.S., Gupta, P.R., Chander, R., 1994a. Diversity of scarabaeid beetles in mid hills of Himachal Pradesh. Himachal J. Agric. Res. 20, 98–101.

Chandel, R.S., Gupta, P.R., Chander, R., 1995. Behaviour and biology of the defoliating beetle, *Brahmina coriacea* (Hope) (Coleoptera: Scarabaeidae) in Himachal Pradesh. J. Soil Biol. Ecol. 15, 82–89.

Chandel, R.S., Chander, R., Gupta, P.R., Verma, T.D., 1994b. Relative efficacy of some insecticides against apple defoliating beetle *Brahmina coriacea* (Hope) on apple. Indian J. Plant Prot. 22, 45–49.

Chandel, R.S., Gupta, P.R., Thakur, J.R., 1997. Host preference and seasonal abundance of defoliating beetles infesting fruit trees in mid hills of Himachal Pradesh. J. Soil Biol. Ecol. 17, 140–146.

Chandel, R.S., Kumar, R., Kashyap, N.P., 2000. Saving potato from tuber moth. Indian Farming 50 (9), 28.

Chandel, R.S., Kumar, R., Kashyap, N.P., 2001. Bioecology of potato tuber moth, *Phthorimaea operculella* Zeller in mid hills of Himachal Pradesh. J. Entomol. Res. 25, 195–203.

Chandel, R.S., Chandla, V.K., Sharma, A., 2003. Population dynamics of potato white grubs in Shimla hills. J. Indian Potato Assoc. 30, 151–152.

Chandel, R.S., Chandla, V.K., Singh, B.P., 2005. Potato tuber moth – *Phthorimaea operculella* (Zeller). Tech. Bull. (No. 65) CPRI, Shimla, India.

Chandel, R.S., Dhiman, K.R., Chandla, V.K., Kumar, V., 2007. Integrated pest management in potato. In: Jain, P.C., Bhargava, M.C. (Eds.), Entomology: Novel Approaches in Entomology, New India Publishing Agency, New Delhi, India, pp. 377–398.

Chandel, R.S., Dhiman, K.R., Chandla, V.K., Desh, R., 2008a. Insect pests of potato – I: Root and tuber eating pests. Pestology 32, 39–46.

Chandel, R.S., Chandla, V.K., Garg, I.D., 2008b. Integrated pest management of potato tuber moth in India. Trop. Agric. 20, 129–140.

Chandel, R.S., Dhiman, K.R., Garg, I.D., Desh, R., 2008c. Insect pests of potato – II: Sap feeding pests. Pestology 32, 47–53.

Chandel, R.S., Banyal, D.K., Singh, B.P., Malik, K., Lakra, B.S., 2010. Integrated management of whitefly, *Bemisia tabaci* (Gennadius) and potato apical leaf curl virus in India. Potato Res. 53, 129–139.

Chandel, R.S., Sharma, P.C., Verma, K.S., Mehta, P.K., Vinod, K., 2011. Insect pests of potato – III: Leaf eating and defoliating insects. Pestology 35, 60–66.

Chandla, V.K. 1985. Potato pests and their management. Indian Farming 34, 31–32.

Chandla, V.K., Misra, S.S., Bist, S.S., Bhalla, O.P., Thakur, J.R., 1988. Whitegrub *Brahmina coriacea* (Hope) infesting potato in Shimla Hills. Seeds & Farms 14, 12–13.

Chandla, V.K., Khurana, S.M.P., Garg, I.D., 2004. Aphids, their importance, monitoring and management in seed potato crop. Tech. Bull. (No. 61) CPRI, Shimla, India.

Chandla, V.K., Verma, K.D. 2000. Potato Aphids. In: Khurana Paul, S.M. (Ed.), Diseases and Pests of Potato – A Mannual. CPRI, Shimla, India, pp. 48–52.

Das, B.B., Ram, G., 1988. Incidence, damage and carryover of cutworms (*Agrotis ipsilon*) attacking potato (*Solanum tuberosum*) crop in Bihar. Indian J. Agric. Sci. 58, 650–651.

Chandla, P.C., 2000. Potato in India. Kalyani Publishers, New Delhi, India.

Dhawan, P., Mandal, R.B., 2008. Effect of apical leaf curl begomovirus disease on growth and yield parameters of potato. In: Proceedings, Global Potato Conference – Opportunities and Challenges in the New Millennium, pp. 159–160, 9–12 Dec. 2008, New Delhi, India.

Dhawan, A.K., Butter, N.S., Narula, A.M., 2007. The cotton whitefly *Bemisia tabaci* (Gennadius). Tech. Bull., Department of Entomology, PAU Ludhiana, India.

Fletcher, T.B., 1919. Life histories of Gelechiidae. Mem. Dep. Agric. India 6, 75–76.

Garg, I.D., Khurana, S.M.P., Kumar, Shiv, Lakra, B.S., 2001. Association of a geminivirus with potato apical leaf curl in India and its immunized electron microscopic detection. J. Indian Potato Assoc. 28, 227–232.

Haq, A., 1962. Notes on the Bionomics of *Lachnosterna longipennis* B1. (Melolonthinae: Coleoptera). Indian J. Entomol. 24, 220–221.

Khurana, S.M.P., Bhale, Usha, Garg, I.D., 2001. Stem Necrosis disease of potato. Tech. Bull. (No. 54) CPRI, Shimla, India.

Kishore, R., Misra, S.S., 1988. Impact of different planting dates on the incidence of cut worm on potato crop. Bull. Entomol. 29, 223–225.

Kishore, R., Misra, S.S., Singh, L., 1993. Incidence of red ant, *Dorylus orientalis*, on different potato genotypes. J. Indian Potato Assoc. 20, 62.

Konar, A., Paul, S., Basu, A., Chattri, M., 2005. Integrated management of mole cricket attacking potato in Eastern Gangetic plains of West Bengal. Potato J. 32, 250.

Kumar, R., Nirula, K.K., 1967. Control of potato tuber moth in the field. Indian J. Agric. Sci. 37, 553–554.

Kumar, S., Singh, P.H., Garg, I.D., Khurana, S.M.P., 2003. Integrated management of potato diseases. Indian Hortic 48, 14.

Lakra, B.S., 2002. Leaf curl: A threat to potato crop in Haryana. Journal Mycol. Plant Pathol. 32, 367.

Lakra, B.S., 2003. Effect of date of planting on white fly population, leaf curl incidence and yield of potato cultivars. J. Indian Potato Assoc. 30, 115–116.

Lal, K.B., 1945. Prevention of damage to stored potatoes by PTM. Curr. Sci. 14, 131.

Lefroy, H.M., 1907. The potato tuber moth. Indian Agric. J. 2, 294–295.

Lingappa, S., Giraddi, R.S., 1995. Insect pests of potato and their management. In: Research Highlights on potato, Division of Horticulture UAS Dharwad, Karnataka, India, pp. 19–22.

Malik, K., Chandel, R.S., Singh, B.P., Chandla, V.K., 2005. Studies on potato apical leaf curl virus disease and its whitefly vector *Bemisia tabaci*. In: Proceedings, Annual Meeting of Indian Society of Plant Pathologists and Centenary Symposium on Plant Pathology, 7–8 April 2005, India. CPRI, Shimla, India.

Mathur, Y.S., Bhatnagar, A., Singh, S., 2010. Bioecology and management of phytophagous whitegrubs of India. Technical Bulletin 4. All India Network Project on Whitegrubs and Other Soil Arthropods. Agriculture Research Station Durgapura, Jaipur, India.

Mehta, P.K., Chandel, R.S., Mathur, Y.S., 2008. Phytophagous white grubs of Himachal Pradesh. Tech. Bull. CSK HPKV, Palampur, India.

Mehta, P.K., Chandel, R.S., Mathur, Y.S., 2010. Status of whitegrubs in north western Himalaya. J. Insect Sci. 23, 1–14.

Mishra, P.N., 2001. Scarab Fauna of Himalayan Region and Their Management. In: Sharma, G., Mathur, Y.S., Gupta, R.B.L. (Eds.), Indian Phytophagous Scarabs and their Management Present Status and Future Strategy, Agrobios (India), Jodhpur, India, pp. 74–85.

Mishra, P.N., Singh, M.P., 1993. Field biology of white grubs, *Holotrichia longipennis* on potatoes in UP hills. J. Indian Potato Assoc. 20, 249–251.

Misra, S.S., 1995. Potato pests and their management. Tech. Bull. (No. 45) CPRI, Shimla, India.

Misra, S.S., Chandel, R.S., 2003. Potato white grubs in India. Tech. Bull. (No. 60) CPRI, Shimla, India.

Misra, S.S., Chandla, V.K., 1989. White grubs infesting potatoes and their management. J. Indian Potato Assoc. 16, 29–33.

Misra, S.S., Chandla, V.K., Singh, A.K., 1995. Potato pests and their management. Tech. Bull. 45 CPRI, Shimla, India.

Misra, S.S., Singh, D.B., Chandla, V.K., Raj, B.T., 2003. Potato pests and their management. In: Khurana, S.M.P., Minhas, J.S., Pandey, S.K. (Eds.), The Potato – Production and Utilization in Subtropics, Mehta Publishers, New Delhi, India, pp. 252–269.

Musthak Ali, T.M., 2001. Biosystematics of Phytophagous Scarabacidae – An Indian Overview. In: Sharma, G., Mathur, Y.S., Gupta, R.B.L. (Eds.), Indian Phytophagous Scarabs and their Management Present Status and Future Strategy, Agrobios (India), Jodhpur, India, pp. 5–37.

Nair, M.R.G.K., 1975. Insect and mites of crops in India. Indian Council of Agricultural Research, New Delhi, India.

Nirula, K.K., 1960. Control of potato tuber moth. Indian Potato J. 2, 47–51.

Pandey, R.P., 2002. The Potato. Kalyani Publishers, New Delhi, India.

Pandey, S.K., Kang, G.S., 2003. Ecological and varietal improvement. In: Khurana, S.M.P., Minhas, J.S., Pandey, S.K. (Eds.), The Potato – Production and Utilization in Subtropics, Mehta Publishers, New Delhi, India, pp. 48–60.

Puri, S.N., Baranwal, V.K., Surender, K., 1995. Cotton leaf curl disease and its management. NCIPM Extension Folder 1, National Centre for Integrated pest management, New Delhi, India.

Rahman, K.A., 1944. Prevention of damage to stored potatoes by the potato tuber moth. Curr. Sci. 13, 133–134.

Rai, K.M., Joshi, R., 1988. Control of white grubs (Kurmula) damaging potato crop in U.P. hills. Prog. Hortic. 20, 333–334.

Raj, B.T., Parihar, S.B.S., 1993. Red ant damage on potato in western Uttar Pradesh. J. Indian Potato Assoc. 20, 61–62.

Rana, R.K., 2011. The Indian potato processing industry: global comparison and business prospects. Outlook Agric. 40, 237–243.

Rao, N.V., Reddy, A.S., 1989. Seasonal influence on the developmental duration of whitefly (*Bemisia tabaci*) in upland cotton (*Gossypium hirsutum*). Indian J. Agric. Sci. 59, 283–285.

Reddy, A.S., Rao, N.V., 1989. Cotton whitefly (*Bemisia tabaci Genn.*). Indian J. Plant Prot. 17, 171–179.

Regupathy, A., Palanisamy, S., Chandramohan, N., Gunathilagaraj, K., 1997. A Guide on Crop Pests. Sooriya Desktop Publishers, Coimbatore, India.

Roonwal, M.L., 1976. Plant pest status of root eating ants, *Dorylus orientalis* with notes on taxonomy, distribution and habits (Insects: Hymenoptera). J. Bombay Nat. Hist. Soc. 72, 305–313.

Saini, H.K., 1998. Effect of synthetic pyrethroids on biology of whitefly *Bemisia tabaci* (Gennadius) on *Gossypium hirsutum* (Linn.). MS thesis, Punjab Agricultural University, Ludhiana, India.

Shah, N.K., Shah, L., 1990. Bionomics of *Holotrichia longipennis* (Coleptera: Melolonthinae) in western Himalayas. Indian J. For. 13, 234–237.

Shankar, U., Priya, S., Kumar, D., 2008. Vegetable pest management: guide for farmers. International Book Distributing Company, Lucknow, India.

Sharma, P., Rishi, N., 2004. Population build-up of the cotton whitefly, *Bemisia tabaci* in relation to weather factors at Hissar, Haryana. Pest Manage. Econ. Zool. 12, 33–38.

Sharma, P.L., Attri, B.S., Aggarwal, S.C., 1969. Beetles causing damage to pome and stone fruits in Himachal Pradesh and their control. Indian J. Entomol. 31, 377–379.

Singh, M.B., Bhagat, R.M., Sharma, D.C., 1990. Life history and host range of potato tuber moth (*Phthorimaea operculella* Zeller). Himachal J. Agric. Res. 16, 59–62.

Singh, R.B., Khurana, S.M.P., Pandey, S.K., Srivastava, K.K., 1997. Screening germplasm for potato stem necrosis resistance. Indian Phytopathol. 51, 222–224.

Singh, R.B., Khurana, S.M.P., Pandey, S.K., Srivastava, K.K., 2000. Tuber treatment with imidacloprid is effective for control of potato stem necrosis disease. Indian Phytopathol. 53, 142–145.

Singh, S.P., 1987. Studies on some aspects of biology – ecology of potato cutworms in India. J. Soil Biol. Ecol. 7, 135–143.

Srivastava, K.P., Butani, D.K., 1998. Pest management in vegetables. Part I Periodicals and Book Publishing House, El Paso, Texas.

Sushil, S.N., Mohan, M., Selvakumar, G., Bhatt, J.C., 2006. Relative abundance and host preference of white grubs (Coleoptera: Scarabaeidae) in Kumaon hills of Indian Himalayas. Indian J. Agric. Sci. 76, 338–339.

Trivedi, T.P., Rajagopal, D., 1999. Integrated Pest Management in Potato. In: Upadhayay, R.K., Mukeriji, K.G., Dubey, O.P. (Eds.), IPM System in Agriculture, Cash Crops, Vol. 6, Aditya Books Pvt. Ltd., New Delhi, India, pp. 299–313.

Trivedi, T.P., Rajagopal, D., Tandon, P.L., 1994. Environmental correlates of the potato tuber moth, *Phthorimaea operculella* (Zeller) (Lepidoptera:Gelechiidae). Intl J. Pest Manage 40, 305–308.

Veersh, G.K., Kumar, A.R.V., Musthak Ali, T.M., 1991. Biogeography of pest species of white grubs of Karnataka. In: Veersh, G.K., Rajagopal, D., Viraktamath, C.A. (Eds.), Advances in management and conservation of soil fauna Oxford and IBP, Publishing Company Pvt Ltd., Bangalore, India, pp. 191–198.

Verma, K.D., Chandla, V.K., 1999. Potato aphids and their management. Tech. Bull. (No. 26) CPRI, Shimla, India.

Woodhouse, E.J., 1912. Potato moth in Bengal. Indian Agric. J. 7, 264–271.

Yadava, C.P.S., Sharma, G., 1995. Indian white grubs and their management. Tech. Bull. (No. 2) All India Coordinated Research Project on White Grubs, Jaipu, India.

Part III

The Potato Field as a Managed Ecosystem

Chapter 9

Spud Web: Species Interactions and Biodiversity in Potatoes

Christine A. Lynch, David W. Crowder, Randa Jabbour, and William E. Snyder
Department of Entomology, Washington State University, Pullman, WA, USA

INTRODUCTION

Much early work in biological control focused on interactions between particular natural enemy species and their pestiferous prey (DeBach and Rosen 1991). This approach likely reflected the many successes of classical biological control, where natural enemies of an introduced pest were released into the invasive range and dramatically reduced pest densities. Classical biological control agents were sought that had a high degree of prey specificity and a reproductive rate as high as the target pest (DeBach and Rosen 1991). Such highly specialized natural enemies are tightly linked to particular herbivores, and have the clear ability to exert density-dependent regulation of pests (Hawkins *et al.* 1997, 1999). However, recent years have seen growing interest in the community ecology of biological control, where interactions among more than two species are considered (e.g., Vandermeer 1995, Matson *et al.* 1997). First, classical biological introductions have become more difficult and expensive because of growing societal concerns about potential unintended negative impacts of biocontrol agents on native species (Howarth 1991). Furthermore, this approach is only possible when pests are non-native exotics. Instead, there is increasing interest in the conservation of native natural enemies, including generalist predators, which often interact with many other species in addition to any single target pest (Landis *et al.* 2000, Symondson *et al.* 2002). Second, consumer concerns about possible negative human- and environmental-health effects of pesticides have increased the adoption of organic agriculture and other pesticide-reduction schemes (Crowder *et al.* 2010). With a reduction in broad-spectrum pesticide applications by growers, the abundance and diversity of natural enemies generally increases (Bengtsson *et al.* 2005, Hole *et al.* 2005, Straub *et al.* 2008). This inevitably increases the complexity of agricultural food webs (e.g., Tylianakis *et al.* 2007).

The relationship between biodiversity and biocontrol has long been a topic of interest to basic and applied ecologists. Early agroecologists thought that the

Insect Pests of Potato. http://dx.doi.org/10.1016/B978-0-12-386895-4.00009-0

relative ecological simplicity of agricultural monocultures rendered them more prone to herbivore outbreaks than natural systems (Pimentel 1961, Root 1973, van Emden and Williams 1974). If so, then the restoration of more natural levels of biodiversity would improve natural pest control (Straub *et al.* 2008). This viewpoint is consistent with the emerging sub-discipline of "biodiversity-ecosystem function" research, which suggests that ecosystem health is maximized only within species-rich communities (Ives *et al.* 2005, Hooper et al. 2006). Other viewpoints also exist. Simplest is the "green-world hypothesis" of Hairston *et al.* (1960), who proposed that the world's ecosystems break down into three simple trophic levels: plants, herbivores, and predators. Predators generally regulate herbivore densities, so that plants grow and proliferate largely unscathed. Biodiversity is never explicitly discussed in the 1960 paper, but it can be inferred that predators (and other natural enemies) form a cohesive third trophic level that acts consistently regardless of the number of species present (Hairston and Hairston 1993, 1997). In contrast, the "trophic-level omnivory hypothesis" suggests that complex feeding relationships among species blur the formation of distinct trophic levels; without trophic levels, cascading predator effects on herbivores and then plants cannot occur (Strong 1992, Polis and Strong 1996). For example, ecological communities typically include many species of generalist predators, which often feed on plants, detritivores, and other predators in addition to herbivores (Polis 1991). Data from particular studies in agricultural crops variously support each of these hypotheses (Straub *et al.* 2008).

Potato crops often contain diverse communities of herbivores and their natural enemies (Walsh and Riley 1868, Hough-Goldstein *et al.* 1993, Hilbeck and Kennedy 1996, Hilbeck *et al.* 1997, Koss *et al.* 2005), providing an ideal model system to study the ecological issues discussed above. Traditionally, potatoes have been heavily treated with insecticides, due to the high value of the crop and the severe damage caused by some insect and other pests (Hare 1990). However, recent years have seen growing pressure to reduce insecticide use on this food crop, leading to a growing organic sector and the adoption of pesticide-reduction schemes in conventionally managed fields (Koss *et al.* 2005, Werling and Gratton 2010). Here, we explore species interactions among herbivores and natural enemies occurring in potato crops, and how these interactions are influenced by increasing biodiversity. We separately consider how biodiversity effects might operate at the plant, herbivore, and predator trophic levels.

PLANT BIODIVERSITY

Root (1973) presented two hypotheses for how increasing plant diversity in agricultural fields might decrease pest problems. First, specialist herbivores might have a more difficult time finding their host plant against a background of other plant species. Second, diverse plantings might support a more diverse community of non-pest prey and other foods, increasing the abundance and diversity of natural enemies. Indeed, diversified plantings often house fewer pests than do agricultural monocultures (Russell 1989, Andow 1991). However, differentiating between the two above-described hypotheses as the root cause of this diversity effect has been problematic (Bommarco and Banks 2003). Increasingly, agroecologists have examined crop diversification plans as a means to conserve natural enemies (Landis *et al.* 2000). Plant diversification can be accomplished either at local scales, with resource-providing plants installed at field edges or within the crop (Landis *et al.* 2000), or at the scale of landscapes, where regions including more crop species and more non-crop habitats might provide diverse resources for highly-mobile natural enemies (Tscharntke *et al.* 2005).

In-Field Plant Diversity

Perhaps the most common approach to diversifying the in-field habitat of potato crops has been the application of straw mulch. Straw mulch can benefit predators by establishing more benign environmental conditions relative to bare-ground fields, by providing protection from the predators' own natural enemies (including intraguild predators), and/or by providing food for detritivores that act as supplemental, non-pest prey (Settle *et al.* 1996, Halaj and Wise 2002, Langellotto and Denno 2004). Brust (1994) reported that potato plantings receiving straw mulch housed larger populations of generalist predators, and also lower densities of Colorado potato beetles. With fewer potato beetles, potato plants suffered less feeding damage. He also found no difference in potato beetle movement between mulched and unmulched plants, suggesting that the benefits of mulch were not due to potato beetles avoiding the colonization of mulched potatoes (Brust 1994). Several other studies confirmed that potato beetle densities decrease in straw-mulched plots (Zehnder and Hough-Goldstein 1990, Stoner 1993, Johnson *et al.* 2004). Often, the application of straw mulch has been associated with higher potato yields (Zehnder and Hough-Goldstein 1990, Stoner 1993), although it is not always clear whether this effect is due to reduced potato beetle densities, better moisture retention in the soil, or both (Stoner *et al.* 1996).

Nonetheless, increased structural complexity is not universally beneficial. Szendrei and Weber (2009) found that potato crops planted into rye stubble attracted higher densities of predatory *C. maculata* lady beetles but lower densities of the predatory carabid beetle *Lebia grandis*. Mulching decreased overall beetle suppression by predators because the carabid was the most voracious predator of potato beetles (Szendrei and Weber 2009). Indeed, molecular gut-content analysis revealed that predators were more likely to have fed on potato beetles in unmulched than in mulched plots (Szendrei *et al.* 2010).

Predators of aphid and potato beetle in potatoes, such as the lady beetle *Coleomegilla maculata*, are often highly mobile, suggesting they may be affected by the mixture of potato and other plant habitats within and around farms. For example, corn crops seem to be most attractive to predatory lady beetles, such that lady beetle densities in potato fields planted near corn can be too low to exert significant impacts on potato pests (Groden *et al.* 1990, Nault and Kennedy 2000). However, if the attractiveness of potato crops could be increased by, for example, planting flowering plants at field edges or between crop rows, potatoes might be more attractive to lady beetles and other natural enemies (e.g., Patt *et al.* 1997). This could increase biological control impacts in potatoes. For example, Groden *et al.* (1990) suggested timed cutting of adjacent alfalfa crops as a strategy to encourage *C. maculata* to move from alfalfa to potatoes as potato-pest densities grow. Idoine and Ferro (1990) indicated that adult females of the Colorado potato beetle egg parasitoid *Edovum puttleri* benefited from feeding on aphid honeydew, suggesting that providing a sugar source for the wasps could benefit the control of potato beetles. Such approaches to predator conservation are fraught with risk, however. For example, Baggen and Gurr (1998) found that planting flowering plants near potatoes provided food for adults of the parasitoid *Copidosoma koehleri*, an egg parasitoid of the pestiferous potato moth *Phthorimaea operculella*. These "floral resources" extended parasitoid longevity and increased fecundity, both of which would benefit potato moth biocontrol. However, in field trials, providing flowers actually increased pest densities in potatoes and resulting crop damage; this occurred because potato moths also benefitted from the use of floral resources (Baggen and Gurr 1998). Subsequent work showed that flower species could be selected which benefitted only the parasitoid and not the moth (Baggen *et al.* 1999), such that the unintended benefits to the pest could be eliminated and the parasitoid selectively promoted (Baggen *et al.* 1999).

Landscape-Scale Plant Diversity

Plant diversity can also be beneficial at scales larger than single potato fields. Werling and Gratton (2010) examined how landscape structure and diversity impacted predation of green peach aphid (*Myzus persicae*) and Colorado potato beetle (*Leptinotarsa decimlineata*) pests, and found that control of the two pests was impacted at different spatial scales. Green peach aphid predation was highest in field margins adjacent to potato fields that were imbedded within landscapes containing diverse non-crop habitats within 1.5 km; however, this effect of landscape diversity did not alter predation of green peach aphids within the potato fields themselves (Werling and Gratton 2010). In contrast, predation of Colorado potato beetle eggs was unaffected by landscape diversity at larger scales, but instead increased as the ratio of field margin to crop increased. Thus, potato beetle predation was greatest in smaller potato fields where relatively diverse edge habitats were always close by (Werling and Gratton 2010). Ground beetle predation of weed seeds was greater in field margins compared to adjacent potato crops (Gaines and Gratton 2010), again suggesting that greater plant diversity in field margins increased predation relative to what was seen in the plant-species-poor potato crop. In conclusion, the presence and size of adjacent non-crop habitat can have important implications for predation of herbivores in potato fields, but there is limited evidence for the role of landscape complexity at larger scales.

HERBIVORE BIODIVERSITY

While pest management decisions in potatoes may often focus on just one or two pests perceived to be of the greatest economic importance, these crops typically house a diverse community of many herbivore species (Lynch *et al.* 2006). Multiple species of herbivores have the potential to indirectly impact one another's densities, with these interactions generally transmitted indirectly by the host plant (Karban and Baldwin 1997, Lill and Marquis 2001, Ohgushi 2005, Rodriguez-Saona and Thaler 2005). These multi-herbivore effects can be either harmful or beneficial to particular key potato pests. Likewise, the presence of multiple herbivore species can either heighten or weaken the impact of natural enemies on a particular species of pest (e.g., Eubanks and Denno 2000). Below, we review research literature examining how herbivore diversity impacts interactions among herbivore species, and among predators and their herbivore prey, on potatoes.

Herbivore-Herbivore Interactions Mediated by Plants

Plant-mediated indirect interactions among herbivores generally fall into three categories: (1) those due to resource depletion, where one herbivore species consumes plant resources that are then unavailable to a second herbivore species; (2) those due to induced plant defenses, where feeding by one herbivore species triggers the plant to deploy defensive strategies that harm a second, often later-arriving, herbivore species; and (3) those due to alterations in plant chemistry following feeding by one herbivore species that render a plant more susceptible and/or attractive to a second herbivore species. Interactions of all types have been described in potato crops, or are likely to occur.

The Colorado potato beetle is often on the losing end of plant-mediated competition with other herbivore species (e.g., Tomlin and Sears 1992a, 1992b, Wise 2002, Lynch *et al.* 2006, Kaplan *et al.* 2007). This might be because the Colorado potato beetle generally does its greatest damage relatively late in the growing season, such that its feeding is influenced by changes in plant chemistry/quality that have been induced by earlier-arriving herbivores. For example, early-season feeding by both flea beetles (Wise and Weinberg 2002) and potato leafhoppers (Lynch *et al.* 2007) causes female potato beetles to avoid ovipositing on affected potato plants, and slows development of any larvae that do hatch. Slower larval development not only makes it more difficult for Colorado potato beetles to complete multiple generations each year, but also heightens the risk of beetle larvae falling victim to predators; predation risk is heightened because the beetles are forced to spend more time in vulnerable smaller stages (Kaplan *et al.* 2007).

Although the precise mechanism through which earlier feeding by flea beetles or leafhoppers harms potato beetles has not been demonstrated, potato plants defend themselves through some combination of altered nutritional quality, allelochemical defenses, and morphological alterations (Tomlin and Sears 1992a, 1992b; Hlywka *et al.* 1994, Bolter and Jongsma 1995, Pelletier *et al.* 1999). It is likely that one or more of these defenses are involved in induced resistance to Colorado potato beetle. The indirect effect of leafhoppers on potato beetles depends on the density of the leafhoppers, unlike chemical defenses that can be triggered to high levels by relatively little herbivore feeding (Heil and Kost 2006), and leafhoppers are known to alter the amino acid profile of potato foliage (Tomlin and Sears 1992a). This implies that leafhoppers harm potato beetles, at least in part, by reducing the nutritional value of potato plants (Lynch *et al.* 2006).

It is worth noting that previous feeding by other herbivore species does not invariably deter subsequent potato beetle attack. For example, in one study, potato plants that had cabbage looper regurgitant applied to wounds in the foliage, simulating caterpillar attack, attracted more potato beetle adults than did undamaged plants (Landolt *et al.* 1999). Similarly, potato plants attacked by beet armyworm larvae were more attractive to colonizing potato beetles (Bolter *et al.* 1997). In the Bolter *et al.* (1997) study, previous feeding by Colorado potato beetles also heightened attractiveness of damaged plants to later-arriving potato beetles, suggesting that plant damage by any chewing herbivore rendered potato plants more attractive.

Herbivore-Herbivore Interactions Mediated by Shared Predators

Herbivore species can also indirectly impact one another by changing the behavior of shared predators. For example, "apparent competition" occurs when the presence of one herbivore species draws in larger numbers of natural enemies than might otherwise be found. When these enemies switch to feeding also on a second herbivore species, the second herbivore species is harmed (Holt 1977, Harmon and Andow 2004). Thus, one herbivore species harms another by increasing the second species' risk of predation. Apparent completion among potato herbivores has not been directly demonstrated, but there is good circumstantial evidence to suggest it might occur. For example, green peach aphid and other aphid pests of potato can attract large numbers of aphid-associated lady beetles and other generalist predators (Koss *et al.* 2005), which likely opportunistically feed on potato beetle eggs and other vulnerable herbivore species (e.g., Chang and Snyder 2004). Similarly, potato plants damaged by Colorado potato beetles are more attractive to the predatory pentatomid *Perillus bioculatus* than are undamaged plants (Weissbecker *et al.* 1999, 2000); presumably, once these generalist predators are drawn to a crop they would feed also on prey other than potato beetles.

However, the presence of one herbivore species will not necessarily lead to higher predator attack rates on a second, co-occurring herbivore species. In fact, the presence of a preferred prey might draw predator attacks away from a second, unpalatable or less-preferred herbivore species (Harmon and Andow 2004). For example, in laboratory arenas *Coleomegilla maculata* lady beetles eat fewer Colorado potato beetle eggs when aphids are available as alternative prey (Groden *et al.* 1990, Hazzard and Ferro 1991, Mallampalli *et al.* 2005). Similarly, in field cages where predators cannot aggregate at aphid infestations, the presence of green peach aphids protects Colorado potato beetles from attack by a diverse guild of spider and predatory bug generalist predators (Koss and Snyder 2005). Disruption of potato beetle predation in the presence of aphids apparently occurs because aphids are more attractive prey for most predator species. Consistent with this interpretation, the presence of Colorado potato beetles as prey has no impact on these same predators' likelihood of attacking aphids (Koss *et al.* 2004).

NATURAL ENEMY BIODIVERSITY

Potato crops often house a remarkably high diversity of predator, pathogen, and parasitoid natural enemies (Walsh and Riley 1868, Hough-Goldstein *et al.* 1993, Alyokhin and Sewell 2004, Straub and Snyder 2006, Ramirez and Snyder 2009). For example, in North America, Colorado potato beetles are attacked by a diverse guild of generalist egg and larval predators, egg and larval parasitoids, and entompathogenic nematodes and fungi (Lopez *et al.* 1993, Hilbeck and Kennedy 1996, Berry *et al.* 1997, Koss *et al.* 2005, Crowder *et al.* 2010). Indeed, both observational studies and predator surveys using molecular gut content analysis have found a vast array of predators feeding on Colorado potato beetles under entirely natural, open-field conditions (Chang and Snyder 2004, Greenstone *et al.* 2010, Szendrei *et al.* 2010).

As we have seen for plants and herbivores, increasing biodiversity among natural enemies can have either positive or negative consequences for particular potato pests. On the one hand, as more enemy species are added to a community this increases the risk that one enemy will feed on another. Such "intraguild" predation has the potential to greatly disrupt biological control (Polis *et al.* 1989, Rosenheim *et al.* 1995, Ives *et al.* 2005) (Fig. 9.1). On the other hand,

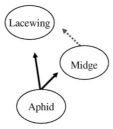

FIGURE 9.1 Lucas *et al.* (1998) found that lacewing larvae fed not only on potato aphids, but also on larvae of predatory *Aphidoletes* midges. "Intraguild" predation of this type has the potential to disrupt aphid control, as lacewing larvae consume midge larvae that otherwise might have eaten many aphids. Arrows denote energy flow and so point from prey to predator; solid arrows indicate predation of the herbivore by predators, while the dotted arrow indicates predation of one predator by another.

combining natural enemies that fill different ecological niches can lead to complementary impacts on a pest species, where different enemy species eliminate spatial or temporal refuges from predation that the pest might otherwise enjoy (Wilby and Thomas 2002, Casula *et al.* 2006, Straub *et al.* 2008) (Fig. 9.2, Plate 9.3). Furthermore, natural enemies sometimes facilitate one another's prey capture. For example, a predator may chase a prey species from one habitat into the waiting clutches of a second predator species located somewhere else in the environment (e.g., Losey and Denno 1998). In these cases, pest control is strongest where several natural enemy species co-occur. We next discuss examples of negative, and then positive, predator-predator interactions that have been found to impact biological control in potatoes.

Negative Predator-Predator Interactions

Natural enemies often feed upon one another, and this intraguild predation has the potential to greatly limit biological control (Rosenheim *et al.* 1995). Some evidence for this comes from potato crops. For example, Mallampali et al. (2002) showed that the spined soldier bug (*Podisus maculiventris*) fed heavily on larvae of the 12-spotted lady beetle (*Coleomegilla maculata*), which disrupted predation of Colorado potato beetle eggs by that lady beetle. Similarly, in laboratory feeding trials the predatory bug *Anthocoris nemorum* was as likely

FIGURE 9.2   In potato foliage, *Nabis alternatus* bugs and *Hippodamia convergens* beetles exert complementary impacts on *Myzus persicae* aphids (Straub and Snyder 2008). The lady beetles forage primarily at leaf edges, whereas the predatory bugs also forage at leaf centers. Because of these differences in space use, aphids face heavy predation pressure everywhere on the plant only when both predator species occur together.

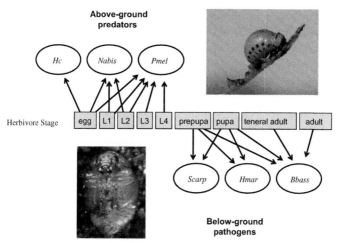

FIGURE 9.3   In Washington potato crops, predators and pathogens exert complementary impacts on Colorado potato beetles (Ramirez and Snyder 2009). Potato beetle eggs and larvae in the foliage are eaten by a diverse group of predatory *Hippodamia convergens* (*Hc*) and *Pterostichus melanarius* (*Pmel*) beetles, and *Nabis alternatus* (*Nabis*) bugs. Once the beetles enter the soil to pupate they are infected by entomopathogenic *Steinernema carpocapsae* (*Scarp*) and *Heterorhabditis marelatus* (*Hmar*) nematodes, and *Beauveria bassiana* (*Bbass*) fungi. Because of the spatiotemporal separation of predators and pathogens, potato beetles face attack throughout their life cycle only when both classes of natural enemy are present. See also Plate 9.3.

to feed on green peach aphids parasitized by the wasp *Aphidius colemani* as on unparasitized aphids (Meyling *et al.* 2004), and *Beauveria bassiana* fungi pathogenic to green peach aphids and Colorado potato beetles also attacked *C. maculata* lady beetles (Todorova *et al.* 2000). If these interactions also occur in the field, they could disrupt biological control.

One of the more detailed series of studies on intraguild predation among aphid predators examined interactions among predatory midges (*Aphidoletes aphidomyza*), lacewings (*Chrysoperla rufilabris*), and lady beetles (*C. maculate*) attacking potato aphids (*Macrosiphum euphorbiae*) (Fig. 9.1). The midges are highly susceptible to intraguild predation by the lacewing and lady beetle; lacewing and lady beetle larvae also attack one another (Lucas *et al.* 1998). Apparently in response to its high risk of falling victim to intraguild predation, the midge has developed several strategies to reduce this danger. First, midge larvae will "hide" at the center of aphid aggregations, so that aphids at the colony periphery will absorb most predator attacks (Lucas and Brodeur 2001). Second, midge eggs are deposited on plants with high trichome density, which makes them relatively immune from predation by *C. maculata* predators (Lucas and Brodeur 1999).

The parasitoid wasp *Aphidius nigripes* attacks green peach aphids (*Myzus persicae*) in potato crops in North America, where it faces attack by hyperparasitoids (Brodeur and McNeil 1992). As the parasitoid larva nears the end of its development, it somehow alters the host aphid's behavior, causing the doomed aphid to walk to the top of the potato plant before being killed by the parasitoid. The wasp then pupates beneath its former host's exoskeleton. Brodeur and McNeil (1992) found that parasitoid pupae located at the tips of plants were unlikely to fall victim to hyperparasitoids. In alfalfa crops, this placement of parasitoid pupae was found to make the wasps less accessible for ground-foraging, predatory carabid beetles that occasionally climb onto plants to forage (Snyder and Ives 2001). Altogether, this suggests that the selective pressures exerted by intraguild predators have led the parasitoid to develop its ability to change host behavior, in order to be delivered to a pupation site largely out of reach of the parasitoid's own natural enemies (Brodeur and McNeil 1992). Similarly, the lady beetle *C. maculata* leaves aphid-infested potato plants to pupate elsewhere, apparently in order to avoid intraguild predation by the many predators that aphids attract (Lucas *et al.* 2000).

Positive Predator-Predator Interactions

There is growing evidence that species that occupy different niches consume more resources than any single species can consume (Cardinale *et al.* 2006, Hooper et al. 2006). The same may hold true when herbivorous agricultural pests are the "resource", and predator, parasitoid, and pathogen natural enemies are the "consumers" (Straub *et al.* 2008). After all, natural enemy species differ from one another in where they hunt in the environment, the time of day or year that they are active, and the particular hunting style they use, such that pests are likely to face a broad-based attack only when several natural enemy species co-occur (Wilby and Thomas 2002, Snyder 2009). In this sense, natural enemy species are likely to complement one another, and some of the best evidence of this comes from work in potato crops.

Working with potato plants enclosed in large field cages, Straub and Snyder (2006, 2008) compared the effects of single natural enemy species on green peach aphid prey to those of diverse mixes of three or four enemy species (Fig. 9.2). The biocontrol agents considered in this work were a diverse group comprised of spiders, ladybeetles, ground beetles, *Nabis* and *Geocoris* true bugs, and parasitoid wasps. All experiments manipulated predators following a substitutive design, in which total predator density (or, in some experiments, predator biomass) was held constant across species richness levels. Such designs isolate the impacts of changes in species number by eliminating differences among treatments in species abundance (Loreau and Hector 2001). Initial experiments showed that the different predator species strongly differed in how effective they were at killing aphids, but that diverse mixes of predator species killed only slightly more aphids than did single enemy species (Straub and Snyder 2006). This suggested that predator species did not strongly complement one another. However, subsequent work painted a more complex picture. When predator diversity was manipulated simultaneously on both collard (*Brassica oleracea*) and potato plants colonized by the green peach aphids – in that experiment, plants of the two species were present in different cages – diverse predator communities killed more aphids than any single enemy species on plants of both species (Straub and Snyder 2008). Predators complemented one another quite strongly on collard plants, where diverse predator communities killed 200 more aphids than did single species, but quite weakly on potatoes, where diverse predator communities killed just 6 more aphids than did single predator species (Straub and Snyder 2008). Behavioral observations indicated that lady beetles foraged largely at the edges of leaves, while predatory bugs and parasitoids foraged at leaf centers (Fig. 9.2). Because collard leaves are larger, these space-use niche differences, and thus predator-predator complementarity, were greater on collards than potatoes (Straub and Snyder 2008).

Natural enemy facilitation occurs when the presence of one natural enemy species indirectly heightens prey capture by a second enemy species. In one well-known example of this phenomenon, lady beetles foraging in alfalfa foliage cause aphids to drop to the ground; once on the ground, aphids are readily eaten by ground beetles on the soil surface (Losey and Denno 1998). Something analogous has been reported for Colorado potato beetle prey, where predators facilitate prey infection by pathogens (Ramirez and Snyder 2009) (Fig. 9.3). During their development, the beetles move from the foliage where they feed to the soil where they pupate, and as they do so they transition between two quite distinct communities of natural enemies (Fig. 9.3). Above ground, the beetles are attacked by lady beetles, ground beetles, and *Nabis* true bugs. Below ground, they face infection by entomopathogenic nematodes (*Heterorhabditis* spp. and *Steinernema* spp.), and *Beauveria bassiana* fungi. Ramirez and Snyder (2009) manipulated species number among these predator and pathogen natural enemies, and then measured potato beetle survival from egg to adult in large field cages. Enemy species number was again manipulated following a substitutive design, so that natural enemy densities did not differ as species richness was changed. They found that potato beetle mortality increased as predator/pathogen biodiversity increased. This happened because predators and pathogens were increasingly likely to co-occur at higher species-richness levels, and predator–pathogen pairings were particularly lethal to potato beetles. The authors suggest that energetically costly anti-predator defenses of the potato beetle larvae, deployed to escape from predators early in potato beetle development, weakened the beetles' later ability to fight off pathogen infection (Ramirez and Snyder 2009, Ramirez *et al.* 2010). Thus, exposure to predators indirectly weakened beetle immune function. A similar type of facilitation appears to occur even within entomopathogen communities, as exposure to entomopathogenic fungi increases potato beetle susceptibility to entomopathogenic nematodes (Jabbour *et al.* 2011).

GETTING EVEN WITH PESTS: NATURAL BALANCE AND BIOCONTROL

The term "biodiversity" is often used to indicate the number of species present, but this is just one component of biodiversity that is more correctly labeled "species richness". Ecologists have long suspected that ecosystem function might also improve when species' relative abundances are evenly matched, a second component of biodiversity known as "evenness" (Hillebrand *et al.* 2008). Indeed, most biodiversity indices include some measure of evenness, in addition to richness, in their calculation. It is thought that communities with just a few very common species, typical for example of highly-disturbed areas dominated by weedy and/or invasive species, are less healthy than communities with many similarly abundant species. Unfortunately, ecologists' intuition has, until recently, been backed by relatively little empirical evidence that more-even communities are in fact more stable and productive than their uneven counterparts (Hillebrand *et al.* 2008).

Interestingly, some of the best evidence for the importance of greater evenness comes from potato crops. Crowder *et al.* (2010) examined how the transition from conventional farming practices, dominated by intense insecticide use, to organic farming, with its greater reliance on natural processes, influenced richness and evenness among natural enemies of insects in Washington potato fields. As in the Ramirez and Snyder (2009) study discussed previously, natural enemies of insects in Washington potatoes are dominated by insect generalist predators and nematode and fungal pathogens. In a regional survey of predators and pathogens spanning a broad geographic area and several years, Crowder *et al.* (2010) found, surprisingly, no increase in predator/pathogen species richness in fields using organic practices. However, natural enemy evenness was significantly higher in organic than conventional potato fields. This meant that just one or two natural enemy species were abundant in conventional fields, but many enemy species were similarly common in organic fields. A meta-analysis of predator surveys for crops worldwide, not just potatoes, showed that evenness among natural enemy communities was generally greater in organic than conventional fields. When the authors constructed natural enemy communities that ranged from very even to very uneven, they found that control of potato beetle pests was significantly stronger when predators and/or pathogens were evenly abundant. Indeed, greater enemy evenness translated into significantly larger potato plants, and thus presumably higher potato yields (although yields were not measured). Thus, balance among natural enemies may be as important for strong pest suppression as having a large number of natural enemy species (Crowder *et al.* 2010).

SUMMARY AND FUTURE DIRECTIONS

Agroecologists often suggest that greater biodiversity is the key to reducing pest problems. Work in potato crops, however, paints a more nuanced picture.

Increasing plant structural diversity in the crop by mulching often appears to augment natural enemy populations and increase enemy impacts on pests (Brust 1994). Likewise, more diverse landscapes can foster greater impacts of natural enemies (Werling and Gratton 2010), as can the addition of flowering plants that provide resources to natural enemies (Baggen *et al.* 1999). However, poorly chosen flowering plants can feed potato pests in addition to predators and parasitoids, worsening pest problems (Baggen and Gurr 1998).

Increasing herbivore diversity has similarly complex effects. Herbivores such as flea beetles and leafhoppers that attack early in the growing season trigger induced resistance in potato plants that renders plants less attractive and/or nutritious to later-occurring herbivores like Colorado potato beetles (e.g., Lynch *et al.* 2007). As a result, early-season herbivory can dampen later herbivory by other species. However, the presence of highly attractive prey like aphids might draw predator attacks away from less desirable prey like potato beetles, such that aphids indirectly protect potato beetles (e.g., Koss and Snyder 2005). Thus, increasing herbivore diversity can either harm or benefit particular pest species.

The impacts of predator diversity are equally multifaceted. In a few cases, adding predators that mostly eat other predators disrupts overall biological control (e.g., Mallampalli *et al.* 2002) (Fig. 9.1). More often, however, predators, parasitoids, and pathogens complement one another, attacking pests in different habitats and/or during different life stages. As a result, biological control is most effective when several enemy species are present (Straub and Snyder 2008, Ramirez and Snyder 2009) (Figs. 9.2, 9.3). In summary, while greater diversity sometimes makes pest problems worse, it appears that increasing biodiversity within potato crops is more likely to make pest problems less frequent.

While entomologists have made great progress in delineating interactions among potato arthropods, several topics, in our opinion, are still worthy of further exploration:

1. **The role of behavioral interactions among species.** Species interact not only by killing one another (or by outcompeting one another for food), but also by changing one another's behavior (Lima and Dill 1990). For example, predators that chase herbivores from preferred feeding sites can protect plants from damage even when the pest is not killed (Schmitz *et al.* 1997, 2004, Werner and Peacor 2003, Preisser *et al.* 2005). Such behavior-mediated or "non-trophic" interactions have been shown to be important in other cropping systems, but relatively little attention has been paid to them in potatoes. Similarly, interference among predator species occurs not only when predators actually eat one another, but also when a predator flees a particular habitat to avoid being eaten by another predator (Moran and Hurd 1994); whether a predator is truly killed or simply frightened away, herbivore suppression can be equally disrupted (Schmitz 2008, Steffan and Snyder 2010). The obvious anti-predator behaviors of potato-pests such as the Colorado potato beetle (e.g., Ramirez *et al.* 2010), and existing good

evidence for natural enemies in potato behaviorally avoiding one another (e.g., Lucas and Brodeur 2001), together suggest that non-trophic interactions might be an important way that species interact in potatoes.

2. **Greater attention to biodiversity aspects other than species richness.** Ecologists interested in biodiversity's effects most often manipulate the number of species present, or species richness (Cardinale *et al.* 2006, Hooper *et al.* 2006). However, as discussed above, biodiversity includes a second component: species' relative abundances or evenness (Hillebrand *et al.* 2008). Future work should build upon the intriguing results of Crowder *et al.* (2010) to see if the benefits of greater natural enemy evenness found in potatoes extend to other trophic groups and cropping systems. Furthermore, the impacts of in-field and landscape factors on promoting or reducing evenness warrants further consideration.

3. **The mechanisms of induced plant defenses in potatoes.** Robert Denno and his colleagues have clearly shown that early-season feeding by leaf-hopper herbivores renders potato plants less susceptible to Colorado potato beetles later in the growing season (Lynch *et al.* 2006, Kaplan *et al.* 2007). Potatoes are known to have many possible defenses against these and other herbivores, but it is not always clear precisely which defenses are activated against which herbivore species, and whether chewing and sucking herbivores are battled in the same ways. Increasing knowledge about the molecular bases of defenses within closely related plant species, such as tomato, should facilitate our ability to learn about the specific operations of anti-herbivore defenses in potato (Mueller *et al.* 2005). Such information would increase our understanding of tradeoffs for the plant in defending against one herbivore species versus another (e.g., Kaplan *et al.* 2009).

ACKNOWLEDGMENTS

The authors were supported during the preparation of this book chapter by the National Research Initiative of the USDA National Institute of Food and Agriculture (NIFA), grant #2007-02244.

REFERENCES

Alyokhin, A., Sewell, G., 2004. Changes in lady beetle community following the establishment of three alien species. Biol. Invasions 6, 463–471.

Andow, D.A., 1991. Vegetational diversity and arthropod population response. Annu. Rev. Entomol. 36, 561–586.

Baggen, L.R., Gurr, G.M., 1998. The influence of food on *Copidosoma koehleri* (Hymenoptera: Encyrtidae), and the use of flowering plants to enhance biological control of potato moth, *Phthorimaea operculella* (Lepidoptera: Gelechiidae). Biol. Control 11, 9–17.

Baggen, L.R., Gurr, G.M., Meats, A., 1999. Flowers in tri-trophic systems: mechanisms allowing selective exploitation by insect natural enemies for conservation biological control. Entomol. Exp. Appl. 91, 155–161.

Bengtsson, J., Ahnström, J., Weibull, A., 2005. The effects of organic agriculture on biodiversity and abundance: A meta-analysis. J. Appl. Ecol. 42, 261–269.

Berry, R.E., Liu, J., Reed, G., 1997. Comparison of endemic and exotic entomopathogenic nematode species for control of Colorado potato beetle (Coleoptera: Chrysomelidae). J. Econ. Entomol. 90, 1528–1533.

Bolter, C.J., Jongsma, M.A., 1995. Colorado potato beetles (*Leptinotarsa decemlineata*) adapt to proteinase inhibitors induced in potato leaves by methyl jasmonate. J. Insect Physiol. 41, 1071–1078.

Bolter, C.J., Dicke, J.M., Van Loop, J.J., Visser, J.H., Posthumus, M.A., 1997. Attraction of Colorado potato beetle to herbivore-damaged plants during herbivory and after its termination. J. Chem. Ecol. 23, 1003–1023.

Bommarco, R., Banks, J.E., 2003. Scale as a modifier in vegetation diversity experiments: effects on herbivores and predators. Oikos 102, 440–448.

Brodeur, J., McNeil, J.M., 1992. Host behavior-modification by the endoparasitoid *Aphidius nigripes* – a strategy to reduce hyperparasitism. Ecol. Entomol. 17, 97–104.

Brust, G.E., 1994. Natural enemies in straw-mulch reduce Colorado potato beetle populations and damage in potato. Biol. Control 4, 163–169.

Cardinale, B.J., Srivastava, D.S., Duffy, J.E., Wright, J.P., Downing, A.L., Sankaran, M., Jouseau, C., 2006. Effects of biodiversity on the functioning of trophic groups and ecosystems. Nature 443, 989–992.

Casula, P., Wilby, A., Thomas, M.B., 2006. Understanding biodiversity effects on prey in multi-enemy systems. Ecol. Lett. 9, 995–1004.

Chang, G.C., Snyder, W.E., 2004. The relationship between predator density, community composition, and field predation on Colorado potato beetle eggs. Biol. Control 31, 453–461.

Crowder, D.W., Northfield, T.D., Strand, M.R., Snyder, W.E., 2010. Organic agriculture promotes evenness and natural pest control. Nature 466, 109–112.

Debach, P., Rosen, D., 1991. Biological control by natural enemies. Cambridge University Press, New York, NY.

Eubanks, M.D., Denno, R.F., 2000. Host plants mediate omnivore–herbivore interaction and influence prey suppression. Ecology 81, 936–947.

Gaines, H.R., Gratton, C., 2010. Seed predation increases with ground beetle diversity in a Wisconsin (USA) potato agroecosystem. Agric. Ecosyst. Environ. 137, 329–336.

Greenstone, M.H., Szendrei, Z., Payton, M.E., Rowley, D.L., Coudron, T.C., Weber, D.C., 2010. Choosing natural enemies for conservation biological control: Use of the prey detectability half-life to rank key predators of Colorado potato beetle. Entomol. Exp. Appl. 136, 97–107.

Groden, E., Drummond, F.A., Casagrande, R.A., Haynes, D.L., 1990. *Coleomegilla maculata* (Coleoptera: Coccinellidae): its predation upon the Colorado potato beetle (Coleoptera: Chrysomelidae) and its incidence in potatoes and surrounding crops. J. Econ. Entomol. 83, 1306–1315.

Hairston Jr., N.G., Hairson Sr., N.G., 1993. Cause-effect relationships in energy flow, trophic structure, and interspecies interactions. Am. Nat. 142, 379–411.

Hairston Jr., N.G., Hairson Sr., N.G., 1997. Does food-web complexity eliminate trophic level dynamics? Am. Nat. 149, 1001–1007.

Hairston, N.G., Smith, F.E., Slobotkin, L.B., 1960. Community structure, population control and competition. Am. Nat. 94, 421–425.

Halaj, J., Wise, D.H., 2002. Impact of a detrital subsidy on trophic cascades in a terrestrial grazing food web. Ecology 83, 3141–3151.

Hare, D., 1990. Ecology and management of the Colorado potato beetle. Annu. Rev. Entomol. 35, 81–100.

Harmon, J.P., Andow, D.A., 2004. Indirect effects between shared prey: Predictions for biological control. Biocontrol 49, 605–626.

Hawkins, B.A., Cornell, H.V., Hochberg, M.E., 1997. Predators, parasitoids, and pathogens as mortality agents in phytophagous insect populations. Ecology 78, 2145–2152.

Hawkins, B.A., Mills, N.J., Jervis, M.A., Price, P.W., 1999. Is the biological control of insects a natural phenomenon? Oikos 86, 493–506.

Hazzard, R.V., Ferro, D.N., 1991. Feeding response of adult *Coleomegilla maculata* (Coleoptera: Coccinellidae) to eggs of Colorado potato beetle (Coleoptera: Chrysomelidae) and green peach aphids (Homoptera: Aphididae). Environ. Entomol. 20, 644–651.

Heil, M., Kost, C., 2006. Priming of indirect defenses. Ecol. Lett. 9, 813–817.

Hilbeck, A., Kennedy, G.G., 1996. Predators feeding on the Colorado potato beetle in insecticide-free plots and insecticide-treated commercial potato fields in eastern North Carolina. Biol. Control 6, 272–282.

Hilbeck, A., Eckel, C., Kennedy, G.G., 1997. Predation on Colorado potato beetle eggs by generalist predators in research and commercial potato plantings. Biol. Control 8, 191–196.

Hillebrand, H., Bennett, D.M., Cadotte, M.W., 2008. Consequences of dominance: a review of evenness effects on local and regional ecosystem processes. Ecology 89, 1510–1520.

Hlywka, J.J., Stephenson, G.R., Sears, M.K., Yada, R.Y., 1994. Effects of insect damage on glycoalkaloid content in potatoes (*Solanum tuberosum*). J. Agri. Food Chem. 42, 2545–2550.

Hole, D.G., Perkins, A.J., Wilson, J.D., Alexander, I.H., Grice, F., Evans, A.D., 2005. Does organic farming benefit biodiversity? Biol. Conserv. 122, 113–139.

Holt, R.D., 1977. Predation, apparent competition, and the structure of prey communities. Theor. Pop. Biol. 12, 197–229.

Hooper, D.U., Chapin, F.S., Ewel, J.J., Hector, A., Inchausti, P., Lavorel, S., Lawton, J.H., Lodge, D.M., Loreau, M., Naeem, S., Schmid, B., Setala, H., Symstad, A.J., Vandermeer, J., Wardle, D.A., 2005. Effects of biodiversity on ecosystem functioning: A consensus of current knowledge. Ecol. Monogr. 75, 3–35.

Hough-Goldstein, J.A., Heimpel, G.E., Bechmann, H.E., Mason, C.E., 1993. Arthropod natural enemies of the Colorado potato beetle. Crop Prot. 12, 324–334.

Howarth, F.G., 1991. Environmental impacts of classical biological control. Annu. Rev. Entomol. 36, 485–509.

Idoine, K., Ferro, D.N., 1990. Persistence of *Edovum putleri* (Hymenoptera: Eulophidae) on potato plants and parasitism of *Leptinotarsa decemlineata* (Coleoptera: Chrysomelidae): Effects of resource availability and weather. Environ. Entomol. 19, 1732–1737.

Ives, A.R., Cardinale, B.J., Snyder, W.E., 2005. A synthesis of subdisciplines: predator–prey interactions, and biodiversity and ecosystem functioning. Ecol. Lett. 8, 102–116.

Jabbour, R., Crowder, D.W., Aultman, E.A., Snyder, W.E., 2011. Entomopathogen biodiversity increases host mortality. Biol. Control 59, 277–283.

Johnson, J.M., Hough-Goldstein, J.A., Vangessel, M.J., 2004. Effects of straw mulch on pest insects, predators, and weeds in watermelons and potatoes. Environ. Entomol. 33, 1632–1643.

Kaplan, I., Lynch, M.E., Dively, G.P., Denno, R.F., 2007. Leafhopper-induced plant resistance enhances predation risk in a phytophagous beetle. Oecologia 152, 665–675.

Kaplan, I., Dively, G.P., Denno, R.F., 2009. The costs of anti-herbivore traits in agricultural crop plants: a case study involving leafhoppers and trichomes. Ecol. Appl. 19, 864–872.

Karban, R., Baldwin, I.T., 1997. Induced responses to herbivory. University of Chicago Press, Chicago, IL.

Koss, A.M., Snyder, W.E., 2005. Alternative prey disrupt biocontrol by a guild of generalist predators. Biol. Control 32, 243–251.

Koss, A.M., Chang, G.C., Snyder, W.E., 2004. Predation of green peach aphids by generalist predators in the presence of alternative, Colorado potato beetle egg prey. Biol. Control 31, 237–244.

Koss, A.M., Jensen, A.S., Schreiber, A., Pike, K.S., Snyder, W.E., 2005. A comparison of predator and pest communities in Washington potato fields treated with broad-spectrum, selective or organic insecticides. Environ. Entomol. 34, 87–95.

Landis, D.A., Wratten, S.D., Gurr, G.M., 2000. Habitat management to conserve natural enemies of arthropod pests in agriculture. Annu. Rev. Entomol. 45, 175–201.

Landolt, P.J., Tumlinson, J.H., Alborn, D.H., 1999. Attraction of Colorado potato beetle (Coleoptera: Chrysomelidae) to damaged and chemically induced potato plants. Environ. Entomol. 28, 973–978.

Langellotto, G.A., Denno, R.F., 2004. Responses of invertebrate natural enemies to complex-structured habitats: a meta-analytical synthesis. Oecologia 139, 1–10.

Levin, D.A., 1976. The chemical defenses of plants to pathogens and herbivores. Annu. Rev. Ecol. Syst. 7, 121–159.

Lill, J.T., Marquis, R.J., 2001. The effects of leaf quality on herbivore performance and attack from natural enemies. Oecologia 126, 418–428.

Lima, S.L., Dill, L.M., 1990. Behavioral decisions made under the risk of predation: a review and prospectus. Can. J. Zool. 68, 619–640.

Lopez, E.R., Ferro, D.N., Van Driesche, R.G., 1993. Direct measurement of host and parasitoid recruitment for assessment of total losses due to parasitism in the Colorado potato beetle *Leptinotarsa decemlineata* (Say) (Coleoptera: Chrysomelidae) and *Myiopharus doryphorae* (Riley) (Diptera: Tachinidae). Biol. Control 3, 85–92.

Loreau, M., Hector, A., 2001. Partitioning selection and complementarity in biodiversity experiments. Nature 412, 72–76.

Losey, J.E., Denno, R.F., 1998. Positive predator-predator interactions: enhanced predation rates and synergistic suppression of aphid populations. Ecology 79, 2143–2152.

Lucas, E., Brodeur, J., 1999. Oviposition site selection by the predatory midge *Aphidoletes aphidimyza* (Diptera: Cecidomyiidae). Environ. Entomol. 28, 622–627.

Lucas, E., Brodeur, J., 2001. A fox in sheep's clothing: Furtive predators benefit from the communal defense of their prey. Ecology 82, 3246–3250.

Lucas, E., Coderre, D., Brodeur, J., 1998. Intraguild predation among aphid predators: Characterization and influence of extraguild prey density. Ecology 79, 1084–1092.

Lucas, E., Coderre, D., Brodeur, J., 2000. Selection of molting sites by *Coleomegilla maculata* (Coleoptera: Coccinellidae): Avoidance of intraguild predation. Environ. Entomol. 29, 454–459.

Lynch, M.E., Kaplan, I., Dively, G.P., Denno, R.F., 2006. Host-plant-mediated competition via induced resistance: interactions between pest herbivores on potatoes. Ecol. Appl. 16, 855–864.

Mallampalli, N., Castellanos, I., Barbosa, P., 2002. Evidence for intraguild predation by *Podisus maculiventris* on a ladybeetle, *Coleomegilla maculata*: Implications for biological control of Colorado potato beetle, *Leptinotarsa decemlineata*. BioControl 47, 387–398.

Mallampalli, N., Gould, F., Barbosa, P., 2005. Predation of Colorado potato beetle eggs by a polyphagous lady beetle in the presence of alternative prey: potential impact on resistance evolution. Entomol. Exp. Appl. 114, 47–54.

Matson, P.A., Parton, W.J., Power, A.G., Swift, M.J., 1997. Agricultural intensification and ecosystem properties. Science 277, 504–509.

Meyling, N.V., Enkegaard, A., Brodsgaard, H., 2004. Intraguild predation by Anthocoris nemorum (Heteroptera: Anthocoridae) on the aphid parasitoid *Aphidius colemani*. Biocontrol Sci. Tech. 14, 627–630.

Moran, M.D., Hurd, L.E., 1994. Short-term responses to elevated predator densities: noncompetitive intraguild interactions and behaviour. Oecologia 98, 269–273.

Mueller, L.A., Solow, T.H., Taylor, N., Skwarecki, B., Buels, R., Binns, J., Lin, C., Wright, M.H., Ahrens, R., Wang, Y., Herbst, E.V., Keyder, E.R., Menda, N., Zamir, D., Tanksley, S.D., 2005. The SOL genomics network: a comparative resource for Solanaceae biology and beyond. Plant Physiol. 138, 1310–1317.

Nault, B.A., Kennedy, G.C., 2000. Seasonal changes in habitat preference by *Coleomegilla maculata*: Implications for Colorado potato beetle management in potato. Biol. Control 17, 164–173.

Ohgushi, T., 2005. Indirect interaction webs: herbivore-induced effects through trait change in plants. Annu. Rev. Ecol. Evol. Syst. 36, 81–105.

Patt, J.M., Lashomb, J.H., Hamilton, G.C., 1997. Impact of strip-insectary interplanting with flowers on conservation biological control of the Colorado potato beetle. Adv. Hort. Sci. 11, 175–181.

Pelletier, Y., Grondin, G., Maltais, P., 1999. Mechanism of resistance to the Colorado potato beetle in wild *Solanum* species. J. Econ. Entomol. 92, 708 713.

Pimentel, D., 1961. Species diversity and insect population outbreaks. Ann. Entomol. Soc. Am. 54, 76–86.

Polis, G.A., 1991. Complex trophic interactions in deserts: An empirical critique of food-web theory. Am. Nat. 138, 123–155.

Polis, G.A., Strong, D.R., 1996. Food web complexity and community dynamics. Am. Nat. 147, 813–846.

Polis, G.A., Myers, C.A., Holt, R.D., 1989. The ecology and evolution of intraguild predation: potential competitors that eat each other. Annu. Rev. Ecol. Syst. 20, 297–330.

Preisser, E.L., Bolnick, D.I., Benard, M.F., 2005. Scared to death? The effects of intimidation and consumption on predator-prey interactions. Ecology 86, 501–509.

Ramirez, R.A., Snyder, W.E., 2009. Scared sick? Predator-pathogen facilitation enhances the exploitation of a shared resource. Ecology 90, 2832–2839.

Ramirez, R.A., Crowder, D.W., Snyder, G.B., Strand, M.R., Snyder, W.E., 2010. Antipredator behavior of Colorado potato beetle larvae differs by instar and attacking predator. Biol. Control 53, 230–237.

Rodriguez-Saona, C., Thaler, J.S., 2005. The jasmonate pathway alters herbivore feeding behaviour: consequences for plant defenses. Entomol. Exp. Appl. 115, 125–134.

Root, R.B., 1973. Organization of a plant-arthropod association in simple and diverse habitats: the fauna of collards. Ecol. Monogr. 43, 95–124.

Rosenheim, J.A., Kaya, H.K., Ehler, L.E., Marois, J.J., Jaffee, B.A., 1995. Intraguild predation among biological-control agents: theory and practice. Biol. Control 5, 303–335.

Russell, E.P., 1989. Enemies hypothesis: a review of the effect of vegetational diversity on predatory insects and parasitoids. Environ. Entomol. 18, 590–599.

Schmitz, O.J., 2008. Predators avoiding predation. Proc. Natl Acad. Sci. USA 39, 14749–14750.

Schmitz, O.J., Beckerman, A.P., O'Brien, K.M., 1997. Behaviorally mediated trophic cascades: effects of predation risk on food web interactions. Ecology 78, 1388–1399.

Schmitz, O.J., Krivan, V., Ovadia, O., 2004. Trophic cascades: the primacy of trait-mediated interactions. Ecology Letters 7, 153–163.

Settle, W.H., Ariawan, H., Astuti, E.T., Cahyana, W., Hakim, A.L., Hindayana, D., Lestari, A.S., Pajarningsih, S., 1996. Managing tropical rice pests through conservation of generalist natural enemies and alternative prey. Ecology 77, 1975–1988.

Snyder, W.E., 2009. Coccinellids in diverse communities: which niche fits? Biol. Control 51, 323–335.

Snyder, W.E., Ives, A.R., 2001. Generalist predators disrupt biological control by a specialist parasitoid. Ecology 82, 705–716.

Steffan, S.A., Snyder, W.E., 2010. Cascading diversity effects transmitted exclusively by behavioral interactions. Ecology 91, 2242–2252.

Stoner, K.A., 1993. Effects of straw and leaf mulches and trickle irrigation on the abundance of Colorado potato beetles (Coleoptera: Chrysomelidae) on potato in Connecticut. J. Entomol. Sci. 28, 393–403.

Stoner, K.A., Ferrandino, F.J., Gent, M.P.N., Elmer, W.H., Lamonida, J.A., 1996. Effects of straw mulch, spent mushroom compost, and fumigation on the density of Colorado potato beetle (Coleoptera: Chyrsomelidae) in potatoes. J. Econ. Entomol. 89, 1267–1280.

Straub, C.S., Snyder, W.E., 2006. Species identity dominates the relationship between predator biodiversity and herbivore suppression. Ecology 87, 277–282.

Straub, C.S., Snyder, W.E., 2008. Increasing enemy biodiversity strengthens herbivore suppression on two plant species. Ecology 89, 1605–1615.

Straub, C.S., Finke, D.L., Snyder, W.E., 2008. Are the conservation of natural enemy biodiversity and biological control compatible goals? Biol. Control 45, 225–237.

Strong, D.R., 1992. Are trophic cascades all wet? Differentiation and donor-control in speciose systems. Ecology 73, 747–754.

Symondson, W.O.C., Sunderland, K.D., Greenstone, M.H., 2002. Can generalist predators be effective biocontrol agents? Annu. Rev. Entomol. 47, 561–594.

Szendrei, S., Weber, D.C., 2009. Response of predators to habitat manipulation in potato fields. Biol. Control 50, 123–128.

Szendrei, S., Greenstone, M.H., Payton, M.E., Weber, D.C., 2010. Molecular gut-content analysis of predator assemblage reveals the effect of habitat manipulation on biological control in the field. Basic Appl. Ecol. 11, 153–161.

Todorova, S.I., Coderre, D., Cote, J.C., 2000. Pathogenicity of *Beauveria bassiana* isolates toward *Leptinotarsa decemlineata* (Coleoptera: Chrysomelidae), *Myzus persicae* (Homoptera: Aphididae) and their predator *Coleomegilla maculata lengi* (Coleoptera: Coccinellidae). Phytoprotection 81, 15–22.

Tomlin, E.S., Sears, M.K., 1992a. Indirect competition between the Colorado potato beetle (Coleoptera: Chrysomelidae) and the potato leafhopper (Homoptera: Cicadellidae) on potato: laboratory study. Environ. Entomol. 21, 787–792.

Tomlin, E.S., Sears, M.K., 1992b. Effects of Colorado potato beetle and potato leafhopper on amino acid profile of potato foliage. J. Chem. Ecol. 18, 481–488.

Tscharntke, T., Klein, A.M., Kruess, A., Steffan-Dewenter, I., Thies, C., 2005. Landscape perspectives on agricultural intensification and biodiversity-ecosystem service management. Ecol. Lett. 8, 857–874.

Tylianakis, J.M., Tscharntke, T., Lewis, O.T., 2007. Habitat modification alters the structure of tropical host-parasitoid food webs. Nature 445, 202–205.

van Emden, H.F., Williams, G.F., 1974. Insect stability and diversity in agro-ecosystems. Annu. Rev. Entomol. 19, 455–475.

van Loon, J.J.A., de Vos, E.W., Dicke, M., 2000. Orientation behavior of the predatory hemipteran *Perillus bioculatus* to plant and prey odours. Entomol. Exp. Appl. 96, 51–58.

Vandermeer, J., 1995. The ecological basis of alternative agriculture. Annu. Rev. Ecol. Syst. 26, 201–224.

Walsh, B.D., Riley, C.V., 1868. Potato bugs. Am. Entomol. 1, 21–49.

Weissbecker, B., van Loon, J.J.A., Dicke, M., 1999. Electroantennogram responses of a predator, *Perillus bioculatus*, and its prey, *Leptinotarsa decemlineata*, to plant volatiles. J. Chem. Ecol. 25, 2313–2325.

Weissbecker, B., van Loon, J.J.A., Posthumus, M.A., Bouwmeester, H.J., Dicke, M., 2000. Identification of volatile potato sesquiterpenoids and their olfactory detection by the two-spotted stinkbug *Perillus bioculatus*. J. Chem. Ecol. 26, 1433–1445.

Werling, B.P., Gratton, C., 2010. Local and landscape structure differentially impact predation of two potato pests. Ecol. Appl. 20, 1114–1125.

Werner, E.E., Peacor, S.D., 2003. A review of trait-mediated indirect interactions in ecological communities. Ecology 84, 1083–1100.

Wilby, A., Thomas, M.B., 2002. Natural enemy diversity and pest control: patterns of pest emergence with agricultural intensification. Ecol. Lett. 5, 353–360.

Wise, M.J., Weinberg, A.M., 2002. Prior flea beetle herbivory affects oviposition preferences and larval performance of a potato beetle on their shared host plant. Ecol. Entomol. 27, 115–122.

Zehnder, G.W., Hough-Goldstein, J., 1990. Colorado potato beetle (Coleoptera: Chrysomelidae) population development and effects on yield of potatoes with and without straw mulch. J. Econ. Entomol. 83, 1982–1987.

Chapter 10

Interactions among Organic Soil Amendments, Plants, and Insect Herbivores

Andrei Alyokhin[1], and Serena Gross[2]

[1]*School of Biology and Ecology, University of Maine, Orono, ME, USA,* [2]*Department of Entomology, Purdue University, West Lafayette, IN, USA*

INTRODUCTION

Development of sustainable agricultural technologies is impossible without regarding cultivated fields as agricultural ecosystems and not as random associations of plants, insects, soils, and manufactured inputs. While the importance of proper soil management for improving crop yields is widely recognized by scientists and farmers alike, its impact on the populations of herbivorous insect pests is still poorly understood. In this chapter, we review existing information on the effects of soil management on insect pests of potatoes and discuss its implications for integrated pest management.

INSECT PEST PRESSURE IN ORGANIC AGRICULTURE

One of the foundations of organic farming is an assumption that organic production systems create a generally unfavorable environment for pest populations (Oelhaf 1978, Beanland *et al.* 2003, Zehnder *et al.* 2006). Indeed, insect pressure has often been comparable between organic and conventional farms, despite the fact that organically certified insecticides are typically less effective than their conventional counterparts (Feber *et al.* 1997, Gallandt *et al.* 1998, Letourneau and Goldstein 2001, Delate *et al.* 2003). Although the claim that healthy soils produce healthy plants has been enthusiastically embraced by proponents of organic agriculture, historically it was supported mainly by indirect evidence and/or anecdotal observations.

Increasing interest in organic and sustainable agriculture has boosted, among other things, research on the effects of organic soil amendments on insect herbivores. As a result, there is a mounting body of quantitative evidence

Insect Pests of Potato. http://dx.doi.org/10.1016/B978-0-12-386895-4.00010-7

generally supporting the idea of lower herbivorous populations on plants grown on organically amended soils.

Eigenbrode and Pimentel (1988) reported lower populations of flea beetles, *Phyllotreta crucifera* and *P. striolata*, on collard plants fertilized with manure compared to plants fertilized with synthetic fertilizer. Corn treated with organic fertilizer hosted fewer corn leaf aphids, *Rhopalosiphum maidis*, than corn receiving synthetic fertilizer in the study by Morales *et al.* (2001). European corn borers (*Ostrinia nubilalis*) preferred to lay eggs on corn plants grown in conventionally managed soil rather than in organically managed soil (Phelan *et al.* 1995). Hsu *et al.* (2009) found that white cabbage butterfly (*Pieris rapae crucivora*) larvae grew faster on cabbage plants grown using synthetic, rather than organic, fertilizer. Adults also preferred ovipositing on synthetically fertilized plants. Similarly, the diamondback moth, *Plutella xylostella*, was more abundant on synthetically fertilized cabbage plants and preferred to oviposit on those plants compared to organically fertilized plants (Staley *et al.* 2010). The green peach aphid, *Myzus persicae*, also had higher populations on synthetically fertilized plants, although the opposite was true for the cabbage aphid, *Brevicoryne brassicae*, which was tested in the same study (Staley *et al.* 2010). Findings by Cardoza (2011) show that applying vermicompost to *Arabidopsis thaliana* plants resulted in higher mortality, lower weight gain, and slower development of the generalist herbivore *Helicoverpa zea*.

The most immediate explanation for the observed reduction in insect numbers is that organic fertilization regimes are inferior to synthetic fertilization regimes and thus result in poor-quality small plants incapable of supporting large herbivorous populations. While sometimes this may indeed be the case (e.g., Ponti *et al.* 2007), none of the aforementioned studies found such an association. On the contrary, plant biomass and yields were often higher on the organically amended soils (Phelan *et al.* 1996, Alyokhin *et al.* 2005, Hsu *et al.* 2009).

In many situations, a single application of an organic amendment might be sufficient to cause a negative effect on insect herbivores (Eigenbrode and Pimentel 1988, Hsu *et al.* 2009, Staley *et al.* 2010, Cardoza 2011). However, sometimes multi-year organic management may be required for soils to acquire their ability to affect insect herbivores (Phelan *et al.* 1995). Therefore, soil history might be a more important factor than the current-season amendment applications (Phelan *et al.* 1995, Alyokhin and Atlihan 2005).

Insect populations are not universally lower on plants grown in manure-amended soils. A number of studies either did not detect any fertilizer effect on the numbers of herbivorous insects (Costello and Altieri 1995, Letourneau *et al.* 1996, Bengtsson *et al.* 2005) or found lower populations under synthetic fertilizer treatments (Culliney and Pimentel 1986, Staley *et al.* 2010). Also, different types of organic amendments may have different effects on insect herbivores (Eigenbrode and Pimentel 1988, Garratt *et al.* 2011), and there could be significant interactions between soil amendments and plant cultivars (Hsu *et al.*

2009). However, reduction in the numbers of insect herbivores on plants grown in organically fertilized soils has been reported from a considerable number of diverse insect-plant systems located in a variety of geographic regions. Therefore, we are confident that it is a real and fairly widespread phenomenon that should be considered in devising crop management approaches.

EVIDENCE FROM POTATOES

Potato crops cause intensive soil disturbance and contribute very little organic matter back to the soil (Porter and McBurnie 1996). Because of this, soil on potato fields is often susceptible to compaction and erosion. Therefore, agronomic techniques that ameliorate these problems are likely to be generally beneficial to potato plants. Adding organic matter to the soil through manure or compost applications can substantially increase soil aggregation. Well-aggregated soil allows for proper water flow; this in turn causes the soil to retain more moisture and thus reduces erosion. Aggregated soil also improves root penetration, which enables plants to have access to soil moisture further down into the soil and allows for healthier stands (Killham 1994). Increasing soil organic matter also provides a better food source for plants and beneficial soil organisms (Altieri 1999), allowing for healthier, stronger plants.

Alyokhin *et al.* (2005) reported results of a multi-year field study investigating the effects of soil amendment practices on Colorado potato beetle (*Leptinotarsa decemlineata*) populations in potato fields. The study was a part of the multidisciplinary Potato Ecosystem Project and was conducted during the 1999–2003 growing seasons at relatively large (41.0 m long and 14.6 m wide) field plots located on the University of Maine's Aroostook Research Farm. The land used for the study had a long history of commercial and research potato production. Weekly counts of all Colorado potato beetle life stages were conducted on each plot.

Colorado potato beetle densities were almost always lower in plots that had received, over the course of a decade, manure soil amendments in combination with reduced amounts of synthetic fertilizers compared to plots that had received full rates of synthetic fertilizers, but no manure (Alyokhin *et al.* 2005), for the same period of time. Unlike beetle abundance, plant height and canopy cover were comparable between plots receiving manure and synthetic fertilizer. Furthermore, tuber yields were higher in manure-amended plots. Thus, the difference in beetle density was unlikely to be explained simply by poor plant vigor in the absence of synthetic fertilizers. An additional 4 years of observations conducted on the same plots during the 2004–2008 growing seasons (Alyokhin, unpubl. data) have confirmed the trend towards lower Colorado potato beetle abundance on manure-amended plots (Fig. 10.1).

Subsequent field-cage and laboratory experiments (Alyokhin and Atlihan 2005) confirmed that potato plants grown on manure-amended plots of the Potato Ecosystem Project were indeed inferior Colorado potato beetle hosts

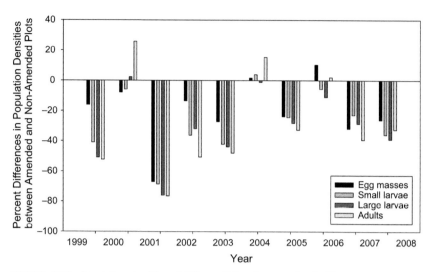

FIGURE 10.1   Population densities of different Colorado potato beetle life stages on manure-amended plots relatively to non-amended plots. Densities on non-amended plots are equal to 100%. Small larvae are the first and the second instars combined, large larvae are the third and the fourth instars combined. Data collected as part of the Potato Ecosystem Project (Alyokhin *et al.* 2005, Alyokhin unpubl. data).

compared to plants grown in synthetically fertilized soil. The observed negative effects were broad in scope. Female fecundity was lower in field cages set up on manure-amended plots early in the season, although it later became comparable between the treatments. Fewer larvae survived past the first instar, and the development of immature stages was slowed on manure-amended plots. In the laboratory, first instars also consumed less foliage excised from plants grown in manure-amended soils (Alyokhin and Atlihan 2005) (Fig. 10.2A).

Another large-scale field study comparing Colorado potato beetle populations on organically-amended and non-amended soils was conducted during the 2007–2009 growing seasons on the Aroostook Research Farm and on a nearby organically managed commercial potato farm (Gross 2010, Bernard *et al.* 2011). Again, it was a part of a large multidisciplinary project investigating the effects of compost and biological control agents on soil-borne diseases and insect pests of potato. Due to operational constraints, field plots in this study were much smaller (7.6 m long by 5.5 m wide) than in the Potato Ecosystem Project described above. Half of the plots on each of the two farms were amended with compost containing lignocellulosic substrates. Weekly counts of each of the Colorado potato beetle life stages and of potato-colonizing aphids (potato aphid (*Macrosiphum euphorbiae*), green peach aphid (*Myzus persicae*), and buckthorn aphid (*Aphis nasturii*)) were conducted on each plot.

Results of the compost study were more variable and less promising from an economic standpoint than the results of the Potato Ecosystem Project.

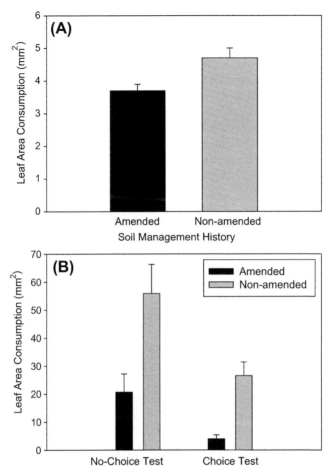

FIGURE 10.2   Consumption of potato foliage grown on manure-amended and non-amended soils by the Colorado potato beetles. (A) First instars in no-choice tests (Alyokhin and Atlihan 2005); (B) Summer-generation adults (Boiteau *et al.* 2008).

Nevertheless, they confirmed the same phenomenon. Unlike the previous study, compost amendment increased the numbers of early-season colonizing Colorado potato beetle adults at the conventional Aroostook Research Farm, but not on the organic farm. The numbers of egg masses were also higher in composted plots early in the season at the conventional farm for 2 out of the 3 years, but no such difference was detected on the organic farm (Gross 2010). Post-diapause beetle aggregation on compost-amended plots was likely explained by earlier plant emergence, which is fairly common for plants grown on composted soils (McCallum *et al.* 1998, Hahm 2000, Willekens *et al.* 2008). Despite the adult build-up, larval populations were more often lower on the composted plots at both farms, suggesting that potato plants grown on compost-amended soils are less suitable for Colorado potato beetle larval development. As a result, a considerable 21% overall increase in adult numbers observed in the study was followed by a small (2–7%) decrease in the numbers of immature stages (Gross 2010). More pronounced compost effects on the conventionally managed farm than on the organically managed farm were probably due to the latter receiving organic amendments for many years before the beginning of the present study. Therefore, soil on the organic farm may have previously acquired the capacity to affect Colorado potato beetles.

Boiteau *et al.* (2008) conducted laboratory experiments that measured Colorado potato beetle performance on foliage excised from synthetically fertilized potato plots and on potato plots fertilized with several rates of dehydrated and pelletized poultry manure. Larval mortality between the first and the end of the third instar was similar, regardless of the fertilizer treatment. However, larval development took 1.4 times longer on organically fertilized plants compared to plants fertilized by an equivalent (based on kg N/ha) amount of synthetic fertilizer. Furthermore, adult beetles consumed 6.6 times more synthetically fertilized foliage when given a choice between the two fertilization regimes, and 2.7 times more synthetically fertilized foliage in no-choice tests (Fig. 10.2B).

The previously discussed studies strongly support the idea that organic soil amendments make potato plants less suitable as Colorado potato beetle hosts. This is highly consistent with the aforementioned evidence obtained from other systems (Eigenbrode and Pimentel 1988, Phelan *et al.* 1995, 1996, Morales *et al.* 2001, Hsu *et al.* 2009, Staley *et al.* 2010, Cardoza 2011). Less information is available for other insect pests affecting this crop. The only currently available data were provided by Gross (2010), who did not find any compost effect on populations of potato-colonizing aphids. Further investigations are needed to determine the possible negative effects of soil amendments on insects other than the Colorado potato beetle. However, even if no such effects are found, selecting a proper soil management approach may still be a valuable pest management tool because of the immense threat posed by the Colorado potato beetle in most potato-growing areas of the world.

POSSIBLE MECHANISMS

While the possibility of organic soil amendments having unfavorable impact on insect herbivores can be regarded as an established fact, its mechanisms still need further elucidation. Observed effects are often variable and not very strong in comparison to other factors affecting insect populations. Therefore, a good understanding of their underlying mechanisms is essential for their successful utilization in pest management programs. It is possible to achieve good pest suppression using an insecticide with an unknown mode of action. However, such an approach is unlikely to work with a more sophisticated, ecologically based technique.

Currently, there are three major hypotheses attempting to explain lower populations of insect herbivores on organically amended soils. The mineral balance hypothesis and induced defense hypothesis attribute this phenomenon to enhanced plant resistance, while the natural enemy hypothesis suggests a build-up in the populations of natural enemies on amended plots. Existing data certainly support the idea of plant-mediated effects, as lower herbivore fitness has been repeatedly demonstrated under laboratory and greenhouse conditions in the absence of natural enemies and other confounding factors that may have affected insects in the field (Phelan *et al.* 1995, 1996, Alyokhin and Atlihan 2005, Boiteau *et al.* 2008, Hsu *et al.* 2009, Staley *et al.* 2010, Cardoza 2011). However, the natural enemy hypothesis also has some support (Morales *et al.* 2001) and should not be ruled out.

Overall, it is doubtful that any single hypothesis provides a universal explanation that is applicable in all situations. Exact mechanisms are most certain to vary among the different agricultural ecosystems. Furthermore, the three hypotheses are not mutually exclusive, and at least in some cases, different mechanisms may be complementary to each other.

The Mineral Balance Hypothesis

The chemical composition of host plants is known to affect both behavioral and developmental responses in herbivorous insects (Jansson and Smilowitz 1985, Clancy 1992, Phelan *et al.* 1995, 1996, Busch and Phelan 1999, Beanland *et al.* 2003). The mineral balance hypothesis suggests that the organic matter and microbial activity associated with organically managed soils affords a buffering capability to maintain nutrient balance in plants (Phelan *et al.* 1996, Phelan 1997). An optimal nutrient balance results in good plant growth. It also results in resistance to herbivory through production of primary or secondary compounds needed for protection from herbivores and/or for healing of wounds inflicted by herbivore feeding. In contrast, plants growing in soils without these biologically based buffering capabilities are more likely to take up either excess or insufficient levels of certain nutrients. In some instances, a resulting imbalance in the ratio of certain mineral nutrients may result in rapid plant growth. However, affected plants may have their primary and/or secondary metabolisms impaired, thus compromising their abilities to resist or tolerate insect damage. Furthermore, biochemical pathways in such plants may operate with reduced efficiency, leading to the accumulation of simple sugars and free amino acids and peptides and thus providing an enriched diet for arthropod herbivores (Phelan *et al.* 1996).

To test the mineral balance hypothesis, Phelan *et al.* (1996), Busch and Phelan (1999), and Beanland *et al.* (2003) conducted a series of laboratory and greenhouse studies using corn and soybean pests as model systems. In their experiments, they measured insect responses to and performance on plants grown in soil collected from both organic and conventional farms. They also artificially manipulated plant mineral content and used statistical mixture models to determine the effects of different nutrient ratios on insect performance. Similar to the earlier data reported by Clancy et al. (1988) and Clancy (1992) on the western spruce budworm, their results provided strong support for the formulated hypothesis.

Laboratory manipulation of the nutrient content of hydroponic solutions revealed that resulting changes in the chemical composition of soybean leaves affected the development of an array of phytophagous arthropods, including the soybean looper (*Pseudoplusia includens*), Mexican bean beetle (*Epilachna varivestis*), velvetbean caterpillar (*Ancarsia gemmatalis*), and two-spotted spider mite (*Tetranychus urticae*) (Busch and Phelan 1999, Beanland et al. 2003). The actual responses displayed by these herbivores to individual nutrients depended not only on a nutrient's concentration but also on its proportion relative to other nutrients in the solution. These proportions highlighted the importance of the interactive effects. Despite extremely valuable insights provided by those experiments, we still do not know the exact secondary metabolites and/or biochemical pathways of their synthesis that might be responsible for the observed phenomena.

Laboratory experiments by Alyokhin and Atlihan (2005) and Boiteau et al. (2008) confirmed that deleterious effects of organic soil-amendments on Colorado potato beetles were at least partially plant-mediated. Consistent with the mineral balance hypothesis, the mineral content of potato leaves explained 40–57% of the variation in Colorado potato beetle populations observed among the field plots in the study by Alyokhin et al. (2005). There was also a dramatic dissimilarity in the mineral composition of potato leaves collected from manure-amended and synthetic fertilizer-treated plots. That included significant differences in concentrations of nitrogen, calcium, magnesium, phosphorus, aluminum, boron, copper, iron, manganese, and zinc. Among these elements, boron was the most dramatically affected by soil amendment (Alyokhin et al. 2005). Its concentration was typically about two-fold higher in the foliage of plants grown on manure-amended soil (Fig. 10.3).

A similar increase in boron concentrations in plants grown on organically amended soils was reported by a number of other authors (Warman and Havard 1997, Sharma et al. 1999, Warman and Cooper 2000). Furthermore, Alyokhin et al. (2005) showed that the boron concentration of potato foliage had a strong negative correlation with all Colorado potato beetle stages except for overwintered adults. This was consistent with the findings by Beanland et al. (2003), who detected deleterious effects of elevated boron concentrations in plant tissues on the soybean looper, Mexican bean beetle, and velvetbean caterpillar. Therefore, Alyokhin et al. (2005) hypothesized that elevated boron concentrations may have been at least partially responsible for the observed reduction in Colorado potato beetle populations. However, a subsequent laboratory experiment, which created a gradient of boron concentrations in hydroponically grown potatoes (Alyokhin, unpubl. data), failed to establish any relation

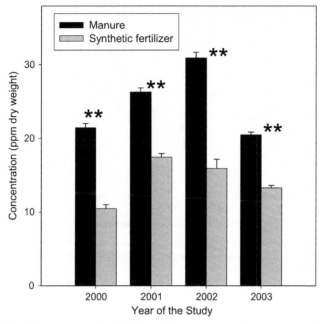

FIGURE 10.3   Boron concentration in foliage of potatoes grown in manure-amended and non-amended soils (Alyokhin *et al.* 2005). **, statistically significant difference (ANOVA, $P < 0.01$).

between larval survivorship and the boron contents of potato foliage (Fig. 10.4). Field sprays with a solution of boron on potato plots also did not show any effects (Alyokhin, unpubl. data). Although these results exclude boron as a sole factor affecting the Colorado potato beetle on foliage from manure-amended plots, they do not necessarily rule out the importance of its interactions with other minerals (Beanland *et al.* 2003).

Induced Defense Hypothesis

The induced defense hypothesis suggests that organic amendments enhance populations of naturally-occurring soil microorganisms. Being related to plant parasites (or even facultatively parasitic themselves), these microorganisms provide general physical or chemical stimuli that induce innate plant defenses. For example, rhizobacteria and fungi have both been found to be abundant in composts and induce systemic resistance to plant pathogens (Leeman *et al.* 1995, Liu *et al.* 1995, Zhang *et al.* 1998). Activated defenses, in turn, make the affected plants less suitable for development of insect herbivores by increasing their toxicity, causing antifeedant effects, delaying larval development, or promoting attack by parasitoids (Baldwin and Preston 1999, Stotz *et al.* 2000, Vallad and Goodman 2004).

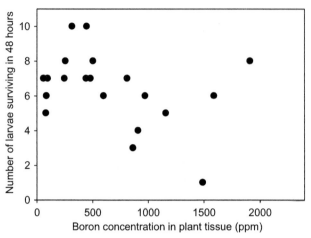

FIGURE 10.4   Apparent lack of correlation between boron content of potato foliage and survivorship of the Colorado potato beetle larvae (Spearman Rank Correlations, $r_s = -0.31$, $P = 0.189$) (Alyokhin unpubl. data).

The expression of induced plant defenses is mediated by complex signaling networks that are regulated by the plant hormones jasmonic acid (JA) and salicylic acid (SA). Induction of the JA pathway may result in the production of proteinase inhibitors, defense-related volatile compounds, secondary metabolites (e.g., nicotine), active phenolics and phytoalexins (Balbi and Devoto 2008), insect repellents (De Morales *et al.* 2001), and natural enemy attractants (Turlings *et al.* 1990). Induction of the SA pathway is followed by the rapid death of host-plant cells in the area of initial infection, which kills the pathogen through starvation and limits the spread of infection. It may also trigger the expression of suites of pathogen resistance genes that code for a number of defense chemicals (Smith *et al.* 2009).

JA pathways are usually implicated in the regulation of defenses against insect herbivores (particularly, but not exclusively, against insects with chewing mouthparts), while the SA pathway is associated with defense against pathogens (Smith *et al.* 2009). However, this dichotomy is far from absolute. JA-mediated responses have been shown to be triggered by bacteria, while SA-mediated responses trigger plant defenses against phloem-feeding insects (e.g., aphids) and cell-content-feeding insects (e.g., thrips) that do not cause extensive cellular damage (Walling 2000, Brader *et al.* 2001, Smith *et al.* 2009). Furthermore, interactions between the JA and SA pathways may play important roles in fine-tuning defense responses. Shared defense pathways allow the possibility of the immune response triggered by microorganisms to be effective against insects, and *vice versa*.

Karban *et al.* (1987) demonstrated that previous exposure of cotton seedlings to the spider mite *Tetranychus urticae* reduced the probability of infection and severity of symptoms caused by the vascular wilt fungus, *Verticillium dahlia*. Similarly, rice plants that had been fed upon by the white-backed planthopper (*Sogatella furcifera*) were found to have induced systemic resistance to the rice blast fungus, *Pyricularia grisea* (Kanno and Fujita 2003, Kanno *et al.* 2005). Planthopper feeding induced the expression of two genes coding for the defense protein β-1,3-gluconase (Kanno *et al.* 2005) that is associated with the SA-mediated defense pathway (Reymond and Farmer 1998).

Cardoza (2011) observed a two-fold reduction in survival of *H. zea* caterpillars on *Arabidopsis* plants fertilized with unsterile vermicompost. However, no such effect was detected when vermicompost was sterilized by freezing or autoclaving in the same experiment. Spraying the tea prepared from the same vermicompost also failed to have any effect on the tested insects, indicating that entomopathogens were not likely the main mortality factor. While, theoretically, autoclaving could have compromised the chemical composition of the substrate, freezing would not have such an effect. The chemical composition of frozen vermicompost was comparable with that of the intact substrate (Cardoza 2011). This strongly suggests that microbial activity associated with the vermicompost was responsible for inducing plant resistance to pests in plants grown in vermicompost-amended soil.

Hsu *et al.* (2009) and Staley *et al.* (2010) found higher glucosinolate concentrations in plants grown in organically amended soils. Eigenbrode and Pimentel (1988) reported a significant negative effect of the same group of compounds on insect herbivores but did not detect any difference between organically and synthetically fertilized plots. Glucosinolates are important defensive compounds that are known to be induced through both SA (Kiddle *et al.* 1994) and JA pathways (Fritz *et al.* 2010). Therefore, observations by Hsu *et al* (2009) and Staley *et al.* (2010) provide some indirect support to the induced plant defense hypothesis. However, higher glucosinolate production might have also been explained by a better nutrient balance in plants grown on organically amended soils. As discussed above, the two mechanisms are not mutually exclusive and can possibly complement each other.

Arbuscular mycorrhizal fungi are among the most important below-ground organisms that have significant effects on plant growth and fitness. They are also known to affect plant defenses against insect herbivores, both through changing the availability of resources used by crop plants to manufacture defenses against pests and to compensate for pest damage as well as through inducing plant defense signaling pathways (Vannette and Hunter 2009). Higher production of secondary defensive compounds in the presence of arbuscular mycorrhizae has been reported for a number of plant species (Crush 1974, Gange and West 1994, Subhashini and Krishnamurty 1995, Abu-Zeyad *et al.* 1999). Furthermore, mycorrhizal symbiosis often "primes" systemic plant defenses by increasing JA concentrations in pathogen-infected plants (Hause *et al.* 2002, Meixner *et al.* 2005, Conrath *et al.* 2006). Such priming increases the speed and magnitude of induced responses to insect or pathogen attack (Vannette and

Hunter 2009). Organic soil amendments can be expected to benefit arbuscular mycorrhizal fungi (Mosse 1959, Hepper and Warner 1983, St John *et al.* 1983, Ishac *et al.* 1986, Tarkalson *et al.* 1998), while synthetic fertilization often has a negative effect (Ellis *et al.* 1992). In potatoes, arbuscular mycorrhizae have been shown to promote plant growth and yield (Duffy and Cassells 2000, Davies *et al.* 2005, Douds *et al.* 2007) and to induce production of defensive compounds (McArthur and Knowles 1992).

Bernard *et al.* (2012) observed an increase in culturable microbes on field potato plots amended with compost as well as shifts in the soil community. It is conceivable that the reported effect could induce plant resistance which would affect the performance of the Colorado potato beetle. However, experimental studies of this issue have yet to be conducted.

Natural Enemy Hypothesis

An increase in populations of natural enemies on organically amended soils could provide yet another explanation for the decreased densities of insect herbivores. Clearly, natural enemies cannot be the only factor responsible. Decreased herbivorous performance on organically fertilized plants has been repeatedly demonstrated in the absence of natural enemies under confined laboratory conditions (see above). Nevertheless, higher organic matter in the soil is likely to provide more moisture and detritivorous alternative prey for soil-dwelling natural enemies. The numbers of predatory arthropods, such as spiders and carabid beetles, have been shown to increase due to organic amendments (Morris 1922, Reichert and Bishop 1990, Mathews *et al.* 2002, Brown and Tworkoski 2004, Garratt *et al.* 2011). Similarly, larger lady beetle (Coleoptera: Coccinellidae) populations were seen by Morales *et al.* (2001) when organic fertilizer amendments were added for the first time to a corn field, even though more aphids were encountered on plants grown with synthetic fertilizer. Also, Ponti *et al.* (2007) found that parasitism of the cabbage aphid, *Brevicoryne brassicae*, was greater on broccoli plants grown in compost-amended soil compared to those grown in synthetically fertilized soil.

Brust (1994, 1996) reported significantly lower Colorado potato beetle numbers and defoliation on potato plots that received applications of straw mulch. This coincided with a significant increase in populations of generalist predators on the mulched plots approximately 2–3 weeks after straw was placed in the field (Brust 1994). In the first half of the season, soil-dwelling carabid beetles comprised the majority of the predators found on potato foliage. The community became dominated by lady beetles, *Coleomegilla maculata* and *Hippodamia convergence*; green lacewings, *Chrysopa carnea*; and predatory stinkbugs, *Perillus bioculatus*. Predators were observed feeding on eggs and early instars of the Colorado potato beetle, with soil-dwelling carabid beetles climbing on plants in search of the prey (Brust 1994). To the contrary, in the study by Szendrei *et al.* (2010) predator species abundance and diversity were not influenced by planting living mulches of winter rye (*Secale cereal*) and hairy vetch

(*Vicia villosa*), even though prey density was also highest in plots without mulch as reported by Brust (1994, 1996). Gut-content analysis revealed that the highest incidence of predators positive for the Colorado potato beetle DNA was in plots without mulch, indicating that the lower prey abundance in mulched plots was not due to predation.

Results of those studies cannot be directly extrapolated onto the systems discussed in this chapter because of the obvious difference between the modes of action of organic soil-incorporated amendments and mulches. However, they highlight the importance that generalist predators may play in controlling the Colorado potato beetle. If organic soil amendments increase populations of generalist natural enemies, we can expect a significant negative impact on the Colorado potato beetle and, potentially to a larger degree, on potato-colonizing aphids.

Another potentially important insect natural enemy in potato ecosystems is the entomopathogenic fungus *Beauveria bassiana*. *B. bassiana* infects a wide range of insect species, including the Colorado potato beetle. It is probably the most common commercially-used natural enemy of the Colorado potato beetle, with readily available formulations that can be applied using a regular pesticide sprayer. Applications of *B. bassiana* have been shown to reduce beetle populations by up to 75% (Cantwell *et al.* 1986). The effectiveness of this biocontrol agent can also be affected by organic soil amendments, but the direction of that impact depends on particular circumstances. On one hand, this fungus is capable of surviving in soil as a saprophyte; as a result, the addition of organic matter may increase its survival and growth rates (Rosin *et al.* 1996). On the other hand, increased microbial activity associated with organic amendments is known to have a negative effect on entomopathogenic fungi, including *B. bassiana*, through competition for resources and active inhibition by producing antibiotic compounds (Lingg and Donaldson 1981, Quesaga-Morada *et al.* 2007, Jabbour and Barbercheck 2009).

Organic soil amendments may also have a negative impact on predatory and parasitic arthropods. Increased plant defenses, which are to be expected in plants grown in organically amended soils, may affect natural enemies both directly and indirectly through changes in herbivore size or quality (Price *et al.* 1980). Staley *et al.* (2011) found that the parasitoid wasp *Cotesia vestalis* parasitized more diamondback moth (*Plutella xylostella*) caterpillars on cabbage plants that had received either a synthetic ammonium nitrate fertilizer or were unfertilized compared to the plants receiving a composite organic fertilizer containing hoof and horn, limestone, superphosphate, and potassium sulfate. Parasitism was intermediate on plants fertilized with organically produced animal manure. The parasitoids also showed a preference for unfertilized plants in olfactometer experiments. Similarly, the parasitoid wasp *Diaeretiella rapae* attacked fewer cabbage aphids on cabbage plants fertilized with chicken manure compared to plants fertilized with the hoof- and horn-containing fertilized described above (Pope *et al.* 2011). Thus, there is a possibility that, under some circumstances, natural enemy populations may actually decrease on organically amended plots.

INTEGRATED APPROACH

Integrated pest management (IPM) is the systems approach to reducing pest damage to tolerable levels by using a diverse array of chemical and non-chemical control methods and relying on a variety of decision-making paradigms (Rajotte 1993, Zalom 1993). IPM seeks to find a balance between purchased inputs and resulting outputs, with concomitant goals of improving economic, social, and environmental conditions (Rajotte 1993). It emphasizes knowledge-based integration of non-pesticidal methods, such as host-plant resistance and biological, cultural, and behavioral control, with the judicious use of pesticides.

Prokopy (1993) compared the progress towards the implementation of IPM in its ideal form to climbing a stepladder. The first step involves having multiple ecologically sound management methods for a single class of pests, such as insects and other arthropods, microbes, weeds, or vertebrates. On the second step, management practices for all classes of pests are integrated. The third step integrates combined pest management procedures with the entire system of crop production, and the fourth step blends the concerns of all groups having stakes in pest management. As one ascends this imaginary ladder, the distance between steps and their thicknesses increase, signifying the greater difficulty of attaining the next level. It is important to try and reach higher levels of IPM to optimize pest control and to be ecologically and economically sound. However, as one moves up through the levels of IPM, it becomes more difficult to coordinate management methods and alleviate the concerns of all those involved.

Proper soil management lays the very foundation of an effective and sustainable system of agricultural production. Therefore, proceeding onto the third level of IPM is impossible without a good understanding of soil effects on insect pests. The complexity of the system makes it a rather challenging task. However, even initial advances discussed in this chapter provide grounds for optimism. Potato ecosystems look particularly promising in this regard because conventional potato production requires intensive soil disturbance and contributes very little organic matter back to the soil. Therefore, there is already a strong impetus to amend soils with manure and composts, with a possible reduction in herbivorous pressure serving as the "icing on the cake".

It is highly unlikely that soil management can be sufficient in keeping potato pests below economically damaging levels. Nevertheless, they do create a less favorable environment for Colorado potato beetles (and possibly other species that have not yet been studied in potatoes; evidence from other systems indicates that they may also be affected). Higher death rates of early instars may complement other mortality sources, thus decreasing the amount of chemicals necessary to suppress beetle populations. Slower larval development may allow plants more opportunity to compensate for damage, especially in combination with lower leaf consumption by adult beetles. It may also extend a window of opportunity for using biorational insecticides, which are most efficient against early instars. Furthermore, a slower rate of development may allow for more predation and parasitism by natural enemies. While organic soil amendments are not going to provide a "silver-bullet" solution for insect problems in potatoes, they can be useful for designing fully integrated, ecologically sound crop management systems.

REFERENCES

Abu-Zeyad, R., Khan, A.G., Khoo, C., 1999. Occurrence of arbuscular mycorrhiza in *Castanospermum australe* A. Cunn. & C. Fraser and effects on growth and production of castanospermine. Mycorrhiza 9, 111–117.

Altieri, M.A., 1999. The ecological role of biodiversity in agroecosystems. Agr. Ecosyst. Environ. 74, 19–31.

Alyokhin, A., Atlihan, R., 2005. Reduced fitness of the Colorado potato beetle (Coleoptera: Chrysomelidae) on potato plants grown in manure-amended soil. Environ. Entomol. 34, 963–968.

Alyokhin, A., Porter, G., Groden, E., Drummond, F., 2005. Colorado potato beetle response to soil amendments: A case in support of the mineral balance hypothesis? Agr. Ecosyst. Environ. 109, 234–244.

Balbi, V., Devoto, A., 2008. Jasmonate signalling network in *Arabidopsis thaliana*: crucial regulatory nodes and new physiological scenarios. New Phytol. 177, 301–318.

Baldwin, I.T., Preston, C.A., 1999. The eco-physiological complexity of plant responses to insect herbivores. Planta 208, 137–145.

Beanland, L., Phelan, P.L., Salminen, S., 2003. Micronutrient interactions on soybean growth and the developmental performance of three insect herbivores. Environ. Entomol. 32, 641–651.

Bengtsson, J., Ahnstrom, J., Weibull, A.C., 2005. The effects of organic agriculture on biodiversity and abundance: A meta-analysis. J. Appl. Ecol. 42, 261–269.

Bernard, E., Larkin, R.P., Tavantzis, S., Erich, M.S., Alyokhin, A., Sewell, G., Lannan, A., Gross, S.D., 2011. Compost, rapeseed rotation, and biocontrol agents significantly impact soil microbial communities in organic and conventional potato production systems. Appl. Soil Ecol. 52, 29–41.

Boiteau, G., Lynch, D.H., Martin, R.C., 2008. Influence of fertilization on the Colorado potato beetle, *Leptinotarsa decemlineata*, in organic potato production. Environ. Entomol. 37, 575–585.

Brader, G., Tas, E., Palva, E.T., 2001. Jasmonate dependent induction of indole glucosinolates in Arabidopsis by culture filtrates of the nonspecific pathogen *Erwinia carotovora*. Plant Physiol. 126, 849–860.

Brown, M.W., Tworkoski, T., 2004. Pest management benefits of compost mulch in apple orchards. Agr. Ecosyst. Environ. 103, 465–472.

Brust, G.E., 1994. Natural enemies in straw-mulch reduce Colorado potato beetle populations and damage in potato. Biological Control 4, 163–169.

Brust, G.E., 1996. Interaction of mulch and *Bacillus thuringiensis* subsp. *tenebrionis* on Colorado potato beetle (Coleoptera: Chrysolllelidae) populations and damage in potato. J. Econ. Entomol. 89, 467–474.

Busch, J.W., Phelan, P.L., 1999. Mixture models of soybean growth and herbivore performance in response to nitrogen–sulphur–phosphorous nutrient interactions. Ecol. Entomol. 24, 132–145.

Cantwell, G.E., Cantelo, W.W., Schroder, R.F.W., 1986. Effect of *Beauveria bassiana* on underground stages of the Colorado potato beetle, *Leptinotarsa decemlineata* (Coleoptera: Chrysomelidae). Great Lakes Entomol. 19, 81–84.

Cardoza, Y.J., 2011. *Arabidopsis thaliana* resistance to insects, mediated by an earthworm-produced organic soil amendment. Pest Manage. Sci. 67, 233–238.

Clancy, K.M., 1992. Response of Western spruce budworm (Lepidoptera: Tortricidae) to increased nitrogen in artificial diets. Environ. Entomol. 21, 331–344.

Clancy, K.M., Wagner, M.R., Tinus, R.W., 1988. Variation in host foliage nutrient concentrations in relation to western spruce budworm herbivory. Can. J. Forest Res. 18, 530–539.

Conrath, U., Conrath, G.F., Beckers, M., Flors, V., Garcia-Agustin, P., Jakab, G., Mauch, F., Newman, M.A., Pieterse, C.M.J., Poinssot, B., Pozo, M.J., Pugin, A., Schaffrath, U., Ton, J., Wendehenne, D., Zimmerli, L., Mauch-Mani, B., 2006. Priming: getting ready for battle.. Mol. Plant–Microbe In. 19, 1062–1071.

Costello, M.J., Altieri, M.A., 1995. Abundance, growth rate and parasitism of *Brevicoryne brassicae* and *Myzus persicae* (Homoptera, Aphididae) on broccoli grown in living mulches. Agric. Ecosyst. Environ. 52, 187–196.

Crush, J.R., 1974. Plant-growth responses to vesicular-arbuscular mycorrhizae. 7. Growth and nodulation of some herbage legumes. New Phytol. 73, 743–747.

Culliney, T.W., Pimentel, D., 1986. Ecological effects of organic agricultural practices on insect populations. Agric. Ecosyst. Environ. 15, 253–266.

Davies Jr., F.T., Calderón, C.M., Huainan, Z., 2005. Influence of arbuscular mycorrhizae indigenous to Peru and a flavonoid on growth, yield, and leaf elemental concentration of "Yungay" potatoes. HortScience 40, 381–385.

De Morales, C.M., Mescher, M.C., Tumlinson, J.H., 2001. Caterpillar-induced nocturnal plant volatiles repel conspecific females. Nature 410, 577–580.

Delate, K., Friedrich, H., Lawson, V., 2003. Organic pepper production systems using compost and cover crops. Biol. Agric. Hort. 21, 131–150.

Douds. Nagahashi Jr., D.D.G., . Nagahashi, G., Reider, C., Hepperly, P.R., 2007. Mycorrhizal fungi increases the yield of potatoes in a high P soil. Biol. Agric. Hortic. 25, 67–78.

Duffy, E.M., Cassells, A.C., 2000. The effect of inoculation of potato (*Solanum tuberosum* L.) microplants with arbuscular mycorrhizal fungi on tuber yield and tuber size distribution. Appl. Soil Ecol. 15, 137–144.

Eigenbrode, S.D., Pimentel, D., 1988. Effects of manure and chemical fertilizers on insect pest populations on collards. Agric. Ecosyst. Environ. 20, 109–125.

Ellis, J.R., Roder, W., Mason, S.C., 1992. Grain sorghum-soybean rotation and fertilization influence in vesicular-arbuscular mycorrhizal fungi. Soil Sci. Soc. Am. J. 56, 789–794.

Feber, R.E., Firbank, L.G., Johnson, P.J., McDonald, D.W., 1997. The effects of organic farming on pest and non-pest butterfly abundance. Agric. Ecosyst. Environ. 64, 133–139.

Fritz, V.A., Justen, V.L., Bode, A.M., Schuster, T., Wang, M., 2010. Glucosinolate enhancement in cabbage induced by jasmonic acid application. HortScience 45, 1188–1191.

Gallandt, E.R., Mallory, E.B., Alford, A.R., Drummond, F.A., Groden, E., Leibman, M., Marra, M.C., McBurnie, J.C., Porter, G.A., 1998. Comparison of alternative pest and soil management strategies for Maine potato production systems. Am. J. Altern. Agric. 13, 146–161.

Gange, A.C., West, H.M., 1994. Interactions between arbuscular mycorrhizal fungi and foliar-feeding insects in *Plantago lanceolata*. L. New Phytol. 128, 79–87.

Garratt, M.P.D., Wright, D.J., Leather, S.R., 2011. The effects of farming system and fertilisers on pests and natural enemies: A synthesis of current research. Agric. Ecosyst. Environ. 141, 261–270.

Gross, S.D., 2010. Effect of soil amendments on pest insects in potatoes. MS thesis. University of Maine, Orono, ME.

Hahm, J.M., 2000. The effects of topically-applied municipal biosolids on seedling emergence and early seedling growth. MS thesis. Texas Tech University, Lubbock, TX.

Hause, B., Maier, W., Miersch, O., Kramell, R., Strack, D., 2002. Induction of jasmonate biosynthesis in arbuscular mycorrhizal barley roots. Plant Physiol. 130, 1213–1220.

Hepper, C.M., Warner, A., 1983. Role of organic matter in growth of a vesiculararbuscular mycorrhizal fungi in soil. Trans. Br. Mycol. Soc. 81, 155–156.

Hsu, Y., Shen, T., Hwang, S., 2009. Soil fertility management and pest responses: A comparison of organic and synthetic fertilization. Ecotoxicology 102, 160–169.

Ishac, Y.Z., El-Haddad, M.E., Daft, M.J., Ramadan, E.M., El-Demerdash, M.E., 1986. Effect of seed inoculation, mycorrhizal infection and organic amendment on wheat growth. Plant Soil. 90, 373–382.

Jabbour, R., Barbercheck, M.E., 2009. Soil management effects on entomopathogenic fungi during the transition to organic agriculture in a feed grain rotation. Biol. Control 51, 435–443.

Jansson, R.K., Smilowitz, Z., 1985. Influence of nitrogen on population parameters of potato insects: Abundance, development, and damage of the Colorado potato beetle, *Leptinotarsa decemlineata* (Coleoptera: Chrysomelidae). Environ. Entomol. 14, 500–506.

Kanno, H., Fujita, Y., 2003. Induced systemic resistance to rice blast fungus in rice plants infested with white-backed planthopper. Entomol. Exp. Appl. 107, 155–158.

Kanno, H., Satoh, M., Kimura, T., Fujita, Y., 2005. Some aspects of induced resistance to rice blast fungus, *Magnaporthe grisea*, in rice plant infested by white-backed planthopper, *Sogatella furcifera*. Appl. Entomol. Zool. 40, 91–97.

Karban, R., Adamchak, R., Schnathorst, W.C., 1987. Induced resistance and interspecific competition between spider mites and a vascular wilt fungus. Science 235, 678–680.

Kiddle, G.A., Doughty, K.J., Wallsgrove, R.M., 1994. Salicylic acid-induced accumulation of glucosinolates in oilseed rape (*Brassica napus* L.) leaves. J. Exp. Botany 45, 1343–1346.

Killham, K., 1994. Structural aspects of the soil habitat. In: Killham, K. (Ed.), Soil Ecology, Cambridge University Press, Cambridge, UK, pp. 7–12.

Leeman, M., van Pelt, J.A., Hendrickx, M.J., Scheffer, R.J., Bakker, P.A.H.M., Schippers, B., 1995. Biocontrol of Fusarium wilt of radish in commercial greenhouse trials by seed treatment with *Pseudomonas fluorescens* WCS374. Phytopathology 85, 1301–1305.

Letourneau, D.K., Goldstein, B., 2001. Pest damage and arthropod community structure in organic vs conventional tomato production in California. J. Appl. Ecol. 38, 557–570.

Letourneau, D.K., Drinkwater, L.E., Shennan, C., 1996. Effects of soil management on crop nitrogen and insect damage in organic vs conventional tomato fields. Agric. Ecosyst. Environ. 57, 179–187.

Lingg, A.J., Donaldson, M.D., 1981. Biotic and abiotic factors affecting stability of *Beauveria bassiana* conidia in soil. J. Invert. Path. 38, 191–200.

Liu, L., Kloepper, J.W., Tuzun, S., 1995. Induction of systemic resistance in cucumber by plant growth-promoting rhizobacteria: Duration of protection and effect of host resistance on protection and root colonization. Phytopathology 85, 1064–1068.

Mathews, C.R., Bottrell, D.G., Brown, M.W., 2002. A comparison of conventional and alternative understory management practices for apple production: Multi-trophic effects. J. Appl. Soil Ecol. 21, 221–231.

McArthur, D.A.J., Knowles, N.R., 1992. Resistance responses of potato to vesicular-arbuscular mycorrhizal fungi under varying abiotic phosphorus levels. Plant Physiol. 100, 341–351.

McCallum, K.R., Keeling, A.A., Beckwith, C.P., Kettlewell, P.S., 1998. Effects of green waste compost on spring wheat (*Triticum aestivum* L. CV Avans) emergence and early growth. Acta Hort 469, 313–318.

Meixner, C., Ludwig-Muller, J., Miersch, O., Gresshoff, P., Staehelin, C., Vierheilig, H., 2005. Lack of mycorrhizal autoregulation and phytohormonal changes in the supernodulating soybean mutant nts1007. Planta 222, 709–715.

Morales, H., Perfecto, I., Ferguson, B., 2001. Traditional fertilization and its effect on corn insect populations in the Guatemalan highlands. Agr. Ecosyst. Environ. 84, 145–155.

Morris, H.M., 1922. The insect and other invertebrate fauna of arable land at. Rothamsted. Ann. Appl. Biol. 9, 282–305.

Mosse, B., 1959. Observations on the extra-matrical mycelium of a vesicular-arbuscular endophyte. Trans. Brit. Mycol. Soc. 42, 439–448.

Oelhaf, R.C., 1978. Organic Farming: Economic and Ecological Comparisons with Conventional Methods. Allanheld & Osmun, Montclair, NJ.

Phelan, L.P., Mason, J.F., Stinner, B.R., 1995. Soil-fertility management and host preference by European corn borer, *Ostrinia nubilalis* (Hubner), on *Zea mays* L.: A comparison of organic and conventional chemical farming. Agric. Ecosyst. Environ. 56, 1–8.

Phelan, L.P., Norris, K.H., Mason, J.F., 1996. Soil management history and host preference by *Ostrinia nubilalis*: evidence for plant mineral balance mediating insect–plant interactions. Environ. Entomol. 25, 1329–1336.

Phelan, P.L., 1997. Soil-management history and the role of plant mineral balance as a determinant of maize susceptibility to the European corn borer. Biol. Agric. Hortic. 15, 25–34.

Ponti, L., Altieri, M.A., Gutierrez, A.P., 2007. Effects of crop diversification levels and fertilization regimes on abundance of *Brevicoryne brassicae* (L.) and its parasitization by *Diaeretiella rapae* (M'Intosh) in broccoli. Agric. Forest Entomol. 9, 209–214.

Pope, T.W., Girling, R.D., Staley, J.T., Trigodet, B., Wright, D.J., Leather, S.R., van Emden, H.F., Poppy, G.M., 2012. Effects of organic and conventional fertilizer treatments on host selection by the aphid parasitoid *Diaeretiella rapae*. J. Appl. Entomol. 136: 445–455.

Porter, G.A., McBurnie, J.C., 1996. Crop and Soil Research. In: Marra, M.C. (Ed.), The Ecology, Economics, and Management of Potato Cropping Systems: A Report of the First Four Years of the Maine Potato Ecosystem Project, Maine Agric. Forest Experiment Station Bull. 843, pp. 8–62.

Price, P.W., Bouton, C.E., Gross, P., McPheron, B.A., Thompson, J.N., Weis, A.E., 1980. Interactions among three trophic levels – influence of plants on interactions between insect herbivores and natural enemies. Annu. Rev. Ecol. Syst. 11, 41–65.

Prokopy, R.J., 1993. Stepwise progress towards IPM and sustainable agriculture. IPM Practitioner 15, 1–4.

Quesada-Moraga, E., Navas-Cortés, J.A., Maranhao, E.A.A., Ortiz-Urquiza, A., Santiago-Álvarez, C., 2007. Factors affecting the occurrence and distribution of entomopathogenic fungi in natural and cultivated soils. Mycol. Res. 111, 947–966.

Rajotte, E.G., 1993. From profitability to food safety and the environment: Shifting the objectives of IPM. Plant Dis. 77, 296–299.

Reichert, S.E., Bishop, L., 1990. Prey control by an assemblage of generalist predators: Spiders in garden test systems. Ecology 71, 1441–1450.

Reymond, P., Farmer, E.E., 1998. Jasmonate and salicylate as global signals for defense gene expression. Curr. Opin. Plant Biol. 1, 404–411.

Rosin, F., Shapiro, D.I., Lewis, L.C., 1996. Effect of fertilizers on the survival of *Beauveria bassiana*. J. Invert. Pathol. 68, 194–195.

Sharma, K.R., Srivastava, P.C., Ghosh, D., Gangwar, M.S., 1999. Effect of boron and farmyard manure application on growth, yields, and boron nutrition of sunflower. J. Plant Nutr. 22, 633–640.

Smith, J.L., Moraes, C.M.D., Mescher, M.C., 2009. Jasmonate- and salicylate-mediated plant defense responses to insect herbivores, pathogens and parasitic plants. Pest Manage. Sci. 65, 497–503.

St John, T.V., Coleman, D.C., Reid, C.P.P., 1983. Association of vesicular-arbuscular mycorrhizal hyphae with soil organic particles. Ecology 64, 957–959.

Staley, J.T., Stewart-Jones, A., Pope, T.W., Wright, D.J., Leather, S.R., Hadley, P., Rossiter, J.T., van Emden, H.F., Poppy, G.M., 2010. Varying responses of insect herbivores to altered plant chemistry under organic and conventional treatments. Proc. R. Soc. B 277, 779–786.

Staley, J.T., Girling, R.D., Stewart-Jones, A., Poppy, G.M., Leather, S.R., Wright, D.J., 2011. Organic and conventional fertilizer effects on a tritrophic interaction: parasitism, performance and preference of *Cotesia vestalis*. J. Appl. Entomol. 135: 658–665.

Stotz, H.U., Pittendrigh, B.R., Kroymann, J., Weniger, K., Fritsche, J., Bauke, A., Mitchell-Olds, T., 2000. Induced plant defense responses against chewing insects. Ethylene signaling reduces resistance of *Arabidopsis* against Egyptian cotton worm but not diamondback moth. Plant Physiol. 124, 1007–1017.

Subhashini, D.V., Krishnamurty, V., 1995. Influence of vesiculararbuscular mycorrhiza on phosphorus economy, yield and quality of flue-cured Virginia tobacco in rain-fed Alfisols: biofertilizers for the future. In: Proceedings, 3rd National Conference on Mycorrhiza, pp. 328–330, 13–15 March 1995, New Delhi, India.

Szendrei, Z., H.Greenstone, M., Payton, M.E., Weber, D.C., 2010. Molecular gut-content analysis of a predator assemblage reveals the effect of habitat manipulation on biological control in the field. Basic Appl. Ecol. 11, 153–161.

Tarkalson, D.D., Jolley, V.D., Robbins, C.W., Terry, R.E., 1998. Mycorrhizal colonization and nutrition of wheat and sweet corn grown in manure-treated and untreated topsoil and subsoil. J. Plant Nutr. 21, 1985–1999.

Turlings, T.C.J., Tumlinson, J.H., Lewis, W.J., 1990. Exploitation of herbivore induced plant odors by host-seeking parasitic wasps. Science 250, 1251–1253.

Vallad, G.E., Goodman, R.M., 2004. Systemic acquired resistance and induced systemic resistance in conventional agriculture. Crop Sci. 44, 1920–1934.

Vannette, R.L., Hunter, M.D., 2009. Mycorrhizal fungi as mediators of defense against insect pests in agricultural systems. Agric. Forest Entomol. 11, 351–358.

Walling, L.L., 2000. The myriad plant responses to herbivores. J. Plant Growth Regul. 19, 195–216.

Warman, P.R., Cooper, J.M., 2000. Fertilization of a mixed forage crop with fresh and composted chicken manure and NPK fertilizer: Effects on soil and tissue Ca, Mg, S, B, Cu, Fe, Mn, and Zn. Can. J. Soil Sci. 80, 345–352.

Warman, P.R., Havard, K.A., 1997. Yield, vitamin and mineral contents of organically and conventionally grown carrots and cabbage. Agric. Ecosyst. Environ. 61, 155–162.

Willekens, K., De Vliegher, A., Vandecasteele, B., Carlier, L., June 2008. 2008. Effect of compost versus animal manure fertilization on crop development, yield and nitrogen residue in the organic cultivation of potatoes. In: Cultivating the future based on science: 2nd Conference of the International Society of Organic Agriculture Research ISOFAR. Italy, Modena 18–20.

Zalom, F.G., 1993. Reorganizing to facilitate the development and use of integrated pest management. Agric. Ecosyst. Env. 46, 245–256.

Zehnder, G., Gurr, G.M., Kühne, S., R.Wade, M., D.Wratten, S., Wyss, E., 2006. Arthropod Pest management in organic crops. Annu. Rev. Entomol. 52, 57–80.

Zhang, W., Han, D.Y., Dick, W.A., Davis, K.R., Hoitink, H.A.J., 1998. Compost and compost water extract-induced systemic acquired resistance in cucumber and arabidopsis. Phytopathology 88, 450–455.

Chapter 11

Aphid-Borne Virus Dynamics in the Potato-Weed Pathosystem

Rajagopalbabu Srinivasan[1], Felix A. Cervantes[2], and Juan M. Alvarez[3]

[1]Department of Entomology, University of Georgia, Tifton, GA, USA, [2]Entomology and Nematology Department, University of Florida, Gainesville, FL, USA, [3]DuPont Crop Protection, Stine Haskell Research Center, Newark, DE, USA

INTRODUCTION TO THE POTATO-VIRAL PATHOSYSTEM

Aphids can affect potato plants by inducing feeding injuries, as well as by transmitting plant viruses. In potato pathosystems, aphids are known to transmit a suite of persistent and non-persistent viruses capable of causing severe economic losses. As a consequence, potato production is severely hampered by numerous aphid-borne viruses worldwide. The management of aphids and aphid-transmitted viruses is of primary concern to growers, particularly seed growers. To complicate matters, potato-infesting aphids are highly polyphagous and potato-infecting viruses can also infect numerous alternate hosts across plant families.

The role of alternate hosts or weeds during the occurrence and spread of plant viral diseases is an integral part of viral epidemiology (Duffus 1971). Weeds or alternate hosts affect virus transmission by acting as reservoirs of vectors or viruses. In some cases, weeds act as reservoirs for both vector and virus and significantly hasten viral spread in an agroecosystem (Duffus 1971, Norris and Kogan 2005). The potato ecosystem has numerous alternate weedy hosts that can serve as reservoirs for colonizing aphids, as well as for persistently and non-persistently transmitted viruses. Evidence indicates that a single alternate host can serve as a reservoir for several colonizing aphid species and harbor numerous potato viruses simultaneously. Thus, a simple tricomponent pathosystem (potato–aphid–virus) can become a complicated pathoweb due to the presence of alternate hosts. The goal of this chapter is to characterize the intricate interactions in the potato pathosystem as influenced by alternate hosts.

Insect Pests of Potato. http://dx.doi.org/10.1016/B978-0-12-386895-4.00011-9

COMPONENTS OF THE POTATO-VIRAL PATHOSYSTEM

Potato

The cultivated potato, *Solanum tuberosum* (L.), is susceptible to aphids and the viruses that they transmit. However, its wild relatives are known to display varying levels of resistance to both aphids and viruses. Glandular trichomes in *Solanum berthaultii* (Hawkes) are known to negatively affect aphid biology and behavior (Avé *et al.* 1987). Other plants in the genus *Solanum*, such as *S. circaeifolium* spp. *capsicibaccatum* (Cãrdenas), *S. polyadenium* (Greenm.), *S. tarijense* (Hawkes), and *S. trifidum* (Correll) have been found to adversely affect the biology of green peach aphid, *Myzus persicae* (Sulzer), and potato aphid, *Macrosiphum euphorbiae* (Thomas) (Pelletier *et al.* 2010). Similarly, resistance to potato viruses has been identified in wild species such as *S. tuberosum* ssp. *andigena* (L.), *S. brevidens* (Phil.), *S. hougasii* (Correll), *S. stoloniferum* (Schltdl. & Bouché), and *S. chacoense* (Bitter) (Valkonen *et al.* 1991, Heldak *et al.* 2007).

So far, it has been extremely difficult to introgress resistance conferring genes to cultivated potatoes due to ploidy-level differences. Most wild *Solanum* spp. are non-tuberizing diploids, while the cultivated potato is a tetraploid. Despite this, breeding efforts have concentrated heavily on introgressing resistance from wild plants. As a result, a few cultivars with viral resistance have been released (Whitworth *et al.* 2010). However, there are virtually no resistant cultivars that possess a high degree of consumer acceptability and broad-spectrum resistance against aphid-borne viruses. In the United States and elsewhere, potato is propagated clonally. Therefore, stringent management tactics are required to minimize virus infection in seed production.

Aphids

A number of aphid species (Aphididae: Homoptera), including *M. persicae* and *M. euphorbiae*, are known to colonize potato plants. These aphid species are typically polyphagous and are known to feed on hundreds of host plants in multiple plant families (Blackman and Eastop 1984), including both cultivated and alternate weed hosts. Under temperate conditions, some potato-colonizing aphids utilize woody perennials such as peach (*Prunus persica* (L.) Batsch) as their primary hosts in the fall and winter seasons. During this period, their reproduction switches from parthenogenetic mode to sexual mode. In the spring and summer seasons, aphids typically use herbaceous hosts, including crops and weeds, and reproduce parthenogenetically. Under tropical conditions, it is not unusual for aphids to utilize herbaceous and other available hosts throughout the year and reproduce through parthenogenetic means. On a suitable host plant, the parthenogenetic mode of reproduction aids in the quick build-up of populations and efficient host utilization.

Colonizing aphids are phloem feeders. Aphids typically reach the phloem through their stylets. The feeding process causes negligible damage to the sieve tube and to other cells in the vicinity of the stylet path. While feeding on the phloem, aphids also salivate. Salivation assists in indirectly transmitting circulative and phloem-restricted viruses to plants. Colonizing aphids transmit viruses both non-persistently as well as persistently. However, non-colonizing aphids only probe the epidermal surfaces, and do not feed on the phloem. Hence, non-colonizing aphids can only transmit non-persistent viruses. Besides exhibiting differences in modes of reproduction, aphids can also have both winged and wingless morphs. These abilities are dictated by diet and environmental factors, as well as by their innate characteristics (Müller *et al.* 2001). These morphological variations can play an important role in colonizing host plants, as well as in initiating viral epidemics.

Potato Viruses

Aphid-borne potato viruses are exclusively single-stranded RNA viruses. Most of the aphid-transmitted potato viruses are found in four viral families and in one floating genus (Brunt and Loebenstein 2001, Hull 2001). *Potato leafroll virus* (genus *Polerovirus*) is found in the family *Luteoviridae* (Brunt and Loebenstein 2001). Three potyviruses (potato viruses A, Y, and V) are found in the genus *Potyvirus* and family *Potyviridae* (Brunt and Loebenstein 2001). *Cucumber mosaic virus*, *Alfalfa mosaic virus*, and *Potato yellowing virus* are found in *Bromoviridae*. A *Nucleorhabdovirus* is found in *Rhabdoviridae*. Potato viruses M and S, and *Potato latent virus*, are found in the genus *Carlavirus*. Another relatively new and unassigned *potato* virus is *Potato virus P* (Brunt and Loebenstein 2001).

Of these aphid-borne viruses, only *Potato leafroll virus* (PLRV) and *Potato virus Y* (PVY) are considered to be economically important in North America. With the exception of PLRV, aphids transmit almost all other viruses non-persistently; PLRV is transmitted by aphids in a persistent and circulative manner. In addition, there are several strains of the same virus that are known to differentially affect potato. For example, in the case of PVY, strain differences are identifiable serologically, as well as based on the symptoms expressed in host plants. In North America, three PVY strains are commonly found. The ordinary strain (PVY^O) is the predominant strain, but strains PVY^{NTN} and PVY$^{N:O}$ are also regularly found in potato (Lorenzen *et al.* 2006, Singh *et al.* 2008). Aphid-borne potato viruses can induce a suite of symptoms on potato foliage and on tubers and often cause severe economic losses. Besides infecting potato, these viruses have been known to infect hundreds of alternate hosts.

Alternate Hosts (Weeds)

Weeds can play an important role in the epidemiology of potato viruses. They can serve as vector reservoirs or as virus reservoirs, and some can function as both. In order to influence the epidemiology of a persistent virus, weed hosts have to serve as reservoirs of the vector and harbor the virus. Otherwise, alternate hosts would only serve as dead-end hosts.

Numerous weed species have been identified as hosts of PLRV. These hosts are found not only within Solanaceae but also in other families, including Amaranthaceae, Brassicaceae, and Portulaceae. Aphids such as *M. persicae* and *M. euphorbiae* are known to colonize members of Solanaceae and other plant families, including Brassicaceae, Amaranthaceae, Compositae, Chenopodiaceae, and Cucurbitaceae. Non-persistently transmitted viruses, such as potyviruses and carlaviruses, are also known to infect alternate hosts from several plant families, including Solanaceae. These weeds are often ubiquitously found in the ecosystem and can produce enormous quantities of seeds that are equipped to deal with adverse weather conditions.

Upon infection with potato viruses, some weed hosts such as *Physalis floridana* (Rydb.), *Solanum sarrachoides* (Sendtner), and *Datura stramonium* L. exhibit symptoms, but symptoms are not obvious in most weed species. Most weed species that are aphid and virus reservoirs are herbaceous annuals. The role of weeds as inoculum sources in non-temperate regions is widely acknowledged (Souza-Diás *et al.* 1993, Hanafi *et al.* 1995, Ragsdale *et al.* 2001). However, in temperate regions the role of alternate weed hosts has been largely discounted because annuals do not survive winter and transmission of important aphid-borne potato viruses does not occur via true seeds. Unfortunately, this is not always true. Research has shown that annual-weed hosts can survive the winter in temperate conditions in ditches, near buildings and greenhouses, and close to hot springs (Duffus 1971, Alvarez *et al.* 2003), although not in significant numbers. Also, weeds can survive within potato fields longer than the crop itself.

The emergence of weed hosts early in the spring season in temperate regions typically coincides with the emergence of volunteer potatoes that sprout from tubers inadvertently left from the previous season. The interactions among volunteers, weed hosts, and aphids provide opportunities for virus spread within a season. The presence of weed hosts in the potato ecosystem in the Pacific Northwest has been shown to enhance virus spread efficiently (Srinivasan *et al.* 2008), reiterating that alternate weed hosts can significantly impact virus incidence in potato even in temperate ecosystems.

APHID–VIRUS INTERACTIONS

Virus Transmission by Aphids

Aphid-transmitted potato viruses have evolved complex interactions with their vectors. Thus, viruses can be transmitted in different ways. The mechanism by which a virus is transmitted defines the mode of transmission. The amount of time required for the vector to acquire the virus from an infected host and to inoculate it to a healthy host and the amount of time the virus is retained in the body of the vector influence the mechanism of virus transmission.

Persistent Virus Transmission by Aphids

Persistently transmitted viruses are generally restricted to sieve elements and companion cells. These viruses can either replicate (propagative viruses) or not (non-propagative viruses) within the aphid. As the virus is restricted to the phloem tissue, the aphid needs to feed on the host plant and ingest phloem sap to acquire the virus (Prado and Tjallingii 1994). Sustained feeding is imperative for virus acquisition.

PLRV, a *Luteovirus*, is a good example of a persistent aphid-transmitted virus. Potato-colonizing aphids, such as *M. persicae*, need to feed on PLRV-infected plants for a few minutes to 5 hours in order to acquire PLRV (acquisition access period) (Smith 1931, Tanaka and Shiota 1970, Leonard and Holbrook 1978), and for about 10 minutes to 4 hours to inoculate the virus (inoculation access period) (Kirkpatrick and Ross 1952). However, once aphids acquire the virus there is a latent period for the virus to enter the aphid heomocele, and subsequently the salivary glands, before it can be released and inoculated to a non-infected host. The latent period for PLRV has been reported to last from 8 hours (Tanaka and Shiota 1970) to as long as 4 days. About 13 species of aphids have been documented to transmit PLRV, but some transmit more efficiently than others. *M. persicae* and *Aphis gossypii* (Glover) are known to be better PLRV vectors than other species, with transmission rates of up to 84% and 75%, respectively (Halbert *et al.* 1995, Woodford *et al.* 1995).

Non-Persistent Virus Transmission by Aphids

Unlike persistent viruses, non-persistently transmitted viruses tend to be distributed in multiple tissues of their host plants, including epidermal cells. Non-persistent viruses can be readily transmitted mechanically. Aphids acquire them during short stylet penetrations of epidermal cells at the beginning of a probe. Only a few seconds are required for an optimal acquisition access period. There is no latent period, and the virus particles can be inoculated into a non-infected host, also in a few seconds, during the short stylet penetration. Most potyviruses and carlaviruses that infect potato are transmitted in this manner. Due to the rapid acquisition and inoculation periods, aphids do not have to colonize the potato plant.

Potyviruses, such as PVY, are transmitted in a non-persistent manner and can be carried by both potato-colonizing and non-colonizing aphid species (DiFonzo *et al.* 1996, Cervantes 2008, Cervantes and Alvarez 2011). However, aphid species differ in their ability to transmit a potyvirus. For instance, *M. persicae* is considered to be the most efficient vector of PVY, and its efficiency varies between 4.7% and 71.1%. Transmission efficiency for *M. euphorbiae* varies between 4.0% and 29.0% (Ragsdale *et al.* 2001). Non-colonizing aphids such as *Rhopalosiphum padi* (L.) are also capable of transmitting PVY, albeit at a lower efficiency of 0.5–11.5% (van Hoof 1980, Sigvald 1984, Harrington and Gibson 1989, Cervantes and Alvarez 2011).

Effect of Viruses on Potato

Among aphid-transmitted viruses that affect potato, PLRV and PVY are the most important. Hence our discussion will be limited to these two viruses. PLRV infection can induce upward rolling of leaves, especially at the base. The edges of leaflets of some cultivars may also develop reddening. Other symptoms include stunting and marginal yellowing of a leaf at the apical portion. Necrosis may develop in the phloem tissue of stems and petioles, as well as in tubers. Infected tubers may produce internal net necrosis, and occasionally thin sprouts ("spindling sprouts") develop from infected tubers (Quanjer 1916, Rodriguez and Jones 1978, Brunt and Loebenstein 2001).

PVY has high genetic variability, with several distinct strain groups infecting the potato crop (De Bokx and Huttinga 1981, Kerlan *et al.* 1999). The strains are recognized by the symptoms that they produce on naturally infected potato and tobacco (*Nicotiana tabacum* L.) (De Bokx and Huttinga 1981, Kerlan *et al.* 1999). The most common strain is PVY^O, which induces mild to severe mosaic and leaf-drop streaks in potato and systemic mottle on tobacco. PVY^N, the tobacco veinal necrosis strain, induces very mild mottling on most potato cultivars, with occasional necrotic leaves on some cultivars (Chachulska *et al.* 1997, Kerlan *et al.* 1999). However, it induces severe systemic necrosis of leaf veins and petioles on tobacco (De Bokx and Huttinga 1981). PVY^{NTN}, a strain belonging to the necrotic group of PVY^N, causes potato tuber ring necrotic disease (PTNRD) in potato tubers (Kus 1992). It induces chlorotic mottle to mosaic symptoms on potato plants and superficial to deeply sunken necrotic rings on the tuber (Le Romancer *et al.* 1994). $PVY^{N:O}$, a new recombinant strain between O and N strains, is serologically similar to PVY^O, but produces necrosis on tobacco (Nie and Singh 2002, Crosslin *et al.* 2005) and necrotic rings on potato tubers of some cultivars (Piche *et al.* 2004).

Effect of Virus-Induced Changes in Potato on Aphid Behavior and Biology

Upon infection, potato viruses induce drastic alterations to potato plant morphology and physiology. These changes are known to affect the behavior and fitness of vectors, and, subsequently, virus spread. Potato viral infections are known to modify host-plant preference of aphids (Castle and Berger 1993, Castle *et al.* 1998, Srinivasan *et al.* 2006, Alvarez *et al.* 2009). The symptoms produced by PLRV-infected potato plants, such as yellowing and stunting, and ensuing physiological changes, arrested winged and wingless aphids longer on PLRV-infected plants than on non-infected potato plants (Castle and Berger 1993, Castle *et al.* 1998, Eigenbrode *et al.* 2002, Srinivasan *et al.* 2006, Alvarez *et al.* 2009). Studies also have indicated that besides visual and gustatory cues, olfactory cues are associated with aphid preference and/or settling. Volatiles trapped from the headspace of potato plants infected with PLRV more strongly arrested

M. persicae nymphs and adults than volatiles trapped from non-infected potato plants (Eigenbrode *et al.* 2002, Srinivasan *et al.* 2006, Alvarez *et al.* 2009).

The settling of aphids on virus-infected plants is more commonly associated with persistent viruses and potato-colonizing aphids. Two potato-colonizing species (*M. persicae* and *M. euphorbiae*) consistently settled on PLRV-infected plants more often than on PVY-infected plants. This effect was more prominent when PLRV-infected plants also were infected with PVY (Fig. 11.1). Upon mixed infection, potato plants typically expressed more severe symptoms compared to plants infected with one virus (Srinivasan and Alvarez 2007). It is not uncommon to find plants simultaneously infected with PLRV and PVY in potato ecosystems. The prolonged arrestment of aphids on PLRV-infected plants may provide aphids with a longer acquisition access period, which in turn may aid in successful acquisition and inoculation of the virus.

On the contrary, the effect of non-persistent virus infections on aphid settling behavior has been negligible or minimal. Our studies indicated that winged and wingless morphs of *M. persicae* and *M. euphorbiae* did not prefer potato plants infected with PVY to non-infected plants. Unlike persistent viruses, non-persistent viruses are acquired and transmitted by both colonizing and non-colonizing aphids' in short epidermal probes. Shorter durations of aphid arrestment may favor the spread of non-persistent virus, as their acquisition and

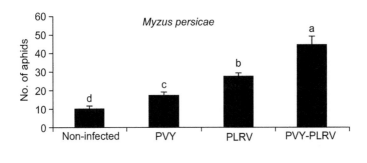

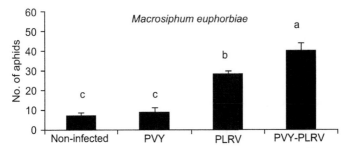

FIGURE 11.1   Settling of *Myzus persicae* and *Macrosiphum euphorbiae* on *Potato leafroll virus* (PLRV)-infected and *Potato virus Y* (PVY)-infected potato plants. Alphabets (a–d) indicate differences among treatments. Treatments or bars with the same alphabet are not different from each other. *(With permission: Srinivasan and Alvarez 2007: J. Econ. Entomol. 100: 646–655.)*

inoculation periods typically last only a few seconds. Research conducted on vector behavior as influenced by non-persistent viruses in the potato system and in other systems indicates that virus-infected plants may be arresting aphids for a short period of time to facilitate their spread (Mauck *et al.* 2010). However, these effects seem to vary between aphid species.

Electrical penetration graph (EPG) studies have shed more light on the effect of non-persistent virus infection on feeding behavior of colonizing and non-colonizing aphids. Boquel *et al.* (2012) compared feeding behavioral differences of *M. euphorbiae* on PVY-infected vs. non-infected potato foliage and observed that the duration of the E1 and E2 phases (associated with salivation in the phloem and phloem sap ingestion) decreased and the number of probes increased with PVY infection. *M. euphorbiae* is a very mobile aphid (Alyokhin and Sewell 2003); therefore, more probing and less time spent in the phloem ingestion phase could likely increase the probability of PVY acquisition and transmission.

In contrast, *M. persicae*, the most efficient vector of PVY, started probing sooner on PVY-infected plants than on healthy plants. The duration of E2 increased in PVY-infected plants vs non-infected plants. Acquisition and inoculation of a non-persistent virus are associated with subphases II-1 and II-3, respectively, of the potential drop that occurs during a stylet cell puncture (Powell *et al.* 1995). When comparing the effect of PVY infection on feeding behavior of colonizing vs. non-colonizing aphids, Boquel *et al.* (2012) observed that the mean duration of the potential drop and that of subphase II-3 increased in PVY-infected plants for *M. persicae*. On the contrary, for the non-colonizing aphid species *Sitobion avenae* Fabricius, the mean duration of subphases II-1 and II-3 was reduced on PVY-infected plants.

Besides influencing the behavior of aphids, potato virus infections have also been known to alter the biology of their aphid vectors. As in the case of behavior, the effects induced by a persistent virus seem to positively influence the fitness of colonizing aphids. Vector life history studies conducted with potato or other Solanaceae hosts infected with PLRV revealed that aphids multiplied faster on infected plants than on their non-infected counterparts (Ponsen 1969,Castle and Berger 1993). *M. persicae* and *M. euphorbiae* nymphal survival rates in general were greater on PLRV-infected plants than on non-infected plants. The fecundity and intrinsic rate of increase of colonizing aphids were higher on PLRV-infected plants than on non-infected plants (Fig. 11.2) (Srinivasan *et al.* 2008). Ponsen (1969) also found similar results; the daily fecundity of *M. persicae* was greater on PLRV-infected *P. floridana* than on non-infected plants. The reproductive period of *M. persicae* and *M. euphorbiae* on PLRV-infected potato plants was longer than on non-infected potato plants (Srinivasan *et al.* 2008). The exact physiological basis of the responses observed in these studies is unknown, but it is widely speculated that the improved nutritional quality, reduced concentrations of deterrents, and the presence and absence of feeding stimulants and toxicants, respectively, in the phloem of PLRV-infected plants could have positively influenced the fitness of aphids (Guntner *et al.* 1997, Karley 2002). The interactions between persistently transmitted luteoviruses,

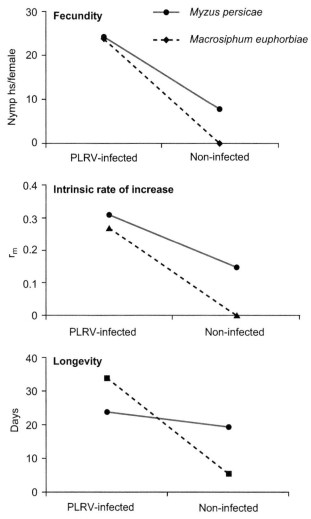

FIGURE 11.2 Fecundity, intrinsic rate of increase and longevity of *Myzus persicae* and *Macrosiphum euphorbiae* on *Potato leafroll virus* (PLRV)-infected and non-infected potato plants.

such as PLRV, and their vectors are postulated to be mutualistic and may have co-evolved, leading to increased fitness of the vector and propagation of the virus (Castle and Berger 1993, Srinivasan *et al.* 2008).

On the contrary, positive effects of non-persistent viruses on vector fitness are not consistent. The fecundity of *M. persicae* was greater on PVY-infected plants than on non-infected potato plants. However, the fecundity on PVY-infected plants was less than on PLRV-infected plants (Fig. 11.3) (Srinivasan and Alvarez 2007). In the case of *M. euphorbiae*, no difference in fecundity was observed between PVY-infected and non-infected potato plants. However, significant increase in fecundity was observed on PLRV-infected plants

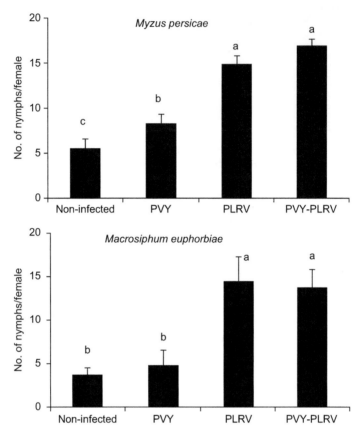

FIGURE 11.3 Fecundity of *Myzus persicae* and *Macrosiphum euphorbiae* on *Potato leafroll virus* (PLRV)-infected and *Potato virus Y* (PVY)-infected potato plants. Alphabets (a–d) indicate differences among treatments. Treatments or bars with the same alphabet are not different from each other. *(With permission: Srinivasan and Alvarez 2007: J. Econ. Entomol. 100: 646–655.)*

compared to PVY-infected or non-infected potato plants (Fig. 11.3) (Srinivasan and Alvarez 2007).

WEED HOSTS AND APHID–VIRUS INTERACTIONS

Phenology and Competition with Potato

Most weed hosts are herbaceous annuals. Weed hosts are common and abundant in temperate and non-temperate potato ecosystems. The importance of these weed hosts as vector and viral reservoirs is clearly documented in non-temperate regions, where they have previously been reported to be associated with PLRV transmission. Souza-Diás *et al.* (1993) identified a complex of solanaceous weeds as efficient inoculum sources of PLRV in São Paulo, Brazil. These weeds were found to be the cause for high PLRV-infection levels (20–80%) in potato grown from high-quality seed tubers containing less than 1% PLRV infection. Jimsonweed, *D. stramonium*, a solanaceous plant that develops interveinal necrosis upon infection, is considered to be the principal source of inoculum for PLRV in Morocco (Hanafi *et al.* 1995).

The importance of these weeds in viral epidemiology in temperate regions has been largely discounted. However, numerous studies have shown that some of these weeds can function as vector and virus reservoirs throughout the year. Even in temperate conditions, weeds such as *S. sarrachoides* have been documented to survive the winter in canals and ditches, near hot springs, and adjacent to heated buildings in the Pacific Northwest. These weeds were found to support viviparous aphid populations that were likely to be viruliferous (Wallis 1967a,1967b, Duffus 1971, Alvarez *et al.* 2003). Winter survival of weed hosts presents a continuum of host availability for the virus and its vectors. Even if overwintering *S. sarrachoides* is rare enough to be an inconsequential source of viral inoculum, preferred alightment and colonization of aphids on *S. sarrachoides* early in the spring (Alvarez *et al.* 2003, Srinivasan and Alvarez 2011) could enhance the build-up of aphid populations in the fields and aid subsequent dispersal to the crop.

Annual weeds have been known to act as bridges for virus and vectors before the emergence of crops. Weeds such as *S. sarrachoides* and *P. floridana* are available longer in potato ecosystems (from March to December) than the crop itself. This extended availability provides an avenue for build-up of vector populations prior to the emergence of potato plants. Short-range dispersal of aphids between volunteer potato plants and weed hosts can result in virus infection in weed hosts. Besides serving as reservoirs of aphids and viruses, several weed species can directly compete with the potato crop for resources. Recent research indicated that season-long *S. sarrachoides* competition with the potato crop for water and other essential nutrients could reduce yield and tuber quality by up to 30% (Hutchinson *et al.* 2011). However, some varieties were less affected by competition with hairy nightshade than others. For example, "Russet Burbank" is a more competitive variety than "Russet Norkotah" due to a vegetative canopy which closes over the rows for a longer period during the season, allowing much less light to reach weeds growing below the canopy. Russet Burbank yields were reduced by only 11% in plots with a hairy nightshade density of three plants per meter row. Yield reduction in Russet Norkotah plots with the same hairy nightshade density was as high as 27% (Hutchinson *et al.* 2011).

Solanaceous annual weeds such as *S. nigrum* L. and *S. sarrachoides* are capable of growing 30–60 cm tall, with flowers arranged in clusters resembling those of potato and tomato, and developing into green berries containing numerous tiny seeds (Whitson *et al.* 1996). These weeds also possess tremendous seed production capabilities. For instance, *S. sarrachoides* can produce over 45,000 seeds per plant at low infestation densities (Blackshaw 1991), and a maximum hairy nightshade seed production of over 300,000 seeds per square meter was observed at a density of 30 plants per meter row (Blackshaw 1991). The seeds can remain viable in soil for 5–10 years. Even a mere 1% seed survival in a field could lead to establishment of thousands of *S. sarrachoides* plants every year. *S. sarrachoides* plants emerging as late as 6–7 weeks before a killing frost can still produce mature and viable seeds. *Solanum* weeds are extremely difficult to control as they are too closely related to potato to permit selective herbicide control in the field (Quackenbush and Anderson 1984, 1985, Eberlein *et al.* 1992). Besides serving as reservoirs of aphids, these weed species also can serve as hosts of other insect pests such as the Colorado potato beetle, as well as nematodes and fungal pathogens (including *Alternaria* sp. and *Phytophthora* sp.).

Weed Hosts as Reservoirs of Aphids

Aphids that typically colonize potato plants, such as *M. persicae* and *M. euphorbiae*, have also been found colonizing solanaceous weed hosts. Aphids utilize some as their host plants better than others. Tamaki and Olsen (1979) evaluated a number of weeds, and their research indicated the greatest rate of increase of *M. persicae* on *S. sarrachoides* in comparison with other common orchard weeds in the Pacific Northwest. In our study, the average number of nymphs produced by one *M. persicae* in 48 hours was 3.56 and 4.41 for potato and *S. sarrachoides*, respectively (Alvarez and Srinivasan 2005). Besides, aphids preferentially settled on *S. sarrachoides* rather than on potato when given a choice (Srinivasan *et al.* 2006). Field observations clearly indicated that *M. persicae* and *M. euphorbiae*

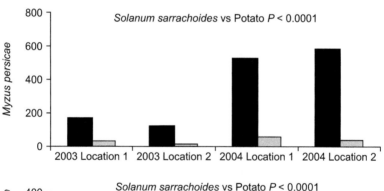

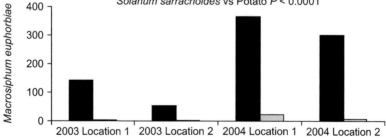

FIGURE 11.4   Counts of *Myzus persicae* and *Macrosiphum euphorbiae* on *Solanum sarrachoides* (black bars) and potato over 2 years in two locations. Counts were taken at weekly intervals for 6 weeks in the growing season. Total counts are represented. *(With permission: Alvarez and Srinivasan 2005: J. Econ. Entomol. 98: 1101–1108; Srinivasan and Alvarez 2011. Environ. Entomol. 40: 350–356.)*

preferred *S. sarrachoides* plants over adjacently occurring potato plants (Fig. 11.4) (Thomas 2002, Alvarez and Srinivasan 2005, Srinivasan and Alvarez 2011). Furthermore, the intrinsic rate of increase and longevity of both aphid species was greater on *S. sarrachoides* than on potato (Fig. 11.5) (Srinivasan *et al.* 2008).

These findings indicate that some weed hosts could be nutritionally superior and more suitable for aphid growth and development than potato. However, not all solanaceous weeds are equally suitable for aphids. Some wild *Solanum* spp., such as *S. berthaultii*, are known to possess glandular trichomes and can adversely affect aphid utilization of the host plant (Avé *et al.* 1987), whereas other species, such as *S. sarrachoides*, do not possess glandular terpenoid-secreting trichomes and seem to positively influence the fitness of aphids.

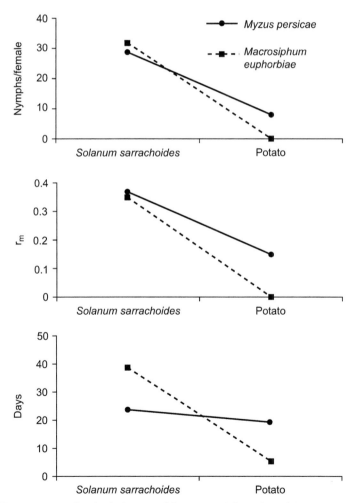

FIGURE 11.5   Fecundity, intrinsic rate of increase and longevity of *Myzus persicae* and *Macrosiphum euphorbiae* on *Solanum sarrachoides* and potato plants.

The ubiquitous presence of wild solanums in potato ecosystems, as well as their ability to persist in the landscape for a longer duration, has led to unique interactions with aphids. For instance, we documented the presence of *S. sarrachoides*-specific biotype of *M. euphorbiae* in Pacific Northwest's potato ecosystems (Srinivasan and Alvarez 2011). Though polyphagous in nature, this particular biotype has resigned to monophagy. The monophagous biotype outcompeted *M. persciae* on *S. sarrachoides* (Fig. 11.6), but could not reproduce on potato. The *S. sarrachoides* biotype produced more winged adults than *M. persicae*. *S. sarrachoides* is ubiquitous in potato ecosystems and is present longer in a temperate ecosystem than the crop itself. Thus, specializing on *S. sarrachoides* could be beneficial to *M. euphorbiae* from an evolutionary standpoint. The ability of *S. sarrachoides* specific biotype to outcompete *M. persicae* on that host and its ability to produce more winged aphids might be strategies that the aphid biotype employs to better utilize a nutritionally superior host such as *S. sarrachoides* (Srinivasan and Alvarez 2011). Careful field observations over the past few years in Idaho's potato ecosystems clearly indicated that *M. persicae* were more commonly observed colonizing potato plants than *M. euphorbiae* (Srinivasan and Alvarez 2011). This once again reiterates that the host utilization patterns of *M. euphorbiae* have shifted to take advantage of a nutritionally superior host. Thus, the presence of alternate weed hosts such as *S. sarrachoides* in the ecosystem can result in a substantial population increase of colonizing aphids. The obvious question is whether these aphids

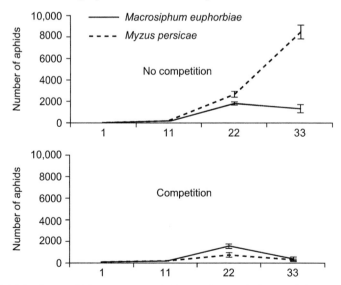

FIGURE 11.6 Counts of *Myzus persicae* and *Macrosiphum euphorbiae* on *Solanum sarrachoides* in the absence and presence of competition. Experiments were conducted by individually caging *S. sarrachoides* plants and releasing aphids either alone or in combination (*M. persicae* and *M. euphorbiae* together). Counts were taken at 10-day time intervals for a period of 1 month. *(With permission: Srinivasan and Alvarez 2011. Environ. Entomol. 40: 350–356.)*

disperse to potato plants due to overcrowding or for other reasons. Field studies conducted by Srinivasan *et al.* (unpubl. data) showed that the presence of *S. sarrachoides* in the vicinity of a potato field led to increase in aphid populations on potato plants.

Weed Hosts as Virus Inoculum Sources

Solanum weeds can become infected with PLRV and PVY. Some weed species, such as *D. stramonium* and *P. floridana*, always display characteristic symptoms upon infection with potato viruses, while others may or may not display symptoms. The presence of solanaceous weeds for a longer period of time during the growing season than the crop itself results in very high rates of virus infection, and enhances their chances of serving as inoculum sources. A survey conducted by Souza-Diáz *et al.* (1993) found a complex of solanaceous weeds comprising *S. lycocarpum* St. Hill, *S. erianthum* D. Don, *S. paniculatum* L., and *S. variabile* Mart. to be infected with PLRV at very high rates, and also documented that these weeds can serve as efficient inoculum sources of PLRV in São Paulo, Brazil. The incidence of PLRV increased up to 80% in the presence of *Solanum* spp. weeds on potato fields planted with high-quality (<1% infection) seed materials. High PLRV infection rates were also found on *S. sarrachoides* plants in potato ecosystems in the Pacific Northwest (Thomas 2002, Alvarez *et al.* 2003, Alvarez and Srinivasan 2005). In Morocco, *D. stramonium* along with volunteer potatoes was implicated as a principal source of viruliferous aphids and virus inoculum (Hanafi *et al.* 1995).

Recent studies have indicated that *Solanum* weeds, such as *S. sarrachoides*, are not only being infected with the virus but also have a potential to accumulate greater virus titers. Thomas (2002) reported greater PLRV titers in *S. sarrachoides* than in potato. However, another study indicated that PLRV titers were lower in *S. sarrachoides* than in potato (Alvarez and Srinivasan 2005). Cervantes and Alvarez (2010, 2011), through greenhouse experiments, found that *S. sarrachoides* could be infected with necrotic and non-necrotic strains of PVY (PVY^O, PVY^{NTN}, and PVY$^{N:O}$). Also, the titers of PVY in general were comparable with PVY titers in one of the most susceptible potato cultivars, Russet Burbank. There were no differences in titers between potato and *S. sarrachoides* for PVY^O and PVY$^{N:O}$. However, in both plant species, the titers of PVY^{NTN} were greater than the titers of other two strains. The titer of PVY^{NTN} also was greater in *S. sarrachoides* than in potato (Fig. 11.7) (Cervantes and Alvarez 2011). PVY^{NTN} can induce tuber necrosis and is capable of causing serious economic losses.

Weed Host-Induced Effects on Vector–Virus Interactions

Potato viruses can be transmitted by aphids from potato to weed hosts and *vice versa*, emphasizing the ability of *Solanum* weeds to serve as inoculum sources.

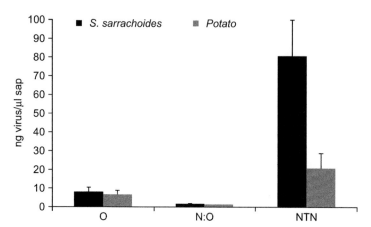

FIGURE 11.7 Titers of three *Potato virus Y* strains in a microliter of plant sap of *Solanum sarrachoides* and potato.

For example, PLRV was transmitted to and from *S. sarrachoides* by *M. persciae*. The transmission rates were at least four times greater when *S. sarrachoides* was used as an inoculum source (Alvarez and Srinivasan 2005), despite PLRV titers being lower in *S. sarrachoides* when compared with potato.

Recent research has focused on examining the effects of PLRV infection in weeds on aphid biology and the mechanisms or cues that potentially mediate such effects. Both winged and wingless morphs of *M. persicae* settled on PLRV-infected *S. sarrachoides* more readily than on non-infected *S. sarrachoides*. These experiments indicated that the ability of virus-infected weed hosts, such as *S. sarrachoides*, to serve as virus reservoirs could increase substantially. Furthermore, choice tests conducted under light and dark conditions indicated that besides visual and gustatory cues, olfactory cues could also be involved (Srinivasan *et al.* 2006). Examination of headspace volatile organic compounds from *S. sarrachoides* and potato plants with and without PLRV infection, followed by gas chromatographic analysis and mass spectroscopy, led to the identification of various green leaf volatiles or 6-C containing compounds such as 2-hexenal, 2-hexen-1-ol, 3-hexen-1-ol, 2-hexen-1-ol-acetate, and 3-hexen-1-ol-acetate in *S. sarrachoides* and potato. In addition, an aldehyde, nonanal, and methyl salicylate were identified in *S. sarrachoides*. One monoterpene and 12 sesquiterpene compounds (both identified and unidentififed) were detected in potato but not in *S. sarrachoides* (Srinivasan 2006).

PLRV-infected *S. sarrachoides* plants released more volatile organic compounds than non-infected plants. Conversely, most of the volatile organic compound fractions detected in potato were in greater concentrations from non-infected than from PLRV-infected plants. However, concentrations of a few compounds such as 2-hexen-1-ol, β-myrcene, α-caryophyllene, and an unknown sesquiterpene (204 mol. wt) were greater in PLRV-infected plants than in non-infected potato plants (Fig. 11.8) (Srinivasan 2006). These results

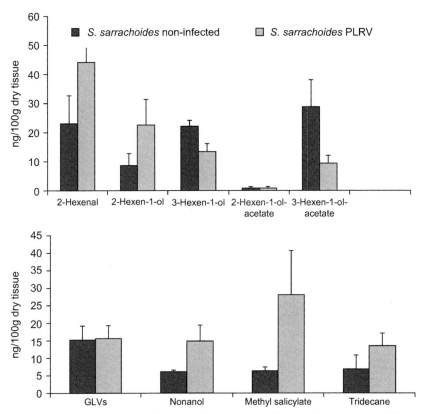

FIGURE 11.8   Headspace volatile fractions in *Potato leafroll virus*-infected and non-infected *Solanum sarrachoides* plants.

clearly indicate variations in volatile organic compounds in two hosts of the same genus, and that pathogen infection could alter volatile organic-compounds emission in host plants. The roles of these volatile compounds in vector-virus interactions are yet to be elucidated in the *S. sarrachoides* PLRV pathosystem. However, many of these chemicals have already been identified as affecting aphid-virus interactions in potato and cereal *Luteovirus* pathosytems (Eigenbrode *et al.* 2002, Ngumbi *et al.* 2007).

The fecundity of *M. euphorbiae* was greater on PLRV-infected *S. sarrachoides* plants than on non-infected plants (Srinivasan and Alvarez 2007, Srinivasan *et al.* 2008). No such effects were observed with *M. persicae*. The intrinsic rate of increase on *S. sarrachoides* was not influenced by PLRV infection. The longevity of both aphid species was greater on PLRV-infected plants than on non-infected plants (Fig. 11.9). Typically, PLRV-infected potato plants exhibited severe foliar yellowing and upward curling of leaves, which in turn could have led to increased accumulation of soluble sugars and free amino acids. Such

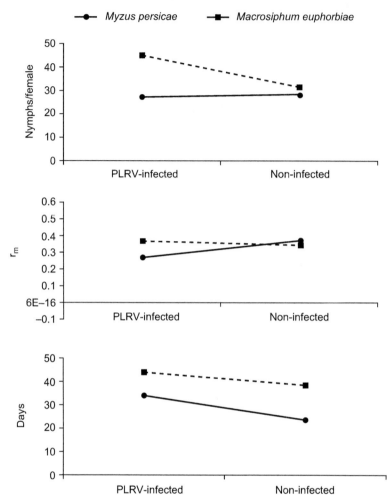

FIGURE 11.9   Fecundity, intrinsic rate of increase and longevity of *Myzus persicae* and *Macrosiphum euphorbiae* on *Potato leafroll virus* (PLRV)-infected and non-infected *Solanum sarrachoides* plants.

alterations were not always associated with *S. sarrachoides*, and hence positive effects of *Luteovirus* infection on fitness of aphids were subdued.

Weed Host-Induced Complexity in the Potato-Viral Pathosystem

The presence of alternate weed hosts in potato ecosystems that serve as vector and/or virus reservoirs can lead to a number of component interactions that are normally absent in a simple tricomponent (aphid–vector–virus) pathosystem. The number of component interactions (multicomponent interactions) will in turn increase the complexity of the pathosystem (Fig. 11.10).

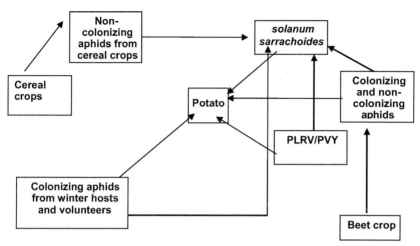

FIGURE 11.10   Potato viral pathosystem in the Pacific Northwest characterized by a solanaceous weed, hairy nightshade (*Solanum sarrachoides*).

Non-potato colonizing aphid species such as the black bean aphid *Aphis fabae* (Scopoli) are known to colonize *S. sarrachoides* plants in the Pacific Northwest potato ecosystems. Aphids from other crop species, such as beet, can also disperse and colonize weed hosts. It should be taken into account that these *Solanum* weed hosts can be infected with almost all typical potato-infecting viruses. Their ability to serve as virus inoculum sources and their ubiquitous presence can aid in the transmission of non-persistent viruses, following the dispersal of non-potato colonizing aphids. Thus, a single alternate weed host can increase the complexity of the potato pathosystem; it is not uncommon to find a potato system characterized by more than one solanaceous weed host.

Impact of Weed Hosts on Virus Epidemiology

Virus-infected weed hosts can be responsible for both the primary and the secondary spread of the virus within the cropping system. Studies conducted by establishing an inoculum focus using PLRV-infected *S. sarrachoides* plants revealed that the numbers of *M. persicae* were greater both on potato plants in plots with an inoculum focus, as well as on plots with non-infected *S. sarrachoides* plants (Srinivasan and Alvarez 2008). Also, by tagging potato plants, the spatial as well as the temporal spread of PLRV was monitored in the plots with *S. sarrachoides*. Results indicated that the presence of virus-infected weeds significantly influenced PLRV spread. PLRV infection was initially noticed at distances closer to inoculum sources (placed in the middle of the experimental plot), but it then spread further away (Fig. 11.11). That sequential spatial spread indicated that the spread was predominantly aided by the movement of *S. sarrachoides* colonizing wingless aphids as opposed to a random in-field virus

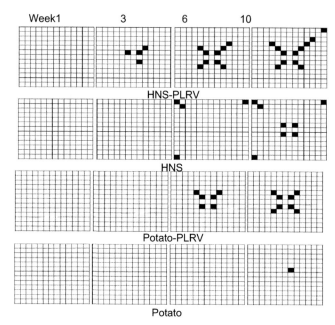

FIGURE 11.11   Spatial and temporal spread of *Potato leafroll virus* (PLRV) from inoculum foci characterized by PLRV-infected *Solanum sarrachoides* and potato, respectively and placed in the middle of the plot. Each grid represents a potato plant. Dark grids represent PLRV-infected plants. *HNS is an abbreviation for hairy nightshade (*S. sarrachoides*).

spread by winged aphids. However, increased PLRV incidence was observed on the edges of plots with non-infected *S. sarrachoides* and potatoes. These results are consistent with PLRV spread from an inoculum focus, as well as with the background of inflight of viruliferous aphids (Srinivasan and Alvarez 2008).

Similarly, experiments were conducted with microcosms (made up of aphid-proof mesh cages) to assess the effect of *S. sararchoides* on the spread of PVY. Two potato-colonizing (*M. persicae* and *M. euphorbiae*) and one non-colonizing (*R. padi*) aphid species were tested for PVY spread. The infection rates in cages with PVY-infected *S. sarrachoides* were twice as high as the infection rates in cages with PVY-infected potato as an inoculum source. Also, the infection rates in microcosms with colonizing aphids were three to six times greater than the infection rates in microcosms with non-colonizing aphids (Cervantes 2008).

Although non-colonizing aphids are usually implicated as being responsible for most PVY transmission, colonizing aphids can significantly impact PVY spread. The presence of alternate weed hosts, such as *S. sarrachoides*, may hasten the spread. Within the microcosms with colonizing aphid species, infection rates were typically higher in cages with *M. euphorbiae* than with *M. persicae*. Increased PVY incidence could have been brought about by the fact that

M. euphorbiae is more mobile and has a greater intrinsic ability to produce winged morphs (Alyokhin and Sewell 2003, Srinivasan and Alvarez 2011).

WEED MANAGEMENT AND POTATO VIRUS EPIDEMICS

Management Tactics

In this chapter, it has been illustrated that the presence of an alternate weed host can increase the complexity of the potato viral pathosystem. The complexity may increase multifold if more than one weed host is found in the potato ecosystem. In order to successfully manage aphid-transmitted viruses in such ecosystems, management tactics should concentrate on all the potential contributing factors.

A virus management plan should include strategies that reduce vectors, viruses, and alternate weed hosts in the potato ecosystem. A holistic approach that is inclusive of aphid and virus management in potato, weed hosts, and other crop hosts is essential. Besides planting high-quality seed tubers, other inoculum sources such as *Solanum* weeds and volunteer potato plants within potato fields and in their vicinities should be kept at a minimum. In addition, aphid management should include the potato crop, cereal crops, and other crop hosts, as well as alternate weed hosts.

Aphid management as currently practiced by potato farmers predominantly relies on insecticides, which are generally effective in preventing PLRV transmission. However, due to its non-persistent nature, no single insecticide is capable of preventing PVY transmission. Still, growers can reduce PVY infection in their fields by reducing the densities of colonizing aphids. Currently available management plans do not incorporate the management of aphids in weed hosts or in other crops. However, even the presence of non-infected *Solanum* weeds adjacent to potato plots can contribute to aphid dispersal on potato and increased virus incidence in potato (Srinivasan 2006). Thus, insecticides targeting weed hosts in field edges can prevent aphid dispersal into potato fields. Care should be taken to ensure that aphids are killed rapidly, otherwise they can become agitated. This, in turn, may lead to increased transmission of non-persistent viruses through increased probing.

Although desirable, eradication of all *Solanum* weeds in and near the potato fields may be impractical. Nevertheless, it is necessary to advocate intense weed control measures in potato fields before weed seed accumulates in the soil. Research trials in the Pacific Northwest have shown that satisfactory weed control of solanaceous weeds has been achieved by a timely combination of cultivation and hilling operations, followed by herbicide application performed just prior to potato emergence. Since solanaceous weeds are in the same plant family as potato, and have similar germination and emergence times, selective control without risk of potato crop injury can be difficult. Very few herbicides currently labeled for use in potato have activity on emerged solanaceous weeds.

Nevertheless, available pre-emergence herbicides can provide 90% or greater season-long control of *Solanum* weeds. Following the pre-emergence herbicide application with an early post-emergence application can also provide good control.

Crop rotation is probably a good tactic to manage solanaceous weeds in a potato cropping system, as there are more options for control of *Solanum* weeds in non-solanaceous crops. Unfortunately, the seeds of some solanaceous weeds can be dormant in the soil for years. In such cases, crop rotation may only offer partial relief.

Management of aphids in other crop hosts, such as beet and cereals, should also be given due importance to reduce the incidence of viruses in potato. In the Pacific Northwest, suction trap data indicated an increase in the number of cereal aphids coinciding with the maturity of wheat, corn, other small grain crops, and beet (Srinivasan, unpubl. data). One or two applications of systemic foliar insecticides on infested cereal fields or beet fields adjacent to seed potato fields before harvest can facilitate the reduction of cereal aphid movement into seed potato fields and consequently suppress virus transmission.

Pitfalls and Future Research

Although some weeds, such as *D. stramonium* and *P. floridana*, display prominent symptoms of viral infections, other alternate weed hosts are often infected at very high rates but do not display virus-associated symptoms. As a result, it is difficult to assess the amount of inoculum that is available in the cropping system. *Solanum* weeds harbor PLRV and various strains of PVY. Therefore, identification of viruses in weed hosts would require serological and/or molecular testing, which is cost-prohibitive on a commercial scale. Eliminating weeds from the ecosystem is also logistically impossible. Thus, aphid management remains the most practical option.

Aphid management on potato is commonly practiced by commercial potato growers, but aphids emigrating from weed hosts and other crop hosts are not targeted. The ubiquitous presence of weeds and aphid migrations from adjacent non-potato fields can severely impact virus epidemics in potato. An area-wide management program can provide some relief. However, there are numerous logistical issues associated with implementing an area-wide management program. It is not economical for either cereal or beet growers to treat their crops for aphids, but these fields are often interspersed with potato fields in a production landscape. Primary and secondary spread of the viruses initiated by immigrating aphids can present considerable issues to growers, particularly seed growers.

Aphid management is also constrained by the availability of insecticides and their modes of action, particularly when it pertains to the transmission of non-persistent viruses. Aphicides with a quick knockdown can slightly reduce PVY transmission, but not to satisfactory levels. Also, only a few such insecticides are available. Rotation of insecticide classes is necessary to prevent insecticide resistance development. Potato-colonizing aphids such as *M. persicae* can develop insecticide resistance rapidly (Ragsdale *et al.* 2001). In addition, the blanket application of insecticides will lead to adverse non-target effects, such as eliminating natural enemies in the ecosystem. The absence of natural enemies would further favor aphid population growth.

Some alternate weed hosts are known to provide refuge to natural enemies. However, planting refuge species that also are hosts of potato viruses may lead to increased virus incidence. It may be possible to encourage competition among weed species in the ecosystem by planting desirable natural enemy refuge species; therefore, wild solanaceous weeds could be outcompeted. However, these management options are time consuming and it may take years before desirable results can be attained. Such an approach may not be possible in intensive potato production systems, particularly when there are stringent levels of allowable infection rates in seed potato production.

Given the complexity associated with the potato pathosystem and the advent of tuber necrotic non-persistently transmitted viruses and/or their strains, very few available management options are effective. Planting resistant cultivars is probably the most important management option. Unfortunately, due to difficulties associated with conventional breeding, the availability of an ideal potato cultivar with broad-spectrum resistance against potato viruses and/or their strains is a remote possibility. Transgenic potato cultivars are effective, but currently have poor consumer acceptability. Until that attitude changes, the management of vectors and viruses in the potato crop and in alternate weed hosts is the sole alternative.

Research on vector-pathogen interactions as influenced by alternate hosts has provided a platform to study intricate component interactions in the pathosystem and the factors that drive such interactions. Understanding vector behavior as influenced by virus infection and by alternate hosts can contribute to a greater understanding of the pathosystem, which may lead to the development of novel management options. For instance, recent research has indicated that the volatile organic compounds in alternate hosts and in virus-infected plants may serve as olfactory cues that mediate aphid host preferences. The exact role of all the volatile organic compounds in determining vector preference is not known. Information on the volatile organic compounds influencing vector behavior could increase understanding of the chemical ecology of an ecosystem, and could be used to develop efficient aphid monitoring and management tactics in the future.

REFERENCES

Alvarez, J.M., Srinivasan, R., 2005. Evaluation of hairy nightshade as an inoculum source for aphid-mediated transmission of potato leafroll virus. J. Econ. Entomol. 98, 1101–1108.

Alvarez, J.M., Stoltz, R.L., Baird, C.R., Sandvol, L.E., 2003. Potato insects and their management. In: Love, S., Stark, J. (Eds.), Potato Production Systems, Vol. 2. University of Idaho Agricultural Communications, pp. 204–239.

Alvarez, J.M., Srinivasan, R., Cervantes, F., 2009. Potato viral infections affect the biology and behaviour of aphid vectors. Redia 92, 169–170.

Alyokhin, A., Sewell, G., 2003. On-soil movement and plant colonization by walking wingless morphs of three aphid species (Homoptera: Aphididae) in greenhouse arenas. Environ. Entomol. 32, 1393–1398.

Avé, D.A., Gregory, P., Tingey, W.M., 1987. Aphid repellent sesquiterpenes in glandular trichomes of *Solanum berthaultii* and *S. tuberosum*. Entomol. Exp. Appl. 44, 131–138.

Blackman, R.L., Eastop, V.F., 1984. Aphids on the world's crops: An identification guide. Wiley, New York, NY.

Blackshaw, R.E., 1991. Hairy nightshade (*Solanum sarrachoides*) interference in dry beans (*Phaseolus vulgaris*). Weed Sci. 39, 48–53.

Boquel, S., Delayen, C., Couty, A., Giordanengo, P., Ameline, A., 2010. Modulation of aphid vector activity by Potato virus Y on *in vitro* potato plants. Plant Dis. 96, 82–86.

Brunt, A.A., Loebenstein, G., 2001. Potato leafroll virus (PLRV; genus Polerovirus; family Luteoviridae). pp. 65–134. In: Loebenstein, G., Berger, P.H., Brunt, A.A., Lawson, R.H. (Eds.), Virus and virus-like diseases of potatoes and production of seed potatoes, Kluwer Academic Publisher, Dordrecht, The Netherlands.

Castle, S.J., Berger, P.H., 1993. Rates of growth and increase of *Myzus persicae* on virus-infected potatoes according to type of virus vector–vector relationship. Entomol. Exp. Appl. 69, 51–60.

Castle, S.J., Mowry, T.M., Berger, P.H., 1998. Differential settling by *Myzus persicae* (Homoptera: Aphididae) on various virus infected host plants. Ann. Entomol. Soc. Am. 91, 661–667.

Cervantes, F., 2008. Role of hairy nightshade Solanum sarrachoides (Sendtner) in the transmission of Potato virus Y (PVY) strains by aphids and characterization of the PVY strain reactions on Solanum tuberosum (L.). Ph.D. dissertation. University of Idaho, Idaho.

Cervantes, F.A., Alvarez, J.M., 2010. Role of hairy nightshade in the transmission of different Potato virus Y strains on *Solanum tuberosum* (L.). Plant Health Progress. doi:10.1094/PHP-2010-0526-05-RS.

Cervantes, F.A., Alvarez, J.M., 2011. Within plant distribution of Potato virus Y in hairy nightshade (*Solanum sarrachoides*): An inoculum source affecting PVY aphid transmission. Virus Res. 159, 194–200.

Chachulska, A.M., Chrzanowska, M., Robadlia, C., Zagorski, W., 1997. Tobacco veinal necrosis determinants are unlikely to be located within the 5′ and 3′-terminal sequences of the Potato virus Y genome. Arch. Virol. 142, 765–779.

Crosslin, J.M., Hamm, P.B., Shiel, P.J., Hane, D.C., Brown, C.R., Berger, P.H., 2005. Serological and molecular detection of tobacco veinal necrosis isolates of Potato virus Y (PVYN) from potatoes grown in the western United States. Am. J. Potato Res. 82, 263–269.

DeBokx, J.A., Huttinga, H., 1981. Potato virus Y. Descriptions of plant viruses, No. 242. Common Mycol Inst./Assoc. Appl. Boil., Kew, England.

DiFonzo, C.D., Ragsdale, D.W., Radcliffe, E.B., 1996. Integrated management of PLRV and PVY in seed potato. With emphasis on the Red River Valley of Minnesota and North Dakota. URL: http://ipmworld.umn.edu In: Radcliffe, E.B., Hutchison, W.D. (Eds.), Radcliffe's IPM World Textbook, University of Minnesota, St Paul, MN.

Duffus, J.E., 1971. Role of weeds in the incidence of virus diseases. Annu. Rev. Phytopathol. 9, 319–340.

Eberlein, C.V., Guttieri, M.J., Schaffers, W.C., 1992. Hairy nightshade (Solanum sarrachoides) control in potatoes (Solanum tuberosum) with bentazon plus additives. Weed Technol. 6, 85–90.

Eigenbrode, S.D., Ding, H., Shiel, P., Berger, P.H., 2002. Volatiles from potato plants infected with Potato leafroll virus attract and arrest the virus vector, *Myzus persicae* (Homoptera: Aphididae). Proc. R. Soc. B Biol. Sci. 269, 455–460.

Guntner, D., Gonzales, A., Dos-Reis, R., Gonzales, G., Vazquez, A., Ferreira, F., Moyna, P., 1997. Effect of Solanum glycoalkaloids on potato aphid, Macrosiphum euphorbiae. J. Chem. Ecol. 23, 1651–1659.

Halbert, S.E., Castle, S.J., Mowry, T.M., 1995. Do Myzus (Nectarosiphon) species other than M. persicae pose a threat to the Idaho potato crop? Amer. Potato J. 72, 85–97.

Hanafi, A., Radcliffe, E.B., Ragsdale, D.W., 1995. Spread and control of potato leafroll virus in the Souss Valley of Morocco. Crop Prot. 14, 145–153.

Harrington, R., Gibson, R.W., 1989. Transmission of Potato virus Y by aphids trapped in potato crops in southern England. Potato Res. 32, 167–174.

Heldak, J., Bezo, M., Stefunova, V., Gallikova, A., 2007. Selection of DNA markers for the extreme resistance of Potato virus Y in tetraploid potato (Solanum tuberosum L.) F_1 progenies. Czech. J. Genet. Plant Breed 43, 125–134.

Hull, R., 2002. Mathew's plant virology (4 edn). Academic Press, San Diego, CA.

Hutchinson, P., Beutler, B., Farr, J., 2011. Hairy nightshade (Solanum sarrachoides) competition with two potato varieties. Weed Sci. 59, 37–42.

Karley, A.J., Douglas, A.E., Parker, W.E., 2002. Amino acid composition and nutritional quality of potato leaf phloem sap for aphids. J. Exp. Biol. 205, 3009–3018.

Kerlan, C., Tribodet, M., Glais, L., Guillet, M., 1999. Variability of Potato virus Y in potato crops in France. J. Phytopathol. 147, 643–651.

Kirkpatrick, H.C., Ross, A.F., 1952. Aphid-transmission of potato leaf roll virus to solanaceous species. Phytopathology 42, 540–546.

Kus, M., 1992. Potato tuber necrotic ringspot disease. Varietal differences in appearance of ringspot necrosis symptoms on tubers, pp. 81–83. In Proceedings of the EAPR meeting virology section. 29 June–3 July 1992 Vitoria-Gasteiz, Spain.

Le Romancer, M., Kerlan, C., Nedellec, M., 1994. Biological characterization of various geographical isolates of Potato virus Y inducing superficial necrosis on potato tubers. Plant Pathol. 43, 138–144.

Leonard, S.H., Holbrook, F.R., 1978. Minimum acquisition and transmission times for potato leaf roll virus by the green peach aphid. Ann. Entomol. Soc. Am. 71, 493–495.

Lorenzen, J.H., Piche, L.M., Gudmestad, N.C., Meacham, T., Shiel, P., 2006. A multiplex PCR assay to characterize Potato virus Y isolates and identify strain mixtures. Plant Dis. 90, 935–940.

Mauck, K.E., De Moraes, C.M., Mescher, M.M., 2010. Deceptive chemical signals induced by a plant virus attract insect vectors to inferior hosts. PNAS 107, 3600–3605.

Müller, C.B., Williams, I.S., Hardie, J., 2001. The role of nutrition, crowding, and interspecific interactions in the development of winged aphids. Ecol. Entomol. 26, 330–340.

Ngumbi, E., Eigenbrode, S.D., Bosque-Perez, N.A., Ding, H., Rodriguez, A., 2007. Myzus persicae is arrested more by blends than by individual compounds elevated in headspace of PLRV-infected potato. J. Chem. Ecol. 33, 1733–1747.

Nie, X., Singh, R.P., 2002. A new approach for the simultaneous differentiation of biological and geographical strains of Potato virus Y by uniplex and multiples RT-PCR. J. Virol. Methods 104, 41–54.

Norris, R.F., Kogan, M., 2005. Ecology of interactions between weeds and arthropods. Annu. Rev. Entomol. 50, 479–503.

Pelletier, Y., Pompon, J., Dexter, P., Quiring, D., 2010. Biological performance of Myzus persicae and Macrosiphum euphorbiae (Homoptera: Aphididae) on seven wild Solanum species. Ann. Appl. Biol. 165, 329–336.

Piche, L.M., Singh, R.P., Nie, X., Gudmestad, N.C., 2004. Diversity among Potato virus Y isolates obtained from potatoes grown in the United States. Phytopathology 94, 1368–1375.

Ponsen, M.B., 1969. The effect of potato leafroll virus on the biology of *Myzus persicae*. Netherlands J. Pl. Pathol. 75, 360–368.

Powell, G., Pirone, T., Hardie, J., 1995. Aphid stylet activities during potyvirus acquisition from plants and an *in vitro* system that correlate with subsequent transmission. Eur. J. Plant. Pathol. 101, 411–420.

Prado, E., Tjallingii, W.F., 1994. Aphid activities during sieve element punctures. Entomol. Exp. Appl. 72, 157–165.

Quackenbush, L.S., Andersen, R.N., 1984. Effect of soybean (*Glycine max*) interference on eastern black nightshade (*Solanum ptycanthum*). Weed Sci. 32, 638–645.

Quackenbush, L.S., Andersen, R.N., 1985. Susceptibility of five species of *Solanum nigrum* complex to herbicides. Weed Sci. 33, 386–390.

Quanjer, H.M., van der Lek, H.A., Oortwijn Botjes, J.G., 1916. Aard. Verpreidingswijze en bestrijding van pheromonecrose (bladrol) en verwante ziekten, o.a. Meded LamdbHogesch 10, 1–138.

Ragsdale, D.V., Radcliffe, E.B., DiFonzo, C.D., 2001. Epidemiology and field control of PVY and PLRV, pp. 195–225. In: Loebenstein, G., Berger, P.H., Brunt, A.A., Lawson, R.H. (Eds.), Virus and virus-like diseases of potatoes and production of seed potatoes, Kluwer Academic Publishers, Dordrecht, The Netherlands.

Rodriguez, A., Jones, R.A.C., 1978. Enanismo amarillo disease of *Solanum andigena* potatoes is caused by Potato leafroll virus. Phytopathology 68, 39–43.

Sigvald, R., 1984. The relative efficiency of some aphid species as vectors of Potato virus Y^O (PVY^O). Potato Res. 27, 285–290.

Singh, R.P., Valkonen, J.P.T., Gray, S.M., Boonham, N., C.Jones, R.A., Kerlan, C., Schubert, J., 2008. Discussion paper: The naming of Potato virus Y strains infecting potato. Arch. Virol. 153, 1–13.

Smith, K.M., 1931. Studies on potato virus disease. IX. Some further experiments on the insect transmission of potato leafroll. Ann. Appl. Biol. 18, 141–157.

Souza-Diás, J.A.C., Costa, A.S., Nardin, A.M., 1993. Potato leafroll virus in solanaceous weeds in Brazil explains severe outbreaks of the disease in absence of known potato donor sources. Summa Phytopathol. 19, 80–85.

Srinivasan, R., 2006. Influence of hairy nightshade, *Solanum sarrachoides* (Sendtner) on the Potato leafroll virus pathosystem). PhD dissertation University of Idaho, ID.

Srinivasan, R., Alvarez, J.M., 2007. Effect of mixed viral infections (Potato virus Y-Potato leafroll virus) on biology and preference of vectors *Myzus persicae* and *Macrosiphum euphorbiae* (Hemiptera: Aphididae). J. Econ. Entomol. 100, 646–655.

Srinivasan, R., Alvarez, J.M., 2008. Hairy nightshade as a potential Potato leafroll virus inoculum source in Pacific Northwest potato ecosystems. Phytopathology 98, 985–991.

Srinivasan, R., Alvarez, J.M., 2011. Specialized host utilization of *Macrosiphum euphorbiae* on a nonnative weed host, *Solanum sarrachoides*, and competition with *Myzus persicae*. Environ. Entomol. 40, 350–356.

Srinivasan, R., Alvarez, J.M., Eigenbrode, S.D., Bosque-Perez, N.A., 2006. Influence of hairy nightshade *Solanum sarrachoides* (Sendtner) and Potato leafroll virus (Luteoviridae: *Polerovirus*) on the host preference of *Myzus persicae* (Sulzer) (Homoptera: Aphididae). Environ. Entomol. 35, 546–553.

Srinivasan, R., Alvarez, J.M., Eigenbrode, S.D., Bosque-Pérez, N.A., Novy, R., 2008. Effect of an alternate weed host, hairy nightshade, *Solanum sarrachoides* (Sendtner), on the biology of the two important Potato leafroll virus (Luteoviridae: *Polerovirus*) vectors, *Myzus persicae* (Sulzer) and *Macrosiphum euphorbiae* (Thomas) (Homoptera: Aphididae). Environ. Entomol. 37, 592–600.

Tamaki, G., Olsen, D., 1979. Evaluation of orchard weed hosts of green peach aphid and the production of winged migrants. Environ. Entomol. 8, 314–317.

Tanaka, S.H., Shiota, H., 1970. Latent period of potato leafroll virus in the green peach aphid (*Myzus persicae* Sulzer). Ann. Phytopathol. Soc. Jpn. 36, 106–111.

Thomas, P.E., 2002. First report of *Solanum sarrachoides* (hairy nightshade) as an important host of potato leafroll virus. Plant Dis. 86, 559.

Valkonen, J.P.T., 1994. Natural genes and mechanisms for resistance to viruses in cultivated and wild potato species (*Solanum* spp.). Pl. Breed.112, 1–16.

Valkonen, J.P.T., Pehu, E., Jones, M.G.K., Gibson, R.W., 1991. Resistance in *Solanum brevidens* to both Potato virus Y and Potato virus X may be associated with a slow cell-to-cell spread. J. Gen. Virol. 72, 231–236.

Van Hoof, H.A., 1980. Aphid vectors of Potato virus Y. Neth. J. Plant Pathol. 86, 159–162.

Wallis, R.L., 1967a. Green peach aphids and spread of beet western yellows in the Northwest. J. Econ. Entomol. 60, 313–315.

Wallis, R.L., 1967b. Some host plants of the green peach aphid and beet western yellows in the Pacific Northwest. J. Econ. Entomol. 60, 904–907.

Whitson, T.D., Burril, L.C., Dewey, S.A., Cudney, D.W., Nelson, B.E., Lee, R.D., Parker, R., 1996. Weeds of the West. pp. 578–579 University of Wyoming, Pioneer of Jackson Hole, Jackson, WY.

Whitworth, J.L., Novy, R.G., Stark, J.C., Pavek, J.J., Corsini, D.L., Vales, M.I., Mosley, A.R., James, S.R., Hane, D.C., Shock, C.C., Charlton, B.A., Knowles, N.R., Pavek, M.J., 2010. Yukon Gem: A yellow-fleshed potato cultivar suitable for fresh-pack and processing with resistances to PVYO and late blight. Amer.. J. Potato Res. 87, 327–336.

Woodford, J.A.T., Jolly, C.A., Aveyard, C.S., 1995. Biological factors influencing the transmission of potato leafroll virus by different aphid species. Potato Res. 38, 133–141.

Chapter 12

Successional and Invasive Colonization of the Potato Crop by the Colorado Potato Beetle: Managing Spread

Gilles Boiteau[1] and Jaakko Heikkilä[2]

[1]Agriculture and Agri-Food Canada Potato Research Centre, Fredericton, NB, Canada, [2]MTT Agrifood Research Finland, Economic Research, Helsinki, Finland

INTRODUCTION

The Colorado potato beetle (CPB), *Leptinotarsa decemlineata* (Say), has emerged as the dominant insect pest of the potato crop in much of the world and is continuing its conquest of the rest of the globe. In the past 170 years, the CPB has traveled from its native host plants in Mexico to the cultivated potato fields of China, and its voyage continues. Our intent has been to review the research on its colonization of the world, one potato patch at a time, with a particular emphasis on the management of colonization events.

A MOBILE PEST WITH A DUAL PERSONALITY

While North American growers, extension specialists, and researchers focus their attention on the management of crop colonization by a well-established resident CPB, growers, extension specialists, and researchers of countries such as the British Isles, Norway, and Finland have their attention focused on the prevention or early management of colonization events by an alien invasive CPB. The beetle takes a different personality depending on one's perspective. In this chapter we review the literature on CPB crop colonization to try and determine how lessons learned by those managing seasonal recolonizers and by those managing invasive CPB can be useful to both groups.

The separation of the beetles into "colonizers" and "invaders" seems to be the consequence of a generally inconsistent and imprecise use of terminology. The two personalities of the CPB are in fact the opposite ends

Insect Pests of Potato. http://dx.doi.org/10.1016/B978-0-12-386895-4.00012-0

of the same gradient of colonization. An increasing occurrence of invasive events around the globe has generated an expanding research effort on the management of invasive insect species. In parallel, the definition of an invasive species has been expanding (see Davis and Thompson 2000) to the point that it is now applied to a wide range of colonization events that occur when any group of individuals has been introduced into a new area in which they have established themselves, increased in number, and spread geographically (e.g., Guillemaud *et al.* 2011). By opposition, the original definition of invasion required spread into new ecological conditions and necessarily resulted in negative effects on the invaded ecosystem. The definition of invasive species has become so broad (Davis and Thompson 2000) that it now overlaps that of colonization.

There seems to be no advantage to continuing to separate research related to successional colonization (seasonal recolonization) from that related to invasive colonization. Such isolation can only slow down research and progress in the development of appropriate management strategies for each type (see Davis and Thompson 2000). In this chapter, invasion will be viewed as a specific case of the more general process of colonization. Fundamentally similar ecological processes are involved in the different types of colonization events (Davis and Thompson 2000). The terms "invaders" and "invasion" will be reserved for novel colonizers that have an impact, usually undesirable, on a new environment (Davis and Thompson 2000). In this context, the Colorado potato beetle can be said to become an invasive colonizer when it moves to or is transported to a new location outside of its distribution range, in which it establishes a population sufficient to prevent natural extinction and to spread within the area (see Liebhold and Tobin 2008).

The CPB is included in many lists of invasive species because it has inevitably become a pest directly threatening agriculture in every new location that it has colonized (see CABI 2011). The Colorado potato beetle makes the Wikipedia European invasive species list, probably in part because of its other personality (as a resident pest), but not the Wikipedia 100 top invasive species of the world. The economic and social consequences of colonization by the CPB, regardless of type, are high: production loss, decreases in the value of the potato tubers, and increased management costs related to monitoring, population control, quarantine, and eradication (see, for example, Heikkila and Peltola 2003).

It is interesting to note that although CABI (2011) includes the CPB in its invasive species list, its factsheet on the insect pest states that "spread could not be called invasive" in the strict sense "because it occurred in an introduced crop planted over large areas as a monoculture". The species has incrementally expanded its range by colonizing new regions where the potato crop had been introduced. It has not affected the area of the crop grown, and has no direct effects on the novel environment. There are no indications that it affects wild plants in the natural environment to any significant extent (CABI 2011). It is not invasive in an environmental sense (Beenen and Roques 2010).

At a continental scale, the CPB has shown a "classic pattern of geographical spread" (CABI 2011) within North America and then Europe and Western Asia after it colonized that continent. The voyage of the beetle (Lu and Lazell 1996) from its native Mexico to conquer potato fields and other Solanaceae throughout most of the globe has been extensively documented by Gibson *et al.* (1925), Wegorek (1959), Hurst (1975), Hsiao (1988), Jacques (1988), and Jolivet (1991). Within the North American continent its geographical spread is almost complete, but range expansion opportunities northward or into Newfoundland, California and Nevada (Capinera 2001) may arise out of global warming. The CPB is established in some Central American countries, many Asian countries and European countries, except for Britain, Ireland, Norway, Sweden, Finland, and some Spanish and Portuguese islands (EPPO 2006, Heikkila and Peltola 2007), where it threatens the potato industry. Northern Africa should have been colonized but has not been, so far, perhaps more by chance than for any other reasons (Jolivet 1991). On all these continents, successional colonization (seasonal recolonization) of the crop maintains the species range of distribution and produces the colonizers required for continued geographical spread.

COLONIZATION TYPES

Colonization, whether it is within or beyond the distribution range of an insect, is, by definition, the process by which dispersing individuals establish viable populations on host plants. In the case of the CPB the host plants are Solanaceae, and mostly the cultivated potato, *Solanum tuberosum* L. Successional colonization of fields, small range expansionist colonization at the border of the distribution range, and invasive colonization of isolated or fragmented areas outside the distribution range operate at different scales but share a similar process (Table 12.1). There are, however, differences among factors promoting and regulating colonization between the different scales that will affect management options (see next sections).

For the purpose of this review, like CABI (2011), we have adopted the nomenclature of Davis and Thompson (2000) that recognizes eight different types of colonization. Four of the eight types are immediately applicable to the CPB (Table 12.2). Invasive CPB colonizers of areas separate or isolated from the existing distribution range correspond to Type 8, usually characterized by a need for long-distance dispersal by the founding beetles and the potential to add important economic hardship to the existing potato industry. Successional colonizers (within the distribution range) (Type 2) as well as expansionist colonizers across the border of the distribution range (Type 3) or in areas surrounding an initial invasive colonization (Type 4) are characterized by relatively short dispersal distances. Pushing the borders of a distribution range (Type 3) can be expected to have less impact on an industry experienced at managing the pest than does the spread of the distribution into areas free from the pest as a result of landscape barriers or others (Type 4).

TABLE 12.1 Colonization Phases and Corresponding Management Considerations for Different Types of Colonization Events by Adult Colorado Potato Beetles

	Successional colonization	Expansionist colonization – border of range	Expansionist colonization – founder's range	Invasive colonization
Arrival and spread	Active dispersal; if aerial, landscape randomly seeded with propagules	Active, wind-aided or human-aided dispersal (if aerial, landscape randomly seeded with propagules)	Active, wind-aided or human-aided dispersal (if aerial, landscape randomly seeded with propagules)	Active, wind-aided or human-aided dispersal (if aerial, landscape randomly seeded with propagules)
Establishment	Host availability Superdiapause	Host and habitat availability (required for colonizer survival) Reproduction Superdiapause	Host and habitat availability (required for colonizer survival) Reproduction Superdiapause	Host and habitat availability (required for colonizer survival)
				Overwintering success (required for colonizer success – requires adequate temperature, humidity and soil conditions in the area) Superdiapause
Management considerations	Preventive: e.g., rotation	Preventive	Preventive	Preventive: domestic or international quarantine
	Monitoring	Monitoring	Monitoring	Monitoring
	Curative: Chemical Biological Cultural	Curative Containment Chemical Biological Cultural	Curative Containment Chemical Biological Cultural	Curative Containment Eradication

POTATO COLONIZATION

Original Source Population

The CPB is native to southern Mexico, and Hsiao (1985) suggests that it originally had a distribution range extending to the foothills of the Rocky Mountains.

TABLE 12.2 The Four Types of Colonization in which the Adult Colorado Potato Beetle Engages

	Successional colonization	Expansionist colonization – border of range	Expansionist colonization – founder's range	Invasive colonization
Colonization	Type 2	Type 3	Type 4	Type 8
Dispersal distance	Short	Short	Short	Long
Uniqueness to site	Common (resident)	Novel	Novel	Novel
Impact	Chronic	Additive	Additive with long-term consequences	Additive with long-term consequences
Description	Colonization or recolonization of an area where the CPB is a resident	Incremental range expansion usually without significant human assistance	Incremental range expansion usually without significant human assistance	Colonization of an area where the CPB was not a resident
	Diffusion colonizer	Diffusion colonizer	Diffusion with occasional saltation	Saltation colonizers often with human facilitation

Adapted from Davis and Thompson (2000).

It was a relatively rare and scattered insect, feeding on buffalo burr and other wild Solanaceae (Metcalf and Metcalf 1993). The range of the beetle matched that of its natural hosts (Hsiao 1985, Casagrande 1985). The CPB may be one of the best examples, in agriculture, of the key role played by human intervention in determining whether or not a particular insect species becomes an important pest (Metcalf and Metcalf 1993). After pioneers started planting the cultivated potato, the CPB deserted its weed hosts for this new and much more available food source and started its march from potato patch to potato patch (Metcalf 1999). The host range expansion for *L. decemlineata* has not been limited to the cultivated potato (Horton and Capinera 1990), but this expansion has had the most consequences, for biological and economic reasons. *S. tuberosum* is a preferred host that effectively recruits foraging CPBs (Weber *et al.* 1995) and is the world's fourth largest food crop after rice, wheat, and maize. Geographical range expansion has been a continuous process ever since (Jolivet 1991, 1994).

Cultivated Potato Source Population

Guillemaud *et al.* (2011) suggest that the desertion of the weed hosts by the CPB resulted from a single acquisition of invasive characteristics that occurred within or close to the original source population distribution range. This important evolutionary shift may have turned it into a primary pest of the potato crop. Our understanding of the fitness differences between the original CPB population and the CPB population adapted to the cultivated potato remains limited (Hsiao 1981, Lu and Logan 1994). According to Guillemaud *et al.* (2011), the spread of the beetle throughout the world originated from this single bridgehead population of invasive CPB colonizers adapted to the cultivated potato rather than from multiple introductions (arrivals) that would have each, independently, acquired the necessary invasive traits.

The lack of "interest" of the Mexican CPB population regarding the cultivated potato could be temporary and part of a latent colonization period (see below), but this is unlikely since the population has had as much time as populations of nearby regions to adapt. Obviously other factors such as, perhaps, dominance of alternate hosts and genetic differences help maintain the distinctive Mexican population. The true dual personality of the CPB may reside in this coexistence of a geographically restricted original source population adapted to buffalo burr and a bridgehead population adapted to the cultivated potato spreading worldwide.

Arrival and Spread in Potato Crop

General Colonization Scenario

Although the differences in the colonization fitness of the original Mexican CPB population and that of the CPB population adapted to the cultivated potato have yet to be clearly established, it is clear that the intrinsic dispersal abilities of this more aggressive CPB combined with a range of favorable environmental and agronomic factors did create opportunities for the colonization of host plants near and afar, and continue to do so (Weber and Ferro 1994). Each colonization event begins as a small isolated focus made up of one or more founding beetles resulting from human intervention (Type 8) (Metcalf 1999), massive flights from contaminated regions (Type 8), or diffusion of poorly controlled populations (Types 2, 3, 4). The establishment of the population is completed with the confluence of foci over an entire field or region. Feytaud (1930) compared successional and border range expansion colonization to the coalescence of the droplets in an oil slick, and invasive colonization to "splashing" of the occasional oil droplet away from the main oil film – the "splashing" being equivalent to human intervention or unusual weather events.

Dispersal of the first three types of colonizers is somewhat predictable, at least in range. It can be modeled by combining exponential growth and diffusive movement (Liebhold and Tobin 2008). In the case of successional colonization, the emerging overwintered CPB population must, every spring, redistribute itself among nearby host sites. Within a farm, the lag time between the emergence of the overwintered beetles and the colonization of field borders can be very short. This is partially explained by the presence of a reservoir of colonizers (estimated at 35% of the season-long colonizing population) which emerge before potato plants break the ground, and are ready to colonize these plants (Boiteau *et al.* 2008). Then, colonizing CPBs spend only a few days on field borders (Boiteau 2005) before rapidly spreading over the total field area (Blom 2001, 2002, Boiteau 2005). Increasing the distance between overwintering CPBs and the potato crop delays the onset and reduces the period of colonization.

Natural and Anthropogenic Regulators of Spread
Intrinsic Dispersal Ability
The intrinsic walking and flight dispersal abilities of the adult beetle are well suited for local dispersal (Boiteau 2003). They favor the spread of successional colonizers but can constrain the dispersal of invasive founding populations. Only a small proportion of beetles disperse long distances from their natal or overwintering patches. During a potato field season, the adult beetle disperses by a combination of walks and flights within a relatively short mean distance range of 1.5–2.0 km (Follett *et al.* 1996, Weisz *et al.* 1996, Boiteau *et al.* 2001, Sexton and Wyman 2005), reaching a maximum distance of 5 km (Johnson 1965). If this takes place in the middle of a potato farm, the population level is maintained or increased; if it occurs in an area where potato fields are scattered, there is a higher probability of individuals ending up in areas without host plants. All of this is in agreement with a CPB residency strategy (Gui *et al.* 2011) that prioritizes dispersal locally, where there is a high probability of host, over wide-range dispersal. The strategy is characterized by a relatively long residency time on the potato plant (Bach 1982, Boiteau, unpubl. data), a tendency to remain within the borders of a host patch (Boiteau, unpubl. data), and an unrushed random walking track within arable land free of hosts (Gui *et al.* 2011). There is no indication that adult CPBs can determine where they land or where they walk to when they are engaged in long-distance dispersal; this would make reliance on long-range dispersal very risky and therefore local dispersal and long host residency periods more likely (Sandeson *et al.* 2002, Gui *et al.* 2011).

The orientation of dispersing successional colonizers is probably towards the sun in response to the adult CPB's strong phototactic behavior. Also, the orientation range would be likely limited to the position of the sun in the warm part of the day because of the relatively high temperature thresholds for walking and, especially, flight. The orientation of invasive colonizers is difficult to predict because the events are rare and can be triggered or favored by a range of causes. However, their orientation is more likely to be in the direction of the dominant wind (Feytaud 1938, Wiktelius 1981) than in the direction of the sun. Historically, CPB spread has proceeded faster eastward than southwards (Hurst 1975). It has been suggested that eastward spread was downwind and southward spread more difficult because often against the wind. Also, especially in Europe, eastward spread followed mountainous barriers rather than crossing them.

In spite of a number of studies (Boiteau et al. 2003), the orientation of walking and flying CPBs remains an unresolved complex phenomenon, at all spatial scales. In the absence of hard data, there is a tendency for modelers to allocate randomly the total number of beetles dispersing between the four cardinal points (e.g., Valosaari et al. 2008). New technology such as harmonic radar tracking is only beginning to improve our understanding of CPB orientation (e.g., Boiteau and Colpitts 2002, Boiteau et al. 2011, Gui et al. 2011).

Host-Plant Availability

It is unclear if the absence of host plants in the area where the CPB emerges, walks, or lands will stimulate residency or dispersal. Some studies have concluded that CPB flight is strongly encouraged by the absence of food (Caprio and Grafius 1990, Ferro et al. 1991, Weber and Ferro 1996) (in Baker et al. 2002), but others have reported no increase in frequency of flight take-off or walking bouts in response to an increasing period of starvation (e.g., Gui and Boiteau 2010, Gui et al. 2011). The latter would be in agreement with the CPB's remarkable ability to survive starvation (e.g., Gui et al. 2010). Although the need for food and its impact on dispersal remains open for debate, the need of teneral beetles and overwintered beetles coming out of diapause for water is well established and critical to their dispersal and survival (MacQuarrie and Boiteau 2003). The diapause switch that takes place at the end of the summer season is the best known trigger of CPB long walks and flights (Weber and Ferro 1994, Noronha and Cloutier 1999, Boiteau et al. 2003).

Anthropogenic Contribution

The greatest human contribution to the success of the CPB was no doubt the introduction of the cultivated potato and its rapid distribution throughout much of the globe. This provided the beetle not only with an additional source of food but also a worldwide "homogenized" colonizer's habitat (Guillemaud et al. 2011). The usual need for the successive waves of founding colonizers to quickly evolve adaptations to the local ecological conditions to ensure the success of colonization events was thereby greatly attenuated (Guillemaud et al. 2011). However, the expansion of the distribution of the potato would not have been sufficient to cause the range expansion of the CPB (Feytaud 1930). One did not follow the other. The wide distribution of the potato plant simply provides a geographically wider habitat suitable for colonization.

The intrinsic CPB dispersal ability (mostly local) allows for diffusion of colonizers between fields (Type 2) or in areas immediately outside the current range (Type 3), but Types 4 and 8 colonization require anthropogenic transport or dispersal assisted by unusual meteorological events (saltation) (Valosaari *et al.* 2008). It is likely that the dispersal of the CPB within each continent has been and continues to be stratified – that is, results from a combination of short- and long-distance movement, and population-driven and accidental dispersal. It is not clear if incremental colonization of areas outside the distribution range by human transport is as common as by walking and flight. In the case of long distances between the source population and the new location, or of geographically isolated areas protected from population diffusion by mountain ranges or bodies of water, it is likely that CPB spreads more by accidental transport than by flight. Certainly, the beetle would have been confined to America were it not for accidental transport by ship, aircraft, etc. (Hurst 1975).

Regardless of the cause of the long-distance events, accidental or not, they are important accelerators of the spread of the species. The jump from North America to Europe is an example (Jolivet 1991, 1994). Otherwise, according to Liebhold and Tobin (2008), the rate of spread of colonizers through diffusion is normally constant and therefore any sudden change in the rate could be presumed to have necessitated saltation. Feytaud (1938) estimated the annual spread of the CPB at a minimum of 20 km and reaching 150 km, presumably with the assistance of exceptional air currents. It averaged 80 km/year and 40 km/year in North America and Europe, respectively (Hurst 1975), but, in spite of the excellent data records, no one seems to have calculated the occurrence of saltation events. The application of a similar exercise to the Asian spread might be beneficial to our understanding of the beetle's future spread throughout China.

Meteorological Conditions

Weather and related abiotic factors are key regulating factors of CPB movement within and between the potato fields of its distribution range. Air temperature and wind determine population growth and dispersal activity of beetles. For example, overwintered beetles do not start flying until they accumulate 150–200 DD (Baker *et al.* 2002). Temperature regulates a number of life history parameters necessary to the survival and dispersal, such as walking and flight take-off threshold (Boiteau *et al.* 2003). Temperature imposes limits on the beetles' activities, but the absence of a statistical link between monthly temperature and CPB geographical distribution (Ulrichs and Hopper 2008), for example, confirms that temperature is but one of the factors determining geographical distribution.

In the case of invasive colonization events, next to anthropogenic transport, meteorological conditions are probably the most important factor in facilitating the introduction of the beetle outside its distribution area. The efficient flight ability of the CPB (Voss and Ferro 1990a, 1990b, Boiteau 2001, Valosaari *et al.* 2008) must be combined with particular weather conditions such as the presence of strong air currents (Wiktelius 1981, Ferro *et al.* 1985, Voss and Ferro 1990b, Grapputo *et al.* 2005) for long distance dispersal. Under favorable meteorological conditions, beetles have, for example, flown over 100 km across the Baltic Sea to Scandinavia (Wiktelius 1981). Wind-borne long-distance migration from Russia or Estonia was the most likely route of spread to Finland in the early 2000s (Heikkila and Peltola 2007, Ooperi and Jolma 2009). On the other hand, winds associated with storms can significantly dwarf the distances flown (Feytaud 1930). Locally, winds as low in speed as 3.4–7.0 m/s have been shown to limit the frequency of CPB flight take-off at exposed sites (Boiteau *et al.* 2010). Cool and humid climates decrease locomotory activities and slow down the spread (Wegorek 1959). There is evidence that CPB populations can adapt to summer climatic conditions cooler than those in their current distribution range (Boman *et al.* 2007), and can therefore, given time, colonize a geographic area much wider than their current niche (expansionist colonization) (EPPO 2006). However, the northern range of the insect seems limited by the low heritability of diapause traits (Piiroinen *et al.* 2011). The resulting lack of adaptability to winter climate would explain why the CPB has not spread further north (Piiroinen *et al.* 2011) in countries such as Canada in spite of having been long-established residents. On the other hand, global warming could directly influence the abundance of the CPB (e.g., Smatas *et al.* 2008) and indirectly improve the overwintering survival (Smatas *et al.* 2008) necessary to the establishment of invasive colonizers.

Habitat

As mentioned earlier, the physical structure of the landscape can play an important role in determining whether or not beetles can disperse beyond their current distribution. Mountains, oceans, cliffs, hills, and tree borders constitute barriers of various importance against dispersal, but perhaps more in the case of invasive and expansionist than successional colonization. Whenever mountains, valleys, and rivers block the dispersing beetles, the latter may also reorient their dispersal and thus such features become highways for CPB range expansion (Feytaud 1930). Early on, large water features such as Lake Michigan and high mountains were considered insurmountable barriers to the beetle's flight (Chittenden 1914, Hurst 1975). The low temperatures at high altitude could prevent flight (Chittenden 1914), while temperatures above large bodies of water may not support flight or the water may extend further than the flight duration of the beetle. It has been suggested that most of the beetles landing on water are not likely to survive (Hurst 1975), but this is contradicted by the number of beetles "entering" ports in Great Britain every year (Bartlett 1980) and observations of CPBs crossing bodies of water in the USA (Riley 1876) and Canada (Peters 1954).

The presence of the crop in the habitat is of course necessary, and can arrest the spread of the colonizers or bridge a series of local dispersal events. It is noteworthy that the impact of host crop connectivity on the dispersal and spread of the CPB remains essentially unknown both for successional and for invasive colonization. Crop age and condition at the time of the colonization can also influence the spread of the colonizers (e.g., Boiteau 1986, Mbungu and Boiteau 2008). The low quality of the crop through blight or wilting in very humid or very dry years could have an indirect effect on the spread of the invasion (Feytaud 1930).

Establishment

Colonization Sites

For colonization to be successful, foci must be initiated where (1) the host plants are present, (2) the soil is suitable for pupation and overwintering, and (3) climatic conditions are appropriate for beetle multiplication and overwintering. These requirements determine the probability of establishment following an invasive colonization event, and explain much of the relative stability of the distribution range elsewhere. The CPB colonizers are typically at greater risk of extinction in invasive than in successional colonization. In the case of successional colonization and some cases of expansionist colonization, host-plant absence is the factor most likely to limit establishment. Climatic conditions are suitable, and beetles can escape freezing temperatures and a prolonged absence of host plant by entering into diapause and burrowing into the soil. In the case of invasive colonization, the establishment of the CPB also depends on the ability to develop from eggs to adults during the growing season and an acceptable winter survival rate (Valosaari et al. 2008).

Whenever all the requirements for establishment of the founder population are present but marginal, foci may become extinct (see Allee effects, below). For example, foci may establish under poor soil conditions, but they will not thrive and have a high probability of dying out (Feytaud 1930). The soil type can have a major impact on overwintering survival – for example, studies have shown that beetles may survive up to 10 times more in loamy sand than in clay loam (Hiiesaar et al. 2006). It is likely that the ability of the insects to dig in the soil and the drainage capacity of the soil have a large impact on the survival. Essentially, even if the temperature range is suitable for the resident population or the colonizing population, the local variations in soil type, drainage, and rainfall may very well limit the probability of establishment or range expansion.

With the CPB already present in most of the countries where potatoes are grown, it may not be surprising to find a rather low rate of successful invasive colonization as the species reaches the margins of the suitable geographical areas. For example, in Europe the northernmost populations can be found in Russia, near 62° N (EPPO 2006, Piiroinen et al. 2011), where high overwintering mortality is likely to have stopped its invasive colonization (Valosaari et al. 2008,

Piiroinen *et al.* 2011). Furthermore, Piiroinen *et al.* (2011) have shown a low heritability of diapause behavior and physiological traits associated with diapause. The CPB may not have enough genetic adaptive potential to respond to the selection exerted by harsh winter conditions and expand its range further north. However, partial adult survival even under extreme conditions, the existence of superdiapause, and the high fecundity of females will continue to generate sporadic northern expansions. These colonization events combined with the northern movement of the temperature cline under global climate change (increased winter survival) may result, in the near future, in permanent localized range expansions.

Attention is usually focused on range expansion, but range diminution can occur. Any change in the ability of colonized sites to provide the required conditions can lead to retreat of the beetle from the area. This could explain the occasional fall-back of the distribution range. For example, the CPB originally colonized the potato fields of British Columbia, eventually retreated and then made a re-entry (Peters 1954).

Founding Populations

The likelihood of successful establishment by founder beetles depends on the size of the introduction (Memmott *et al.* 2005; Liebhold and Tobin 2008) and whether there are one or more introductions from one or more populations (Guillemaud *et al.* 2011), as well as on the timing of the introductions. Repeated introductions of founding beetles generated by areas with two consecutive generations of CPB per season may also have an increased probability of encountering conditions suitable for establishment. Resistance to some types of insecticides may play a key role in a successful establishment, whether during the recolonization of a field or the invasion of a new country where the insecticide is the main line of defense against the colonizer (Guillemaud *et al.* 2011).

The number of colonizing individuals is a predictable parameter in successional colonizations but an essentially unpredictable one in invasive colonizations. This is not to say that CPB populations have to be at very high level of abundance to represent a source of expansion. This was demonstrated in the 1970s, when family fields in rural areas such as those in Normandy continued to threaten the Anglo Normand islands (Vergnaud 1991) even though the CPB populations in commercial fields in France were generally very low. The ability of a single mated female CPB to overwinter with fertile sperm and produce up to 500 eggs in the spring favors the establishment of the species, even when present in small numbers.

Allee Effects

Liebhold and Tobin (2008) have discussed how Allee effects and inimical demographic and environmental random variation can bring colonizing or recently established low-density insect populations to extinction. It is generally unknown whether the establishment of CPB colonizers outside their distribution range is influenced by Allee effects. The common types of Allee effects, such as decreased ability to locate males, inbreeding, or reduced predator establishment, are not likely to have a large impact on the CPB. The majority of overwintering female CPBs carry sperm in their spermatheca and do not have to find a male in the spring to fertilize their eggs. The number of predators is small, and their impact limited by thanatosis and the production of deterrent chemicals. Furthermore, males and females are likely to find each other even at low colonizing densities since they share a very limited host-plant range; the inability to locate the host plant would be a more likely cause of colonization failure, especially in an invasive colonization event. It is unknown if inbreeding could have an impact.

The information available suggests that the Allee threshold for the CPB is high and therefore offers a limited potential as a contributor to the design and implementation of a management strategy against the establishment or spread of the beetle.

Establishment Lag Phase

Most invasive colonization events, with insects and other organisms, are characterized by a lag phase between the introduction phase and complete establishment. This is especially common where the number of colonizers at any foci may be small and may not coincide with the presence of a host crop. This is also the case with the CPB, even though genetic adaptation is not a factor since the cultivated potato is genetically similar throughout the world. The CPB was first recorded in Europe in the 1870s (Hsiao 1985) but did not become a serious pest for another 40 years or so, even though a potato-adapted strain was introduced. The genetic reduction accompanying the invasion of new host countries could have been a determinant factor slowing down the successful establishment and growth of the new arrivals. However, recent evidence suggests that the genetic breadth of these new arrivals is much greater than had been expected (Boman *et al.* 2007) and, as indicated above, the host crop is remarkably homogeneous. The latent period is therefore most likely to result from factors such as low numbers of colonizers, limited crop acreage, and marginally suitable climatic or habitat factors.

In the case of successional colonization, the CPB has the advantage of being already established, but individuals still need, season after season, to redistribute themselves within the farm or the region according to host availability.

MANAGEMENT OF SUCCESSIONAL, EXPANSIONIST AND INVASIVE COLONIZATION

Introduction

The management procedures applicable vary according to each colonization type (Table 12.2) and each colonization step or phase within it (arrival, establishment, spread). With Type 2 and 3 colonizers, management focuses on local prevention tools such as rotation followed by abundance monitoring. Curative methods such as chemical, biological, or cultural controls are then applied according to perceived need. Unfortunately, perhaps because real-time population monitoring combined with effective curative methods has proven effective at managing population abundance (if not spread), the use of preventive control methods based on population estimates from predictive models is rarely adopted. With Types 4 and 8, where the colonizer's presence and abundance have to be predicted, management is necessarily focused on prevention tools such as international and domestic quarantine followed by monitoring at potential points of colonization. Eradication follows, if necessary.

Preventative Management

Monitoring

The detection of colonizing CPB is challenging regardless of colonization type. The high probability that a colonizing individual or group will escape detection efforts is believed to explain, at least in part, the spread latency observed in many cases of past invasive colonization events. Whether the latency is real (and caused by the slow build-up of populations) or the consequence of insufficient investment in detection, or results from a combination of both, the successful management of the colonizers, especially in terms of containment or eradication, is proportional to how early the population was detected. A CPB colonizing focus may be as small as the progeny of one mated female and may not be detected for some generations or until many such small populations have coalesced. In a case of successional colonization, delayed detection can make it more difficult to control population abundance; however, in a case of invasive colonization, delayed detection could jeopardize quarantine efforts or the successful application of an eradication program.

The "Achille's heel" of monitoring methods and forecasting methods available to measure the risk of CPB colonization (Zehnder *et al.* 1994, Hoy *et al.* 2008) is the unpredictable nature of the number of colonizers at any foci of colonization (Valosaari *et al.* 2008). Information on the different physical barriers and distances from existing distribution ranges of the CPB that might help determine the probability of accidental beetle introduction is complex and difficult to gather. So-called gravity models (Liebhold and Tobin 2008) have been suggested. In the case of the CPB, Ooperi and Jolma (2009) used human population density per predetermined area to estimate the probability of accidental beetle transportation in vehicles in the case of invasive colonization. Although invasive colonization models have grown in sophistication, they remain dependent on that information and a still very limited understanding of the dispersal ecology of the beetle and the almost random occurrence of introductions.

Predictive Tools

The French/British program to keep the Channel Islands free of the CPB provides a good example of real-time monitoring-based preventative CPB management. The program used monitoring cages located in France on sites where massive invasive flights could take place towards the Jersey islands. The cages were used for short-term prediction of the invasive colonization events. Although these sea invasions are, according to some, the most dangerous type of accidental introductions (compared to aircraft, ships, and cargoes), this program proved that they can usually be detected with proper monitoring and, if action is taken, eradicated effectively (Thomas and Wood 1980). The monitoring system of cages has proven effective in providing a real-time warning of colonizing flights for over 30 years (Vergnaud 1991, Thomas and Wood 1980). There was a strong incentive to maintain the beetle-free status of the Channel Islands after the invasion of the continent by the insect because of the economic importance of the export market to the British Isles (Thomas and Wood 1980). Cages set up in Normandy were provided with newly emerged overwintered adult CPBs and flight frequency towards the outside of the containment area was recorded daily. A 40% flight rate accompanied by high daily temperatures and appropriate wind direction triggered a warning to intensify monitoring on the islands for potential invasive colonization flights from the mainland. The objective of the original project was to eradicate the CPB from a whole region, but monitoring cages could also be considered as a powerful tool to modulate the use of curative control methods in regions with well-established resident CPB populations. Cages can also be used in parallel to their monitoring role, or independently, to research the role of different environmental parameters on CPB dispersal (e.g., Boiteau et al. 2010). Cages are a monitoring tool for early detection of possible invasions and early intervention that could be considered complementary to predictive models of invasive colonization.

Predictive models are becoming essential tools for the development or analysis of CPB protection strategies where beetle colonization is a serious threat to potato production. Spatial and climate-based simulation models have been developed to predict CPB invasive colonization. A spatially realistic model of invasive colonization developed by Valosaari et al. (2008) provides useful insights for the selection of appropriate management strategies, but the authors themselves suggested that the addition of a climate model such as CLIMEX would be required to improve its predictive power. Two climate-based modeling approaches have been used to forecast invasive colonization by CPB: (1) measuring beetle response to the climate, and (2) determining the climatic characteristics of the regions where the beetle is already present. The underlying assumption of the latter is that if an organism does not occur in an area from which it is not separated by physical barriers, it is because environmental conditions are unfavorable (Ulrichs et al. 2008). Also, when estimating climate response parameters based on the observed distribution of a pest, it is uncertain whether its geographical distribution is an expression of its climatic tolerance, or whether it is only an expression of how far it has had time to spread or adapt (Rafoss and Saethe 2003). There is evidence that CPB populations can adapt to summer climatic conditions cooler than those in their current distribution range

(Boman *et al.* 2007) and can therefore, given time, colonize a geographic area much wider than their current niche (expansionist colonization) (EPPO 2006). However, the northern range of the insect seems limited by the low heritability of diapause traits (Piiroinen *et al.* 2011). The resulting lack of adaptability to winter climate would explain why the CPB has not spread further north (Piiroinen *et al.* 2011) in countries such as Canada in spite of being a long-established resident.

The CLIMEX software has been used to determine the overall suitability of existing and anticipated climate conditions for the establishment and long-term presence of a CPB population in Norway (Rafoss and Saethe 2003) and the different regions of the Czech Republic and Austria (Kocmánková *et al.* 2010). The CLIMEX model uses weekly and annual indices that describe the responses of the insect to temperature and moisture to produce an ecoclimatic index. The index takes into account population growth under favorable conditions, survival during unfavorable periods, and stress interactions. Like any model, it depends on the availability of development thresholds (see, for example, Table 12.1 in Kocmánková *et al.* 2010) for the nearby populations in the case of expansionist colonization or for likely founder populations in the case of invasive colonization. Although the model makes it possible to measure the potential for geographical distribution range expansion between now and 2050 (Kocmánková *et al.* 2010) and the increased area with more than one generation per year, it does not address the concurrent implications for dispersal into the additional areas suitable for CPB development. Jarvis and Baker (2001) used a more complex model of insect development and temperature, including estimates of pest activity and flight potential, to estimate the risk of CPB becoming established in England and Wales. It could be said that where models such as CLIMEX using averaged monthly data assess the impact of climate on pest risk, models such as the one used by Jarvis and Baker (2001) using daily data assess the impact of weather on pest risk. The scale of the data or data interpolation (e.g., 1 or 2 km in Jarvis and Baker 2001, and 5 km in Ooperi and Jolma 2009) has an important impact on the sensitivity or reliability of the model. A model specific to the location of a threatened crop may provide a more objective assessment of risk (Jarvis and Baker 2001).

A spatially explicit model of successional colonization has been initiated by Blom *et al.* (2004), but its application is limited to diffusive short-distance dispersal. Further work will be required to better take into account the geographical scales of expansionist and invasive colonization events. At these scales, the probability of colonization decreases with distance and becomes less predictable than with the diffusive short-distance dispersal.

The applicability of the results from these predictive models will grow as the extent and accuracy of the information on the response parameters of the beetle, as well as habitat and environmental indices, improves. For example, on the positive side, it is likely that the winter survival requirement for an invasive CPB founder population to successfully establish is within range of the 20–30% survival rate suggested by Heikkila and Peltola (2007) as a reasonable critical factor for making prevention measures economically sensible. On the other hand, there is difficulty in determining what the intrinsic or basic (environmental) parameters for the development and survival of the species are. For example, the temperature range favorable to beetle development and population growth used in a Finland model (15–30°C) (Valosaari et al. 2008) was estimated more narrowly than in a Norway model (12–35°C) (Rafoss and Saethre 2003). The duration of the warm period favorable to CPB development in Finland was initially believed to be too short for CPB larvae to mature, but Boman et al. (2008) have shown that the average development time of the beetles from first instar larvae to adult stage at 17°C, which corresponds to mean temperature in June and July in Finland, is 46.4 days. This result suggested that there was enough time for the development of a new beetle generation that is capable of overwintering if the immigration takes place early enough during the season. This is also now known to be possible since successfully overwintered adults have been found in Finland (Valosaari et al. 2008). Climatic models for Norway suggest that the CPB cannot complete development under existing climatic conditions (e.g. Rafoss and Saethre 2003, also mentioned in Kocmánková et al. 2010), but the growing season duration and temperatures in January and July at many locations compare well with locations in Canada where the beetle is a major pest. For example, Tonsberg, Norway, located in the potato-growing area, has a climate very similar to that in Fredericton in New Brunswick, Canada (Table 12.3), where the CPB is well established. Both locations are warmer, drier, and have generally longer growing seasons than St John's, the warmest region of Newfoundland – the only Canadian province free of CPB.

Once the abiotic requirements have been more firmly established, modelers will still be left with having to take into account founder population size and degree of synchronization with the crop. For example, in a study in the Ukraine, CPB winter mortality ranged between 30% and 83%, depending on the opportunity for proper development during the summer (EPPO 2006). It is also difficult to determine how these parameters combine to form the northern

TABLE 12.3 Comparison of Mean Temperature, Precipitation, and Growing Season in Selected Cities from Norway, New Brunswick (Canada), and Newfoundland and Labrador (Canada)

| City (Country) | Temperature (°C) | | Precipitation (mm/year) | Growing season (days) |
	Jan	July		
Tensberg (Norway)	−3.2	16.8	930	194
Fredericton (Canada)	−9	19	1150	190
St John's (Canada)	−3	12 to 15	1300–1500	170–190

and southern temperature limits (for growth in summer and survival in winter, respectively) of a suitable distribution range. Wegorek (1959) estimated these limits for North America at mean annual isotherms of 0° and 20°C. Also, a climate change scenario similar to that outlined by Kocmánková *et al.* (2010) could increase temperature sufficiently to shift the conditions in countries such as Norway from unfavorable to favorable for CPB.

Spatial simulation models can also be used to test the probability of eradication after invasive flights (Ooperi and Jolma 2009). Considering the difficulty of predicting invasive colonization events, it is surprising that so little attention has been given to models that test the probability of eradication success after massive inflights of invasive CPB have been detected in a new colonization area. The spatially explicit population model of Ooperi and Jolma (2009) provides a framework to test whether a spatially targeted eradication strategy would perform better than a strategy without prioritizing. It is not clear whether the targeted approach would actually improve the level of control, but it should reduce costs by focusing surveillance and eradication measures to a smaller geographical area. Although the model was developed for the eradication of CPB founding populations in the process of establishment into new territory, it could provide an interesting tool to manage expansionist colonization as well as successional colonization. The model is developed around the "oil splatter" spread typical of the CPB, rather than the frontline spread typical of many other species. The model also takes into account dispersal in the form of an active flight index and a logistic-aided dispersal index. Of course, having a model to help focus the attention on the key clusters of founder beetles scattered over a wide geographical area is only a first step in developing the tools required to successfully eradicate it. The eradication process itself remains the real challenge.

The historical record for the spread of the CPB over continents is remarkably good, and could be extended to estimate the most probable sites for range expansion or invasion. Many methods are available to do so, as outlined in Liebhold and Tobin (2008). The reconstruction of routes of colonization is an additional tool to assist in developing appropriate management strategies. Knowledge of the geographic origin of the potential colonizers (overwintering sites for successional colonization; airports, harbors, or others for invasive colonization) makes it possible to develop monitoring plans, barriers, rotation programs, quarantine programs, etc. to prevent colonization or determine where to apply chemical, biological, or cultural control methods. In England, for example, eradication or containment applied shortly after the arrival of the quarantined pest at harbors (Bartlett 1980) has been successful at controlling the threat to UK crops after more than 163 outbreaks (Jarvis and Baker 2001). The choice of an appropriate predator or parasite or effective insecticide may also depend on appropriate knowledge of the origin of the population colonizing, and whether it is for field management or for the protection of a country's crop (Guillemaud *et al.* 2011).

Quarantine

Domestic and international quarantine are the most effective methods available to prevent invasive colonization. The cost of monitoring low-density populations can be high, although costs for the CPB should be modest compared to those for many other species because only sites with potatoes or, in some cases, with patches of solanaceous weeds have to be considered. Once a quarantine program has been initiated, the most difficult aspect of the strategy may become that of determining when the benefits of not having the pest outweigh the costs of surveillance, labeling, import restrictions, eradication, and post-monitoring (Heikkila and Peltola 2007).

Sustained international quarantine has kept Great Britain free of the CPB in spite of repeated sea introductions (Bartlett 1980). The Canadian province of Newfoundland and Labrador provides a good example of effective domestic quarantine. Potatoes shipped to this province must be free of the pest (CFIA 2009). The program is believed to have kept the province free of the CPB in spite of its abundance in nearby provinces. The short cool seasons, the isolation (island province), and the relatively small acreage of potato in the province are all important supporting factors of this successful quarantine program.

To prevent the spread of quarantine organisms such as the CPB within the European Union (EU), plant health legislation oversees protected zones (ZP) where the organism must be eradicated if detected (Heikkila and Peltola 2007). The ZP status is voluntary for EU member states, which can apply for such a status if the CPB does not exist in their country. However, if the ZP status is given, the member state needs to report the pest and instigate eradication whenever the species is encountered. Compensation to farmers may be needed to ensure compliance with the regulations. There have been five important episodes of invasive colonization in Finland to date: localized and short-lived in 1983, 1998, and 2002; and confirmed cases of winter survival in 2004 and 2011 (Heikkila and Peltola 2007). Although the CPB is a quarantined pest in Europe (CABI 2011), the waiver of formerly routine phytosanitary inspections on goods transported within the European Union may be weakening the protection and thus allowing the insect to move freely throughout the Union (Beenen and Roques 2010).

The second function of quarantine is to contain the spread of the CPB once it has colonized an area. In Finland, for example, the authorities attempt to eradicate any focus of colonization by quarantining for 1 year any fields where a beetle is found. A small amount of potato is left as a trap crop on the quarantine field for the next year to catch and prevent the spread of the potentially overwintered individuals (Valosaari et al. 2008). Widely spaced, late seeded potatoes are also used to continue trap-cropping later. All trap plants are then inspected daily for CPB. So far, these control methods have been sufficient to prevent the establishment of pest populations (Valosaari et al. 2008). The ability to enforce a combination of rotation/quarantine whenever beetles occur reinforces the effectiveness of the rotation and trap-crop practices. Once the CPB is officially recognized as established, rotation and trap-crop practices can no longer be enforced but only recommended. In the absence of quarantine and the sporadic application of control practices, reinfestation sites ensure the maintenance of the CPB population.

CPB-Resistant Potato

CPB-resistant potato crops could be major contributors to the prevention or containment of colonization pressure. Beetles landing in or spreading into beetle-resistant potato crops such as *Bt* potato would have no or a low probability of survival compared to those landing in conventional potato fields. Valosaari *et al.* (2008) used a simulation model to investigate the potential of plant resistance in a CPB eradication plan. *Bt*-potato cultivation in areas where CPB colonizers are detected seemed to be the most efficient control method and did not need to be integrated with other control methods.

Unfortunately, there are essentially no resistant cultivars available to manage CPB population abundance or spread. Conventional plant-breeding efforts are underway, but no varieties have yet reached the market. CPB resistance has been achieved by transferring the *Bacillus thuringiensis* (*Bt*) subsp. *tenebrionis cry3* toxin gene to potato plants (Shelton *et al.* 2002). The genetically modified plants have been successfully field tested, but they are not available commercially because of concerns regarding human health and non-target effects.

Host-Plant Control

The elimination of the potato crop and other hosts from sites under CPB colonization pressure is, at the limit, the ideal preventative measure. The elimination of the host crop is usually not an economical consideration, but managing the crop and its agroecosystem to marginalize its ability to meet the requirements of colonizing beetles may have potential, and should be considered in any control strategy.

Some level of weed control might seem a more feasible option. However, the use of weed control as a strategy to delay the establishment of the beetle in expansion zones, for example, is not likely to contribute significantly to the management of this pest. The CPB feeds essentially on the cultivated potato throughout its distribution range except in Mexico and southern USA, where it continues to feed on its original host plants, *Solanum rostratum* (Dunal) and *S. angustifolium* (Miller) (Hsiao 1978, Casagrande 1985, Hare 1990). The larval and adult stages will occasionally feed on other wild Solanaceae, but the suitability of these plants is marginal and varies between beetle populations (Hsiao 1978, Tscheulin *et al.* 2009). If weeds are called to play a role in the spread of the CPB it would be mostly in areas where the cultivated potato is absent and the weeds could act as a bridge between areas with cultivated potato (e.g., Tscheulin *et al.* 2009). Except in regions where a potential weed host is widely distributed (e.g., *S. elaeagnifolium* in Greece; Tscheulin *et al.* 2009), less preferred crops such as tomato and eggplant are more likely to serve as colonization bridges than are weeds.

Curative Management

Curative management strategies have generally been developed to lower the abundance of beetles in cases of successional or expansionist colonization, but the same strategies are often applied, although more intensively, to eradicate or reduce founding populations of invasive colonizers. With Type 2 and 3 colonizers, the risk of colonization is related to the overwintering success, and the proximity of overwintering sites from last year and the current year crop. Studies have clearly demonstrated that a new year crop within 1.5 km of the previous year's crop or overwintering sites is at high risk of colonization and will require monitoring to determine if and when to apply curative controls (Follett *et al.* 1996, Weisz *et al.* 1996, Sexton and Wyman 2005, Boiteau *et al.* 2008). Beyond this radius, the risk of colonization is low and somewhat similar to that for Type 4 and 8 colonizers. The value of the potato crop and the risk of colonization forecast by available models will determine the intensity of monitoring required and economically justifiable to detect potential colonizers and measure economic action thresholds. Prediction of (non-invasive) spread can be further complicated by the occasional mixing of long-time resident adapted beetles and beetles migrated from other regions (Hiisaar *et al.* 2006).

Chemical Control

Chemical control is regularly called upon whenever there is a need to eliminate invading or new crop season colonizers because of the relative lack of other CPB control tactics that are effective at low densities. The excellent crop coverage and high level of efficacy of insecticides are ideal to control low densities of colonizers. In the case of successional colonization, the repeated use of insecticides against low-density colonizers can lead to the development of insecticide resistance and is usually not recommended. However, the large-area adoption, in the past two decades, of the systemic insecticide Admire® to prevent the establishment of the overwintered colonizer in the current year crop has been credited with a massive reduction in the abundance (propagule pressure) of the CPB in the USA and parts of Canada.

In the case of expansionist or invasive colonization, insecticides can play a substantial role in preventing the establishment of colonizing foci. A systematic program of insecticide spraying played a large role in the control of the 1972 invasive colonization of Sweden (Wiktelius 1981). Where high propagule pressure exists, even in the absence of a confirmed presence, it may be appropriate to consider the use of insecticides to prevent the establishment of colonizers. This was part of the strategy used to protect the Channel Islands from potential colonization from the mainland (Thomas and Wood 1980).

In cases of successional colonization, selective insecticide application to crop borders would have obvious advantages in terms of protection of the beneficial fauna and of environmental health, but its effectiveness at reducing colonization is limited by the need to maintain the border toxic. The lack of persistent products, the need to multiply the applications of non-persistent insecticides (Ferro 1996), and the rapid dispersal of the adult beetles over the whole crop field as they emerge (Blom *et al.* 2002, 2004, Boiteau 2005) have resulted in a low adoption of this control method.

State-coordinated prevention/eradication can aid in controlling for the development of pesticide resistance. The state authorities can come up with a control plan that aims at eradication and which simultaneously tries to control the development of resistance. For instance, the Finnish plant protection authorities took development of resistance specifically into account in their control strategy. For further discussion on the management of insecticide resistance and its prevention, see, for example, Baker *et al.* (2001).

Biological Control

Natural enemies can reduce the growth of a CPB population under particular conditions, and could seemingly enhance Allee effects on spread (Liebhold and Tobin 2008). However, the impact of natural populations of biological control agents on CPB population growth is limited (Cloutier *et al.* 2002). Mass releases of biological control agents to manage peak populations have potential, but there are very few natural enemies that could be mass reared (Cloutier *et al.* 2002). The general predator *Coleomegilla maculaae* DeGeer, the Pentatomidae *Podisus maculiventris* (Say) and *Perillus bioculatus* (F.), the egg parasitoid *Edovum puttleri* Grissell, and the fungus *Beauveria bassiana* (Balsamo) Vuillemin have shown promise, but there are no mass rearing facilities available for these (Cloutier *et al.* 2002). Even if available, their release is better aimed at control than eradication. Also, their release to control invading colonizers would represent exotic introductions in many areas or countries susceptible to CPB invasion.

Cultural and Physical Control

Rotation

Rotation is the most efficient strategy to prevent or delay successional colonization and thereby lower population build-up. As outlined earlier, the dispersal radius of the adult CPB within potato-growing areas in the course of a season is limited. As a result, rotation of the current year potato crop away from the location of the previous year crop or from the sites where beetles overwintered decreases the probability of any individual finding a host, and increases the travel time to a host (Follett *et al.* 1996, Weisz *et al.* 1996, Baker *et al.* 2001). The probability of CPB colonizers locating hosts is a negative function of the rotation distance (Boiteau *et al.* 2008). Rotation of the current year's potato crop away from the arrival sites of invasive colonizers can also be an effective method to control the spread and establishment of invasive founding populations (e.g., as in Finland).

Barriers

Because CPB overwinters predominantly in grassy and woody habitats sprinkled throughout the farm landscape, strategically placed vegetative or physical barriers can be effective at reducing the size of successional colonizing populations in the spring (Boiteau et al. 1994, Ferro 1995) and overwintering site colonization in the fall (Boiteau et al. 1994). Plastic-lined trenches and aboveground extruded plastic barriers (Boiteau and Vernon 2001) set up around current year crop fields or experimental plots have reduced potato colonization by at least 50% (Boiteau et al. 1994, Ferro 1995). The cumulative negative impact of these barriers on crop and overwintering sites should assist with long-term population reduction, but has not been quantified. Although plastic-lined trenches have been designed to manage successional colonization, because the barriers keep trapping beetles over the duration of their period of active dispersal, these traps could play an important role in managing invasive colonizers in the early stages of establishment.

The density of the vegetation making up the field borders, hedges, and meadows of the farm landscape can have a substantial impact on the ability of the CPB to disperse by walking. For example, the walking rate of the adult CPB can be five times greater in grass than on arable soil (Boiteau et al. 2011).

Trap-crop barriers consisting of a few rows of potato planted before the main crop should in theory intercept the beetles moving from the overwintering sites into the main crop. The accumulated beetles could then be destroyed using appropriate insecticides. Effective against some other pests, trap crops have unfortunately been largely ineffective against the CPB because of the lack of a sufficient time interval between adult emergence and main crop emergence to insert a trap crop in most potato growing areas. In areas where it would be possible to set up trap crops, such measures are not economically justifiable in the case of successional colonization, where the objective is only to maintain a low population density. However, the acceptable management cost threshold is lower for invasive colonization than for successional colonization. As introduced above, potato trap-crops have been used effectively to contain the spread of invasive CPB in the quarantine Channel Islands (Thomas and Wood 1980).

A better understanding of the connectivity of the farm landscape and its impact on CPB dispersal might make it possible to manage the rate of successional potato crop colonization using crop border barriers.

Delayed Crop Planting

Late planting is used successfully in many crops to reduce spring colonization pressure by insect pests. However, with potato, the predominance of long-season varieties, the low temperature preferences of the crop, and the relatively short growing season of most potato growing countries combine against this approach. Also, late planted fields could become sinks for beetles emigrating from earlier harvested fields looking for feeding and overwintering sites

(Boiteau 1986, Baker *et al.* 2001, Mbungu and Boiteau 2008). However, in the case of invasive colonization, late planting could be considered as a temporary measure that would allow continued potato production while assisting with the reduction of the abundance of the founder group. Overwintering survival of late diapausing individuals tends to be lower than that of early diapausing individuals (e.g., Piiroinen *et al.* 2011).

Under a climate change scenario, increasing temperatures could mean earlier termination of diapause in the current range of the beetle. This would create an opportunity to make use of late planting to reduce survival among the overwintered population looking for host plants; however, on the negative side, this could lead to an increasing presence of multiple generations (Kocmánková *et al.* 2010).

Shelterbelts

Weather conditions and prevailing wind directions will determine the importance of inflights in the colonization of potato fields between regions and between years. These can only be controlled indirectly by ensuring low-density populations on the crop to minimize potential flights. However, because of the importance of wind in the dispersal of the adult CPB, the local use of shelterbelts to protect fields from inflights has been suggested. Although woodland-surrounded potato fields may be protected to some extent from colonizing inflights, the higher temperature and low wind speeds within the field area have the potential to facilitate the spread of the local population throughout the protected potato fields (Boiteau *et al.* 2010). Although large air masses may carry beetles over long distances once beetles have become airborne, wind acts as a physical barrier to flight take-off (Boiteau *et al.* 2010) and could therefore, indirectly, encourage spread within fields and within the farm rather than over a region. It is not clear from the information available if shelterbelt-protected potato fields are at a lesser or greater risk of invasive colonization than wind exposed potato fields.

Companion Planting

The role of plant biodiversity in colonization by CPB has received remarkably little attention. May and Ahmad (1983) suggested that the range of plants attractive to the CPB is wider than the range of plants acceptable for feeding. The orientation mechanism of the beetle may have evolved in such a manner that the beetle moves towards a wide range of plants that are not necessarily all suitable food sources. Because the beetle is thought to respond to its host by quantitative comparisons of common chemicals, habitat diversity would be detrimental to long-range orientation to the host (May and Ahmad 1983). Emission of some of the key chemicals by non-host plants in the diverse habitat would increase the background noise and make host-finding by walking or flight more difficult. If the CPB evolved such a mechanism for long-range orientation to hosts in its native habitat, the system would remain quite effective in low-diversity agricultural systems (May and Ahmad 1983). It has been suggested that companion planting, the agronomic pendant of plant biodiversity, reduces successional CPB colonization of the potato crop, especially in organic farming operations, but research on the impact of this control method remains inconclusive (Moreau *et al.* 2006).

Eradication

Eradication of CPB breeding colonies can be very effective, but is attempted only to prevent establishment of invasive colonizing foci, because it is a difficult and expensive undertaking. Thomas and Wood (1980) provide an interesting review of the successful but extensive and determined efforts that were required to repeatedly eradicate the beetle from the Channel Islands and maintain its beetle-free status, presenting a fascinating example of the requirements for successful CPB eradication. It is interesting to note that even at this time of high technology, an eradication program cannot overlook any control method and may even include a systematic program of handpicking, as was the case in the 1972 invasion of Sweden (Wiktelius 1981).

Successful eradication requires prompt detection and action to minimize the area over which eradication measures have to be applied, especially outside of contained host areas such as islands. For the purpose of eradicating founding populations, the initial maximum unaided dispersal distance of 5 km (Johnson 1969) provides a convenient buffer area (Ooperi and Jolma 2009) around the mean 1.5- to 2.0-km seasonal dispersal area observed around potato fields (Follett *et al.* 1996, Weisz *et al.* 1996, Boiteau *et al.* 2001, Sexton and Wyman 2005). Successful eradication also requires that control measures be repeatedly applied to all founding populations. First control measures often reduce abundance sharply, but eradication will rarely be achieved unless control is maintained over many years. Even at very low population levels, volunteer potato plants and solanaceous garden plants can generate survivors for the next season. Even in the absence of any evidence of a CPB population on insecticide-treated fields, Boiteau *et al.* (2008) were able to scout and quantify overwintered beetles walking to the field edges of the next year crop, thereby confirming the continued presence of successional colonization pressure even at low CPB density.

Management or Not? Economic and Social Consequences

The CPB/potato system should, in principle, be a relatively simple one to manage, with few conflicting interests. However, because of the CPB's high adaptability to control attempts, the insect remains an economically significant pest. Beyond the biology or the ecology of the pest, the provision of appropriate incentives for the growers is the next greatest challenge to successful management. The impact of actions taken (or not taken) by one grower extends beyond the boundaries of his or her own fields. Each strategy carries a different set of costs, and these costs do not necessarily devolve to the same parties. Governments, for instance, are likely to pay a larger proportion of costs related to prevention than of costs related to control or adaptation. Sound implementation of the chosen strategies needs the support of the growers and all other parties involved. Economic incentives – for example, to switch to resistant potato varieties – will likely be required, especially in invasive cases.

Whether successional or invasive, the challenge with CPB colonization management is how to set priorities for action. In the case of invasive colonization, plant protection organizations have the mandate to determine the domestic and international risks of expansionist and invasive colonization. One of their key tools for pest control is quarantine. The social and economic consequences of quarantine on domestic or international trade can be substantial, and must therefore be justified by thorough risk analyses (e.g., Jarvis and Baker 2001, Heikkila and Peltola 2007). The probability of the CPB dispersing into a previously non-colonized field, region, or country is but one aspect of the risk assessment procedures. The reliability of the assessment depends on the extent of knowledge on the biology of the insect and its interaction with both the crop production environment and the natural environment.

For a coordinated management program to have a probability of success, quarantine legislation requires incentives as well as penalties. The incentives given to the commercial and hobby producers may use the stick (such as penalties, fines, denied compensation) or the carrot (compensation, subsidies, information), or a combination of both (Heikkila and Peltola 2007). Even if the invasive colonizers do not establish in the area, the invasion may still have an economic impact if it occurs at a critical time. For example, some massive invasive flights of the CPB from Normandy to the Channel Islands at the time of crop harvest prevented their export as a beetle-free crop or forced a postponement of the harvest (Thomas and Wood 1980). In the EU, protected zones restrict the import of host plants to other protected zones while allowing export anywhere. However, loss of the protected-zone status basically ends the right to export to other protected zones (Heikkila and Peltola 2007).

The costs of pre-emptive management may seem greater than those of reactive management because of the annual fixed costs. In the case of invasive colonization, pre-emptive management of potential CPB colonization includes the appropriate infrastructure for regular monitoring as fixed costs, and pest eradication and potential financial compensation for the producers as variable costs (e.g., Heikkila and Peltola 2007). Reactive management has no fixed costs, only variable costs associated to potential price changes, pest control, and value of lost production. A related reason which may make pre-emptive costs seem larger than reactive costs is the certainty of those costs and the uncertainty of the benefits acquired (Finnoff et al. 2007). Management decisions made in one region or one country can affect, positively or negatively, the risk of colonization in neighboring regions or countries. For example, decisions made in Finland (e.g., protection zones) or France could result in these countries providing buffer zones for neighboring Norway/Sweden and the Channel Islands, respectively (Heikkila and Peltola 2007, Thomas and Wood 1980).

The effectiveness of management strategies at the national level will vary depending on whether they are part of an authority-driven and coordinated protection system, or whether they are carried out independently by producers outside of any coordinated plan. The choice of the management strategies and their cost-effectiveness at the national level also determine to what extent it is the taxpayers, the producers, or the consumers that are affected by the strategy choice.

In geographical regions without a resident CPB population, determination of the need for management is largely dependent on the tools available to estimate colonization risk. Areas of high value and at low risk of colonization will have priority for management action because the success rate will be high and the crop will have an economic return that justifies management costs. Areas with a high risk of colonization are exposed to failure of the protection system and additional costs of eradication programs for founding populations. Bioeconomic research can help determine when the probability of prevention success no longer justifies continuing preventative actions.

SUMMARY AND FUTURE DIRECTIONS

Management of CPB spread is obviously linked to the management of CPB dispersal, but the link to population growth should not be overlooked. The sustained application of an IPM strategy that minimizes the abundance of resident CPB populations on the crop should actually be the first step in reducing the likelihood of introductions in current year crop fields or in areas outside the distribution range. Each potato grower within and outside the CPB distribution range depends on careful monitoring and management practices of neighbors.

The extent of our knowledge of CPB ecology is now such that there is less and less uncertainty as to whether or not the pest will recolonize the current year's potato crop in areas where it is already established. The remaining uncertainty has more to do with its level of abundance at different locations and throughout the season than with its presence or absence. Rotation and monitoring followed by appropriate curative management is the line of defense against recolonization.

Dispersal models are most useful at predicting the potential for range expansion, but marginal differences in life history traits and behavior combined with changes in the climate make it difficult to predict occurrence (e.g., Valosaari *et al.* 2008). Predictions can be improved by combining spatial and climate models and the information used to guide the development of sound management strategies against CPB colonization. However, it is important not to lose sight, as Henderson-Sellers (1996) and Jarvis and Baker (2001) have wisely pointed out, that, however useful the risk assessments can be, they remain "sketching images of the future" providing "vague contours of the plausible".

The expansion of the distribution range of the CPB is ongoing and likely to continue wherever potatoes are grown; such expansion of the range is, however, likely to proceed slowly. Expansionist colonization is essentially driven by rare and unpredictable anthropogenic and weather-related transportation of CPB colonizers outside their current distribution range. Furthermore, the immediate survival of these early colonizers is only possible if landing takes place in a host patch. Given the high level of adaptation of the CPB to the potato crop, its establishment in the new area could proceed rapidly, but a number of other factors are still required for survival and long-term establishment. Successful establishment also requires that one or more of the arrivals occur at an appropriate time of the season under appropriate temperatures and where the soil type is suitable for overwintering success. With expansionist and invasive colonization, climate and spatial models are helping to measure the risk of CPB colonization in new areas; however, uncertainty remains, with climate change and anthropogenic dispersal as wild cards. Monitoring and quarantine measures are the first line of defense against invasive colonizers to try and prevent them from finding a host and establishing a focus. Preventative management is far more effective and economically justifiable than eradication of already widespread beetles (Heikkila and Peltola 2003, Boman et al. 2007). Eradication of colonization foci is difficult and likely eventually to fail where invasion events are repeated spatially or over time.

The current management approach to contain expansionist or invasive CPB colonizers is to attempt to accomplish eradication and, if eradication fails, to slow down the colonization. Theoretical analysis has shown the empirical approach to be near optimal even under severe uncertainty (Carrasco et al. 2009). The switching point between eradication and slowing down of colonization was identified as the time when small invaded areas coalesced or spread into large areas economically and physically too large for the application of an eradication strategy (Carrasco et al. 2009). Eventually, the management of the invasive colonization stops when it is no longer effective and reverts to options suitable for the management of successional colonization.

Advances in the management of all types of CPB colonization are likely to come from a wider adoption of IPM practices and a better understanding of the landscape ecology of the CPB.

ACKNOWLEDGMENTS

We wish to thank Pam MacKinley for her assistance with the manuscript.

REFERENCES

Bach, C.E., 1982. The influence of plant dispersion on movement patterns of the Colorado potato beetle, *Leptinotarsa decemlineata* (Coleoptera:Chrysomelidae). Great Lakes Entomologist 15, 247–252.

Baker, M.B., Ferro, D.N., Porter, A.H., 2001. Invasions on large and small scales: Management of a well-established crop pest, the Colorado potato beetle. Biol. Invasions 3, 295–306.

Bartlett, P.W., 1980. Interception and eradication of Colorado beetle in England and Wales, 1958–1977. Bulletin OEPP/EPPO Bulletin 10, 481–490.

Beenen, R., Roques, A., 2010. Leaf and Seed Beetles (Coleoptera: Chrysomelidae). In: Roques, A., Kenis, M., Lees, D., Lopez-Vaamonde, C., Rabitsch, W., Rasplus, J.-Y., Roy, D.B. (Eds.), Alien Terrestrial Arthropods of Europe. Biorisk, Vol. 4. Pensoft Publishers, Sofia, Bulgaria, pp. 267–292.

Blom, P.E., Fleischer, S., 2001. Dynamics in the spatial structure of the Colorado potato beetle, *Leptinotarsa decemlineata* (Say) (Coleoptera: Chrysomelidae). Environ. Entomol. 30, 350–364.

Blom, P.E., Fleischer, S.J., Smilowitz, Z., 2002. Spatial and temporal dynamics of Colorado potato beetle (Coleoptera: Chrysomelidae) in fields with perimeter and spatially targeted insecticides. Environ. Entomol. 31, 149–159.

Blom, P.E., Fleischer, S.J., Harding, C.L., 2004. Modeling colonization of overwintered immigrant *Leptinotarsa decemlineata* (Coleoptera: Chrysomelidae). Environ. Entomol. 33, 267–274.

Boiteau, G., 1986. Effect of planting date and plant spacing on field colonization by Colorado potato beetles, *Leptinotarsa decemlineata* (Say), in New Brunswick. Environ. Entomol. 15, 311–315.

Boiteau, G., 2001. Recruitment by flight and walking in a one-generation Colorado potato beetle (Coleoptera: Chrysomelidae) environment. Environ. Entomol. 30, 306–317.

Boiteau, G., 2005. Within-field spatial structure of Colorado potato beetle (Coleoptera: Chrysomelidae) populations in New Brunswick. Environ. Entomol. 34, 446–456.

Boiteau, G., Colpitts, B., 2001. Electronic tags for the tracking of insects in flight: Effect of weight on flight performance of adult Colorado potato beetles. Entomol. Exp. Appl. 100, 187–193.

Boiteau, G., Vernon, R.S., 2001. Physical barriers for the control of insect pests. pp. 61–73 In: C Vincent, B.P., Fleurat-Lessard, F. (Eds.), Physical Control Methods in Plant Protection, Springer-Verlag, Berlin, Germany.

Boiteau, G., Pelletier, Y., Misener, G.C., Bernard, G., 1994. Development and evaluation of a plastic trench barrier for protection of potato from walking adult Colorado potato beetles (Coleoptera: Chrysomelidae). J. Econ. Entomol. 87, 1325–1331.

Boiteau, G., Alyokhin, A., Ferro, D.N., 2003. The Colorado potato beetle in movement. CP Alexander Review. Can. Entomol. 135, 1–22.

Boiteau, G., Picka, J.D., Watmough, J., 2008. Potato field colonization by low-density populations of Colorado potato beetle as a function of crop rotation distance. J. Econ. Entomol. 101, 1575–1583.

Boiteau, G., McCarthy, P.C., MacKinley, P.D., 2010. Wind as an abiotic factor of Colorado potato beetle (Coleoptera: Chrysomelidae) flight take-off activity under field conditions. J. Econ. Entomol. 103, 1613–1620.

Boiteau, G., Vincent, C., Meloche, F., Leskey, T.C., Colpitts, B.G., 2011. Evaluation of tag entanglement as a factor in harmonic radar studies of insect dispersal. Environ. Entomol. 40, 94–102.

Boman, S., Grapputo, A., Lindstrom, L., Lyytinen, A., Mappes, J., 2008. Quantitative genetic approach for assessing invasiveness: geographic and genetic variation in life-history traits. Biol. Invasions, 1–11.

CABI (Commonwealth Agricultural Bureau International), 2011. Invasive Species Compendium. CABI Publishing, Wallingford, UK. www.cabi.org/ISC.

Capinera, J.L., 2001. Handbook of Vegetable Pests. Academic Press, New York, NY.

Caprio, M.A., Grafius, E.J., 1990. Effects of light, temperature, and feeding status on flight initiation in post-diapause Colorado potato beetles (Coleoptera: Chrysomelidae). Environ. Entomol. 19, 281–285.

Carrasco, L.R., Baker, R., MacLeod, A., Knight, J.D., Mumford, J.D., 2010. Optimal and robust control of invasive alien species spreading in homogeneous landscapes. J. R. Soc. Interface 7, 529–540.

Casagrande, R.A., 1985. The "Iowa" potato beetle, its discovery and spread to potatoes. B. Entomol. Soc. Am. 31, 27–29.

CFIA (Canadian Food Inspection Agency), 2009. Import Requirements for Seed Potatoes and Other Potato Propagative Material, 7th revision. D-98–01 CFIA, Ottawa, ON.

Chittenden, F.H., 1914. The Colorado potato beetle migrating to the Pacific coast. J. Econ. Entomol. 7, 15.

Cloutier, C., Boiteau, G., Goettel, M.S., 2002. Leptinotarsa decemlineata (Say), Colorado potato beetle (Coleoptera: Chrysomelidae). In: Mason, P.G., Huber, J.T. (Eds.), Biological Control Programmes in Canada 1981–2000, CABI, Wallingford, UK, pp. 145–152.

Davis, M.A., Thompson, K., 2000. Eight ways to be a colonizer; Two ways to be an invader. B. Ecol. Soc. Am. 81, 226–230.

EPPO, 2006. Data sheets on quarantine pests: Leptinotarsa decemlineata. European and Mediterranean Plant Protection Organization. www.eppo.org/QUARANTINE/insects/Leptinotarsa_decemlineata/LPTNDE_ds.pdf.

Ferro, D.N., Logan, J.A., Voss, R.H., Elkinton, J.S., 1985. Colorado potato beetle (Coleoptera: Chrysomelidae) temperature-dependent growth and feeding rates. Environ. Entomol. 14, 343–348.

Ferro, D.N., Tuttle, A.F., Weber, D.C., 1991. Ovipositional and flight behavior of overwintered Colorado potato beetle (Coleoptera: Chrysomelidae). Environ. Entomol. 20, 1309–1314.

Ferro, D.N., 1995. Mechanical and physical control of the Colorado potato beetle and aphids, pp. 53–68. In: R.-M. Duschesne, and G. Boiteau (eds.), Proceedings of the Potato Insect Pest Control Symposium, 31 July–1 August 1995, Quebec City.

Feytaud, J., 1930. Recherches sur Leptinotarsa decemlineata Say. 1. Observations biologiques. Ann. Epiph. 16, 303–390.

Feytaud, J., 1938. Le rôle des facteurs naturels dans la dissémination du Doryphore en Europe. Intl Congr. Entomol. 4, 2655–2659.

Finnoff, D., Shogren, J.F., Leung, B., Lodge, D., 2007. Take a risk: Preferring prevention over control of biological invaders. Ecol. Econ. 62, 216–222.

Follett, P.A., Cantelo, W.W., Roderick, G.K., 1996. Local dispersal of overwintered Colorado potato beetle (Chrysomelidae: Coleoptera) determined by mark and recapture. Environ. Entomol. 25, 1304–1311.

Gibson, A., Gorham, R.P., Hudson, H.F., Flock, J.A., 1925. The Colorado potato beetle (Leptinotarsa decemlineata) in Canada. Canadian Department of Agriculture, Bulletin 52.

Grapputo, A., Boman, S., Lindström, L., Lyytinen, A., Mappes, J., 2005. The voyage of an invasive species across continents: Genetic diversity of North American and European Colorado potato beetle populations. Mol. Ecol. 14, 4207–4219.

Gui, L.Y., Boiteau, G., 2010. Effect of food deprivation on the ambulatory movement of the Colorado potato beetle, Leptinotarsa decemlineata. Entomol. Exp. Appl. 134, 138–145.

Gui, L.Y., Boiteau, G., Colpitts, B.G., MacKinley, P., McCarthy, P.C., 2011. Random movement pattern of fed and unfed adult Colorado potato beetles in bare-ground habitat. Agr. Forest Entomol. 14, 59–68.

Guillemaud, T., Ciosi, M., Lombaert, E., Estoup, A., 2011. Biological invasions in agricultural settings: insights form evolutionary biology and population genetics. C. R. Biol. 334, 237–246.

Hare, J.D., 1990. Ecology and management of the Colorado potato beetle. Annu. Rev. Entomol. 35, 81–100.

Heikkila, J., Peltola, J., 2003. Conceptualising the economics of plant health protection against invasive pests. Agr. Food Sci. Finland 12, 67–81.

Heikkila, J., Peltola, J., 2004. Analysis of the Colorado potato beetle protection system in Finland. Agr. Econ. 31, 343–352.

Heikkila, J., Peltola, J., 2007. Phytosanitary measures under uncertainty. A cost-benefit analysis of the Colorado potato beetle in Finland. In: Oude Lansink, A.G.J.M. (Ed.), New Approaches to the Economics of Plant Health, Springer, Dordrecht, The Netherlands, pp. 147–161.

Henderson-Sellers, A., 1996. Can we integrate climate modeling and assessment?. Environ. Model. Assess 1, 59–70.

Hiiesaar, K., Metspalu, L., Jõudu, J., Jõgar, K., 2006. Over-wintering of the Colorado potato beetle (*Leptinotarsa decemlineata* Say) in field conditions and factors affecting its population density in Estonia. Agron. Res. 4, 21–30.

Horton, D.R., Capinera, J.L., 1990. Host utilization by the Colorado potato beetle (Coleoptera: Chrysomelidae) in a potato/weed (*Solanum sarrachoides* Sendt.) system. Can. Entomol. 122, 113–121.

Hoy, C.W., Boiteau, G., Alyokhin, A., Dively, G., Alvarez, J.M., 2008. Managing insect and mite pests. In: Johnson, D.A. (Ed.), Potato Health Management, The American Phytopathological Society, St Paul, MN, pp. 133–147.

Hsiao, T.H., 1978. Host plant adaptations among geographic populations of the Colorado potato beetle. Entomol. Exp. Appl. 24, 237–247.

Hsiao, T., 1981. Ecophysiological adaptations among geographic populations of the Colorado potato beetle in North America. In: Lashcomb, J.H., Casagrande, R. (Eds.), Advances in potato pest management, Hutchinson Ross Publishing Co, Stroudsburg, PA, pp. 69–85.

Hsiao, T.H., 1985. Ecophysiological and genetic aspects of geographic variations of the Colorado potato beetle. In: Ferro, D.N., Voss, R.H. (Eds.), Proceedings of the Symposium on the Colorado Potato Beetle, 17th International Congress of Entomology. Massachusetts Experiment Station, University of Massachusetts, Amherst, MA, pp. 63–78.

Hsiao, T.H., 1988. Host specificity, seasonality and bionomics of Leptinotarsa beetles. In: Jolivet, P., Petitpierre, E., Hsiao, T.H. (Eds.), Biology of Chrysomelidae, Kluwer Academic Publishers, Dordrecht, the Netherlands, pp. 581–599.

Hurst, G.W., 1975. Meteorology and the Colorado potato beetle. Secretariat of the world meteorological organization, Geneva, Switzerland.

Jacques, R.L., 1988. The Potato Beetles: The Genus Leptinotarsa in North America (Coleoptera, Chrysomelidae). CRC Press, Boca Raton, FL.

Jarvis, C.H., Baker, R.H.A., 2001. Risk assessment for nonindigenous pests. I. Mapping the outputs of phenology models to assess the likelihood of establishment. Divers Distrib. 7, 223–235.

Johnson, C.G., 1969. Migration and dispersal of insects by flight. Methuen, London, UK.

Jolivet, P., 1991. The Colorado beetle menaces Asia (*Leptinotarsa decemlineata* Say) (Coleoptera: Chrysomelidae). Entomol. 47, 29–48.

Jolivet, P., 1994. Dernières nouvelles de la progression du Doryphore: *Leptinotarsa decemlineata* (Say, 1824) (Coleoptera: Chrysomelidae). Entomol. 50, 105–111.

Kocmánková, E., Trnka, M., Eitzinger, J., Formayer, H., Dubrovský, M., Semerádová, D., Ăalud, Z., Juroch, J., Možný, M., 2010. Estimating the impact of climate change on the occurrence of selected pests in the central European region. Climate Res. 44, 95–105.

Liebhold, A.M., Tobin, P.C., 2008. Population ecology of insect invasions and their management. Annu. Rev. Entomol. 53, 387–408.

Lu, W., Lazell, J., 1996. The voyage of the beetle. Nat. Hist 105, 36–39.

Lu, W., Logan, P., 1994. Genetic variation oviposition between and within populations of *Leptinotarsa decemlineata* (Coleoptera: Chrysomelidae). Ann. Entomol. Soc. Am. 87, 634–640.

MacQuarrie, C.J.K., Boiteau, G., 2003. Effect of diet and feeding history on flight of Colorado potato beetle, *Leptinotarsa decemlineata*. Entomol. Exp. Appl. 107, 207–213.

May, M.L., Ahmad, S., 1983. Host location in the Colorado potato beetle: searching mechanisms in relation to oligophagy. In: Ahmad, S. (Ed.), Herbivorous insects: host-seeking behavior and mechanisms, Academic Press, New York, NY, pp. 173–199.

Mbungu, N.T., Boiteau, G., 2008. Flight take-off performance of Colorado potato beetle in relation to potato phenology. J. Econ. Entomol 101 (1): 56–60.

Memmott, J., Craze, P.G., Harman, H.M., Syrett, P., Fowler, S.V., 2005. The effect of propagule size on the invasion of an alien insect. J. Animal Ecol. 74, 50–62.

Metcalf, R.L., Metcalf, R.A., 1993. Destructive and Useful Insects: Their habits and control. McGraw Hill, New York, NY.

Moreau, T., Warman, P.R., Hoyle, J., 2006. An evaluation of companion planting and botanical extracts as alternative pest controls for the Colorado potato beetle. Biol. Agr. Hort 23, 351–370.

Noronha, C., Cloutier, C., 1999. Ground and aerial movement of adult Colorado potato beetle (Coleoptera: Chrysomelidae) in a univoltine population. Can. Entomol. 131, 521–538.

Ooperi, S., Jolma, A., 2009. Modeling invasion dynamics of Colorado potato beetle to test spatially targeted management strategy. 18th World IMACS/ MODSIM Congress, Cairns, Australia.

Peters, E.H., 1954. The Colorado potato beetle in Canada 1870–1952. 36th Report of Quebec Society for Protection of Plants, pp. 38–43.

Piiroinen, S., Ketola, T., Lyytinen, A., Lindström, L., 2011. Energy use, diapause behaviour and northern range expansion potential in the invasive Colorado potato beetle. Func. Ecol. 25, 527–536.

Rafoss, T., Saethre, M.G., 2003. Spatial and temporal distribution of bioclimatic potential for the Codling moth and the Colorado potato beetle in Norway: Model predictions versus climate and field data from the 1990s. Agr. Forest Entomol. 5, 75–85.

Riley, C.V., 1876. Eighth Annual Report on the Noxious, Beneficial and Other Insects of the State of Missouri St. Bd. Agric. Regan and Carter, Jefferson City, MO.

Sandeson, P.D., Boiteau, G., Le Blanc, J.P.R., 2002. Adult density and the rate of Colorado potato beetle (Coleoptera: Chrysomelidae) flight take-off. Environ. Entomol. 31, 533–537.

Sexton, D.L., Wyman, J.A., 2005. Effect of crop rotation distance on populations of Colorado potato beetle (Coleoptera:Chrysomelidae): Development of area wide Colorado potato beetle pest management strategies. J. Econ. Entomol. 98, 716–724.

Smatas, R., Semaskiene, R., Lazauskas, S., 2008. The impact of climate conditions on the occurrence of the Colorado potato beetle (*Leptinotarsa decemlineata*). Zemdirbyste 95, 235–241.

Shelton, A., Zhao, J., Roush, R., 2002. Economic, ecological, food safety, and social consequences of the deployment of Bt transgenic plants. Ann. Rev. Entomol 47, 845–881.

Thomas, G., Wood, F., 1980. Colorado beetle in the Channel Islands. Bulletin OEPP/EPPO Bulletin 10, 491–498.

Tscheulin, T., Petanidou, T., Settele, J., 2009. Invasive weed facilitates incidence of Colorado potato beetle on potato crop. Int. J. Pest Manage 55, 165–173.

Ulrichs, C., Hopper, K.R., 2008. Predicting insect distributions from climate and habitat data. Bio-Control 53, 881–894.

Ulrichs, C., Mucha-Pelzer, T., Scobel, E., Kretschmer, L., Bauer, R., Bauer, E., Mewis, I., 2008. Plant protection with silica particles: Electrostatic application and impact of particle layer density on the insecticidal efficacy [Silikate im Pflanzenschutz: Elektrostatische Applikation und Abhängigkeit der Wirksamkeit von der Schichtdicke]. Gesunden Pflanz 60, 29–34.

Valosaari, K.-R., Aikio, S., Kaitala, V., 2008. Spatial simulation model to predict the Colorado potato beetle invasion under different management strategies. Ann. Zool. Fenn. 45, 1–14.

Vergnaud, A., 1991. Lutte contre le doryphore de la pomme de terre (*Leptinotarsa decemlineata*) en Normandie (France). Bulletin OEPP/EPPO Bulletin 21, 17–21.

Voss, R.H., Ferro, D.N., 1990a. Ecology of migrating Colorado potato beetles (Coleoptera: Chryso-melidae) in western Massachusetts. Environ. Entomol. 19, 123–129.

Voss, R.H., Ferro, D.N., 1990b. Phenology of flight and walking by Colorado potato beetle (Coleoptera: Chrysomelidae) in western Massachusetts. Am. Potato J. 69, 473–482.

Weber, D.C., Ferro, D.N., 1994. Colorado potato beetle: diverse life history poses challenge to management. In: Zehnder, G.W., Powelson, M.L., Jansson, R.K., Raman, K.V. (Eds.), Advances in Potato Pest Biology and Management, American Phytopathological Society, St Paul, MN pp. 54–65 .

Weber, D.C., Ferro, D.N., 1996. Flight and fecundity of Colorado potato beetles (Coleoptera: Chrysomelidae) fed on different diets. Ann. Entomol. Soc. Am. 89, 297–306.

Weber, D.C., Drummond, F.A., Ferro, D.N., 1995. Recruitment of Colorado potato beetles (Coleoptera: Chrysomelidae) to solanaceous hosts in the field. Environ. Entomol. 24, 608–622.

Wegorek, W., 1959. The Colorado potato beetle (Leptinotarsa decemlineata Say). Prace Naukowe Instytutu Ochrony Roslin, Warszawa, Poland.

Weisz, R., Smilowitz, Z., Fleischer, S., 1996. Evaluating risk of Colorado potato beetle (Coleoptera:Chrysomelidae) infestation as a function of migratory distance. J. Econ. Entomol. 89, 435–441.

Wiktelius, S., 1981. Wind dispersal of insects Aphids, Leptinotarsa decemlineata, Colorado potato beetle, crop pests, weather situation as dispersal factor. Grana 20, 205–207.

Zehnder, G.W., Powelson, M.L., Jansson, R.K., Raman, K.V. (Eds.), 1994. Advances in Potato Pest Biology and Management., American Phytopathological Society, St Paul, MN.

Part IV

Management Approaches

Chapter 13

Chemical Control of Potato Pests

Thomas P. Kuhar, Katherine Kamminga, Christopher Philips, Anna Wallingford, and Adam Wimer

Department of Entomology, Virginia Tech, Blacksburg, Virginia, USA

INTRODUCTION

For more than a century, chemical control has been one of the most widely used pest management tactics in potato production. Although environmental and human safety concerns have influenced the registration status of many insecticides around the world, and effective non-chemical strategies have been identified for most pests (see other chapters in this book), chemical control still remains one of the most widely used strategies for eliminating crop damage by arthropod pests, and will likely remain the basis of pest management for the foreseeable future (Alyokhin 2009). In this chapter we will review where we've been and where we are in the present day with the use of insecticides in potato production.

EARLY HISTORY OF CHEMICAL CONTROL IN POTATOES

In North America and Europe, chemical control in potatoes has largely been driven by the pest management challenges brought on by the Colorado potato beetle, *Leptinotarsa decemlineata* (Say) (Coleoptera: Chrysomelidae). This insect species greatly impacted the history of insecticide use in potatoes, and agriculture in general. Gauthier *et al.* (1981) provides a very good review of the early history of chemical control in potatoes. In the 19th century, the aceto-arsenite of copper called "Paris Green" was first used to control Colorado potato beetle (Riley, 1871), and other types of arsenical compounds such as lead arsenate and calcium arsenate would continue to be used for its control into the 1940s (Gauthier *et al.* 1981). However, arsenical insecticides were difficult to mix, difficult to apply effectively, did not have a long residual life on plants, and sometimes caused phytotoxicity. Thus, alternatives to the use of arsenicals in potatoes were sought throughout the early 1900s.

Botanical insecticides such as veratrine alkaloids from Sabadilla, ryania extract, and rotenone were evaluated for the control of Colorado potato beetle

Insect Pests of Potato. http://dx.doi.org/10.1016/B978-0-12-386895-4.00013-2

(Brown 1951). Although rotenone demonstrated sufficient efficacy against this pest, the focus on botanical insecticides as a replacement for arsenicals would soon be overshadowed as the "Age of Pesticides" began in the 1940s (Metcalf 1980).

THE PESTICIDE TREADMILL

Chlorinated Hydrocarbons

In the 1940s, Colorado potato beetle was one of the first agricultural targets for the chlorinated hydrocarbon insecticide, DDT, and it also became one of the first pests to develop resistance to the chemical in the US during the 1950s (Hofmaster, 1956). Other chlorinated hydrocarbons (mostly cyclodienes), including aldrin, dieldrin, endrin, heptachlor, methoxychlor, endosulfan, and others, would be widely used on potatoes, but the Colorado potato beetle quickly developed resistance to those insecticides as well (Gauthier *et al.* 1981, Alyokhin *et al.* 2008a). Nonetheless, the long residual activity of these insecticides in the soil made them ideal for controlling subterranean pests of potatoes such as wireworms (Coleoptera: Elateridae) (Merrill 1952, Gunning and Forrester 1984, Parker and Howard 2001). Thus, cyclodienes would continue to be used on potatoes until most agricultural uses would eventually be canceled (in the US, by 1980) because of the persistence of these compounds in the environment, resistance that developed in several insect pests, and biomagnification in some wildlife food chains (Ware and Whitacre 2004).

Organophosphates and Carbamates

Gradually, carbamates and organophosphates would replace the chlorinated hydrocarbons in potato production. These cholinesterase-inhibiting neurotoxins have broad-spectrum activity against most insect pests attacking potatoes. In addition, the systemic activity of many of these compounds, including aldicarb, disulfoton, fensulfothion, carbofuran, phorate, and oxamyl, made possible the introduction of the insecticide into growing plant parts via soil application. This enabled longer residual activity of the chemical and, in theory, the need for fewer foliar insecticide sprays (Gauthier *et al.* 1981). However, Colorado potato beetle would eventually develop resistance and cross-resistance to carbamates and organophosphates, rendering virtually all insecticides in these two classes practically useless against this pest (Casagrande 1987). Nonetheless, a few carbamates and organophosphates are still widely used today in commercial potato production; oxamyl, aldicarb, ethoprop, and methamidophos are listed among the top five insecticides in total amount of active ingredient applied to potatoes in the US (United States Department of Agriculture – National Agricultural Statistics Service 2010). These insecticides are generally not applied for Colorado potato beetle control, but rather for control of plant parasitic nematodes and wireworms in the soil, and for control of aphids, potato psyllids, and potato tuberworms.

Pyrethroids

In the 1970s the first synthetic pyrethroid insecticides, fenvalerate and perme-thrin, were registered on potatoes in the US. Modeled after the plant-derived pyrethrins, these insecticides offered a similar mode of action to that of DDT, modulating the sodium channel on neuronal membranes (Ware and Whitacre 2004). Throughout the 1970s until the 1990s, a number of pyrethroid insecti-cides would be registered throughout the world (Table 13.1). These insecticides were shown to be active on a broad range of insect pests, were efficacious at extremely low use rates, and typically did not break down quickly in sunlight like natural pyrethrins. However, because the mode of action of pyrethroids is not that different from that of DDT (Ware and Whitacre 2004), it is not surpris-ing that Colorado potato beetle quickly developed resistance to this insecticide class (Casagrande 1987, Tisler and Zehnder 1990, Alyokhin *et al.* 2008a).

Thus, over a 30- to 40-year span in the US, Colorado potato beetle would develop resistance to all of the aforementioned classes of insecticides; as a result, many potato growers had to change insecticides every few years as well as tank-mix multiple compounds for a single spray application (Casagrande 1987). By 1990, it was reported that Colorado potato beetle had developed resistance to over 25 insecticides from all major classes (Forgash 1985, Harris and Turnball 1986, Roush *et al.* 1990, Tisler and Zehnder 1990, French *et al.* 1992, Grafius and Bishop 1996). In certain potato-producing areas of the eastern US, growers were running out of effective insecticide options for Colorado potato beetle. A number of different mechanisms of resistance have been identified in Colorado potato beetle, including target site insensitivity, enhanced metabolic enzyme activity of the target organism, reduced insecticide penetration, and increased excretion (Rose and Brindley 1985, Ioannidis *et al.* 1991, Argentine *et al.* 1994, Wierenga and Hollingworth 1994, Alyokhin *et al.* 2008a).

In addition to resistance problems in Colorado potato beetle, the frequent applications of these broad-spectrum insecticides resulted in pest resurgences and outbreaks of secondary pests because they destroyed natural enemy popula-tions in fields (Metcalf 1980). Moreover, insecticide resistance to chlorinated hydrocarbons, carbamates, organophosphates, and pyrethroids also occurred in other potato pests, including the green peach aphid, *Myzus persicae* (Sulzer) (Hemiptera: Aphididae), twospotted spider mite, *Tetranychus urticae* (Koch) (Acari: Tetranychidae), and beet armyworm, *Spodoptera frugiperda* (J.E. Smith) (Lepidoptera: Noctuidae), among others (Penman and Chapman 1988, Brewer and Trumble 1994, Kerns *et al.* 1998, Foster *et al.* 2007, Castañeda *et al.* 2011).

The mistakes and risks of indiscriminate and excessive use of insecticides could not have been more clearly demonstrated than in the case of the Colorado potato beetle in the US by 1990 (Casagrande 1987). Nonetheless, the conven-tional control strategy today for the pest has not changed. Chemical control is the primary tool used, and very little regard is given to integrated pest manage-ment (Alyokhin 2009). Over the past two decades there has been a major shift in

TABLE 13.1 Insecticides and Miticides Currently Registered for Use on Potatoes in the US as of 2011

IRAC Classification Number[a]	Group	Mode of Action	Insecticide(s)[b]
1A	Carbamate	Acetylcholine esterase inhibitor (reversible)	Aldicarb** Carbaryl Carbofuran** Methomyl Oxamyl
1B	Organophosphate	Acetylcholine esterase inhibitor (irreversible)	Azinphosmethyl** Dimethoate Disulfoton** Ethoprop Malathion Methamidophos Phorate Phosmet
2A	Organochlorines	GABA-gated chloride channel antagonist	Endosulfan**
2B	Phenylpyrazoles	GABA-gated chloride channel antagonist	Fipronil
3	Pyrethroids	Sodium channel modulator	Beta-cyfluthrin Bifenthrin Cyfluthrin Esfenvalerate Lambda-cyhalothrin Permethrin Zeta-cypermethrin
4	Neonicotinoid	Nicotinic acetylcholine receptor agonist	Acetamiprid Clothianidin Dinotefuran Imidacloprid Thiamethoxam
5	Spinosyns	Nicotinic acetylcholine receptor allosteric activator	Spinetoram Spinosad
6	Avermectins	Chloride channel activator	Abamectin
9	None	Selective homopteran feeding blocker	Flonicamid Pymetrozine
10	None	Mite growth inhibitor	Hexythiazox

TABLE 13.1 Insecticides and Miticides Currently Registered for Use on Potatoes in the US as of 2011—*Cont'd*

IRAC Classification Number[a]	Group	Mode of Action	Insecticide(s)[b]
11	None	Microbial disruptors of insect midgut membrane	*Bacillus thuringiensis tenebrionensis* *Bacillus thuringiensis kurstaki*
12C	None	Inhibitors of mitochondrial ATP synthesis	Propargite
15	Benzoylureas	Inhibitors of chitin biosynthesis, type 0	Novaluron
17	None	Molting disruptor	Cyromazine
18B	Azadirachtins	Unknown (UN) – Multiple effects on insects including antifeedancy and insect growth disruption.	Azadirachtin
22A	None	Voltage-dependent sodium channel blockers	Indoxacarb
23	Tetronic and tetramic acid derivaties	Inhibitors of acetyl CoA carboxylase – Lipid synthesis, growth regulation	Spiromesifen Spirotetramat
28	Diamide None None None	Ryanodine receptor modulator Unknown (UN) Unknown (UN) Unknown (UN)	Chlorantraniliprole Bifenazate Chenopodium extract Cryolite

[a]*Insecticide Resistance Action Committee (IRAC) mode of action classification is the definitive global authority on the target site of insecticides.*
[b]*All products are not registered for use in all states.*
**, Insecticides targeted for cancelation by the US Environmental Protection Agency.

insecticide development to more targeted or narrow-spectrum insecticides with novel modes of action. These new insecticides are often less toxic to mammals and the environment and have reduced impacts on natural enemies compared to carbamates, organophosphates, and pyrethroids. In the US there are over 30 insecticides from at least 15 different classes currently registered for Colorado potato beetle control on potatoes (Alyokhin *et al.* 2008a), and even more are in development or are currently in the registration process (Table 13.1).

Neonicotinoids

Neonicotinoids are also referred to as nitro-quanidines, neonicotinyls, nicotinoids, chloronicotines, and chloronicotinyls. The first neonicotinoid insecticide, imidacloprid, was introduced in Europe and Japan in 1990, and was registered for use on potatoes in the US in 1996. Other neonicotinoids, including thiamethoxam, acetamiprid, dinotefuran, and clothianidin, were registered a few years later. Since that period these chemicals have been the most commonly used insecticides on potatoes for control of Colorado potato beetle as well as other insect pests, including leafhoppers, potato psyllids, aphids, and flea beetles. Neonicotinoids are neurotoxins that target the nicotinic acetylcholine receptor acting as agonists (Maienfisch et al. 2001). Although they are effective as contact insecticides, it is the ability of these chemicals to translocate from the soil into leaves as systemic insecticides that has been one of the primary reasons for their popularity. Most commercial potato growers apply these chemicals in the seed furrow at planting or as a pre-planting treatment to seed pieces. Both application methods provide long-term systemic protection to the potato plant against Colorado potato beetle (Boiteau et al. 1997, Kuhar et al. 2003a, 2007, Kuhar and Speese 2005a, 2005b) and sucking pests such as leafhoppers, aphids, and psyllids (Boiteau et al. 1997, Pavlista 2002, Kuhar and Speese 2005a, 2005b). Neonicotinoids also provide efficacy against wireworms in the soil (Kuhar et al. 2003b, Kuhar and Alvarez 2008). Currently, neonicotinoid insecticides represent the foundation for insect control in most potato-growing regions.

Field dissipation rates for imidacloprid and thiamethoxam are variable. The half-life for imidacloprid has been reported to be as short as 60 days (Liu et al. 2011) or as long as 280 days in field soils (Saran and Kamble 2008). The half-life for thiamethoxam is much shorter, and ranges from 9 days in field soils (Karmakar and Kulshrestha 2009) to up to 75 days in lab soils (Maienfisch et al. 2001). Field efficacy trials conducted on sandy loam soils in Virginia (USA) showed that imidacloprid, thiamethoxam, and clothianidin applied to potato seed pieces at planting provided effective control of both Colorado potato beetle and potato leafhopper, Empoasca fabae (Harris) (Hemiptera: Cicadellidae), for more than 60 days after planting (Table 13.2). Soil type, pH, groundcover, cultivation (i.e., exposure to sunlight), moisture, temperature, and microbial communities present all play a role in the residual life of an insecticide in the soil. Both imidacloprid and thiamethoxam are stable in neutral and acidic water, although these compounds will slowly degrade in basic solutions (Liu et al. 2006). Soil-dwelling microorganisms have been described that degrade imidacloprid and thiamethoxam (Anhalt et al. 2007, Pandey et al. 2009). Bare-ground soils will see longer half-lives for imidacloprid than soils with groundcover (Scholz and Spiteller 1992), likely due to higher populations of those microorganisms in soils with growing vegetation. Higher levels of organic matter in the soil make for a longer half-life, as sorption of imidacloprid increases as organic

TABLE 13.2 Counts of Colorado Potato Beetle (CPB) Larvae at 64 Days after Planting (DAP) and Potato Leafhopper (PLH) Nymphs at 79 DAP on Potatoes with Seed (ST) or In-Furrow (IF) Treatments of Various Neonicotinoid Insecticides in a Small-Plot Field Experiment Conducted in Painter, Virginia (USA), April–July 2003

Treatment	Rate (kg AI/ha)	No. CPB larvae/ 10 stems	No. PLH nymphs/ 10 leaves[a]
Untreated control		66.0 a	51.0 a
Thiamethoxam (IF)	0.11	0.0 b	0.0 b
Imidacloprid (IF)	0.28	0.0 b	0.0 b
Imidacloprid (ST)	0.14	0.0 b	1.0 b
Imidacloprid (ST)	0.28	0.0 b	0.0 b
Clothianidin (ST)	0.17	0.0 b	0.0 b
Clothianidin (ST)	0.22	0.0 b	0.0 b

[a]*Numbers within a column with a letter in common are not significantly different according to analysis of variance followed by Fisher's protected LSD to separate means.*

carbon content increases (Cox *et al.* 1997, 1998), thereby decreasing the bio-availability to microorganisms that degrade the compound.

Researchers have documented non-target effects of neonicotinoid applications to the soil. There is evidence that imidacloprid and thiamethoxam induce systemic acquired resistance (SAR) in citrus (Graham and Myers 2011) as well as increase plant vigor by inducing salicylic acid-associated plant responses (Ford *et al.* 2010). On a negative note, imidacloprid can affect the health of earthworms and impair their burrowing behavior (Dittbrenner *et al.* 2011a, 2011b). Repeated annual use of neonicotinoid insecticides also can have negative effects on ground beetles (Peck 2009, Kunkel *et al.* 2001). In general, imidacloprid is considered safer than older classes of insecticides to use near bodies of water because of its soil binding properties and low leaching potential (Oi 1999, Churchel *et al.* 2011). Environmental levels of imidacloprid found in agricultural fields are rarely reported to reach concentrations that are acutely toxic to model organisms, such as *Daphnia magna* (Straus) (Cladocera: Daphniidae), *Vibrio fischeri* (Beirjerinck) (Vibrionales: Vibrionaceae), and *Desmodesmus subspicatus* (Chodat) (Sphaeropleales: Scenedesmaceae) (Jemec *et al.* 2007, Tisler *et al.* 2009). Further, thiamethoxam is reported to have very low toxicity to fish, *Daphnia*, mollusks, and earthworms (Maienfisch *et al.* 2001).

Unfortunately, as had been the case for virtually all other insecticides that preceded it, Colorado potato beetle would develop resistance to imidacloprid, with greater than 100-fold resistance levels detected in populations from Long

Island, NY in 1997 (Olson *et al.* 2000, Zhao *et al.* 2000). Since that time, several other populations of the beetle in North America have developed resistance to imidacloprid as well as cross-resistance to thiamethoxam (Mota-Sanchez *et al.* 2000, 2006, Tolman *et al.* 2005, Alyokhin *et al.* 2006, 2008a).

Entomologists have attempted to be proactive with slowing the rate of resistance development to this important class of insecticides. In the interest of insecticide resistance management (IRM) it is strongly recommended that growers explore non-chemical control options such as crop rotation, avoid using neonicotinoids where beetle populations have demonstrated resistance, avoid foliar applications of neonicotinoids if at-planting systemic applications were used in a field, use economic thresholds, only spray if necessary, leave untreated refuge areas in fields, and apply full rates of products (Sexson *et al.* 2005, Alyokhin 2011). Moreover, neonicotinoid resistance monitoring of Colorado potato beetle populations from across North America is conducted annually, and alternative insecticide mode of actions are recommended to potato growers when resistance is suspected. Insecticide rotation is strongly encouraged in general as a sound IRM practice in agriculture.

A PLETHORA OF CHEMICAL CONTROL OPTIONS FOR COLORADO POTATO BEETLE

Today there is a wide range of effective insecticides for control of Colorado potato beetle and other insect pests (Table 13.1). These include some biologically-derived products as well as synthetic compounds with novel modes of action. For most of these foliar spray products, monitoring potato fields for beetle eggs and larvae allows growers to accurately target needed sprays. This strategy keeps the number of sprays to a minimum and avoids excessive selection of resistant beetles and other pests (Sexson *et al.* 2005). Several alternative insecticides and novel classes of insecticides for control of Colorado potato beetle are discussed below.

Cryolite is an inorganic fluoride insecticide that was used for control of insecticide-resistant Colorado potato beetles in the 1990s. The fluoride ion inhibits many enzymes that contain iron, calcium, and magnesium. Several of these enzymes are involved in energy production in cells, as in the case of phosphatases and phosphorylases (Ware and Whitacre 2004). Cryolite has been used as a relatively safe fruit and vegetable insecticide in integrated pest management programs. It provides effective control of Colorado potato beetle (Noetzel and Holder 1996, Sorensen and Holloway 1997).

Avermectins are macrocyclic lactone derivatives from the fermentation of *Streptomyces avermitilis*, a soil actinomycete (Campbell 1989) (see Chapter 16 for additional information). The insecticide **abamectin** is a mixture of avermectins containing more than 80% avermectin B1a and less than 20% avermectin B1b. Abamectin blocks the transmittance of electrical activity in nerves and muscle cells by stimulating the release and binding of gamma-aminobutyric acid

(GABA) at nerve endings, which causes an influx of chloride ions into the cells, leading to hyperpolarization and subsequent paralysis of the neuromuscular system (Bloomquist 1996, 2003). Abamectin is toxic to a wide range of insects and mites, and has been shown to be highly effective at controlling Colorado potato beetle larvae and adults (Nault and Speese 1999a, Kuhar et al. 2006a, Marčića et al. 2009, Sewell and Alyokhin 2010b). Prior to the introduction of neonicotinoid insecticides, abamectin was one of the top insecticides used for control of insecticide-resistant Colorado potato beetles. However, avermectin is expensive to manufacture, and its major use on crops today is for control of mites.

Azadirachtin is a tetranortriterpenoid (limonoid) found in the seeds of the neem tree (*Azadirachta indica*). This compound has been shown to cause antifeedancy and insect growth disruption in insects by blocking the release of the morphogenic peptide hormone (Mordue and Blackwell 1993, Seymour *et al.* 1995, Abudulai *et al.* 2003). It has been shown to be effective on a wide range of insects, including lepidopteran pests and Colorado potato beetle. In general, azadirachtin is most effective as a growth regulator on eggs and small larvae (Trysyono and Whalon 1999, Kowalska 2007), and therefore application timing is important for successful control, particularly when targeting Colorado potato beetle. The insecticide has demonstrated moderate efficacy in the field for Colorado potato beetle control (Zehnder and Warthen 1988, Marčića *et al.* 2009). Because azadirachtins are biologically derived, they can be found in insecticide formulations that are approved for use in organic agriculture.

Bacillus thuringiensis* subspecies *tenebrionis is a bacterium that produces delta-endotoxins that target midgut cells and are toxic to certain beetles (see Chapter 16 for additional information). If the endotoxins are ingested, they form an ion channel that causes shrinking or swelling in the epithelium cells, leading to cell lysis and eventual death of the insect (Slaney *et al.* 1992). *Bacillus thuringiensis* (*Bt*) subsp. *tenebrionis* applications are most effective against small larvae of Colorado potato beetle, and thus, as with azadirachtin, application timing is critical for effective control in the field (Ghidiu and Zehnder 1993). Even with proper application timing, the efficacy of this insecticide against Colorado potato beetle has been moderate at best (Sewell and Alyokhin 2009). Moreover, resistance to *Bt* subsp. *tenebrionis* was reported in isolated populations of Colorado potato beetle in the early 1990s (Whalon *et al.* 1993). This insecticide is not widely used today.

Spinosyns are a group of insecticidal macrocyclic lactones derived from the fermentation of *Saccharapolyspora spinosa*, which is a soil actinomycete (Thompson *et al.* 1995, 2000). The insecticide **spinosad** is a mixture of spinosyns A and D. Spinosad disrupts binding of acetylcholine in nicotinic acetylcholine receptors at the postsynaptic cell, and acts upon the insect by exciting neurons in the central nervous system. This causes tremors and spontaneous muscle contractions, leading to paralysis and loss of body fluids (Salgado 1998). The insecticide is active against most lepidopterans, thysanopterans, dipterans, and some coleopterans (Thompson *et al.* 2000). Foliar applications of spinosad provide excellent control of Colorado potato beetle larvae (Byrne *et al.* 2006, Kuhar and Doughty 2009, Sewell and Alyokhin 2009, 2010a). Recently, however, some populations of Colorado potato beetle that are resistant to neonicotinoids have demonstrated cross-resistance (or inherent reduced susceptibility) to spinosad (Mota-Sanchez *et al.* 2006) (see Chapter 16 for additional information). **Spinetoram** is a more recent spinosyn insecticide that was derived from spinosyns J and L, which have been chemically modified to produce a semi-synthetic insecticide (Sparks *et al.* 2008). Spinetoram is active against the same pest groups as spinosad, and has shown excellent control against most lepidopteran pests as well as Colorado potato beetle (Sewell and Alyokhin 2007, Kuhar and Doughty 2009, Groves *et al.* 2011a).

Novaluron is an insect growth regulator that belongs to the benzoylphenyl urea (or benzoylurea) class of chemicals (IRAC Group 15). These insecticides target and disrupt chitin biosynthesis on the larval stages of many insects (Ishaaya *et al.* 2003, Ware and Whitacre 2004). Novaluron is very effective at controlling the larval stage of Colorado potato beetle (Cutler *et al.* 2007), but can also cause egg mortality (Alyokhin *et al.* 2008b) and a decrease in reproductive viability of adult females when ingested (Alyokhin *et al.* 2010). Two foliar applications of novaluron will provide effective control of Colorado potato beetle (Kuhar *et al.* 2006b, Kuhar and Doughty 2009, Sewell and Alyokhin 2009, 2010a). Novaluron will also control European corn borer, Ostrinia nubilalis Huebner, (Kuhar *et al.* 2006b).

Cyromazine, a triazine, is also a potent chitin synthesis inhibitor (Ware and Whitacre 2004). It is selective toward dipterous insects, and is used for the control of leafminers and root maggots. However, the insecticide has also been shown to provide effective control of Colorado potato beetle larvae (Sirota and Grafius 1994, Linduska *et al.* 1996).

Indoxacarb is a broad-spectrum insecticide belonging to the oxadiazine class of insecticides. Indoxacarb is a voltage-dependent sodium channel blocker (Wing *et al.* 2000) that is efficacious against most lepidopteran pests. Indoxacarb alone provides moderate control of Colorado potato beetle as well as potato leafhopper (Linduska et al. 2002, Davis et al. 2003, Kuhar and Speese 2005c), but when tank-mixed with the synergist piperonyl butoxide it is highly efficacious against those pests (Linduska et al. 2002, Sewell and Alyokhin 2003, 2006).

Metaflumizone also belongs to the oxadiazine class of insecticides. It is currently not registered for use in the US. Field efficacy trials have shown that it provides excellent control of Colorado potato beetle (Kuhar *et al.* 2006a, Sewell and Alyokhin 2009, 2010b).

Chlorantraniliprole is a novel insecticide from a relatively new class of chemistry, the anthranilic diamides, which activate the insect ryanodine receptors affecting calcium release during muscle contraction (Cordova *et al.* 2006). Insects treated with chlorantraniliprole exhibit rapid feeding cessation, lethargy, regurgitation, muscle paralysis, and ultimately death (Hannig *et al.* 2009). Chlorantraniliprole has demonstrated tremendous efficacy against a variety of lepidopteran pests, whiteflies, and beetles. It has demonstrated excellent efficacy against Colorado potato beetle in the field (Kuhar and Doughty 2009, 2010, Sewell and Alyokhin 2009). Chlorantraniliprole is also xylem-mobile for root uptake, providing systemic control of Colorado potato beetle (Groves *et al.* 2011b, Sewell and Alyokhin 2011).

Research is also currently being conducted on another diamide insecticide, cyantraniliprole, which has also shown tremendous efficacy against Colorado potato beetle as both a foliar and a pre-plant or at-planting (systemic) application (Sewell and Alyokhin 2009, 2011, Kuhar and Doughty 2010, Groves *et al.* 2011b).

In 1996 the pyrazole (or phenoxybenzylamide) insecticide **tolfenpyrad** was discovered by Mitsubishi Chemical Corporation (now the Nihon Nohyaku Co. Ltd), and was registered as a broad-spectrum insecticide in Japan in 2002. Pyrazole pesticides (IRAC Group 21) are respiratory poisons that inhibit mitochondrial electron transport at the NADH-CoQ reductase site, leading to the disruption of adenosine triphosphate (ATP) formation (Ware and Whitacre 2004). Until recently, there was little knowledge or development of the insecticide in the US. Nichino America is currently developing tolfenpyrad for use in agricultural markets in the US. In Virginia (USA), leaf-dip bioassays indicated that tolfenpyrad was very toxic to Colorado potato beetle larvae and adults, and that significant control of this pest in the field could be achieved with rates as low as 0.15 kg ai/ha (TPK, unpublished data). Sewell and Alyokhin (2011) also achieved effective control of Colorado potato beetle in the field with tolfenpyrad. The insecticide is also efficacious against potato leafhopper and potato aphid. If registered, tolfenpyrad should offer yet another new mode of action for IRM of Colorado potato beetle and other pests.

CHEMICAL CONTROL OF HEMIPTERAN PESTS

Potatoes are attacked by several phloem-feeding hemipteran pests. In North America, the major hemipteran pests include the potato psyllid, *Bactericerca cockerelli* (Sulc) (Hemiptera: Triozidae), potato leafhopper, and various aphids (Hemiptera: Aphidae), including the green peach aphid, *Myzus persicae*, potato aphid, *Macrosiphum euphorbiae* (Thomas), and buckthorn aphid, *Aphis nasturtii* (Kaltenbach). Moreover, in other parts of the world potatoes are also attacked by the cotton aphid, *Aphis gossypii* (Glover) and the leafhopper, *Amrasca biguttula biguttula* (Ishida) (Kumar *et al.* 2011).

Feeding by potato psyllids causes yellowing of the leaves, referred to as "psyllid yellows", stunting, leaf curling, and yield loss (Wallis 1995, Gharalari *et al.* 2009). In addition, this species transmits the pathogen causing zebra chip syndrome (Muyaneza et al. 2007 Gao et al. 2009, Secor et al. 2009). Potato leafhopper nymphs and adults feed on leaves and stems and secrete a salivary toxin into the plant, which causes cellular abnormalities that result in "hopperburn" and subsequent yield loss (Backus and Hunter 1989). Aphids are considered important pests of potato primarily because of their role as vectors of viruses to seed potatoes; the two most important viruses transmitted to potato by aphids are potato leafroll virus (PLRV) and Potato virus Y (PVY or Mosaic) (see Chapter 10 for more details).

Often the insecticides that are applied to potato for other pests, such as the Colorado potato beetle will also provide control of most hemipteran pests. Most insecticides targeted specifically at hemipteran pests are applied in an effort to control or reduce viral infections. Because these viruses spread rapidly, prevention by use of insecticides is difficult. Nevertheless, some insecticides will help to slow the spread of viruses (Collar *et al.* 1997). While thresholds exist for most hemipteran pests (Sexson *et al.* 2005, Goolsby *et al.* 2007), the concept is not widely used for these pests, particularly in potato seed production, where most growers apply a systemic insecticide at planting (Sexson *et al.* 2005). For decades, carbamates or organophosphates such as aldicarb, phorate, and disulfoton have provided effective early season control of hemipteran pests, and they continue to be used today, particularly in the Pacific Northwest potato-growing region of the US (Gerhardt and Turley 1961, Harding 1962, Gerhardt 1966, Cranshaw 1997). Neonicotinoids such as imidacloprid, thiamethoxam, and clothianidin also provide excellent control of hemipteran pests as systemic or contact insecticides (Boiteau *et al.* 1997, Pavlista 2002, Kuhar *et al.* 2003a, Liu and Trumble 2005, Kund *et al.* 2006, Kuhar and Doughty 2010, Groves *et al.* 2011b). More recently, Kumar *et al.* (2011) developed controlled-release formulations of carbofuran and imidacloprid that were found to control the aphid, *A. gossypii*, and leafhopper, *A. biguttula biguttula*, better than commercial formulations.

If foliar insecticide applications are needed, many pyrethroid, organophosphate, and carbamate insecticides will provide rapid knockdown of hemipteran pests (Sexson *et al.* 2005, Berry *et al.* 2009) and can reduce probing by aphids, which will slow the spread of viruses (Dewar 2007). However, because insecticide resistance to carbamates, organophosphates, and pyrethroids has emerged in several of the key hemipteran pest species, most notably green peach aphid (Foster *et al.* 2007, Castañeda *et al.* 2011), combining or rotating insecticides that have different modes of action is strongly recommended. Berry *et al.* (2009) showed that abamectin, azadirachtin, and thiacloprid were highly toxic to potato psyllid, resulting in almost 100% mortality after 48 hours. Gharalari *et al.* (2009) also showed that abamectin was highly efficacious against potato psyllid and suggested that the translaminar activity of the insecticide enables it to perform better against sucking insects that may be difficult to reach with spray on undersides of leaves. Numerous insecticides have been proven to work in controlling aphid pests as well. Nevertheless, the push for safer insecticides has stimulated the development of several novel insecticides specifically designed to control hemipteran pests. These novel insecticides include pymetrozine, flonicamid, and spirotetramat.

Pymetrozine belongs to the pyridine-azomethine class of chemicals. This insecticide is active on a number of different hemipteran pests, such as aphids, whiteflies, and leafhoppers. The mode of action of pymetrozine interferes with the regulatory mechanism of food intake (Kristinsson 1995). It has shown a high efficacy against aphids (Sewell and Alyokhin 2010b), but has provided only marginal control of potato psyllid in field trials (Russell *et al.* 2001, Liu and Trumble 2005).

Flonicamid is an insecticide belonging to the pyridinecarboxamide class of chemicals. Flonicamid is a novel systemic compound with activity on hempiteran pests such as aphids, whiteflies, and thysanopteran pests. The mode of action of this compound is that it inhibits the feeding capabilities of sucking insects; this inability to feed is observed until death (Morita *et al.* 2007).

Spirotetremat is a novel insecticide derived from spirocyclic tetramic acid, and inhibits lipid biosynthesis. This compound is a two-way systemic (ambimobile), meaning it can be applied to the root zone or foliage and move systemically to offer complete plant protection. It has activity on a number of pests, such as aphids, psyllids, scale insects, mealy bugs, and whiteflies (Bretschneider *et al.* 2007, Nauen *et al.* 2008, Sewell and Alyokhin 2010b).

CHEMICAL CONTROL OF WIREWORMS

Many species of wireworms (Coleoptera: Elateridae) can be serious pests of potato throughout the world, and chemical control of these subterranean pests is difficult (Hancock *et al.* 1986, Kwon *et al.* 1999, Parker and Howard 2001, Kuhar *et al.* 2003b). Unlike the chemical control advancements that have been made for many of the other pests mentioned in this chapter, relatively little progress has been made in recent years in the control of wireworms. To be most effective, it is recommended that insecticides be incorporated into the soil prior to planting to reach their target (Thomas *et al.* 1983), but also be very persistent in the soil to ensure adequate protection of tubers late in the season (Parker and Howard 2001). In the 1950s, soil-applied organochlorine insecticides such as DDT and aldrin became the standard treatment for wireworms in many parts of the world (Merrill 1952, Gunning and Forrester 1984, Parker and Howard 2001). However, widespread concerns over the environmental impact of organochlorine insecticides led to the removal of these chemicals from agricultural use (Ware and Whitacre 2004). Organophosphate and carbamate insecticides have served as replacements for the organochlorines in potatoes for more than 30 years in the US and Europe (Edwards and Thompson 1971, Parker *et al.* 1990, Kuhar *et al.* 2003b). These chemicals, which include aldicarb, bendiocarb, carbofuran, carbosulfan, chlorpyrifos, diazinon, disulfoton, ethoprophos, fonofos, phorate, and phosmet, have not always provided effective or consistent wireworm control in potatoes (Hancock *et al.* 1986, Toba 1987, Jansson *et al.* 1988, Noetzel and Ricard 1988, Parker *et al.* 1990, Sorensen and Kidd 1991, Pavlista 1997, Shamiyeh *et al.* 1999, Nault and Speese 2000, Kuhar *et al.* 2003b). All of the aforementioned insecticides are relatively toxic to humans and the environment, and consequently several are no longer registered or used on potatoes in the US or other countries.

Fipronil is a phenylpyrazole insecticide that was registered for use on potatoes in the US in the mid-2000s. Fipronil blocks the gamma-aminobutyric acid (GABA)-regulated chloride channel in neurons antagonizing the "calming" effects of GABA, similar to the action of the cyclodienes. Fipronil is a systemic material with contact and stomach activity. Although it has been shown to be efficacious as a foliar insecticide on Colorado potato beetle (Moffat 1993, Noetzel and Holder 1996, Linduska *et al.* 1999, Nault and Speese 1999b) and as a systemic material for control of European corn borer (Nault and Speese 1999a, Kuhar *et al.* 2010), the primary targets for this insecticide in agriculture are soil pests such as wireworms. Kuhar and Alvarez (2008) showed that fipronil, the pyrethroid bifenthrin, and the neonicotinoids imidacloprid and thiamethoxam applied to the soil at-planting provided similar control of wireworms (50–80%) to that of the organophosphate standards of phorate and ethoprop. In that study, combinations of imidacloprid or thiamethoxam with fipronil or bifenthrin did not enhance the efficacy of any one compound used alone. Other researchers have also shown effective wireworm control in potatoes with these insecticides (DeVries and Wright 2005). The aforementioned products are all currently registered for use on potatoes, and provide much needed alternative insecticides for wireworm control in potato.

CHEMICAL CONTROL OF POTATO TUBERWORM

Potato tuberworm, *Phthorimaea operculella* (Zeller) (Lepidoptera: Gelechiidae), is the primary pest of potato throughout tropical and subtropical regions of the world. It is also a pest in the Pacific Northwest of the US. This pest is discussed further in Chapter 6, and an excellent review of its biology, ecology, and management can be found in Rondon (2010).

Chemical control of the potato tuberworm has posed a challenge for potato growers because eggs can be deposited on tubers after they are harvested (Rondon 2010) and because insecticide efficacy on this pest has been unpredictable (von Arx *et al.* 1987, Berlinger 1992). In the US, resistance to the pyrethroid esfenvalerate and the phenylpyrazole fipronil was documented in 2005 from field-collected potato tuberworms from the Columbia Basin in the Pacific Northwest. Resistance to the organophosphate methamidophos was not detected in these strains (Doframaci and Tingey 2007).

Recently, Clough et al. (2010) found that rotations of esfenvalerate and indoxacarb applications before and at vine kill were effective at reducing potato tuberworm damage. Those application timings are critical for effective control (Clough *et al.* 2008, 2010, Rondon 2010). Also, in addition to chemical control, immediate harvest of potatoes after vine kill is strongly recommended because the risk of potato tuberworm damage increases if the tubers are left in the field (Rondon 2010).

Researchers have determined that during the daytime the adult potato tuberworm moths rest on the bottom of potato leaves, becoming more active during the evening (Gubbaiah and Thontadarya 1977). Therefore, insecticide applications should coordinate with the evening peak of insect activity.

Recently, the insecticides chlorantraniliprole and spinetoram have offered a more targeted and IPM-friendly option for lepidopteran control in potatoes. They have shown very good efficacy on potato tuberworm (Lawrence 2009, Dobie 2010). These insecticides can also be efficacious at controlling other lepidopteran pests, such as beet armyworm, *Spodoptera exigua* (Hubner) (Kund *et al.* 2011, Natwick and Lopes 2011, Palumbo 2011), and European corn borer (Sewell and Alyokhin 2007, Kuhar *et al.* 2011). These new narrow-spectrum insecticides are safer for the environment and less disruptive to natural enemies.

FINAL THOUGHTS

Chemical control remains the most widely used strategy for eliminating potato damage by pests, and will likely remain the base of pest management for the foreseeable future (Alyokhin 2009). However, a more sustainable and responsible approach to chemical control is possible; one that avoids past mistakes and uses insecticides efficiently, with a better understanding of the pest's biology, and as part of an integrated pest management program. A number of novel insecticide chemistries have been registered in recent years and many more are in development. These insecticides will undoubtedly be safer for the user, have less of an impact on beneficial insects, and therefore be able to fit better into potato IPM programs. If chemical control is truly needed for a given pest situation on potatoes, it should be the job of entomologists and crop consultants to recommend the use of these new chemical tools over the more disruptive, broad-spectrum insecticides. Also, frequent rotation of insecticide classes should minimize insecticide resistance development in Colorado potato beetle and other pests.

REFERENCES

Abudulai, M., Shepard, B.M., Mitchell, P.L., 2003. Antifeedant and toxic effects of a neem (*Azadirachta indica* A. Juss) based formulation Neemix against *Nezara viridula* (L.) (Hemiptera: Pentatomidae). J. Entomol. Sci. 38, 398–408.

Alyokhin, A., 2009. Colorado potato beetle management on potatoes: Current challenges and future prospects. Fruit. Vegetable and Cereal Science and Biotechnology 3 (Special Issue 1), 10–19.

Alyokhin, A., 2011. Insecticide resistance in the Colorado potato beetle. PotatoBeetle.Org. Retrieved December 10, 2011, from http://resistance.potatobeetle.org/management.html.

Alyokhin, A., Dively, G., Patterson, M., Mahoney, M., Rogers, D., Wollam, J., 2006. Susceptibility of imidacloprid-resistant Colorado potato beetle to non-neoneonicotinoid insecticides in the laboratory and field trials. Am.. J. Potato Res. 83, 485–494.

Alyokhin, A., Baker, M., Mota-Sanchez, D., Dively, G., Grafius, E., 2008a. Colorado potato beetle resistance to insecticides. Am.. J. Potato Res. 85, 395–413.

Alyokhin, A., Sewell, G., Choban, R., 2008b. Reduced viability of Colorado potato beetle, Leptinotarsa decemlineata, eggs exposed to novaluron. Pest Manag. Sci. 64, 94–99.

Alyokhin, A., Guillemette, R., Choban, R., 2010. Stimulatory and suppressive effects of novaluron on the Colorado potato beetle reproduction. J. Econ. Entomol. 102, 2078–2083.

Anhalt, J., Moorman, T.B., Koskinen, W.C., 2007. Biodegradation of imidacloprid by an isolated soil microorganism. J. Environ. Sci. Health 42, 509–514.

Argentine, J.A., Zhu, K.Y., Lee, S.H., Clark, J.M., 1994. Biochemical mechanisms of azinphos-methyl resistance in isogenic strains of Colorado potato beetle. Pestic. Biochem. Physiol. 48, 63–78.

Backus, E.A., Hunter, W.B., 1989. Comparison of feeding behavior of the potato leafhopper, *Empoasca fabae* (Homoptera: Cicadellidae) on alfalfa and broad beans leaves. Environ. Entomol. 18, 473–480.

Berlinger, M.J., 1992. Pests of processing tomatoes in Israel and suggested IPM model. Acta Hort 301, 185–192.

Berry, N.A., Walker, M.K., Butler, R.C., 2009. Laboratory studies to determine the efficacy of selected insecticides on tomato/potato psyllid. NZ Plant Prot. 62, 145–151.

Bloomquist, J.R., 1996. Ion channels as targets for insecticides. Annu. Rev. Entomol. 41, 163–190.

Bloomquist, J.R., 2003. Chloride channels as tools for developing selective insecticides. Arch. Insect Biochem. Physiol. 54, 145–146.

Boiteau, G., Osborn, W.P.L., Drew, M.E., 1997. Residual activity of imidacloprid controlling Colorado potato beetle (Coleoptera: Chrysomelidae) and three species of potato colonizing aphids (Homoptera: Aphidae). J. Econ. Entomol. 90, 309–319.

Bretschneider, T., Fischer, R., Nauen, R., 2007. Inhibitors of lipid synthesis (acetyl-CoA-carboxylase inhibitors). In: Kramer, W., Schirmer, U. (Eds.), Modern Crop Protection Compounds, Wiley-VCH, Weinheim, Germany, pp. 909–925.

Brewer, M.J., Trumble, J.T., 1994. Beet armyworm resistance to fenvalerate and methomyl: resistance variation and insecticide synergism. J. Agric. Entomol. 11, 291–300.

Brown, A.W.A., 1951. Chemical control of insects feeding on plants. Insect Control by Chemicals, John Wiley & Sons, New York, NY, pp. 574–667.

Byrne, A.M., Grafius, E., Pett, W., 2006. Colorado potato beetle control 2005. Arthropod Manage. Tests 31, E54.

Campbell, W.C., 1989. "Ivermectin and Abamectin." Springer-Verlag, New York, NY.

Casagrande, R.A., 1987. The Colorado potato beetle: 125 years of mismanagement. Bull. Entomol. Soc. Am. 33, 142–150.

Castañeda, L.E., Barrientos, K., Cortes, P.A., Figueroa, C.C., Fuentes-Contreras, E., Luna-Rudloff, M., Silva, A.X., Bacigalupe, L.D., 2011. Evaluating reproductive fitness and metabolic costs for insecticide resistance in *Myzus persicae* from Chile. Physiol. Entomol. 36, 253–260.

Churchel, M.A., Hanula, J.L., Berisford, C.W., Vose, J.M., Dalusky, M.J., 2011. Impact of imidacloprid for control of hemlock wooly adlegid on nearby aquatic macroinvertebrate assemblages. Journal of Applied Forestry 35, 26–32.

Clough, G., DeBano, S., Rondon, S., David, N., Hamm, P., 2008. Use of cultural and chemical practices to reduce tuber damage from the potato tuberworm in the Columbia Basin. Hortscience 43, 1159–1160.

Clough, G.H., Rondon, S.I., DeBano, S.J., David, N., Hamm, P.B., 2010. Reducing tuber damage by Potato tuberworm (Lepidoptera: Gelechiidae) with cultural practices and insecticides. J. Econ. Entomol. 103, 1306–1311.

Collar, J.L., Avilla, C., Duque, M., Fereres, A., 1997. Behavioral response and virus vector ability of *Myzus persicae* (Homoptera: Aphididae) probing on pepper plants treated with aphicides. J. Econ. Entomol. 90, 1628–1634.

Cordova, D., Benner, E.A., Sacher, M.D., Rauh, J.J., Sopa, J.S., Lahm, G.P., 2006. Anthranilic diamides: a new class of insecticides with a novel mode of action, ryanodine receptor activation. Pestic. Biochem. Physiol. 84, 196–214.

Cox, L., Koskinen, W.C., Yen, P.Y., 1997. Sorption–desorption of imidacloprid and its metabolites in soils. J. Agric. Food Chem. 45, 1468–1472.

Cox, L., Koskinen, W.C., Yen, P.Y., 1998. Changes in sorption of imidacloprid with incubation time. Soil Sci. Soc. Am. J. 62, 342–347.

Cranshaw, W.S., 1997. The potato (tomato) psyllid, *Paratrioza cackerelli* (Sulc) as a pest of potatoes. 83–95 In: Zehnder, G.W., Powelson, M.L., Jansson, R.K., Raman, K.V. (Eds.), Advances in potato pest management, American Phytopathological Society, St. Paul, MN.

Cutler, G.C., Scott-Dupree, C.D., Tolman, J.H., Harris, C.R., 2007. Field efficacy of novaluron for control of Colorado potato beetle (Coleoptera: Chrysomelidae) on potato. Crop Prot. 26, 760–767.

Davis, J.A., Radcliff, E.B., Ragsdale, D.W., 2003. Control of potato leafhopper on potatoes using foliar insecticides 2002. Arthropod Manage. Tests 28, E52.

DeVries, T., Wright, R.J., 2005. Larval wireworm control 2004. Arthropod Manage. Tests 29, F17.

Dewar, A.M., 2007. Chemical control. 391–422 In: van Emden, H.F., Harrington, R. (Eds.), Aphids as Crop Pests, CABI, Wallingford, UK.

Dittbrenner, N., Isabelle, M., Triebskorn, R., Capowiez, Y., 2011a. Assessment of short and long-term effects of imidacloprid on the burrowing behaviour of two earthworm species (Aporrectodea caliginosa and Lumbricus terrestris) by using 2D and 3D post-exposure techniques. Chemosphere 84, 1349–1355.

Dittbrenner, N., Schmitt, H., Capowiez, Y., Triebskorn, R., 2011b. Sensitivity of *Eisenia fetida* in comparison to *Aporrectodea caliginosa* and *Lumbricus terrestris* after imidacloprid exposure, Body mass change and histopathology. J. Soils Sediments 11, 1000–1010.

Dobie, C.H., 2010. Pesticide susceptibility of potato tuberworm in the Pacific Northwest. MS Thesis. Washington State University, Pullman, WA. www.dissertations.wsu.edu/Thesis/Spring2010/C_Dobie_122109.pdf.

Doframaci, M., Tingey, W.M., 2007. Comparison of insecticide resistance in a North American field population and a laboratory colony of potato tuberworm (Lepidoptera: Gelechiidae). J. Pest Sci. 81, 17–22.

Edwards, C.A., Thompson, A.R., 1971. Control of wireworms with organophosphorus and carbamate insecticides. Pestic. Sci. 2, 185–189.

Ford, K.A., Casida, J.E., Chandran, D., 2010. Neonicotinoid insecticides induce salicylate-associated plant defense responses. Proc. Natl. Acad. Sci. 107, 12527–17532.

Forgash, A.J., 1985. Insecticide resistance in the Colorado potato beetle. Research Bulletin Massachussets Agricultural Experiment Station, 33–52.

Foster, S.P., Devine, G., Devonshire, A.L., 2007. Insecticide resistance. In: van Emden, H.F., Harrington, R. (Eds.), Aphids as Crop Pests, CABI, Wallingford, UK, pp. 261–285.

French II, N.M., Heim, D.C., Kennedy, G.G., 1992. Insecticide resistance patterns among Colorado potato beetle, *Leptinotarsa decemlineata* (Say) (Coleoptera: Chrysomelidae), populations in North Carolina. Pestic. Sci. 36, 1091.

Gao, F., Jifon, J., Yang, X., Liu, T., 2009. Zebra chip disease incidence on potato is influenced by timing of potato psyllid infestation, but not by the host plants on which they were reared. Insect Sci. 16, 399–408.

Gauthier, N.L., Hofmaster, R.N., Semel, M., 1981. History of Colorado potato beetle control. In: Lashomb, J.H., Casagrande, R.A. (Eds.), Advances in potato pest management, Hutchinson Ross, Stroudsburg, PA, pp. 13–23.

Gerhardt, P.D., 1966. Potato psyllid and green peach aphid control on Kennebec potatoes with Temik and other insecticides. J. Econ. Entomol. 59, 9–11.

Gerhardt, P.D., Turley, D.L., 1961. Control of certain potato insects in Arizona with soil applications of granulated phorate. J. Econ. Entomol. 54, 1217–1221.

Gharalari, A.H., Nansen, C., Lawson, D.S., Gilley, J., Munyaneza, J.E., Vaughn, K., 2009. Knockdown mortality, repellency, and residual effects of insecticides for control of adult *Bactericera cockerelli* (Hemiptera: Psyllidae). J. Econ. Entomol. 102, 1032–1038.

Ghidiu, G.M., Zehnder, G.W., 1993. Timing of the initial spray application of *Bacillus thuringiensis* for control of the Colorado potato beetle (Coleoptera: Chrysomelidae) in potatoes. Biol. Control 3, 348–352.

Goolsby, J.A., Adamczyk, J., Bextine, B., Lin, D., Munyaneza, J.E., Bester, G., 2007. Development of an IPM program for management of the potato psyllid to reduce incidence of zebra chip disorder in potatoes. Subtrop. Plant Sci. 59, 85–94.

Grafius, E.J., Bishop, B.A., 1996. Resistance to imidacloprid in Colorado potato beetles from Michigan. Resistant Pest Management Newsletter 8, 21–25.

Graham, J.H., Myers, M.E., 2011. Soil application of SAR inducers imidacloprid, thiamethoxam, and acibenzolar-S-methyl for citrus canker control in young grapefruit trees. Plant Dis. 95, 725–728.

Groves, R.L., Chapman, S., Schramm, S., 2011a. Registered and experimental foliar insecticides to control Colorado potato beetle and potato leafhopper in potato 2010. Arthropod Manage. Tests 36, E60.

Groves, R.L., Chapman, S., Schramm, S., 2011b. Evaluations of systemic insecticides for the control of Colorado potato beetle, potato leafhopper and aphids in potato 2010. Arthropod Manage. Tests 36, E61.

Gubbaiah, Thontadarya, T.S., 1977. Bionomics of potato tuberworm, *Gnorimoschema Operculella* Zeller (Lepidoptera Gelechiidae) in Karnataka. Mysore. J. Agric. Sci. 11, 380–386.

Gunning, R.V., Forrester, N.W., 1984. Cyclodiene lindane resistance in *Agrypnus variabilis* (Candeze) (Coleoptera: Elateridae) in northern New South Wales. J. Aust. Entomol. Soc. 23, 247–248.

Hancock, M., Green, D., Lane, A., Mathias, P.L., Port, C.M., Tones, S.J., 1986. Evaluation of insecticides to replace aldrin for the control of wireworms on potatoes. Ann. Appl. Biol. 108 (Suppl), 28–29.

Hannig, G.T., Ziegler, M., Marcon, P.G., 2009. Feeding cessation effects of chlorantraniliprole, a new anthranilic diamide insecticide, in comparison with several insecticides in distinct chemical classes and mode-of-action groups. Pest Manag. Sci. 65, 969–974.

Harding, J.A., 1962. Tests with systemic insecticides for control of insects and certain diseases on potatoes. J. Econ. Entomol. 55, 62–64.

Harris, C.R., Turnbull, S.A., 1986. Contact toxicity of some pyrethroid insecticides, alone and in combination with piperonyl butoxide, to insecticide-susceptibile and pyrethroid-resistant strains of the Colorado potato beetle (Coleoptera: Chrysomelidae). Can. Entomol. 118, 1173–1176.

Hofmaster, R.N., 1956. Resistance of the Colorado potato beetle to DDT. Vegetable Growers News 10, 3–4.

Ioannidis, P.M., Grafius, E., Whalon, M.E., 1991. Patterns of insecticide resistance to azinphosmethyl, carbofuran, and permethrin in the Colorado potato beetle (Coleoptera: Chrysomelidae). J. Econ. Entomol. 84, 1417–1423.

Ishaaya, I., Kontsedalov, S., Horowitz, A.R., 2003. Novaluron (Rimon), a novel IGR: Potency and cross-resistance. Arch. Insect Biochem. Physiol. 54, 157–164.

Jansson, R.K., Lecrone, S.H., Tyson, R., Daigle Jr., C.F., 1988. Wireworm control on potato, 1987. Insecticide and Acaricide Tests 13, 145.

Jemec, A., Tisler, T., Drobne, D., Sepcic, K., Fournier, D., Trebse, P., 2007. Comparative toxicity of imidacloprid, of its commercial liquid formulation and of diazinon to a non-target arthropod, the microcrustacean Daphnia magna. Chemosphere 68, 1408–1418.

Karmakar, R., Kulshrstha, G., 2009. Persistence, metabolism and safety evaluation of thiamethoxam in tomato crop. Pest Manag. Sci. 65, 931–937.

Kerns, D.L., Palumbo, J.C., Tellez, T., 1998. Resistance of field strains of beet armyworm (Lepidoptera: Noctuidae) from Arizona and California to carbamate insecticides. J. Econ. Entomol. 91, 1038–1043.

Kowalska, J., 2007. Azadirachtin as a product for control of Colorado beetles. J. Res. Appl. Agric. Eng. 52, 78–81.

Kristinsson, H., 1995. Pyridine-azomethines – a novel class of insecticides with a new mode of action. Opportunities for pymetrozine. Agro Food Industry Hi-Tech. 6, 21–26.

Kuhar, T.P., Alvarez, J.M., 2008. Timing of injury and efficacy of soil-applied insecticides against wireworms on potato in Virginia. Crop Prot. 27, 792–798.

Kuhar, T.P., Doughty, H.B., 2009. Evaluation of foliar insecticides for the control of Colorado potato beetle in potato 2008. Arthropod Manage. Tests 34, E71.

Kuhar, T.P., Doughty, H.B., 2010. Evaluation of foliar insecticides for the control of Colorado potato beetle and potato leafhopper in potatoes in Virginia 2009. Arthropod Manage. Tests 35, E17.

Kuhar, T.P., Speese, J., 2005a. Evaluation of seed-piece treatment insecticides in potatoes 2004. Arthropod Manage. Tests 30, E64.

Kuhar, T.P., Speese, J., 2005b. Evaluation of soil-applied insecticides in potatoes 2004. Arthropod Manage. Tests 30, E65.

Kuhar, T.P., Speese, J., 2005c. Evaluation of foliar insecticides in potatoes 2004. Arthropod Manage. Tests 30, E63.

Kuhar, T.P., Speese, J., Barlow, V.M., Cordero, R.J., Venkata, R.Y., 2003a. Evaluation of in-furrow and seed piece insecticides for controlling insects in potato 2002. Arthropod Manage. Tests 28, E58.

Kuhar, T.P., Speese, J., Whalen, J., Alvarez, J.M., Alyokhin, A., Ghidiu, G., Spellman, M.R., 2003b. Current status of insecticidal control of wireworms in potatoes. Pesticide Outlook 14, 265–267.

Kuhar, T.P., Hitchner, E.M., Chapman, A.V., 2006a. Evaluation of foliar insecticides on potatoes 2005. Arthropod Manage. Tests 31, E56.

Kuhar, T.P., Hitchner, E.M., Chapman, A.V., 2006b. Evaluation of Rimon 0.83EC for control of insect pests on potatoes 2005. Arthropod Manage. Tests 31, E57.

Kuhar, T.P., Doughty, H., Hitchner, E., Chapman, A., Cassell, M., 2007. Evaluation of seed-piece insecticide treatments on potatoes 2006. Arthropod Manage. Tests 32, E37.

Kuhar, T.P., Ghidiu, G., Doughty, H.B., 2010. Decline of European corn borer as a pest of potatoes. Online. Plant Health Progress. doi:10.1094/PHP-2010-0129-01-PS.

Kuhar, T.P., Schultz, P., Doughty, H., Wimer, A., Andrews, H., Philips, C., Cassel, M., Jenrette, J., 2011. Evaluation of foliar insecticides for the control of lepidopteran larvae in bell peppers in Virginia 2010. Arthropod Manage. Tests 36, E54.

Kumar, J., Shakil, N.A., Kahn, M.A., Malik, K., Walia, S., 2011. Development of controlled release formulations of carbofuran and imidacloprid and their bioefficacy evaluation against aphid, *Aphis gossypii* and leafhopper, *Amrasca biguttula* biguttula Ishida on potato crop. J. Environ. Sci. Health, Part B. 46, 678–682.

Kund, G.S., Carson, W.G., Trumble, J.T., 2006. Effect of insecticides on pepper insects 2005. Arthropod Manage. Tests 31, E44.

Kund, G.S., Carson, W.G., Trumble, J.T., 2011. Effect of insecticides on celery insects 2009. Arthropod Manage. Tests 36, E27.

Kunkel, B.A., Held, D.W., Potter, D.A., 2001. Lethal and sublethal effects of bendiocarb, halfenozide, and imidacloprid on *Harpalus pennsylvanicus* (Coleoptera: Carabidae) following different modes of exposure in turfgrass. J. Econ. Entomol. 94, 60–67.

Kwon, M.Y., Hahm, I., Shin, K.Y., Ahn, Y.J., 1999. Evaluation of various potato cultivars for resistance to wireworms (Coleoptera: Elateridae). Am. J. Potato Res. 76, 317–319.

Lawrence, J.L., 2009. Damage relationships and control of the Tobacco splitworm (Gelechiidae: Phthorimaea operculella) in flue-cured tobacco. MS thesis. North Carolina State University, Raleigh, NC. http://repository.lib.ncsu.edu/ir/bitstream/1840.16/784/1/etd.pdf.

Linduska, J.J., Ross, M., Mulford, K., Baumann, D., 1996. Colorado potato beetle control on potatoes with foliar insecticide sprays, 1995. Arthropod Manage. Tests 21, E83.

Linduska, J.J., Ross, M., Abbott, B., Steele, S., Ross, E., Eastman, R., 2002. Colorado potato beetle control on potatoes 2001. Arthropod Manage. Tests 27, E69.

Liu, D., Trumble, J.T., 2005. Interactions of plant resistance and insecticides on the development and survival of *Bactericerca cockerelli* [Sulc](Homoptera: Psyllidae). Crop Prot. 24, 111–117.

Liu, W.P., Zheng, W., Ma, Y., Liu, K.K., 2006. Sorption and degradation of imidacloprid in soil and water. J. Environ. Sci. Health, Part B. 41, 623–634.

Liu, Z., Dai, Y., Huang, G., Gu, Y., Ni, J., Wei, H., Yuan, S., 2011. Soil microbial degradation of neonicotinoid insecticides imidacloprid, acetamiprid, thiacloprid and imidaclothiz and its effect on the persistence of bioefficacy against horsebean aphid *Aphis craccivora* Koch after soil application. Pest Manag. Sci. 67, 1245–1252.

Maienfisch, P., Angst, M., Brandl, F., Fischer, W., Hofer, D., Kayser, H., Kobel, W., Rindlisbacher, A., Senn, R., Steinemann, A., Widmer, H., 2001. Chemistry and biology of thiamethoxam: a second generation neoneonicotinoid. Pest Manag. Sci. 57, 906–913.

Marčića, D., Perić, P., Krasteva, L., Panayotov, N., 2009. Field evaluation of natural and synthetic insecticides against *Leptinotarsa decemlineata* Say. Acta Hortic 830, 391–396.

Merrill, L.G., 1952. Reduction of wireworm damage to potatoes. J. Econ. Entomol. 45, 548–549.

Metcalf, R.L., 1980. Changing role of insecticides in crop protection. Annu. Rev. Entomol. 25, 219–256.

Moffat, A.S., 1993. New chemicals seek to outwit insect pests. Science 5121, 550–551.

Mordue, A.J., Blackwell, A., 1993. Azadirachtin: an Update. J. Insect Physiol. 39, 903–924.

Morita, M., Ueda, T., Yoneda, T., Koyanagi, T., Haga, T., 2007. Flonicamid, a novel insecticide with a rapid inhibitory effect on aphid feeding. Pest Manag. Sci. 10, 969–973.

Mota-Sanchez, D., Whalon, M., Grafius, E., Hollingworth, R., 2000. Resistance of Colorado potato beetle to imidacloprid. Resistant Pest Management Newsletter 11, 31–34.

Mota-Sanchez, D., Hollingworth, R.M., Grafius, E.J., Moyer, D.D., 2006. Resistance and cross-resistance to neonicotinoid insecticides and spinosad in the Colorado potato beetle, *Leptinotarsa decemlineata* (Say) (Coleoptera: Chrysomelidae). Pest Manag. Sci. 62, 30–37.

Munyaneza, J.E., Crosslin, J.M., Upton, J.E., 2007. Association of *Bactericera cockerelli* (Homoptera: Psyllidae) with "zebra chip", a new potato disease in southwestern United States and Mexico. J. Econ. Entomol. 100, 656–663.

Natwick, E.T., Lopes, M.I., 2011. Anthranilic diamide worm control in romaine lettuce 2010. Arthropod Manage. Tests 36, E34.

Nauen, R., Reckmann, U., Thomzik, J., Thielert, W., 2008. Biological profile of spirotetramat (Movento) – a new two-way systemic (ambimobile) insecticide against sucking pest species. Bayer CropSci. J. 61, 245–278.

Nault, B.A., Speese, J., 1999a. Evaluation of Agri-mek for CPB control in potatoes, 1998. Arthropod Manage. Tests 24, E74.

Nault, B.A., Speese, J., 1999b. Evaluation of Agenda for control of ECB in potatoes, 1998. Arthropod Manage. Tests 24, E75.

Nault, B.A., Speese, J., 2000. Evaluation of insecticides to control soil insect pests on potatoes, 1999. Arthropod Manage. Tests 25, 148.

Noetzel, D.M., Holder, B., 1996. Control of resistant Colorado potato beetles Andover, MN 1994. Arthropod Manage. Tests 21, E87.

Noetzel, D., Ricard, M., 1988. Wireworm and white grub control in potato, 1987. Arthropod Manage. Tests 13, 160–161.

Oi, M., 1999. Time-dependent sorption of imidacloprid in two different soils. J. Agric. Food Chem. 47, 327–332.

Olson, E.R., Dively, G.P., Nelson, J.O., 2000. Baseline susceptibility to imidacloprid and cross resistance patterns in Colorado potato beetle (Coleoptera: Chrysomelidae) populations. J. Econ. Entomol. 93, 447–458.

Palumbo, J.C., 2011. Evaluation of new insecticides for control of lepidopterous larvae on head lettuce 2010. Arthropod Manage. Tests 36, E43.

Pandey, G., Dorrian, S.J., Russell, R.J., Oakeshott, J.G., 2009. Biotransformation of the neonicotinoid insecticides imidacloprid and thiamethoxam by Pseudomonas sp. 1G. Biochem. Biophys. Res. Commun. 380, 710–714.

Parker, W.E., Howard, J.J., 2001. The biology and management of wireworms (Agriotes spp.) on potato with particular reference to the UK. Agric. Forest Entomol. 3, 85–98.

Parker, W.E., Clarke, A., Ellis, S.A., Oakley, J.N., 1990. Evaluation of insecticides for control of wireworms (Agriotes spp.) on potato. Ann. Appl. Biol. 116 (Suppl), 28–29.

Pavlista, A.D., 1997. Mocap and Aztec on reducing wireworm damage to potato tubers, 1995. Arthropod Manag. Tests 22, 158.

Pavlista, A.D., 2002. Comparison of systemic insecticides applied in-furrow or as a seed dust for potato psyllid and leafhopper control on potato 2001. Arthropod Manag. Tests 27, E75.

Peck, D.C., 2009. Long-term effects of imidacloprid on the abundance of surface- and soil-active nontarget fauna in turf. Agric. Forest Entomol. 11, 405–419.

Penman, D.R., Chapman, R.B., 1988. Pesticide-induced mite outbreaks: pyrethroids and spider mites. Exp. Appl. Acarol. 4, 265–276.

Riley, C.V., 1871. Third annual report on the noxious, beneficial, and other insects of the state of Missouri. Horace Wilcox, Jefferson City, MO.

Rondon, S.I., 2010. The potato tuberworm: A literature review of its biology, ecology, and control. Am.. J. Potato Res. 87, 149–166.

Rose, R.L., Brindley, W.A., 1985. An evaluation of the role of oxidative enzymes in Colorado potato beetle resistance to carbamate insecticides. Pestic. Biochem. Physiol. 23, 74–84.

Roush, R.T., Hoy, C.W., Ferro, D.N., Tingey, W.M., 1990. Insecticide resistance in the Colorado potato beetle (Coleoptera: Chyrsomelidae): influence of crop rotation and insecticide use. J. Econ. Entomol. 83, 315–319.

Russell, J.S., Grichar, J., Besler, B., Brewer, K., 2001. Evaluation of selected insecticides against potato psyllids on potatoes 2000. Arthropod Manage. Tests 26, E62.

Salgado, V.L., 1998. Studies on the mode of action of spinosad: Insect symptoms and physiological correlates. Pestic. Biochem. Physiol. 2, 91–102.

Saran, R.J., Kamble, S.T., 2008. Concentration-dependent degradation of three termiticides in soil under laboratory conditions and their bioavailability to eastern subterranean termites (Isoptera: Rhinotermitidae). J. Econ. Entomol. 101, 1373–1383.

Scholz, K., Spiteller, M., 1992. Influence of ground cover on the degradation of 14C-imidacloprid in soil. Brighton Crop Prot. Conf. Pests Dis. 2, 883–888.

Secor, G.A., Rivera, V.V., Abad, J.A., Lee, I.M., Clover, G.R.G., 2009. Association of *"Candidatus Liberibacter solanacearum"* with zebra chip disease of potato established by graft and psyllid transmission, electron microscopy, and PCR. Plant Dis. 93, 574–583.

Sewell, G.H., Alyokhin, A., 2007. Control of European corn borer on potato 2006. Arthropod Manage. Tests 32, E42.

Sewell, G.H., Alyokhin, A., 2009. Control of Colorado potato beetle on potato 2008. Arthropod Manage. Tests 34, E52.

Sewell, G.H., Alyokhin, A., 2010a. Control of Colorado potato beetle on potato 2009. Arthropod Manage. Tests 35, E19.

Sewell, G.H., Alyokhin, A., 2010b. Control of aphids on Irish potato 2009. Arthropod Manage. Tests 35, E18.

Sexson, D.L., Wyman, J.A., Radcliffe, E.B., Hoy, C.W., Ragsdale, D.W., Dively, G., 2005. Chapter 5: Potato. In: Foster, R., Flood, B.R. (Eds.), Vegetable Insect Management, Meister Media Worldwide, Willoughby, OH, pp. 93–106.

Seymour, J., Bowman, G., Crouch, M., 1995. Effects of neem seed extract on feeding frequency of *Nezara viridula* L. (Hemiptera, Pentatomidae) on pecan nuts. J. Aust. Entomol. Soc. 34, 221–223.

Shamiyeh, N.B., Pereira, R., Straw, R.A., Follum, R.A., 1999. Control of wireworms in potatoes, 1998. Arthropod Manage. Tests 24, 164–165.

Sirota, J.M., Grafius, E., 1994. Effects of cyromazine on larval survival, pupation, and adult emergence of Colorado potato beetle (Coleoptera: Chrysomelidae). J. Econ. Entomol. 87, 577–582.

Slaney, A.C., Robbins, H.L., English, L., 1992. Mode of action of *Bacillus thuringiensis* toxin CryIIIA: An analysis of toxicity in *Leptinotarsa decemlineata* (Say) and *Diabrotica undecimpunctata* Howardi Barber. Insect Biochem. Mol. Biol. 22, 9–18.

Sorensen, K.A., Holloway, C.W., 1997. Colorado potato beetle and European corn borer control with insecticides, 1996. Arthropod Manage. Tests 22, E88.

Sorensen, K.A., Kidd, K.A., 1991. Wireworm control, Currituck County, 1989. Arthropod Manage. Tests 16, E106.

Sparks, T.C., Crouse, G.D., Dripps, J.E., Anzeveno, P., Martynow, J., DeAmicis, C.V., Thetford, L.C., 1993. Trigard a new tool for resistance management of the Colorado potato beetle. Resistant Pest Management Newsletter 5, 18–19.

Thomas, G.W., Keaster, A.J., Grundler, J.A., 1983. Control of wireworms and other corn soil insects. Agricultural Guide, 4154.

Thompson, G.D., Busacca, J.D., Jantz, O.K., Kirst, H.A., Larson, L.L., Sparks, T.C., 1995. An overview of new natural insect management systems. Proc. Beltwide Cotton Conf., 1039–1043 Spinosyns.

Thompson, G.D., Dutton, R., Sparks, T.C., 2000. Spinosad – a case study: an example from a natural products discovery programme. Pest Manag. Sci. 56, 696–702.

Tisler, A.M., Zehnder, G.W., 1990. Insecticide resistance in the Colorado potato beetle (Coleoptera: Chrysomelidae) on the eastern shore of Virginia. J. Econ. Entomol. 83, 666–671.

Tisler, A.M., Jemec, T.A., Mozetic, B., Trebse, P., 2009. Hazard identification of imidacloprid to aquatic environments. Chemosphere 76, 907–914.

Toba, H.H., 1987. Treatment regimens for insecticidal control of wireworms on potato. J. Agric. Entomol. 4, 207–212.

Tolman, J.H., Hilton, S.A., Whistlecraft, J.W., McNeil, J.R., 2005. Susceptibility to insecticides in representative Canadian populations of Colorado potato beetle, *Leptinotarsa decemlineata* (Say). Resistance Pest Management Newsletter 15, 22–25.

Trisyono, A., Whalon, M.E., 1999. Toxicity of neem applied alone and in combinations with *Bacillus thuringiensis* to Colorado potato beetle (Coleoptera: Chrysomelidae). J. Econ. Entomol. 92, 1281–1288.

United States Department of Agriculture–National Agricultural Statistics Service. (2010). Multistate insecticide applications on potatoes 2010. USDA-NASS Quick Stats Database. Retrieved December 10 2011, from http://quickstats.nass.usda.gov/

von Arx, R., Goueder, J., Cheikh, M., Bentemime, A., 1987. Integrated control of potato tubermoth *Phthorimaea operculella* (Zeller) in Tunisia. Insect Sci. Appl. 8, 989–994.

Wallis, R.L., 1995. Ecological studies on the potato psyllid as a pest of potatoes. USDA Technical Bulletin No. 1107. USDA, p. 25.

Ware, G.W., Whitacre, D.M., 2004. The Pesticide Book, 6th ed. Meister Media Worldwide, Willoughby, OH, p. 496.

Whalon, M.E., Miller, D.L., Hollingworth, R.M., Grafius, E.J., Miller, J.R., 1993. Selection of a Colorado potato beetle (Coleoptera: Chrysomelidae) strain resistant to *Bacillus thuringiensis*. J. Econ. Entomol. 86, 226–233.

Wierenga, J.M., Hollingworth, R.M., 1994. The role of metabolic enzymes in insecticide- resistant Colorado potato beetles. Pestic. Sci. 40, 259–264.

Wing, K.D., Sacher, M., Kagaya, Y., Tsurubuchi, Y., Mulderig, L., Connair, M., Schnee, M., 2000. Bioactivation and mode of action of the oxadiazine indoxacarb in insects. Crop Prot. 19, 537–545.

Zehnder, G., Warthen, D., 1988. Feeding inhibition and mortality effects of neem-seed extract on the Colorado potato beetle (Coleoptera: Chrysomelidae). J. Econ. Entomol. 81, 1040–1044.

Zhao, J.Z., Bishop, B.A., Grafius, E.J., 2000. Inheritance and synergism of resistance to imidacloprid in the Colorado potato beetle (Coleoptera: Chrysomelidae). J. Econ. Entomol. 93, 1508–1514.

Chapter 14

Biological Control of Potato Insect Pests

Donald C. Weber

Invasive Insect Biocontrol and Behavior Laboratory, USDA Agricultural Research Service, Beltsville, MD, USA

INTRODUCTION

Predators and parasitoids which actively seek out the pest have an enormous potential to suppress potato insect pests in the context of a truly integrated pest management approach, locally adapted to include essential cultural controls, pest thresholds, and a variety of compatible intervention tactics such as biopesticides, pheromone-based technologies, trap cropping, and selective insecticides. Here we discuss arthropod predators and parasitoids of major potato pests, with an emphasis on those pests more or less specific to the potato crop. This excludes specific discussion of such important groups as leafhoppers, scarab grubs, and wireworms, which are, nevertheless, the key pests in many potato fields. Microbial control agents, including viruses, bacteria, fungi, and entomopathogenic nematodes, are described in Chapter 16, as are natural products such as botanical and semiochemical preparations.

NATURAL ENEMIES OF MAJOR POTATO PESTS

Colorado Potato Beetle (Coleoptera: Chrysomelidae)

The Colorado potato beetle (CPB; *Leptinotarsa decemlineata* (Say)) now ranges throughout most of the North Temperate Zone, with the exception of the British Isles, most of China, and South and Southeast Asia. Since it is native to northern Mexico and the southwestern USA, almost all of its current range represents an expansion, and all of its cultivated hosts – principally potato, eggplant, and tomato – are also novel (Weber 2003). In addition to studies of natural enemies in North America and Eurasia, a number of classical and one neoclassical biological control introductions have been attempted against CPB.

Since the first review of natural enemies of Colorado potato beetle by Walsh and Riley (1868), only one major species of native predator in North America

Insect Pests of Potato. http://dx.doi.org/10.1016/B978-0-12-386895-4.00014-4

has been added to that early list. This section reviews primarily the research published since Ferro's (1994) review of the arthropod and microbial natural enemies. Soon after that date, the introduction of imidacloprid, which is a highly effective systemic neonicotinoid for control of CPB, decreased interest in its biological control in North America and elsewhere. This occurred despite pesticide resistance remaining an ever-present risk with this insect (see Chapter 2), and the repeated lesson that resistance risk is mitigated by a suite of alternative tactics such as crop rotation and other cultural practices, as well as microbial and biological controls (Alyokhin 2009). CPB resistance to neonicotinoids has since appeared in several growing areas; however, growers now have a number of other novel insecticide classes with which to suppress CPB (see Chapter 13).

The recent growth of organic potato culture has increased interest in non-chemical controls as part of sustainable CPB management. For instance, in the USA, certified organic potato plantings almost tripled to just over 1% of total potato acreage between 1997 and 2007, while the overall acreage decreased by 13% (ERS 2012). Furthermore, only a portion of those acres grown using organic techniques are certified organic. This trend, as well as the expansion of CPB to the east in Asia, where it is threatening to impact the crops of the two leading potato-growing nations, China and India, has also heightened interest in possibilities for classical biological control with the three species of CPB parasitoids commonly attacking the pest in North America.

CPB populations in North America are commonly preyed upon by several species of native and exotic lady beetles, a specialized carabid which is also a pupal parasitoid, a few species of asopine stink bugs, two species of tachinid parasitoids, and a variety of generalist arthropod predators, including predatory bugs of the genera *Orius*, *Geocoris*, and *Nabis*, Carabidae, Cantharidae, and Opiliones (Heimpel and Hough-Goldstein 1992, Ferro 1994, Hilbeck and Kennedy 1996).

Egg Parasitoids

Egg parasitoids are rare in North America. However, the eulophid *Edovum puttleri* Grissell (1981) was introduced each growing season to eggplant and also potato, from New Jersey Department of Agriculture insectary rearings, during the period 1981 to 1997 (Schroder and Athanas 1989, Tipping 1999). The population, derived from field collections in Colombia (Medellin and Palmira [Valle], between 1000 and 1500 m elevation), where it parasitizes a related species *L. undecimlineata* (Stål) (Grissell 1981), is not winter-hardy in North America (Schroder and Athanas 1989) or in Europe (Pucci and Dominici 1988). It is also ineffective in cooler weather in which potatoes are grown (Cloutier *et al.* 2002).

Van Driesche and Ferro (letter to B. Puttler, 1987), Acosta and O'Neil (1999), and O'Neil *et al.* (2005) made several explorations for CPB natural enemies, including cold-hardy *E. puttleri* in cooler habitats in Central America and Colombia. Also, Cappaert *et al.* (1991) reported *E. puttleri* on native CPB populations in the Mexican state of Morelos. These efforts did not discover strains better adapted to North American conditions. However, the higher elevations of Colombia are still not thoroughly explored.

E. puttleri is no longer available from any source in North America or Europe. Hamilton and Lashomb (1996) demonstrated that a program of biointensive pest management for New Jersey eggplant was economically viable, as it was apparently for the crop in Italy (Colazza and Bin 1992). Unfortunately, the lower value per acre of potato, as well as the parasitoid's poorer performance on this crop (Colazza and Bin 1992), led to its abandonment for commercial potatoes well before its demise in eggplant. In the warmer latitudes where it might better prosper in the potato crop, CPB is not much of a pest and/or potato is grown in the cooler portion of the year. For *E. puttleri* (at least the Colombian strain) there is a crop-parasitoid mismatch: the potato crop is grown under cooler conditions compared to eggplant, and the parasitoid may also, for other reasons, prefer eggplant to potato (Vasquez *et al.* 1997). The wild host from which it was collected, *Solanum torvum* Swartz, is very closely related to eggplant, as reflected in their interfertility, architecture, and foliar texture (Daunay 2008).

Parasitoids: Myiopharus *spp.*

Myiopharus aberrans and *M. doryphorae* are common tachinid parasitoids of CPB in North America. They both appear to be specialists on CPB. *M. doryphorae* was also recorded from sunflower leaf beetle, *Zygogramma exclamationis* (Fitch), in Canada, even though *M. macellus* (Reinhard) is much more common on this host beetle (Charlet 2003). López and Ferro (1990) and López *et al.* (1992) discovered that both of these species overwinter within the adult host, and emerge the subsequent spring. Both species larviposit into the CPB larvae, and many of the larvae laid by *M. doryphorae* from the second week of August onwards remain within the host, probably as a first instar. They develop within the overwintering adult and emerge in the spring (López *et al.* 1997a). Second and third CPB instars are the preferred hosts.

CPB larvae have a broad repertoire of defensive behaviors against the two tachinids (López *et al.* 1997b), suggesting co-evolution of CPB with this natural enemy. Ramirez *et al.* (2010) also found some similar, and some different, defensive responses to generalist predators in Washington state larval populations. Both tachinid species guard the parasitized host following successful larviposition during certain times, usually in late summer, when the preferred hosts are in short supply. This is the same period during which superparasitism may occur, even though, for most of the year, larvipositing flies do not parasitize previously parasitized hosts (López *et al.* 1995).

Unlike *M. doryphorae*, *M. aberrans* females shift in late season to larvipositing directly into adult CPB by rapidly gaining access to their vulnerable abdominal dorsum when the host raises its wings to fly. This larviposition behavior appears again in *M. aberrans* on early-season post-diapause CPB (López *et al.*

1997b). With this early- and late-season parasitization of adult CPB, the two *Myiopharus* apparently have a significant differentiation of niche relative to the common beetle host. The late emergence of *M. doryphorae* does not allow a build-up of its numbers early in the season to suppress CPB numbers enough to prevent crop damage even though, later on, parasitism may reach 50–80% (e.g., Tamaki *et al.* 1983). Horton and Capinera (1987) found high rates of parasitism by *M. doryphorae* in Colorado in early season both on potato and on the weed *Solanum sarrachoides* Sendtner. Interestingly, this was an area in which CPB, though present on potato, was not an economic pest (Horton and Capinera 1987).

Carabid Predator–Parasitoid, Lebia grandis *Hentz*

CPB was not the original host of *Lebia grandis*; instead, it was the so-called false potato beetle, *L. juncta* Germar. This *Leptinotarsa* species is native to the southeastern USA. Only after the CPB invaded its range was *L. grandis* discovered as a predator on the CPB (Weber *et al.* 2006). The carabid's larval life cycle was not known until a classical biological control program exported the carabid to France, by boat, in the 1930s. Chaboussou (1939) discovered that first-instar *L. grandis* larvae sought out CPB prepupae or pupae soon after they dug into the soil, and he reared them successfully for introduction into France. This effort ultimately failed, probably due to climatic mismatch, with the target area having too dry summers. *L. grandis* is not present north of southern Maine in eastern USA (Alyokhin, pers. comm.), and is less able to complete larval development at low temperatures (<20°C) than is its host (Weber, unpubl. data). Groden (1989) and Weber *et al.* (2006) considered this species to be the most promising CPB natural enemy, because of its faithfulness to the host and voracity of feeding. In cage studies, Szendrei and Weber (2009) showed the per capita suppression by *L. grandis* to be far superior to that of *C. maculata*. Field studies using molecular gut analysis to assess predation also showed that *Lebia* had the highest predation rate, along with the pentatomid *P. maculiventris* (Szendrei *et al.* 2010). Although *L. grandis* is difficult to rear in the lab, it is worthy of conservation biocontrol in the field, especially by use of selective insecticides, and of further study to determine its limiting factors and possibilities for in-field augmentation.

Candidates for Classical Biological Control

Because of their host specificity and the lack of native *Leptinotarsa* and related genera in Eurasia, *Lebia grandis* and *Myiopharus* spp. are candidates for classical biocontrol in warm humid areas of the Palearctic. Introduction of *Myiopharus* (apparently *M. doryphorae*) was attempted but failed in France, along with shipments of *Lebia grandis* (Trouvelot 1931). Another attempt with 11 shipments from Canada in 1958–1963 to the far-western USSR (Zakarpatsia region of Ukraine) also failed, probably due to lack of knowledge of overwintering habits

(Sikura and Smetnik 1967). Current improved knowledge of all three species bodes well for future introductions. However, classical weed biocontrol programs in Eurasia may possibly give rise to non-target concerns. *Leptinotarsa* species are being considered to target exotic weedy *Solanum eleaegnifolium* Cavanilles in the Mediterranean (Hoffmann *et al.* 1998, Sforza and Jones 2007). Furthermore, *Zygogramma suturalis* (F.) is already established to suppress exotic ragweed *Ambrosia* (Reznik 1991), with other *Zygogramma* under discussion (Gerber *et al.* 2011). Given that all species involved (weeds, their introduced herbivores, CPB, and its natural enemies) are exotic to Eurasia, potential conflicts require a balancing of the various pest and damage concerns amongst the exotic species.

Predatory Stink Bugs: *Perillus* and *Podisus*

Many studies have tested augmentation of the North American native asopine pentatomids *Perillus bioculatus* and/or *Podisus maculiventris* in the USA, Canada, and southern Europe (reviewed in Hough-Goldstein 1998). Both species have been introduced to Europe (Jermy 1980, Stamopoulos and Chloridis 1994, Manole *et al.* 2002). Colazza *et al.* (1996) found *P. maculiventris* was superior to the South American *Perillus connexivus* Bergroth in field tests with potato in central Italy. *P. maculiventris* is a generalist predator, whereas *P. bioculatus* is thought to be more of a specialist on CPB, although it is capable of attacking a wider prey range (Saint-Cyr and Cloutier 1996). Nymphs of *P. bioculatus* benefit from group feeding, and a number of factors influence the success of nymphs and their dispersal in the field (Hough-Goldstein *et al.* 1996, LaChance and Cloutier 1997). Adults and nymphs of both species are attracted to their respective aggregation pheromones (Aldrich 1999), and are also sensitive to plant volatiles (Sant'ana and Dickens 1998, Weissbecker *et al.* 1999).

P. bioculatus releases are clearly capable of drastically reducing CPB egg numbers in the field (Cloutier and Baudin 1995, Hough-Goldstein 1998). However, mass-rearing them for augmentative release in potatoes is difficult to justify economically (Tipping *et al.* 1999), even with refinements in diet for rearing (Rojas *et al.* 2000). An alternative source for *Podisus* is early-season field-trapped insects. For small plantings, they can be attracted with pheromone and set up in a "field nursery", probably situated in an early-planted trap crop. This allows reproduction of the enclosed adults and subsequent dispersal of nymphs out of the coarse-screened field enclosures and into the surrounding crop (Aldrich and Cantelo 1999). Such an approach relies on a background of conservation biocontrol to support the endemic predatory bug population (Perdikis *et al.* 2011); namely, having a diversified ecosystem with alternate prey for this generalist predator to thrive in the rest of the season, and to generate sufficient numbers for capture and emplacement in the potatoes in the early spring. A commercial system to attract *P. maculiventris* (Predator Rescue) was marketed by the Sterling Company in Spokane, Washington, USA, but is now not marketed for retail.

Coleomegilla maculata (DeGeer) and other Coccinellidae

Throughout its range, the CPB is preyed upon by a variety of coccinellid species (see, for example, Snyder and Clevenger 2004, Szendrei *et al.* 2010). Of those, the most studied is *Coleomegilla maculata*, a North American omnivorous species capable of developing on a variety of pure and mixed diets (Weber and Lundgren 2009). In North America it is most abundant in late summer in corn (maize) fields, feeding on corn pollen, Lepidoptera larvae, and aphids (Nault and Kennedy 2000, Lundgren *et al.* 2004). Adults overwinter adjacent to cornfields and other late-season habitats. Both larvae and adults consume CPB eggs and early-instar larvae, and their phenology may be well-synchronized with reproduction of overwintered CPB populations colonizing potato crops. During this period, their predation was found to be an important source of mortality for CPB eggs in North Carolina (Hilbeck *et al.* 1997) and in Massachusetts (Hazzard *et al.* 1991). Potato fields following corn in crop rotation were heavily colonized in Massachusetts (Hazzard *et al.* 1991). However, in North Carolina overwintered coccinellids dispersing in spring often moved to small grain fields before potatoes had emerged, and thus were less abundant in potatoes (Nault and Kennedy 2000).

In mulched and tilled potato field plots in Maryland, Szendrei and Weber (2009) found that *C. maculata* preferred rye-straw mulched plots in spite of the higher density of CPB in the tilled plots, and that in field cages *Lebia grandis* adults were far more effective per capita than *C. maculata* adults at reducing CPB numbers *ab ovo*. PCR-based gut content analysis showed that, in the field, of the four major predators, *C. maculata* adults were least likely to consume CPB, even when adjustments were made for its more rapid digestion of CPB marker DNA (Szendrei *et al.* 2010). However, *C. maculata* may in many situations be several times more abundant than either *L. grandis* or pentatomid predators (e.g., Hilbeck and Kennedy 1996, Szendrei and Weber 2009). Depending on their numbers, as well as on prevalence of eggs and small instar CPBs, and alternate prey, *C. maculata* may or may not be an important predator for pest suppression. This is consistent with Ferro's (1994) mixed conclusion regarding the importance of this species for CPB biological control.

Because of their abundance, fecundity, and ease of rearing, several authors have considered augmentative or inundative releases of *C. maculata*. Giroux *et al.* (1995) concluded that third-instar larvae would be the optimal stage for inundative releases because of their ability to prey on CPB eggs and early-stage larvae, and their inability to disperse from target fields by flight.

Cultural Effects: Mulching and Native Vegetation

Mulching with grain straw suppresses Colorado potato beetle populations; however, it is unclear whether this is due to a positive benefit for natural enemies (as Brust 1994 concluded), a behavioral effect on the pest insect, and/or other factors (see Chapter 10 and references therein).

Interaction of CPB Natural Enemies

Interaction of CPB natural enemies has been investigated both for practical and for theoretical reasons (Moldenke and Berry 1999). Lynch and colleagues (see Chapter 9) discuss the interactions of these and other potato pests and natural enemies. In potatoes, using green peach aphid and CPB respectively as focal pests, Straub and Snyder (2006) and Szendrei and Weber (2009) both emphasized that the identity of the natural enemies is more important than simply their diversity or evenness. Identifying key predators requires an objective assessment not just of their presence but also of their predation (both day and night) on the focal pest (Weber *et al.* 2008). Molecular-based assessment of predation of CPB in Maryland potatoes (Greenstone *et al.* 2007, 2010, Szendrei *et al.* 2010) has indicated that, among the four most abundant predators sampled, *P. maculiventris* adults and *L. grandis* were highest ranked in terms of per capita predation, and that *C. maculata* larvae and adults, and other stinkbug lifestages, including *P. bioculatus* nymphs and adults, were less effective predators of CPB. This information can guide future choices about what species should be targeted for conservation or augmentation in the potato system.

Potato Tuberworms (Lepidoptera: Gelechiidae)

Three species of Lepidoptera (Gelechiidae) are known as potato tuberworms because the larvae bore into potato tubers in the field and in potato stores. The potato tuberworm (PTW, *Phthorimaea operculella* (Zeller)) is by far the most widespread. The PTW originated in the Andes, but is now present in all potato-growing regions with the exception of colder north temperate areas; in many regions this is the insect responsible for the largest potato losses (Rondon 2010). *Tecnia solanivora* (Povolný) (Guatemalan potato tuberworm, GPTW) has in the past 20 years spread from central America into Venezuela, Colombia, and Ecuador, and more recently to southern Mexico (Cruz Roblero *et al.* 2011) and the Canary Islands, thus threatening invasion of Africa and Eurasia (Povolný and Hula 2004). The Andean potato tuberworm (*Symmetrischema tangolias* (Gyen)) originated in Peru and Bolivia, but is now also found in Ecuador and Colombia, typically at above 2800–3000 m elevation (Palacios *et al.* 1999, Dangles *et al.* 2008). This species has also spread to Australia including Tasmania, New Zealand, and reportedly Indonesia (see Fig. 6.1). It is important to realize that while some authors lump these species together as "potato tuberworm complex", they differ in many aspects of biology and damage (Dangles *et al.* 2008). Thus, appropriate pest management approaches, including specialist natural enemies, may be species-specific. The vast majority of ecological and management experience has been with *P. operculella*, including that with native and introduced biological controls.

Potato Tuberworm, Phthorimaea operculella

The PTW is by preference a leafminer; the ovipositing female prefers foliage over tubers, attacking eggplant (aubergine, brinjal), tomato, tobacco, and a number of other non-tuberous solanaceous crops, weeds, and wild plants (Rondon 2010). It may also bore in above-ground stems. Foliar damage to potatoes is usually not economically important (Rondon 2010). However, as foliar quality declines due to natural or chemically-induced senescence near the end of the potato-growing cycle, tubers come under attack by larvae moving down from the canopy, and by females ovipositing directly on exposed tubers or through cracks in the soil (Kroschel 2006). During the pre-harvest period, when tubers are still in the ground, degree and duration of tuber exposure is a critical factor in PTW infestation of the harvested crop (Alvarez *et al.* 2005). This in turn depends on soil type, moisture, tillage (particularly hilling), cultivar-dependent tuber depth, and, of course, population of PTW. If tubers are left above ground in the process of harvest, additional infestation is invited (Von Arx *et al.* 1990, and other studies).

Growers with access to controlled temperature storage typically arrest any PTW infestation there. However, many regions, especially in developing countries where potatoes are a staple, have few such storage facilities, and store the harvest in rustic conditions – under trees, in sheds, or simply in piles – where their exposure to further PTW development and reproduction can result in huge losses, often up to 100%. In rustic storage, farmers often cover the tuber stores with plant remains, including potato haulms from the harvested crop. Use of PTW-infested halms is particularly detrimental to stores because the remaining larvae have direct access to the tubers (Kroschel 2006).

Native Natural Enemies

A variety of natural enemies attack PTW in all growing regions where it occurs. Particularly prevalent are larval and egg-larval parasitoids in the hymenopteran families Braconidae and Ichneumonidae. A well-studied example is the situation in South Africa (reviewed by Kfir [2003]), in which five species of native parasitoid wasps were found to attack PTW, in each case allowing larvae to develop to just before pupation before killing the host. Rates of parasitism often ranged around 70%, but unacceptable tuber infestation still often resulted. These native species have a variety of other hosts, since the crop is native to South America. Also, a variety of coccinellids, chrysopid larvae, predatory Heteroptera, carabid and staphylinid beetles, and earwigs prey on PTW eggs and larvae (Kfir 2003).

The range of native parasitoids and predators present in South Africa is typical of that in other regions; however, parasitism rates may vary. For example, in Israel, Coll *et al.* (2000) found that five indigenous braconids and ichneumonids attacked PTW larvae. However, their overall level of parasitism was below 10% in commercial fields, with the exception of up to 40% larvae parasitized on volunteer plants that harbored high densities of PTW. That study was particularly notable for its examination of the predator complex and evaluation of its suppressive effects on PTW using emplaced eggs with and without fine-mesh exclosures. The most important predators in that system were *Coccinella septempunctata*,

Chrysoperla carnea, Orius spp., and four identified ant species, causing about 80% mortality in the field. To the best of our knowledge, that has been the only quantitative assessment of predation in the field for PTW, although many other studies have noted species of parasitoids and predators, and nominal rates of parasitism. Kroschel and Koch (1994) found in Yemen that the native larval parasitoid ichneumonid *Diadegma mollipla* Holmgren and the egg parasitoid *Chelonus phthorimaeae* Gahan (Braconidae), along with native predators, helped suppress potato tuberworm. Unfortunately, poor cultural practices (particularly the planting of infested tubers) frequently resulted in high levels of damage.

Classical Biological Control of Potato Tuberworm

Beginning in the mid-20th century, the Commonwealth Institute of Biological Control (CIBC) undertook a classical biological control effort against PTW. That resulted in the introduction of several hymenopterous parasitoids, native to South America, into South Africa and many other potato-growing areas. *Copidosoma koehleri* Blanchard (Encyrtidae), an egg-larval polyembryonic parasitoid, and the larval solitary koinobiont *Apanteles subandinus* Blanchard (Braconidae), established in South Africa and greatly increased the levels of parasitism in larvae of PTW in the field (Kfir 2003). Subsequently, PTW damage was significantly and permanently reduced in South Africa, Zimbabwe, Zambia, and other countries of southern Africa (Kfir 2003). In Zimbabwe, PTW was eliminated as a pest (Mitchell 1978). In Australia, introductions of the braconids *Orgilus lepidus* Muesbeck and *A. subandinus*, and the egg parasitoid *C. koehleri*, remain a major factor controlling PTW in potato fields without insecticides (Horne 1990). *O. lepidus*, which was introduced but did not establish in South Africa, is the most important parasitoid, and effectively seeks out the host using volatiles produced by PTW damage to foliage (Keller and Horne 1993). Natural enemies form the basis in Australia for a biointensive IPM program including all potato pests such as PTW, aphids, and noctuid and geometrid larvae (Horne 1990, Horne and Page 2008). In New Zealand, *A. subandinus* is the most important parasitoid, reaching high levels of PTW parasitism especially on foliar populations (Herman 2008) where insecticide treatments were reduced or absent. Another braconid, *Diadegma semiclausum* (Hellén), which was introduced to New Zealand to suppress diamondback moth, reached up to 24% parasitism. All three species of parasitoids have been established in India as well (Kroschel 2006), increasing modest parasitism by native species (Chandla and Verma 2000). The International Potato Center has rearings of these species available for future classical biological control introductions against PTW (CIP 2012), and is assessing target areas worldwide for suitability of introductions (Sporleder *et al.* 2011).

Guatemalan Potato Tuberworm

The GPTW (*T. solanivora*) differs from PTW in its apparent preference for tubers over the above-ground plant parts (Torres *et al.* 1997), and also in its thermal response (Dangles *et al.* 2008). Similarly, study of the natural enemies in the field, mainly since its invasion of northwestern South America, has revealed that the natural enemy complex also differs (Osorio *et al.* 2001). In Colombia, two anthocorid bugs are important predators, along with lycosid and salticid spiders, Carabidae, Coccinellidae, Staphylinidae, and Tenebrionidae (Osorio *et al.* 2001). Two unidentified parasitoid wasps, *Apanteles* sp. (Braconidae) and *Trichogramma* sp. (Trichogrammatidae), were commonly associated with GPTW. Intriguingly, an unidentified tachinid fly was associated with GPTW as well (Osorio *et al.* 2001). *C. koehleri*, native to the Andes, seems to be specialized on PTW and not GPTW. Báez and Gallegos (2011) showed that 36% and 25% of eggs of PTW were parasitized when offered separately and in a mixture with the other two tuberworm species, respectively. In a similar setting, the parasitism of GPTW was only 1% and 2%, respectively. This suggests that one of the principal natural enemies of PTW is not effective on GPTW eggs.

Andean Potato Tuberworm

APTW (*S. tangolias*) is even less-known than GPTW, and less widespread, being cited as a pest in Peru and Bolivia (Palacios *et al.* 1998) although the three tuberworm species are now sympatric in parts of Colombia and Ecuador, where they have been observed to increase crop damage when two or three species occur together (Dangles *et al.* 2009). Infestation is primarily in stems in the field, but tubers are infested both in the field and in storage (Palacios *et al.* 1998). In Australia and New Zealand, it is known as tomato stem borer, and also attacks other fruiting *Solanum* crops. Little information specific to APTW parasitoids and predators is yet available (see Chapter 6). However, Sánchez-Aguirre and Palacios (1996) and Báez and Gallegos (2011) showed that APTW was not a preferred host of *C. koehleri*.

Inundative Biological Control

Several efforts at inundative parasitoid releases have resulted in variable, often unsatisfactory, suppression of PTW in the field and in storage. Inundative releases of exotic parasitoids, particularly *C. koehleri*, have been employed against PTW in the field (e.g., Pokharkar *et al.* 2002, in India; Pucci *et al.* 2003, in Italy) with mixed results, probably due to the sensitivity of the species to desiccation and hot temperatures (Kfir 2003). Inundative approaches with *C. koehleri* and *Trichogramma* species have also been tried against PTW in rustic storage (Keasar and Sadeh 2007, Mandour *et al.* 2012). Rubio *et al.* (2004) evaluated *Trichogramma lopezandinensis* for control of *T. solanivora* (GPTW) in Colombian potato stores and determined that young female parasites, released at 3-day intervals, were necessary for best results. For stored potatoes, a combination of microbial controls (PTW granulosis virus and/or *Bt*), proper sanitation, and storage, as well as use of repellant botanicals, has proven more reliable than parasitoids (Kroschel and Koch 1996, Chandla and Verma 2000, Hanafi 2005, Mandour *et al.* 2012; see also Chapter 16, this volume).

Summary of Biological Control of Tuberworm Species

Additional research is critical for biological control; in many regions PTW bio-control is poorly understood, as are the biocontrol possibilities for GTPW and ATPW. In the field, predators are consistently present, and at least sometimes are a major source of mortality (Coll *et al.* 2000); these must be researched, recognized, and promoted wherever possible. In the field, native parasitoids play a variable role, often not sufficient to adequately suppress PTW (e.g., Kfir 2003). However, introduced parasitoids from the presumed native range of PTW, and also possibly from outside the native range, have the potential to play a major role in suppression of PTW in the crop, depending on climate and potato culture (Kroschel 2006). Classical biological control may also play a role in the future of GPTW and APTW. Introduced and native parasitoids, as well as predators, are expected to be more successful when supported by practices of pesticide reduction and selection, along with conservation biocontrol (Baggen *et al.* 1999, Horne and Page 2008), which provide resources such as nectar in such a way as to selectively favor natural enemies over the pest, as well as protection from desiccation and from temperature extremes.

Pheromones are known for all three species of potato tuberworms; phero-mone-based technologies such as attract-and-kill schemes (Kroschel and Zegarra 2010) are powerful and promising tools which should be selective and compatible with natural enemies. Cultural controls must provide the foundation for protection of tubers from exposure to tuberworm infestation, and continued protection of the tubers during and after harvest. Otherwise, the benefits of biocontrol during the growing season are lost (Kroschel 2006). Simple techniques such as solarization (Gallegos *et al.* 2005) should also be pursued as appropriate. Storage practices can easily be the downfall of potatoes by PTW infestation. A combination of temperature control, sanitation, and exclusion at the outset of storage and over time, and safe biopesticides such as PoGV and/or *Bt*, are most promising where secure refrigeration is not available. Inundative biocontrol with parasitoids is not the most efficient means of control of PTW in potato stores; biopesticides are more reliable and economical in this environment (see Chapter 16).

Hadda Beetle and Potato Lady Beetle: *Epilachna* spp. (Coleoptera: Coccinellidae)

Epilachna species form a large genus of phytophagous coccinellids which feed chiefly on foliage of the Solanaceae, Cucurbitaceae, Fabaceae, and Asteraceae. The systematics of this group is confused; as a result, many specific names in the literature are not correct (Richards and Filewood 1990, CABI 2010). The most important species attacking potato and other solanaceous plants is the hadda beetle, or 28-spotted lady beetle, *Epilachna vigintoctopunctatum* (F.) (also placed in the genus *Henosepilachna*, and taken here to include *E.sparsa* (Herbst); nomenclature follows CABI (2010)). The geographic range of this species includes South, Southeast, and East Asia, Oceania including Australia

(CABI 2010), and, recently, New Zealand (MAFBNZ 2010, David Yard (MAFBNZ), pers. comm. 2012). This species was also detected in Brazil by Schroder *et al.* (1993), has spread to Argentina (Folcia *et al.* 1996), and persists on wild nightshades in Brazil (Araujo-Siqueira and Massutti 2004). The other important solanaceous feeder is the potato lady beetle (also known as the large 28-spotted lady beetle), *Epilachna vigintoctomaculatum* Motschulsky. It is the most damaging potato pest in China (most notably in northern China; see Chapter 7), and a pest of potatoes in eastern Russia (Ivanova 1962; AgroAtlas 2012), Korea (Lee *et al.* 1988), and Japan (Nakamura 1987).

Hadda Beetle, Epilachna vigintoctopunctatum

Throughout much of its range, hadda beetle is the principal foliar feeder on potatoes and eggplant, and second only to potato tuberworm among potato pests. Schaefer (1983) reported at least 15 species of parasitoids and 4 predator species. Cannibalism is also significant (Patalappa and ChannaBasavanna 1979; Nakamura *et al.* 2004). In India and Pakistan, hadda beetle is attacked by three predatory bugs: *Rhynocoris fuscipes* (F.) (Reduviidae), *Cantheconidea furcellata* (Wolff) (Pentatomidae), and *Geocoris tricolor* F. (Lygaeidae) (Patalappa and ChannaBasavanna 1979, Schaefer 1983, CABI 2010). In southeastern China, Tu and Wang (2010) reported that *Campylomma chinensis* Schuh (Hemiptera: Miridae) was an important predator. Published observations of generalist predators, such as other coccinellids, have not been detailed, but unattributed egg and larval mortality of hadda beetle suggests that predators, many of them nocturnal, can be major mortality factors (e.g., Nakamura 1976, Nakamura *et al.* 1988). Where predator observations have been made on *Epilachna*, as with *E. varivestis* in USA and Mexico, a large number of species of coccinellids, carabids, reduviids, pentatomids, and chrysopids have been found as predators (Schaefer 1983).

For hadda beetle, the most important of the parasitoids are the gregarious larval-pupal eulophids *Pediobius foveolatus* (Crawford) and *P. epilachnae* (Rohwer), the eulophid egg parasitoid *Oomyzus ovulorum* (Ferriere) (formerly *Tetrastichus*), and chalcidids *Uga* spp. These three genera are also recorded from *Epilachna ocellata* Redtenbacher, a closely-related potato pest of northern India (Chandel *et al.* 2007, CABI 2010). Parasitism, particularly by *Pediobius*, frequently reaches 75% on wild nightshades, as well as on potato and eggplant (Patalappa and ChannaBasavanna 1979, Rajagopal and Trivedi 1989; Nakamura *et al.* 2004). Sheng and Wang (1992) found a peak of 28.5% parasitism on potato, and later a peak of 64.5% on *Solanum nigrum*, followed by a precipitous decline of both host and parasitoid due to high temperatures in mid-July. Other studies in India (see Rajagopal and Trivedi 1989), Bangladesh (FAO 2003), and Thailand (Kernasa *et al.* 2002) confirm the importance of *P. foveolatus* as a natural enemy of hadda beetle. There are at least 11 generations per year reported in China (Sheng and Wang 1992), and as many as 18 generations in

India (Rajagopal and Trivedi 1989). Females oviposit into larvae, mainly of the third and fourth instars, and between 10 and 30 adults usually emerge from the resulting larval "mummies" (Lall 1961). Adult longevity is greatly increased by feeding on dilute honey in the laboratory (Sheng and Wang 1992). As early as 1954 Puttarudriah and Krishnamurti observed the seasonally high parasitism on both potato and eggplant, and tested the effects of early chlorinated hydrocarbon and calcium arsenate insecticides on *P. foveolatus*. They urged growers to withhold sprays against hadda beetle when parasitism reached about 70%, and demonstrated the negative effects of the DDT, BHC, and toxaphene, but not the arsenical, on the parasitoid. Fenvalerate, abamectin, and particularly dichlorvos severely affected larval survival inside the host in lab tests, especially within 7 days of oviposition by *P. foveolatus* (Wang *et al.* 1998). Schaefer *et al.* (1986) reported *Tetrastichus* sp. to be hyperparasitic on *Pediobius* at several locations in China.

Potato Lady Beetle, Epilachna vigintoctomaculatum

For the potato lady beetle, *P. foveolatus* is not as important a parasitoid as for the hadda beetle, but appears to be the same wide-ranging species (Peng and Bao 1988). Lee *et al.* (1988) studied potato lady beetle on potato and eggplant crops, and reported that, in South Korea, *Nothoserphus afissae* (Watanabe) (Proctotrupidae) was the dominant parasitoid. *Uga menoni* (Kerrich) and *P. foveolatus* were also present, the latter only near the end of the season, except in southern Korea, indicating lack of winter hardiness. The authors advocated leaving the late-season wild plant host *Solanum nigrum* intact as a refuge for *E. vigintoctomaculatum* and its parasitoids. Zhuang and Sun (2009) constructed life tables for this pest, and considered predation and rainfall to be the most important mortality factors in Daqing, in far northeastern China. They mentioned spiders and lady beetles (not identified) as predators, and an average of 22% of pupae as being parasitized (parasitoids not specified). Three species of Tachinidae, *Medina collaris* Fallen, *M. separata* (Meigen), and *Bessa parallela* (Meigen), have also been reported from Japan (Schaefer 1983), but their impact on the target species was not quantified.

Annual Inoculation of Pediobius against Epilachna in USA

P. foveolatus was exported from India (original host, hadda beetle) and is the basis for a 40-year history of annual inoculative releases against *Epilachna varivestis* Mulsant (Mexican bean beetle) in eastern North America (Kogan 1999, Robbins *et al.* 2010). The parasitoid is also available through four commercial vendors in the USA (White and Johnson 2010). Releases are made each year because it does not overwinter. In New Jersey, where the state Department of Agriculture rears and releases *P. foveolatus*, area-wide control has now suppressed *E. varivestis* numbers on bean and soybean crops by ~95% compared to densities in the 1980s (Robbins *et al.* 2010). A trap crop is typically employed to attract the overwintering hosts, which functions as an in-field nursery for the lab-reared parasitoids (Robbins *et al.* 2010). This strategy could also be employed with Asian *Epilachna* species.

Summary of Role of Biological Control for Epilachna Potato Pests

To make optimal use of natural enemies in suppressing *Epilachna* pests of pota-toes, overwintering adult aggregations (see Chapter 7) should be targeted in temperate regions for destruction to reduce immigrating pest numbers, which otherwise commonly outstrip predator and parasitoid suppression. Staging and managing trap crops as in-field nurseries for *Pediobius* parasitoids, using the model of the New Jersey scheme for Mexican bean beetle, would likely result in early-season biological control establishment and effectiveness against Asian *Epilachna* spp. There is a need for more selective insecticides for *Epilachna* control which would conserve the oft-abundant natural enemies. Although *Bacillus thuringiensis tenebrionis* (*Btt*) is not toxic to *Epilachna* (specifically, *E. varivestis*; Krieg *et al.* 1987), there have been reports of strains toxic to coc-cinellids (e.g., Peña *et al.* 2006), and these, if selected and deployed carefully to avoid harming predatory coccinellids, could function in concert, rather than against, the beneficial insect guild.

Andean Potato Weevil (Coleoptera: Curculionidae)

There are at least 14 species which make up the Andean potato weevil species complex: *Premnotrypes latithorax* (Pierce), *P. suturicallus* Kuschel, *P. vorax* (Hustache), and 9 other *Premnotrypes* spp., as well as *Rhigopsidius piercei* Heller and one other species in this genus. The four named species are the most important, according to Kühne *et al.* (2007). They are distributed from Andean Venezuela south to northern Chile and Argentina, mostly at between 2800 and 4500 m in altitude; they are often the only important pest above 3800 m (Kühne *et al.* 2007, Kroschel *et al.* 2009).

All Andean potato weevils are specialists on tuberous *Solanum* spp. including all potato species. They are univoltine in Peru and Bolivia, corresponding to a single potato season, but two to three generations occur to the north, where rainfall allows longer or even continuous potato crops. Adults of all species are flightless, and colonize new hosts by walking. Emergence from dormancy is variable by species and growing region, and *P. suturicallus* and *P. vorax* are reported to survive a few months if starved. On potatoes, they feed modestly during the first half of the night, copulating and ovipositing numerous eggs at the base of the plant. Larvae seek out the tuber and enter it to complete their development in four instars. In the soil, almost all larvae of *Premnotrypes* spp. typically emerge from the tuber to pupate, but *Rhigopsidius* spp., in contrast, pupate within the tuber (Kühne *et al.* 2007).

In traditional communal potato culture, a system of "sectoral fallow", still practiced in parts of Peru and Bolivia, limited potato plantings to blocks of

~100 ha, which were rotated each year amongst 7–10 similar blocks. Long fallow periods of several years were designated for grazing rather than crops. Under such a system, the Andean potato weevil typically damaged less than 10% of tubers, and these would typically be fed to animals or made into *chuños* (traditional freeze-dried potato meal). Within the past several decades, the communal land tenure, and along with it the rotational system of sectoral fallow, has been dismantled. Small plots of potatoes and other crops raised on private plots in uncoordinated individual rotations accompanied the ascendancy of Andean potato weevil to key pest status throughout most of the upper-elevation growing areas of South America (Parsa 2010, Parsa *et al.* 2011, Rios and Kroschel 2011).

Natural Enemies

The natural enemy complex appears to be lacking in specific natural enemies. No parasitoids have been discovered despite prolonged and intensive efforts (Parsa 2010). Known predators are six genera of carabids: *Harpalus turmalinus* Erichson, *Hylitus, Meotachys, Metius, Notiobia* including *N. schnusei* Van Emden, and *Blennidus,* a predatory tenebrionid (*Metius* sp.) (Alcalá and Alcázar 1976, Alcázar and Cisneros 1999, Kaya *et al.* 1999), and a predatory ant found in potato storage (Garmendia 1961). Kroschel *et al.* (2009) caught carabids in abundance along with weevil adults when plastic barriers were combined with pitfall traps. The authors stated that high carabid numbers, and their association with lower weevil populations, indicated an "important role in improved natural control of Andean potato weevils". They also suggested two methods of favoring carabids over weevils: removal of barriers after ~95% of weevil captures, to allow the later-colonizing carabids to move into new potato plantings, and selective release by growers of carabids captured in traps along with the weevils. The carabids identified are thought to be generalists (Parsa 2010), although there is no quantitative data on their diets; they readily accept weevil eggs as food, and some prefer early-instar larvae (Alcázar and Cisneros 1999, Loza and Apaza 2001, Yábar *et al.* 2006). Wild birds and toads also prey on weevil adults, likely with minimal impact, but chickens at harvest time may be useful (Alcázar and Cisneros 1999). Predators may have been negatively impacted across the region with adoption of broad-spectrum synthetic pesticides (Parsa 2010), of which the most commonly used are carbofuran and methamidiphos (Kühne *et al.* 2007). Both these chemicals are extremely toxic to most invertebrates and vertebrates, including humans (Orozco *et al.* 2009, Pradel *et al.* 2009).

Biological Control in Context of IPM for Andean Potato Weevil

Within this changing spatiotemporal context, the IPM strategies that are considered most promising are crop rotation away from sources of emerging weevils (Rios and Kroschel 2011), development of practical barriers to weevil movement (Kroschel *et al.* 2009), use of natural barriers such as streams as well as ditches (Parsa *et al.* 2011), trap crops and baits prior to planting (Gallegos and Castillo 2011), border treatments of newly immigrated weevil concentrations, manipulation of harvest timing and storage (Kühne *et al.* 2007), and use of entomopathogens, including *Beauveria*, and particularly cold-adapted and virulent *Heterorhabditis* sp. nematodes (Kaya *et al.* 2009; see also Chapter 16, this volume). It is likely that predation on Andean potato weevils has a useful supporting role in the field when cultural controls are in place, and chemical control is judicious and selective in nature. Molecular approaches to identifying and quantifying subterranean predation, as used by Juen and Traugott (2007) for wireworms and Lundgren *et al.* (2009) for western corn rootworm larvae, could yield surprising information regarding unseen trophic relationships. Given the dearth of knowledge on the predator fauna, more research on carabids and possible other arthropods as natural enemies of Andean potato weevil complex is much needed.

Potato Psyllid (Hemiptera: Triozidae)

The potato psyllid (*Bactericera cockerelli* Sulc) (Hemiptera: Triozidae), native to the Rocky Mountain region of the USA and northern Mexico, has recently expanded its range to the entire western USA, including California, Oregon, and Washington, southern Canada, including Ontario, as well as Mexico and Central America, and New Zealand, with major impacts on potatoes, tomatoes, peppers, and other solanaceous crops (Crosslin *et al.* 2010, Munyaneza 2010, Rehman *et al.* 2010). The damage, including zebra chip disease of potatoes, is now unequivocally associated with the transmission by potato psyllid of the recently-discovered *Candidatus* Liberibacter psyllaurous/solanacearum (see Chapter 4). In response to increasing potato damage, "At present, application of insecticide targeted against the potato psyllid [is] the only way to effectively manage zebra chip" (Munyaneza 2010). In California, recommendations call for monitoring by yellow sticky cards and plant sampling, coupled with insecticide applications (UC IPM Online 2009).

Predators in Rocky Mountain States

Natural enemies identified as preying on the potato psyllid include predatory bugs in the genera *Nabis*, *Geocoris*, *Orius*, *Anthocoris*, and *Deraeocoris*, and also adult and larval *Hippodamia* (Coccinellidae), larval Chrysopidae, and Syrphidae (Knowlton 1933a, 1933b, 1934a, 1934b, Knowlton and Allen 1936). Romney (1939) noted that "numerous predators (coccinellids and chrysopids) reduce the numbers of eggs and nymphs to a varying degree from year to year," but it is not clear whether this was on wild hosts in Arizona or on crops in Colorado. Among those early works, field observations on predation were generally not detailed. However, Pletsch (1947) noted "an exceedingly heavy and aggressive population of [coccinellid] adults and larvae feeding on psyllids" on tomato in Bozeman, Montana, especially *Hippodamia parenthesis* (Say), *H. convergens* Guerin, *H. quinquesignata* Kirby, *Coccinella novemnotata* Herbst, and *C. transversoguttata*

Faldermann. He also considered lacewings (Chrysopidae) among the most important predators, and ascertained that newly eclosed lacewings as well as later instars fed readily on potato psyllid eggs.

Al-Jabr (1999) found that two species of Chrysopidae (*Chrysoperla carnea* and *C. rufilabris*) had promise for potato psyllid suppression on greenhouse tomatoes, consuming (under laboratory conditions) a mean of 24.4 and 17.5 nymphs per day during a larval development period of 12 and 8 days, respectively. Development and survival was comparatively worse on green peach aphid prey, with *C. rufilabris* not completing larval development. In laboratory choice trials, both species consumed both prey species without preference. A field trial of augmentation of *C. carnea* eggs in Colorado potatoes failed to show suppression of either potato psyllid or green peach aphid populations at 7 and 14 days post-release (Al-Jabr 1999).

Early Work with Parasitoids, and Changing Context

A parasitoid "*Tetrastichus* sp." was mentioned by Romney (1939), and described later by Burks (1943) as *Tetrastichus triozae*. Pletsch (1947) in Montana and Johnson (1971) in Colorado concluded that *Tetrastichus* (now *Tamarixia*) *triozae* was not a promising biocontrol agent for potato psyllid. However, that assessment was based largely on the parasitoid's localized occurrence and late phenology, which does not pertain in mild climate areas where infestations are worst now (California, Texas, Mexico, Central America and New Zealand). Johnson (1971) also noted high pupal mortality in the field. Lab studies by Pletsch (1947) showed that oviposition was only in fourth- and fifth-instar nymphs.

Natural Enemies of Invasive Psyllid Populations in California

Butler *et al.* (2010) examined endemic biological controls on potatoes grown in southern California, and on a common weed in the crop environment, *Solanum americanum* P. Miller, which also supported the psyllid. Over 90% of natural enemies were in the following groups: hymenopterous parasitoids *Tamarixia triozae* (Eulophidae) and *Metaphycus psyllidis* Compere (Encyrtidae), *Chrysopa* (Chrysopidae), spiders (Araneae), predaceous bugs *Orius tristicolor* (White) (Anthocoridae), *Geocoris*, *Nabis*, and various Miridae, the coccinellids *Hippodamia convergens, Coccinella septempunctata*, and *Harmonia axyridis*. Cage exclusion studies with potato and *S. americanum* showed similar and significant suppression of nymphal populations (Butler *et al.* 2010). The recorded parasitism rates by *T. triozae* were less than 20% (Butler and Trumble, 2012).

Introduction to New Zealand, and Natural Enemies There

Following the discovery of potato psyllid on the North Island of New Zealand in 2006, and its rapid spread to most areas of both principal islands (Teulon *et al.* 2009), chemical control was the main response for severely affected greenhouse crops, including tomato and peppers, and for field-grown potatoes (Walker *et al.* 2011). The natural enemy complex in unsprayed psyllid-infested potatoes on North Island includes many of the same higher taxa found to be important in California, with the widespread Australian native lacewing *Micromus tasmaniae* (Walker) (Neuroptera: Hemerobiidae) and the endemic *Melanostoma fasciatum* (Macquart) (Diptera: Syrphidae) judged in preliminary studies to be most important (Walker *et al.* 2011). These two species, as well as *Coccinella undecimpunctata* L., *Harmonia conformis* (Boisduval), *Nabis kingergii* Reuter (Hemiptera: Nabidae), and Linyphiidae (Araneae), were common in unsprayed potato fields, and all consumed potato psyllid nymphs and adults in laboratory assays (MacDonald *et al.* 2010, Larsen *et al.* 2011,Walker *et al.* 2011). The native psyllid *Trioza vitreoradiata* (Maskell), which feeds on native woody plants, has a predator community including *M. tasmaniae* and *Drepanacra binocula* (Newman) (Hemerobiidae); coccinellids *Harmonia conformis*, *Adalia bipunctata* (L.), and *Halmus chalybeus* (Boisduval); and the predatory mirid *Sejanus albisignata* (Knight). All six of these species preyed on potato psyllid in lab no-choice tests, with *D. binocula* and the first two coccinellids considered the most promising predators of potato psyllid except in tomato crops, where the coccinellids avoided the crop (Gardner-Gee 2011). No parasitoids or entomopathogens were found in the field study by Walker *et al.* (2011), but an undescribed *Tamarixia* was discovered in 1997 in New Zealand (Workman and Whiteman 2009, Workman 2009) and is now in culture (Gardner-Gee 2011).

Candidates for Classical Biological Control

The North American native eulophid *Tamarixia triozae* has been studied in Mexico (Luna Cruz 2010, Rojas Rojas 2010) and then exported from Koppert Mexico to quarantine in New Zealand (Workman and Whiteman 2009, Workman 2009). *T. triozae* prefers fourth- and fifth-instar nymphs for oviposition, but undertakes host-feeding particularly on younger nymphs (this feeding was not quantified) (Rojas Rojas 2010). Both females and males benefit enormously from non-prey food. In the laboratory, honey increased the lifespan of females from a mean of less than 2 days with water only, to more than 46 days. Females in the lab averaged 143 parasitized hosts in their lifetime; the population doubling time appears to be significantly less than that of potato psyllid (Rojas Rojas 2010). Its reported host range is restricted to the psyllid families Calophyidae, Psyllidae, and Triozidae, for a total of 7 genera in 13 species, including the potato psyllid (Jensen 1957, Zuparko *et al.* 2011). It is quite sensitive to commonly used insecticides (Luna Cruz 2010). A single unidentified sphegigastrine pteromalid was reared from *T. triozae* by Pletsch (1947), and two hyperparasitoid *Encarsia* species have been documented in California on tomato and pepper crops at totals between 5.3% and 6.9% (Butler and Trumble 2011).

  Greenhouse infestations of potato psyllid in Ontario on tomatoes and peppers have been suppressed with releases of *T. triozae* (OMAFRA 2012, Workman and

Whiteman 2009), and this application may become more widespread (including in New Zealand, if the parasitoid is released from quarantine). Koppert Mexico (2012) sells the species commercially, and permission to import to Canada has been issued by the Canadian Food Inspection Agency (2011).

New Zealand researchers have begun screening of *T. triozae* against the diverse native psyllid fauna (83 described species), as well as to compare its performance against potato psyllid with the newly-discovered New Zealand *Tamarixia* sp. No-choice tests against four native psyllid species showed only one species partially accepted, without adult emergence (Gardner-Gee 2011).

Compere (1943) described the encyrtid *Metaphycus psyllidis* from *B. cockerelli* on peppers in Southern California. Until the report of Butler *et al.* (2010), there were no subsequent published observations. It is one of only three species of *Metaphycus*, all in the New World, which parasitize psyllids (Guerrieri and Noyes 2000).

Spread of Liberibacter and the Context for Psyllid Microflora

As molecular tools are used to discover more information about the occurrence and transmission of the pathogenic bacterium *Ca.* L. psyllaurous/solanacearum, there are major implications for the management both of the potato psyllid and the pathogen. Studies of potato psyllids in New Zealand (Berry *et al.* 2010) have shown a large proportion of insects carry the Liberibacter; furthermore, in New Zealand field studies, PCR shows putative presence in plants of four plant families other than Solanaceae. This in turn raises the possibility that additional reservoirs and pathogen-vector relationships may develop. Molecular confirmation of a distinct haplotype of *Ca.* L. psyllaurous/solanacearum in carrot psyllid and symptomatic carrots in Finland (Munyaneza *et al.* 2010, Nelson *et al.* 2011), and the report of zebra chip symptomatic potatoes infested with *Bactericera nigricornis* (Forster) in northern Iran (Fathi 2011), suggest wider occurrence of *Ca.* L. psyllaurous/solanacearum. Molecular studies also reveal that the pathogen is universally present in first-instar nymphs through adults on potato (Hansen *et al.* 2008), and that pathogen transmission by potato psyllid is very rapid, putting a premium on management tools providing rapid death and/or cessation of feeding and/or repellency (Buchman *et al.* 2010, Yang *et al.* 2010, Peng *et al.* 2011).

The symbiotic flora of potato psyllid may also yield possibilities for management of the pathogen, including biological control within the symbiont community. Psyllid bacteriocytes harbor the obligate symbiont *Candidatus* Carsonella ruddii which is maternally transmitted (reviewed in Baumann 2005), and may also contain secondary symbionts. All potato psyllids studied contained Carsonella; many contained Liberibacter; some also contained *Wollbachia*, *Acinetobacter*, and/or *Methylibium* (Alvarado *et al.* 2010, Nachappa *et al.* 2011). Furthermore, there are indications that parasitoids can promote secondary symbiont transmission in psyllids (Hansen *et al.* 2007). Together with the potential for oral ingestion of RNAi constructs for potato psyllid control (Wuriyanghan *et al.* 2011), one can imagine using the same technology to achieve elimination of *Ca.* L. psyllaurous/solanacearum from its psyllid host, or transgenerational passage of a secondary symbiont which is antagonistic to the Liberibacter, or even transmission of Liberobactor-antagonistic symbionts into potato psyllid by its parasitoid.

Overall Context for Sustainable Potato Psyllid Management Including Biocontrol

Returning to the agroecological level, it is essential to recognize the importance and potential value of population suppression of potato psyllid by native and introduced natural enemies, in, near, and distant from agroecosystems. This population suppression works to reduce many losses and risks: crop quantity and quality loss, spread of Liberibacter, spread of potato psyllid on local to global scales, insecticide resistance, provocation of secondary pest outbreaks, and other deleterious effects of pesticides to non-target organisms including humans. In combination with more selective chemical controls, semiochemical tactics (see Guédot *et al.* 2010), cultural manipulations, and crop resistance, biological control must play an important role in sustainable management of potato psyllid.

Aphids (Hemiptera: Aphididae)

Several species of aphids feed on potatoes throughout the world; the most important are green peach aphid (*Myzus persicae* Sulzer), potato aphid (*Macrosiphum euphorbiae* (Thomas)), buckthorn aphid (*Aphis nasturtii* Kaltenbach), and foxglove aphid (*Aulacorthum solani* (Kaltenbach)); they may vector several damaging viruses in a persistent or non-persistent manner (Ragsdale *et al.* 1994; Flint 2006). Virus transmission is of highest concern for growing seed potatoes, which are used to plant subsequent crops (Nakata 1995, Flint 2006). In potatoes, aphids rarely reach populations which lower potato yields by their feeding alone, due to natural enemy complexes typically including Coccinellidae, predatory bugs in genera *Orius*, *Nabis*, and *Geocoris*, lacewings, spiders, syrphid fly larvae, and/or predatory gall midge larvae (Cecidomyiidae), as well as aphid-specific parasitoids, typically solitary koinobionts in family Aphidiidae (Hautier *et al.* 2006, Straub and Snyder 2006). Alyokhin *et al.* (2005, 2011), using a 34-year record of aphid and natural enemy populations on potatoes in northern Maine, concluded that pest populations were regulated in a density-dependent manner. Predators suppressed buckthorn and potato aphids, potato aphids were also suppressed by entomopathogenic fungi, green peach aphids were negatively affected by interactions with the former two species, and parasitoids did not significantly affect any of the three aphid populations. Management implications of this study were that chemical controls should minimize impacts on predators and entomopathogenic fungi, and that non-damaging populations of buckthorn and potato aphids should be allowed to develop, because these might help prevent population increase in green peach aphid, which is recognized as a more effect virus vector in potatoes (Alyokhin *et al.* 2011). Insecticide applications often only temporarily suppress aphid populations, resulting in decimation of natural enemies and subsequent resurgence of aphids within a few weeks (Ito *et al.* 2005, Horne and Page 2008). Evolution of insecticide resistance in aphids aggravates this phenomenon (Hardin *et al.* 1995). Even insecticides applied specifically against aphids (e.g., pirimicarb, Jansen *et al.* 2011) and those allowed in organic farming (Jansen *et al.* 2010) are often damaging to aphid antagonists. However, there are several more-or-less specific aphidicides which minimally impact aphid predators and parasitoids, avoiding pest resurgence (Hautier *et al.* 2006). Border treatments may reduce chemical quantities by over 90% while achieving similar aphid population suppression, due to concentration of colonizing alates in field edges (Carroll *et al.* 2009). Sampling of alate aphids has recently shown that they carry a wide variety of entomopathogens and also larval parasitoids, which inoculate colonizing populations with these natural enemies (Feng *et al.* 2007, Huang *et al.* 2008). In some areas, if potatoes are grown using certified seed (reducing concerns of virus transmission), avoidance of all insecticide applications may be practical (e.g., Ito *et al.* 2005, Ito and Furukawa 2009). Lack of knowledge of widely varying aphid and virus levels has encouraged prophylactic treatment with systemic insecticides and/or calendar foliar treatments which are often not needed (Ragsdale *et al.* 1994).

BIOLOGICAL CONTROL INTERACTION WITH OTHER MANAGEMENT METHODS

Interaction with Chemical Control

There are countless instances of non-selective insecticides impairing natural enemy function, up to and including decimating their numbers and causing resurgence both of the target pests, and of other herbivores in the absence of their respective natural enemies. In potato systems, there is no shortage of examples. Horne and Page (2008) show resurgence of potato aphid (*M. euphorbiae*) populations approximately 1 month after pyrethroid application against lepidoptera larvae. The pyrethroid esfenvalerate was highly toxic to *C. maculata* and *Chysoperla carnea* (Stephens), natural enemies of CPB (Hamilton and Lashomb 1997), as was fenvalerate and methamidiphos to *E. puttleri* (Obrycki *et al.* 1986), endosulfan, methamidophos, and permethrin to PTW parasitoid *Orgilus lepidus* (Symington 2003), and imidacloprid to *C. maculata* (Lucas *et al.* 2004). All of these insecticides are commonly used in potato applications. Insecticides specific to targeted key pests, such as *Bt* strains for lepidoptera or CPB, and the aphidicides pymetrozine and flonicamide, are highly compatible with arthropod natural enemies (Ferro 1994, Jansen *et al.* 2011). Deployment of border treatments, even with relatively non-selective materials, also has the potential for large cost savings (Carroll *et al.* 2009); additional delayed benefits also accrue, by conservation of predators and parasitoids. Conversely, effectiveness of natural enemies serves to avoid pesticide applications, thereby mitigating or delaying insecticide resistance on the part of target pests (Alyokhin *et al.* 2008). Interactions of predators with multiple prey may heighten this effect (Mallampalli *et al.* 2005).

Interaction with Biopesticides and Nematodes

Microbial biopesticides can be extremely selective in favor of natural enemies; for instance, *Bacillus thuringiensis tenebrionis* (*Btt*) does no harm to coccinellids regardless of application method (Lucas *et al.* 2004, Kühne 2010). Ramirez and Snyder (2009) found a fascinating synergistic effect on Colorado potato beetle between above-ground generalist predators and below-ground pathogens (*Beauveria* and the nematodes *Heterorhabditis marelatus* and *Steinernema carpocapsae*). Negative interaction between arthropod natural enemies and entomopathogens is of concern. For example, *Beauveria* strains varied from harmless to quite pathogenic against *C. maculata*, making at least 6 of 10 strains tested unacceptable for these non-target effects (Todorova *et al.* 2000).

Interaction with Cultural Controls

If adequate cultural controls are not in place, biological control will often not succeed either through excessive pest populations or through frequent insecticide applications, or through both. For instance, for potato tuber moth, covering tubers with intact (not cracked) soil during the growing season restricts their infestation of the potato plant to the canopy, both avoiding tuber damage and providing access to hosts for a suite of parasitoids and predators which can attack them there (Von Arx *et al.* 1990, Kroschel 2006). Crop rotation is necessary to provide a foundation for biological control as part of an integrated, sustainable approach to Colorado potato beetle management (Alyokhin *et al.* 2008); it is also essential for Andean potato weevil management, and possibly beneficial to hadda beetle and potato lady beetle management (see above).

Interaction with Crop Resistance

Plant cultivar affects the success of natural enemies through a variety of mechanisms, including access to the host over time (e.g., positive effect on *C. maculata* predation on CPB feeding on tomatoes, Lu *et al.* 1996). Studying potato and tomato host plants, Baggen and Gurr (1995) found a negative lethal effect of glandular trichomes of tomato, and Gooderham *et al.* (1998) found a negative non-lethal effect of non-glandular foliar pubescence on the potato tuberworm parasitoid *Copidosoma koehleri*. Therefore, plant resistance to pests can be favorable or detrimental to natural enemies, depending on the mechanism involved and the pest and the beneficial species. Arpaia *et al.* (1997) proposed that predation could in theory slow or accelerate evolution of CPB resistance to *Bt*-transgenic potatoes, and showed that *C. maculata* could decrease this undesirable selection – a process which could also work to slow pest resistance to non-transgenic crop resistance.

CURRENT AND FUTURE RESEARCH NEEDS

The potato crop is surprisingly tolerant of foliar damage that is inflicted by many of the important potato pests (Alyokhin 2009). Pest damage at certain growth stages can even result in overcompensation, as demonstrated for Guatemalan potato tuberworm by Poveda *et al.* (2010). Yet pest management practices rarely take this compensatory capacity, or any scientifically established pest thresholds, into account (Alyokhin 2009). Chemical control is the dominant pest management tactic for potato pests (see Chapter 13), often applied too frequently and/or in excess. This has negative consequences for pest management. In the short run, target pests and other species resurge because their natural enemies are no long present. Also, especially in developing countries, pesticide exposure reaches dangerous levels (Giri *et al.* 2009, Orozco *et al.* 2009). In the long term, pesticide resistance causes more pest damage, and increased expense in attempting to control it. In order for biological controls to contribute to management of potato insect pests, pesticide applications must be based on rational decision-making, be more selective for the target pest(s), and take account of these short- and long-term risks. For this to happen, potato growers must be more knowledgeable regarding both the risks and the benefits of more sustainable strategies (Kroschel *et al.* 2012).

Conservation Biocontrol

There are ample opportunities for enhancing the performance of both native and introduced natural enemies using reduced applications of selective insecticides or bioinsecticides only when necessary (e.g., Horne and Page 2008). Moreover, provision of food resources for parasitoids and predators based on scientific knowledge of their requirements, which is quite poor at the present moment, would further benefit natural enemies. Companion plantings may provide some biocontrol benefits (e.g., Patt *et al.* 1997) or fail altogether (Moreau *et al.* 2006) if their multiple benefits and risks are not known. The example of parasitoid-selective flowering plants offered by Baggen and Gurr (1998) and Baggen *et al.* (1999) is both directly relevant to potato pest biocontrol, and a powerful paradigm for sorely-needed practical research to develop conservation biocontrol tools. Some very easy steps to provide in-field resources for natural enemies could, a number of studies suggest, increase the longevity and fecundity of parasitoid adults from 2-fold to 20-fold.

Predation impact on pest populations is poorly known, and is often more effective than we realize. Furthermore, it is probably easy to increase, once we have a better understanding of predator biology and ecology. The few studies where exclusion cages have tested the magnitude of suppression by predators (e.g., Coll *et al.* 2000, Butler and Trumble 2012) show that surprises are in store both for the existing power of insect predation, and for the increased potential of predation through conservation.

Augmentative and Inundative Biocontrol

Potato value as a crop makes inundative biocontrol using parasitoids or predators an economically dubious proposition. Inundative control using egg parasitoids to protect potatoes from potato tuberworm has only been attempted on a research basis, and even then has not been promising (Horne and Page 2008). Inundative control of Colorado potato beetle with the exotic egg parasitoid *Edovum puttleri* has only been economically viable in eggplant, a much more valuable crop, when registered pesticides were not effective on CPB (Ferro 1994, Hamilton and Lashomb 1996). Similar practical considerations pertain to possible inundative control with predatory stink bugs in the potato crop.

Augmentative control, on the other hand, has a place where the crop environment is supportive of natural enemies but they are not present early enough in the crop cycle, or in high enough numbers, to provide adequate control. Rearing costs may still discourage large releases, but these are not always necessary if in-field conditions (including in-field nurseries) can be established which allow natural enemies optimal conditions to reproduce, as with wild-caught *Podisus maculiventris* against CPB (Aldrich and Cantelo 1999) and lab-reared *Pediobius foveolatus* against *Epilachna varivestis* in beans (Robbins *et al.* 2010). Candidates for augmentation include *P. foveolatus* against *Epilachna* beetles, especially in temperate Asia where it overwinters poorly or not at all (as in Korea, where it only became a mortality factor late in the growing season (Lee *et al.* 1988)); *Copidosoma koehleri* where there are low populations at some times of the season but climatic conditions allow successful survival in the field; and possibly the potato psyllid parasitoid *Tamarixia triozae*.

Classical and Neoclassical Biocontrol

A number of the major potato pests, including Colorado potato beetle, potato tuberworm, Guatemalan potato tuberworm, and potato psyllid, have invaded large new areas of potatoes and related crops. Because of non-target risks, classical biological control programs consider only specialized natural enemies, and they must continue to take this approach. However, there are still many opportunities, within this framework, to export natural enemies from areas of pest origin, or from areas of origin of closely related species (a so-called neoclassical approach), creating associations of target pests and novel natural enemies. Colorado potato beetle parasitoids, the carabid beetle *Lebia grandis*, and two species of specialist tachinid parasitoids in the genus *Myiopharus*, could benefit potato, eggplant, and tomato agroecosystems in the Old World. The last time such an introduction was attempted was almost 50 years ago, in the USSR (Sikura and Smetnik 1967), before much of the biology, including the overwintering habit of the tachinids, was known. Climatic matching today would yield a broader range of opportunities than for early efforts, when the beetle was restricted to France and adjacent western Europe. For CPB, there is also still the possibility that an egg parasitoid or other effective natural enemy may be found on other *Leptinotarsa* species. Unfortunately, this seems unlikely given the fruitless effort already expended.

Classical biocontrol programs for potato tuberworm have yielded immense returns in crop yield and pest control savings, particularly in southern Africa and Australia. For the three parasitoids most commonly introduced, each region has had a unique combination of relative successes and failures. The failure of some species to establish or to play a significant role in pest suppression may in the future yield to an understanding of biotypes along with more sophisticated climatic models and matching. With the expansion of the Guatemalan potato tuberworm may come opportunities to suppress invasive populations in South America, or in potentially invaded areas such as Africa, with natural enemies yet to be determined.

A potato psyllid classical biological control program for New Zealand is already underway. However, it is not clear that candidate species will ultimately be introduced in this newly invaded territory, both because of non-target concerns, and because native psyllid predators that are already present may be capable, with management, of adequately suppressing the invasive pest.

The Epilachninae are an extremely diverse group offering many possibilities for natural enemy matching. The importation of *Pediobius foveolatus* from India to combat a Mexican pest, *Epilachna varivestis*, in temperate USA, is an example of successful biocontrol in this group of pests. In contrast, for Asian *Epilachna* pests of potatoes and related crops, because potentially powerful endemic natural enemy complexes including parasitoids are already present, the best approach is to research and implement suppression of these pests through conservation and augmentation, combined with appropriate cultural controls and application of monitoring and thresholds with selective chemical or microbial treatments if needed.

Interaction with Future Technologies, Practices, and Growing Regions

Potatoes are grown under an increasingly extreme diverse range of locations, conditions, and pest pressures (see Chapter 1). If our expectation is for this ancient crop to provide even as large a proportion of our food needs tomorrow as it does today, we will have to employ a much more diverse set of tactics, and to combine them in a much more sophisticated manner. These tactics will not employ themselves; they must be brought to and made accessible to the potato grower – who lives and farms under a vast range of economic and environmental conditions. This will require a practical and knowledge-intensive participatory approach (e.g., FAO 2003, Ortiz *et al.* 2004, Dangles *et al.* 2010), which researchers have, to date, rarely taken. Furthermore, IPM must be locally adapted and take into account all pests impinging on the crop, otherwise inappropriate management practices for novel pests – even minor pests or non-pests – may disrupt the entire crop system (Horne and Page 2008, 2009; Kroschel *et al.* 2012). Researchers and growers will have to answer a myriad of questions and will need to apply the answers in a judicious manner, or suffer the consequences of poor pest management and wasted resources. Successful tactics of the future may range from apparently simple ones like intercropping potatoes with onions, which is reported to greatly reduce hadda beetle (Potts 1990), or nursery crops for parasitoids of pest *Epilachna* beetles, to the high technology of detecting insect predation using quantitative PCR, or applying RNA interference to cure insect vectors of their potato huanglongbing. Only if these tactics respect the beneficial organisms in and surrounding potato agroecosystems around the globe are they likely to succeed in providing the opportunity for sustainable management of potato insect pests.

ACKNOWLEDGMENTS

Andrei Alyokhin provided valuable editing and commentary, and also translation of an important publication in Russian. Jian Jun Duan provided translation and commentary on several Chinese publications, which improved the section on *Epilachna* significantly. Several authors provided hard-to-locate publications and guidance on research progress. Michael Athanas and Tony DiMeglio helped obtain numerous publications from the National Agricultural Library.

REFERENCES

Acosta, N.M., O'Neil, R.J., 1999. Life history characteristics of three populations of *Edovum puttleri* Grissell (Hymenoptera: Eulophidae) at three temperatures. Biol. Control 16, 81–87.

AgroAtlas, 2012. Interactive Agricultural Ecological Atlas of Russia and Neighboring Countries: Economic Plants and their Diseases, Pests and Weeds. Entry for *Epilachna vigintioctomaculata*. www.agroatlas.ru/en/content/pests/Epilachna_vigintioctomaculata.

Alcalá, C.P., Alcázar, J., 1976. Biologia y comportamiento de *Premnotrypes suturicallus* Kuschel (Col.: Curculionidae). Rev. Peru. Entomol. 19, 49–52.

Alcázar, J., Cisneros, F., 1999. Taxonomy and bionomics of the Andean potato weevil complex: *Premnotrypes* spp. and related genera. Centro Internacional de la Papa, Program Report 1997–1998, pp. 141–151.

Aldrich, J.R., 1999. Predators, In: Hardie, J., Minks, A.K. (Eds.), Pheromones of non-lepidopteran insects associated with agricultural plants, CABI Publishing, Wallingford, UK, pp. 357–381.

Aldrich, J.R., Cantelo, W.W., 1999. Suppression of Colorado potato beetle infestation by pheromone-mediated augmentation of the predatory spined soldier bug, *Podisus maculiventris* (Say) (Heteroptera: Pentatomidae). Agric. Forest Entomol. 1, 209–217.

Al-Jabr, A.M., 1999. Integrated Pest Management of tomato/potato psyllid, *Paratrioza cockerelli* (Sulc) (Homoptera: Psyllidae) with emphasis on its importance in greenhouse grown tomatoes. PhD dissertation. Colorado State University, Fort Collins, CO.

Alvarado, V.Y., Duncan, O., Odokonyero, D., Scholthof, H.B., 2010. Quantifying "*Candidatus* Liberibacter" and other microbes in psyllids in relation to insect colony behavior and ZC disease. In: Workneh, F., Rush, C.M. (Eds.), Proceedings, 10th Annual Zebra Chip Reporting Session (Specialty Crop Research Initiative), 7–10 November 2010, Dallas, Texas, pp. 24–28.

Alvarez, J.M., Dotseth, E., Nolte, P., 2005. Potato tuberworm: A threat for Idaho potatoes. University of Idaho Extension Leaflet CIS-1125, 4pp.

Alyokhin, A., 2009. Colorado potato beetle management on potatoes: current challenges and future prospects. Fruit Veg. Cereal Sci. Biotechnol. 3, 10–19.

Alyokhin, A., Drummond, F.A., Sewell, G., 2005. Density-dependent regulation in populations of potato-colonizing aphids. Popul. Ecol. 47, 257–266.

Alyokhin, A., Baker, M., Mota-Sanchez, D., Dively, G., Grafius, E., 2008. Colorado Potato beetle resistance to insecticides. Am.. J. Potato Res. 85, 395–413.

Alyokhin, A., Drummond, F.A., Sewell, G., Storch, R.H., 2011. Differential effects of weather and natural enemies on coexisting aphid populations. Environ. Entomol. 40, 570–580.

Araujo-Siqueira, M., Massutti de Almeida, L., 2004. Comportamento e ciclo de vida de *Epilachna vigintioctopunctata* (Fabricius) (Coleoptera, Coccinellidae) em *Lycopersicum esculentum* Mill. (Solanaceae). Rev. Bras. Zool. 21, 543–550.

Arpaia, S., Gould, F., Kennedy, G., 1997. Potential impact of *Coleomegilla maculata* predation on adaptation of *Leptinotarsa decemlineata* to Bt-transgenic potatoes. Entomol. Exp. Appl. 82, 91–100.

Báez, F., Gallegos, P., 2011. Prospección y eficiencia de parasitoides nativos de las polillas de la papa *Tecia solanivora* Povolny, *Symmetrischema tangolias* Gyen y *Phthorimaea operculella* Zeller (Lepidoptera: Gelechiidae) en el Ecuador. Memorias Completas del 4˚ Congreso Ecuatoriano de la Papa, Guaranda, Ecuador, 28–30, June 2011. www.quito.cipotato.org/ivcongresopapa_esp.htm.

Baggen, L.R., Gurr, G.M., 1995. Lethal effects of foliar pubescence of solanaceous plants on the biological control agent *Copidosoma koehleri* Blanchard (Hymenoptera: Encyrtidae). Plant Prot. Q. 10, 116–118.

Baggen, L.R., Gurr, G.M., 1998. The influence of food on *Copidosoma koehleri* (Hymenoptera: Encyrtidae), and the use of flowering plants as a habitat management tool to enhance biological control of potato moth, *Phthorimaea operculella* (Lepidotera: Gelechiidae). Biol. Control 11, 9–17.

Baggen, L.R., Gurr, G.M., Meats, A., 1999. Flowers in tri-trophic systems: mechanisms allowing selective exploitation by insect natural enemies for conservation biological control. Entomol. Exp. Appl. 91, 155–161.

Baumann, P., 2005. Biology of bacteriocyte-associated endosymbionts of plant sap-sucking insects. Annu. Rev. Microbiol. 59, 155–189.

Berry, N.A., Scott, I., Thompson, S., Beard, S., 2010. Detection of *Candidatus* Liberibacter solanacearum in trapped insects and non-crop plants in New Zealand. In: Workneh, F., Rush, C.M. (Eds.), Proceedings of the 10th Annual Zebra Chip Reporting Session (Specialty Crop Research Initiative), Dallas, Texas, pp. 159–163.

Brust, G.E., 1994. Natural enemies in straw-mulch reduce Colorado potato beetle populations and damage in potato. Biol. Control 4, 163–169.

Buchman, J.L., Sengoda, V.G., Munyaneza, J.E., 2010. Potato psyllid density and feeding duration required to cause zebra chip. In: Workneh, F., Rush, C.M. (Eds.), Proceedings of the 10th Annual Zebra Chip Reporting Session (Specialty Crop Research Initiative), Dallas, Texas, pp. 187–191.

Burks, B.D., 1943. The North American parasitic wasps of the genus *Tetrastichus*. Proc. US Natl. Mus. 93, 505–608.

Butler, C.D., Trumble, J.T., 2010. Distribution and phenology of the potato psyllid in southern California. In: Workneh, F., Rush, C.M. (Eds.), Proceedings of the 10th Annual Zebra Chip Reporting Session (Specialty Crop Research Initiative), Dallas, Texas, pp. 37–41.

Butler, C.D., Trumble, J.T., 2011. New records of hyperparasitism of *Tamarixia triozae* (Burks) (Hymenoptera: Eulophidae) by *Encarsia* spp. (Hymenoptera: Aphelinidae) in California. Pan-Pac. Entomol. 87, 130–133.

Butler, C.D., Trumble, J.T., 2012. The potato psyllid, *Bactericera cockerelli* (Sulc) (Hemiptera: Triozidae): life history, relationship to plant diseases, and management strategies. Terr. Arthropod Rev. 5, 87–111.

Butler, C.D., Novy, R.G., Miller, J.C., Trumble, J.T., 2010. In: Alternative strategies: Plant resistance and biological control. Workneh, F., Rush, C.M. (Eds.), Proceedings of the 10th Annual Zebra Chip Reporting Session (Specialty Crop Research Initiative), Dallas, Texas, pp. 69–73.

CABI (Centre for Agriculture and Biosciences International), 2010. The Crop Protection Compendium, 2010 Edition, CABI Publishing, Wallingford, UK, www.cabi.org/cpc2010.

Canadian Food Inspection Agency, 2011. List of approved arthropod biological control agents from commercial sources that do not fall under the mandate of the Plant Protection Act. www.inspection.gc.ca/english/plaveg/protect/dir/biocontrole.shtml.

Cappaert, D.L., Drummond, F.A., Logan, P.A., 1991. Incidence of natural enemies of the Colorado potato beetle, *Leptinotarsa decemlineata* (Coleoptera: Chrysomelidae) on a native host in Mexico. Entomophaga 36, 369–378.

Carroll, M.W., Radcliffe, E.B., MacRae, I.V., Ragsdale, D.W., Olson, K.D., Badibanga, T., 2009. Border treatment to reduce insecticide use in seed potato production: Biological, economic, and managerial analysis. Am.. J. Potato Res. 86, 31–37.

Chaboussou, F., 1939. Contribution à l'étude biologique de *Lebia grandis* Hentz, prédateur américain du doryphore. Ann. Épiphyt. Phytogénet. 5, 387–433.

Chandel, R.S., Dhiman, K.R., Chandla, V.K., Kumar, V., 2007. Integrated pest management in potato.In: Jain, P.C., Bhargava, M.C. (Eds.), Entomology: Novel Approaches, New India Publishing Agency, New Delhi, India, pp. 377–398.

Chandla, V.K., Verma, K.D., 2000. Present status of potato tuber moth and its management. In: Khurana, S.M., Shekhawat, G.S., Singh, B.P., Pandey, S.K. (Eds.), Potato, global research and development: Proceedings of the Global Conference on Potato, 6–11 December 1999, Indian Potato Association, New Delhi, India, pp. 363–369.

Charlet, L.D., 2003. Incidence of sunflower beetle (Coleoptera: Chrysomelidae) and parasitism of its larvae by *Myiopharus macellus* (Reinhard) (Diptera: Tachinidae) in native sunflowers in North Dakota and Minnesota. J. Kans. Entomol. Soc. 76, 436–441.

CIP (Centro Internacional de la Papa), 2012. Potential IPM options ready for implementation: Inoculative biological control. http://cipotato.org/research/crop-management-production-systems.

Cloutier, C., Bauduin, F., 1995. Biological control of the Colorado potato beetle *Leptinotarsa decemlineata* (Coleoptera: Chrysomelidae) in Quebec by augmentative releases of the two-spotted stinkbug *Perillus bioculatus* (Hemiptera: Pentatomidae). Can. Entomol. 127, 195–212.

Cloutier, C., Boiteau, G., Goettel, M.S., 2002. *Leptinotarsa decemlineata* (Say), Colorado potato beetle (Coleoptera: Chrysomelidae). In: Mason, P.G., Huber, J.T. (Eds.), Biological Control Programmes in Canada, CABI Publishing, Wallingford, UK, pp. 145–152.

Colazza, S., Bin, F., 1992. Introduction of the oophage *Edovum puttleri* Griss. (Hymenoptera: Eulophidae) in Italy for the biological control of the Colorado potato beetle. Redia 75, 203–225.

Colazza, S., Czepak, C., Prosperi, G., 1996. Prove in campo di controllo biologico della dorifora della patata con due preditori esotici, *Podisus maculiventris* and *P. connexivus* (Eterotteri: Pentatomidi). Inf. Fitopatol. 46, 44–48.

Coll, M., Gavish, S., Dori, I., 2000. Population biology of the potato tuber moth, *Phthorimaea operculella* (Lepidoptera: Gelechiidae), in two potato cropping systems in Israel. Bull. Entomol. Res. 90, 309–315.

Compere, H., 1943. A new species of *Metaphycus* parasite on psyllids. Pan-Pac. Entomol. 19, 71–73.

Crosslin, J.M., Munyaneza, J.E., Brown, J.K., Liefting, L.W., 2010. A history in the making: Potato zebra chip disease associated with a new psyllid-borne bacterium – a tale of striped potatoes. Online APSnet Features, 2010–0110. http://www.apsnet.org/publications/apsnetfeatures/Pages/PotatoZebraChip.aspx

CruzRoblero, E.N., Castillo Vera, A., Malo, E.A., 2011. First report of *Tecia solanivora* (Lepidoptera: Gelechiidae) Attacking the potato *Solanum tuberosum* in Mexico. Fla. Entomol. 94, 1055–1056.

Dangles, O., Carpio, C., Barragan, A.R., Zeddam, J., Silvain, J., 2008. Temperature as a key driver of ecological sorting among invasive pest species in the tropical Andes. Ecol. Appl. 18, 1795–1809.

Dangles, O., Mesias, V., Crespo-Perez, V., Silvain, J., 2009. Crop damage increases with pest species diversity: evidence from potato tuber moths in the tropical Andes. J. Appl. Ecol. 46, 1115–1121.

Dangles, O., Carpio, F.C., Villares, M., Yumisaca, F., Liger, B., Rebaudo, F., Silvain, J.F., 2010. Community-based participatory research helps farmers and scientists to manage invasive pests in the Ecuadorian Andes. Ambio 39, 325–335.

Daunay, M.C., 2008. Eggplant, Vegetables II In: Prohens, J., Nuez, F. (Eds.), Fabaceae, Liliaceae, Solanaceae, and Umbelliferae (Handbook of Plant Breeding), Springer, New York, New York, pp. 163–220.

ERS (Economic Research Service), 2012. US Department of Agriculture, US Potato Statistics, www.ers.usda.gov2012 (91011 and "Organic Production").

FAO Inter-country Programme for IPM in Vegetables in South and Southeast Asia., 2003. Eggplant Integrated Pest Management. An Ecological Guide. FAO Regional Office for Asia and the Pacific, Bangkok, Thailand.

Fathi, S.A., 2011. Population density and life-history parameters of the psyllid *Bactericera nigricornis* (Forster) on four commercial cultivars of potato. Crop Prot. 30, 844–848.

Feng, M., Chen, C., Shang, S., Ying, S.H., Shen, Z., Chen, X., 2007. Aphid dispersal flight disseminates fungal pathogens and parasitoids as natural control agents of aphids. Ecol. Entomol. 32, 97–104.

Ferro, D.N., 1994. Biological control of the Colorado potato beetle, In: Zehnder, G.W., Jansson, R.K., Powelson, M.L., Raman, K.V. (Eds.), Advances in Potato Pest Biology and Management, American Phytopathological Society Press, St Paul, Minnesota, pp. 357–375.

Flint, M.L., 2006. Integrated pest management for potatoes in the western United States, 2nd Edition. Statewide Integrated Pest Management Program University of California Division of Agriculture and Natural Resources Publication, 3316.

Folcia, A.M., Rodríguez, S.M., Russo, S., 1996. Aspectos morfológicos, biológicos y de preferencia de *Epilachna vigintioctopunctata* Fabr. (Coleóptera: Coccinellidae). Boletín de Sanidad Vegetal, Plagas 22, 773–780.

Gallegos, P., Castillo, C., 2011. Manejo integrado de gusano blanco en papa. Memorias Completas del 4° Congreso Ecuatoriano de la Papa. Guaranda, Ecuador, 28–30 June 2011. www.quito.cipotato.org/ivcongresopapa_esp.htm.

Gallegos, P., Asasquibay, C., Chamorro, F., Rodríguez, P., Williams, R., 2005. Asolacion de los tuberculos de semilla de papa como método de control para la polilla, Una tecnología tradicional para la solución a un nuevo problema *Tecia solanivora*. Instituto Nacional Autónomo de Investigaciones Agropecuarias (INIAP), Quito, Ecuador.

Gardner-Gee, R., 2011. Progress towards biological control of *Bactericera cockerelli* in covered crops in New Zealand. IOBC/WPRS Bull. 68, 41–45.

Garmendia, A., 1961. Observaciones sobre un posible método de control biológico de la gusanera de la papa depositada en almacén. Rev. Peru. Entomol. Agríc 4, 76–77.

Gerber, E., Schaffner, U., Gassmann, A., Hinz, H.L., Seier, M., Müller-Schärer, H., 2011. Prospects for biological control of *Ambrosia artemisiifolia* in Europe: learning from the past. Weed Res. 51, 559–573.

Giri, Y.P., Maharjan, R., Sporleder, M., Kroschel, J., 2009. Pesticide use practices and awareness among potato growers in Nepal. In: Papers of 15th Triennial Symposium of the International Society for Tropical Root Crops (ISTRC), 2–6 November, Lima, Peru, pp. 21–30.

Giroux, S., Duchesne, R., Coderre, D., 1995. Predation of *Leptinotarsa decemlineata* (Coleoptera: Chrysomelidae) by *Coleomegilla maculata* (Coleoptera: Coccinellidae): comparative effectiveness of predator developmental stages and effect of temperature. Environ. Entomol. 24, 748–754.

Gooderham, J., Bailey, P.C., Gurr, G.M., Baggen, L.R., 1998. Sub-lethal effects of foliar pubescence on the egg parasitoid *Copidosoma koehleri* and influence on parasitism of potato moth, *Phthorimaea operculella*. Entomol. Exp. Appl. 87, 115–118.

Greenstone, M.H., Rowley, D.L., Weber, D.C., Payton, M.E., Hawthorne, D.J., 2007. Feeding mode and prey detectability half-lives in molecular gut-content analysis: an example with two predators of the Colorado potato beetle. Bull. Entomol. Res. 97, 201–209.

Greenstone, M.H., Szendrei, Z., Payton, M.E., Rowley, D.L., Coudron, T.A., Weber, D.C., 2010. Choosing natural enemies for conservation biological control: use of the prey detectability half-life to rank key predators of Colorado potato beetle. Entomol. Exp. Appl. 136, 97–107.

Grissell, E.E., 1991. *Edovum puttleri*, n.g., n. sp. (Hymenoptera: Eulophidae), an egg parasite of the Colorado potato beetle (Chrysomelidae). Proc. Entomol. Soc. Wash. 83, 790–796.

Groden, E., 1989. Natural mortality of the Colorado potato beetle. Leptinotarsa decemlineata (Say). PhD dissertation. Michigan State University, East Lansing, MI.

Guédot, C., Horton, D.R., Landolt, P.J., 2010. Sex attraction in *Bactericera cockerelli* (Hemiptera: Triozidae). Environ. Entomol. 39, 1302–1308.

Guerrieri, E., Noyes, J.S., 2000. Revision of European species of genus *Metaphycus* Mercet Hymenoptera: Chalcidoidea: Encyrtidae), parasitoids of scale insects (Homoptera: Coccoidea). Syst. Entomol. 25, 147–222.

Hamilton, G.C., Lashomb, J., 1996. Comparison of conventional and biological control intensive pest management programs on eggplant in New Jersey. Fla. Entomol. 79, 488–496.

Hamilton, G.C., Lashomb, J.H., 1997. Effect of insecticides on two predators of the Colorado potato beetle (Coleoptera: Chrysomelidae). Fla. Entomol. 80, 10–23.

Hanafi, A., 2005. Integrated management of potato tuber moth in field and storage. In: Haverkort, A.J., Struik, P.C. (Eds.), Potato in Progress: Science Meets Practice, Wageningen Academic Publishers, Wageningen, The Netherlands, pp. 203–210.

Hansen, A.K., Jeong, G., Paine, T.D., Stouthamer, R., 2007. Frequency of secondary symbiont infection in an invasive psyllid relates to parasitism pressure on a geographic scale in California. Appl. Environ. Microbiol. 73, 7531–7535.

Hansen, A.K., Trumble, J.T., Stouthamer, R., Paine, T.D., 2008. A new huanglongbing species, "*Candidatus* Liberibacter psyllaurous", found to infect tomato and potato, is vectored by the psyllid *Bactericera cockerelli* (Sulc). Appl. Environ. Microbiol. 74, 5862–5865.

Hardin, M.R., Benrey, B., Coll, M., Lamp, W.O., Roderick, G.K., Barbosa, P., 1995. Arthropod pest resurgence: An overview of potential mechanisms. Crop Prot. 14, 3–18.

Hautier, L., Jansen, J.P., Mabon, N., Schiffers, B., 2006. Building a selectivity list of plant protection products on beneficial arthropods in open field: A clear example with potato crop. IOBC–WPRS Bull. 29, 21–32.

Hazzard, R.V., Ferro, D.N., Van Driesche, R., Tuttle, A.F., 1991. Mortality of eggs of Colorado potato beetle (Coleoptera: Chrysomelidae) from predation by *Coleomegilla maculata* (Coleoptera: Coccinellidae). Environ. Entomol. 20, 841–848.

Heimpel, G.E., Hough-Goldstein, J.A., 1992. A survey of arthropod predators of *Leptinotarsa decemlineata* (Say) in Delaware potato fields. J. Agric. Entomol. 9, 137–142.

Herman, T.J.B., 2008. Biological control of potato tuber moth, by *Apanteles subandinus* Blanchard in New Zealand. In: Kroschel, J., Lacey, L. (Eds.), Integrated Pest Management for the Potato Tuber Moth, *Phthorimaea operculella* Zeller – a Potato Pest of Global Importance. Tropical Agriculture 20, Advances in Crop Research 10, Margraf Publishers, Weikersheim, Germany, pp. 73–80.

Hilbeck, A., Kennedy, G.G., 1996. Predators feeding on the Colorado potato beetle in insecticide-free plots and insecticide-treated commercial potato fields in eastern North Carolina. Biol. Control 6, 273–282.

Hilbeck, A., Eckel, C., Kennedy, G.G., 1997. Predation on Colorado potato beetle eggs by generalist predators in research and commercial potato plantings. Biol. Control 8, 191–196.

Hoffmann, J.H., Moran, V.C., Impson, F.A.C., 1998. Promising results from the first biological control programme against a solanaceous weed (*Solanum elaeagnifolium*). Agric. Ecosyst. Environ. 70, 145–150.

Horne, P.A., 1990. The influence of introduced parasitoids on the potato moth, *Phthorimaea operculella* (Lepidoptera: Gelechiidae) in Victoria, Australia. Bull. Entomol. Res. 80, 159–164.

Horne, P., Page, J., 2008. Integrated Pest Management dealing with potato tuber moth and all other pests in Australian potato crops. In: Kroschel, J., Lacey, L. (Eds.), Integrated Pest Management for the Potato Tuber Moth, *Phthorimaea operculella* Zeller – a Potato Pest of Global Importance. Tropical Agriculture 20, Advances in Crop Research 10, Margraf Publishers, Weikersheim, Germany, pp. 111–117.

Horne, P., Page, J., 2009. IPM in Australian potato crops and the threat from potato psyllid. In: Nelson, W. (Ed.), 7th World Potato Congress, Solanaceous crops, Psyllids and Liberibacter, Proceedings of the Workshop, Christchurch, New Zealand, pp. 73–78.

Horton, D.R., Capinera, J.L., 1987. Seasonal and host plant effects on parasitism of Colorado potato beetle by *Myiopharus doryphorae* (Riley) (Diptera: Tachinidae). Can. Entomol. 119, 729–734.

Hough-Goldstein, J., 1998. Use of predatory pentatomids in integrated management of the Colorado potato beetle (Coleoptera: Chrysomelidae). In: Coll, M., Ruberson, J.R.E. (Eds.), Predatory Heteroptera: Their ecology and use in biological control. Thomas Say Publications in Entomology Proceedings, Entomological Society of America, Lanham, Maryland, pp. 209–223.

Hough-Goldstein, J., Janis, J.A., Ellers, C.D., 1996. Release methods for *Perillus bioculatus* (F.), a predator of the Colorado potato beetle. Biol. Control 6, 114–122.

Huang, Z., Feng, M., Chen, X., Liu, S., 2008. Pathogenic fungi and parasitoids of aphids present in air captures of migratory alates in the low-latitude plateau of Yunnan, China. Environ. Entomol. 37, 1264–1271.

Ito, K., Furukawa, K., 2009. Effects of insecticide-free potato cultivation on the occurrence of aphids, their predators and tuber yield. Jpn. J. Appl. Entomol. Zool. 53, 45–51.

Ito, K., Furukawa, K., Okubo, T., 2005. Conservation biological control of aphids in potato fields with reduced use of insecticides in Hokkaido, Japan. Jpn. J. Appl. Entomol. Zool. 49, 11–22.

Ivanova, A.N., 1962. The Potato Ladybird in the Far East. Primorskii Publishing House, Vladivostok, USSR. 53 pp.

Jansen, J.P., Defrance, T., Warnier, A.M., 2010. Effects of organic-farming-compatible insecticides on four aphid natural enemy species. Pest Manage. Sci. 66, 650–656.

Jansen, J.P., Defrance, T., Warnier, A.M., 2011. Side effects of flonicamide and pymetrozine on five aphid natural enemy species. BioControl 56, 759–770.

Jensen, D.D., 1957. Parasites of the Psyllidae. Hilgardia 27, 71–99.

Jermy, T., 1980. The introduction of *Perillus bioculatus* into Europe to control the Colorado beetle. EPPO Bull. 10, 475–479.

Johnson, S.A., 1971. The effectiveness of *Tetrastichus triozae* Burks (Hymenoptera: Eulophidae) as a biological control agent of *Paratrioza cockerelli* (Sulc.) (Homoptera: Psyllidae) in north central Colorado. MS Thesis. Colorado State University, Fort Collins, CO.

Juen, A., Traugott, M., 2007. Revealing species-specific trophic links in soil food webs: molecular identification of scarab predators. Mol. Ecol. 16, 1545–1557.

Kaya, H.K., Alcazar, J., Parsa, S., Kroschel, J., 2009. Microbial control of the Andean potato weevil complex. Fruit Veg. Cereal Sci. Biotechnol. 3, 39–45.

Keasar, T., Sadeh, A., 2007. The parasitoid *Copidosoma koehleri* provides limited control of the potato tuber moth, *Phthorimaea operculella*, in stored potatoes. Biol. Control 42, 55–60.

Keller, M.A., Horne, P.A., 1993. Source of host-location cues for the parasitic wasp *Orgilus lepidus* (Braconidae). Aust. J. Zool. 41, 335–341.

Kernasa, O., Noynonmuang, K., Suasa-ard, W., 2002. Biological study of 28-spotted lady beetle, *Henosepilachna vigintioctopunctata* (F.) (Coleoptera: Coccinellidae) and its natural enemies. Annual Symposium of the National Biological Control Research Center, Ubon Ratchathani, Thailand, pp. 21–22.

Kfir, R., 2003. Biological control of the potato tuber moth *Phthorimaea operculella* in Africa. In: Neuenschwander, P., Borgemeister, C., Langewald, J. (Eds.), Biological Control in IPM Systems in Africa, CABI Publishing, Wallingford, UK, pp. 77–85.

Knowlton, G.F., 1933a. Aphis lion predators of the potato psyllid. J. Econ. Entomol. 26, 977.

Knowlton, G.F., 1933b. Ladybird beetles as predators of the potato psyllid. Can. Entomol. 65, 241–243.

Knowlton, G.F., 1934a. Potato psyllid investigations. Proc. Utah Acad. Sci. 11, 261–265.

Knowlton, G.F., 1934b. A big-eyed bug predator of the potato psyllid. Fla. Entomol. 18, 40–43.

Knowlton, G.F., Allen, M., 1936. Three hemipterous predators of the potato psyllid. Proc. Utah Acad. Sci. 13, 293–294.

Kogan, M., Gerling, D., Maddox, J.V., 1999. Enhancement of biological control in annual agricultural environments. In: Bellows, T.S., Fisher, T.W. (Eds.), Handbook of Biological Control, Academic Press, New York, NY, pp. 789–818.

Koppert, México, 2012. Código de Productos Koppert. www.koppert.com.mx/lista_productos.htm.

Krieg, A., Schnetter, W., Huger, A.M., Langenbruch, G.A., 1987. *Bacillus thuringiensis* subsp. *tenebrionis*, strain BI256–82: a third pathotype within the H-serotype 8a8b. Syst. Appl. Microbiol. 9, 138–141.

Kroschel, J., 2006. Management of the potato tuber moth, an invasive pest of global proportions. 6th World Potato Congress, 20–26 August 2006, Boise, Idaho, USA.

Kroschel, J., Koch, W., 1994. Studies on the population dynamics of the potato tuber moth (*Phthorimaea operculella* Zell. (Lep., Gelechiidae)) in the Republic of Yemen. J. Appl. Entomol. 118, 327–341.

Kroschel, J., Koch, W., 1996. Studies on the use of chemicals, botanicals and *Bacillus thuringiensis* in the management of the potato tuber moth in potato stores. Crop Prot. 15, 197–203.

Kroschel, J., Zegarra, O., 2010. Attract-and-kill: a new strategy for the management of the potato tuber moths *Phthorimaea operculella* (Zeller) and *Symmetrischema tangolias* (Gyen) in potato: laboratory experiments towards optimising pheromone and insecticide concentration. Pest Manage. Sci. 66, 490–496.

Kroschel, J., Alcázar, J., Poma, P., 2009. Potential of plastic barriers to control Andean potato weevil *Premnotrypes suturicallus* Kuschel. Crop Prot. 28, 466–476.

Kroschel, J., Mujica, N., Alcazar, J., Canedo, V., Zegarra, O., 2012. Developing integrated pest management for potato: Experiences and lessons from two distinct potato production systems of Peru. In: He, Z., Larkin, R., Honeycutt, W. (Eds.), Sustainable Potato Production: Global Case Studies, Springer Publishing, New York, NY, pp. 419–450.

Kühne, M., Alcázar, J., Vidal, S., Jung, K., 2007. Biology and management of the Andean potato weevil – A review. Ph.D. dissertation. Georg-August-Universität, Göttingen, Germany.

Kühne, S., 2010. Regulierung des Kartoffelkäfers (*Leptinotarsa decemlineata* Say) mit biologischen Pflanzenschutzmitteln (Azadirachtin, B.t.t., Pyrethrum, Spinosad) und deren Nebenwirkungen auf Blattlausprädatoren im ökologischen Kartoffelanbau. J. Kulturpflanz. 62, 331–340.

LaChance, S., Cloutier, C., 1997. Factors affecting dispersal of *Perillus bioculatus* (Hemiptera: Pentatomidae), a predator of the Colorado potato beetle (Coleoptera: Chrysomelidae). Environ. Entomol. 26, 946–954.

Lall, B.S., 1961. On the biology of *Pediobius foveolatus* (Eulophidae: Hymenoptera). Indian J. Entomol. 23, 268–273.

Larsen, N.J., MacDonald, F.H., Connolly, P.G., Walker, G.P., 2011. Could *Harmonia conformis* be an important predator of *Bactericera cockerelli*? N. Z. Plant Prot. 64, 293.

Lee, J., Reed, D.K., Lee, H., Carlson, R.W., 1988. Parasitoids of *Henosepilachna vigintioctomaculata* (Moschulsky) (Coleoptera: Coccinellidae) in Kyonggido area, Korea. Korean J. Appl. Entomol. 27, 28–34.

López, E.R., Ferro, D.N., 1990. *Myiopharus aberrans*, a tachinid parasitoid of the Colorado potato beetle: foraging behavior under field conditions. Oral presentation. Entomological Society of America Annual Meetings 3 December 1990, New Orleans, LA.

López, R., Ferro, D.N., van Driesche, R.G., 1992. Overwintering biology of *Myiopharus aberrans* and *Myiopharus doryphorae* (Dipt.: Tachinidae) larval parasitoids of the Colorado potato beetle (Col. : Chrysomelidae). Entomophaga. 37, 311–315.

López, R., Ferro, D.N., van Driesche, R.G., 1995. Two tachinid species discriminate between parasitized and non-parasitized hosts. Entomol. Exp. Appl. 74, 37–45.

López, R., Ferro, D.N., Elkinton, J., 1997a. Temperature-dependent development rate of *Myiopharus doryphorae* (Diptera: Tachinidae) within its host, the Colorado potato beetle (Coleoptera: Chrysomelidae). Environ. Entomol. 26, 655–660.

López, R., Roth, L.C., Ferro, D.N., Hosmer, D., Mafra-Neto, A., 1997b. Behavioral ecology of *Myiopharus doryphorae* (Riley) and *M. aberrans* (Townsend), tachinid parasitoids of the Colorado potato beetle. J. Insect Behav. 10, 49–78.

Loza, A.L., Apaza, A., 2001. Amplitud depredadora y preferencia de presa en tres especies de carábidos (Coleoptera) del altiplano de Puno, Perú. Rev. Peru. Entomol. 42, 73–78.

Lu, W., Kennedy, G.G., Gould, F., 1996. Differential predation by *Coleomegilla maculata* on Colorado potato beetle strains that vary in growth on tomato. Entomol. Exp. Appl. 81, 7–14.

Lucas, E., Giroux, S., Demougeot, S., Duchesne, R.M., Coderre, D., 2004. Compatibility of a natural enemy, *Coleomegilla maculata lengi* (Col., Coccinellidae) and four insecticides used against the Colorado potato beetle (Col., Chrysomelidae). J. Appl. Entomol. 128, 233–239.

Luna Cruz, A., 2010. Toxicidad de cuatro insecticidas sobre *Tamarixia triozae* (Burks) (Hymenoptera: Eulophidae) y su hospedero *Bactericera cockerelli* (Sulc) (Hemíptera: Psyllidae). M. S. thesis,

Colegio de Postgraduados. Institución de Enseñanza e Investigación en Ciencias Agrícolas, Campus Montecillo, Estado de México, México.

Lundgren, J.G., Razzak, A.A., Wiedenmann, R.N., 2004. Population responses and food consumption by predators *Coleomegilla maculata* and *Harmonia axyridis* (Coleoptera: Coccinellidae) during anthesis in an Illinois cornfield. Environ. Entomol. 33, 958–963.

Lundgren, J.G., Ellsbury, M.E., Prischmann, D.A., 2009. Analysis of the predator community of a subterranean herbivorous insect based on polymerase chain reaction. Ecol. Appl. 19, 2157–2166.

MacDonald, F.H., P Walker, G., Larsen, N.J., Wallace, A.R., 2010. Naturally occurring predators of *Bactericera cockerelli* in potatoes. N. Z. Plant Prot. 63, 275.

MAFBNZ, 2010. Hadda beetle established in Auckland. Ministry of Agirculture and Food, Biosecurity, New Zealand (Wellington). Press release of 25 March 2012. www.biosecurity.govt.nz/media/25–;03–10/hadda-beetle-auckland.

Mallampalli, N., Gould, F., Barbosa, P., 2005. Predation of Colorado potato beetle eggs by a polyphagous ladybeetle in the presence of alternate prey: potential impact on resistance evolution. Entomol. Exp. Appl. 114, 47–54.

Mandour, N.S., Sarhan, A.A., Atwa, D.H., 2012. The integration between *Trichogramma evanescens* West. (Hymenoptera: Trichogrammatidae) and selected bioinsecticides for controlling the potato tuber moth *Phthorimaea operculella* (Zell.) (Lepidoptera: Gelechiidae) of stored potatoes. J. Plant Prot. Res. 52, 40–46.

Manole, T., Iamandei, M., Teodorescu, I., 2002. Biological control of Colorado potato beetle in Romania by inundative releases of *Podisus maculiventris* Say (Heteroptera: Pentatomidae). Rev. Roum. Biol. Ser. Biol. Anim. 47, 117–121.

Mitchell, B.L., 1978. The biological control of potato tuber moth *Phthorimaea operculella* (Zeller) in Rhodesia. Rhod. Agric. J. 75, 55–58.

Moldenke, A.F., Berry, R.E., 1999. Biological control of Colorado potato beetle, *Leptinotarsa decemlineata* (Say) (Chrysomelidae: Chrysomelinae). In: Cox, M.L. (Ed.), Advances in Chrysomelidae Biology 1, Backhuys Publishers, Leiden, The Netherlands, pp. 169–183.

Moreau, T.L., Warman, P.R., Hoyle, J., 2006. An evaluation of companion planting and botanical extracts as alternative pest controls for the Colorado potato beetle. Biol. Agric. Hortic. 23, 351–370.

Munyaneza, J.E., 2010. Psyllids as vectors of emerging bacterial diseases of annual crops. Southwest. Entomol. 35, 471–477.

Munyaneza, J.E., Fisher, T.W., Sengoda, V.G., Garczynski, S.F., Nissinen, A., Lemmetty, A., 2010. Association of "*Candidatus* Liberibacter solanacearum" with the psyllid, *Trioza apicalis* (Hemiptera: Triozidae) in Europe. J. Econ. Entomol. 103, 1060–1070.

Nachappa, P., Levy, J., Pierson, E., Tamborindeguy, C., 2011. Diversity of endosymbionts in the potato psyllid, *Bactericera cockerelli* (Triozidae), vector of zebra chip disease of potato. Curr. Microbiol. 62, 1510–1520.

Nakamura, K., 1976. Studies on the population dynamics of the 28-spotted lady beetle, *Henosepilachna vigintioctopunctata* F., I. Analysis of life tables and mortality process in the field population. Jpn. J. Ecol. 26, 49–59.

Nakamura, K., 1987. Population study of the large 28-spotted ladybird *Epilachna vigintioctomaculata* (Coleoptera: Coccinellidae). Res. Popul. Ecol. 29, 215–228.

Nakamura, K., Abbas, I., Hasyim, A., 1988. Population dynamics of the phytophagous lady beetle, *Epilachna vigintioctopunctata* in an eggplant field in Sumatra. Res. Popul. Ecol. 30, 25–42.

Nakamura, K., Hasan, N., Abbas, I., Godfray, H.C., Bonsall, M.B., 2004. Generation cycles in Indonesian lady beetle populations may occur as a result of cannibalism. Proceedings of the Royal Society Biological Sciences Series B 271 (Suppl 6), S 501–S 504.

Nakata, T., 1995. Population fluctuations of aphids and their natural enemies on potato in Hokkaido, Japan. Appl. Entomol. Zool. 30, 129–138.

Nault, B.A., Kennedy, G.G., 2000. Seasonal changes in habitat preference by *Coleomegilla maculata*: Implications for Colorado potato beetle management in potato. Biol. Control 17, 164–173.

Nelson, W.R., Fisher, T.W., Munyaneza, J.E., 2011. Haplotypes of "*Candidatus* Liberibacter solanacearum" suggest long-standing separation. Eur. J. Plant Pathol. 130, 5–12.

Obrycki, J.J., Tauber, M.J., Tingey, W.M., 1986. Comparative toxicity of pesticides to *Edovum puttleri* (Hymenoptera: Eulophidae), an egg parasitoid of the Colorado potato beetle (Coleoptera: Chrysomelidae). J. Econ. Entomol. 79, 948–951.

OMAFRA, 2012. Potato psyllid – A pest in greenhouse tomatoes and peppers. Ontario Ministry of Agriculture Food and Rural Affairs, www.omafra.gov.on.ca/english/crops/facts/potato_psyllid.htm.

O'Neil, R.J., Cañas, L.A., Obrycki, J.J., 2005. Foreign exploration for natural enemies of the Colorado potato beetle in Central and South America. Biol. Control 33, 1–8.

Orozco, F.A., Cole, D.C., Forbes, G., Kroschel, J., Wanigaratne, S., Arica, D., 2009. Monitoring adherence to the International Code of Conduct – highly hazardous pesticides in central Andean agriculture and farmers' rights to health. Int. J. Occup. Environ. Health 15, 255–268.

Ortiz, O., Garret, K.A., Heath, J.J., Orrego, R., Nelson, R.J., 2004. Management of potato late blight in the Peruvian highlands: Evaluating the benefits of farmer field schools and farmer participatory research. Plant Dis. 88, 565–571.

Osorio, M.P.A., Espitia, E.M., Luque, E.Z., 2001. Reconocimiento de enemigos naturales de *Tecia solanivora* (Lepidoptera: Gechiidae) en localidades productoras de papa en Colombia. Rev. Colomb. Entomol. 27, 177–185.

Palacios, M., Tenorio, J., Vera, M., Zevallos, F., Lagnaoui, A., 1999. Population dynamics of the Andean potato tuber moth, *Symmetrischema tangolias* (Gyen), in three different agroecosystems in Peru, pp. 153–160. International Potato Center Program Report 1997–1998, Peru, Lima.

Parsa, S., 2010. Native herbivore becomes key pest after dismantlement of a traditional farming system. Am. Entomol. 56, 242–251.

Parsa, S., Ccanto, R., Rosenheim, J.A., 2011. Resource concentration dilutes a key pest in indigenous potato agriculture. Ecol. Appl. 21, 539–546.

Patalappa, G., Basavanna, P.C., 1979. Seasonal incidence and life history of *Pediobius foveolatus* (Hymenoptera: Eulophidae), a parasite of *Henosepilachna vigintioctopunctata* (Fabricius) (Coleoptera: Coccinellidae). Mysore J. Agric. Sci. 13, 191–196.

Patt, J.M., Hamilton, G.C., Lashomb, J.H., 1997. Impact of strip-insectary intercropping with flowers on conservation biological control of the Colorado potato beetle. Adv. Hortic. Sci. 11, 175–181.

Peña, G., Miranda-Rios, J., de la Riva, G., Pardo-López, L., Soberón, M., Bravo, A., 2006. A *Bacillus thuringiensis* S-Layer Protein involved in toxicity against *Epilachna varivestis* (Coleoptera: Coccinellidae). Appl. Environ. Microbiol. 72, 353–360.

Peng, H., Bao, J.Z., 1988. Studies on the taxonomic status of a *Pediobius* species (Hym.: Eulophidae), a parasitoid of *Henosepilachna vigintioctomaculata* (Col.: Coccinellidae) in Beijing. Chin. J. Biol. Control 4, 123–126.

Peng, L., Trumble, J.T., Munyaneza, J.E., Liu, T., 2011. Repellency of a kaolin particle film to potato psyllid, *Bactericera cockerelli* (Hemiptera: Psyllidae), on tomato under laboratory and field conditions. Pest Manage. Sci. 67, 815–824.

Perdikis, D., Fantinou, A., Lykouressis, D., 2011. Enhancing pest control in annual crops by conservation of predatory Heteroptera. Biol. Control 59, 13–21.

Pletsch, D.J., 1947. The potato psyllid. *Paratrioza cockerelli* (Sulc), its biology and control Mont. Agric. Stn. Tech. Bull. 446.

Pokharkar, D.S., Ghorpade, S.A., Bade, B.A., 2003. Development of mass release technique of parasitoids, *Copidosoma koehleri* Blanchard and *Chelonus blackburni* Cameron against tuber worm, *Phthorimaea operculella* (Zeller) on potato. In: Tandon, P.L., Ballal, C.R., Jalali, S.K., Rabindra, R.J. (Eds.), Proceedings of the Symposium on Biological Control of Lepidopteran Pests, Society for Biocontrol Advancement, Bangalore, India, 319–323 .

Potts, M.J., 1990. Influence of intercropping in warm climates on pests and diseases of potato, with special reference to their control. Field Crops Res. 25, 133–144.

Poveda, K., Gomez Jimenez, M.I., Kessler, A., 2010. The enemy as ally: herbivore-induced increase in crop yield. Ecol. Appl. 20, 1787–1793.

Povolný, D., Hula, V., 2004. Ein neuer, nach Südwesteuropa eindringender Kartoffelschädling, die Grosse Kartoffel-Knollenmotte, *Scrobipalpopsis solanivora* (Lepidoptera: Gelechiidae). Entomol. Gen. 27, 155–168.

Pradel, W., Forbes, G.A., Ortiz, O., Cole, D., Wanigaratne, S., Maldonado, L., 2009. Use of the environmental impact quotient to estimate impacts of pesticide usage in three Peruvian potato production areas. Integrated Crop Management Division Working Paper No. 2009–2. Centro Internacional de la Papa, Lima.

Pucci, C., Dominici, M., 1988. Field evaluation of *Edovum puttleri* Grissel (Hym., Eulophidae) on eggs of *Leptinotarsa decemlineata* Say (Col., Chrysomelidae) in central Italy. J. Appl. Entomol. 106, 465–472.

Pucci, C., Spanedda, A.F., Minutoli, E., 2003. Field study of parasitism caused by endemic parasitoids and by the exotic parasitoid *Copidosoma koehleri* on *Phthorimaea operculella* in Central Italy. Bull. Insect. 56, 221–224.

Puttarudriah, M., Krishnamurti, B., 1954. Problem of *Epilachna* control in Mysore: Insecticidal control found inadvisable when natural incidence of parasites is high. Indian J. Entomol. 16, 137–141.

Ragsdale, D.W., Radcliffe, E.B., Difonzo, C.D., Connelly, M.S., 1994. Action thresholds for an aphid vector of potato leafroll virus. In: Zehnder, G.W., Jansson, R.K., Powelson, M.L., Raman, K.V. (Eds.), Advances in Potato Pest Biology and Management, American Phytopathological Society Press, St Paul, MN, pp. 99–110.

Rajagopal, D., Trivedi, T.P., 1989. Status, bioecology and management of Epilachna beetle, *Epilachna vigintioctopunctata* (Fab.) (Coleoptera: Coccinellidae) on potato in India: a review. Trop. Pest Manage. 35, 410–413.

Ramirez, R.A., Snyder, W.E., 2009. Scared sick? Predator–pathogen facilitation enhances exploitation of a shared resource. Ecology 90, 2832–2839.

Ramirez, R.A., Crowder, D.W., Snyder, G.B., Strand, M.R., Snyder, W.E., 2010. Antipredator behavior of Colorado potato beetle larvae differs by instar and attacking predator. Biol. Control 52, 230–237.

Rehman, M., Melgar, J.C., Rivera, C.J., Idris, A.M., Brown, J.K., 2010. First report of "*Candidatus* Liberibacter psyllaurous" or "*Ca.* Liberibacter solanacearum" associated with severe foliar chlorosis, curling, and necrosis and tuber discoloration of potato plants in Honduras. Plant Dis. 94, 376–377.

Reznik, S.Y., 2011. Host plant population density and distribution pattern as factors limiting geographic distribution of the ragweed leaf beetle *Zygogramma suturalis* F. (Coleoptera, Chrysomelidae) Entomol. Rev. 91, 292–300.

Richards, A.M., Filewood, L.W., 1990. Feeding behaviour and food preferences of the pest species comprising the *Epilachna vigintioctopunctata* (F.) complex (Col., Coccinellidae). J. Appl. Entomol. 110, 501–515.

Rios, A.A., Kroschel, J., 2011. Evaluation and implications of Andean potato weevil infestation sources for its management in the Andean region. J. Appl. Entomol. 135, 738–748.

Robbins, G., Hudson, W., Dorsey, T., Mayer, M., Bronhard, L., 2010. Biological control of the Mexican bean beetle *Epilachna varivestis* (Coleoptera: Coccinellidae) using the parasitic wasp *Pediobius foveolatus* (Hymenoptera: Eulophidae). New Jersey Department of Agriculture, www.state.nj.us/agriculture/divisions/pi/pdf/mexicanbeanbeetle.pdf.

Rojas, M.G., Morales-Ramos, J.A., King, E.G., 2000. Two meridic diets for *Perillus bioculatus* (Heteroptera: Pentatomidae), a predator of *Leptinotarsa decemlineata* (Coleoptera: Chrysomelidae). Biol. Control 17, 92–99.

Rojas Rojas, P., 2010. Biología de *Tamarixia triozae* (Burks) (Hymenoptera: Eulophidae) parasitoide de *Bactericera cockerelli* (Sulc) (Hemíptera: Triozidae). MS thesis. Colegio de Postgraduados. Institución de Enseñanza e Investigación en Ciencias Agrícolas, Campus Montecillo, Estado de México, México.

Romney, V.E., 1939. Breeding areas of the tomato psyllid, *Paratrioza cockerelli* (Sulc). J. Econ. Entomol. 32, 150–151.

Rondon, S.I., 2010. The potato tuberworm: A literature review of its biology, ecology, and control. Am.. J. Potato Res. 87, 149–166.

Rubio, S.A, B. I. Vargas, B.I., López-Ávila, A., 2004. Evaluación de la eficiencia de *Trichogramma lopezandinensis* (Hymenoptera: Irichogrammatidae) para el control de *Tecia solanivora* (Lepidoptera: Gelechiidae) en papa almacenada. Rev. Colomb. Entomol. 30, 107–114.

Saint-Cyr, J.F., Cloutier, C., 1996. Prey preference by the stinkbug *Perillus bioculatus*, a predator of the Colorado potato beetle. Biol. Control 7, 251–258.

Sánchez-Aguirre, R., Palacios, M., 1996. Eficacia del parasitismo de *Copidosoma koehleri* en el complejo polilla de la papa Rev. Peru. Entomol. 58, 59–62.

Sant'ana, J., Dickens, J.C., 1998. Comparative electrophysiological studies of olfaction in predaceous bugs, *Podisus maculiventris* and *P. nigrispinus*. J. Chem. Ecol. 24, 965–984.

Schaefer, P.W., 1983. Natural enemies and host plants of species in the Epilachninae. (Coleoptera: Coccinellidae) – a world list. Bull. Agric. Exp. Stn. Univ. Del. No. 445, 42 pp.

Schaefer, P.W., Peng, H., Gou, X., 1986. Preliminary survey of parasites of *Epilachna* sp. in China. Chin. J. Biol. Control 2, 112–115.

Schroder, R.F.W., Athanas, M.M., 1989. Potential for the biological control of *Leptinotarsa decemlineata* Coleoptera Chrysomelidae by the egg parasite *Edovum puttleri* Hymenoptera Eulophidae in Maryland USA 1981–1984. Entomophaga 34, 135–141.

Schroder, R.F.W., Athanas, M.M., Pavan, C., 1993. *Epilachna vigintioctopunctata* (Coleoptera: Coccinellidae), new record for Western Hemisphere, with a review of host plants. Entomol. News 104, 111–112.

Sforza, R., Jones, W.A., 2007. Potential for classical biocontrol of silverleaf nightshade in the Mediterranean Basin. EPPO Bull. 37, 156–162.

Sheng, J.K., Wang, G.H., 1992. Studies on the biology of *Pediobius foveolatus* (Hym.: Pteromalidae) in the Nanchang area. Chin. J. Biol. Control 8, 110–114.

Sikura, A.I., Smetnik, A.I., 1967. Results of the acclimatization entomophages of Colorado beetle and American white butterfly in the Zakarpatskaya region USSR. Sb. Karantinu Rast 19, 114–127.

Snyder, W.E., Clevenger, G.M., 2004. Negative dietary effects of Colorado potato beetle eggs for the larvae of native and introduced ladybird beetles. Biol. Control 31, 353–361.

Sporleder, M., Tonnang, E.Z.H., Carhuapoma, P., Gonzales, J.C., Juarez, H., Simon, R., Kroschel, J., 2011. ILCYM – Insect Life Cycle Modeling. A software package for developing temperature-based insect phenology models with applications for regional and global pest risk assessments and mapping. International Potato Center, Lima, Peru, 68 pp.

Stamopoulos, D.C., Chloridis, A., 1994. Predation rates, survivorship and development of *Podisus maculiventris* (Het.: Pentatomidae) on larvae of *Leptinotarsa decemlineata* (Col.: Chrysomelidae) and *Pieris brassicae* (Lep.: Pieridae), under field conditions. Entomophaga 39, 3–9.

Straub, C.S., Snyder, W.E., 2006. Species identity dominates the relationship between predator biodiversity and herbivore suppression. Ecology 87, 277–282.

Symington, C.A., 2003. Lethal and sublethal effects of pesticides on the potato tuber moth, *Phthorimaea operculella* (Zeller) (Lepidoptera: Gelechiidae) and its parasitoid *Orgilus lepidus* Muesebeck (Hymenoptera:Braconidae). Crop Prot. 22, 513–519.

Szendrei, Z., Weber, D.C., 2009. Response of predators to habitat manipulation in potato fields. Biol. Control 50, 123–128.

Szendrei, Z., Greenstone, M.H., Payton, M.E., Weber, D.C., 2010. Molecular gut-content analysis of a predator assemblage reveals the effect of habitat manipulation on biological control in the field. Basic Appl. Ecol. 11, 153–161.

Tamaki, G., Chauvin, R.L., Burditt Jr., A.K., 1983. Field evaluation of *Doryphorophaga doryphorae* (Diptera: Tachinidae), a parasite, and its host the Colorado potato beetle (Coleoptera:Chrysomelidae). Environ. Entomol. 12, 386–389.

Teulon, D.A., Workman, P.J., L.Thomas, K., Nielsen, M.C., Zydenbos, S.M., 2009. Bactericera cockerelli Incursion, dispersal and current distribution on vegetable crops in New Zealand. N. Z. Plant Prot. 62, 136–144.

Tipping, P.W., Holko, C.A., Abdul-Baki, A.A., Aldrich, J.R., 1999. Evaluating *Edovum puttleri* Grissell and *Podisus maculiventris* (Say) for augmentative biological control of Colorado potato beetle in tomatoes. Biol. Control 16, 35–42.

Todorova, S.I., Coderre, D., Cote, J.C., 2000. Pathogenicity of *Beauveria bassiana* isolates toward *Leptinotarsa decemlineata* (Coleoptera: Chrysomelidae), *Myzus persicae* (Homoptera: Aphididae) and their predator *Coleomegilla maculata lengi* (Coleoptera: Coccinellidae). Phytoprotection 81, 15–22.

Torres, F.W., Notz, A., Valencia, L., 1997. Cicio de vida y otros aspectos de la biología de la polilla de la papa *Tecia solanivora* (Povolny) (Lepidoptera: Gelechiidae) en el estado Tachira, Venezuela. Bol. Entomol. Venez. 12, 81–94.

Trouvelot, B., 1931. Recherches sur les parasites et predateurs attaquant le doryphore en Amerique du Nord. Ann. Épiphyt. 17, 408–445.

Tu, X., Wang, G., 2010. Research progress of bio-control of *Henosepilachna vigintioctopunctata*. Chin. Plant Protec. 30, 13–16.

UC IPM Online, 2009. Potato: Potato Psyllid UC Pest Management Guidelines. University of California ANR Statewide IPM Program, www.ipm.ucdavis.edu/PMG/r607300811.html.

Vasquez, B., Lashomb, J.H., Hamilton, G., 1997. Alightment of *Edovum puttleri* Grissell (Hymenoptera: Eulophidae) in response to colors mimicking three hosts of Colorado potato beetle (Coleoptera: Chrysomelidae). J. Entomol. Sci. 32, 386–397.

Von Arx, R., Roux, O., Baumgartner, J., 1990. Tuber infestation by potato tubermoth, *Phthorimaea operculella* (Zeller), at potato harvest in relation to farmers' practices. Agric. Ecosyst. Environ. 31, 277–292.

Walker, G.P., MacDonald, F.H., Larsen, N.J., Wallace, A.R., 2011. Monitoring *Bactericera cockerelli* and associated insect populations in potatoes in South Auckland. NZ Plant Prot. 64, 269–275.

Walsh, B.D., Riley, C.V., 1868. Potato Bugs: the Colorado potato-bug, its past history and future progress. Am. Entomol. 1, 41–49.

Wang, G., Li, H., Sheng, J., 1998. Effects of insecticides on *Henosepilachna vigintioctopunctata* and its larval parasite, *Pediobius foveolatus*. Nat. Enemies Insects 20, 164–168.

Weber, D.C., 2003. Colorado beetle: pest on the move. Pestic. Outlook 14, 256–259.

Weber, D.C., Lundgren, J.G., 2009. Assessing the trophic ecology of the Coccinellidae: Their roles as predators and as prey. Biol. Control 51, 199–214.

Weber, D.C., Rowley, D.L., Greenstone, M.H., Athanas, M.M., 2006. Prey preference and host suitability of the predatory and parasitoid carabid beetle, *Lebia grandis*, for several species of Leptinotarsa beetles. J. Insect Sci. 6, 09.

Weber, D.C., Pfannenstiel, R.S., Lundgren, J.G., 2008. Diel predation pattern assessment and exploitation of sentinel prey: New interpretations of community and individual behaviors. In: Mason, P.G., Gillespie, D.R., Vincent, C. (Eds.), Proceedings of the Third International Symposium on Biological Control of Arthropods, 8–13 February 2009, Christchurch, New Zealand. USDA Forest Service Publication FHTET-2008-06, Morgantown, WV, USA, pp. 485–494.

Weissbecker, B., Van Loon, J.J., Dicke, M., 1999. Electroantennogram responses of a predator, *Perillus bioculatus*, and its prey, *Leptinotarsa decemlineata*, to plant volatiles. J. Chem. Ecol. 25, 2313–2325.

White, J., Johnson, D., 2010. Vendors of Beneficial Organisms in North America. University of Kentucky Cooperative Extension ENTFACT–125, 7 pp.

Workman, P.J., 2009. Current research program-entomology. In: Nelson, W. (Ed.), 7th World Potato Congress, Solanaceous crops, Psyllids & Liberibacter, pp. 65–69. Proceedings of the Workshop held 26 March 2009, Christchurch, New Zealand. New Zealand Plant & Food Research Ltd., Wellington, NZ.

Workman, P.J., Whiteman, S.A., 2009. Importing *Tamarixia triozae* into containment in New Zealand. N. Z. Plant Protec. 62, 412.

Wuriyanghan, H., C. Rosa, and B. W. Falk, 2011. Oral delivery of double-stranded RNAs and siRNAs induces RNAi effects in the potato/tomato psyllid, *Bactericerca cockerelli*. PLoS One 6: e27736.

Yábar, E., Castro, E., Meló, L., Gianoli, E., 2006. Predación de *Bembidion* sp., *Notiobia peruviana* (Dejean) y *Metius* sp. (Coleoptera: Carabidae) sobre huevos de *Premnotrypes latithorax* (Pierce) (Coleoptera: Curculionidae) en condiciones de laboratorio. Rev. Peru. Entomol. 45, 91–94.

Yang, X., Zhang, Y., Hua, L., Peng, L., Munyaneza, J.E., Trumble, J.T., Liu, T., 2010. Repellency of selected biorational insecticides to potato psyllid, *Bactericera cockerelli* (Hemiptera: Psyllidae). Crop Prot. 29, 1320–1324.

Zhuang, H.D., Sun, Q., 2009. Life tables of natural population of *Henosepilachna vigintioctomaculata* (Mots.). Jiangxi Plant Prot. 32, 103–106.

Zuparko, R.L., De Queiroz, D.L., La Salle, J., 2011. Two new species of *Tamarixia* (Hymenoptera: Eulophidae) from Chile and Australia, established as biological control agents of invasive psyllids (Hemiptera: Calophyidae, Triozidae) in California. Zootaxa 2921, 13–27.

Chapter 15

Potato Resistance Against Insect Herbivores: Resources and Opportunities

Yvan Pelletier[1], Finbarr G. Horgan[2], and Julien Pompon[13]

[1]Potato Research Centre, Agriculture and Agri-Food Canada, Fredericton, NB, Canada, [2]Crop and Environmental Sciences Division, International Rice Research Institute, Metro Manila, The Philippines, [3]Department of Biology, University of New Brunswick, Fredericton, NB, Canada

INTRODUCTION

Beginning in the 1600s, the potato (*Solanum tuberosum tuberosum* L.), originally from South America, has been introduced to most of the world's regions and ecologies to become the fourth most important crop worldwide (Jansky *et al.* 2009). Many of the most destructive pests of potato have also been introduced from regions where potatoes and their relatives are native. These include potato cyst nematodes (*Globodera* spp.), the Colorado potato beetle (*Leptinotarsa decemlineata* L.) (CPB), potato tuber moths (*Phthorimaea operculella* [Zell.], *Tecia solanivora* Polovny, *Symmetrischema tangolias* [Gyen]), and potato weevils (*Premnotrypes* spp.) (Wale *et al.* 2008). These pests can cause considerable losses to production and often lead to high pesticide use in potato fields. Despite advances in chemical formulations, pesticides continue to be toxic, dangerous to the environment, hazardous to human health, and subject to adaptation by insect herbivores that leads to increasingly higher application concentrations and increasing pesticide inefficiency (Crissman *et al.* 1994, Whalon *et al.* 2008, Juraske *et al.* 2011).

Integrated Pest Management (IPM) combines a series of knowledge-based elements to reduce the use of insecticides and other pesticides while achieving greater crop protection. One IPM component that is gaining increasing attention is the use of host-plant resistance (HPR) to reduce insect damage (Fig. 15.1). Host-plant resistance uses natural variability in crop plants, introgressed traits from wild crop ancestors, or, more recently, introduced traits from other animal, plant, or bacterial species, to protect potatoes from herbivore attack. However, in its broadest sense HPR also includes knowledge of insect-plant interactions at

Insect Pests of Potato. http://dx.doi.org/10.1016/B978-0-12-386895-4.00015-6

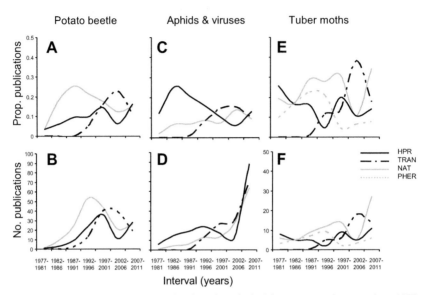

FIGURE 15.1   Trends in research related to the principal insect pests of potato since 1977. Publications were retrieved using the ISI Web of Science and categorized according to research focus. Trends indicate an increasing emphasis on molecular approaches to host-plant resistance (HPR) and particularly transgenic approaches (TRAN) since the early 1990s. Together, these research areas greatly exceed output from other research areas including natural enemies (NAT) and pheromone technology (PHER). The success of granulosis viruses in controlling tuber moths in stored potatoes has produced a recent peak in attention to viral natural enemies for biological control.

landscape (spatial) and generation (temporal) scales (Smith 2005). In this chapter, we divide HPR into two main categories: (1) conventional resistance (resistance obtained from within the species or by introgressing traits from near relatives through conventional breeding and somatic hybridization), and (2) engineered resistance, which introduces novel traits to the plant from the same species, or from closely related species, or from distant species via bacterial carriers. Separation of HPR approaches into these categories is based on the major scientific disciplines that underlie their development and the crop management implications related to each. In this chapter we concentrate on sources of potato resistance, the mechanisms underlying this resistance, and the opportunities and constraints in applying HPR for sustainable potato crop protection. Our discussion is limited to the most destructive insect pests of potato, including the CPB, tuber moths, and the most economically important aphids (the potato aphid, *Macrosiphum euphorbiae* [Thomas], and the green peach aphid, *Myzus persicae* [Sulzer]).

Trends in Potato HPR Research

Host-plant resistance is not new to agriculture, but has been applied for at least 200 years. Among the earliest successes were resistance to Hessian fly, *Mayetiola destructor* (Say), in wheat, *Triticum* spp. (1792); resistance to woolly apple aphid, *Eriosoma lanigerum* (Hausmann), in apple, *Malus* spp (1831); and resistance to grape phylloxera, *Daktulospharia vitifoliae* (Fitch), in grapes, *Vitis* spp. (ca 1860) (Smith 2005, and references therein). Host-plant resistance is likely to become increasingly important given the continuing homogenization of agricultural landscapes. Until the advent of the green revolution, crops were diverse and often hidden in heterogeneous landscapes that consisted of trees, woodland, or other natural and derived habitats. Monocultures have eroded landscape diversity and resulted in a loss of habitat for natural enemies, which often increases the vulnerability of crops (Meehan *et al.* 2011; but see also Parsa *et al.* 2011). Agricultural expansion, declining agricultural diversity, and declining crop genetic diversity have demanded a greater focus on intrinsic plant protection. This has coincided with an increasing emphasis on plant molecular biology since the 1990s, largely driven by advances in the molecular tools available to science, including molecular breeding and transgenic technologies. Fig. 15.1 indicates that for CPB, potato aphids, and tuber moths, the number and proportion of scientific publications related to HPR and insect-plant interactions have generally increased over the past 35 years. During the 2000s, research into conventional HPR and transgenic approaches to plant protection dominated all other management approaches (CPB 31%, tuber moths 38%, and aphids/viruses 24% of all papers published from 2000 to 2011), including all research on pest natural enemies and biological control. For tuber moths, significant advances have been made in the use of pheromones and granulosis viruses for protection of stored potatoes (Fig. 15.1e, 15.1f), but for field crops HPR remains the main focus of research. Interestingly, Fig. 15.1 suggests that research into conventional and transgenic resistance may be mutually competitive in terms of attention or funding opportunities. In the early 2000s, research into transgenics reached its peak; however, more recently there has been a shift to research of natural induced responses to insect attack and plant biochemical defenses.

CONVENTIONAL RESISTANCE

Native and Crop Potatoes as Sources of Resistance

Together with *S.t. tuberosum* (henceforth crop potato), eight other diploid and tetraploid potato species (henceforth native potatoes) are domesticated throughout the world, and especially in Andean regions. These include *Solanum pureja* Juz. & Bukasov, *Solanum goniocalyx* Juz. & Bukasov, and *Solanum tuberosum andigena*, the latter of which is the suggested predecessor of the crop potato (National Research Council 1989). Most edible potato varieties are susceptible to potato pests (Jansky *et al.* 2009). However, there is often considerable variability in insect performance on different crop and native potato cultivars, and many varieties show clear potential to reduce pest damage in the field and in storage. For example, in a study by Ghassemi-Kahrizeh *et al.* (2010) CPB larval survival ranged from 13% to 91% across 20 Iranian potato cultivars. The intrinsic rate of increase of the two main potato aphid pests also varied widely across 49 commercial cultivars (Davis *et al.* 2007). Horgan *et al.* (2012) found high resistance (up to 55%) to two tuber moth species in native purple potatoes (*S.t. andigena*) in Peru, indicating that the resistance mechanisms are likely general and broad-spectrum. With over 2500 known varieties, native potatoes hold promise as sources of pest resistance that is often easily introgressed with crop potato. Furthermore, even where resistance is low in breeding materials, careful attention to resistance traits during breeding programs can improve resistance levels. In a study by Fisher *et al.* (2002), after three rounds of recurrent selection initiated with moderately resistant potato varieties grown in Iowa, estimates of resistance against the CPB increased almost three-fold and exceeded levels observed in a highly resistant variety.

Wild Potatoes as Sources of Resistance

Insect- and disease-resistance traits can be found in a group of approximately 230 wild *Solanum* species (*Solanaceae*, Section Petota) that produce tubers (Flanders *et al.* 1999). Wild potatoes are native to the area from the southwestern United States, southward through Mexico and Central America, along the Andes of South America, and into the plains of Argentina, Paraguay, and Uruguay. Most species have a narrow range defined by a specific set of ecogeographic characteristics, but some are more widely distributed. Several important collections of wild *Solanum* species make material available to potato breeders. These collections also record and maintain information on different traits of interest for the development of commercial potato varieties (Flanders *et al.* 1999).

The genetics of the wild *Solanum* species is complicated by the fact that most species (approximately 64%) are diploid ($2 \times = 24$) but the remaining species are either tetraploid ($4 \times = 48$) or hexaploid ($6 \times = 72$) (Jansky *et al.* 2009). The balance of genetic factors contributed by gametes to the endosperm (the endosperm balance number) is key to determining the crossability between species (Jansky *et al.* 2009). However, modern cellular techniques such as somatic hybridization can be used to overcome the reproductive barriers between incompatible species and crop potato (Chen *et al.* 2008).

The pedigrees of several modern crop potato varieties include wild potato species incorporated into breeding programs specifically to increase resistance against insect pests. Two notable wild species that have been extensively studied as sources of resistance are *Solanum berthaultii* Hawkes and *Solanum chacoense* Bitt. (see below). Their modes of action are predominantly based on the possession of glandular trichomes and defensive glycoalkaloids, respectively. However, evidence suggests that several other resistance mechanisms may also underlie defenses in wild potatoes and their interspecific hybrids.

Mechanisms of Potato Resistance

Glycoalkaloids

Glycoalkaloids form a family of compounds present in a few genera of the Solanaceae, including the genus *Solanum* (Osman 1983). They are constructed from an alkaloid derived from cholesterol, and a sugar. Changes in the nature of the alkaloid or the sugar produce different glycoalkaloids with slightly different activities. These are generally considered toxic for human consumption, and are well known for their fungicidal activity. Glycoalkaloids are bitter – a characteristic that is avoided in the development of modern potato varieties; nevertheless, the bioactive properties of glycoalkaloids make them obvious candidates for insect resistance.

The presence and amounts of glycoalkaloids in tuber-bearing *Solanum* species often do not correlate with resistance against the main insect pests (Tingey and Sinden 1982, Sinden *et al.* 1991, Lyytinen *et al.* 2007), but the nature of the specific glycoalkaloids is often significant. Of eight species containing the glycoalkaloid tomatine, six are resistant to at least one insect (Flanders *et al.* 1992). For example, tomatine inhibits both CPB larval development and adult feeding at concentrations that occur naturally in the foliage of some wild potato species (Sinden *et al.* 1991). Commersonine and dehydrocommersonine have also been linked to resistance against the CPB (Tingey 1984).

The use of glycoalkaloids for breeding is limited since plant production of the chemicals is influenced by environmental factors, and, more importantly, because foliar glycoalkaloid content is correlated with tuber content – a negative trait for modern potato cultivars (Tingey 1984). However, a group of glycoalkaloids, the leptines, does not show such a correlation. Leptines are found only in a few accessions of *S. chacoense*, and differ from commonly occurring *Solanum* glycoalkaloids such as solanine and chaconine only in the substitution of an acetyl on carbon-23 of the steroid aglycone (Tingey and Yencho 1994). In a study of hybrid populations of *S. chacoense* × *S. tuberosum*, the level of leptines, but not of other glycoalkaloids, was correlated with CPB resistance (Lorenzen *et al.* 2001). Further breeding produced parental lines with a significant reduction of defoliation in the field (Cooper *et al.* 2007). This material had an antibiotic effect on the CPB, slowing the development and reducing the survival of larvae, reducing feeding, and reducing oviposition and survival in adults (Lorenzen *et al.* 2001). Recently a potato variety, Dakota Diamond, with high levels of foliar leptine, and consequently low levels of herbivory, has been released for commercialization (Thompson *et al.* 2008).

Trichomes

Trichomes are uni- or multicellular structures that originate from epidermal cells of above-ground plant tissues. Trichomes can be classified morphologically as either non-glandular (hairs) or glandular. They can take diverse forms but only three types from the tuber-bearing *Solanum* have received attention: (1) the tall,

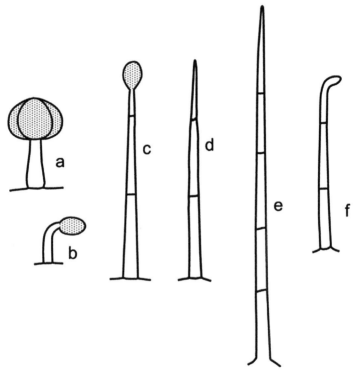

FIGURE 15.2　Several types of trichome occur on the leaf surface of *S. berthaultii:* Short (120–210 μm) type-A glandular trichomes (a) occur on the leaf blade and club-shaped glandular trichomes (b) along leaf veins. Longer (650–950 μm) type B glandular trichomes (c) also occur on some accessions but are largely absent from *S. berthaultii (tarijense),* which bears morphologically similar leaf hairs (d) instead. Occasional long (1.5–2.0 mm) (e) or hooked trichomes (f) occur on some *S. berthaultii* plants (i.e., accessions HHA 6548 and HHA 6661a, respectively).

non-glandular hairs, (2) the short, four-lobed type A glandular trichomes containing phenolic compounds, and (3) the hair-like type B trichomes that exude droplets containing sucrose esters of carbocylic acids from their tips (Flanders *et al.* 1992) (Fig. 15.2). Wild potatoes can bear all, some, or none of these types of trichomes. Trichomes occurring together on the leaf surface may function in several different ways. For example, type A and B trichomes defend potato foliage against ovipositing potato tuber moths, whereas non-glandular leaf hairs may actually increase egg-laying by moths (Malakar and Tingey 2000, Horgan *et al.* 2007a).

Glandular trichomes are present in moderate to high densities on the leaves of at least 13 tuber-bearing *Solanum* species (Tingey *et al.* 1981). Three tuber-bearing species with glandular trichomes exhibit resistance to CPB, aphids, tuber moths, and other insects. These are *S. polyadenium* Greenm., *S. berthaultii* (which includes two distinct morphotypes that were previously regarded as separate species and are referred to in this text as *S. berthaultii* and *S.berthaultii (tarijense)* [Spooner *et al.* 2007], the latter generally lacking type B trichomes), and *S. neocardenasii* Hawkes (Gibson 1976, Tingey *et al.* 1982, Dimock *et al.* 1985, 1986, Lapointe and Tingey 1986, Sanford and Cantelo 1989, Hanzlik *et al.* 1997, Flanders *et al.* 1999, Pelletier and Tai 2001, Horgan *et al.* 2007a). Resistance levels to aphids and tuber moths have been positively correlated with trichome densities (Tingey and Laubengayer 1981, Horgan *et al.* 2007a); however, resistance is significantly improved where the two types of glandular trichome, type A and type B, are present together (Tingey and Sinden 1982). A genetic study of a population of hybrids segregating for both resistance and trichome densities has suggested that trichome abundance does not entirely explain the level of resistance to the CPB. Both resistance and trichome characteristics pointed to the same quantitative trait loci (QTL), but other QTLs were linked with resistance only (Yencho *et al.* 1996).

Several classical manipulative studies have shown that the removal of trichomes from leaves increases feeding and fitness of the CPB and tuber moths on potato foliage, whereas the mechanical transfer of exudate to the leaves of susceptible potato varieties reduces feeding (Yencho and Tingey 1994, Malakar and Tingey 2000, Pelletier and Dutheil 2006). Furthermore, disruption of selection behavior by aphids of *S. polyadenium* and *S. berthaultii* as host plants disappears when trichomes are removed (Alvarez *et al.* 2006). The mechanism of trichome-mediated resistance involves several steps (Gregory *et al.* 1986): type B trichome exudate forms an adhesive coating on the tarsi of attacking insects, and also elicits an antifeedant behavior through its sesquiterpene content; the insect struggles during attempts to escape and breaks the type A trichome heads; and α- and/or β-polyphenol oxidases are released from the broken heads of type A trichomes and react with a phenolic substrate (chlorogenic acid) to initiate an enzymatic oxidative process leading to the formation of quinones. The exudate gradually accumulates on the appendages to the point where insect behavior is impaired (often known as tarsal gumming). Inhibition of locomotion through tarsal gumming appears to be significant for small insects only (Dimock and Tingey 1987). However, the level of polyphenol oxidase on the leaves of *S. berthaultii* is correlated positively with CPB larval mortality and development time, and negatively with adult fecundity (Castañera *et al.* 1996). Type B trichome exudate from *S. berthaultii* is also a known feeding deterrent for CPB larvae (Dimock and Tingey 1988) and adults (Yencho and Tingey 1994).

Trichome-based resistance has disadvantages. Trichomes are not present on buds or flowers, enabling aphids, for example, to colonize these plant parts (Ashouri *et al.* 2001). Under field conditions, depletion of type A trichomes through ware, or dust accumulation on the exudates (Tingey and Sinden 1982), reduces resistance levels. Furthermore, the production of trichome B exudate is strongly associated with unfavorable agronomic traits in *S. berthaultii* × *S. tuberosum* hybrids (Kalazich and Plaisted 1991) and possibly has tradeoffs associated with tuber resistance (Horgan *et al.* 2009). Glandular trichomes can also cause increased mortality of natural enemies (Obrycki 1986). For example, the parasitoid *Copidosoma koehleri* (Blanchard) makes fewer and shorter visits to plants with high trichome densities, resulting in lower parasitism of tuber moths in potato fields (Baggen and Gurr 1995, Gooderham *et al.* 1998). Furthermore, aphid predation by *Coleomegilla maculata* (DeGeer) (Coleoptera: Coccinellidae) is lower on *S. berthaultii (tarijense)* compared to *S. tuberosum*, but this effect has not been observed with another Coccinellidae, *Harmonia axyridis* (Pallas) (Fréchette *et al.* 2012). In spite of these disadvantages, trichome-based resistance derived from wild potatoes has generally proved to be a useful, broad-spectrum, and popular resistance source.

Undefined Resistance from Wild Solanum Species

Field evaluations of defoliation and laboratory or greenhouse screening studies were first used to identify candidate *Solanum* species for resistance against insects (Carter 1987, Flanders *et al.* 1999). Grafting experiments have since provided evidence that several wild *Solanum* species produce chemicals that are translocated throughout the plant and reduce food consumption by insects such as the CPB (Pelletier and Clark 2004). Furthermore, a volatile fraction from leaf rinse of *S. berthaultii* and *S. berthaultii (tarijense)* had a feeding-deterrent effect on adults of the CPB (Yencho *et al.* 1994). However, the specific nature of the factors underlying these observations remains unknown.

Multivariate analysis techniques can be used with field population data to develop hypotheses on the mode of resistance in term of antibiosis (a plant's ability to reduce the fitness of an attacking herbivore) or antixenosis (a plant's ability to deter herbivore attack). For example, using this approach, *S. berthaultii*, *Solanum circaeifolium* Bitt. *(capsicibaccatum)* (now considered to be the same species as *S. circaeifolium*, and referred to as such here), *Solanum jamesii* Torr., *Solanum pinnatisectum* Dun, and *Solanum trifidum* Corr. demonstrate both antixenosis and antibiosis against the CPB but express different levels of resistance (Pelletier and Tai 2001). The mode of resistance of *S. polyadenium* seems to be antibiosis, and that of *S. berthaultii (tarijense)* antixenosis (Pelletier and Tai 2001). Field and laboratory evaluations of CPB performance support conclusions from field analyses (Pelletier *et al.* 1999). A similar approach was used to demonstrate the antixenosis effect of *Solanum oplocense* Hawkes, *S. okadae* Hawkes and Hjerting, and *S. berthaultii (tarijense)* (Pelletier *et al.* 2001). Laboratory evaluation of CPB performance in the same study confirmed the antixenosis effect of *S. berthaultii (tarijense)*, but the repellent effect was limited to adult CPB. The reduction of oviposition by adult CPB can also be interpreted as resulting from antixenotic factors present in *S. berthaultii* (Casagrande 1982). However, at least part of the reduced fecundity could be explained by a slower ovarian development and a negative impact on the digestive tract when feeding on *S. berthaultii* foliage, and is thus linked to antibiosis factors (Franca and Tingey 1994).

In aphids, resistance mechanisms have mainly been studied through analysis of host-plant selection (Powell *et al.* 2006), potentially resulting in a biased emphasis on certain mechanisms – antibiosis normally being underestimated. Electrical penetration graph (EPG) devices enable researchers to record the activity of aphid stylets inside the leaf, and to detect and quantify several behavioral parameters (Tjallingii 1995). A multivariate approach is best suited to analyze these behavioral datasets (Pompon *et al.* 2010a). However, in-depth analysis of precise plant compounds or physical plant characteristics is still required to determine resistance mechanisms. Such resistance mechanisms appear rather specific to aphid species.

Repellent plant factors in the superficial layers of *S. polyadenium*, *S. berthaultii*, *S. berthaultii (tarijense)*, *Solanum spegazzinii* Bitt., *S. circaeifolium (capsicibaccatum)*, and *Solanum stoloniferum* Schltdl. & Bouché may be responsible for delayed stylet insertion by *M. persicae* (Alvarez *et al.* 2006, Pompon *et al.* 2010a). Physical and/or chemical characteristics of the mesophyll are thought to trigger stylet penetration impairment in *S. stoloniferum*, resulting in stylet derailment, and an extended probing time as on *Solanum chomatophilum* Bitt. for *M. persicae* (Pompon *et al.* 2010a). When aphids insert their stylets into vascular tissues, the plants mount a wound response resulting in the coagulation of phloem sap. Such a mechanism is suspected to be responsible for the lower performance of *M. persicae* on *S. stoloniferum* (Alvarez *et al.* 2006), and of *M. euphorbiae* on *S. oplocense*, *S. stoloniferum* (Pompon *et al.* 2010a), and *S. trifidum* (Le Roux *et al.* 2008). Aphids eject watery saliva to prevent the sieve elements sealing (Tjallingii 2006), but this does not always stop the coagulation process.

Phloem sap composition (Pescod *et al.* 2007) and the presence of toxic compounds (Golawska 2007) can also affect penetration behavior at the level of vascular tissues. Salivation in phloem vessels increases when aphids face nutritionally unbalanced (Ponder *et al.* 2000) or toxic phloem sap (Ramirez and Niemeyer 1999), both of which are suggested to underlie resistance in *S. pinnatisectum* to *M. euphorbiae* (Pompon *et al.* 2010a). Nutritionally unbalanced or toxic phloem sap may also be responsible for difficulties in maintaining phloem sap ingestion, as observed for *M. persicae* on *Solanum multiinterruptum* Bitt. (Alvarez *et al.* 2006), and for *M. euphorbiae* on *S. polyadenium* and *S. berthaultii (tarijense)* (Pompon *et al.* 2010a). Aphids occasionally consume xylem sap to regulate the osmotic pressure of the gut content, but the precise causes of xylem sap consumption on resistant accessions of *S. chomatophilum*, *S. oplocense*, and *S. pinnatisectum* by *M. euphorbiae* remain to be elucidated (Le Roux *et al.* 2008, Pompon *et al.* 2010a, 2010b, 2011).

The Periderm and Tuber Defenses

The principal barrier to insect attack on potato tubers is the periderm. The periderm consists of numerous layers of cork (Cutter 1992). The outer phellem layers contain wax and subarins that form an effective barrier to water. Periderm thickness and chemical composition vary according to potato species, variety, and environmental conditions during tuber development (Cutter 1992, Tyner *et al.* 1997). The periderm of wild potatoes can contribute from 0% to 100% of resistance against tuber moths, but more typically is only a part of the tuber defense, with the remainder related to unidentified cortex properties (Horgan *et al.* 2007b, 2010). Because the periderm is considerably thinner near the tuber eyes, most neonate tuber moths enter tubers (both commercial varieties and wild potato species) through the eyes (Malakar-Kuenen and Tingey 2006, Horgan *et al.* 2007b).

Tuber resistance generally breaks down as tubers age and sprout (Horgan *et al.* 2007b, 2010). Lower resistance against tuber moths on sprouting tubers suggests that resistance is related to tuber dormancy (Horgan *et al.* 2007b). Dormancy break is governed by a series of interacting hormones, including phenolics located in the periderm that inhibit plant growth (Suttle 2004). During dormancy, food availability and quality may be lower for developing insects (Cutter 1992, Coleman 2000). Dormant tubers also have reduced water and gas permeability (Cutter 1992), suggesting that larval penetration of the periderm is also inhibited. Genetically determined dormancy potential may be linked to the resistance of recently formed (new), non-sprouting tubers, because tubers that are programmed to undergo extended dormancy will have an increased number and intensity of physiological adaptations that prevent tuber decay. Therefore, tuber resistance may generally be a secondary feature of dormancy governed by adaptations to climatic conditions rather than to herbivore pressure in the potato's native habitat.

ENGINEERED RESISTANCE

Using Genetic Transformation Methods

Developments in the fields of tissue culture, genetic transformation, and genomics have contributed to the emergence of novel pest-control strategies based on the use of resistant "transgenic" plants. These strategies consist of integrating new genes encoding for resistance factors into the genome of the potato plant. These novel genes can come from diverse sources, including wild *Solanum* species (Sagredo *et al.* 2006), other plants (Lecardonnel *et al.* 1999), bacteria (Mohammed *et al.* 2000), sea urchins (Outchkourov *et al.* 2003), and, potentially, spider and scorpion venom (Pham Trung *et al.* 2006, Hernández-Campuzano *et al.* 2009). Furthermore, green-tissue specific or tuber-specific promoters can be used to restrict transgene expression to above- or below-ground plant tissues, respectively, to reduce costs to the plant, improve direct resistance factors, and avoid consumer concerns about possible negative effects of the novel proteins (Meiyalaghan *et al.* 2006, Kumar *et al.* 2010). Therefore, while traditional breeding methods are restricted to introgression of genes from closely related species, genetic transformation greatly increases the potential to find new sources of resistance.

To date only a few genes coding for resistance factors against insects have been introgressed into the potato crop. Many resistance factors are polygenic in nature, which complicates their use in transgenic methods. Laboratories in several countries have developed insect-resistant transformed potatoes, and some countries are in the process of evaluating such cultivars for national registration; however, there is currently no commercial use of transgenic potatoes; although *Bt*-transgenic potatoes (Monsanto Newleaf® potatoes) were sold in the US and Canada between 1995 and 2001, they have since been withdrawn from the market (Thornton 2003, Clive 2010).

Transformation Events and their Effects

Bacillus thuringiensis endotoxin gene

Bacillus thuringiensis (*Bt*) produces a protein (Cry) that interacts with specific protein receptors present in the epithelial cells of the insect midgut, causing death of the target pest (Pelletier and Michaud 1995). Different forms of the Cry protein have been isolated and have different levels of toxicity, depending on the group of insects.

The *cry3A* gene has been used to provide resistance against the CPB to several potato varieties in field and cage experiments (Coombs *et al.* 2005, Kamenova *et al.* 2008). Mortality of CPB larvae fed on transformed potato likely results from greatly reduced ingestion of plant material (Altre *et al.* 1996). Adult CPBs make shorter flights when fed *Bt*-transformed potato foliage (Alyokhin *et al.* 1999), but residency time is shorter in *Bt*-transformed potato field plots compared to non-transformed plots (Pelletier *et al.* 2000).

A modified *Bt-cry1Ia1* gene (often referred to as *cry5)* has also been used to transform potato for resistance to both the CPB and lepidopterans (Douches *et al.* 2002). In field and laboratory evaluations, *cry3A* was more effective than *cry1Ia1* in reducing CPB damage, even when combined with other, natural resistance factors (Coombs *et al.* 2003). Nevertheless, the *cry1Ia1* gene significantly reduces damage to potatoes by the potato tuber moth in field trials compared to non-transgenic control plants, and in stored potato tubers (Mohammed *et al.* 2000, Cooper *et al.* 2007). Cry1Ia1 potato lines with increased resistance to CPB have been achieved through a *cry1Ba/cry1Ia* hybrid gene *(SN19)* encoding a protein consisting of domains I and III of Cry1Ba and domain II of Cry1Ia (Naimov *et al.* 2006).

Transformation with *Bt* genes can be applied to practically any potato germplasm, including CPB-resistant clones expressing leptines or trichomes (Jansky *et al.* 1999). A CPB-resistant potato clone derived from *S. chacoense* expressing leptine glycoalkaloids was transformed to produce Bt toxin and was found to be more resistant to the CPB, including *Bt*-resistant CPB populations, than when only one resistance factor was present in the potato foliage (Jansky *et al.* 1999, Coombs *et al.* 2005, Cooper *et al.* 2004). Similar results have been found with the potato tuber moth (Douches *et al.* 1998). The approach has also been used to transform trichome producing germplasm with the *Bt-cry11a1* gene, but with little success in terms of field resistance to the CPB (Coombs *et al.* 2003).

Aphids are not highly sensitive to Cry toxins (Cloutier *et al.* 2008). Moderate toxicity against the pea aphid, *Acyrthosiphon pisum* (Harris), can be observed for four types of Cry toxin (*Cry3A, Cry4A, Cry11Aa*, and *Cry1Aa*) fed through artificial diets at high doses (Porcar *et al.* 2009). *Macrosiphum euphorbiae* performance was also slightly negatively affected on *Bt*-transformed potato plants, although this effect might be attributed to disturbance of the plant physiology by the transgene (Ashouri *et al.* 2001). Interestingly, the flight incidence of aphids is higher on *Bt* varieties, suggesting a low quality of phloem sap (Ashouri *et al.* 2001, Muller *et al.* 2001). Cry effects on aphids appear to be species-specific, and to depend on other unknown factors. *Myzus persicae* populations were higher on Newlcaf® *Bt* potato plants than on non-transformed plants in almost half of the impact studies conducted during development of the variety (Cloutier *et al.* 2008).

Protease inhibitors

Protease inhibitors (PIs) from several sources have been evaluated in transgenic potatoes against insect pests. Protease inhibitors of the "cysteine" class, such as the cystatins, inhibit a major part of CPB digestive proteases (Pelletier and Michaud 1995). Ingestion of protease inhibitors by small CPB larvae results in reduced growth and compromises survival. A barley cystatin, HvCPI-1, and one of its variants, HvCPI-1 C68 fi G, have been used in the transformation of potato to produce a significant increase in the development time of CBP larvae (Álvarez-Alfageme *et al.* 2007). Furthermore, PIs from a locust hemolymph (SGTI and SGCI), inhibiting both trypsin and chymotrypsin, produced a slight increase of the development time of CPB larvae on transformed potatoes (Kutas *et al.* 2004, Kondrák *et al.* 2005).

Transformed potato lines expressing a cathepsin D inhibitor (CDI) from tomato have been developed and evaluated for their effect on larvae of the CPB (Brunelle *et al.* 2004). Leaf consumption and relative growth rates were slightly reduced during the first 12 hours for third-instar larvae fed CDI-transformed foliage, but no significant differences were observed thereafter. Monitoring of the digestive proteases showed that the CPB is able to compensate for the loss of cathepsin D activity by modulating its digestive protease complement in response to aspartate-type inhibitors in the diet (Brunelle *et al.* 2004).

CPB larvae feeding on potato transformed with the rice cystatin I (oryzacystatin I: OCI) consumed leaf material 14% faster, gained weight 28% faster, and weighed 20% more at the end of the L3 stage, leading to a faster development time compared to controls (Cloutier *et al.* 1999). However, a different transformation using the same transgene produced effectively resistant germplasm (Lecardonnel *et al.* 1999). Further variability in the efficiency of protease inhibitors to reduce the biological performance of CPB larvae was noted during the development of resistant transformed potato germplasm. Resistance varied with the age of the larvae but also with the exposure time to the resistant foliage (Cloutier *et al.* 1999). The CPB has the ability to rapidly (within a few days) change its complement of proteases and switch to proteases insensitive to the inhibitor (Cloutier *et al.* 2000). This is a major inconvenience for the long-term use of protease inhibitors as a resistance factor against the CPB and other insects.

Immunological labeling of OCI indicated that the protease inhibitor was present along the digestive gut of aphids, and associated with bacteriocytes in *M. persicae* fed on transgenic oilseed rape (Rahbé *et al.* 2003). OCI may affect fitness through digestive tract targets, but also by reaching the haemolymph, thereby inhibiting extra-digestive proteolytic cascades and interacting with functions related to reproduction. Accordingly, OCI fed through artificial diets moderately but significantly reduced growth of several aphid species, including *M. persicae* (Rahbé *et al.* 2003).

Protease inhibitors impact aphid performance in a dose-dependent manner. A transgenic potato, expressing chicken egg-white cystatin (CEWc) and developed to control nematodes, was tested in field settings and by restricting *M. persicae* to leaves using clip cages (Cowgill *et al.* 2002). The toxin did not affect aphid density or fitness. However, when aphids were fed on artificial diets containing CEWc, the survival and growth of larvae was inhibited (Cowgill *et al.* 2002). Lack of effect in the plant may result from a lower concentration of PI in the phloem than in the artificial diets. The use of promoters other than the commonly used CaMV 35S could increase expression levels. Notably, in such a dose-dependent situation, the assessment method, and whether or not it is possible to regulate the PI concentration, directly influence the outcome of the study.

Avidin

Biotin is a water-soluble vitamin produced by plants, bacteria, and some fungi, and required by all forms of life. Avidin is a biotin-binding protein produced in the oviducts of birds, reptiles, and amphibians, and deposited in the whites of their eggs. Because of the universal dependence on biotin, avidin is effective against a broad range of plant pests, including Diptera, Lepidoptera, and Coleoptera (Cooper *et al.* 2006, 2009a). Potato can be transformed to produce avidin (chicken avidin), causing a significant decrease in biological performance of CPB larvae; however, the CPB tends to be less sensitive to avidin than are other insects (Cooper *et al.* 2009a). Tobacco plants transformed with the pPLA2 (chicken) avidin and pSAV2α streptavidin (from *Streptomyces avidinii*) genes produce significant mortality (93%) of tuber moth larvae (Markwick *et al.* 2003). However, in a study of transgenic potatoes, despite considerably higher avidin expression than in the tobacco study, tuber moth mortality was not higher than in susceptible, control plants. By combining avidin with natural defenses derived from *S. chacoense*, 98% mortality of larvae could be achieved (Cooper *et al.* 2009b).

Lectins

Lectins are sugar-binding proteins involved in many biological processes. Several plant lectins play a role in defense against insect pests, and can affect Coleoptera, Lepidoptera, Hemiptera, and Diptera. However, lectins are generally harmful to mammals (Licastro *et al.* 1993), and care should be taken to assess the potential effects on humans if these proteins are to be used in crop protection.

Lectins are particularly attractive for controlling Hemiptera, which are not highly sensitive to *Bt* or protease inhibitors. The snowdrop lectin (*Galanthus nivalis* L. agglutinin, GNA) is a mannose-binding lectin and seems to be non-toxic to mammals (Pusztai *et al.* 1993). GNA inhibits fecundity and growth development in the foxglove aphid, *Aulacorthum solani* (Kaltenbach), and in *M. persicae* when delivered in artificial diets and transgenic potato (Sauvion *et al.* 1993; Down *et al.* 1996, Gatehouse *et al.* 1996). A delay in the onset of reproduction for adults released on a GNA-containing diet suggests that GNA has an antifeedant effect (Down *et al.* 1996).

Concanavalin A (ConA) is a glucose/mannose-specific lectin from jack-bean (*Canavalia ensiformis* [L.]). ConA has a detrimental effect on *M. persicae* fecundity when aphids ingest the toxin from an artificial diet or from transgenic potatoes (Gatehouse *et al.* 1999). The expression level of the transgene in the plants can be lower than expected, indicating that transposition of the gene to another species does not always result in a functional protein in the modified plant (Gatehouse *et al.* 1999).

CONSTRAINTS ON HPR

Tradeoffs, Yield Penalties and Insect Adaptation

Two of the principal constraints to adopting HPR are inherent tradeoffs associated with the resistance mechanisms and the ability of insect pests to overcome the resistance. These constraints apply to both transgenic and conventionally-bred plants. Tradeoffs arise because of restrictions in resource allocations in plants to growth, reproduction, and defense (Herms and Mattson 1992). High costs of resistance are thought to deplete resources allocated to other plant functions. Until the advent of marker-aided selection and transgenic technology, it was difficult to detect such tradeoffs; however, there is increasing evidence that HPR can sometimes cause yield penalties (Kalazich and Plaisted 1991, Chakrabarti *et al.* 2006, Xia *et al.* 2010). Avoidance of such penalties can be achieved through careful selection of elite lines during conventional breeding, and through tissue-specific or other promoters during transgenic development. Penalties may be small when one major gene governs resistance; however, yield penalties could represent a major constraint when several major resistance genes are combined in a single plant.

Adaptation by insects to varieties with major resistance genes is a major concern of HPR. Virulent insect populations have been associated with the large-scale deployment of several resistant crops (Gould 1991, 1998). For example quantitative genetic and artificial selection studies indicate that the CPB can overcome *S. berthaultii* resistance within several generations (Groden and Casagrande 1986, Pelletier and Smilowitz 1991, Franca *et al.* 1994), and Dipel (*Bt*) resistant tuber moths have been found in the wild (Rico *et al.* 1998). The rate at which a pest overcomes resistance depends on several factors, including the area and duration of planting of the resistant cultivar, the size of the pest population, the strength of the resistance, the pest's diet range and availability of alternate host-plants, the pest's reproductive strategy, and the efficiency of pest regulation by natural enemies (Gould 1998). Crop deployment and management strategies aimed at preventing the development of virulent pest populations should limit the scale of deployment (in the case of *Bt* genes, limiting deployment of all *Bt* crops, not only potato, that are potential hosts for the insect) and avoid resurgence pesticides that disrupt natural regulation of pest populations.

Specific Issues with Transgenic Resistance

Transgenic crops are "new" in evolutionary time. It is only since the early 1990s, when transgenic crops were first publically marketed, that scientists, farmers and consumers have been forced to evaluate and predict the potential positive and/or negative effects of these crops. Difficulties in reaching agreement between proponents of transgenic technologies and those who prefer greater caution makes transgenic resistance considerably less attractive than conventionally bred resistance, especially where the conventional resistance is already effective (i.e., Coombs *et al.* 2003). Furthermore, transgenic potatoes are not suitable for farms in regions where they could potentially contaminate native tuber-bearing or weedy Solanacea (Celis *et al.* 2004). Public distrust, whether justified or not, is largely responsible for poor adoption of transgenic potatoes (Thornton 2003). Of particular interest to IPM are issues of transgenic compatibility with biological control and natural pest regulation. Optimally, resistant crops, whether transgenic or conventionally bred, should work together with natural enemies to reduce pest densities and cause greater mortality to pests than when either is working alone (van Emden 1986, Bell *et al.* 2001). As discussed above, even conventional resistance can sometimes reduce the efficiency of natural enemies.

Several studies have examined the effects of transgenic crops on natural enemies (Lövei and Arpaia 2005, Marvier *et al.* 2007, Shelton *et al.* 2009, Gatehouse *et al.* 2011). Many of these studies have found no negative effects of insecticidal proteins (*Bt* toxins) or protease inhibitors on non-target herbivores or natural enemies; however, the results of risk assessment studies depend on the choice of natural enemy and the evaluation method. For example, the effects of protease inhibitors on natural enemies, as mediated through the target herbivore, depend on the nature of the inhibitor expressed by the crop and whether the same or different proteases are used by the herbivore and its natural enemy (Gatehouse *et al.* 2011). A number of studies indicate that transgenic crops reduce farm applications of insecticides and have positive effects on certain predators compared to chemical-based pest control methods (Gatehouse *et al.* 2011). However, more meaningful comparisons might include farms adopting ecologically based pest management methods (Marvier *et al.* 2007). Such comparative studies will also need to more carefully address mechanisms. Concerns should only arise where clear associations have been drawn between biodiversity loss and any toxic effects of transgenic-crops on non-target beneficial species.

FUTURE DIRECTIONS

Host-plant resistance has been a prominent area in IPM research in potatoes for the past three decades, and is gaining increased attention. Whereas some of this research has ultimately led to the release of new, resistant potato varieties, there is still a need to apply newer technologies and knowledge to increase the strength and stability of resistance, to diversify resistance sources (and therefore enhance resistance durability at landscape levels), and to better integrate HPR into current IPM practices. There are rich sources of resistance on which to base future research. This chapter has given some indication of the wealth of these sources, which include over 230 wild potato species, some 2500 native potato varieties, and several moderately resistant commercial cultivars whose resistance can be augmented through recurrent breeding programs. Furthermore, transgenic technologies have produced plants that derive their resistance from sources as diverse as bacterial toxins, plant-derived protease inhibitors, and biotin-binding avidins.

This diversity of resistance sources represents an array of resistance mechanisms, some of which are highly species-specific (i.e., certain secondary chemicals) and some of which are broad-spectrum (i.e., trichomes). Research has mainly focused on identifying these sources, on understanding their underlying mechanisms, and on developing tools to integrate the resistance into modern potato varieties. However, there is a need to look more holistically at potato resistance, particularly as regards the evolution of virulent (adapted) insect populations. Researchers will need to focus on potato-herbivore interactions as part of a dynamic system. For example, the current trend of storing herbivore populations as closed, inbred, laboratory colonies for screening and testing of resistance cannot possibly address this dynamic nature, and needs to be reviewed. The observation that increased resistance can be gained from existing varieties through recurrent selection may relate directly to this dynamic nature of insect-plant interactions and the co-evolution of plants and their insect pests. Regionally, herbivore populations are likely to be best adapted to the most commonly grown, local varieties, such that temporal and spatial variability in varieties is expected to decrease crop vulnerability.

Host-plant resistance needs to be compatible with other IPM methods and with consumer demands, and it has to avoid associated yield and quality costs. These requirements can limit the application of new technologies and the application of seemingly highly effective resistance sources or mechanisms. Where effective and accepted by farmers and consumers, HPR must reduce insecticide use, which requires an associated, informed agricultural extension system. In effect, this is the up-scaling component of HPR development, but it is rarely regarded as part of integrated HPR breeding and development programs. Therefore, although several sources of resistance to potato pests have been identified, and although the search for new resistance sources should continue, further efforts are needed to apply HPR successfully in the field. The release of a few modern resistant potato varieties, especially those developed through crossing with wild potato species such as *S. berthaultii* and *S. chacoense*, indicates the large potential for future successes of HPR in modern potato farms.

REFERENCES

Altre, J.A., Grafius, E.J., Whalon, M.E., 1996. Feeding behavior of cryIIIA-resistant and susceptible Colorado potato beetle (Coleoptera: Chrysomelidae) larvae on *Bacillus thuringiensis* tenebrionis-transgenic *cryIIIA*-treated and untreated potato foliage. J. Econ. Entomol. 89, 311–317.

Alvarez, A.E., Tjallingii, W.F., Garzo, E., Vleeshouwers, V., Dicke, M., Vosman, B., 2006. Location of resistance factors in the leaves of potato and wild tuber-bearing *Solanum* species to the aphid. Myzus persicae. Entomol. Exp. Appl. 121, 145–157.

Álvarez-Alfageme, F., Martínez, M., Pascual-Ruiz, S., Castañera, P., Diaz, I., Ortego, F., 2007. Effects of potato plants expressing a barley cystatin on the predatory bug *Podisus maculiventris* via herbivorous prey feeding on the plant. Transgenic Res. 16, 1–13.

Alyokhin, A., Ferro, D.N., Hoy, C.W., Graham, H., 1999. Laboratory assessment of flight activity displayed by Colorado potato beetles (Coleoptera: Chrysomelidae) fed on transgenic and *Cry3a* toxin-treated potato foliage. J. Econ. Entomol. 92, 115–120.

Ashouri, A., Michaud, D., Cloutier, C., 2001. Unexpected effects of different potato resistance factors to the Colorado potato beetle (Coleoptera: Chrysomelidae) on the potato aphid (Homoptera: Aphididae). Environ. Entomol. 30, 524–532.

Baggen, L.R., Gurr, G.M., 1995. Lethal effects of foliar pubescence of solanaceous plants on the biological control agent *Copidosoma koehleri* Blanchard (Hymenoptera: Encyrtidae). Plant Protec. Quart 10, 116–118.

Bell, H.A., Fitches, E.C., Down, R.E., Ford, L., Marris, G.C., Edwards, J.P., Gatehouse, J.A., Gatehouse, A.M.R., 2001. Effects of dietary cowpea trypsin inhibitor (*CpTI*) on the growth and development of tomato moth *Lacanobia oleracea* (Lepidoptera: Noctuidae) and on the success of the gregarious ectoparasitoid *Eulophus pennicornis* (Hymenoptera: Eulophidae). Pest Manag. Science 57, 57–65.

Brunelle, F., Cloutier, C., Michaud, D., 2004. Colorado potato beetles compensate for tomato cathepsin D inhibitor expressed in transgenic potato. Arch. Insect Biochem. Physiol. 55, 103–113.

Carter, C.D., 1987. Screening *Solanum* germplasm for resistance to Colorado potato beetle. Am. Potato J. 64, 563–568.

Casagrande, R.A., 1982. Colorado potato beetle resistance in wild potato. Solanum berthaultii. J. Econ. Entomol. 75, 368–372.

Castañera, P., Steffens, J.C., Tingey, W.M., 1996. Biological performance of Colorado potato beetle larvae on potato genotypes with differing levels of polyphenol oxidase. J. Chem. Ecol. 22, 91–101.

Celis, C., Scurrah, M., Cowgill, S., Chumbiauca, S., Green, J., Franco, J., Main, G., Kiezebrink, D., Visser, R.G.F., Atkinson, H.J., 2004. Environmental biosafety and transgenic potato in a centre of diversity for this crop. Nature 432, 222–225.

Chakrabarti, S.K., Lutz, K.A., Lertwiriyawong, B., Svab, Z., Maliga, P., 2006. Expression of the cry9Aa2 *B.t.* gene in tobacco chloroplasts confers resistance to potato tuber moth. Transgenic Res. 15, 481–488.

Chen, Q., Li, H.Y., Shi, Y.Z., Beasley, D., Bizimungu, B., Goettel, M.S., 2008. Development of an effective protoplast fusion system for production of new potatoes with disease and insect resistance using Mexican wild potato species as gene pools. Can. J. Plant Sci. 88, 611–619.

Clive, J., 2010. Global status of commercialized biotech/GM crops: 2010, ISAAA *Brief No. 42*, Ithaca, NY.

Cloutier, C., Fournier, M., Jean, C., Yelle, S., Michaud, D., 1999. Growth compensation and faster development of Colorado potato beetle (Coleoptera: Chrysomelidae) feeding on potato foliage expressing oryzacystatin I. Arch. Insect Biochem. Physiol. 40, 69–79.

Cloutier, C., Jean, C., Fournier, M., Yelle, S., Michaud, D., 2000. Adult Colorado potato beetles, *Leptinotarsa decemlineata* compensate for nutritional stress on oryzacystatin I-transgenic potato plants by hypertrophic behavior and over-production of insensitive proteases. Arch. Insect Biochem. Physiol. 44, 69–81.

Cloutier, C., Boudreault, S., Michaud, D., 2008. Impact de pommes de terre résistantes au doryphore sur les arthropodes non visés: une méta-analyse des facteurs possiblement en cause dans l'échec d'une plante transgénique Bt. Cah. Agric. 17, 388–394.

Coleman, W.K., 2000. Physiological ageing of potato tubers: A review. Ann. Appl. Biol. 137, 189–199.

Coombs, J.J., Douches, D.S., Li, W., Grafius, E.J., Pett, W.L., 2003. Field evaluation of natural, engineered, and combined resistance mechanisms in potato for control of Colorado potato beetle. J. Am. Soc. Hort. Sci. 128, 219–224.

Coombs, J.J., Douches, D.S., Cooper, S.G., Grafius, E.J., Pett, W.L., Moyer, D.D., 2005. Combining natural and engineered host plant resistance mechanisms in potato for Colorado potato beetle: Choice and no-choice field studies. J. Am. Soc. Hort. Sci. 130, 857–864.

Cooper, S.G., Douches, D.S., Grafius, E.J., 2004. Combining genetic engineering and traditional breeding to provide elevated resistance in potatoes to Colorado potato beetle. Entomol. Exp. Appl. 112, 37–46.

Cooper, S.G., Douches, D.S., Grafius, E.J., 2006. Insecticidal activity of avidin combined with genetically engineered and traditional host plant resistance against Colorado potato beetle (Coleoptera: Chrysomelidae) larvae. J. Econ. Entomol. 99, 527–536.

Cooper, S.G., Douches, D.S., Coombs, J.J., Grafius, E.J., 2007. Evaluation of natural and engineered resistance mechanisms in potato against Colorado potato beetle in a no-choice field study. J. Econ. Entomol. 100, 573–579.

Cooper, S.G., Douches, D.S., Grafius, E.J., 2009a. Combining engineered resistance, avidin, and natural resistance derived from *Solanum chacoense* Bitter to control Colorado potato beetle (Coleoptera: Chrysomelidae). J. Econ. Entomol. 102, 1270–1280.

Cooper, S.G., Douches, D.S., Zarka, K., Grafius, E.J., 2009b. Enhanced resistance to control potato tuberworm by combining engineered resistance, avidin, and natural resistance derived from Solanum chacoense. Am. J. Potato Res. 86, 24–30.

Cowgill, S.E., Wright, C., Atkinson, H.J., 2002. Transgenic potatoes with enhanced levels of nematode resistance do not have altered susceptibility to nontarget aphids. Mol. Ecol. 11, 821–827.

Crissman, C.C., Cole, D.C., Carpio, F., 1994. Pesticide use and farm worker health in Ecuadorian potato production. Am. J. Agric. Econ. 76, 593–597.

Cutter, E.G., 1992. Structure and development of the potato plant. In: Harris, P. (Ed.), The potato crop: The scientific basis for improvement, second ed. Chapman and Hall, London, UK, pp. 65–146.

Davis, J.A., Radcliffe, E.B., Ragsdale, D.W., 2007. Resistance to green peach aphid, *Myzus persicae* (Sulzer), and potato aphid, *Macrosiphum euphorbiae* (Thomas), in potato cultivar. Am. J Potato Res. 84, 259–269.

Dimock, M.B., Tingey, W.M., 1987. Mechanical interaction between larvae of the Colorado potato beetle and glancular trichomes of *Solanum berthaultii*. Am. Potato J. 64, 507–516.

Dimock, M.B., Tingey, W.M., 1988. Host acceptance behavior of Colorado potato beetle larvae influenced by potato glandular trichomes. Physiol. Entomol. 13, 399–406.

Dimock, M.B., Lapointe, S.L., Tingey, W.M., 1985. Resistance in Solanum spp. to the Colorado potato beetle: Mechanisms, genetic resources and potential. In: Ferro, D.N., Voss, R.H. (Eds.), Proceedings of the Symposium on the Colorado potato beetle, XVIIth International Congress of Entomolgy, Mass. Agric. Exp. Sta. Bull. 704.

Dimock, M.B., Lapointe, S.L., Tingey, W.M., 1986. *Solanum cardenasii*: A new source of potato resistance to the Colorado potato beetle (Coleoptera: Chrysomelidae). J Econ Entomol. 79, 1269–1275.

Douches, D.S., Westedt, A.L., Zarka, K., Schroeter, B., 1998. Potato transformation to combine natural and engineered resistance for controlling tuber moth. HortScience 33, 1053–1056.

Douches, D.S., Li, W., Zarka, K., Coombs, J., Pett, W., Grafius, E., El-Nasr, T., 2002. Development of Bt-cry5 insect resistant potato lines "Spunta-G2" and "Spunta-G3". HortScience 37, 1103–1107.

Down, R.E., Gatehouse, A.M.R., Hamilton, W.D.O., Gatehouse, J.A., 1996. Snowdrop lectin inhibits development and decreases fecundity of the glasshouse potato aphid (*Aulacorthum solani*) when administered in vitro and via transgenic plants both in laboratory and glasshouse trials. J. Insect Physiol. 42, 1035–1045.

Fisher, D.G., Deahl, K.L., Rainforth, M.V., 2002. Horizontal resistance in Solanum tuberosum to Colorado potato beetle (*Leptinotarsa decemlineata* Say). Am. J. Potato Res. 79, 281–293.

Flanders, K.L., Hawkes, J.G., Radcliffe, E.B., Lauer, F.I., 1992. Insect resistance in potatoes: sources, evolutionary relationship, morphological and chemical defenses, and ecogeographical associations. Euphytica 61, 83–111.

Flanders, K.L., Arnone, S., Radcliffe, E.B., 1999. The potato: genetic resources and insect resistance. In: Global Plant Genetic Resources for Insect-Resistance Crops (S. L. Clement, and S. S. Quinsberry), CRC Press, Boca Raton, FL, pp. 207–239.

Franca, F.H., Tingey, W.M., 1994. *Solanum berthaultii* Hawkes affects the digestive system, fat body and ovaries of the Colorado potato beetle. Am. Potato J. 71, 405–411.

Franca, F.H., Plaisted, R.L., Roush, R.T., Via, S., Tingey, W.M., 1994. Selection response of the Colorado potato beetle for adaptation to the resistant potato. *Solanum berthaultii*. Entomol. Exp. Appl. 73, 101–109.

Fréchette, B., Vincent, C., Giordanengo, P., Pelletier, Y., Lucas, É, 2012. *In:* Do resistant plants provide an enemy-free space to aphids? Eur. J. Entomol. 109, 135–137.

Gatehouse, A.M.R., Down, R.E., Powell, K.S., Sauvion, N., Rahbé, Y., Newell, C.A., Merryweather, A., Hamilton, W.D., Gatehouse, J.A., 1996. Transgenic potato plants with enhanced resistance to the peach-potato aphid. *Myzus persicae*. Entomol. Exp. Appl. 79, 295–307.

Gatehouse, A.M.R., Davison, G.M., Stewart, J.N., Galehouse, L.N., Kumar, A., Geoghegan, I.E., Birch, A.N.E., Gatehouse, J.A., 1999. Concanavalin A inhibits development of tomato moth (*Lacanobia oleracea*) and peach-potato aphid (*Myzus persicae*) when expressed in transgenic potato plants. Mol. Breed 5, 153–165.

Gatehouse, A.M.R., Ferry, N., Edwards, M.G., Bell, H.A., 2011. Insect-resistant biotech crops and their impacts on beneficial arthropods. Philo. Trans. R. Soc. B-Biol. Sci. 1569, 1438–1452.

Ghassemi-Kahrizeh, A., Nouri-Ganbalani, G., Shayesteh, N., Bernousi, I., 2010. Antibiosis effects of 20 potato cultivars to the Colorado potato beetle, Leptinotarsa decemlineata (Say)(Col. Chrysomelidae). J. Food Agric. Env. 8, 795–799.

Gibson, R.W., 1976. Glandular hairs on Solanum polyadenium lessen damage by the Colorado beetle. Ann. Appl. Biol. 82, 147–150.

Golawska, S., 2007. Deterrence and toxicity of plant saponins for the pea aphid Acyrthosiphon pisum Harris. J. Chem. Ecol. 33, 1598–1606.

Gooderham, J., Bailey, P.C.E., Gurr, G.M., Baggen, L.R., 1998. Sub-lethal effects of foliar pubescence on the egg parasitoid Copidosoma koehleri and influence on parasitism of potato tuber moth. Phthorimaea operculella. Entomol. Exp. Appl. 87, 115–118.

Gould, F., 1991. The evolutionary potential of crop pests. Am. Sci. 79, 496–507.

Gould, F., 1998. Sustainability of transgenic insecticidal cultivars: integrating pest genetics and ecology. Annu. Rev. Entomol. 43, 701–726.

Gregory, P., Tingey, W.M., Ave, D.A., Bouthyette, P.Y., 1986. Potato glandular trichomes: A physicochemical defense mechanism against insects. ACS Symp. Series 296, 160–167.

Groden, E., Casagrande, R.A., 1986. Population dynamics of the Colorado potato beetle, Leptinotarsa decemlineata (Coleoptera: Chrysomelidae), on Solanum berthaultii. J. Econ. Entomol. 79, 91–97.

Hanzlik, M.W., Kennedy, G.G., Sanders, D.C., Monks, D.W., 1997. Response of European corn borer (Ostrinia nubilalis, Hubner) to two potato hybrids selected for resistance to Colorado potato beetle. Crop Protect 16, 487–490.

Herms, D.A., Mattson, W.J., 1992. The dilemma of plants: to grow or defend. Q. Rev. Biol. 67, 283–335.

Hernández-Campuzano, B., Suárez, R., Lina, L., Hernández, V., Villegas, E., Corzon, G., Iturriaga, G., 2009. Expression of a spider venom peptide in transgenic tobacco confers insect resistance. Toxicon 53, 122–128.

Horgan, F.G., Quiring, D.T., Lagnaoui, A., Pelletier, Y., 2007a. Variable responses of tuber moth to the leaf-trichomes of wild potatoes. Entomol. Exp. Appl. 125, 1–12.

Horgan, F.G., Quiring, D.T., Lagnaoui, A., Salas, A.R., Pelletier, Y., 2007b. Periderm- and cortex-based resistance to tuber-feeding Phthorimaea operculella in two wild potato species. Entomol. Exp. Appl. 125, 249–258.

Horgan, F.G., Quiring, D.T., Lagnaoui, A., Pelletier, Y., 2009. Trade-off between foliage and tuber resistance to Phthorimaea operculella in wild potatoes. Entomol. Exp. Appl. 131, 130–137.

Horgan, F.G., Quiring, D.T., Lagnaoui, A., Salas, A.R., Pelletier, Y., 2010. Variations in resistance against Phthorimaea operculella in wild potato tubers. Entomol. Exp. Appl. 137, 269–279.

Horgan, F.G., Quiring, D.T., Lagnaoui, A., Pelletier, Y., 2012. Life-histories and fitness of two tuber moth species feeding on native Andean potatoes. Neotrop. Entomol. in press.

Jansky, S., Austin-Phillips, S., McCarthy, C., 1999. Colorado potato beetle resistance in somatic hybrids of diploid interspecific Solanum clones. HortScience 34, 922–927.

Jansky, S.H., Simon, R., Spooner, D.M., 2009. A test of taxonomic predictivity: Resistance to the Colorado Potato Beetle in wild relatives of cultivated potato. J. Econ. Entomol. 102, 422–431.

Juraske, R., Mosquera-Vivas, C.S., Erazo-Velásquez, A., García-Santos, G., Berdugo-Moreno, M.B., Diaz-Gomez, J., Binder, C.R., Hellweg, S., Guerrero-Dallos, J.A., 2011. Pesticide uptake in potatoes: model and field experiments. Environ. Sci. Technol. 45, 651–657.

Kalazich, J.C., Plaisted, R.L., 1991. Association between trichome characters and agronomic traits in Solanum tuberosum (L.) × S. berthaultii (Hawkes) hybrids. Am. Potato J. 68, 833–847.

Kamenova, I., Batchvarova, R., Flasinski, S., Dimitrova, L., Christova, P., Slavov, S., Atanassov, A., Kalushkov, P., Kaniewski, W., 2008. Transgenic resistance of Bulgarian potato cultivars to the Colorado potato beetle based on Bt technology. Agro. Sustain. Dev. 28, 481–488.

Kondrák, M., Kutas, J., Szenthe, B., Patthy, A., Bánfalvi, Z., Nádasy, M., Gráf, L., Asbóth, B., 2005. Inhibition of Colorado potato beetle larvae by a locust proteinase inhibitor peptide expressed in potato. Biotech. Letters 27, 829–834.

Kumar, M., Chimote, V., Singh, R., Mishra, G.P., Nail, P.S., Pandey, S.K., Chakrabarti, S.K., 2010. Development of Bt transgenic potatoes for effective control of potato tuber moth by using cry1Ab gene regulated by GBSS promoter. Crop Prot. 29, 121–127.

Kutas, J., Kondrak, M., Szenthe, B., Patthy, A., Banfalvi, Z., Nadasy, M., Graf, L., Asboth, B., 2004. Colorado potato beetle larvae on potato plants expressing a locust proteinase inhibitor. Com. Agric. Appl. Biol. Sci. 69, 281–287.

Lapointe, S.L., Tingey, W.M., 1986. Glandular trichomes of Solanum neocardenasii confer resistance to green peach aphid (Homoptera: Aphididae). J. Econ. Entomol. 79, 1264–1268.

Le Roux, V., Dugravot, S., Campan, E., Dubois, F., Vincent, C., Giordanengo, P., 2008. Wild Solanum resistance to aphids: Antixenosis or antibiosis? J. Econ. Entomol. 101, 584–591.

Lecardonnel, A., Chauvin, L., Jouanin, L., Beaujean, A., Prévost, G., Sangwann-Norreel, B., 1999. Effects of rice cystatin I espression in trasngenic potato on Colorado potato beetle larvae. Plant Sci. 140, 71–79.

Licastro, F., Davis, L.J., Morini, M.C., 1993. Lectins and superantigens – Membrane interactions of these compounds with T-lymphocyes affect immue responses. Int. J. Biochem. 25, 845–852.

Lorenzen, J.H., Balbyshev, N.F., Lafta, A.M., Casper, H., Tian, X., Sagredo, B., 2001. Resistant potato selections contain leptine and inhibit development of the Colorado potato beetle (Coleoptera: Chrysomelidae). J. Econ. Entomol. 94, 1260–1267.

Lövei, G.L., Arpaia, S., 2005. The impact of transgenic plants on natural enemies: A critical review of laboratory studies. Entomol Exp. Appl. 114, 1–14.

Lyytinen, A., Lindstram, L., Mappes, J., Julkunen-Tiitto, R., Fasulati, S.R., Tiilikkala, K., 2007. Variability in host plant chemistry: Behavioural responses and life-history parameters of the Colorado potato beetle (Leptinotarsa decemlineata). Chemoecology 17, 51–56.

Malakar, R., Tingey, W.M., 2000. Glandular trichomes of Solanum berthaultii and its hybrids with potato deter oviposition and impair growth of potato tuber moth. Entomol. Exp. Appl. 94, 249–257.

Malakar-Kuenen, R., Tingey, W.M., 2006. Aspects of tuber resistance in hybrid potatoes to potato tuberworm. Entomol. Exp. Appl. 120, 131–137.

Markwick, N.P., Docherty, L.C., Phung, M.M., Lester, M.T., Murray, C., Yao, J.-L., Mitra, D.S., Cohen, D., Beuning, L.L., Kutty-Amma, S., Christeller, J.T., 2003. Transgenic potato and apple plants expressing biotin-binding proteins are resistant to two cosmopolitan insect pests, potato tuber moth and lightbrown apple moth, respectively. Transgen. Res. 12, 671–681.

Marvier, M., McCreedy, C., Regetz, J., Kareiva, P., 2007. A meta-analysis of effects of Bt cotton and maize on nontarget invertebrates. Sci. 316, 1475–1477.

Meehan, T.D., Werling, B.P., Landis, D.A., Gratton, C., 2011. Agricultural landscape simplification and insecticide use in the Midwestern United States. Proc. Natl Acad. Sci. 108, 11500–11505.

Meiyalaghan, S., Jacobs, J.M.E., Butler, R.C., Wratten, S.D., Conner, A.J., 2006. Expression of cry1Ac9 and cry9Aa2 genes under a potato light-inducible Lhca3 promoter in transgenic potatoes for tuber moth resistance. Euphytica 147, 297–309.

Mohammed, A., Douches, D.S., Pett, W., Grafius, E., Coombs, J., Liswidowati, L., Li, W., Madkour, M.A., 2000. Evaluation of potato tuber moth (Lepidoptera: Gelechiidae) resistance in tubers of Bt-cry5 transgenic potato lines. J. Econ. Entomol. 93, 472–476.

Muller, C.B., Williams, I.S., Hardie, J., 2001. The role of nutrition, crowding and interspecific interactions in the development of winged aphids. Ecol. Entomol. 26, 330–340.

Naimov, S., Zahmanova, G., Boncehva, R., Kostova, M., Minkov, I., Dukiandjiev, S., De Maagd, R., 2006. Expression of synthetic SN 19 hybrid delta-endotoxin encoding gene in transgenic potato. Biotechnol. Biotec. Eq. 20, 38–41.

National Research Council, 1989. The crops of the Incas: little-known plants of the Andes and the promise for worldwide cultivation. National Academy Press, Washington, DC, pp. 415.

Obrycki, J.J., 1986. The influence of foliar pubescence on entomophagous species. In: Boethel, D.J., Eikenbary, R.D. (Eds.), Interactions of plant resistance and parasitoids and predators of insects, John Wiley & Sons, New York, NY, pp. 61–83.

Osman, S.F., 1983. Glycoalkaloids in potatoes. Food Chem. 11, 235–247.

Outchkourov, N.S., Rogelj, B., Strukelj, B., Jongsma, M.A., 2003. Expression of sea anemone equistatin in potato: Effects of plant proteases on heterologous protein production. Plant Physiol. 133, 379–390.

Parsa, S., Canto, R., Rosenheim, J.A., 2011. Resource concentration dilutes a key pest in indigenous potato agriculture. Ecol. Appl. 21, 539–546.

Pelletier, Y., Clark, C., 2004. Use of reciprocal grafts to elucidate mode of resistance to Colorado potato beetle (*Leptinotarsa decemlineata* (Say)) and potato aphid (*Macrosiphum euphorbiae* (Thomas)) in six wild *Solanum* species. Am. J. Potato Res. 81, 341–346.

Pelletier, Y., Dutheil, J., 2006. Behavioural responses of the Colorado potato beetle to trichomes and leaf surface chemicals of *Solanum tarijense*. Entomol. Exp. Appl. 120, 125–130.

Pelletier, Y., Michaud, D., 1995. Insect pest control on potato: genetically-based control. In: Duchesne, R.-M., Boiteau, G. (Eds.), Proceedings Potato Insect Pest Control: Development of a Sustainable Approach, July 31–August 1, 1995, Québec, Canada, pp. 69–79.

Pelletier, Y., Smilowitz, Z., 1991. Biological and genetic study on the utilization of *Solanum berthaultii* Hawkes by the Colorado potato beetle (*Leptinotarsa decemlineata* (Say)). Can. J. Zool. 69, 1280–1288.

Pelletier, Y., Tai, G.C.C., 2001. Genotypic variability and mode of action of Colorado potato beetle (Coleoptera: Chrysomelidae) resistance in seven *Solanum* species. J. Econ. Entomol. 94, 572–578.

Pelletier, Y., Grondin, G., Maltais, P., 1999. Mechanism of resistance to the Colorado potato beetle in wild *Solanum* species. J. Econ. Entomol. 92, 708–713.

Pelletier, Y., Clark, C., Boiteau, G., Feldman-Riebe, J., 2000. The effect of Bt-transgenic potatoes on the movement of the Colorado potato beetle [Coleoptera: Chrysomelidae]. Phytoprotec. 81, 107–114.

Pelletier, Y., Clark, C., Tai, G.C., 2001. Resistance of three wild tuber-bearing potatoes to the Colorado potato beetle. Entomol. Exp. Appl. 100, 31–41.

Pescod, K.V., Quick, W.P., Douglas, A.E., 2007. Aphid responses to plants with genetically manipulated phloem nutrient levels. Physiol. Entomol. 32, 253–258.

Pham Trung, N., Fitches, E., Gatehouse, J.A., 2006. A fusion protein containing a lepidopteran-specific toxin from the South Indian red scorpion (*Mesobuthus tamulus*) and snowdrop lectin shows oral toxicity to target insects. BMC Biotech. 6, 18.

Pompon, J., Quiring, D., Giordanengo, P., Pelletier, Y., 2010a. Role of host plant selection in resistance of wild *Solanum* species to *Macrosiphum euphorbiae* (Thomas) and *Myzus persicae* (Sulzer) (Hemiptera: Aphididae). Entomol. Exp. Appl. 137, 73–85.

Pompon, J., Quiring, D., Giordanengo, P., Pelletier, Y., 2010b. Role of xylem consumption on osmo-regulation in *Macrosiphum euphorbiae* (Thomas). J. Insect Physiol. 56, 610–615.

Pompon, J., Quiring, D., Goyer, C., Giordanengo, P., Pelletier, Y., 2011. A phloem-sap feeder mixes phloem and xylem sap to regulate osmotic potential. J. Insect Physiol. 57, 1317–1322.

Ponder, K.L., Pritchard, J., Harrington, R., Bale, J.S., 2000. Difficulties in location and acceptance of phloem sap combined with reduced concentration of phloem amino acids explain lowered performance of the aphid *Rhopalosiphum padi* on nitrogen deficient barley (*Hordeum vulgare*) seedlings. Entomol. Exp. Appl. 97, 203–210.

Porcar, M., Grenier, A.M., Federici, B., Rahbe, Y., 2009. Effects of *Bacillus thuringiensis* delta-Endotoxins on the pea aphid (*Acyrthosiphon pisum*). Appl. Environ. Microbiol. 75, 4897–4900.

Powell, G., Tosh, C.R., Hardie, J., 2006. Host plant selection by aphids: Behavioral, evolutionary, and applied perspectives. Annu. Rev. Entomol. 51, 309–330.

Pusztai, A., Grant, G., Spencer, R.J., Duguid, T.J., Brown, D.S., Ewen, S.W.B., Peumans, W.J., Van-damme, E.J.M., Bardocz, S., 1993. Kidney bean lectin-induced *Escherichia coli* overgrowth in the small intestine is blocked by GNA, a mannose-specific lectin. J. Appl. Bact. 75, 360–368.

Rahbé, Y., Deraison, C., Bonade-Bottino, M., Girard, C., Nardon, C., Jouanin, L., 2003. Effects of the cysteine protease inhibitor oryzacystatin (OC-I) on different aphids and reduced per-formance of *Myzus persicae* on OC-I expressing transgenic oilseed rape. Plant Sci. 164, 441–450.

Ramirez, C.C., Niemeyer, H.M., 1999. Salivation into sieve elements in relation to plant chemistry: The case of the aphid *Sitobion fragariae* and the wheat,. *Triticum aestivum*. Entomol. Exp. Appl. 91, 111–114.

Rico, E., Ballester, V., Mensua, J.L., 1998. Survival of two strains of *Phthorimaea operculella* (Lepidoptera: Gelechiidae) reared on transgenic potatoes expressing a *Bacillus thuringiensis* crystal protein. Agronomie 18, 151–155.

Sagredo, B., Lafta, A., Casper, H., Lorenzen, J., 2006. Mapping of genes associated with leptine content of tetraploid potato. Theor. Appl. Gen. 114, 131–142.

Sanford, L., Cantelo, W., 1989. Larval development rate and mortality of Colorado potato beetle on detached leaves of wild *Solanum* species. Am. J. Potato Res. 66, 575–582.

Sauvion, N., Rahbe, Y., Peumans, W.J., VanDamme, E.J.M., Gatehouse, J.A., Gatehouse, A.M.R., 1996. Effects of GNA and other mannose binding lectins on development and fecundity of the peach-potato aphid *Myzus persicae*. Entomol. Exp. Appl. 79, 285–293.

Shelton, A.M., Naranjo, S.E., Romeis, J., Hellmich, R.L., Wolt, J.D., Federici, B.A., Albajes, R., Bigler, F., Burgess, E.P.J., Dively, G.P., Gatehouse, A.M.R., Malone, L.A., Roush, R., Sears, M., Sehnal, F., 2009. Setting the record straight: a rebuttal to an erroneous analysis on trans-genic insecticidal crops and natural enemies. Transgenic Res. 18, 317–322.

Sinden, S.L., Cantelo, W.W., Sanford, L.L., Deahl, K.L., 1991. Allelochemically mediated host resis-tance to the Colorado potato beetle, *Leptinotarsa decemlineata* (Say) (Coleoptera: Chrysomeli-dae). Mem. Entomol. Soc. Can. 157, 19–28.

Smith, C.M., 2005. Plant resistance to arthropods: molecular and conventional approaches. Springer, Dordrecht, The Netherlands.

Spooner, D.M., Fajardo, D., Bryan, G.J., 2007. Species limits of *Solanum berthaultii* Hawkes and *S. tarijense* Hawkes and the implications for species boundaries in *Solanum* sect. Petota. Taxon 56, 987–999.

Suttle, J.C., 2004. Physiological regulation of potato tuber dormancy. Am. J. Potato Res. 81, 253–262.

Thompson, A., Farnsworth, B., Gudmestad, N., Secor, G., Preston, D., Sowokinos, J., Glynn, M., Hatterman-Valenti, H., 2008. Dakota Diamond: An Exceptionally High Yielding, Cold Chip-ping Potato Cultivar with Long-Term Storage Potential. Am. J. Potato Res. 85, 171–182.

Thornton, M., 2003. The rise and fall of Newleaf potatoes. Biotechnology: Science and society at a crossroads. Nat. Agric. Biotech. Council Rep. 15, 235–243.

Tingey, W.M., 1984. Glycoalkaloids as pest resistance factors. Am. Potato J. 61, 157–167.

Tingey, W.M., Laubengayer, J.E., 1981. Defense against the green peach aphid and potato leafhopper by glandular trichomes of. *Solanum berthaultii*. J. Econ. Entomol. 74, 721–725.

Tingey, W.M., Sinden, S.L., 1982. Glandular pubescence, glycoalkaloid composition, and resistance to the green peach aphid, potato leafhopper, and potato fleabeetle in *Solanum berthaultii*. Am. Potato J. 59, 95–106.

Tingey, W.M., Yencho, G.C., 1994. Insect resistance in potato: A decade of progress. In: Zehnder, G.W., Powelson, M.L., Jackson, R.K., Raman, K.V. (Eds.), Advances in potato pest biology and management, American Phytopathology Society, St Paul, MN, pp. 405–425.

Tingey, W.M., Mehlenbacher, S.A., Laubengayer, J.E., 1981. Occurrence of glandular trichomes in wild *Solanum* species. Am. J. Potato Res. 58, 81–83.

Tingey, W.M., Plaisted, R.L., Laubengayer, J.E., Mehlenbacher, S.A., 1982. Green peach aphid resistance by glandular trichomes in *Solanum tuberosum* × S. *berthaultii*. Am. Potato J. 59, 241–251.

Tjallingii, W.F., 1995. Aphid–plant interactions: what goes on in the depth of the tissues?. Proc. Sect. Exp. Appl. Entomol. Netherlands Entomol. Soc. 6, 189–200.

Tjallingii, W.F., 2006. Salivary secretions by aphids interacting with proteins of phloem wound responses. J. Exp. Bot. 57, 739–745.

Tyner, D.N., Hocart, M.J., Lennard, J.H., Graham, D.C., 1997. Periderm and lenticle characterization in relation to potato cultivar, soil moisture and maturity. Potato Res. 40, 181–190.

van Emden, H.F., 1986. The interactions of plant resistance and natural enemies: Effects on populations of sucking insects. In: Boethel, D.J., Eikenbary, R.D. (Eds.), Interactions of plant resistance and parasitoids and predators of insects, John Wiley & Sons, New York, NY, pp. 138–150.

Wale, S., Platt, H.W., Cattlin, N., 2008. Diseases, pests and disorders of potatoes. Manson Publishing, London, UK, p. 176.

Whalon, M., Mota-Sanchez, D., Hollingworth, R., 2008. Global pesticide resistance. CABI Publishing, Wallingford, UK, p. 169.

Xia, H., Chen, L., Wang, F., Lu, B.R., 2010. Yield benefit and underlying cost of insect-resistance transgenic rice: Implication in breeding and deploying transgenic crops. Field Crops Res. 118, 215–220.

Yencho, G.C., Tingey, W.M., 1994. Glandular trichomes of *Solanum berthaultii* alter host preference of the Colorado potato beetle. *Leptinotarsa decemlineata*. Entomol. Exp. Appl. 70, 217–225.

Yencho, G.C., Renwick, J.A.A., Steffens, J.C., Tingey, W.M., 1994. Leaf surface extracts of *Solanum berthaultii* Hawkes deter Colorado potato beetle feeding. J. Chem. Ecol. 20, 991–1007.

Yencho, G.C., Bonierbale, M.W., Tingey, W.M., Plaisted, R.L., Tanksley, S.D., 1996. Molecular markers locate genes for resistance to the Colorado potato beetle, *Leptinotarsa decemlineata*, in hybrid *Solanum tuberosum* x S. *berthaultii* potato progenies. Entomol. Exp. Appl. 81, 141–154.

Chapter 16

Biopesticides

Marc Sporleder[1] and Lawrence A. Lacey[2]

[1]*International Potato Center (CIP), ICIMOD Building, Khumaltar, Lalitpur, Nepal,* [2]*IP Consulting International, PO Box 8338, Yakima, WA 98908*

INTRODUCTION

The term *biopesticides* defines compounds that are used to manage agricultural pests by means of specific biological effects rather than as broader chemical pesticides. It refers to products containing biocontrol agents – i.e., natural organisms or substances derived from natural materials (such as animals, plants, bacteria, or certain minerals), including their genes or metabolites, for controlling pests. According to the FAO definition, biopesticides include those biocontrol agents that are passive agents, in contrast to *biocontrol agents* that actively seek out the pest, such as parasitoids, predators, and many species of entomopathogenic nematodes. The latter biocontrol agents used to manage potato pests are described in Chapter 14. Thus biopesticides cover a wide spectrum of potential products that can be classified as follows:

- Microbial pesticides and other entomopathogens: pesticides that contain microorganisms, like bacteria, fungi, or virus, which attack specific pest species, or entomopathogenic nematodes as active ingredients. Although most of these agents attack insect species (called entomopathogens; products referred to as bioinsecticides), there are also microorganisms (i.e., fungi) that control weeds (bioherbicides).
- Plant-Incorporated Protectants (PIPs): these include pesticidal substances that are produced in genetically modified plants/organisms (GMO) (i.e., through the genetic material that has been incorporated into the plant).
- Biochemical pesticides: pesticides based on naturally occurring substances that control pests by non-toxic mechanisms, in contrast to chemical pesticides that contain synthetic molecules that directly kill the pest. Biochemical pesticides fall into different biologically functional classes, including pheromones and other semiochemicals, plant extracts, and natural insect growth regulators.

Insect Pests of Potato. http://dx.doi.org/10.1016/B978-0-12-386895-4.00016-8

Biopesticides generally have several advantages compared to conventional pesticides (Kaya and Lacey 2007, Kaya and Vega 2012). While chemical pesticides are responsible for extensive pollution of the environment, a serious health hazard due to the presence of their residues in food, development of resistance in targeted insect pest populations, a decrease in biodiversity, and outbreaks of secondary pests that are normally controlled by natural enemies, biopesticides, in contrast, are inherently less toxic to humans and the environment, do not leave harmful residues, and are usually more specific to target pests. Often they affect only the target pest and closely related organisms, substantially reducing the impact on non-target species. A further advantage of most microbial pesticides is that they replicate in their target hosts and persist in the environment due to horizontal and vertical transmission, which may cause long-term suppression of pest populations even without repeating the application.

Since the use of the biopesticides is markedly safer for the environment and users, and more sustainable than the application of chemicals, their use as alternatives to chemical pesticides, especially as components in Integrated Pest Management (IPM) strategies, is of growing interest. Several biopesticides of the different classes have proved to be very effective in controlling potato pests; however, there are certain disadvantages associated with their use that have prevented them from being used on a wider basis in potato production today. The very high specificity of the products might be a disadvantage when a complex of pest species needs to be controlled. Since biopesticides often contain living material, the products have reduced shelf lives. Also, their efficacy is often variable due to the influence of various biotic and abiotic factors. For using biopesticides effectively, users need to have good knowledge about managing the particular pests or pest complexes. Due to limited commercial use (niche products) biopesticides often are developed by research institutions rather than by the traditional pesticide industry. While effective active ingredients have been discovered, products might lack appropriate formulation for efficient field use. A broader set of perspectives in the design and launch of a biopesticide would be helpful. Farmers consider biopesticides often as an alternative to a chemical pesticide, in which the active ingredient is thought to be synthetic, having a similar mode of action to the chemical pesticide. But the truth is that biopesticides differ in their modes of action from conventional chemical pesticide considerably; their modes of action are almost always specific. Therefore, using biopesticides efficiently requires specific user knowledge on the agent and the target pest for optimizing application time, field rates, and application intervals. Biopesticides should not be considered as a one-for-one replacement of chemical pesticides.

As pesticides in general, biopesticides need to be approved and registered as such in most countries before they can be used, sold, or supplied. Since biopesticides pose fewer risks than conventional pesticides, authorities generally require fewer data for their registration. For example, the Environmental Protection Agency (EPA) in the USA often registers new biopesticides in less than a year, compared with an average of more than 3 years for conventional pesticides. However, in some cases it is difficult to determine whether a product meets the criteria for classification as a biopesticide, and the decision by local agencies might vary depending on the regulations in each country. There might be specific requirements pertinent to the different categories of biopesticides.

In this chapter, the major biopesticides of potato pests and their potential for integrated pest management are reviewed according to their categories. Because knowledge about the mode of action of each type of active ingredient is crucial, each subsection includes a brief description of the biocontrol agent.

MICROBIAL PESTICIDES AND OTHER ENTOMOPATHOGENS

A diverse spectrum of microscopic and multicellular organisms (bacteria, fungi, viruses, protozoa, and nematodes) parasitizes and kills insect pests of virtually every crop. Several of these agents have been developed as microbial pesticides (Kaya and Lacey 2007, Lacey *et al.* 2001), some of which have been used to control certain insect pests of potato (Lacey *et al.* 2009a, Wraight *et al.* 2009). Substantial effort has gone into the development of certain microbial agents for controlling the potato tuber moth *Phthorimaea operculella* (Zeller) (Lacey and Kroschel 2009). Although most coleopteran and hemipteran potato pests are currently managed using chemical pesticides, there is a growing demand and potential for alternative control options including the use of microbial pesticides (Wraight *et al.* 2009).

Viruses

The most important group of viruses used for biocontrol belongs to the highly host-specific family of Baculoviridae, which are pathogenic for invertebrates (Blissard and Rohrmann 1990). Baculoviruses have been found in over 700 species of invertebrates, mainly Lepidoptera, and are considered an effective and selective means for biological insect control (Hunter-Fujita *et al.* 1998, Moscardi 1999). They have never been found to cause disease in any organism outside the phylum Arthropoda. Baculoviruses have been recently divided into four genera: *Alpha-, Beta-, Gamma-,* and *Deltabaculoviruses* (Jehle *et al.* 2006, Eberle *et al.* 2012). Based on the morphology of their occlusion bodies (OBs), they can be distinguished between nucleopolyhedroviruses (NPVs) and granuloviruses (GVs). The OBs of NPVs typically enclose several to many virions per OB, while the virions contain either multiple (MNPV) or single (SNPV) nucleocapsids. In contrast, the OBs of GVs enclose only one virion that always contains a single nucleocapsid. The OBs are composed of a crystalline protein matrix, mainly consisting of a single protein, the so-called polyhedrin in NPVs and granulin in GVs (referred to as polyhedras and granules in earlier literature). *Alphabaculovirus* comprises lepidopteran-specific NPVs, *Betabaculovirus* lepidopteran-specific GVs, *Gammabaculovirus* hymenopteran-specific NPVs, and *Deltabaculovirus* dipteran-specific NPVs, respectively.

The host range of baculoviruses, especially GVs, is normally restricted to only one species or at best to a few closely related species, although it is well known that differences in virulence exist between strains or isolates within a given host species (Hamm and Styer 1985, Shapiro and Robertson 1991). Like other viruses, baculoviruses are obligate parasites and unable to reproduce without a host.

Baculoviruses must usually be ingested by the host to produce an infection. Once inside the host the OBs dissolve in the alkaline pH of the midgut, liberating the nucleocapsids which pass through the peritrophic membrane and then fuse with the microvilli of midgut epithelium. Infection of these cells is transient without the production of OBs (Federici 1997). Subsequently, baculoviruses invade a variety of host cells and produce hundreds of millions of OBs per larva. The larval fat cells are the predominant site of virus production. Ultimately, infected larvae die and become a source of inoculum for infecting other hosts continuing the cycle.

Since baculoviruses already exist as natural components in the environment (Heimpel *et al.* 1973), their release represents no considerable environmental consequences in contrast to applications of chemical pesticides. Vogel (1986) and Moscardi (1999) suggested that approximately 30% of all agricultural pests could be controlled by baculoviruses.

These viruses are excellent candidates for species-specific, narrow-spectrum insecticidal applications. They have been shown to have no negative impacts on plants, mammals, birds, fishes, or even on non-target insects (Gröner 1986, Moscardi 1999). This is especially desirable when beneficial insects are being conserved to aid in an overall IPM program, or when an ecologically sensitive area is being treated. An advantage of baculoviruses is that they replicate and persist in the environment and may suppress host populations through horizontal transmission long after their application.

The Granulovirus Infecting P. operculella (PhopGV)

The granulovirus attacking the common potato tuber moth (PTM) *P. operculella* (i.e., *Phop*GV) has the potential to play a key role in managing the moth, especially for protecting stored tubers. The granulovirus was first isolated from diseased larvae in Sri Lanka and propagated in Australia (Reed 1969), where it showed high potential for PTM control (Reed and Springett 1971, Matthiessen *et al.* 1978). Later it was isolated in various parts of the world, such as South America (Alcázar *et al.* 1991, 1992a, Mascarin *et al.* 2010), Africa (Broodryk and Pretorius 1974, Laarif *et al.* 2003), the Middle East (Kroschel and Koch 1994), Asia (Setiawati *et al.* 1999, Zeddam *et al.* 1999), Australia (Reed 1969, Briese 1981), and North America (Hunter *et al.* 1975), and it seems that the virus has accompanied the moth from its South American center of origin to most countries where it has become established. Laboratory bioassays on the biological activity of *Phop*GV, including 14 geographical isolates of the virus, revealed a wide range of activity among isolates covering several orders of magnitude

(Sporleder 2003). Restriction endonuclease DNA analysis showed minor genetic differences between *Phop*GV isolates, although certain geographic isolates were distinguishable by minor genetic polymorphisms (Vickers *et al.* 1991, Kroschel *et al.* 1996a, Zeddam *et al.* 1999, Sporleder 2003). In contrast, Léry *et al.* (1998) demonstrated considerable genetic heterogeneity between a Tunisian isolate and isolates of *Phop*GV from other regions. Histopathology studies showed the fat body and epidermis are the main tissues infected by the virus and that the virus morphogenesis is similar to other GVs, with the exception that small vesicles appear between mature granules (Lacey *et al.* 2011a). Infected *P. operculella* larvae can be recognized by their opaque, milky white color, and by their behavior. Infected larvae do not respond vigorously when disturbed. The effect of the virus on the larvae is lethal since they fail to pupate; however, very high dosages of *Phop*GV can cause death by toxicosis within 48 hours.

In 1984, researchers of the International Potato Center (CIP) in Lima, Peru, identified *Phop*GV from a potato store in Lima (Raman and Alcázar 1988) and initiated research on the beneficial role of *Phop*GV in an IPM program (Alcázar *et al.* 1991, 1992b, Alcázar and Raman 1992, Lagnaoui *et al.* 1995). CIP has developed a simple technique for multiplication and formulation of the virus (CIP 1993). A dust formulation, produced by selecting and grinding virus-infected larvae from damaged potato tubers and then mixing them with ordinary talc, has been used at the rate of 5 kg/tonne of stored potatoes (20 infected larvae per kg). Research showed that the granulovirus would reduce damage in stores by 91% and 78%, 30 and 60 days after application (Raman and Alcázar 1990), respectively. The virus, in this dust formulation, has been promoted successfully for protecting farmers' home-stored potatoes in Peru, Bolivia, Ecuador, Tunisia, and Egypt by using low-cost facilities for propagation (Gelernter and Trumble 1999). Good protection of treated tubers in non-refrigerated storage using *Phop*GV products has been reported by several researchers. A substantial amount of successful testing of *Phop*GV has been conducted on stored tubers in the Andean countries (Alcázar *et al.* 1992b, CIP 1992, Zeddam *et al.* 2003a) and in several countries in the Middle East, Northern Africa, and Asia (Amonkar *et al.* 1979, Hamilton and Macdonald 1990, Islam *et al.* 1990, Ali 1991, Das *et al.* 1992, Setiawati *et al.* 1999). Protection of tubers generally lasted several months. Lacey *et al.* (2010) showed that *Phop*GV in a liquid formulation can be used for protecting tubers stored in refrigerated warehouse conditions. Like most granuloviruses, *Phop*GV has a fairly specific host range. Only the common potato tuber moth, *P. operculella*, and certain other species in the same family (Gelechiidae) are infected by the virus. *Tuta (Scrobipalpuloides) absoluta* (Meyrick) and *Tecia solanivora* (Povolny) are susceptible to *Phop*GV, but at lower levels than PTM (Angeles and Alcázar 1995, 1996, Zeddam *et al.* 2003b). Although *Phop*GV could be isolated from the Andean PTM species *Symmetrischema tangolias* (Gyen), it does not appear to affect this species (J. Kroschel, unpubl. data). Since *S. tangolias* is becoming economically more important in the Andean region, it limits the use of *Phop*GV in that region. Pokharkar and Kurhade (1999) reported no infectivity to 11 other lepidopteran species.

Earlier field experiments in different countries, though the results were variable in some cases, showed that virus applications might successfully suppress the host population. In some cases applications gave similar results to chemical insecticide applications. Applications resulted in a large production of secondary inoculum, and the virus could spread extensively to untreated areas (Reed 1969, Reed and Springett 1971, Kroschel *et al.* 1996b, Arthurs *et al.* 2008), and subsequent in-store infestation failed to develop (Ben Salah and Aalbu 1992). However, the high amount of virus-infected larvae needed for field applications is a limiting factor. In addition, studies of PTM field populations have, in some cases, revealed natural *Phop*GV incidence levels as high as 35–40% (Kroschel 1995, Laarif *et al.* 2003). Several authors have shown that the infestation of potato tubers at harvest can be significantly reduced by effectively controlling PTM on the foliage during the growing season (Radcliffe 1982, Kroschel 1995, Arthurs *et al.* 2008). Such observations encourage the investigation of the microbial control potential of *Phop*GV from the standpoint of inundative augmentation in potato crops.

One of the main constraints using *Phop*GV in the field is its rapid inactivation due to solar (ultraviolet, UV) radiation. Kroschel *et al.* (1996a) determined half-life times of 1.3 days for *Phop*GV. However, *Phop*GV degradation does not follow a simple exponential curve (Sporleder *et al.* 2001), as found for several other baculoviruses (Brassel and Benz 1979, Huber and Lüdcke 1996, Jones *et al.* 1993). Initial inactivation can be very fast but the inactivation curve curtails when about 95–99% of the viruses are inactivated, showing a half-life increased about six-fold compared with the initial phases (Sporleder and Kroschel 2008). Therefore, a certain portion of virus particles remains active for long periods of time after application and may contribute to long-term suppression of the pest and further dissemination of the pathogen. A variety of adjuvants that have been used to protect other baculoviruses from UV inactivation were reviewed by Burges and Jones (1998). Sporleder (2003) investigated the use of dyes, optical brighteners, antioxidants, insect-host derived materials, and type of formulation for protection of *Phop*GV from UV inactivation. Optical brightener Tinopal® and certain antioxidants protected the infectivity of irradiated virus; however, *Phop*GV-infected larvae macerated in water were superior to other preparations in protecting the virus from UV irradiation. Sporleder and Kroschel (2008) discuss several aspects that need to be considered using *Phop*GV as a biocontrol agent, and outline implications and possible directions for further improving the strategies using the virus for pest control in the field. Besides virus field stability and possibilities of using UV screens for improving virus stability, the low slope of the virus concentration-host mortality curve, and changing host susceptibility to the virus with larval age, have implications for justifying field dosages. The slope for *Phop*GV and its host *P. operculella* derived from field and laboratory experiments varies around 0.65 (Sporleder *et al.* 2005, 2008, Mascarin *et al.* 2010), which is in contrast to chemical insecticides that show steeper slopes in dose-effect relationships. This implies that a proportionately low increase in mortality rate will be achieved for a given rise in dose so that even a high increase of the dose provides no significant increase in response. Although high mortalities are expected to be difficult to achieve, the low slope provides the advantage that field doses might be decreased without significantly reduced mortality responses. In practice, it will be difficult to achieve mortalities over 95% in neonate larvae because the number of infected larvae/ha would be too high to be economically feasible, while moderate mortalities might be achieved with extremely reduced numbers of *Phop*GV-infected larvae. In addition, *Phop*GV affects the larval stage only and resistance increases significantly with larval age (Sporleder *et al.* 2007). Therefore, *Phop*GV applications must be directed against first-instar larvae as well as against the eggs (emerging larvae may take up a lethal dose with consumption of the egg chorion as long as the virus was deposited on the egg surface). Sporleder and Kroschel (2008) suggest using *Phop*GV at short intervals, depending on temperature and PTM severity, as a relatively inexpensive partial suppression agent in potato fields through the use of low dosages per hectare; however, such an approach has not yet been tested and specific treatment thresholds still need to be determined. Sporleder and Kroschel combined results on *Phop*GV activity with an age-stage structured theoretical temperature-based pest population/phenology model for PTM (Sporleder *et al.* 2004), and suggested such modeling studies to support adjusting virus management strategies taking into account conditions in different agroecological regions. As with other modeling studies on microbial control agents (Anderson *et al.* 1982), Sporleder and Kroschel's modeling results indicated that for long-term control of the pest population and for inoculative augmentation, subsequent applications causing moderate infection in the host population may be better than a single hit with greater virulence.

Other Baculoviruses Attacking Potato Pests

Several baculoviruses, GV and NPV, attacking specific lepidopteran species have the potential to control defoliating caterpillars and loopers attacking potato (Lacey, unpubl. data); for example the cabbage looper, *Trichoplusia ni* (Hübner), can be controlled by nucleopolyhedroviruses (*Tn*SNPV, AcMNPV). Other potential baculoviruses are listed in Table 16.1.

Bacteria

Many naturally occurring bacteria pathogenic to insects have been isolated from insects and soils, but only a few species have been studied intensively. The most widely studied and used as a biopesticide is the gram-positive,

TABLE 16.1 Potential Biopesticides to Control Potato Pests and their Commercial Availability

Name	Spectrum of control (target organism)	Production process	Availability (countries of registration)
Microbial pesticides			
Viruses			
*Phop*GV	PTM (GPTM)	In PTM larvae (in vivo)	Peru, Bolivia u.a. (used by government agencies)
*Ha*SNPV	*Helicoverpa armigera*		Commercially used (China)
*Tn*SNPV AcMNPV	*Trichoplusia ni*		Commercially used
*Sl*NPV	*Spodoptera litura*		Commercial insecticide in China
*To*SNPV	*Thysanoplusia orichalcea*		Not available
MbNPV	(PTM)		Europe
Bacteria			
Bt var. *tenebrionis*	CPB(larvae)	Fermentation	USA, commercialized worldwide
Bt var. *kurstaki*	Most lepidopteran pests		USA, commercialized worldwide
Bt var. aizawai	Most lepidopteran pests (with high gut pH)		USA, commercialized worldwide
Modified *Rahnella aquatilis*	Wireworms (treatment of seed tubers)		n.a.
Fungi			
Beauvaria bassiana	Adults and larvae of many kinds of insects; eggs of lepidopteran pests (APW, PTM, CPB, WG, etc.)	Fermentation	Commercially used in Europe and USA; local isolates available in many countries
Isaria fumoso-rosea (Pae-cilomyces fumosoroseus)	LMF, Potato psyllid *Bactericera (Paratri-oza) cockerelli* [vector of zebra chip]		

TABLE 16.1 Potential Biopesticides to Control Potato Pests and their Commercial Availability—*Cont'd*

Name	Spectrum of control (target organism)	Production process	Availability (countries of registration)
Metharhizium anisopliae	WG, PTM, wire-worms, and others		Local isolates available in many countries
Muscodor albus	PTM		Not available; suggested for stored potatoes
Entomophtho-rales	Aphids		Not available; fungicide applications might limit natural control of aphids
Zoophthora (Erynia) radicans	*Empoasca fabae*		
EPN *Heterorhabditis* sp.	APW larvae and pupae of *Premno-trypes suturicallus*, WG, CPB	In vivo and in vitro	EPN are commercially available in USA and Europe, however, APW-specific strains were isolated in Peru which are not commercially used yet
Steinernema sp.	WG, cutworms		
PIPs Cry3A, Cry3B and Cry3C	CPB		USA
Cry5 (Cry1Ial)	PTM		Tested in Egypt, South Africa
Cry1Ab	PTM, *H. armigera*		India
Semiochemicals			
Botanicals Azadirachtin	Caterpillars, beetles, aphids, LMF, thrips, grasshoppers, leaf-hoppers, etc.	From neem tree seeds, neem oil	Many countries
Pheromones Attract and kill	PTM	Contains cyfluthrin as the contact insecticide	Peru (registration in process) CIP

Continued

TABLE 16.1 Potential Biopesticides to Control Potato Pests and their Commercial Availability—*Cont'd*

Name	Spectrum of control (target organism)	Production process	Availability (countries of registration)
Biochemicals			
Metabolites from microorganisms Avermectins			
Abamectin	LMF, PTM, potato psyllid		Germany, China, USA
Growth regulators			
Cyromazine	LMF	Organic synthesis	China, USA, a.o.Suppliers export to many countries
Diflubenzuron	PTM	Organic synthesis	Many countries
Fenoxycarb	PTM	Organic synthesis	Worldwide
Spinosad	PTM, flea beetles	Organic synthesis	Commercially available

spore-forming bacterium *Bacillus thuringiensis* (*Bt*) Berliner. There are several subspecies which are effective on different insects groups, such as Lepidoptera (*Bt kurstaki* and *Bt aizawai*), Coleoptera (Chrysomelidae) (*Bt tenebronis*), and Diptera (Nematocera) (*Bt israeliensis*). The bacterium produces a parasporal proteinaceus endotoxin crystal during sporulation, which dissolves after ingestion in the host's alkaline gut fluids, releasing toxic polypeptides. Different *Bt* strains produce crystals with slightly different properties, and the crystals from each strain are specific for a small number of related insect species. The variable activity of *Bt* strains is due to the variety of toxins produced by a given *Bt* subspecies. All *Bt* endotoxins target the host midgut epithelium and thus can be considered gut poisons. Effects are generally rapid. When ingested, the crystals disrupt the osmotic balance in the midgut, causing lysis of cells that finally paralyzes the digestive tract of the insect, which stops feeding and often dies within 24–48 hours; however, susceptibility decreases with larval age – late-instar larvae are quite resistant to intoxication, and adults are not susceptible (Beegle and Yamamoto 1992, Tanada and Kaya 1993). *Bt* is the most widely commercialized microbial insecticide produced throughout the world today. There are more than 40 *Bt* products available worldwide for controlling caterpillars (Lepidoptera), beetles (Chrysomelidae), and mosquitoes and related families of Diptera. The total of the *Bt* products makes up about 1% of the world insecticide market.

Colorado Potato Beetle

The Colorado potato beetle (CPB), *Leptinotarsa decemlineata* (Say), is a widespread defoliator of potato and other solanaceous vegetables, and is considered a major pest in the Northern Hemisphere. The discovery of *Bt* var. *tenebrionis* (*Btt*) (Langenbruch *et al.* 1985) and other *Bt* toxins (*cry*3A3B3) with activity against beetles has broadened the options for microbial control of the CPB. The bacterium provides excellent control of larvae, especially when applied against early instars. Timing and frequency of application, amount of inoculum, spray coverage, crop canopy, rainfall, and UV inactivation can have strong influences on the efficacy of *Btt* (Bystrak *et al.* 1994, Lacey *et al.* 1999, Wraight *et al.* 2007). Efficacy of *Btt*-based biopesticides for controlling CPB was demonstrated in research trials (Ferro and Gelernter 1989, Zehnder and Gelernter 1989, Zehnder *et al.* 1992), and commercial products were marketed during late 1980s and early 1990s (Gelernter and Trumble 1999). The influence of biotic and abiotic factors on *Btt* is seen as a major disadvantage of the product. A new class of insecticides, neonicotinoids, which entered the market in the mid-1990s and proved extremely toxic to both CPB adults and larvae, with prolonged residual activity through a strong systemic activity, has meant that the demand for *Btt* products has been drastically reduced; however, *Btt* is still used in organic farming.

Potato Tuber Moth

The only bacterium that has been evaluated for PTM control is *Bt*. *Bt* var. *kurstaki* (*Btk*) is the most commonly used against lepidopterous insects. Natural isolates of *Bt* were found within the PTM's native range in Bolivia (Hernández *et al.* 2005). Several strains were isolated from agricultural soils, warehouses, and tubers infested with PTM. Some of these isolates were shown to have equal or even greater toxicity compared with a standard commercial strain of *Btk*, suggesting more effective indigenous strains of *Bt* could be developed for PTM control.

   Bt has been reported effective for control of PTM infestations under field conditions (Awate and Naik 1979, Broza and Sneh 1994, Kroschel and Koch 1996, Arthurs *et al.* 2008). However, repeated applications have been required because *Bt* is degraded by UV light from the sun, and rain washes it onto the soil (Salama *et al.* 1995a). Three consecutive applications of *Bt* (Bio-T™) at 8-day intervals were required to control PTM in an infested tomato crop in Israel (Broza and Sneh 1994). A high application volume (500 L/ha) was used to bring the active ingredient into the tunnels in the leaves where young larvae were mining. In field plot tests in India, foliar application of *Bt* (Thuricide® at 2–5 kg/ha) at 15-day intervals beginning 60 days after planting was almost as effective at controlling PTM infestations as parathion and carbaryl (Awate and Naik 1979). In the Republic of Yemen, PTM infestations are very high. Kroschel (1995) tested *Bt* (DiPel®) over two seasons at two concentrations (0.2% and 0.3%) with three and four applications per potato season. In the control treatments, PTM leaf infestation reached 26 and 35 mines per plant. Until the plant-yellowing stage, *Bt* application reduced PTM leaf infestation by 41% and 54% and final tuber infestation at harvest by 23% and 10%, respectively, compared to the control treatment. Arthurs *et al.* (2008) reported fairly good control of very high PTM populations with *Btk*, but several applications of 1.12 kg/ha were required throughout the growing season. A *PhopGV/Btk* alternation was significantly more effective than *Btk* alone and as effective as *PhopGV* at 10^{13} OBs/ha. In greenhouse and laboratory studies where *Bt* was applied to the soil to protect seedlings or tubers in pots, it retained its potency for up to 60 days (Amonkar *et al.* 1979).

Bt has also been widely tested to control PTM infestations under laboratory and storage conditions. Under laboratory conditions, PTM larvae are susceptible at differing degrees to various *Bt* subspecies, including *kurstaki*, *thuringiensis*, *tolworthi*, *galleriae*, *kenyae*, and *aizawai*, although the lethal concentration (LC_{50}) required increases with larval age (Salama *et al.* 1995b). For example, *Btk* (Thuricide HP) applied at 200 mg/kg potatoes reduced PTM survival from egg to adult emergence to 0.4%, compared with *PhopGV* (0.8–34.7% depending on dosage) or controls (32.5%) (von Arx and Gebhardt 1990). In other laboratory studies, dust formulations of *Bt* (5000 L/ha IU/mg), along with permethrin (0.1%), prothiofos (1%), and rotenone (2.4%) provided good protection of potato tubers against PTM infestations and were more effective at controlling existing infestations compared with 1% chlorpyrifos (Hamilton and Macdonald 1990). In Egypt, another *Bt* preparation (DiPel® 2X at 0.3% concentration) was also reported to be very effective to protect tubers in stores, eliminating PTM infestation compared with 100% infestation in untreated controls 60 days after treatment (Farrag 1998). In Tunisia, an integrated control approach comprising *Bt* applied at the beginning of the storage period in combination with cultural control (early harvest) eliminated the reliance on parathion sprays (von Arx *et al.* 1987). In cases when tubers had a high initial infestation (over 20%), *Bt* was replaced with a synthetic pyrethroid (permethrin). In tests in Indonesia, tubers treated with *Btk* (Thuricide at 2 g/L) caused 79% larval mortality after 4 months of storage compared with 58% mortality of larvae on foliage in a screenhouse (Setiawati *et al.* 1999). In other studies, *Btt* (0.2% Bactospeine® wettable powder (WP) 16,000 IU/mg) was reported ineffective at protecting tubers in storage, resulting in as much tuber damage as in untreated controls (Das *et al.* 1992).

Formulation of *Bt* with various carriers has been reported by several researches to improve *Bt* activity and/or to reduce product costs. *Btk* mixed with fine sand dust containing quartz provided effective control in tuber storage in the Republic of Yemen (Kroschel and Koch 1996). A very low proportion, 40 g *Btk* mixed with 960 g sand, applied to 1 tonne of stored potatoes proved to be efficacious. This treatment also controlled 96% of larvae that were already inside tubers. In Peru, Raman *et al.* (1987) reported that *Btk* (DiPel) was effective in reducing feeding damage in storage when applied as a dust formulation. Formulation of *Btk* with various diluents was effective against neonate larvae. Arthurs *et al.* (2008) demonstrated that tubers treated with 37.5 mg *Btk* WP mixed in talcum or diatomaceous earth/kg tuber before infestation resulted in 99% PTM larval mortality. Different inert materials alone were tested to determine their capacity for providing additional physical protection against moth attack in stored potatoes (Das and Rahman 1997, Mamani *et al.* 2011). Tubers treated with talc only were better protected against *P. operculella* and *S. tangolias* attack than tubers treated with kaolin, lime, or sand (Mamani *et al.* 2011).

Bt also proved to be very effective in controlling the other species of the potato tuber moth complex, namely the Andean PTM (*S. tangolias*) and the Guatemalan PTM (*T. solanivora*). This is especially important where these species co-exist, as it is the case in the Andes. The Andean PTM is often the most prevalent pest species in potato stores for which *PhopGV* is not effective. A rate of 15 g of the commercial product DiPel 2X mixed with 1 kg of talcum is recommended to protect 200 kg of tubers (Lacey and Kroschel 2009).

Other Potato Pests

Several other lepidopteran pests in the family Noctuidae attacking potato, like armyworms and cutworms, can be controlled by using *Btk* (Broza and Sneh 1994, Salama *et al.* 1999). In reporting the rDNA sequences of 86 bacterial isolates from the gut of wireworm *Limonius canus* (LeConte), Lacey *et al.* (2007) suggested that modified *Rahnella aquatilis* Izard, Gavini, Trinel & Leclerc (Enterobacteriaceae) expressing wireworm-active toxins might be useful for wireworm control by treating seed tubers, particularly with the ability of this bacterium to colonize the rhizosphere.

Fungi

Numerous species of entomopathogenic fungi are effective microbial control agents of several insect pests (Goettel *et al.* 2005, Ekesi and Maniania 2007), including key pests of potato. There are some 700 species of entomopathogenic fungi from about 100 genera, but only 10 species are utilized for insect control. The most important fungi which are commercially used are *Beauveria* spp., *Metarhizium* spp., *Isaria fumosorosea,* and *Lecanicillium* spp., which belong in the Ascomycetes (Order: Hypocreales). These are the easiest fungi to produce *in vitro*, and have a wide range of hosts. They are often naturally occurring in soils and commonly associated with soil-inhibiting insects. Most fungi enter the host through the insect cuticle, via germination and hyphal growth, which eventually emerges into the haemocoel of the insect. The fungi produce a broad range of toxic metabolites and a number of biochemical processes take place in the fungus and insect host, enhancing the infection through suppression of the host immune system and enabling sporulation on the killed host via antibiotic activity against invading saprophytes.

There is limited research on the feasibility of using fungi for PTM control. Laboratory studies on two common Hypocreales, *Metarhizium anisopliae* (Metschnikoff) Sorok and *Beauveria bassiana* (Balsamo) Vuillmen indicate they have potential for control of PTM larvae, particularly younger larvae (Hafez *et al.* 1997, Sewify *et al.* 2000). Hafez *et al.* (1997) also demonstrated activity of *B. bassiana* against prepupae, pupae, and adult PTM. Sewify *et al.* (2000) reported that the combination of *M. anisopliae* and *Phop*GV resulted in synergistic larval control when a high concentration of the fungus was used with a low concentration of the virus.

The endophytic fungus *Muscodor albus* (Xylariales: Xylariaceae) (Worapong *et al.* 2001) produces several volatile compounds (alcohols, esters, ketones, acids, and lipids) that are biocidal for a range of organisms, including plant pathogenic bacteria and fungi, nematodes, and insects (Strobel *et al.* 2001, Worapong *et al.* 2001, Lacey *et al.* 2008, Riga *et al.* 2008). Adulticidal and larvicidal activity of *M. albus* was reported against PTM by Lacey and Neven (2006) and Lacey *et al.* (2008). PTM adults and neonate larvae were exposed to *M. albus* volatiles for 72 hours in hermetically sealed chambers. Mean percent mortalities of adult PTM in chambers with 15 and 30 g of formulated mycelia were 84.6% and 90.6%, respectively. Development to the pupal stage of PTM that were exposed as neonate larvae on tubers to 15 or 30 g *M. albus* formulation was reduced by 61.8% and 72.8%. Lacey *et al.* (2008) observed that the length of exposure to *M. albus* significantly affected mortality of larvae within infested tubers and their development to the adult stage. Exposure durations of 3, 7, or 14 days at 24°C followed by incubation at 27°C until emergence resulted in mortalities of 84.2%, 95.5%, and 99.6%, respectively. Mortality of larvae was significantly reduced at 10 and 15°C.

Beauveria bassiana had been proposed as a potential microbial control agent for the Andean potato weevil (APW) complex (Coleoptera: Curculionidae) management in potato stores of Peru, Colombia, and Bolivia (Ewell *et al.* 1994, Raman 1994). However, it has not been widely made available and used by farmers. Potato tubers become infested by the APW during tuber development, and are therefore already infested when stored after harvest. Here, the application of the fungus to tubers was able to reduce the weevil population but not tuber damage, which was unacceptable for farmers.

Until the mid-1980s, *B. bassiana* was the only microbial control agent used against CPB. Control of CPB ranging from poor to excellent has been reported for the fungus (Hajek *et al.* 1987, Poprawski *et al.* 1997, Lacey *et al.* 1999, Wraight and Ramos 2002, Wraight *et al.* 2007). It offers the advantage of recycling in host cadavers and persisting in the soil beneath potato plants, thereby affecting the survival of subterranean stages of the beetle.

Naturally occurring fungi are important regulators of aphid populations (Latgé and Papierok 1988, Nielsen *et al.* 2007), including aphids on potato in more humid areas (Soper 1981). The use of fungicides to combat late blight of potato, *Phytophthora infestans* (Mont) de Bary, has been correlated with aphid outbreaks due to the suppressing effects of the fungicides on entomophthoralean fungi that normally control the aphid (Lagnaoui and Radcliffe 1998). Although development of fungi as mycoinsecticides of aphids in potato has been studied (Soper 1981), no large-scale implementation of artificially cultured fungus in natural populations of aphids in potato has yet been attempted.

Other Hemiptera have also been responsible for economic losses in potato production. Similar to aphids, as they have piercing and sucking mouthparts, fungi are the only entomopathogens with potential for microbial control. McGuire *et al.* (1987) presented research on factors affecting the success of introductions of *Zoophthora* (*Erynia*) *radicans* (Brefeld) Batko for control of the potato leafhopper *Empoasca fabae* Harris (Cicadellidae). Psyllids can also be significant pests of potato. Potato psyllid, *Bactericera* (*Paratrioza*) *cockerelli* (Sulc), has recently been incriminated in the transmission of a disease-causing agent that results in a condition known as zebra chip (Munyaneza *et al.* 2007). Control of *B. cockerelli* using the fungi *Isaria fumosorosea* (*Paecilomyces fumosoroseus*) (Wize) Brown and Smith (order: Hypocreales) and *M. anisopliae* has been demonstrated in the laboratory and field (Lacey *et al.* 2009b, 2011b).

The larvae of click beetles (Coleoptera: Elateridae), also known as wireworms, can be locally important pests of potato tubers (Jansson and Seal 1994, Vernon *et al.* 2005), especially if potato is rotated with crops preferred by the beetle, such as grains. Relatively little research on the microbial control of these pests has been conducted. The fungus *M. anisopliae* has been reported from wireworms and is currently under development for control of these potato pests. Field trials of *M. anisopliae* by Kabaluk *et al.* (2005) resulted in a 30% reduction in wireworm damage to potato tubers, and in significant in-field infection and mortality of wireworms. The testing of new isolates of the fungus bioassayed against three wireworm species produced LT_{50} values as short as 8 days using 10^6 conidia/g of soil. Adults were also found to be highly susceptible to infection by *M. anisopliae*.

In Nepal, where white grubs are becoming an increasing problem in several potato production zones, *M. anisopliae* and *B. bassiana* have been detected as natural antagonists of these pests. Research has been initiated to use these fungi as a biocontrol agent (Yubac Dhoj 2006, 2009).

The leafminer fly (LMF), *Liriomyza huidobrensis* Blanchard (Diptera, Agromyzidae), is a serious potato pest in coastal Peru and Chile and in certain areas of Brazil, Central America, and other countries when insecticides are used intensively and where particularly susceptible cultivars are planted (Ewell *et al.* 1994, Raman 1994). Natural epizootics of the fungus *I. fumosorosea* on leafminer fly adults have been observed in Peru. Its high pathogenicity was approved under laboratory conditions, and the entomopathogenic fungus is being field tested in Peru (Lacey *et al.* 2009a).

Entomopathogenic Nematodes

Entomopathogenic nematodes (EPNs) are insect-specific parasites in the genera *Steinernema* (Steinernematidae) and *Heterorhabditis* (Heterorhabditidae) that are obligately associated with symbiotic bacteria (*Xenorhabdus* spp. and *Photorhabdis* spp., respectively) which are responsible for rapidly killing host insects. After entering a host insect, the infective juvenile stage of EPNs releases its symbiotic bacteria. In addition to killing the host, the bacteria digest host tissues, and produce antibiotics to protect the host cadaver from saprophytes and scavengers. After two to three reproductive cycles, when host nutrients are depleted, infective juveniles are produced and begin leaving the host insect. This stage is capable of immediately infecting a new host, or may persist for months in the absence of a host. Applied and basic research conducted on EPNs over the past five decades has demonstrated their potential as biological control agents of a wide variety of insect pests (Grewal *et al.* 2005, Georgis *et al.* 2006). They have been commercially developed for control of several economically important insect species; however, research for their use for controlling particular potato pest species has only recently been initiated.

EPNs have demonstrated good activity against weevil species with subterranean larvae (Booth *et al.* 2007). In 2003, an EPN was isolated from infected APW larvae and pupae of *Premnotrypes suturicallus* (Kuschel) in a potato store at 2750 m elevation, Junin, Peru (Alcázar and Kaya 2003), which was identified as a putative new *Heterorhabditis* sp. Its high potential to control APW under controlled (Parsa *et al.* 2006) and field conditions (Alcázar *et al.* 2007) was shown recently.

EPNs have also been proposed as microbial control agents of CPB (Berry *et al.* 1997, Armer *et al.* 2004) and white grubs (Koppenhöfer and Fuzy 2008).

PLANT-INCORPORATED PROTECTANTS (PIPS)

The Cry3A toxin from the bacterium *B. thuringiensis* var. *tenebrionis* was used first to produce transgenic potatoes resistant to attack by CPB (Adang *et al.* 1993, Perlak *et al.* 1993). Later additional *Bt* strains with novel *cry3* genes (*cry3B* and *cry3C*) that encodes toxins for CPB and other chrysomelid beetles were identified. The Cry3B toxin is substantially more toxic against adult CPB than the Cry3A toxin (Johnson *et al.* 1993). Constant feeding by female beetles on transgenic Cry3A or B toxin producing potato plants resulted in nearly complete inhibition of egg production (Perlak *et al.* 1993, Arpaia *et al.* 2000). The discovery of multiple *cry3* genes has enabled development of transconjugant *Bt* strains with both enhanced toxicity toward CPB and broader toxicity spectra. However, transgenic potato has not been widely accepted due to public concerns over genetically modified organisms (Douches *et al.* 2011), and since the introduction of neonicotinoid insecticides for controlling CPB, transgenic potato no longer plays a significant role in CPB management.

Transgenic potatoes producing *Bt* toxin with resistance against the PTM have been also tested (Jansens *et al.* 1995, Douches *et al.* 1998, Davidson *et al.* 2002). Cry5 (revised nomenclature Cry1Ial), specifically toxic to Lepidoptera and Coleoptera, was expressed into cultivar Spunta and tested in Egypt under natural infestations (Douches *et al.* 2002, 2004, Mohammed *et al.* 2000). These lines resulted in 99–100% moth-damage-free tubers in the field, however, in storage these lines remained >90% free of PTM damage for 3 months only (Douches *et al.* 2004). The performance of the transgenic potato "Spunta G2" was tested in South Africa (Douches *et al.* 2010), showing no significant differences to non-transgenic "Spunta". Transgenic potato lines containing *cry1Ab* gene were tested in India with good resistance against *P. operculella* and *Helicoverpa armigera* Hübner (Lepidoptera, Noctuidae) (Chakrabarti *et al.* 2000), and in Peru (Cañedo *et al.* 1999, Lagnaoui *et al.* 2001). In Peru, transgenic lines showed resistance to *P. operculella*, *S. tangolias*, and *T. absoluta*, but had no significant effects on *L. huidobrensis*, *Russelliana solanicola* Tuthill (Hemiptera, Psyllidae), and *Myzus persicae* Sulzer (Hemiptera, Aphididae). Until now no transgenic line has been commercialized. One possible problem associated with the use of transgenic *Bt* crops is that due to the constant exposure to the toxin and evolutionary pressure, target pests might develop resistant to the toxin. Resistance to *Bt* in spray form has been already reported from a diamondback moth, *Plutella xylostella* L. (Lepidoptera, Plutellidae), population (Tabashnik *et al.* 2008), and recently resistance was observed in pink bollworm (*Pectinophora gossypiella* Saunders [Lepidoptera, Gelechiidae]) to *Bt* cotton in India. Rico *et al.* (1998) tested the resistance in transgenic plants expressing Cry1Ab against two *P. operculella* populations; one reported resistant and the other susceptible to DiPel. The resistant moth population proved to have less mortality than the susceptible one, but resistance to Cry1Ab was not total; transgenic potatoes were only partially protected against moth attack. For reducing the risk of insect resistance development, management strategies need to be considered when using transgenic *Bt* crops on a large scale (they are mandatory for planting *Bt* crops in the USA and other countries). One method for resistance management is the establishment of non-*Bt* crop refuges that allow some non-resistant insects to survive and maintain a susceptible population (Cameron *et al.* 2002, 2009). As a second approach, it has been suggested to use a mixture of different toxins for delaying resistance development in insects more effectively (Zhao *et al.* 2003). Meiyalaghan *et al.* (2006) evaluated such pyramiding of pairwise combinations of *cry1Ac9*, *cry9Aa2* and *cry1Ba1* genes in potato against PTM attack. All combinations of the three *cry* genes were largely consistent with additive impacts on PTM larval growth, although the combination of the *cry1Ac9* and *cry9Aa2* was slightly synergistic. The authors suggested that pyramiding these *cry* genes in potato could provide a more effective strategy to control PTM compared to single *cry* gene transgenic plants. However, since all Cry endotoxins have a similar mode of action it can be expected that a PTM population that developed resistance against one particular endotoxin also might develop resistance against a second.

BIOCHEMICAL PESTICIDES

Biochemical pesticides can be seen as closely related to conventional chemical pesticides but are distinct because they occur naturally, and their mode of action is non-toxic and often species-specific; however, the active ingredient in a product may be a synthetic analog to the naturally occurring substance. Biochemical pesticides can be separated according to the functional categories into plant extracts (botanicals), semiochemicals, metabolites from microorganisms, and growth regulators.

Botanicals

Plant extracts contain specific chemicals or mixtures of such that are derived from higher plants. Such products are highly diverse in their composition and mode of action, and generally act less directly and are less specific to target pests then other biocontrol agents. Plant extracts contain several different types of metabolites, including alkaloids, phenolics, terpenoids, and secondary chemicals that the plant species has developed to protect itself from pests, and each plant species has its unique chemical complex structure. These compounds might work as repellents or have lethal effects on insect species; for example, neem (*Azadirachta indica*) oil, which contains azadirachtin as the principal active ingredient but also others, acts as an antifeedant, growth inhibitor, and endocrine disruptor. Thus, the plant kingdom offers an unlimited resource pool of biodegradable, economical, and renewable alternatives to chemical pesticides for controlling insect pests. Depending on the mode of action, not all products derived from plants can be considered biopesticides because the level of toxicity can be high. For example, pyrethrum, which is an extract from chrysanthemum species, paralyzes and kills insects by affecting the nervous system with a mode of action similar to DDT (i.e., NA^+ channel). Although nicotine is extracted from tobacco, it is toxic to bees and can therefore not be considered as a biopesticide.

Many plant extracts have been tested especially to reduce PTM damage in unrefrigerated potato storerooms. Das (1995) reviewed the literature on the use of plant extracts tested to prevent PTM damage either in non-refrigerated storages or in the laboratory published from 1915 to 1993, and revealed that 35 plant species are effective in reducing PTM damage or killing the moth at different stages of the pest. In the reviewed studies, different types of preparations were utilized; some studies used chopped and dried leaves, while others used leaf/seed extracts, fruit peel, bulb, roots, or rhizomes. In developing countries, where farmers often home-store their potatoes in rustic non-refrigerated storerooms, they frequently cover the potato piles using local plants like Muna (*Minthostachys* spp.), Eucalyptus (*Eucalyptus globulus*), Chilca (*Baccharis* spp.), Curry plants, Indian pivets, *Lantana camera*, Pangam leaves, *Chenopodium botrys*, *Mentha arvensis*, *Artemesia vulgaris*, *Lycopersicon hirsutum*, etc., depending on the availability of the plants, to prevent PTM damage. The beneficial effect of such treatments, often based on local traditional knowledge, has been proven in many studies (Kroschel and Koch 1996, Castillo *et al.* 1998, Mariy *et al.* 2000a; Alrubeai *et al.* 2001, Iannacone and Lamas 2003, Ibrahim 2008, and references cited by Das 1995). For example, a 2 cm thick layer of *Ageratum haustonianum* or *Cannabis sativa* protected potatoes for up to 120 days while layers of *Vitex negundo* and *Mentha longifolia* were almost equally effective, showing only 6% infestation after 120 days (Kashyap *et al.* 1992).

Deshpande *et al.* (1990) reported that acetone extracts of *Glycosmis pentaphyllum* (*G. pentaphylla*) combined with equal amounts of extracts of *Catharanthus roseus*, *Salvadora oleododes*, and *Breneya* sp. exhibited a significant increase in ovipositional deterrence activity against *P. operculella* compared with the activity of the individual extracts, and suggested that there are important practical implications of this synergism. Spraying with acetone extracts of *L. camara* resulted in a higher reduction of PTM infestation (97.3%) than with treatments of *PhopGV* or *B. thuringiensis* (Mariy *et al.* 2000b). Certain plant extracts might also have synergistic effects when applied together with other entomopathogens. Sabbour and Ismail (2002) reported synergistic effects of plant extracts of *Solanum nigrum*, *Atropa belladonna*, and *Hyoscyamus niger* when applied in combination with *B. thuringiensis* or *B. bassiana* against PTM.

Commercial products of Azadirachtin (extracts of neem) revealed activity against PTM in laboratory experiments (Chatterjee 2005). Kroschel and Koch (1996) reported high efficacy of a water extract of neem applied in storages. In growing potato fields, light irrigation every 4 days and mulching with neem leaves during the latter 4 weeks before harvest were effective for reducing tuber infestation at harvest (Ali 1993).

Semiochemicals and Attract-and-Kill Approach

Commercial sex pheromones are available for all PTM species, *P. operculella*, *S. tangolias*, and *T. solanivora*. Their use for disrupting mating in these species appears an economically feasible method of control in non-refrigerated potato storerooms, and helps to monitor the pest during storage. Kroschel and Zegarra (2010) developed an attract-and-kill strategy for the PTM species *P. operculella* and *S. tangolias*. The attract-and-kill product (attracticide) consisted of pure pheromones and cyfluthrin as the contact insecticide, formulated with plant oils and ultraviolet screens. The product was applied in droplet sizes of 100μL and resulted, under controlled conditions, in 100% mortality of adult male moths, without reduction in efficacy of the formulation for a period of 36 days. The preliminary field experiments indicated good potential using the attract-and-kill technology in potato (Kroschel and Zegarra 2007). Droplet densities of 1 drop per 4 m^2 reduced the number of daily PTM male catches compared with the untreated control by 83.8%. Such treatment corresponds to an application of 1.25 g cyfluthrin as active ingredient to kill the moths per hectare, which is 32-fold less than the recommended field rate of Baythroid® EC 100 for controlling lepidopteran pests. The use of this product seems appropriate for use in developing countries where plant protection is usually done with knapsack sprayers. The application of single droplets applied by using an appropriate hand disperser requires less manpower than the application of chemical pesticides, which involves transportation of water (about 250–500 L/ha) to the field.

Bosa *et al.* (2005) demonstrated the potential of using pheromones for controlling PTM field populations. A potato field treated with the sex pheromone specific to *T. solanivora* at a rate of 28 g/ha suppressed male moth attraction to synthetic pheromone traps almost completely for 2 months.

Metabolites from Microorganisms

Avermectins

Avermectins are naturally occurring compounds (fermentation products) from the soil actinomycete *Streptomyces avermitilis*. The avermectins stimulate the γ-aminobutyric acid (GABA) receptor in the peripheral nervous system, blocking the electrical transmittance between nerves and muscle cells, which leads to hyperpolarization and subsequent paralysis of the neuromuscular systems in insects and nematodes (Bloomquist 1993, 1996). However, resistance to avermectins has been already reported, including in the CPB (Christiane *et al.* 2003), which suggests its use in moderation (Clark *et al.* 1995). Abamectin is one of last insecticides to which the LMF, *L. huidobrensis*, has not developed resistance. Mujica and Kroschel (2005) recommend management strategies for this pest based on the use of resistant potato cultivars and sticky yellow traps to capture adult flies, and the use of abamectin and growth regulators, like cyromazine, as a last option when required. Weintraub (2001) reported that both abamectin and cyromazine applied at recommended field rates early in the growing season significantly reduced LMF populations; however, cyromazine was significantly more effective than abamectin, while parasitoid populations from abamectin-treated plots recovered sooner than those from cyromazine-treated plots.

In Egypt, abamectin was the most effective control agent in potato fields (followed by profenofos, *Bt*, and *Phop*GV) and storages (followed by fenitrothion, *Bt*, and *Phop*GV) preventing *P. operculella* attack (Abdel-Megeed *et al.* 1998).

After field application, abamectin gave the highest residual half-life followed by profenofos and the *Bt* var. *aizawai* product Xen Tari® (Belal *et al.* 2005). For controlling PTM in potato storerooms, the use of sex pheromone water traps in combination with abamectin (1.8%) and Xen Tari (10.3%) provided better protection than sex pheromone water traps only (Moustafa *et al.* 2005). ZhengYue and QingWen (2005) reported that the application of abamectin for controlling PTM exhibited inhibitory effects on the germination of the fungus *B. bassiana* when applied simultaneously. Kay (2006) reported that abamectin (at 8.1 g a.i./ ha) was, in contrast to Spinosad, effective against neonate PTM larvae in mines.

Spinosad

The efficacy of Spinosad, a product derived from the spinosyn-producing soil actinomycete *Saccharopolyspora spinosa* (Mertz and Yao), has been recently tested against PTM. Spinosad is a commercial product that is a mixture of spinosyn A and spinosyn D. It has a novel mode of action; in insects it activates the nicotinic acetylcholine receptor, but at a different site from nicotine or imidacloprid. It also affects the γ-aminobutyric acid (GABA) receptor, but the role in the overall activity is unclear. Spinosad quickly kills target insects like caterpillars, leafminers, thrips, and foliage-feeding beetles.

In field experiments in Egypt, Spinosad efficiently reduced *P. operculella* larvae infestation by >95%, similar to conventional insecticides (Temerak 2003). In Australia, applications of Spinosad at a field rate of 96 g a.i./ha were effective against neonate larvae of PTM but did not efficiently affect the larvae in mines (Kay 2006). ZhengYue and QingWen (2005) reported that the application of Spinosad for controlling PTM is compatible with the simultaneous use of the fungus *B. bassiana*. In Tunisia, Spinosad was tested in non-refrigerated potato storerooms, where it provided better tuber protection against PTM than did deltamethrine (Nouri and Arfaoui 2008).

Growth Regulators

Applications of the growth regular diflubenzuron to potato tubers inoculated with *P. operculella* eggs prevented the development of larvae once eggs had hatched (Kroschel and Koch 1996, Chatterjee 2005); however, the effect is greater when applied to prepupae and pupae instead of larvae (Reddy and Urs 1991). El-Sheikh *et al.* (1988) obtained similar results with triflumuron and chlorfluazuron, with LC_{50}s for PTM at 10 and 20 ppm, respectively. Application of the insect juvenile hormone methoprene to potato tubers or to a sandy substrate under tubers satisfactorily suppressed PTM development; however, low and high concentrations had more effect than moderate concentrations, and males were affected more than females (Hamdy and Salem 1988).

CONCLUSIONS

There are several methods documented for managing potato pests using biopesticides; some of them are commercialized almost worldwide, as, for example, *Bt*, while others are available only in some countries (e.g., *Phop*GV), and others are still in the initial stages of research and development. Several biopesticides offer good prospects in a variety of situations as alternatives for problematic traditional chemical pesticides and for enhancing the sustainability of potato production. Of special importance is the control of PTM in the non-refrigerated potato storage condition, which is common in many developing countries, due to the fact that chemical pesticides can not be applied to harvested tubers used as a staple.

Certain microbial biopesticides have shown good potential for "narrow-spectrum" management of certain potato pests; however, most of them have not been commercialized on a large scale. These are often "niche products", and the biopesticides are often produced by government organizations and available only locally. Many would benefit from broader testing of their field performance, and for improving their production and formulation. Their use is limited because these products mainly contain living material (with a reduced shelf life compared to chemical pesticides, and vulnerability to abiotic factors), are generally slow acting, and present host specificity. The latter might be a problem when a complex of pest species needs to be controlled; otherwise it must be seen as an advantage. These agents also provide the opportunity to developing countries for development, production, and sale of their own natural biopesticide resources. There is no doubt that formulations might improve the field performance of these agents considerably; however, these formulations might also have disadvantages. For example, certain dyes for improving the field stability of baculoviruses are toxic to the environment (e.g., optical brighteners, Congo red, etc.), and their use in a formulation might render the final product hazardous to the environment. The possibility of developing new ways for applying these pathogens, resulting in better control (e.g., by taking advantage of their reproduction in the field), and how different conditions affect the outcome need to be explored. On the other hand, these pathogens might already exist naturally in many potato production zones affecting pest populations. There are reports on endophytism by *B. bassiana* in potato (Jones 1994) and tomato (Leckie 2002) that might colonize the rhizosphere; however, their use has still not been tested in potato and could raise the issue of toxic metabolites produced by the fungus on or in harvested tubers; this would then need to be addressed.

The concept of genetically modified crops has not been welcomed by the public, and will probably not play an important role in potato insect pest management in the near future.

Botanicals play a significant role in traditional pest management, particularly in non-refrigerated rustic potato storerooms. They provide a good source of biodegradable, economical, and renewable pest control agents; however, there are limited numbers of botanical biopestides that have been commercialized (such as neem products, which are supplied by several companies worldwide). This is mainly due to the failure to characterize fully the original plant material and its pesticidal components.

Attract-and-kill strategies have been developed for certain insect pests by traditional pesticide enterprises, but the strategy is presently generally considered unlikely to be an effective crop-protection tool because it attracts and kills male insects only, and it is not expected that sufficient insects are attracted and killed to suppress egg-laying to that extent that pest populations are significantly contained. However, the attract-and-kill strategy for PTM developed by CIP demonstrated high potential for field and storage application, probably due to a combined effect of suppressing the male PTM population and mating disruption. The product seems highly economic, especially for treatments in potato storages where limited areas need to be treated. The approach seems adequate for farmers in developing countries, where farm families produce potatoes on limited areas of land and lack the monetary resources, rather than the workforce, to increase their production. However, the product will probably remain a "niche product" with limited interest for production by larger companies. Since the product contains a pyrethroid as the killing agent, it might also not be approved for countries or states which have completely banned chemical pesticides (e.g., Sikkim, in India).

Biopesticides based on microorganism-derived metabolites (e.g., abamectins, Spinosad, *Bt* products that contain endotoxins of *Bt* only, etc.) and insect growth regulators provide a further option with new modes of action for managing insect pests. The market for such products is growing. They have come primarily from developed countries like Japan, Germany, and the USA; however, nowadays companies in other countries, such as China, have taken up production, and the prices for such products can be expected to decline considerably in the future, making them highly competitive to traditional chemical pesticides.

Most biopesticides are compatible with other pest control measures. Due to their specific mode of action, they should not be seen as stand-alone replacements for chemical pesticides. In fact, almost every (agro)ecosystem is chaotic; micro-organisms that affect insect pest populations occur naturally, as do parasitoids and predators, and there is considerable interaction between all of them. Rather than developing biopesticides to be applied as components in IPM programs, future research should explore the attributes to manage pest species through better understanding of the ecology of the insect pests and their indigenous natural antagonists. With this knowledge, biopesticides will then offer powerful tools to create sustainable potato-pest management systems in which natural enemies of pests and the self-regulating capacity of the agroecosystem is preserved.

LIST OF ABBREVIATIONS

a.i.	active ingredient
APW	Andean potato weevil
Bt	*Bacillus thuringiensis* Berliner; Btk = Bt kurstaki, Bta = Bt aizawai, Btt = Bt tenebronis)
CIP	International Potato Center (Centro Internacional de la Papa)
CPB	Colorado potato beetle
DDT	dichlorodiphenyltrichoroethane
DNA	desoxyribonucleic acid
EPA	Environmental Protection Agency
EPN	entomopathogenic nematode
FAO	Food and Agricultural Organisation
GABA	γ-Aminobutyric acid
GMO	genetically modified organism
GV	granulovirus
IPM	integrated pest management
IU	International Unit
LC50	median lethal concentration
LMF	leafminer fly
LT50	median lethal time
MNPV	multiple nucleocapsid NPV
NPV	nucleopolyhedrovirus
OB	occlusion bodies
*Phop*GV	granulovirus specific to *Phthorimaea operculella*
PIP	Plant-Incorporated Protectant
ppm	parts per million
PTM	potato tuber moth
SNPV	single nucleocapsid NPV
UV	ultraviolet wave length of the light spectrum
WG	white grubs, larvae of scarab beetles (Coleoptera: Scarabaeidae)
WP	wettable powder

REFERENCES

Abdel-Megeed, M.I., Abbas, M.G., El-Sayes, S.M., Moharam, E.A., 1998. Efficacy of certain biocides against potato tuber moth, *Phthorimaea operculella* under field and storage conditions. Ann. Agric. Sci. (Cairo), 309–317.

Adang, M.J., Brody, M.S., Cardineau, G., Eagan, N., Roush, R.T., Shewmaker, C.K., Jones, A., Oaks, J.V., McBride, K.E., 1993. The reconstruction and expression of a *Bacillus thuringiensis* cryIIIA gene in protoplasts and potato plants. Plant Mol. Biol. 21, 1131–1145.

Alcázar, J., Kaya, H.K., 2003. Hallazgo de un nematodo nativo del Genero *Heterorhabditis,* parasito del Gorgojo de los Andes *Premnotrypes suturicallus* en Huasahiasi, Junin. Resumenes XLV Convencion Nacional de Entomologia, 1–5 December, 2003, Sociedad Entomologica del Peru, Ayacucho, Peru, pp. 158.

Alcázar, J., Raman, K.V., 1992. Control de *Phthorimaea operculella* en almacenes rústicos, empleando virus granulosis en polvo. Revista Peruana de Entomología 35, 117–120.

Alcázar, J., Raman, K.V., Salas, R., 1991. Un virus como agente de control de la polilla de la papa *Phthorimaea operculella*. Revista Peruana de Entomología 34, 101–104.

Alcázar, J., Cervantes, M., Raman, K.V., 1992a. Caracterización y patogennicidad de un virus granulosis de la polilla de la papa *Phthorimaea operculella*. Revista Peruana de Entomología 35, 107–111.

Alcázar, J., Cervantes, M., Raman, K.V., 1992b. Efectividad de un virus granulosis formulado en polvo para controlar *Phthorimaea operculella* en papa almacenada. Revista Peruana de Entomología 35, 113–116.

Alcázar, J., Kroschel, J., Kaya, H.K., 2007. Evaluation of the efficacy of an indigenous Peruvian entomopathogenic nematode *Heterorhabditis* sp. to control the Andean potato weevil *Premnotrypes suturicallus* Kuschel under field conditions. Proceedings of the XVI International Plant Protection Congress (Vol II), 15–18 October 2007, Scotland, UK, Glasgow, pp. 544–545.

Ali, M.A., 1993. Effects of cultural practices on reducing field infestation of potato tuber moth (*Phthorimaea operculella*) and greening of tubers in Sudan. J. Agric. Sci. 121, 187–192.

Ali, M.I., 1991. Efficacy of a granulosis virus on the control of potato tuber moth, *Phthorimaea operculella* (Zeller) (Gelechiidae: Lepidoptera) infesting potatoes in Bangladesh. Bangladesh J. Zool. 19, 141–143.

Alrubeai, H.F., Al-Ani, K.H., Al-Azawi, A.F., 2001. Efficacy of some plant extracts on potato tuber moth, *Phthorimaea operculella* (Zeller) (Lepidoptera: Gelechiidae). Arab J. Plant Protect. 19, 92–96.

Amonkar, S.V., Pal, A.K., Vijayalakshmi, L., Rao, A.S., 1979. Microbial control of potato tuber moth (*Phthorimaea operculella* Zell.). Indian J. Exp. Biol. 17, 1127–1133.

Anderson, T.E., Kennedy, G.G., Stinner, R.E., 1982. Temperature-dependent models of European corn borer (Lepidoptera: Pyralidae) development in North Carolina. Environ. Entomol. 11, 1145–1150.

Angeles, I., Alcázar, J., 1995. Susceptibilidad de la polilla Scrobipalpuloides absoluta al virus de la granulosis de *Phthorimaea operculella* (PoGV). Revista Peruana de Entomologia 38, 65–70.

Angeles, I., Alcázar, J., 1996. Susceptibilidad de la polilla Symmetrischema tangolais al virus de la granulosis de *Phthorimaea operculella* (PoGV). Revista Peruana de Entomologia 39, 7–10.

Armer, C.A., Berry, R.E., Reed, G.L., Jepsen, S.J., 2004. Colorado potato beetle control by applications of the entomopathgenic nematode *Heterorhabditis marelata* and potato plant alkaloid manipulation. Entomol. Exp. Appl. 111, 47–58.

Arpaia, S., De Marzo, L., Di Leo, G.M., Santoro, M.E., Mennella, G., van Loon, J.J.A., 2000. Feeding behaviour and reproductive biology of Colorado potato beetle adults fed transgenic potatoes expressing the *Bacillus thuringiensis* Cry3B endotoxin. Entomol. Exp. Appl. 95, 31–37.

Arthurs, S.P., Lacey, L.A., Pruneda, J.N., Rondon, S.I., 2008. Semi-field evaluation of a granulovirus and *Bacillus thuringiensis* ssp. kurstaki for season-long control of the potato tuber moth, *Phthorimaea operculella*. Entomol. Exp. Appl. 129, 276–285.

Awate, B.G., Naik, L.M., 1979. Efficacies of insecticidal dusts applied to soil surface for controlling potato tuberworm (*Phthorimaea operculella* Zeller) in field. J. Maharashtra Agric. Unis 4, 100.

Beegle, C.C., Yamamoto, T., 1992. History of *Bacillus thuringiensis* Berliner research and development. Can. Entomol. 124, 587–616.

Belal, M.H., Moustafa, O.K., Girgis, N.R., 2005. Effect of different compounds in the management of potato tuber moth infesting potato and tomato plants. Egyptian J. Agric. Res. 83, 1581–1590.

Ben Salah, H., Aalbu, R., 1992. Field use of granulosis virus to reduce initial storage infestation of the potato tuber moth, *Phthorimaea operculella* (Zeller), in North Africa. Agric., Ecosyst. Environ. 38, 119–126.

Berry, R.E., Liu, J., Reed, G., 1997. Comparison of endemic and exotic entomopathogenic nematode species for control of Colorado potato beetle (Coleoptera: Chrysomelidae). J. Econ. Entomol. 90, 1528–1533.

Blissard, G.W., Rohrmann, G.F., 1990. Baculovirus diversity and molecular biology. Annu. Rev. Entomol. 35, 127–155.

Bloomquist, J.R., 1993. Toxicology, mode of action and target site-mediated resistance to insecticides acting on chloride channels. Comp. Biochem. Physiol., Part C: Pharmacol. Toxicol. Endocrinol. 106, 301–314.

Bloomquist, J.R., 1996. Ion channels as targets for insecticides. Annu. Rev. Entomol. 41, 163–190.

Booth, S.R., Drummond, F.A., Groden, E., 2007. Small fruits. In: Lacey, L.A., Kaya, H.K. (Eds.), Field Manual of Techniques in Invertebrate Pathology: Application and Evaluation of Pathogens for Control of Insects and Other Invertebrate Pests, 2nd Edn. Springer, Dordrecht, The Netherlands, pp. 583–598.

Bosa, C.F., Cotes Prado, A.M., Fukumoto, T., Bengtsson, M., Witzgall, P., 2005. Pheromone-mediated communication disruption in Guatemalan potato moth. *Tecia solanivora*. Entomol. Exp. Appl. 114, 137–142.

Brassel, J., Benz, G., 1979. Selection of a strain of the granulosis virus of the codling moth with improved resistance against artificial ultraviolet radiation and sunlight. J. Invertebr. Pathol. 33, 358–363.

Briese, D.T., 1981. The incidence of parasitism and disease in field populations of the potato moth *Phthorimaea operculella* (Zeller) in Australia. J. Aust. Entomol. Soc. 20, 319–326.

Broodryk, S.W., Pretorius, L.M., 1974. Occurrence in South Africa of a granulosis virus attacking potato tuber moth, *Phthorimaea operculella* (Zeller) (Lepidoptera: Gelechiidae). J. Entomol. Soc. South. Afr. 37, 125–128.

Broza, M., Sneh, B., 1994. *Bacillus thuringiensis* spp. *kurstaki* as an effective control agent of lepidopteran pests in tomato fields in Israel. J. Econ. Entomol. 87, 923–928.

Burges, H.D., Jones, K.A., 1998. Formulation of bacteria, viruses and protozoa to control insects. In: Burges, H.D. (Ed.), Formulation of Microbial Biopesticides, Kluwer Academic Publishers, Dordrecht, The Netherlands, pp. 34–127.

Bystrak, P., Sanborn, S., Zehnder, G.W., 1994. Methods for optimizing field performance of *Bacillus thuringiensis* endotoxins against Colorado potato beetle. In: Zehnder, G.W., Powelson, M.L., Jansson, R.K., Raman, K.V. (Eds.), Advances in Potato Pest Biology and Management, The American Phytopathological Society Press, St Paul, MN, pp. 386–402.

Cameron, P.J., Walker, G.P., Penny, G.M., Wigley, P.J., 2002. Movement of potato tuberworm (Lepidoptera: Gelechiidae) within and between crops, and some comparisons with diamondback moth (Lepidoptera: Plutellidae). Environ. Entomol. 31, 65–75.

Cameron, P.J., Wigley, P.J., Elliott, S., Madhusudhan, V.V., Wallace, A.R., Anderson, J.A.D., Walker, G.P., 2009. Dispersal of potato tuber moth estimated using field application of Bt for mark-capture techniques. Entomol. Exp. Appl. 132, 99–109.

Cañedo, V., Benavides, J., Golmirzaie, A., Cisneros, F., Ghislain, M., Lagnaoui, A., 1999. Assessing Bt-transformed potatoes for potato tuber moth, *Phthorimaea operculella* (Zeller), management. Impact on a changing world. International Potato Center Program Report 1997–1998., International Potato Center (Centro Internacional de la Papa) (CIP), Lima, Peru, pp. 161–169.

Castillo, G., Lúque, E., Moreno, B., 1998. Laboratory evaluation of the insecticidal activity of five plant species on *Tecia solanivora* (Lepidoptera: Gelechiidae). Agronomía Colombiana 15, 34–40.

Chakrabarti, S. K., Mandaokar, A. D., Pattanayak, D., Chandla, V. K., Kumar, P. A., Naik, P. S., and Sharma, R. P. (2000). Transgenic potato lines expressing a synthetic cry1Ab gene acquired tolerance to both potato tuber moth and defoliating caterpillar. In "Potato, global research & development. Proceedings of the Global Conference on Potato, New Delhi, India, 6–11 December, 1999: Volume 1" (S. M. P. Khurana, G. S. Shekhawat, B. P. Singh, and S. K. Pandey, Eds.), pp. 249–255. Indian Potato Association, Shimla, India.

Chatterjee, H., 2005. Studies on the synergistic response of some commercial biopesticides with botanicals, growth regulator and conventional organophosphate against neonate larvae of *Phthorimaea operculella* (Zeller). Crop Res. (Hisar) 29, 499–502.

Christiane, NG.-L., Yoon, K.S., Clark, J.M., 2003. Differential susceptibility to abamectin and two bioactive avermectin analogs in abamectin-resistant and -susceptible strains of Colorado potato beetle, *Leptinotarsa decemlineata* (Say) (Coleoptera: Chrysomelidae). Pestic. Biochem. Physiol. 76, 15–23.

CIP., 1992. Biological control of potato tuber moth using Phthorimaea baculovirus. CIP Training Bulletin 2.. International Potato Center (CIP), Lima, Peru.

CIP., 1993. Biological control of the potato tuber moth using Phthorimaea baculovirus.. International Potato Center (CIP), Lima, Peru.

Clark, J.M., Scott, J.G., Campos, F., Bloomquist, J.R., 1995. Resistance to avermectins: extent, mechanisms, and management implications. Annu. Rev. Entomol. 40, 1–30.

Das, G.P., 1995. Plants used in controlling the potato tuber moth, *Phthorimaea operculella* (Zeller). Crop Prot. 14, 631–636.

Das, G.P., Rahman, M.M., 1997. Effect of some inert materials and insecticides against the potato tuber moth, *Phthorimaea operculella* (Zeller), in storage. Intl J. Pest Manage 43, 247–248.

Das, G.P., Magallona, E.D., Raman, K.V., Adalla, C.B., 1992. Effects of different components of IPM in the management of the potato tuber moth, in storage. Agric. Ecosyst. Environ. 41, 321–325.

Davidson, M.M., Jacobs, J.M.E., Reader, J.K., Butler, R.C., Frater, C.M., Markwick, N.P., Wratten, S.D., Conner, A.J., 2002. Development and evaluation of potatoes transgenic for a cry1Ac9 gene conferring resistance to potato tuber moth. J. Am. Soc. Hortic. Sci. 127, 590–596.

Deshpande, S.G., Nagasampagi, B.A., Sharma, R.N., 1990. Synergistic oviposition deterrence activity of extracts of *Glycosmis pentaphyllum* (Rutaceae) and other plants for *Phthorimaea operculella* (Zell) control. Curr. Sci. 59, 932–933.

Douches, D.S., Westedt, A.L., Zarka, K., Schroeter, B., Grafius, E.J., 1998. Potato transformation to combine natural and engineered resistance for controlling tuber moth. HortScience 33, 1053–1056.

Douches, D.S., Li, W., Zarka, K., Coombs, J., Pett, W., Grafius, E., El-Nasr, T., 2002. Development of *Bt-cry5* insect-resistant potato lines "Spunta-G2" and "Spunta-G3. HortScience 37, 1103–1107.

Douches, D.S., Pett, W., Santos, F., Coombs, J., Grafius, E., Li, W., Metry, E.A., El-Din, T.N., Madkour, M., 2004. Field and storage testing Bt potatoes for resistance to potato tuberworm (Lepidoptera: Gelichiidae). J. Econ. Entomol. 97, 1425–1431.

Douches, D., Pett, W., Visser, D., Coombs, J., Zarka, K., Felcher, K., Bothma, G., Brink, J., Koch, M., Quemada, H., 2010. Field and storage evaluations of "SpuntaG2" for resistance to potato tuber moth and agronomic performance. J. Am. Soc. Hortic. Sci. 135, 333–340.

Douches, D.S., Coombs, J., Lacey, L.A., Felcher, K., Pett, W., 2011. Choice and no-choice evaluations of transgenic potatoes for resistance to potato tuber moth (*Phthorimaea operculella* Zeller) in the laboratory and field. Am. J. Potato Res. 88, 91–95.

Eberle, K.E., Wennmann, J.T., Kleespies, R.G., Jehle, J.A., 2012. Basic Techniques in Insect Virology. In: Lacey, L.A. (Ed.), Manual of Techniques in Invertebrate Pathology, second ed. Academic Press, San Diego, CA, pp. 15–74.

Ekesi, S., Maniania, N., 2007. Use of Entomopathogenic fungi in Biological Pest Management. Research Signpost Kerala, Kerala, India.

El-Sheikh, F.M., El-Naby, L.M.A., Farrag, R.M., 1988. Effect of two insect growth regulators on pupae of the potato tuber moth, *Phthorimaea operculella* (Zeller) in laboratory. Bollettino del Laboratorio di Entomologia Agraria "Filippo Silvestri" 45, 9–14.

Ewell, P.T., Fano, H., Raman, K.V., Alcázar, J., Palacios, M., Carhuamaca, J., 1994. Management of potato pests by farmers in Peru. In: Serie de Investigación en Sistemas Alimentarios, Instituto Nacional de Investigación Agraria y Agroindustrial, Centro Internacional de la Papa. International Potato Center (Centro Internacional de la Papa) (CIP), Lima, Peru, p. 72.

Farrag, R.M., 1998. Control of the potato tuber moth, *Phthorimaea operculella* Zeller (Lepidoptera Gelechiidae) at storage. Egyptian J. Agric. Res. 76, 947–952.

Federici, B.A., 1997. Baculovirus pathogenesis. In: Miller, L.K. (Ed.), The Baculoviruses, Plenum Press, New York, NY, pp. 33–59.

Ferro, D.N., Gelernter, W.D., 1989. Toxicity of a new strain of *Bacillus thuringiensis* to Colorado potato beetle (Coleoptera: Chrysomelidae). J. Econ. Entomol. 82, 750–755.

Gelernter, W.D., Trumble, J.T., 1999. Factors in the success and failure of microbial insecticides in vegetable crops. Integr. Pest Manage. Rev. 4, 301–306.

Georgis, R., Koppenhöfer, A.M., Lacey, L.A., Bélair, G., Duncan, L.W., Grewal, P.S., Samish, M., Tan, L., Torr, P., van Tol, R.W.H.M., 2006. Successes and failures in the use of parasitic nematodes for pest control. Biol. Control 38, 103–123.

Goettel, M.S., Ellenberg, J., Glare, T., 2005. Entomopathogenic fungi and their role in regulation of insect populations. In: Gilbert, L.I., Iatrou, K., Gill, S.S. (Eds.), Comprehensive Molecular Insect Science, Elsevier, Amsterdam, The Netherlands, Vol. 6, pp. 361–405.

Grewal, P.S., Ehlers, R.-U., Shapiro-Ilan, D.I., 2005. Nematodes as Biological Control Agents. CABI Publishing, Wallingford, UK.

Gröner, A., 1986. Specificity and safety of baculoviruses. In: Federici, R.R.G.B.A. (Ed.), The Biology of Baculoviruses, Volume I. CRC Press, Boca Raton, FL.

Hafez, M., Zaki, F.N., Moursy, A., Sabbour, M., 1997. Biological effects of the entomopathogenic fungus, *Beauveria bassiana* on the potato tuber moth *Phthorimaea operculella* (Seller). Anzeiger für Schädlingskunde, Pflanzenschutz, Umweltschutz 70, 158–159.

Hajek, A.E., Soper, R.S., Roberts, D.W., Anderson, T.E., Biever, K.D., Ferro, D.N., LeBrun, R.A., Storch, R.H., 1987. Foliar applications of *Beauveria bassiana* (Balsamo) Vuillemin for control of the Colorado potato beetle, *Leptinotarsa decemlineata* (Say) (Coleoptera: Chrysomelidae): An overview of pilot test results from the northern United States. Can. Entomol. 119, 959–974.

Hamdy, M.K., Salem, S.A., 1988. The possible use of the juvenoid methoprene as a control agent against the tuber moth *Phthorimaea operculella* (Zeller) (Lepidoptera: Gelechiidae). Bulletin of the Entomological Society of Egypt, Economic Series, 59–64.

Hamilton, J.T., Macdonald, J.A., 1990. Control of potato moth, *Phthorimaea operculella* (Zeller) in stored seed potatoes. Gen. Appl. Entomol. 22, 3–6.

Hamm, J.J., Styer, E.L., 1985. Comparative pathology of isolates of *Spodoptera* nuclear polyhedrosis virus in *S. frugiperda* and *S. exigua*. J. Gen. Virol. 66, 1249–1261.

Heimpel, A.M., Thomas, E.O., Adams, J.R., Smith, L.J., 1973. The presence of nuclear polyhedrosis viruses of *Trichoplusia ni* on cabbage from the market shelf. Environ. Entomol. 2, 173–178.

Hernández, C.S., Andrew, R., Bel, Y., Ferré, J., 2005. Isolation and toxicity of *Bacillus thuringiensis* from potato-growing areas in Bolivia. J. Invertebr. Pathol. 88, 8–16.

Huber, J., Lüdcke, C., 1996. UV-inactivation of baculovirus: the bisegmented survival curve. IOBC/WPRS Bull. 19, 253–256.

Hunter, D.K., Hoffmann, D.F., Collier, S.J., 1975. Observations on a granulosis virus of the potato tuberworm, *Phthorimaea operculella*. J. Invertebr. Pathol. 26, 397–400.

Hunter-Fujita, F.R., Entwistle, P.E., Evans, H.F., Crook, N.E., 1998. Insect Viruses and Pest Management. John Wiley & Sons, New York, NY.

Iannacone, J., Lamas, G., 2003. Insecticidal effect of four botanical extracts and cartap on the potato tuber moth, *Phthorimaea operculella* (Zeller) (Lepidoptera: Gelechiidae), in Peru. Entomotropica 18, 95–105.

Ibrahim, M.Y., 2008. Study of effect of temperatures on the natural death and the biotic potential of potato tuber moth, *Phthorimaea operculella* (Zeller), (Lepidoptera: Gelechiidae) and used of some plant extracts as insect repellents against potato tuber moth under lab. conditions. Dirasat. Agric. Sci. 35, 1–10.

Islam, M.N., Karim, M.A., Nessa, Z., 1990. Control of the potato tuber moth, *Phthorimaea operculella* (Zeller) (Lepidoptera: Gelechiidae) in the storehouses for seed and ware potatoes in Bangladesh. Bangladesh J. Zool. 18, 41–52.

Jansens, S., Cornelissen, M., Clercq, R.d., Reynaerts, A., Peferoen, M., 1995. *Phthorimaea operculella* (Lepidoptera: Gelechiidae) resistance in potato by expression of the *Bacillus thuringiensis* CryIA(b) insecticidal crystal protein. J. Econ. Entomol. 88, 1469–1476.

Jansson, R.K., Seal, D.R., 1994. Biology and management of wireworms on potato. In: Zehnder, G.W., Powelson, M.L., Jansson, R.K., Raman, K.V. (Eds.), Advances in Potato Pest Biology and Management,, American Phytopathological Society Press, St Paul, MN, pp. 31–53.

Jehle, J.A., Blissard, G.W., Bonning, B.C., Cory, J.S., Herniou, E.A., Rohrmann, G.F., Theilmann, D.A., Thiem, S.M., Vlak, J.M., 2006. On the classification and nomenclature of baculoviruses: a proposal for revision. Arch. Virol. 151, 1257–1266.

Johnson, T.B., Slaney, A.C., Donovan, W.P., Rupar, M.J., 1993. Insecticidal activity of EG4961, a novel strain of *Bacillus thuringiensis* toxic to larvae and adults of southern corn rootworm (Coleoptera: Chrysomelidae) and Colorado potato beetle (Coleoptera: Chrysomelidae). J. Econ. Entomol. 86, 330–333.

Jones, K.A., Moawad, G., McKinley, D.J., Grzywacz, D., 1993. The effects of natural sunlight on *Spodoptera littoralis* nuclear polyhedrosis virus. Biocontr. Sci. Technol. 3, 189–197.

Jones, K.D., 1994. Aspects of the biology and biological control of the European corn borer in North Carolina. PhD thesis. North Carolina State University, Raleigh, North Carolina.

Kabaluk, J.T., Goettel, M.S., Erlandson, M.A., Ericsson, J.D., Duke, G.M., Vernon, R.S., 2005. *Metarhizium anisopliae* as a biological control for wireworms and a report of some other naturally-occurring parasites. IOBC/WPRS Bulletin 28, 109–115.

Kashyap, N.P., Bhagat, R.M., Sharma, D.C., Suri, S.M., 1992. Efficacy of some useful plant leaves for the control of potato tuber moth, *Phthorimaea operculella* Zell. in stores. J. Entomol. Res. 16, 223–227.

Kay, I.R., 2006. Testing insecticides against *Phthorimaea operculella* (Zeller) (Lepidoptera: Gelechiidae) using a tomato plant bioassay. Plant Prot. Q. 21, 20–24.

Kaya, H.K., Lacey, L.A., 2007. Introduction to microbial control. In: Lacey, L.A., Kaya, H.K. (Eds.), Field Manual of Techniques in Invertebrate Pathology: Application and Evaluation of Pathogens for Control of Insects and Other Invertebrate Pests., Springer, Dordrecht, The Netherlands, pp. 3–7.

Kaya, H.K., Vega, F.E., 2012. Scope and basic principles of insect pathology. In: Vega, F.E., Kaya, H.K. (Eds.), Insect Pathology, Academic Press, San Diego,CA, pp. 1–12.

Koppenhöfer, A.M., Fuzy, E.M., 2008. Attraction of four entomopathogenic nematodes to four white grub species. J. Invertebr. Pathol. 99, 227–234.

Kroschel, J., 1995. Integrated pest management in potato production in the Republic of Yemen with special reference to the integrated biological control of the potato tuber moth *Phthorimaea operculella* Zeller).. Margraf Verlag, Weikersheim, Germany.

Kroschel, J., Koch, W., 1994. Studies on the population dynamics of the potato tuber moth (*Phthorimaea operculella* Zell. (Lep., Gelechiidae)) in the Republic of Yemen. J. Appl. Entomol. 118, 327–341.

Kroschel, J., Koch, W., 1996. Studies on the use of chemicals, botanicals and *Bacillus thuringiensis* in the management of the potato tuber moth in potato stores. Crop Prot. 15, 197–203.

Kroschel, J., Zegarra, O., 2007. Development of an attract-and-kill strategy for the potato tuber moth complex *Phthorimaea operculella* Zeller and *Symmetrischema tangolias* (Gyen) in Peru. XVI International Plant Protection Congress, 15–18 October 2007. Vol. 2, Glasgow, Scotland UK, pp. 576–577.

Kroschel, J., Zegarra, O., 2010. Attract-and-kill: a new strategy for the management of the potato tuber moths *Phthorimaea operculella* (Zeller) and *Symmetrischema tangolias* (Gyen) in potato: laboratory experiments towards optimising pheromone and insecticide concentration. Pest Manage. Sci. 66, 490–496.

Kroschel, J., Fritsch, E., Huber, J., 1996a. Biological control of the potato tuber moth (*Phthorimaea operculella* Zeller) in the Republic of Yemen using granulosis virus: biochemical characterization, pathogenicity and stability of the virus. Biocontr. Sci. Technol. 6, 207–216.

Kroschel, J., Kaack, H.J., Fritsch, E., Huber, J., 1996b. Biological control of the potato tuber moth (*Phthorimaea operculella* Zeller) in the Republic of Yemen using granulosis virus: propagation and effectiveness of the virus in field trials. Biocontr. Sci. Technol. 6, 217–226.

Laarif, A., Fattouch, S., Essid, W., Marzouki, N., Salah, H.B., Hammouda, M.H.B., 2003. Epidemiological survey of *Phthorimaea operculella* granulosis virus in Tunisia. Bull. OEPP 33, 335–338.

Lacey, L.A., Kroschel, J., 2009. Microbial control of the potato tuber moth (Lepidoptera: Gelechiidae). Fruit Veg. Cereal Sci. Biotechnol. 3, 46–54.

Lacey, L.A., Neven, L.G., 2006. The potential of the fungus, *Muscodor albus*, as a microbial control agent of potato tuber moth (Lepidoptera: Gelechiidae) in stored potatoes. J. Invertebr. Pathol. 91, 195–198.

Lacey, L.A., Frutos, R., Kaya, H.K., Vail, P., 2001. Insect pathogens as biological control agents: Do they have a future? Biol. Control 21, 230–248.

Lacey, L.A., Horton, D.R., Chauvin, R.L., Stocker, J.M., 1999. Comparative efficacy of *Beauveria bassiana,Bacillus thuringiensis*, and aldicarb for control of Colorado potato beetle in an irrigated desert agroecosystem and their effects on biodiversity. Entomol. Exp. Appl. 93, 189–200.

Lacey, L.A., Unruh, T.R., Simkins, H.S., Thomsen-Archer, K., 2007. Gut bacteria associated with the Pacific Coast wireworm, *Limonius canus,* inferred from 16s rDNA sequences and their implications for control. Phytoparasitica 35, 479–489.

Lacey, L.A., Horton, D.R., Jones, D.C., 2008. The effect of temperature and duration of exposure of potato tuber moth (Lepidoptera: Gelechiidae) in infested tubers to the biofumigant fungus *Muscodor albus.* J. Invertebr. Pathol. 97, 159–164.

Lacey, L.A., Kroschel, J., Wraight, S.P., Kabaluk, T., Goettel, M.S., 2009a. An introduction to microbial control of insect pests of potato. Fruit Veg. Cereal Sci. Biotechnol. 3, 20–24.

Lacey, L.A., de la Rosa, F., Horton, D.R., 2009b. Insecticidal activity of entomopathogenic fungi (Hypocreales) for potato psyllid, *Bactericera cockerelli* (Hemiptera: Triozidae): development of bioassay techniques, effect of fungal species and stage of the psyllid. Biocontr. Sci. Technol. 19, 957–970.

Lacey, L.A., Headrick, H.L., Horton, D.R., Schreiber, A., 2010. Effect of a granulovirus on mortality and dispersal of potato tuber worm (Lepidoptera: Gelechiidae) in refrigerated storage warehouse conditions. Biocontr. Sci. Technol. 20, 437–447.

Lacey, L.A., Hoffmann, D.F., Federici, B.A., 2011a. Histopathology and effect on development of the PhopGV on larvae of the potato tubermoth, *Phthorimaea operculella* (Lepidoptera: Gelechiidae). J. Invertebr. Pathol. 108, 52–55.

Lacey, L.A., Liu, T.-X., Buchman, J.L., Munyaneza, J.E., Goolsby, J.A., Horton, D.R., 2011b. Entomopathogenic fungi (Hypocreales) for control of potato psyllid, *Bactericera cockerelli* (Šulc) (Hemiptera: Triozidae) in an area endemic for zebra chip disease of potato. Biol. Control 36, 271–278.

Lagnaoui, A., Radcliffe, E.B., 1998. Potato fungicides interfere with entomopathogenic fungi impacting population dynamics of green peach aphid. Am. J. Potato Res. 75, 19–25.

Lagnaoui, A., Ben Salah, H., Ben Temime, A., 1995. Potato tuber moth granulosis virus, a prime candidate for integrated pest managemant. Third Triennal Conference of the African Potato Association, Sousse, Tunisia.

Lagnaoui, A., Cañedo, V., Douches, D.S., 2001. Evaluation of Bt-cry1Ia1 (cryV) transgenic potatoes on two species of potato tuber moth, *Phthorimaea operculella* and *Symmetrischema tangolias* (Lepidoptera: Gelechiidae) in Peru. Scientist and farmer: partners in research for the 21st Century. Program Report 1999–2000, International Potato Center (Centro Internacional de la Papa) (CIP), Lima, Peru, pp. 117–121.

Langenbruch, G.A., Krieg, A., Huger, A.M., Schnetter, W., 1985. Erste Feldversuche zur Bekämpfung der Larven des Kartoffelkäfers (*Leptinotarsa decemlineata*) mit *Bacillus thuringiensis* var. *tenebrionis*. Mededel. Fac. Landbouwweitenschap. Rijksuniv. Gent. 50, 441–449.

Latgé, J.P., Papierok, B., 1988. Aphid pathogens. In: Minks, A.K., Harrewijn, P. (Eds.), Aphids: Their Biology, Natural Enemies and Control, (Vol B), Elsevier Science Publishers B.V., Amsterdam, The Netherlands, pp. 323–335.

Leckie, B.M., 2002. Effects of *Beauveria bassiana* mycelia and metabolites incorporated into synthetic diet and fed to larval *Helicoverpa zea,* and detection of endophytic *Beauveria bassiana* in tomato plants using PCR and ITS. M.S. thesis. Department of Entomology, The University of Tennessee.

Léry, X., Abol-Ela, S., Giannotti, J., 1998. Genetic heterogeneity of *Phthorimaea operculella* granulovirus: restriction analysis of wild-type isolates and clones obtained *in vitro*. Acta Virologica 42, 13–21.

Mamani, D., Sporleder, M., Kroschel, J., 2011. Efecto de materiales inertes de fórmulas bioinsecticidas en la protección de tubérculos almacenados contra las polillas de papa (Effect of inert materials of bioinsecticides formula to protect stored tubers of potato against the potato moths). Revista Peruana de Entomologia 46, 43–49.

Mariy, F.M.A., El-Saadany, G.B., El-Wahed, M.S.A., Ibrahim, M.Y., 2000a. The bio-effect of biocides and plants as natural repellents for controlling the potato tuber moth, *Phthorimaea operculella* infestation in storage. Ann. Agric. Sci. (Cairo) 4, 1501–1509.

Mariy, F.M.A., El-Saadany, G.B., Abdel-Wahed, M.S., Ibrahim, M.Y., 2000b. Efficacy of some biocides, plant extract and mass trapping on potato tuber moth. Ann. Agric. Sci. (Cairo) 4, 1511–1519.

Mascarin, G.M., Alves, S.B., Rampelotti-Ferreira, F.T., Urbano, M.R., Borges Demétrio, C.G., Delalibera, I., 2010. Potential of a granulovirus isolate to control *Phthorimaea operculella* (Lepidoptera: Gelechiidae). BioControl 55, 657–671.

Matthiessen, J.N., Christian, R.L., Grace, T.D.C., Filshie, B.K., 1978. Large-scale field propagation and the purification of the granulosis virus of the potato moth, *Phthorimaea operculella* (Zeller) (Lepidoptera: Gelechiidae). Bull. Entomol. Res. 68, 385–391.

McGuire, M.R., Maddox, J.V., Armbrust, E.J., 1987. Effect of temperature on distribution and success of introduction of an *Empoasca fabae* (Homoptera: Cicadellidae) isolate of *Erynia radicans* (Zygomycetes: Entomophthoraceae). J. Invertebr. Pathol. 50, 291–301.

Meiyalaghan, S., Butler, R.C., Wratten, S.D., Conner, A.J., 2006. An experimental approach to simulate transgene pyramiding for the deployment of cry genes to control potato tuber moth (*Phthorimaea operculella*). Ann. Appl. Biol. 148, 231–238.

Mohammed, A., Douches, D.S., Pett, W., Grafius, E., Coombs, J., Liswidowati, Li, W., Madkour, M.A., 2000. Evaluation of potato tuber moth (Lepidoptera: Gelechiidae) resistance in tubers of *Bt-cry5* transgenic potato lines. J. Econ. Entomol. 93, 472–476.

Moscardi, F., 1999. Assessment of the application of baculoviruses for control of Lepidoptera. Annu. Rev. Entomol. 44, 257–289.

Moustafa, O.K., Belal, M.H., Girgis, N.R., 2005. Two developed programs for controlling potato tuber moth. Egyptian J. Agric. Res. 83, 1591–1599.

Mujica, N., Kroschel, J., 2005. Developing IPM components for leafminer fly in the Cañete Valley of Peru. Proceedings of the Second International Conference on Area-Wide Control of Insect Pests: Integrating the Sterile Insect and Related Nuclear and Other Techniques, Vienna, Austria, pp. 164–165.

Munyaneza, J.E., Crosslin, J.M., Upton, J.E., 2007. Association of *Bactericera cockerelli* (Homoptera: Psyllidae) with "zebra chip", a new potato disease in Southwestern United States and Mexico. J. Econ. Entomol. 100, 656–663.

Nielsen, C., Jensen, A.B., Eilenberg, J., 2007. Survival of entomophthoralean fungi infecting aphids and higher flies during unfavorable conditions and implications for conservation biological control. In: Ekesi, S., Maniania, N. (Eds.), Use of Entomopathogenic fungi in Biological Pest Management, Research Signpost Kerala, Kerala, India, pp. 13–38.

Nouri, N., Arfaoui, I., 2008. Spinosad: a new biopesticide for Integrated Pest Management of the potato tuber moth in Tunisia. In: Kroschel, J., Lacey, L.A. (Eds.), Integrated Pest Management for the Potato tuber moth *Phthorimaea operculella* (Zeller) – A potato pest of global importance., Margraf, Weikersheim, Germany, pp. 81–88.

Parsa, S., Alcázar, J., Salazar, J., Kaya, H.K., 2006. An indigenous Peruvian entomopathogenic nematode for suppression of the Andean potato weevil. Biol. Control 39, 171–178.

Perlak, F.J., Stone, T.B., Muskopf, Y.M., Petersen, L.J., Parker, G.B., McPherson, S.A., Wyman, J., Love, S., Reed, G., Biever, D., Fischhoff, D.A., 1993. Genetically improved potatoes: protection from damage by Colorado potato beetles. Plant Mol. Biol. 22, 313–321.

Pokharkar, D.S., Kurhade, V.P., 1999. Cross infectivity and effect of environmental factors on the infectivity of granulosis virus of *Phthorimaea operculella* (Zeller) (Lepidoptera: Gelechiidae). J. Biol. Control 13, 79–84.

Poprawski, T.J., Carruthers, R.I., Speese, J., Vacek, D.C., Wendel, L.E., 1997. Early-season applications of the fungus *Beauveria bassiana* and introduction of the hemipteran predator *Perillus bioculatus* for control of Colorado potato beetle. Biol. Control 10, 48–57.

Radcliffe, E.B., 1982. Insect pests of potato. Annu. Rev. Entomol. 27, 173–204.

Raman, K.V., 1994. Pest management in developing countries. In: Zehnder, G.W., Powelson, M.L., Jansson, R.K., Raman, K.V. (Eds.), Advances in Potato Pest Biology and Management., The American Phytopathological Society Press, St Paul, MN, pp. 583–596.

Raman, K.V., Alcázar, J., 1988. Biological control of the potato tuber moth (*Phthorimaea operculella* Zeller) using granulosis virus in Peru. In: Second Triennal Conference of the Asian Potato Association, Kinming, China.

Raman, K.V., Alcázar, J., 1990. Peruvian virus knocks potato tuber moth. paper presented at the 74th Annual Meeting of Potato Association of America, Quebec, Canada.

Raman, K.V., Booth, R.H., Palacios, M., 1987. Control of potato tuber moth *Phthorimaea operculella* (Zeller) in rustic potato stores. Trop. Sci. 27, 175–194.

Reddy, G.V.P., Urs, K.C.D., 1991. Insect growth regulator diflubenzuron as a reproductive inhibitor in potato tuber moth, *Phthorimaea operculella* (Zeller). J. Insect Sci., 155–156.

Reed, E.M., 1969. A granulosis virus of potato moth. Austr. J. Sci. 31, 300–301.

Reed, E.M., Springett, B.P., 1971. Large-scale field testing of a granulosis virus for the control of the potato moth (*Phthorimaea operculella* (Zell.) (Lep., Gelechiidae)). Bull. Entomol. Res. 61, 223–233.

Rico, E., Ballester, V., Ménsua, J.L., 1998. Survival of two strains of *Phthorimaea operculella* (Lepidoptera: Gelechiidae) reared on transgenic potatoes expressing a *Bacillus thuringiensis* crystal protein. Agronomie 18, 151–155.

Riga, K., Lacey, L.A., Guerra, N., 2008. The potential of the endophytic fungus, *Muscodor albus*, as a biocontrol agent against economically important plant parasitic nematodes of vegetable crops in Washington State. Biol. Control 35, 380–385.

Sabbour, M., Ismail, I.A., 2002. The combined effect of microbial control agents and plant extracts against potato tuber moth *Phthorimaea operculella* Zeller. Bull. Natl Res. Centre (Cairo) 27, 459–467.

Salama, H.S., Zaki, F.N., Ragaei, M., Sabbour, M., 1995a. Persistence and potency of *Bacillus thuringiensis* against *Phthorimaea operculella* (Zell.) (Lep., Gelechiidae) in potato stores. J. Appl. Entomol. 119, 493–494.

Salama, H.S., Ragaei, M., Sabbour, M., 1995b. Larvae of *Phthorimaea operculella* (Zell.) as affected by various strains of *Bacillus thuringiensis*. J. Appl. Entomol. 119, 241–243.

Salama, H.S., Salem, S.A., Zaki, F.N., Abdel-Razek, A., 1999. The use of *Bacillus thuringiensis* to control *Agrotis ypsilon* and *Spodoptera exigua* on potato cultivation in Egypt. Arch. Phytopathol. Plant Prot. 32, 429–435.

Setiawati, W., Soeriaatmadja, R.E., Rubiati, T., Chujoy, E., 1999. Control of potato tubermoth (*Phthorimaea operculella*) using an indigenous granulosis virus in Indonesia. Indonesian. J. Crop Sci. 14, 10–16.

Sewify, G.H., Abol-Ela, S., Eldin, M.S., 2000. Effects of the entomopathogenic fungus *Metarhizium anisopliae* (Metsch.) and granulosis virus (GV) combinations on the potato tuber moth *Phthorimaea operculella* (Zeller) (Lepidoptera: Gelechiidae). Bull. Faculty Agriculture, University of Cairo 51, 95–106.

Shapiro, M., Robertson, J.L., 1991. Natural variability of three geografical isolates of gypsy moth (Lepidoptera: Lymantriidae) nuclear polihedrosis virus. J. Econ. Entomol, 84.

Soper, R.S., 1981. Role of entomophthoran fungi in aphid control for potato integrated pest management. In: Lashomb, J.H., Casagrande, R. (Eds.), Advances in Potato Integrated Pest Management,, Hutchinson Ross Publishing Co., Stroudsburg, PA, pp. 153–177.

Sporleder, M., 2003. The granulosis of the potato tuber moth *Phthorimaea operculella* (Zeller): Characterisation and prospects for effective mass production and pest control.. Margraf Verlag, Weikersheim, Germany.

Sporleder, M., Kroschel, J., 2008. The potato tuber moth granulovirus (PoGV): use, limitations and possibilities for field applications. In: Kroschel, J., Lacey, L.A. (Eds.), Integrated Pest Management for the Potato tuber moth *Phthorimaea operculella* (Zeller) – A potato pest of global importance., Margraf, Weikersheim, Germany, pp. 49–71.

Sporleder, M., Zegarra, O., Kroschel, J., Huber, J., Lagnaoui, A., 2001. Assessment of the inactivation time of *Phthorimaea operculella* granulovirus (PoGV) at different intensities of natural irradiation. Scientist and farmer: partners in research for the 21st Century. Program Report 1999–2000, International Potato Center (Centro Internacional de la Papa) (CIP), Lima, Peru, pp. 123–128.

Sporleder, M., Kroschel, J., GutierrezQuispe, M.R., Lagnaoui, A., 2004. A temperature-based simulation model for the potato tuberworm, *Phthorimaea operculella* Zeller (Lepidoptera; Gelechiidae).. Environ. Entomol. 33, 477–486.

Sporleder, M., Kroschel, J., Huber, J., Lagnaoui, A., 2005. An improved method to determine the biological activity (LC50) of the granulovirus PoGV in its host *Phthorimaea operculella*. Entomol. Exp. Appl. 116, 191–197.

Sporleder, M., Cauti, E.M.R., Huber, J., Kroschel, J., 2007. Susceptibility of *Phthorimaea operculella* Zeller (Lepidoptera; Gelechiidae) to its granulovirus PoGV with larval age. Agricultural and Forest Entomology 9, 271–278.

Sporleder, M., Zegarra, O., RodriguezCauti, E.M., Kroschel, J., 2008. Effects of temperature on the activity and kinetics of the granulovirus infecting the potato tuber moth *Phthorimaea operculella* Zeller (Lepidoptera: Gelechiidae). Biol. Control 44, 286–295.

Strobel, G.A., Dirkse, E., Sears, J., Markworth, C., 2001. Volatile antimicrobials from Muscodor albus, a novel endophytic fungus. Microbiol. Reading 147, 2943–2950.

Tabashnik, B.E., Gassmann, A.J., Crowder, D.W., Carriére, Y., 2008. Insect resistance to Bt crops: evidence versus theory. Nat. Biotechnol. 26, 199–202.

Tanada, Y., Kaya, H.K., 1993. Insect Pathology.. Academic Press, Inc, San Diego CA.

Temerak, S.A., 2003. Spinosad, a new naturally derived potato tuber worm control agent in comparisons to certain conventional insecticides. Assiut J. Agric. Sci. 34, 153–162.

Vernon, R.S., van Herk, W., Tolman, J.H., 2005. European wireworms (*Agriotes* spp.) in North America: distribution, damage, monitoring, and alternative integrated pest management strategies. Bulletin of OILB/SROP 28, 73–79.

Vickers, J.M., Cory, J.S., Entwistle, P.F., 1991. DNA characterization of eight geographic isolates of granulosis virus from the potato tuber moth (*Phthorimaea operculella*) (Lepidoptera, Gelechiidae). J. Invertebr. Pathol. 57, 334–342.

Vogel, S., 1986. Baculo-Viren: Biologische Insektizide – Werkzeuge der Molekularbiologie.. VCH Verlagsgesellschaft, Weinheim.

von Arx, R., Goueder, J., Cheikh, M., Temime, A.B., 1987. Integrated control of potato tubermoth *Phthorimaea operculella* (Zeller) in Tunisia. Insect Sci. Appl. 8, 989–994.

von Arx, R., Gebhardt, F., 1990. Effects of a granulosis virus, and *Bacillus thuringiensis* on life-table parameters of the potato tubermoth *Phthorimaea operculella*. Entomophaga 35, 151–159.

Weintraub, P.G., 2001. Effects of cyromazine and abbamectin on the pea leafminer Liriomyza *huidrobiensis* (Diptera: Agromyzidae) and its parasitoid *Diglyphus isaea* (Hymenoptera: Eulophidae) in potatoes. Crop Prot. 20, 207–213.

Worapong, J., Strobel, G., Ford, E.J., Li, J.Y., Baird, G., Hess, W.M., 2001. *Muscodor albus* anam. gen. et sp. nov., an endophyte from *Cinnamomum zeylanicum*. Mycotaxon 79, 67–79.

Wraight, S.P., Ramos, M.R., 2002. Application parameters affecting field efficacy of *Beauveria bassiana* foliar treatments against Colorado Potato Beetle *Leptinotarsa decemilineata*. Biol. Control 23, 164–178.

Wraight, S.P., Sporleder, M., Poprawski, T.J., Lacey, L.A., 2007. Application and evaluation of entomopathogens in potato. In: L.Lacey, A., Kaya, H.K. (Eds.), Field Manual of Techniques in Invertebrate Pathology: Application and Evaluation of Pathogens for Control of Insects and Other Invertebrate Pests, Springer, Dordrecht, The Netherlands, pp. 329–359.

Wraight, S.P., Lacey, L.A., Kabaluk, J.T., Goettel, M.S., 2009. Potential for microbial biological control of Coleopteran and Hemipteran pests of potato. Fruit Veg. Cereal Sci. Biotechnol. 3, 25–38.

Yubac Dhoj, G.C., 2006. White grubs (Coleoptera: Scarabaeidae) associated with Nepalese agriculture and their control with the indigenous entomopathogenic fungus *Metarhizium anisopliae* (Metsch.) Sorokin. Faculty of Science. University of Basel, Basel.

Yubac Dhoj, G.C., Keller, S., Nagel, P., 2009. Microbial control of white grubs in Nepal: the way forward. J. Agric. Environ. 10, 134–142.

Zeddam, J.L., Pollet, A., Mangoendiharjo, S., Ramadhan, T.H., López Ferber, M., 1999. Occurrence and virulence of a granulosis virus in *Phthorimaea operculella* (Lep., Gelechiidae) populations in Indonesia. J. Invertebr. Pathol. 74, 48–54.

Zeddam, J.L., VasquezSoberon, R.M., Vargas Ramos, Z., Lagnaoui, A., 2003a. Use of a granulovirus for the microbial control of the potato tuber moths *Phthorimaea operculella* and *Tecia solanivora* (Lepidoptera: Gelechiidae). Boletín de Sanidad Vegetal, Plagas 29, 659–667.

Zeddam, J.L., VásquezSoberon, R.M., Vargas Ramos, Z., Lagnaoui, A., 2003b. Quantification of viral production and rates of application of a granulosis virus used for the biological control ofthe potato moths, *Phthorimaea operculella* and *Tecia solanivora* (Lepidoptera: Gelechiidae). Revista Chilena de Entomología 29, 29–36.

Zehnder, G.W., Gelernter, W.D., 1989. Activity of the M-ONE formulation of a new strain of *Bacillus thuringiensis* against the Colorado potato beetle (Coleoptera: Chrysomelidae): relationship between susceptibility and insect life stage. J. Econ. Entomol. 82, 756–761.

Zehnder, G.W., Ghidiu, G.M., Speese, J., 1992. Use of the occurrence of peak Colorado potato beetle (Coleoptera: Chrysomelidae) egg hatch for timing of *Bacillus thuringiensis* spray applications in potatoes. J. Econ. Entomol. 85, 281–288.

Zhao, J.Z., Cao, J., Li, Y., Collins, H.L., Roush, R.T., Earle, E.D., Shelton, A.M., 2003. Transgenic plants expressing two *Bacillus thuringiensis* toxins delay insect resistance evolution. Nat. Biotechnol. 21, 1493–1497.

ZhengYue, L., QingWen, Z., 2005. Relative virulence of isolates of *Beauveria bassiana* to the potato tuber moth [*Phthorimaea operculella*] and their biological compatibility with ten insecticides. Plant Prot. 31, 57–61.

Chapter 17

Physical Control Methods

Phyllis G. Weintraub

Agricultural Research Organization, Gilat Research Center, D.N. Negev, Israel

INTRODUCTION

Potatoes (*Solanum tuberosum*) are native to South America; Spanish explorers brought them to Europe, where they spread across the continent as a relatively inexpensive food. Potatoes are now the most important non-grain crop and the fourth most important food overall in the world, with an annual production of about 300 million tonnes (FAOSTAT 2011). Because potatoes are grown from tropical to temperate regions, they are attacked by a number of different key pests which can be divided into two groups: those causing direct damage to the plant or tuber (and therefore directly affecting yield), and those transmitting plant pathogens (which may or may not reduce yield). The former group contains insects like the Colorado potato beetle and potato tuber moths; the latter group contains aphids, leafhoppers, and psyllids.

Physical control methods are the oldest form of pest control and an area for current innovation. These methods may directly impact and kill the pest (e.g., vacuum) or have an indirect effect (e.g., barriers). The effects of physical control methods are limited spatially – in contrast to chemicals, which may drift and be accumulated in food chains. Unlike pesticides, there is no need for governmental regulation/registration with the concomitant need to spend millions of dollars satisfying environmental and animal toxicology, food safety, and efficacy requirements. Knowledge of the pests' behavior, biology, and population ecology are critical to determine the correct method, because unlike pesticides, which are either sprayed or applied systematically to manage a wide variety of pests, physical control techniques target specific stages and/or behaviors. In general, physical control methods are relatively labor-intensive and time consuming, but are easily incorporated in integrated management strategies. Furthermore, physical control measures require neither the monetary expenditures necessary for the development of new pesticides or genetically modified control measures, nor the worry about the concomitant development of resistance to those measures. Physical control methods have not been developed for all potato pests; the long-lived beetles known as wireworms are an example. This

Insect Pests of Potato. http://dx.doi.org/10.1016/B978-0-12-386895-4.00017-X

chapter will briefly review the pests and damage they cause, followed by a range of effective physical control methods for these pests.

PESTS CAUSING DIRECT DAMAGE

Gelechiid Potato Moths

The potato tuber moth (PTM), *Phthorimaea operculella*, is a worldwide pest, but a key pest of potatoes in tropical and subtropical areas, although it is starting to appear in temperate regions (Medina *et al.* 2010). *Tecia solanivora* is known as the Guatemalan potato moth and, as the name implies, was originally from Central America but has spread southward and crossed the Atlantic Ocean into Europe (Anon. 2005a). Unlike the other key pests of potatoes, these potato moths are both field and storage pests. In the field, the adults lay eggs on the potato foliage and the larvae mine the foliage and the stems. Tubers are entered as the larvae leave infected stems and drop through cracks in the soil, or hatch from eggs oviposited directly on exposed tubers. Adults also gain access to the tubers through cracks in the soil. Larvae remaining in the tubers at harvest start another cycle in storage. Depending on the storage conditions, there may be several generations on stored potatoes. The point of entry into the potato can be a source for pathogen infection, further reducing potato quality. There has been documented resistance to a number of insecticide classes (Richardson and Rose 1967, Foot 1976, Collantes *et al.* 1986), but insecticides, when effective, are only so on the foliage stage and not on the tubers (Clough *et al.* 2010). The biology, ecology, and control of PTM have been recently reviewed (Rondon 2010).

Beetles

There is one primary beetle pest that occurs worldwide, the Colorado potato beetle, and more minor but potentially dangerous potato weevils, in the Andes Mountains, that are a quarantine pest in Europe because of their potential capacity to establish and develop there. The Colorado potato beetle, *Leptinotarsa decemlineata*, is a native North American foliage pest of potatoes. In the late 1800s the beetle was transmitted to Europe, and eventually became a pan-Eurasian pest of potatoes. A comprehensive review was written by Alyokhin (2009) (see also Chapter 2 of this volume). The females are highly prolific, and both the four larval stages and adults feed on leaves and can defoliate fields if left unchecked, causing yield loss (Weber and Ferro 1994). Adults overwinter in the soil, emerging in the spring and usually walking to find suitable host plants (Weber *et al.* 1994), although some also fly (Voss and Ferro 1990). Adults locate plant hosts by a combination of olfactory and visual cues (Visser and Nielsen 1977, Jermy *et al.* 1988, Dickens 2000), and are attracted to yellow (Zehnder and Speese 1987). There is at least a second "summer" generation and, depending on the climate, sometimes more. The larvae are particularly sensitive to disturbance, immediately falling to the ground (Boiteau and Misener 1996); perhaps this behavior enhances distribution on the plant as they crawl back up and arrive on a new leaf. CPB has developed resistance to more than 25 insecticides in the US (Roush *et al.* 1990, Ioannidis *et al.* 1991, Grafius 1997). Therefore, there is the need for alternative management tactics, especially physical control (see review by Khelifi *et al.* 2007).

The Andean potato weevils are a complex of three genera, *Premnotrypes*, *Phyrdenus*, and *Rhigopsidius*, which occur in the mountainous native habitats of wild potatoes. Adults are flightless and walk to potato fields, where they nocturnally feed on foliage and lay eggs on the ground, close to tubers (Kroschel *et al.* 2009). Larvae burrow into the soil and feed on tubers, and pupate either in the soil or in stored potatoes. Thus, like PTM, they are both a field and a storage pest. In mountainous regions there is one generation per year, and in lower elevations, there may be two generations. Because of the similarity in potato cultivation, the EPPO lists these weevils as a quarantine pest (Anon. 2011).

PESTS CAUSING INDIRECT DAMAGE

Pathogens that are retained on the cuticle, either in the mouthparts or in the fore- or hindgut, can only be non-persistently or semi-persistently transmitted until the pathogen load is exhausted or the subsequent molt in which the cuticle is shed. Those pathogens that are able to enter the hemocoel, replicate and invade the salivary glands are circulative and persistent, and are retained for long periods of time. Potato virus Y (PVY) is an example of the former transmission group, and potato leafroll virus (PLRV), beet leafhopper-transmitted virescence agent (BLTVA), potato purple top, *Spiroplasma citri*, and *Candidatus* Liberibacter solanacearum/psyllaurous are examples of the latter. These different transmission modes have implications for the means of control methods that can be employed.

Aphids

Of all of the arthropod vectors, aphids transmit more plant pathogens, particularly viruses, than any other single group (see Chapter 3). During feeding (see review, Backus 1988), aphids often secrete a small amount of gelling saliva which forms a proteinaceous sheath to protect delicate stylets as they move through different cells, penetrating deeper into the plant. Since they have no chemosensory structures on the labia, they secrete enzyme-laden watery saliva to taste the contents of cells and determine if they are feeding on the correct tissue (Miles 1999). In species that transmit non-persistent viruses, it is during these short probing periods that viruses are transmitted. There are more than 40 species known to transmit potato virus Y (PVY) (Kennedy *et al.* 1962, Piron 1986, Sylvester 1989). When aphids have determined that the host plant is acceptable and have chosen to feed, their stylets move between cells until they reach the phloem, on which they may feed continuously for days (Tagu *et al.* 2008). It is during phloem feeding that persistent viruses are transmitted.

The behavior of aphids in a colony often dictates the efficacy of transmission; alate forms have a greater potential as vectors, as they actively move by flight into fields and between plants in a field. The rate of transmission of non-persistent viruses by the aphids drops to near zero within a short time – a matter of hours. Generally, vector acquisition of pathogens increases with time spent feeding on the infected plant source; however, in this case acquisition usually decreases with sustained feeding, indicating that the vector is feeding in vascular tissues (Nault 1997). Therefore, their phenology and behavior greatly affect the types of management measures to be used.

Leafhoppers

Leafhoppers range in size up to about 10 mm long and have five nymphal instars; all stages feed on the aerial parts of the plant, nymphs and adults feeding on the same plants. Leafhoppers can be direct pests, such as the potato leafhopper, *Empoasca fabae*, or indirect pests transmitting phloem-limited bacteria such as phytoplasmas and/or *Spiroplasma citri* (*Circulifer tenellus*) (Munyaneza 2010a).

In North America, *E. fabae* overwinters along the Gulf of Mexico and reinvades the potato-growing areas east of the Rocky Mountains on an annual basis (Maletta *et al.* 2006). As adult leafhoppers arrive in potato fields they start feeding in mesophyll tissues, which causes hopperburn. Symptoms include leaf curling and browning, and can lead to reduced yields. In the past, leafhoppers were generally controlled by the same insecticides that were used for CPB. When growers switched to systemic neonicotinoids, control of leafhoppers diminished and either additional insecticide applications or alternative control measures were needed (Kaplan *et al.* 2008). The situation is exacerbated in organic potato production due to the lack of effective insecticides.

Leafhopper-transmitted bacterial plant pathogens have become increasingly important in the past decade (Munyaneza *et al.* 2008, and references therein). *Circulifer tenellus* is a confirmed vector of phytoplasma and *Spiroplasma citri*, both of which infect potatoes (Weintraub and Beanland 2006, Munyaneza *et al.* 2008). In addition to *C. tenellus*, other leafhoppers have been implicated in transmission of potato purple top in Europe and Asia.

Psyllids

Psyllids are small insects that superficially resemble aphids as adults. *Bactericera* (=*Trioza*) *cockerelli* is known as the potato/tomato psyllid, and, as its name implies, is a pest of both crops. The nymphal stages are usually on the underside of leaves. Adult and nymphal feeding causes leaf curling, which progresses to generalized yellowing and stunts the plant, and a reduction in yield (Richards 1973). That the symptoms cease when the psyllids are removed strongly suggests that the agent is a toxin and not a pathogen. *Bactericera nigricornis* is a generalist psyllid, but can cause severe damage in potatoes, especially in the Middle East (Fathi 2011).

Bactericera cockerelli has been shown to transmit *Candidatus* Liberibacter solanacearum/psyllaurous (Secor *et al.* 2009), also known as zebra chip. This has heightened its status as a severe pest. With the identification of this new behavior as a vector, it is receiving much attention. Its distribution and movement in potatoes fields have revealed that the females are strong fliers, especially against prevailing winds, and they do not show a preference for colonizing plants based on the age of the plants (Henne *et al.* 2010). Unfortunately, little is known about their feeding behavior (Bonani *et al.* 2010); psyllid stylets move between cells to reach the vascular bundle and can feed on both phloem and xylem tissues, although predominantly on the former. At present, the retention time of the pathogen in the psyllid is unknown (Munyaneza 2010b).

TYPES OF PHYSICAL CONTROL TECHNIQUES AND THEIR APPLICATIONS

Barriers

Barriers are often some sort of vertical projection, including fences, of a height appropriate to the behavior of the pest. They can be used against either crawling or low-flying insects. In the case of flying insects, an overhang is critically important to trap them and prevent them from simply walking over the top (Bomford *et al.* 2000). Barriers are relatively easy to erect and deconstruct and can be reused, thus initial costs are amortized over years of use. The drawback to barriers is the difficulty of entry of farm equipment, and adequate entryways must be developed. Smooth plastic barriers of heights ranging from 25 to 100 cm were tested in greenhouses to determine what height was insurmountable by Andean weevils (Kroschel *et al.* 2009). Since the weevils are flightless a barrier can be very effective, but it must be at least 50 cm high and it must be erected well before the emergence and migration of weevils has started. These researchers found that the barriers were most efficacious when a fallow-potato crop rotation system was used, as opposed to potato-potato, and was fully as effective as the normal insecticide applications. Furthermore, with the barrier system there were more predatory carabid beetles found in the plots, which enhanced weevil control and reduced potato tuber damage.

Mounding

Various sorts of tillage practices can be used either to bury pests deeper in the ground or to bring them to the surface where they will be exposed to various predators or adverse abiotic factors. Potato tuberworms gain access to potato tubers through cracks in the soil, which can be reduced by irrigation practices and soil moisture (Foot 1979). When soils are not allowed to dry out, cracks do not form and the moths do not have access to the tubers. One of the most effective means of preventing damage, if soil cracking cannot be avoided, is by plowing the fields to mound soil over the potatoes. However, the ridging or mounding must be done at the appropriate time; if done too early in the season, the effects are lost (Shelton and Wyman 1979). Clough *et al.* (2010) examined light sprinkler irrigation practices coupled with vine burn-off and mounding hills. Reducing the amount of irrigation water and soil sloughing together with mounding at vine-kill denied PTM access to tubers.

Mulching, and Cover and Trap Crops

Mulches are either living (cover crops) or made of natural or synthetic material to control weeds, retain soil water, and deflect insect pests (see also Chapter 18). Natural materials include straw, compost (including decomposed manure), peat moss, bark chips, sawdust, etc. Synthetic materials include various colored plastics and aluminum. Mulch effectiveness varies considerably according to type, and has been reviewed by Weintraub and Berliner (2004). Mulches are frequently used, but there is no uniform application with regard to materials, crops, or pests. The use of cover or trap crops, in general, has been reviewed (see, for example, Hooks and Fereres 2006, Shelton and Badenes-Perez 2006).

It has been demonstrated that there is a linear relationship between the risk of CPB infestation and the migration distance from the potato field of the previous season to the field of the current season (Weisz *et al.* 1996). Since not all farms are large enough to rotate potato fields with a sufficient distance from one another, habitat diversification in the form of cover or alternative crops has been explored (Weber *et al.* 1994, Hoy *et al.* 2000). Attempts at using living mulches to control CPB have met with mixed results due to the behavior of the beetle. An examination of the ability of CPB to move over fallow ground versus wheat revealed that wheat was harder to navigate, and the authors concluded that wheat could be used to reduce movement when it surrounded potato fields (Schmera *et al.* 2006). However, in an examination of living mulches of rye or vetch versus fallow land the authors found that in the presence of cover the beetles more easily crossed from row to row, whereas when there was no cover the beetles tended to disperse along a row and not cross the bare ground (Szendrei *et al.* 2009). It seems that cover crops will reduce initial colonization, but once in the field cover crops may actually aid in inter-row movement.

Trials with straw mulch for the management of CPB have met with mixed results but, generally, inadequate control. Zehnder and Hough-Goldstein (1990) applied a dense layer (6–10 cm) of straw and reported reduced adult colonization and lower CPB populations due to physical obstruction and reduced soil temperature. They found a 2.5- to 5-fold decrease in potato defoliation. Stoner (1993) found fewer CPB larvae and less damage to potatoes versus bare ground for the first generation only, but not the summer generation. When mulch was applied to potato plants before CPB emerged, straw did not affect immigration; however, the second (summer) generation was significantly lower than in control plots (Brust 1994). This reduction in the second generation was attributed to the larger number of predators found in the straw mulch plots. Within 2–3 weeks of the straw being applied there were significantly more soil predators. Johnson *et al.* (2004) applied straw mulch at planting and found mixed results regarding the number of adult CPBs colonizing the plots; one year there were more than the control, and the next no difference between the control and treated plots. However, in both years there were fewer egg masses and larvae. There was an increased presence of predators in the straw-mulched plots, which probably fed on the eggs and larvae.

The use of insecticides actually contributes to enhanced PVY transmission in peppers and potatoes (Budnik *et al.* 1996, Radcliffe and Ragsdale 2002). The exact mechanism is not clear, but apparently aphids can sense the insecticide, which acts as an irritant, causing them to be more flighty and feed on more plants in a field. Aphid vectors of non-persistent viruses acquire and transmit virus in a matter of seconds, and lose infectivity after several probes. Furthermore, aphids detect the contrast between dark soil borders and the green crop (Smith 1969). Considering all of these factors, border crops are an ideal alternative control measure; infective aphids can land there, probe these non-crop or non-susceptible plants, and when they migrate into the field be virus-free.

A number of plants have been used as border crops: soybean, sorghum, wheat, and potato (Difonzo *et al.* 1996, Ragsdale *et al.* 2001). Aphid numbers landing on these crops versus a potato crop with a fallow border were not significantly different; however, the incidence of PVY, as measured by ELISA, between border rows and the middle of the field was significantly different. The border crop effectively reduced the incidence of PVY in the protected center of the plots. In both studies, no significant difference was found between the choices of border plant species. As a result of these studies, growers in the Great Lakes region of the US routinely use soybean (non-PVY host plant) as a crop border to protect seed potatoes (Davis and Radcliffe 2008).

Initial trials with straw mulch showed promise in reducing aphids and PVY infestations early in the crop season (Saucke and Doring 2004). This approach was combined with chitting (pre-sprouting) and seemed to have a synergistic effect. Doring *et al.* (2005) examined the effects of straw mulch on weeds and soil parameters and found that soil erosion was reduced while yield and number of weeds was not affected by mulching. They found that if the field was weeded before mulching, the number of weeds was kept at a moderate level. Better soil coverage was achieved with chopped instead of long straw, and this approach is good for both the soil and virus control in organically grown potatoes.

Unfortunately, mulching with silver plastic, although effective against leafhoppers in other crops, did little to control hopperburn caused by *E. fabae* (Maletta *et al.* 2006). In fact, the populations of the potato leafhopper were even higher than in untreated control plots. Mulching did, however, increase potato quality and yield.

Solarization

Using solar radiation is an effective means of managing potato soil pathogens (Denner *et al.* 2000, Triki *et al.* 2001), weeds, and nematodes, and, occasionally, arthropods. It is accomplished by placing clear plastic sheeting over moistened soil, or by using drip irrigation lines beneath the plastic, and leaving it in place for 6–8 weeks during hot summer months. This method was first developed in Israel (see review by Katan 1981) and California, and it is an economically viable method in areas of abundant sunshine. While this is not a new technology, with the gradual elimination of methyl bromide from 2005–2015 (revised Montreal Protocol; Anon. 2006) the frequency and scale of its use is growing.

Chauvin *et al.* (2008) reported that solarization is an effective replacement for chemical control of the potato cyst nematodes *Globodera rostochiensis* and *G. palida*. These nematodes, originally from the Andes Mountains in South America, are a quarantine pest, and effective control to prevent their spread is a necessity.

Trenching

Digging a trench to catch crawling pests is a very old method of control. The critical factors are the depth and the vertical angle of the trench sides, to prevent captured pests from crawling out. Various forms of synthetic lining, such a smooth plastic, can hinder crawling capabilities. Trenches were developed for CPB by Boiteau *et al.* (1994) in which a trench was dug and lined with rigid plastic by a device named the "Beetle Excluder". The trench must be at least 25 cm deep and have sides at an angle of at least 50°. They reported that about 50% of the overwintered adults were captured, and up to 95% of them were retained in the trench and died (Misener *et al.* 1993). It is important to place the trench away from the potato crop and close to the overwintering sites to catch adults before they take flight. This method has proved to be so effective that a plastic portable trench has been patented in the US (Bomford and Vernon 2005), and explicit instructions on the depth and placement of trenches are available on the Internet (University of Connecticut) and described in a video produced by Cornell University (Grubinger 2011).

Flaming

Flaming, as opposed to burning residual plant material, can be accomplished by a hand-held or tractor-mounted device in which a very hot, localized flame is produced, often by burning propane. While it is usually used for weeds, it can also be applied to pests and has been applied to CPB when incidents of insecticide resistance were increasing and alternative management methods were being sought. Significant reductions in eggs and larvae have been reported in the US and Canada (Duchesne *et al.* 1992, Moyer *et al.* 1992). Propane flamers can be used to manage colonizing CPB in the early season, before plants are 10 cm high and when the canopy area is still small (Khelifi *et al.* 2007), and are usually used on the field edges where the early migrants first arrive. Mortality is due to the rapid rise in temperature as the burner passes; muscle function in the legs is affected and beetles cannot climb back onto the plants (Pelletier *et al.* 1995). The eggs are also highly sensitive to flaming (Duchesne *et al.* 2001). To improve the efficacy, Lague *et al.* (1999) combined vacuuming with flaming fallen adults to produce better control than in insecticide-treated plots.

Insect Exclusion Screens

One does not normally think of growing potatoes under some sort of cover; however, insect exclusion screens are a physical barrier to a host of flying pests. The advantage of mesh screening, as opposed to solid plastic sheets, is that it permits movement of air and helps reduce the humidity, which enhances plant pathogen development. The various forms of IES (woven, knitted, and micro-perforated) have been reviewed extensively (Weintraub and Berlinger 2004). Insect vectors of plant pathogens present a unique set of problems; regardless of whether the pathogen is circulative/propagative in the insect or not, transmission occurs faster than the action of any insecticide. Furthermore, in some instances, such as in the case of aphid vectors of the non-persistent potato virus Y (PVY), insecticides actually enhance the spread of the virus because the aphids are irritated by the insecticide and tend to move more frequently (Radcliffe *et al.* 1993). For seed potatoes this is especially important as, by the end of the season, there is approximately a 10-fold increase over the beginning levels of PVY infection.

Greenhouses and Screening

Due to the abundance of soil pathogens, potatoes must be rotated on a 4- to 7-year basis to prevent development of devastating levels of pathogens or nematodes. While greenhouses, especially European-style glasshouses, are generally considered to be permanent structures, work in Israel and other Mediterranean countries is focusing on easily erectable metal-framed walk-in tunnels, also known as hoop houses. In Bangladesh, trials were run comparing growers' usual practices versus net houses (Karim *et al.* 2010); five varieties of potatoes were checked by DAS-ELISA for evidence of virus infection (PVY and potato leafroll virus) when the seed potatoes were received from Holland and at the end of the trial. Not only were all varieties of potato virus-free, but yield was highest under the net-house production system. In Israel, seed potato production is critical, and trials are underway with walk-in tunnels (6 m wide by 2.5–3 m tall, curved and metal-framed) covered with 50 mesh insect exclusion screens. These structures afforded 100% control of aphid transmitted PVY when seed potatoes are of the highest quality.

Floating Row Covers

Floating row covers (FRC), or fleeces, are made from spun-bonded polyester or polypropylene and are cloth-like in appearance. They are loosely applied after seeding or transplanting to ensure there will be sufficient material volume to float up as plants grow, and they are removed just prior to harvesting. Various forms of these covers have been tested, starting in the 1980s. A 15-mesh UV stabilized polyethylene net and a thin polypropylene sealed fiber mesh were first developed and experimented with in Switzerland and Germany (Bizer 1987), and results were promising.

Initial laboratory work evaluated three types of sheets (Harrewijn *et al.* 1991) using visual observations and electrical recording of stylet penetration to determine whether four species of aphids could transmit virus through the sheets. Additionally, they followed up with a small-scale field test. They found that fleece could prevent stylet penetration, and field studies found that 25% of uncovered potatoes became infected whereas different forms of fleece offered up to 100% protection. The explanation for less than 100% protection was due to a mechanical breakdown of one of the fleece types due to heavy wind, thus allowing aphids access to the potatoes.

In Israel, using FRCs of 17 and 19 g/m², similar problems are encountered although the primary factor is the intensity of the sun, which breaks down the fabric after about 80 days in the fields, at which point a breeze easily causes tearing. Even so, in trials an increase in potato yields was realized with protection again PVY.

Studies of the effects of the fleece on selected quality parameters of potatoes have been conducted. Lachman *et al.* (2003) found that the variety and year/location of cultivation had more influence on the ascorbic acid and carotenoids content and nitrate levels. Unfortunately this means that no generalized conclusions could be drawn, but varieties will have to be tested locally.

Pheromones

Pheromones are chemicals that are secreted or excreted into the environment and cause a specific behavioral response in conspecifics. They have been developed for use in pest management in two particular areas: mating disruption (especially for nocturnal species), and as lures to traps (Foster and Harris 1997).

For CPB, both plant volatiles (Dickens 2002) and an aggregation pheromone have been found (Dickens 2000); the latter, most unusually, was produced by males (Dickens *et al.* 2002). In laboratory trials, it was found that both adults and larvae are attracted by the pheromone (Dickens *et al.* 2002, Hammock *et al.* 2007). As a result of these findings, pheromones were developed (Mori and Tashiro 2004, Babu and Chauhan 2009).

The plant volatiles were synthesized and used as a kairomone blend of (Z)-3-hexenyl acetate,(±)-linalool and methyl salicylate (Dickens 2002). Using kairomone-baited and non-baited pitfall traps, Martel *et al.* (2005) found that there were significantly more adults in the kairomone traps. They also compared kairomone-treated (formulated for slow release) and non-treated border rows of potatoes in a plot to act as trap plants, and compared the number of beetles, and egg masses in the center rows of the plots. There were significantly more CPB adults and egg masses in the kairomone -treated plants, and the plots with treated border rows had higher yield. This suggests the possibility of a new management technique.

Similar field trials were performed to determine the effect of a synthesized pheromone in pitfall traps, and when applied to border rows as a trap crop (Kuhar *et al.* 2006). Pitfall traps baited with pheromone trapped more beetles, but their efficacy was short-lived. When pheromone was applied to border rows to act as a trap crop, significantly more beetles and egg masses were found in the border rows as compared to untreated rows, resulting in less defoliation and greater potato yields. The pheromone performed better than the kairomone in the previous trials, again reinforcing the possibility of new management methods.

It is well established that CPB is attracted to the color yellow, especially at wavelengths between 550 and 580 nm, which potato plants also reflect (Zehnder and Speese 1987, Doring and Skorupski 2007, Otalora-Luna and Dickens 2011a). In laboratory trials to determine the attractiveness of light versus the aggregation pheromone, and the possibility of a synergism between the two cues, Otalora-Luna and Dickens (2011b) provided three choices: pheromone in the dark, pheromone with yellow light, and yellow light alone. Unexpectedly, male and female beetles were attracted to the pheromone in the dark, whereas in the presence of yellow light the pheromone was less attractive. Since the insects are diurnal, the authors suggest that the visual cues are predominant over olfactory, but they do not rule out the possibility of using pheromones to design more attractive traps especially for night catches.

Two pheromones for potato tuber moths *P. operculella* were identified and synthesized, and are primarily used to monitor populations of adults for correct timing of insecticide applications (Herman *et al.* 2005, Larrain *et al.* 2007) and in the trap-and-kill management strategy (Kroschel and Zegarra 2010). The Guatemalan potato moth *Tecia* (=*Scrobipalpopsis*) *solanivora* has had three pheromones identified. Various ratios of the pheromones were tested in the field for mating disruption; efficacy was tested by using a live calling female. There was almost 90% reduction in male captures, which indicated highly successful mating disruption (Ochoa 2005).

Pneumatic Removal

In pneumatic control, insects can be dislodged from plants with negative and/ or positive (blowing) air pressure and then killed by a system of turbines, or collected and killed upstream in a dedicated system of the blower. Reviews on pneumatic control have been published (for example, Khlelifi *et al.* 2001, Vincent and Boiteau 2001, Lacasse *et al.* 2001). Vacuuming provides a short-term control measure, and must be repeated to be effective. With certain pests, like CPB, which invade a field from the edge, vacuuming this area can be an effective control measure and has been recommended (Hoy *et al.* 2000).

In the 1990s, the first work was reported on the efficacy of vacuum machines. Boiteau *et al.* (1992) described a vacuum removal system that collected 40%, 27%, and 48% of small larvae, large larvae, and adults, respectively, and Lacasse *et al.* (1998) reported similar results of 24%, 58%, and 61% removal for first and second, third, and fourth instars, respectively. Khelifi *et al.* (2001) described several commercial machines especially designed to remove CPB, which are essentially large vacuum devices mounted on tractors and intended to remove pests by a combination of blowing to dislodge them then vacuuming to remove them from the plant. In the past decade many of these machines have additionally been used for other pests on crops other then potatoes: the Beetle Eater® is a three-bed device being used on asparagus against asparagus beetles (van den Broek 2010), the Bio-Collector® continues to be used primarily for CPB adults and larvae, and the Bug-Vac® is primarily used for *Lygus* on strawberries (Kuepper and Thomas 2009).

CONCLUSIONS

A wide variety of physical control measures are effectively being applied to protect potatoes from a number of direct and indirect pests. Due to the range of climatic conditions under which potatoes are grown, no one technique can or need be applied in all growing conditions. Unlike chemical management, which is broadly applied, physical control measures are directed at specific character-istics or behaviors of targeted pests: barriers which are insurmountable to pests that invade fields by walking/crawling, cultural practices to bury exposed tubers from invading larvae or adults, pneumatic control of pests that habitually "play dead" and fall to the ground when disturbed, insect exclusion screens which are directed at all pathogen vectors, solarization for certain soil pests, etc.

Seed potatoes present a unique problem because the pathogens discussed in this chapter ultimately accumulate in the potato tuber. It is crucially important to start production with pathogen-free seed material, as vectors are virtually ubiqui-tous and will result in major, uncontrollable outbreaks of various potato diseases. Therefore, seed potatoes should be isolated and grown only in areas where there are low vector populations. It has been definitively demonstrated that insecticides lead to greater transmission of aphid-transmitted non-persistent viruses in potatoes, although the situation *vis-à-vis* leafhoppers and psyllids is unknown.

Since all of the vectors known to date transmit pathogens faster than insecticides can act, the best protection against pathogen vectors is the use of insect exclusion screens, including floating row covers. While this may represent a slightly more expensive management technique, the necessity for pathogen-free seed material warrants its use. As discussed in this chapter, new physical control measures are being adapted and applied. These measures are very important tools in pest management, but more research on these relatively simple techniques could have vast consequences in terms of pesticide reduction and improved tuber yield and quality.

REFERENCES

Alyokhin, A., 2009. Colorado potato beetle management on potatoes: current challenges and future prospects. Fruit Veg. Cereal Sci. Biotechnol. 3, 10–19.

Anonymous, 2005. *Tecia solanivora*. Bull. OEPP.EPPO 35, 399–401.

Anonymous, 2006. Handbook for the Montreal protocol on substances that deplete the ozone layer – 7th edition. United Nations Environment Programme. http://ozone.unep.org/Publications/MP_Handbook/Section_1.1_The_Montreal_Protocol/

Anonymous, 2011. Data sheets on quarantine pests: *Premnotrypes spp.* (Andean), www.eppo.org/QUARANTINE/insects/Premnotrypes_latithorax/PREMSP_ds.pdf.

Babu, B.N., Chauhan, K.R., 2009. Enantioselective synthesis of (S)-3,7-dimethyl-2-oxo-6-octene-1,3-diol: a Colorado potato beetle pheromone. Tetrahedron. Lett. 50, 66–67.

Backus, E.A., 1988. Sensory systems and behaviors which mediate hemipteran plant-feeding: a taxonomic overview. J. Insect. Physiol. 34, 151–165.

Bizer, E., 1987. Sind schutznetze wirtschaftlich? Deutscher Gartenbau 12.

Boiteau, G., Misener, G.C., 1996. Response of Colorado potato beetles on potato leaves to mechanical vibrations. Can. Agric. Engin. 38, 223–227.

Boiteau, G., Misener, G.C., Singh, R.P., Bernard, G., 1992. Evaluation of a vacuum collector for insect pest control in potato. Am. Potato. J. 69, 157–166.

Boiteau, G., Peltier, Y., Misener, G.C., Bernard, G., 1994. Development and evaluation of a plastic trench barrier for protection of potato from walking adult Colorado potato beetle (Coleoptera: Chrysomelidae). J. Econ. Entomol. 87, 1325–1331.

Bomford, M.K., Vernon, R.S., 2005. Root weevil (Coleoptera: Curculionidae) and ground beetle (Coleoptera: Carabidae) immigration into strawberry plots protected by fence or portable trench barriers. Environ. Entomol. 34, 844–849.

Bomford, M.K., Vernon, R.S., Pats, P., 2000. Importance of collection overhangs on efficacy of exclusion fences for managing cabbage flies (Diptera: Anthomyiidae). Environ. Entomol. 29, 795–799.

Bonani, J.P., Fereres, A., Garzo, E., Miranda, M.P., Appezzato-Da-Gloria, B., Lopes, J.R.S., 2010. Characterization of electrical penetration graphs of the Asian citrus psyllid, *Diaphorina citri*, in sweet orange seedlings. Entomol. Exp. Appl. 134, 35–49.

Brust, G.E., 1994. Natural enemies in straw-mulch reduce Colorado potato beetle populations and damage in potato. Biol. Control 4, 163–169.

Budnik, K., Laing, M.D., da Graca, J.V., 1996. Reduction of yield losses in pepper crops caused by potato virus Y in KwaZula-Natal, South Africa using plastic mulch and yellow sticky traps. Phytoparasitica 24, 119–124.

Chauvin, L., Caromek, B., Kerlan, M.-C., Rulliat, E., Fournet, S., Chauvin, J.-E., Grenier, E., Ellisseche, D., Mugniery, D., 2008. Control of potato cyst nematodes *Globodera rostochiensis* and *Globodera pallida*. Cahiers Agric. 17, 368–374.

Clough, G.H., Rondon, S.I., DeBano, S.J., David, N., Hamm, P.B., 2010. Reducing tuber damage by potato tuberworm (Lepidoptera: Gelechiidae) with cultural practices and insecticides. J. Econ. Entomol. 103, 1306–1311.

Collantes, L.G., Raman, K.V., Cisneros, F.H., 1986. Effect of six synthetic pyrethroids on two populations of potato tuber moth, *Phthorimaea operculella* (Zeller) (Lepidoptera: Gelechiidae), in Peru. Crop Prot. 5, 355–357.

Davis, J.A., Radcliffe, E.B., 2008. The importance of an invasive aphid species in vectoring a persistently transmitted potato virus: *Aphis glycines* is a vector of potato leafroll virus. Plant Dis. 92, 1515–1523.

Denner, F.D.N., Millard, C.P., Wehner, F.C., 2000. Effect of soil solarisation and mould board ploughing on black dot of potato, caused by *Colletotrichum coccodes*. Potato Res. 43 295–201.

Dickens, J.C., 2000. Orientation of Colorado potato beetle to natural and synthetic blends of volatiles emitted by potato plants. Agric. Forest Entomol. 2, 167–172.

Dickens, J.C., 2002. Behavioral response of larvae of the Colorado potato beetle, *Leptinotarsa decemlineata* Say (Coleoptera: Chrysomelidae), to host plant volatile blends attractive to adults. Agric. Forest Entomol. 4, 309–314.

Dickens, J.C., Oliver, J.E., Hollister, B., Davis, J.C., Klun, J.A., 2002. Breaking a paradigm: male-produced aggregation pheromone for the Colorado potato beetle. J. Exp. Biol. 4, 309–314.

Difonzo, C.D., Ragsdale, D.W., Radcliffe, E.B., Gudmestad, N.C., Secor, G.A., 1996. Crop borders reduce potato virus Y incidence in seed potato. Ann. Appl. Biol. 129, 289–302.

Doring, T.F., Skorupski, P., 2007. Host and non-host leaves in the colour space of the Colorado potato beetle (Coleoptera: Chrysomelidae). Entomol. Gen. 29, 81–95.

Doring, T.F., Brandt, M., Hess, J., Finckh, M.R., Saucke, H., 2005. Effects of straw mulch on soil nitrate dynamics, weeds, yield and soil erosion in organically grown potatoes. Field Crops Res. 94, 238–249.

Duchesne, R.-M., Bernier, D., Jean, C., 1992. Utilization of a propane flamer in potato fields in Quebec. Am. Pot. J. 69, 578.

Duchesne, R.-M., Lague, C., Khelifi, M., Gill, J., 2001. Thermal control of Colorado potato beetle in. In: Vincent, C., Panneton, B., Fleurat-Lessard, F. (Eds.), Physical Control Methods in Plant Protection, Springer, Berlin, Germany, pp. 61–73.

FAOSTAT, 2011. Food and Agriculture Organisation of the United Nations Statistical Database. 2011. http://faostat.fao.org.

Fathi, S.A.A., 2011. Population density and life-history parameters of the psyllid *Bactericera nigricornis* (Forster) on four commercial cultivars of potato. Crop Protect 30, 844–848.

Foot, M.A., 1976. Laboratoy assessment of several insecticides against the potato tuber moth *Phthorimaea operculella* Zeller (Lepidoptera, Gelechiidae). New Zeal. J. Agric. Res. 29, 117–125.

Foot, M.A., 1979. Bionomics of the potato tuber moth, *Phthorimaea operculella* (Lepidoptera: Gelechiidae) at Pukekohe. NZ. J. Zool. 6, 623–636.

Foster, S.P., Harris, M.O., 1997. Behavioral manipulation methods for insect pest-management. Annu. Rev. Entomol. 42, 123–146.

Grafius, E., 1997. Economic impact of insecticide resistance in the Colorado potato beetle (Coleoptera: chrysomelidae) on the Michigan potato industry. J. Econ. Entomol. 90, 1144–1151.

Grubinger, V., 2011. Trapping Colorado potato beetles with plastic-lined trenches. University of Connecticut. www.hort.uconn.edu/ipm/veg/htms/colpbtl.htm2011.

Hammock, J.A., Vinyard, B., Dickens, J.C., 2007. Response to host plant odors and aggregation pheromone by larvae of the Colorado potato beetle on a servosphere. Arthropod-Plant Interact 1, 27–35.

Harrewijn, P., den Ouden, H., Piron, P.G.M., 1991. Polymer webs to prevent virus transmission by aphids in seed potatoes. Entomol. Exp. Appli. 58, 101–107.

Henne, D.C., Workneh, F., Rush, C.M., 2010. Movement of *Bactericera cockerelli* (Heteroptera: Psyllidae) in relation to potato canopy structure, and effects on potato tuber weights. J. Econ. Entomol. 103, 1524–1530.

Herman, T.J.B., Clearwater, J.R., Triggs, C.M., 2005. Impact of pheromone traps design, placement and pheromone blend on catch of potato tuber moth. NZ. Plant Prot. 58, 219–223.

Hooks, C.R.R., Fereres, A., 2006. Protecting crops from non-persistently aphid-transmitted viruses: A review on the use of barrier plants as a management too. Virus Res. 120, 1–16.

Hoy, C.W., Vaughn, T.T., East, D.A., 2000. Increasing the effectiveness of spring trap crops for *Leptinotarsa decemlineata*. Entomol. Exp. Appl. 96, 193–204.

Ioannidis, P.M., Grafius, E., Whalon, M.E., 1991. Patterns of insecticide resistance to azinphos-methyl, carbofuran, and permethrin in the Colorado potato beetle (Coleoptera: Chrysomelidae). J. Econ. Entomol 84 1471–1423.

Jermy, T., Szentesi, A., Horvath, J., 1988. Host plant finding in phytophagous insects: the case of the Colorado potato beetle. Entomol. Exp. Appl. 49, 83–98.

Johnson, J.M., Hough-Goldstein, J.A., Vangessel, M.J., 2004. Effects of straw mulch on pest insects, predators, and weeds in watermelons and potatoes. Environ. Entomol. 33, 1632–1643.

Kaplan, I., Dively, G.P., Denno, R.F., 2008. Variation in tolerane and resistance to the leafhopper *Empoasca fabae* (Hemiptera: Cicadellidae) among potato cultivars: Implications for action thresholds. J. Econ. Entomol. 101, 959–968.

Karim, M.R., Hanafi, M.M., Shahidulla, S.M., Rahman, A.H.M.A., Akanda, A.M., Khair, A., 2010. Virus free seed potato production through sprout cutting technique under net-house. African J. Biotechnol. 9, 5852–5858.

Katan, J., 1981. Solar heating (solarization of soil for control of soilborne pests). Annu. Rev. Phytopathol. 19, 211–236.

Kennedy, J.S., Day, M.F., Eastop, V.F., 1962. A conspectus of aphids as vectors of plant viruses. Commonwealth Institute of Entomology, London, UK, p. 114.

Khelifi, N., Lague, C., Lacasse, B., 2001. Pneumatic control of insects in plant protection. In: Vincent, C., Panneton, B., Fleurat-Lessard, F. (Eds.), Physical Control Methods in Plant Protection, Springer, Berlin, Germany, pp. 261–269.

Khelifi, N., Lague, C., de, Ladurantaye, Y., 2007. Physical control of Colorado potato beetle: A review. Appl. Engin. Agric. 23, 557–569.

Kroschel, J., Zegarra, O., 2010. Attract-and-kill: a new strategy for the management of the potato tuber moths *Phthorimaea operculella* (Zeller) and *Symmetrischema tangolias* (Gyen) in potato: laboratory experiments towards optimizing pheromone and insecticide concentration. Pest Manag. Sci. 58, 1029–1037.

Kroschel, J., Alcazar, J., Poma, P., 2009. Potential of plastic barriers to control Andean potato weevil *Premnotrypes suturicallus*. Kuschel. Crop. Prot. 28, 466–476.

Kuepper, G., Thoman, R., 2009. Bug vacuums for organic crop protection. National Sustainable Agriculture Information Service. www.agrisk.umn.edu/cache/ARL02954.htm.

Kuhar, T.P., Mori, K., Dickens, J.C., 2006. Potential of a synthetic aggregation pheromone for integrated pest management of Colorado potato beetle. Agric. Forest. Entomol. 8, 77–81.

Lacasse, B., Lague, C., Khelifi, M., Roy, P.-M., 1998. Pneumatic and thermal control of Colorado potato beetle. Can. Agric. Engin. 40, 273–280.

Lacasse, B., Lague, C., Roy, P.-M., Khelifi, M., Bourassa, S., Cloutier, C., 2001. Pneumatic control of Colorado potato beetle. In: Vincent, C., Panneton, B., Fleurat-Lessard, F. (Eds.), Physical Control Methods in Plant Protection, Springer, Berlin, Germany, pp. 282–293.

Lachman, J., Hamouz, K., Hejtmankova, A., Dudjak, J., Orsak, M., Pivec, V., 2003. Effect of white fleece on the selected quality parameters of early potato (*Solanum tuberosum* L.) tubers. Plant Soil Environ. 49, 370–377.

Lague, C., Khelifi, M., Gill, J., Lacasse, B., 1999. Pneumatic and thermal control of Colorado potato beetle. Can. Agric. Engin. 41, 53–57.

Larrain, P.S., Guillon, M., Kalazich, J.B., Grana, F.S., Vasquez, C.R., 2007. Efficacy of different rates of sexual pheromone of *Phthorimaea operculella* (Zeller) (Lepidoptera: Gellechiidae) in males of potato tuber moth captures. Agric. Tec. (Chile) 67, 431–436.

Maletta, M., Henninger, M., Holmstrom, K., 2006. Potato leafhopper control and plastic mulch culture in organic potato production. HortTechnol. 16, 199–204.

Martel, J.W., Alford, A.R., Dickens, J.C., 2005. Synthetic host volatiles increase efficacy of trap cropping for management of Colorado potato beetle, *Leptinotarsa decemlineata* (Say). Agric. Forest Entomol. 7, 71–78.

Medina, R.F., Rondon, S.I., Reyna, M., Dickey, A.M., 2010. Population structure of *Phthorimaea operculella* (Lepidoptera: Gelechiidae) in the United States. Environ. Entomol. 39, 1037–1042.

Miles, P.W., 1999. Aphid saliva. Biol. Rev. 74, 41–85.

Misener, G.C., Boiteau, G., McMillan, L.P., 1993. A plastic-lining trenching device for the control of Colorado potato beetle. Beetle Excluder. Am. Pot. J. 70, 903–908.

Mori, K., Tashiro, T., 2004. Useful reactions in modern pheromone synthesis. Curr. Organic Synthesis 1, 11–29.

Moyer, D.D., Derksen, R.C., McLeod, M.J., 1992. Development of a propane flamer for Colorado potato beetle control. Am. Pot. J. 69, 599.

Munyaneza, J.E., 2010a. Emerging leafhopper-transmitted phytoplasma diseases of potato. Southwestern Entomol. 35, 451–456.

Munyaneza, J.E., 2010b. Psyllids as vectors of emerging bacterial diseases of annual crops. Southwestern Entomol. 35, 471–477.

Munyaneza, J.E., Jensen, A.S., Hamm, P.B., Upton, J.E., 2008. Seasonal occurrence and abundance of beet leafhopper in the potato growing region of Washington and Oregon Columbia Basin and Yakima Valley. Am. J. Pot. Res. 85, 77–84.

Nault, L.R., 1997. Arthropod transmission of plant viruses: a new synthesis. Ann. Entomol. Soc. America 90, 521–541.

Ochoa, C.F.B., 2005. Pheromone mediated communication disruption in Guatemalan potato moth, Tecia solanivora Povolny. Licentiate Thesis Faculty of Landscape Planning, Horticulture and Agricultural Science, Department of Crop Science. Swedish University of Agricultural Sciences, Alnarp, Sweden. http://pub.epsilon.slu.se/1013/1/FelipeBosa.pdf.

Otalora-Luna, F., Dickens, J.C., 2011a. Spectral preference and temporal modulation of photic orientation by Colorado potato beetles on a servosphere. Entomol. Exp. Appl. 138, 93–103.

Otalora-Luna, F., Dickens, J.C., 2011b. Multimodal stimulation of Colorado potato beetle reveals modulation of pheromone response by yellow light. PLoS One 6 (6), 1–e20990. doi:10.1371/journal.pone.0020990.

Pelletier, Y., McLeod, C.D., Bernard, G., 1995. Description of sublethal injuries caused to the Colorado potato beetle (Coleoptera: Chrysomelidae) by propane flamer treatment. J. Econ. Entomol. 88, 1203–1288.

Piron, P.G.M., 1986. New aphid vectors of Potato virus Y^N. Netherlands J. Plant Pathol. 92, 223–229.

Radcliffe, E.B., Ragsdale, D.W., 2002. Aphid-transmitted potato viruses: the importance of understanding vector biology. Am. J. Pot. Res. 79, 353–386.

Radcliffe, E.B., Ragsdale, D.W., Flanders, K.L., 1993. Management of aphids and leafhoppers. In: Rowe, R.C. (Ed.), Potato Health Management, APS Press, St Paul, MN, pp. 117–126.

Ragsdale, D.W., Radcliffe, E.B., Difonzo, C.D., 2001. Epidemiology and field control of PVY and PLRV. In: Loebenstein, G., Berger, P., Brunt, A.A., Lawson, R. (Eds.), Viruses and Virus-like

Diseases of Potatoes and Production of Seed-Potatoes, Kluwer, Wageningen, The Netherlands, pp. 237–270.

Richards, H.L., 1973. Psyllid yellows of the potato. J. Agric. Res. 46, 189–216.

Richardson, M.E., Rose, D.J.W., 1967. Chemical control of potato tuber moth, *Phthorimaea opercullella* (Zell). Rhodesia. Bull. Entomol. Res. 57, 271–278.

Rondon, S.I., 2010. The potato tuberworm: a literature review of its biology, ecology and control. Am. J. Pot. Res. 87, 149–166.

Roush, R.T., Hoy, C.W., Ferro, D.N., Tingey, W.M., 1990. Insecticide resistance in the Colorado potato beetle (Coleoptera: Chrysomelidae): influence of crop rotation and insecticide use. J. Econ. Entomol. 83, 315–319.

Saucke, H., Doring, T.F., 2004. Potato virus Y reduction by straw mulch in organic potatoes. Ann. Appl. Biol. 144, 237–355.

Schmera, D., Szentesi, A., Jermy, T., 2006. Within field movement of overwintered Colorado potato beetle: a patch-based approach. J. Appl. Entomol. 131, 34–39.

Secor, G.A., Rivera, V.V., Abah, J.A., Lee, I.M., Clover, G.R.G., Liefting, L.W., Li, X., de Boer, S.H., 2009. Association of "Candidatus Liberibacter solanacearum" with zebra chip disease of potato established by graft and psyllid transmission, electron microscopy, and PCR. Plant Dis. 93, 574–583.

Shelton, A.M., Badenes-Perez, F.R., 2006. Concepts and applications of trap cropping in pest management. Annu. Rev. Entomol. 51, 285–308.

Shelton, A.M., Wyman, J.A., 1979. Potato tuberworm damage to potato grown under different irrigation and cultural practices. J. Econ. Entomol. 72, 261–264.

Smith, J.G., 1969. Some effects of crop background on populations of aphids and their natural enemies on Brussels sprouts. Ann. Appl. Biol. 63, 326–329.

Stoner, K.A., 1993. Effects of straw and leaf mulches and trickle irrigation of the abundance Colorado potato beetle (Coleoptera: Chrysomelidae) on potato in Connecticut. J. Entomol. Sci. 28, 393–403.

Sylvester, E.S., 1989. Viruses transmitted by aphids. In: Minks, A.K., Harrewijn, P. (Eds.), Aphids: Their biology, Natural Enemies, and Control, Vol. 2C, Elsevier Publishing, New York, NY, pp. 65–87.

Szendrei, Z., Kramer, M., Weber, D.C., 2009. Habitat manipulation in potato affects Colorado potato beetle dispersal. J. Appl. Entomol. 133, 711–719.

Tagu, D., Klingler, J.P., Moya, A., Simon, J.-C., 2008. Early progress in aphid genomics and consequences for plant-aphid interactions studies. Mol. Plant-Microbe Interact. 21, 701–708.

Triki, M.A., Priou, S., El Mahjoub, M., 2001. Effects of soil solarization on soil-borne populations of *Pythium aphanidermatum* and *Fusarium solani* and the potato crop in Tunisia. Pot. Res. 44, 271–279.

Van den Broek, R., 2010. "Beetle Eater" successful in asparagus. Applied Plant Research. Wageningen University and Research Centre, Wageningen, The Netherlands. www.ppo.wur.nl/UK/newsagenda/archive/news/2010/Beetleeater_090410.htm.

Vincent, C., Boiteau, G., 2001. Pneumatic control of agricultural insect pests. In: Vincent, C., Panneton, B., Fleurat-Lessard, F. (Eds.), Physical Control Methods in Plant Protection, Springer, Berlin, Germany, pp. 270–281.

Visser, J.H., Nielsen, J.K., 1977. Specificity in the olfactory orientation of the Colorado beetle, *Leptinotarsa decemlineata*. Entomol. Exp. Appl. 21, 14–22.

Voss, R.H., Ferro, D.N., 1990. Phenology of flight and walking by Colorado potato beetle (Coleoptera: Chrysomelidae) adults in western Massachusetts. Environ. Entomol. 19, 117–122.

Weber, D.C., Ferro, D.N., 1994. Movement of overwintered Colorado potato beetles in the field. J. Agric. Entomol. 11, 17–27.

Weber, D.C., Ferro, D.N., Buonaccorsi, J., Hazzard, V., 1994. Disrupting spring colonization of Colorado potato beetle to non-rotated potato fields. Entomol. Exp. Appli. 73, 39–50.

Weintraub, P.G., Beanland, L., 2006. Insect vectors of phytoplasmas. Annu. Rev. Entomol. 51, 91–111 plus supplemental table.

Weintraub, P.G., Berlinger, M.J., 2004. Physical control in greenhouses and field crops. In: Horowitz, A.R., Ishaaya, I. (Eds.), Novel Approaches to Insect Pest Management, Springer, Berlin, Germany, pp. 301–318.

Weisz, R., Smilowitz, Z., Fleischer, S., 1996. Evaluating risk of Colorado potato beetle (Coleoptera: Chrysomelidae) infestation as a function of migratory distance. J. Econ. Entomol. 89, 435–441.

Zehnder, G., Speese III, J., 1987. Assessment of color response and flight activity of *Leptinotarsa decemlineata* (Say) (Coleoptera: Chrysomelidae) using window flight traps. Environ. Entomol. 16, 1199–1202.

Zehnder, G.W., Hough-Goldstein, J., 1990. Colorado potato beetle (Coleoptera: Chrysomelidae) population development and effect on yield of potatoes with and without straw mulch. J. Econ. Entomol. 83, 1982–1987.

Chapter 18

Cultural Control and Other Non-Chemical Methods

Beata Gabryś[1] and Bożena Kordan[2]

[1]*Department of Botany and Ecology, University of Zielona Góra, Zielona Góra, Poland,*
[2]*Department of Phytopathology and Entomology, University of Warmia and Mazury in Olsztyn, Olsztyn, Poland*

INTRODUCTION

Cultural control is the manipulation of the agroecosystem in order to make the cropping system less friendly to the establishment and proliferation of pest populations (Dufour 2001).

The potato *Solanum tuberosum* L. is one of the principal food crops, and a high level of production must be maintained to meet the growing demand of the world population. Additionally, climate change is likely to affect agricultural pest management (Strand 2000, Haverkort and Verhagen 2008). Global warming favors the development of certain insects on potato fields, especially those that develop in the soil and cause damage to underground parts of potatoes: high temperatures and periods of dry weather that occur during the growing season accelerate the development of Elateridae (Coleoptera), Noctuidae (Lepidoptera), and the Colorado potato beetle *Leptinotarsa decemlineata* Say (Coleoptera: Chrysomelidae), so now more generations a year can develop than could in the past (Kapsa 2008).

The development of insect pest management strategies for potato has long been based on the final substitution of insecticides by alternative methods (Boiteau 2010). In a theoretical model of environmentally- and human-friendly crop production, four phases of pest management can be distinguished (Wyss *et al.* 2005, Kühne 2008). The first two basic phases are cultural practices and vegetation management to enhance natural enemy impact and exert direct effects on pest populations. The third phase requires the release of biological control agents, and the fourth, last-resort phase requires the use of approved insecticides and of mating disruption (Wyss *et al.* 2005, Boiteau 2010). The first and the second phases correspond with the primary strategy of contemporary cultural control, which is maintaining and increasing the biological diversity in the farm

Insect Pests of Potato. http://dx.doi.org/10.1016/B978-0-12-386895-4.00018-1

system by the management of the abiotic and biotic environment of the crop. The manipulation of abiotic conditions includes site selection, soil practices including irrigation and fertilizer management, and the use of mulches, row covers, etc. Manipulation of the biotic environment embraces various aspects of crop rotation, intercropping, trap crops, companion planting, and the use of semiochemicals, including antifeedants.

MANAGEMENT OF ABIOTIC CONDITIONS

Abiotic factors are all non-living chemical and physical components of the environment that affect the survival or reproductive success of living organisms. Abiotic factors that have a bearing on potato growth include temperature, solar radiation, day length, moisture availability, and soil nutrients (MacKerron and Waister 1985, Haverkort and Verhagen 2008). Potato crop development, including sprout growth rate, emergence, and leaf area development, depends on temperature and dry matter accumulation, the latter being a function of the amount of solar radiation intercepted by the crop and dry matter distribution between the various organs. For example, short days and low temperatures reduce branching and the number of leaves per stem, but increase the size of individual leaves; high temperatures increase specific leaf area but reduce photosynthesis; long days and high temperatures delay stolon and tuber initiation, and delay and reduce partitioning of dry matter to the tubers, which results in low harvest indices. However, a delay of tuber formation may stimulate final yield, provided that the growing season is sufficiently long to profit from the increased duration of ground cover (Struik and Ewing 1995). If water stress occurs (i.e., there is less water available than needed for optimal growth), the plants are lower in height and the canopy coverage of soil is reduced due to the diminished leaf area and foliage (Ojala et al. 1990). Tuber yields can be reduced by water stress imposed at any time during the growing season (Adams and Stevenson 1990, Jeferries 1995).

Potato is best grown at places where daily temperatures are above 5°C and below 21°C, with sufficient availability of water (Vos and Haverkort 2007). The variability in meteorological conditions influences the long-term effect of different soil tillage and fertilization regimes on potato yields, with the fertilization-induced yield differences manifested most noticeably in years with favorable growing conditions. A warm spring brings higher yields but precipitation during the same period is negatively correlated with the crop, whereas the positive influence of precipitation is expressed after flowering (Saue et al. 2010).

Abiotic conditions that are associated with the climate in a particular region of the world are difficult to manipulate in the open field environment. Nevertheless, human activities associated with the farming system of potatoes may contribute to the economic optimization of the potato yield not only by improving the plant growth but also by eliminating or restraining the populations of insect pests. For example, farm site selection, crop isolation, manipulation of planting or harvest time, or the use of mulches may make the crop unavailable to pests in space and time, and enhancement of soil quality and fertility may alter the crop's susceptibility to pests (Zehnder *et al.* 2007).

Effect of Site Selection, and Planting and Harvest Times

The location of the potato field should be as unsuitable as possible for insect pests (Boiteau 2009). This can be accomplished by simply modifying the location of the crop in space and time. The spatial separation of the crop may be gained by increasing the distance between crops and sources of colonizing pests, or separating them by various barriers (vegetational or physical), or by avoiding cultivation in areas where a given pest species occurred in abundance during the previous season. Temporal isolation can be achieved by selecting the planting and harvest dates so as to escape heavy losses due to pest feeding.

The setting of the potato field is especially important to prevent aphid-borne virus spread. The minimum separation from virus sources depends on local conditions and the virus species involved. A distance of 400 m to 5 km is probably sufficient for the reduction of potato virus Y (PVY) spread, but a much greater distance (*ca.* 32 km) may be required in the case of potato leafroll virus (PLRV), as PVY is a short-lived, non-persistent virus easily discharged by the vector during probing while PLRV is a persistent, circulative one and remains in the vector organism for all of its life (Radcliffe and Ragsdale 2002, Radcliffe *et al.* 2007; see also Chapter 10 of this volume).

The development and feeding habits of herbivorous insects are precisely synchronized with the development of their host plants, which is one of the aspects of plant–herbivore co-evolution. Therefore, if the planting and harvest time can be modified in relation to the natural situation, the damage to crops may be reduced. However, the potato planting time depends on local climatic and agronomic conditions and on economic factors, which may limit the use of this method (Alyokhin 2009).

In the case of potato tuberworm *Phthorimaea operculella* (Zeller) (Lepidoptera: Gelechiidae), foliar damage to the potato crop usually does not result in significant yield losses, although the tuberworm larvae prefer green foliage to tubers for feeding and oviposition. However, when foliage starts to decline, the caterpillars are forced to go into the ground. Therefore, the greatest risk of tuber damage occurs immediately before harvest while the crop is left in the field prior to digging, and the longer the potatoes are left in the field after the vines die, the greater the likelihood of tuber infestation (Rondon 2007).

Late as well as early planting is considered in management of Colorado potato beetle. Late planting causes late plant emergence, so the early emerging beetles are forced to migrate from the field because of food unavailability. Early planting and harvest might also reduce the impact of the second generation because the crop can be removed before the emergence of larvae. In addition, late-planted fields may act as sinks for beetles emigrating from earlier harvested fields looking for feeding and overwintering sites (Baker *et al.* 2001). The harvest date and tillage at different times between crop production seasons do not affect overwintering Colorado potato beetle survival significantly (Nault *et al.* 1997).

Bringing forward tuber-lifting dates to the middle of August results in significantly lower wireworm- (the larval stages of click beetles (Coleoptera: Elateridae)) induced tuber losses compared with the middle of September. This is probably due to the fact that the incidence of tuber damage increases in the second half of August, irrespective of wireworm abundance (Erlichowski 2010). Indeed, Schepl and Paffrath (2005) found that 4-week acceleration of the harvest may cause a 31–64% reduction in tuber damage. Early harvesting can be recommended if tuber skin is sufficiently suberized and if cooling facilities are available for the tubers (Neuhoff et al. 2007).

Sowing and planting dates may appear very important in the management of aphid (Hemiptera: Aphididae) infestation and especially in the incidence of aphid-borne viruses. Early planting can be a useful strategy if the vector species does not begin colonization until late in the growing season (Radcliffe and Ragsdale 2002). Saucke and Döring (2004) found that the incidence of PVY decreased when the phase of early crop emergence coincided with low aphid spring flight activity. However, this method of prevention has to be considered in relation to local fluctuations in the aphid (mainly the peach potato aphid *Myzus persicae* [Sulz.]) population, especially aphid flight activity (Wratten *et al.* 2007). Moreover, in many northern temperate production areas, the duration of the growing season is the limiting factor.

The combination of spatial and temporal isolation of potato crops can be achieved by **crop rotation**. This routine practice of growing a series of dissimilar types of crops in the same area in sequential seasons has traditionally been used to maintain and improve soil health and fertility (Nelson *et al.* 2009, Boiteau 2010, Mohr *et al.* 2011). Nowadays, crop rotation is also used for the cultural management of pests and diseases that become established in the soil over time. For example, crop rotation is crucial to the control of the Colorado potato beetle, which overwinters as adults in potato field margins or surrounding woodlands; this was shown by Wright (1984), who found that rotation for 1 year to a non-host grain crop (rye or wheat) was sufficient to reduce adult *L. decemlineata* densities by 70–95% in the following year's potato crop. The timing of adult beetle colonization, population densities, and early-season defoliation were related closely to how isolated the fields were from the previous year planting. Even short distances of 0.3–0.9 km between rotated locations were sufficient to reduce Colorado potato beetle densities and the necessity to apply insecticides by 50% (Weisz *et al.* 1994). Weisz *et al.* (1996) concluded that beetle infestation of a new potato planting is negatively correlated with distance to all potato fields from the previous growing season. Rotation may delay the colonization of fields by spring-emerging Colorado potato beetle from 1 to 3 weeks, due to the time needed for the beetles to locate fields after emerging and leaving remote overwintering sites (Baker *et al.* 2001). "Risk maps" can be drawn to show which potato fields should be rotated out of the area where potatoes were normally grown to substantially reduce the risk of infestation (Hoy *et al.* 2000). Finally, rotated fields also require fewer insecticide applications, which delays the evolution of resistance in Colorado potato beetle (Baker *et al.* 2001).

Crop rotation is an important tool in controlling wireworms. Wireworms tend to increase rapidly in red and sweet clover, and small grains (particularly barley and wheat), but a clean stand of alfalfa that is maintained for 3–4 years tends to reduce wireworm numbers because extreme dryness of soil is harmful to most wireworms, and alfalfa serves as a soil-drying crop. Moreover, if alfalfa fields are allowed to dry during the season in which they are out of production, further reduction in wireworm populations can be expected (Berry *et al.* 2000).

Effect of Soil Tillage

The conventional approach to potato farming system uses autumn ploughing *ca.* 20 cm deep. The potato crop is usually grown in rows *ca.* 75 cm apart and on ridges about 20–25 cm above soil surface level. One of the main objectives of tillage is to keep and maintain a high level of clod in soil, so that the roots can penetrate and develop better (Carter and Sanderson 2001, Ghazavi *et al.* 2010). Soil surface configuration, such as ridge tillage, may allow manipulation of soil water content. For example, ridge till technology can not only overcome the constraints of water logging, but can also capture and store water in the furrows during periods of low rainfall. Soil and nutrient losses are reported to be as much as 68% less under ridge tillage than conventional tillage, and ridge tillage in the fall may increase soil temperature early in the growing season and accelerate crop emergence (Essah and Honeycutt 2004). Due to the form of the ridge and the spatial variation of root distribution, both vertical and horizontal movement of water and nutrients occurs in the soil. It was shown that for identical environmental conditions, nitrogen uptake by potatoes was higher in sandy clay loam than in loamy sand, as sandy clay loam has higher water content at the same pressure head (De Willigen *et al.* 1995).

In terms of plant protection, tillage can be beneficial because it may disrupt the life cycle of insect pests and can expose the soil-living stages to predators and the physical environmental factors. However, different tillage practices may have different effects depending on the specificity of the insect biology.

In the case of wireworms, which spend their life as much as 0.3–1.5 m below ground level for 2–5 years (Andrews *et al.* 2008), repeated disturbance of the soil decreases their populations both by direct injury and by exposure to desiccation or attack by birds (Seal *et al.* 1992). Wireworms are very sensitive to soil moisture: drying of the upper soil layers in combination with high temperatures causes the downsoil migration. Therefore, cultivation is likely to be most effective when wireworms are active in the upper layers of the soil profile (i.e., 10–20 cm), which occurs at *ca.* 13°C. In the UK, for example, this means that autumn ploughing followed by disking will have more effect on reducing wireworm populations than cultivation in February or March (Parker and Howard 2001).

In the case of the Colorado potato beetle, the conventionally tilled crop (tomato *Lycopersicon esculentum* Mill.) had a more abundant beetle population than a non-tilled one in both rotated and non-rotated fields, probably because of the earlier colonization of overwintered adults. In conventionally tilled plots, this resulted in higher egg mass densities and subsequent infestation of first-generation larvae and adults. Moreover, in treatments where fenvalerate was applied to control Colorado potato beetle populations above economic thresholds, four spray applications were required in conventionally tilled plots, compared with two applications in non-tilled tomatoes (Zehnder and Linduska 1987). In another experiment, where tomatoes were grown in a reduced tillage system utilizing rye (*Secale cereale* [L.]) as a cover crop, colonization by newly emerged adult Colorado potato beetles in the spring was significantly more rapid in conventionally tilled than in reduced-tillage plots. Conventionally tilled plots had significantly higher densities of egg masses, larvae, and second-generation adult Colorado potato beetles, which was attributed to the presence of rye residue in the reduced-tillage plots. Eventually, the reduced-tillage plots sustained less defoliation than conventionally tilled plots and had higher yields of ripe fruit (Hunt 1998).

The soil provides an environment for a wide diversity of predatory arthropods, mainly the ground beetles (Coleoptera: Carabidae) and spiders (Arachnida). Ploughing the soil may affect their survival directly by causing mortality, and may also have indirect effects by modifying the habitat and the availability of prey. Generally, the larger species are more vulnerable to soil cultivations than the smaller ones. However, the response of individual species varies due to their species-specific characteristics, so the overall abundance of soil predators may not differ in consequence of ploughing although the species spectrum of this group may change (Holland 2004).

Effect of Soil Moisture

Soil moisture management (soil drying, soil flooding, or alternation of these) is the most frequently considered technique among the preventive cultural methods that are carried out before potato is planted, especially against wireworms. Wireworms are highly responsive to soil moisture and temperature (Parker and Howard 2001). However, the effect of these practices depends on the wireworm species, soil type, and temperature. Continuous or alternate flooding appears effective for control of *Melanotus communis* (Gyllenhal) and *Conoderus* sp., with the minimal effective continuous flooding period of 6 weeks (Genung 1970). The dusky wireworm, *Agriotes obscurus* (L.), and the lined click beetle, *A. lineatus* (L.), submerged at high temperatures died more quickly than those submerged at low temperatures, and wireworms in flooded Delta soil died more quickly than those in flooded Agassiz soil. Soil analysis suggests that soil salinity may affect the effectiveness of flooding as a control strategy. Flooding in fall or summer (higher temperatures) would likely provide more effective control of wireworm populations than flooding in winter (van Herk and Vernon 2006). However, it must be kept in mind that potato responds negatively to variations in water supply. Over-irrigation favors disease, and leads to nitrate leaching, sediment, and nutrient losses (Shock *et al.* 2007). Too much water may cause reduced root development and rotting of the newly formed tubers, and infection with late blight *Phytophthora infestans* (Mont) De Bary; excessive variation in soil moisture, especially water after a prolonged drought, may affect tuber quality due to "second growth" (Haverkort 1982).

Soil moisture is also important for potato tuber moth infestation. Female moths prefer dry soil for oviposition, and the survival of larvae increases with decreasing soil moisture content. The density of adults is higher in relatively dry sandy soil than in moist loess soil. Also, tuber moth larvae on foliage in the field margins are more abundant than in the center, probably because plants on the edges of the field are more exposed to wind and solar radiation, leading to drier conditions than in the field center. Moreover, infested tubers in loess may support more larvae than those in sand, possibly because cracks in loess soil make the tubers accessible to more larvae (Coll *et al.* 2000).

Effect of Mulches

In crop rotation systems, potato farming generally uses intensive tillage throughout the cropping period and produces low levels of crop residue in the potato year, both of which are associated with soil degradation processes: erosion and leaching of nitrates (Carter *et al.* 2005). The application of mulches is one of the most effective soil erosion prevention methods. Essentially, a mulch is a protective cover placed over the soil to retain moisture, reduce erosion, provide nutrients, and suppress weed growth. Different materials are applied, including organic residues such as straw of various origins, compost, plastic, gravel, etc. Organic mulches are used especially in organic farming to add organic matter to the soil and to increase soil moisture-holding capacity and reduce soil temperature. A number of studies have investigated the effects of different mulches on soil properties, potato harvest, and the occurrence of diseases. Zehnder and Hough-Goldstein (1990) found that soil temperature and moisture conditions were more favorable for potato plant growth in Virginia under straw mulch than in bare ground (no mulch) plots. Final tuber yields were significantly greater in mulched plots (with and without insecticides) compared with plots without mulch. The use of organic mulches after the potato harvest presented a practical form of conservation tillage for potato rotations (Carter and Sanderson 2000). The risk of undesirable post-harvest nitrogen leaching was significantly reduced due to the immobilization of nitrate after harvest, and soil erosion was reduced by more than 97% in a rain simulation experiment (Döring *et al.* 2005). When soil temperature is insufficient, plastic and straw mulches enhance tuber yield (Kar and Kumar 2007, Wang *et al.* 2011a). During a fallow period, a mulch can reduce soil desiccation (Wang *et al.* 2011b). Plots with straw mulch generally have lower soil temperatures and higher soil moisture than control (weedy, no straw) plots. Moreover, when straw was applied at planting the weeds were suppressed, whereas straw applied 4 weeks after planting had less effect on weeds (Johnson *et al.* 2004).

Studies have shown that the application of mulches can suppress some insect pests (mainly Colorado potato beetle and aphids), probably through a combination of effects involving migration, overwintering, host-finding ability, and increased predation from natural enemies.

In the case of the Colorado potato beetle, the use of mulches has a detrimental effect on various aspects of its biology, especially on survival during the vegetative period and at overwintering sites. In potato fields where wheat straw mulch was placed, the numbers of second, third, and fourth instars of first-generation and all instars of second-generation *L. decemlineata* were significantly lower than in non-mulched plots. This was attributed to a significant increase in the number of soil predators, mainly coccinellids and chrysopids, that began in mulch plots approximately 2–3 weeks after straw was placed in the field. As a result of heavy predation, mulched plots suffered 2.5 times less defoliation than non-mulched plots and, consequently, tuber yield was approximately 35% greater in mulched plots than in non-mulched ones (Brust 1994). Straw mulch reduced the density per square meter of adults and large larvae in plots without beetle management, so defoliation was lower and leaf area and ground cover were increased in mulched subplots (Stoner *et al.* 1996). Mulching with wheat or rye straw may reduce the Colorado potato beetle's ability to locate potato fields, and the mulch creates a microenvironment that favors its predators. In the first half of the season, soil predators – mostly ground beetles – climb potato plants to feed on second- and third-instar larvae of the Colorado potato beetle. In the second half of the season, lady beetles and green lacewings are the predominant predators, feeding on eggs and on first and second instars. Mulched plots supported greater numbers of predators in comparison to non-mulched plots, resulting in significantly less defoliation by Colorado potato beetle; in consequence, the tuber yields were increased by a third (Brust 1994). Barley straw mulch is significantly preferred to birch sawdust, milled peat, and black plastic mulches by the generalist predators *Pterostichus vulgaris* (L.), *P. niger* (Bonelli), *Carabus nemoralis* (Müll.), and *Harpalus pubescens* (Müll.) (Coleoptera: Carabidae) (Arus *et al.* 2011). Interestingly, potatoes with straw at planting had more colonizing Colorado potato beetle adults than non-mulched potatoes, but the subsequent Colorado potato beetle egg masses and larval numbers were not higher in this treatment, possibly because of the higher numbers of predators in these plots as assessed by pitfall trapping (Johnson *et al.* 2004). However, the impact of predators in mulched vs. non-mulched potatoes depends on the predator species. Szendrei and Weber (2009) studied the effect of *Lebia grandis* (Hentz) (Coleoptera: Carabidae) and *Coleomegilla maculata* (DeGeer) (Coleoptera: Coccinellidae) on Colorado potato beetle in potato fields with and without rye mulch. They found that the two predator species responded in opposing manners to the habitat manipulation treatment in potato fields: on average, 35% of all *C. maculata* but 85% of all *L. grandis* collected over two field seasons were found in tilled plots vs. rye mulched plots, but neither predator was influenced significantly by the presence of rye mulch in the field cage experiment. Nevertheless, *C. maculata* eliminated more (but not significantly more) prey in the rye-mulched than in the tilled treatment. *C. maculata* was frequently observed scurrying along rye stalks, so the presence of stalks might have had a positive behavioral or physiological effect (Szendrei and Weber 2009).

Mulching has no significant effect on beetle migration within the potato field either during the vegetative period or before overwintering (Brust 1994, Hoy *et al.* 1996). Generally, the numbers of overwintered adult beetles, egg masses, and larvae are significantly lower in plots with straw mulch compared with those without (Zehnder and Hough-Goldstein 1990), but the mulch depth has no impact on overwintering depth of beetles in the soil or average date of emergence in the spring (Hoy *et al.* 1996). However, what happens to a mulch during winter is important. The removal of mulch covers or snow over a mulch rapidly depresses soil temperatures at all depths. In the 0–15 cm soil strata, where most of the adults overwinter, temperatures may drop from 0 to −11.7°C, whereas in undisturbed plots the temperature may remain close to 0°C. As a result, adult survival may be significantly higher in snow-covered, non-mulched plots and mulched habitats (26%) than in disturbed habitats (7%). Apparently, thermal shock may increase the overwintering mortality of Colorado potato beetle; direct disturbance of overwintering habitats could be achieved with mulching/unmulching (Milner *et al.* 1992).

Finally, in the fields where mulches are used it is possible to reduce the number of insecticide applications, which was the case in the study by Zehnder and Hough-Goldstein (1990): in plots treated with insecticides, six spray applications were required to control Colorado potato beetle populations above economic thresholds in plots without mulch, compared with two applications in plots with mulch.

In the case of aphids, mainly the peach-potato aphid *M. persicae*, the direct effect of mulches is manifested primarily in the disruption of host-plant location by the winged morphs, especially early in the season (Wratten *et al.* 2007). The effectiveness of mulches depends on aphid response to color and light reflectance. According to Žanić (2009), green mulch was found to be the most attractive to *M. persicae*, black and clear mulches alternated in attractiveness for *M. persicae* during the season, while the overall seasonal number of *M. persicae* was lower on black, brown, and clear mulches than on green and white mulches. According to Adlerz and Everett (1968), yellow and orange mulches attracted *M. persicae*, while aluminum and silver mulches repelled green peach aphids.

Aluminum mulch significantly reduced virus transmission by *M. persicae* on tomato, which was attributed to the increased reflectance of UV light by that mulch (Kring and Schuster 1992). On the other hand, significantly greater aphid fecundity was demonstrated on plants grown through aluminum-coated construction paper than on plants grown on bare soil. Higher temperatures and host-plant physiology were factors modified by the mulch, and could have resulted in larger aphid populations on plants grown over a reflective surface as the season progressed (Zalom 1981). The total number of aphids, and especially the number of winged aphids, was reduced and the degree of PVY was distinctly lower in potatoes with straw mulch as compared to the crop without mulch (Heimbach *et al.* 2002).

Effect of Fertilizers and Other Soil Amendments

Potato demands a high level of soil nutrients due to its relatively poorly developed and shallow root system in relation to yield (Elbordiny and Gad 2008). Mineral nutrition, especially with the use of natural fertilizers such as manure, results in good plant growth and condition. For example, tuber yields were higher in manure-amended plots as compared to plots receiving full rates of synthetic fertilizers but no manure (Alyokhin *et al.* 2005).

Organic soil management has been associated with plant characteristics unfavorable for Colorado potato beetle reproduction and development: the beetle population density was lower in plots receiving manure soil and reduced amounts of synthetic fertilizers compared to plots receiving full doses of synthetic fertilizers but no manure. The effect was attributed to distinct differences in concentrations of macro- and micronutrients in potato leaves from manure- and synthetic fertilizer-treated plots. Of all studied minerals, zinc had a consistently positive effect on beetle populations but boron had a strong negative effect on all beetle stages except for the overwintered adults. Also, concentrations of this element were usually about two-fold higher in plants grown on manure-amended soil (Alyokhin *et al.* 2005). Female fecundity was lower in manure-amended plots early in the season, although it later became comparable to that on potatoes grown in synthetically fertilized soil. Fewer larvae survived past the first instar, and development of immature stages was slowed on manure-amended plots. Moreover, in the laboratory, first instars consumed less foliage from plants grown in manure-amended soils (Alyokhin and Atlihan 2005).

An interesting option for soil insect pest control is the application of allelo-pathic plant products to the soil. Allelopathy is a natural ecological phenomenon that occurs through the release, by one plant species, of chemicals which affect other species in its vicinity (Kruse *et al.* 2000, Bogatek and Gniazdowska 2008). The term is generally used to describe inhibitory and stimulatory effects of one plant on another plant, but the effects of secondary compounds on plant-insect interactions are also included (Kruse *et al.* 2000). In field crops, allelopathy can be used following rotation, using cover crops or mulching (Farooq *et al.*

2011). The allelopathic products can be administered in the form of either green manures or plant extracts. For example, brassica (*Brassica nigra* [L.] and *Sinapis alba* [L.]) green manures, used before the planting of potatoes, can produce a trend for lower levels of wireworm damage to potato tubers. The effect is possibly caused by toxic brassica green-manure breakdown products (McCaffrey *et al.* 1995, Frost *et al.* 2002). Similar effects can be gained by the application of wheat, turnip, vetch, and mustard green manures, which are most effective when ploughed in autumn (Schepl and Paffrath 2005). Nevertheless, consideration must be given to whether allelochemicals affect non-target organisms, and whether the allelopathic plant itself has adverse effects in the cultivated field or in natural environments (Kruse *et al.* 2000).

MANAGEMENT OF BIOTIC CONDITIONS

Biotic factors are the living parts of ecosystems. In agroecosystems, the crop, being the producer in the food chain, interacts with other biotic components – directly with phytophagous organisms, and indirectly with predators and parasites. At the same time, the crop is a member of a biological network of interactions, which means that its welfare depends not only on interactions with other trophic levels but also on indirect effects of other biotic components, such as neighboring vegetation, accompanying vegetation (e.g., weeds), history of vegetation (e.g., preceding crops, cover crops), etc. Considering these facts, various strategies of biotic environment manipulation are applied in potato culture to prevent or avoid agricultural pests and pathogens. The crop may be made unacceptable to pests by interfering with oviposition preferences, host-plant discrimination, or host location by intercropping, trap cropping, the use of living mulches, etc. Additionally, pest survival may be reduced by enhancing natural enemies through an increase in crop ecosystem diversity (Zehnder *et al.* 2007). Finally, the use of behavior-modifying chemicals (semiochemicals) is a promising strategy supplementing other cultural methods of pest management (Norin 2007). Semiochemicals are natural products that act as signals and regulate interactions between organisms – for example, plants and insects (Pickett *et al.* 2000). Semiochemicals are divided into pheromones (functioning in intraspecific interactions) and allelochemicals (functioning in interspecific interactions) (Norin 2007). In pest management, semiochemicals are applied mainly in monitoring insect pest populations and preventing agricultural damage by interfering with insect behavior. Various chemical stimuli may be used alone or in combination, which may give different behavioral outputs and often lead to disorientation of the insects (Cook *et al.* 2007).

Intercropping

Intercropping is the practice of simultaneously growing two or more crops in close proximity. Intercropping has a long history in traditional agriculture (Roder *et al.* 1992, Jamshidi *et al.* 2008). In certain areas, such as Bhutan, up to

40% of potato is grown in intercropping systems (Roder *et al.* 1992). The idea of intercropping is to choose two or more crops that vary in time of planting and harvesting as well as in manner of growth and development, which means that they should be complementary to, and not competing with, each other in terms of used resources such as light, water, and nutrients (Jamshidia *et al.* 2008). There are several ways to arrange the crops: (1) in strip intercropping, two or more crops are grown in strips wide enough to permit separate crop production but close enough for the crops to interact; (2) in row intercropping, at least one crop is planted in rows; (3) in mixed intercropping, there is no distinct row or strip arrangement; and (4) in relay intercropping, the crops are planted in succession with a second crop planted into a standing crop at the reproductive stage before harvesting (Knörzer 2009).

The effect of intercropping on potato yield depends on many factors, including the species and proportion of the interplanted crop, the location of the field, and the arrangement of the crops. For example, in Bhutan, variation in planting geometry and maize planting date did not affect potato yield, but the location of the fields appeared of importance: in a field located at an elevation of 2700 m above sea level (a.s.l.) and with 720 mm average rainfall, intercropping did not have any effect on the economic output; however, at an elevation of 1900 m a.s.l. and with 1242 mm average rainfall it did increase economic benefit by 12–15%. Moreover, it was suggested that an additional effect of intercropping in mountainous regions would be a reduction in high erosion risk at the time of potato harvest (Roder *et al.* 1992). In Iran, a maximum potato yield was obtained from a 3 : 1 potato : maize crop ratio (Jamshidi *et al.* 2008). In Pakistan, intercropping with maize and faba beans reduced the overall potato yield, and the reduction was higher when strip intercropping was applied than when the mixed intercropping was used. Interestingly, a correlation with the size of tubers was found: maize and bean plant populations were negatively correlated with big tubers and positively with seed-size tubers, depending on the amount of the intercropped maize (Farooq *et al.* 1996). In Sri Lanka, in relay-cropping combinations using maize or beans (soybean) as companion crops, shading during the first 4 weeks improved tuber yield by 20% whereas shading for up to 6 or 8 weeks after planting the potato reduced potato yields by 25% and 35%, respectively (Kuruppuarachchi 1990). In Peru, when relay-cropped with maize, potato plant population at harvest was superior to that of a sole crop of potato – an effect mediated through faster emergence and achievement of a greater maximum population, and not through differential survival of shaded or sole potato plants (Midmore *et al.* 1988). In southern England, intercropping potato with cabbage significantly reduced the economic yields of both component crops due to competition for nutrients or light (Opoku-Ameyawi and Harris 2001).

The described situations show that there is no universal rule on how to apply intercropping to increase the yield of potato, or the overall economic effect of this crop arrangement. Conversely, there is ample evidence that the use of an intercropping system helps to control pathogens and insect pest populations. For example, in Germany, the foliar late blight *P. infestans* was significantly reduced in potatoes strip-cropped with cereals or a grass-clover mix compared to pure stands of potato; the most important factors contributing to disease reduction were loss of inoculum outside of the plots and barrier effects of neighboring non-potato hosts (Bouws and Finckh 2008). In Ethiopia, 75% garlic with 25% potato (3:1) intercropped plots showed significantly lower late blight development and high tuber yield (Kassa and Sommartya 2006).

In the case of insects, and especially those life stages that are active on the above-ground parts of plants, an intercropping system can contribute to population control mainly by manipulation of their behavior. One of the most sensitive steps in the herbivorous insect life is host-location activity, which has consequences not only for the survival of an individual but also for the reproduction and survival of the species (Bruce *et al.* 2005). Host location by herbivores relies mainly on visual and olfactory cues that derive from the habitat of the host plant and the host plant itself, and act over long and short distances. Therefore, many phytophagous insects, especially the oligophagous ones, can find their hosts more efficiently in monocultures, when no other plants are present to interfere (Strong *et al.* 1984). The olfactory cues are of special importance (Bruce and Pickett 2011). For example, studies on Colorado potato beetle showed that subtle alterations in the original ratio of the green leaf volatiles emitted by potato leaves (E)-3-hexen-1-ol, (E)-2-hexen-1-ol, (Z)-2-hexen-1-ol and (E)-2-hexenal had a significant impact on host location, switching off attraction to the host plant when presented in an unnatural ratio (Bruce et al. 2005). It is not surprising, then, that manipulation of the crop-accompanying vegetation may prove a successful strategy to disorient the foraging herbivore and reduce economic loss due to its feeding. Colorado potato beetle can definitely be disoriented by non-host plant odors. The beetle population on potato plants was reduced by 60–100% when interplanted with tansy *Tanacetum vulgare* (L.) and 58–83% when interplanted with catnip *Nepeta cataria* (L.) as compared to monocultural plantings (Panasiuk 1984). Thiery and Visser (1987) found, in the laboratory, that the attractiveness of potato odor was neutralized by a mixture of potato and the non-host wild tomato *Lycopersicon hirsutum f. glabratum* (C.H. Muell), and suggested that this fact may be used in practical pest control by mixed cropping.

Potato tuber moth infestations were consistently reduced when potatoes were grown in association with certain other crops. Potato–chilli, potato–onion, and potato–pea associations significantly reduced larval infestation compared to potato alone. Similarly, tuber damage was significantly lower in the plots associated with chilli, onion, and pea, at 11%, 11%, and 13%, respectively, compared to 27% in potato alone (Lal 1991).

Significantly fewer aphids *Myzus* spp., leafhoppers *Empoasca* spp., and field crickets *Gryllus* spp. occurred in potato–berseem (*Trifolium alexandrinum* L.)

and potato–radish mix cropping (Jan *et al.* 2002). Intercropping the potato crop with onion or garlic reduced populations of *M. persicae, A. gossypii*, and *Empoasca* spp. when less than 0.75 berseem *Trifolium alexandrinum* L m separated the potato plants and *Allium* spp.; leaf damage to potato by *Henosepilachna sparsa* (Herbst) (Coleoptera: Coccinellidae) was also reduced at this spacing, but populations of *Thrips palmi* (Karny) or *T. parvispinus* (Karny) (Thysanoptera: Trypidae) were increased (Potts and Gunadi 2008).

Trap Crops and Barrier Crops

Andow (1991) reported that although pest injury is less likely to exceed economic damage levels in polycultures than in monocultures, in vegetationally diverse agroecosystems absolute yield benefits occur only rarely – and only when the arthropod pests cause severe yield losses in monocultures, and only if polycultures have lower pest populations than monocultures; even then, it occurs only intermittently. Considering this, and the fact that the cultivation of two or more plant species in the same agricultural field simultaneously can reduce the yield of the main crop due to plant competition, it is disputable whether this method is a prospective pest management strategy in agricultural production (Szendrei *et al.* 2009). Instead, a similar, alternative approach has been developed, the so called "push–pull strategy" (stimulo-deterrent diversionary strategy). A push–pull strategy means that the pests are repelled or deterred from the crop (the "push" part) and simultaneously attracted (the "pull" part) to other areas, such as trap crops or barrier crops (Cook *et al.* 2007). Trap crops are plants grown to attract insects or other organisms to protect target crops from pest attack, preventing the pests from reaching the crop or concentrating them in a certain part of the field where they can be economically destroyed (Shelton and Badenes-Perez 2006). Barrier crops are a type of trap crops used as a border to protect another crop from virus diseases by acting either as a "sink" for non-persistent viruses (infective virus vectors, mainly aphids, lose the viruses while probing on plants of the barrier crop) or as mechanical obstacle that impedes the colonization of the protected crop (Fereres 2000).

The use of trap crops should be preceded by analysis of the particular pest species characteristics, including its biology and behavior. Migratory, host-finding, and reproduction behaviors are especially important, so the behavior-modifying stimuli for use in push–pull strategies may include visual and chemical cues or signals from the crops, which respond to mechanisms underlying differential pest preferences (Cook *et al.* 2007). In addition to the natural characteristics of a particular plant used as a trap crop, insect preference can be altered in time and space to enhance further the effectiveness of a trap crop – for example, by the use of behavior-modifying chemicals such as non-host or host-derived volatiles or other chemicals, pheromones, antifeedants, etc. (Shelton and Badenes-Perez 2006, Cook *et al.* 2007).

Below, there are examples of various approaches to habitat management targeted at various sensitive phases of the potato pest insect biology and ecology.

Examples of Habitat Management

Push–Pull and Trap Crop Strategies

In the case of Colorado potato beetle, push–pull or trap crop strategies explored a variety of possibilities and were aimed mainly against overwintered adult beetles colonizing potatoes, and adult beetles dispersing within the field later in the season. Weisz *et al.* (1994) reported that winter-wheat and hay buffers significantly delayed overwintered adult colonization. Hoy *et al.* (2000) investigated the effectiveness of spring trap crops, which were host plants placed between the overwintering site and a new potato field, and barriers beyond them that were intended to retain and concentrate overwintered adult beetles and keep them out of the field. They found that planting date affects the pattern of potato beetle infestation by enhancing and maintaining adult Colorado potato beetle at the edges of potato fields. Physical barriers (dense interplanting of rye) had a greater impact than chemical barriers (tansy *T. vulgare* oil, wormwood *Artemisia absinthium* [L.] oil, piperonyl butoxide applied to outer rows of potatoes) on adult beetle movement from a potato trap crop to the protected potatoes beyond the barrier. Barrier treatments reduced beetle numbers in and just beyond the barrier, but the effects were localized and no significant reduction of beetles was observed further into the field, probably due to increased flight from trap or barrier areas or decreased sensitivity to host plants by walking beetles after passing through the barrier (Hoy *et al.* 2000). Martel *et al.* (2005) found that more postdiapausing, colonizing adults, egg masses, and small larvae were present in synthetic host volatile attractant-treated trap crops than in untreated trap crops, and although the yields for conventionally managed plots and plots bordered by attractant treated trap crops did not differ, 44% less insecticide was applied to plots bordered by attractant-treated trap crops. Additionally, the traditional application of pheromones for monitoring purposes may be broadened for a more general field use. The male aggregation pheromone of Colorado potato beetle [(*S*)-3,7-dimethyl-2-oxo-oct-6-ene-1,3-diol] may increase the preventative role of trap crops: more colonizing adults were present in pheromone-treated peripheral rows of potatoes compared with untreated middle rows, and significantly fewer egg masses and larvae were found in potato plots that were bordered by pheromone-treated rows (Kuhar *et al.* 2006). Host-plant chemicals may alter the response of insects to semiochemicals: orientation of Colorado potato beetle males can be disrupted by a combination of male-produced aggregation pheromone, (*S*)-3,7-dimethyl-2-oxo-oct-6-ene-1,3-diol and the three-component plant attractant blend (comprised of (*Z*)-3-hexenyl acetate + (±) linalool + methyl salicylate), which was preferred over the plant attractant alone (Dickens 2006).

Interestingly, potato may be used as a trap crop to protect other crops, such as tomatoes, against Colorado potato beetle: in Canada, tomato plots had significantly fewer adult beetles and significantly higher tomato yields (61–87% higher) when a potato trap crop was present (Hunt and Whitfield 1996). Similar effects were found by Gilboa and Podoler (1994) in Israel.

Wireworms *Agriotes sordidus* (Illiger) orientate towards a blend of volatiles emitted by chopped roots of barley. This finding underlines the importance of the identification of these compounds and their role assessment alone or combined, as for their effect on wireworms. Such compounds could be used in IPM strategies (Barsics *et al.* 2011). Indeed, the maize/wheat mixture bait is very effective in trapping wireworm larvae for monitoring purposes in potato fields (Parker 1994, Brunner *et al.* 2005).

Fereres (2000) found that the use of barrier crops of sorghum *Sorghum* spp. and vetch *Vicia* spp. can be an effective crop management strategy to protect against PVY infection. A 1 m wide barrier of oats was effective in reducing PVY spread (Radcliffe *et al.* 2007). Additionally, if a barrier is sown earlier than the target crop, some immigrating aphids can be filtered out due to the height difference of plants (Wratten *et al.* 2007). Barrier crops should have a fallow outside border with no space between the barrier crop and the potato field, since winged aphids usually alight at the border of bare ground and green crop (Radcliffe *et al.* 2007). Nevertheless, the species of border crop to be used as a virus sink should be selected carefully because it could act as a natural host for either the virus or the vector (Fereres 2000).

Cover-Crop Residues

Habitat vegetation management also includes the treatment of cover-crop residues. Cover crops are grasses, legumes, or small grains grown between regular crop production periods. A cover crop is not intended to be harvested for feed or sale, and its main purpose is to benefit the soil and/or other crops. Cover crops can interfere with the capacity of pests to colonize hosts by imposing physical barriers, disrupting olfactory and visual cues, and creating diversions to non-crop hosts. For example, hairy vetch residue reduced the rate of colonization by the Colorado potato beetle (Teasdale 2004). Szendrei *et al.* (2009) found that the movement of marked Colorado potato beetles into tilled plots was significantly higher than into vetch or rye cover treatments. Interestingly, the marked beetles released inside the potato field tended to move along the release row rather than across rows, and this pattern was stronger for the tilled treatment than for the two mulch cover treatments.

Antifeedants

Once on the plant, a herbivore can be discouraged from feeding by the application of feeding deterrents of natural or synthetic origin (see Chapter 16). Antifeedants can also be part of the push–pull strategy (the "push" part). Basically, antifeedants are behavior-modifying substances that deter feeding through a direct action on peripheral sensilla (i.e., taste organs) in insects (Isman 2002). However, Frazier and Chyb (1995) suggested that insect feeding can be inhibited at three levels: pre-ingestional (immediate effect associated with host finding and host selection processes involving gustatory receptors), ingestional (related to food transport and production, release, and digestion by salivary enzymes), and post-ingestional (long-term effects involving various aspects of digestion and absorption of food). Consequently, the reduced feeding may cause the rejection of a plant, may affect the development and longevity of the insect, or may lead to its death (Wawrzeńczyk *et al.* 2005, Wieczorek *et al.* 2005, Gabryś *et al.* 2006).

The application of antifeedants as crop protectants has attracted a lot of attention and, as a result, a vast literature on laboratory and field trials has been accumulated. The studies concentrate on various aspects of antifeedant use: structure/activity relationships, insect chemoreception mechanisms, insect feeding habits/application method relationships, mode of action at cellular, organismal and ecosystem levels, etc. (Koul 2005). For example, extracts of *Asclepias tuberosa* (L.) and *Hedera helix* (L.), exhibited exceptional levels of feeding deterrency towards wireworms, and in a field trial using an X-ray technique it was found that although the wireworms burrowed indiscriminately between soil containing either of these extracts and surrounding, untreated soil, they were found more frequently in the untreated areas (Villani and Gould 1985). Antifeedant activity towards Colorado potato beetles and their larvae was noted for *Pelargonium* × *hortorum* (Bailey) and *Geranium sanquineum* (L.) extracts. *P.* × *hortorum* extract added to food showed an unfavorable effect on the development of female reproductive organs and significantly inhibited the number of eggs laid; however, it showed no effect on either the period of winter diapause or spring emergence of beetles. The highest effectiveness under field conditions was recorded for an extract from *Erodium cicutarium* (L.). Potato leaves covered with *P.* × *hortorum* extract showed an unfavorable effect on the development of reproductive organs in females, significantly reducing the number of eggs laid; however, they showed no effect on either the period of winter diapause or spring beetle emergence (Lamparski and Wawrzyniak 2004). Pulegone and its derivatives, silphinene and its derivatives, and many others were efficient antifeedants against Colorado potato beetle in the laboratory (Gonzales-Coloma *et al.* 2002, Szczepanik *et al.* 2005). High ovicidal and oviposition-deterrent effects of *Lavandula gibsonii* J. Graham extracts were exhibited against *P. operculella* (Sharma 1981).

A survey of literature on the plants used for the control of the potato tuber moth has revealed that preparations from 35 plant species were effective against the pest either in storage (non-refrigerated) or in the laboratory. In some studies chopped and dried leaves were used, while in others leaf/seed extracts, fruit peel, bulb, root, and rhizome were used. Plant preparations were effective in reducing the pest damage or killing the pest at different stages (Das 1995). Extracts of garlic, wormwood, and tansy deterred the settling of the peach-potato aphid (Dancewicz and Gabryś 2008). A number of natural terpenoids and their synthetic analogs were also feeding-deterrent to *M. persicae* (Gabryś *et al.* 2005, 2006, Dancewicz *et al.* 2008).

Unfortunately, the commercial use of antifeedants in field crop production systems is still very limited. Many factors contribute to such situation: antifeedants are not lethal to the target organism; natural antifeedants are difficult to apply on a large scale because of their low content in plants; laboratory synthesis is often complicated and economically unjustified as insects are extremely sensitive to the spatial structure of chemical compounds; sometimes the activity is stage-specific, etc. (Szczepanik *et al.* 2005, Gabryś *et al.* 2006, Alyokhin 2009). Therefore, the search for effective antifeedants should be concentrated on natural sources or the synthesis of natural antifeedant analogs. Such compounds will be very selective (species-specific) and easily biodegradable in the environment (Koul 2005, Wawrzeńczyk *et al.* 2005, Wieczorek *et al.* 2005, Dancewicz *et al.* 2008, Grudniewska *et al.* 2011).

CONCLUDING REMARKS

Cultural practices are among the oldest techniques used for pest control, and many of the protective procedures used in agriculture today have their roots in traditional crop growing (Altieri 1999, Zehnder *et al.* 2007). Many of these traditional ways are compatible with natural processes (Morales 2002). Nowadays, and in the future, this compatibility should be of the highest priority for consideration in early stages of crop management strategies as indirect, precautionary measures. It is especially important in the situation where organic farming is one of the fastest growing segments of agriculture. Globally, organic agricultural land is estimated to cover 37.2 million hectares in approximately 160 countries; the world organic market is worth *ca.* US$54.9 billion, and organic *per capita* consumption per year is *ca.* US$8 billion (The World of Organic Agriculture 2011, IFOAM and FiBL). In the USA alone, a 19% annual growth rate in *per capita* organic potato consumption to 2013 has been predicted (Greenway *et al.* 2011).

One must keep in mind, though, that there is no universal cultural method that will significantly reduce all insect pests and increase the crop yield concurrently. Moreover, the protected crop is a part of the network of environmental interactions. The simultaneous application of various cultural management techniques in correspondence with other supplementary methods (biological, chemical, physical, behavior-modifying) should finally contribute to the increase in biodiversity, which is crucial for the integrity, stability, and sustainability of agroecosystems (Altieri 1999, Zehnder *et al.* 2007).

REFERENCES

Adams, S.S., Stevenson, W.R., 1990. Water management, disease development, and potato production 1990. Am. J. Potato Res. 67, 3–11.

Adlerz, W.C., Everett, P.H., 1968. Aluminium foil and white polyethylene mulches to repel aphids and control watermelon mosaic. J. Econ. Entomol. 61, 1276–1279.

Altieri, M.A., 1999. The ecological role of biodiversity in agroecosystems. Agr. Ecosyst. Environ. 74, 19–31.

Alyokhin, A., 2009. Colorado potato beetle management on potatoes, current challenges and future prospects. Fruit, Veg. Cereal Sci. Biotech. 3 (Special Issue 1), 10–19.

Alyokhin, A., Atlihan, R., 2005. Reduced fitness of the Colorado potato beetle (Coleoptera, Chrysomelidae) on potato plants grown in manure-amended soil. Environ. Entomol. 34 (4), 963–968.

Alyokhin, A., Porter, G., Groden, E., Drummond, F., 2005. Colorado potato beetle response to soil amendments, A case in support of the mineral balance hypothesis? Agr. Ecosyst. Environ. 109, 234–244.

Andow, D.A., 1991. Yield loss to arthropods in vegetationally diverse agroecosystems. Environ. Entomol 20 (5), 1228–1235.

Andrews, N., Ambrosino, M., Fisher, G., Rondon, S.I., 2008. Wireworm biology and nonchemical management in potatoes in the Pacific Northwest. A Pacific Northwest Extension Publication 607. Oregon State University, Corvallis, OR.

Arus, L., Luik, A., Monikainen, M., Kikas, A., 2011. Does mulching influence potential predators of raspberry beetle?. Acta. Agric. Scand., Sect. B 61 (3), 220–227.

Baker, M.B., Ferro, D.N., Porter, A.H., 2001. Invasions on large and small scales, management of a well-established crop pest, the Colorado potato beetle. Biol. Invasions 3, 295–306.

Barsics, F., Haubruge, E., Verheggen, F.J., 2011. Attraction of wireworms to root-emitted volatile organic compounds of barley. Proceedings of the 13th European Meeting "Biological Control in IPM Systems", Innsbruck, Austria, 19–23 June 2011 IOBC WPRS Bull 66, 475–478.

Berry, R.E., Reed, G.L., Coop, L.B., 2000. Identification & management of major pest & beneficial insects in potato [online]. Publication No. IPPC E.04-00-1 Oregon State University, Department of Entomology and Integrated Plant Protection Center, Corvallis, OR (published 24 April 2000). Available at: http://ippc2.orst.edu/potato.

Bogatek, R., Gniazdowska, A., 2007. ROS and phytohormones in plant–plant allelopathic interaction. Plant Sig. Behav. 2 (4), 317–318.

Boiteau, G., 2010. Insect pest control on potato, harmonization of alternative and conventional control methods. Am. J. Potato Res. 87, 412–419.

Bouws, H., Finckh, M.R., 2008. Effects of strip intercropping of potatoes with non-hosts on late blight severity and tuber yield in organic production. Plant Pathol. 57 (5), 916–927.

Bruce, T.J.A., Pickett, J.A., 2011. Perception of plant volatile blends by herbivorous insects – Finding the right mix. Phytochemistry 72, 1605–1611.

Bruce, T.J.A., Wadhams, L.J., Woodcock, C.M., 2005. Insect host location, a volatile situation. Trends. Plant. Sci. 10 (6), 269–274.

Brunner, N., Kromp, B., Meindl, P., Pázmándi, C., Traugott, M., 2005. Evaluation of different sampling techniques for wireworms (Coleoptera, Elateridae) in arable land. IOBC/WPRS. Bull. 28 (2), 117–122.

Brust, G.E., 1994. Natural Enemies in Straw-Mulch Reduce Colorado Potato Beetle Populations and Damage in Potato. Biol. Control 4 (2), 163–169.

Carter, M.R., Sanderson, J.B., 2001. Influence of conservation tillage and rotation length on potato productivity, tuber disease and soil quality parameters on a fine sandy loam in eastern Canada. Soil. Till. Res. 63, 1–13.

Carter, M.R., Holmstrom, D., Sanderson, J.B., Ivany, J., DeHaan, R., 2005. Comparison of conservation with conventional tillage for potato production in Atlantic Canada, crop productivity, soil physical properties and weed control. Can. J. Soil Sci. 85, 453–461.

Coll, M., Gavish, S., Dori, I., 2000. Population biology of the potato tuber moth, *Phthorimaea operculella* (Lepidoptera, Gelechiidae), in two potato cropping systems in Israel. Bull. Entomol. Res. 90, 309–315.

Cook, S.M., Khan, Z.R., Pickett, J.A., 2007. The use of push-pull strategies in integrated pest management. Annu. Rev. Entomol. 52, 375–400.

Dancewicz, K., Gabryś, B., Dams, I., Wawrzeńczyk, C., 2008. Enantiospecific effect of pulegone and pulegone–derived lactones on settling and feeding of *Myzus persicae* (Sulz.). J. Chem. Ecol. 34, 530–538.

Dancewicz, K., Gabryś, B., 2008. Effect of extracts of garlic (*Allium sativum* L.), wormwood (*Artemisia absinthium* L.) and tansy (*Tanaceum vulgare* L.) on the behaviour of the peach potato aphid *Myzus persicae* (Sulz.) during the settling on plants. Pesticides 3–4, 93–99.

Das, G.P., 1995. Plants used in controlling the potato tuber moth, *Phthorimaea operculella* (Zeller). Crop. Prot. 14 (8), 631–636.

De Willigen, P., Heinen, M., Van Der Broek, B.J., 1995. Modelling water and nitrogen uptake of a potato crop growing on a ridge. In: Potato ecology and modelling of crops under conditions limiting growth. In: Haverkort, A.J., Mac Kerron, D.K.L. (Eds.), Kluwer Academic Publishers, Dordrecht, The Netherlands, pp. 75–88.

Dickens, J.C., 2006. Plant volatiles moderate response to aggregation pheromone in Colorado potato beetle. J. Appl. Entomol. 130, 26–31.

Döring, T.F., Brandt, M., Heß, J., Finckh, M.R., Saucke, H., 2005. Effects of straw mulch on soil nitrate dynamics, weeds, yield and soil erosion in organically grown potatoes. Field. Crop. Res. 94, 238–249.

Dufour, R., 2001. Biointensive integrated pest management. [Online]. Available. at www.attra.org/attrapub/ipm.html.

Elbordiny, M.M., Gad, N., 2008. Response of potatoes cultivated in the alluvial soil for potassium chloride and cobalt. J. Appl. Sci. Res. 4 (7), 847–856.

Erlichowski, T., 2010. Susceptibility of potato cultivars to the wireworms damage and its significance in organic and integrated farming. Prog. Plant Prot. 50 (3), 1230–1234.

Essah, S.Y.C., Honeycutt, C., 2004. Tillage and Seed-Sprouting Strategies to Improve Potato Yield and Quality in Short Season Climates. Am. J. Potato Res. 81, 177–186.

Farooq, K., Hussain, A., Jan, N., Bajwa, K.A., Mahmood, M.M., 1996. Effect of maize, beans and potato intercropping on yield and economic benefits. Research and development of potato production in Pakistan. Proceedings of the National Seminar held at NARC. 23–25 April, 1995 Pakistan, Islamabad, pp. 211–219.

Farooq, M., Jabran, K., Cheema, Z.A., Wahid, A., Siddique, K.H.M., 2011. The role of allelopathy in agricultural pest management. Pest. Manag. Sci. 5 (67), 493–506.

Fereres, A., 2000. Barrier crops as a cultural control measure of non-persistently transmitted aphid-borne viruses. Virus. Res. 71, 221–231.

Frazier, J.L., Chyb, S., 1995. Use of feeding inhibitors in insect control. In: Chapman, R.F., de Boer G, G. (Eds.), Regulatory Mechanisms in Insect Feeding, Chapman & Hall, New York, NY, pp. 364–381.

Frost, D., Clarke, A., McLean, B.M.L., 2002. Wireworm control using fodder rape and mustard – evaluating the use of brassica green manures for the control of wireworm (*Agriotes* spp.) in organic crops. ADAS, Pwllpeiran. [Online]. Available at: www.organic.aber.ac.uk/library/Wireworm%20control%20and%20brassica%20green%20manures.htm.

Gabryś, B., Dancewicz, K., Halarewicz-Pacan, A., Janusz, E., 2005. Effect of natural monoterpenes on the behaviour of the peach potato aphid *Myzus persicae* (Sulz.). IOBC/WPRS Bulletin 28, 29–34.

Gabryś, B., Szczepanik, M., Dancewicz, K., Szumny, A., Wawrzeńczyk, Cz, 2006. Environmentally safe insect control, Feeding deterrent activity of alkyl-substituted γ- and δ-lactones to peach potato aphid (*Myzus persicae* [Sulz.]) and Colorado potato beetle (*Leptinotarsa decemlineata* [Say]. Pol. J. Environ. Stud. 15 (4), 549–556.

Genung, W.G., 1970. Flooding experiments for control of wireworms attacking vegetable crops in the everglades. Fla. Entomol. 53 (2), 55–63.

Ghazavi, M.A., Hosseinzadeh, B., Lotfalian, A., 2010. Evaluating physical properties of potato by a combined tillage machine. Nature and Science 8 (11), 66–70.

Gilboa, S., Podoler, H., 1994. Population dynamics of the potato tuber moth on processing tomatoes in Israel. Entomol. Exp. Appl. 72, 197–206.

Gonzales-Coloma, A., Valencia, F., Martin, N., Hoffmann, J.J., Hutter, L., Marco, J.A., Reina, M., 2002. Silphinene sesquiterpenes as model insect antifeedants. J. Chem. Ecol. 28 (1), 117–129.

Greenway, G.A., Guenthner, J.F., Makus, L.D., Pavek, M.J., 2011. An analysis of organic potato demand in the U.S. Am. J. Potato. Res. 88, 184–189.

Grudniewska, A., Dancewicz, K., Białońska, A., Ciunik, Z., Gabryś, B., Wawrzeńczyk, C., 2011. Synthesis of piperitone-derived halogenated lactones and their effect on aphid probing, feeding, and settling behavior. RSC Advances 1, 498–510.

Haverkort, A.J., 1982. Water management in potato production. Technical. Information Bulletin. 15. International Potato Center, Lima. Peru, p 22.

Haverkort, A.J., Verhagen, A., 2008. Climate change and its repercussions for the potato supply chain. Potato. Res. 51, 223–237.

Heimbach, U., Eggers, C., Thieme, T., 2002. Fewer aphids caused by mulching? Gesunde. Pflanz. 54 (304), 119–125.

Holland, J.M., 2004. The environmental consequences of adopting conservation tillage in Europe, reviewing the evidence. Agr. Ecosyst. Environ. 103, 1–25.

Hoy, C.W., Wyman, J.A., Vaughn, T.T., East, D.A., Kaufman, P., 1996. Food, ground cover, and Colorado potato beetle (Coleoptera, Chrysomelidae) dispersal in late summer. J. Econ. Entomol. 89, 963–969.

Hoy, C.W., Vaughn, T.T., East, D.A., 2000. Increasing the effectiveness of spring trap crops for Leptinotarsa decemlineata. Entomol. Exp. Appl 96, 193–204.

Hunt, D.W.A., Whitfield, G., 1996. Potato trap crops for control of Colorado potato beetle (Coleoptera, Chrysomelidae) in tomatoes. Can. Entomol. 128, 407–412.

Hunt, D.W.A., 1998. Reduced tillage practices for managing the Colorado potato beetle in processing tomato production. Hortic. Sci. 33, 279–282.

Isman, M., 2002. Insect antifeedants. Pesticide Outlook 13, 152–157.

Jamshidia, K., Mazaherib, D., Sabac, J., 2008. An evaluation of yield in intercropping of maize and potato. Desert 12, 105–111.

Jan, M.T., Naeem, M., Khan, M.I., Mehmood, R., 2002. Management of insect pests of autumn potato crop in diverse culture in NWFP (Peshawar). Asian. J. Plant Sci. 1, 577–578.

Jefferies, R.A., 1995. Physiology of crop in response to drought. In: Haverkort, A.J., Mac Kerron, D.K.L. (Eds.), Potato ecology and modelling of crops under conditions limiting growth, Kluwer Academic Publishers, Dordrecht, The Netherlands, pp. 61–74.

Johnson, J.M., Hough-Goldstein, J.A., Vangessel, M.J., 2004. Effects of straw mulch on pest insects, predators, and weeds in watermelons and potatoes. Environ. Entomol. 33, 1632–1643.

Kapsa, J.S., 2008. Important threats in potato production and integrated pathogen/pest management. Potato. Res. 51, 385–401.

Kar, G., Kumar, A., 2007. Effects of irrigation and straw mulch on water use and tuber yield of potato in eastern India. Agric. Water Manage 94, 109–116.

Kassa, B., Sommartya, T., 2006. Effect of intercropping on Potato late blight, *Phytophthora infestans* (Mont.) de Bary development and potato tuber yield in Ethiopia. Kasetsart. J. Nat. Sci. 40, 914–924.

Khan, Z.R., James, D.G., Midega, C.A.O., Pickett, J.A., 2008. Chemical ecology and conservation biological control. Biol. Control 45, 210–224.

Knörzer, H., Graeff-Hönninger, S., Guo, B., Wang, P., Claupein, W., 2009. The rediscovery of inter-cropping in China, a traditional cropping system for future chinese agriculture – a review. In: Lichtfouse, E. (Ed.), Climate change, intercropping, pest control and beneficial microorganisms, Sustainable Agriculture Reviews, Vol. 2. Springer Science+Business Media B.V, Dordrecht, The Netherlands, pp. 13–44.

Koul, O., 2005. Insect antifeedants. CRC Press LLC, Boca Raton USA..

Kring, J.B., Schuster, D.J., 1992. Management of insects on pepper and tomato with UV-reflective mulches. Fla. Entomol. 75, 119–129.

Kruse, M., Strandberg, M., Strandberg, B., 2000. Ecological effects of allelopathic plants – a review. NERI Technical Report No. 315. National Environmental Research Institute, Silkeborg, Denmark.

Kuhar, T.P., Mori, K., Dickens, J.C., 2006. Potential of a synthetic aggregation pheromone for inte-grated pest management of Colorado potato beetle. Leptinotarsa decemlineata (Say). Agric. Forest Entomol. 8, 77–81.

Kühne, S., 2008. Prospects and limits of botanical insecticides in organic farming. Prog. Plant Prot. 48, 1309–1313.

Kuruppuarachchi, D.S.P., 1990. Intercropped potato (*Solanum* spp.). Effect of shade on growth and tuber yield in the northwestern regosol belt of Sri Lanka. Field Crop. Res. 25, 61–72.

Lal, L., 1991. Effect of inter-cropping on the incidence of potato tuber moth. *Phthorimaea opercu-lella* (Zeller). Agr. Ecosyst. Environ. 36, 185–190.

Lamparski, R., Wawrzyniak, M., 2004. Effect of water extracts from geraniaceae (*Geraniaceae*) plants on feeding and development of Colorado potato beetle (*Leptinotarsa decemlineata* Say). EJPAU 7 (2), [Online]. Available. at: www.ejpau.media.pl/volume7/issue2/agronomy/art-01. html.

MacKerron, D.K.L., Waister, P.D., 1985. A simple model of potato growth and yield. Part I. Model development and sensitivity analysis. Agric. Forest. Meteorol. 34, 241–252.

Martel, J.W., Alford, A.R., Dickens, J.C., 2005. Synthetic host volatiles increase efficacy of trap cropping for management of Colorado potato beetle, Leptinotarsa decemlineata (Say). Agric. For. Entomol. 7, 79–86.

McCaffrey, J.P., Williams III, L., Borek, V., Brown, P.D., Morra, M.J., 1995. Toxicity of Ionic Thiocyanate-Amended Soil to the Wireworm Limonius californicus (Coleoptera: Elateridae) Source. J. Econ. Entomol. 88, 793–797.

Midmore, D.J., Berrios, D., Roca, J., 1988. Potato (*Solanum* spp.) in the hot tropics V. Intercropping with maize and the influence of shade on tuber yields. Field. Crop. Res. 18, 159–176.

Milner, M., Kung, K.-J.S., Wyman, J.A., Feldman, J., Nordheim, E., 1992. Enhancing overwin-tering mortality of Colorado potato beetle (Coleoptera, Chrysomelidae) by manipulating the temperature of its diapause habitat. J. Econ. Entomol. 85, 1701–1708.

Mohr, R.M., Volkmar, K., Derksen, D.A., Irvine, R.B., Khakbazan, M., McLaren, D.L., Monreal, M.A., Moulin, A.P., Tomasiewicz, D.J., 2011. Effect of rotation on crop yield and quality in an irrigated potato system. Am. J. Potato Res 88, 346–359.

Morales, H., 2002. Pest management in traditional tropical agroecosystems, lessons for pest prevention research and extension. Integrated Pest Management Reviews 7, 145–163.

Nault, B.A., Hanzlik, M.W., Kennedy, G.G., 1997. Location and abundance of adult Colorado potato beetles (Coleoptera, Chrysomelidae) following potato harvest. Crop. Prot. 16, 511–518.

Nelson, K.L., Lynch, D.H., Boiteau, G., 2009. Assessment of changes in soil health throughout organic potato rotation sequences. Agr. Ecosyst. Environ. 131, 220–228.

Neuhoff, D., Christen, C., Paffrath, A., Schepl, U., 2007. Approaches to wireworm control in organic potato production. IOBC/WPRS. Bull. 30, 65–68.

Norin, T., 2007. Semiochemicals for insect pest management. Appl. Chem. 79, 2129–2136.

Ojala, J.C., Stark, J.C., Kleinkopf, G.E., 1990. Influence of irrigation and nitrogen management on potato yield and quality. Am. J. Potato. Res. 67, 29–43.

Opoku-Ameyawi, K., Harris, P.M., 2001. Intercropping potatoes in early spring in a temperate climate. 1. Yield and intercropping advantages. Potato Res. 44, 53–61.

Panasiuk, O., 1984. Response of Colorado potato beetles, *Leptinotarsa decemlineata* (Say), to volatile components of tansy, *Tanacetum vulgare*. J. Chem. Ecol. 10, 1325–1333.

Parker, W.E., 1994. Evaluation of the use of food baits for detecting wireworms (*Agriotes* spp., Coleoptera, Elateridae) in fields intended for arable crop production. Crop. Prot. 13, 271–276.

Parker, W.E., Howard, J.J., 2001. The biology and management of wireworms (*Agriotes* spp.) on potato with particular reference to the U.K. Agr. Forest. Entomol. 3, 85–98.

Pickett, J.A., Wadhams, L.J., Pye, B.J., Smart, L.E., Wolfe, M., 2000. Studies on exploiting semio-chemicals for pest management in organic farming systems OF0188. Institute of Arable Crops Research, Biological Chemistry Division. http://orgprints.org/9957/1/Studies_on_exploiting_semiochemicals_for_pest_management_in_organic_farming_systems_OF0188.pdf.

Potts, M.J., Gunadi, N., 2008. The influence of intercropping with *Allium* on some insect populations in potato (*Solanum tuberosum*). Ann. App. Biol. 119, 207–213.

Powell, W., Pickett, J.A., 2003. Manipulation of parasitoids for aphid pest management, progress and prospects. Pest. Manag. Sci. 59, 149–155.

Radcliffe, E.B., Ragsdale, D.W., 2002. Aphid-transmitted potato viruses, the importance of understanding vector biology. Am. J. Potato. Res. 79, 353–386.

Radcliffe, E.B., Ragsdale, D.W., Suranyi, R.A., 2007. IPM case studies, seed potato. In: van Emden, H., Harrington, R. (Eds.), Aphids as crop pests, CABI Publishing, Wallingford, UK, pp. 613–625.

Raman, K.V., 1988. Control of potato tuber moth *Phthorimaea operculella* with sex pheromones in Peru. Agr. Ecosyst. Environ. 21, 85–99.

Roder, W., Anderhalden, E., Gurung, P., Dukpa, P., 1992. Potato intercropping systems with maize and faba bean. Am. Potato. J. 69, 195–202.

Rondon, S.I., DeBano, S.J., Clough, G.H., Hamm, P.B., Jensen, A., Schreiber, A., Alvarez, J.M., Thornton, M., Barbour, J., Dogramaci, M., 2007. Biology and management of the Potato tuberworm in the Pacific Northwest. PNW 594. April 2007. A Pacific Northwest Extension publication, Oregon State University, University of Idaho, and Washington State University.

Saucke, H., Döring, T.F., 2004. Potato virus Y reduction by straw mulch in organic potatoes. Ann. Appl. Biol. 144, 347–355.

Saue, T., Viil, P., Kadaja, J., 2010. Do different tillage and fertilization methods influence weather risks on potato yield? Agronomy. Res. 8 (Special Issue II), 427–432.

Schepl, U., Paffrath, A., 2005. Strategies to regulate the infestation of wireworms (*Agriotes* spp.) in organic potato farming: Results. Insect Pathogens and Insect Parasitic Nematodes. Melolontha. IOBC/WPRS Bulletin 28, 101–104.

Seal, D.R., Chalfant, R.B., Hall, M.R., 1992. Effects of cultural practices and rotational crops on abundance of wireworms (Coleoptera, Elateridae) affecting sweet potato in Georgia. Environ. Entomol. 21, 969–974.

Sharma, R.N., Bhosale, A.S., Joshi, V.N., Hebbalkar, D.S., Tungikar, V.B., Gupta, A.S., Patwardhan, S.A., 1981. *Lavandula gibsonii*. A plant with insectistatic potential. Phytoparasitica 9, 101–109.

Shelton, A.M., Badenes-Perez, F.R., 2006. Concepts and applications of trap cropping in pest management. Annu. Rev. Entomol. 51, 285–308.

Shock, C.C., Pereira, A.B., Eldredge, E.P., 2007. Irrigation best management practices for potato. Am. J. Potato Res. 84, 29–37.

Stoner, K.A., Ferrandino, F.J., Gent, M.P.N., Elmer, W.H., Lamondia, J.A., 1996. Effects of straw mulch, spent mushroom compost, and fumigation on the density of Colorado potato beetles (Coleoptera, Chrysomelidae) in potatoes. J. Econ. Entomol. 89, 1267–1280.

Strand, J.F., 2000. Some agrometeorological aspects of pest and disease management for the 21st century. Agr. Forest. Meteorol. 103, 73–82.

Strong, D.R., Lawton, J.H., Southwood, R., 1984. Insects on plants. Community patterns and mechanisms. Harvard University Press, Cambridge, MA.

Struik, P.C., Ewing, E.E., 1995. Crop physiology of potato (*Solanum tuberosum*), responses to photoperiod and temperature relevant to crop modelling. In: Haverkort, A.J., Mac Kerron, D.K.L. (Eds.), Potato ecology and modelling of crops under conditions limiting growth, Kluwer Academic Publishers, Dordrecht, The Netherlands, pp. 19–40.

Szczepanik, M., Dams, I., Wawrzeńczyk, C., 2005. Feeding deterrent activity of terpenoid lactones with the *p*-menthane system against the Colorado potato beetle (Coleoptera, Chrysomelidae). Environ. Entomol. 34, 1433–1440.

Szendrei, Z., Weber, D.C., 2009. Response of predators to habitat manipulation in potato fields. Biol. Control 50, 123–128.

Szendrei, Z., Kramer, M., Weber, D.C., 2009. Habitat manipulation in potato affects Colorado potato beetle dispersal. J. Appl. Entomol. 133, 711–719.

Teasdale, J.R., Abdul-Baki, A.A., Mills, D.J., Thorpe, K.W., 2004. Enhanced pest management with cover crop mulches. Proc. XXVI IHC – Sustainability of horticultural systems (L. Bertschinger and J.D. Anderson, Eds.) Acta Hort 638, 135–140.

The World of Organic Agriculture, 2011. IFOAM and FiBL, www.ifoam.org.

Thiery, D., Visser, J.H., 1987. Misleading the Colorado potato beetle with an odor blend. J. Chem. Ecol. 13, 1139–1146.

van Herk, W.G., Vernon, R.S., 2006. Effect of temperature and soil on the control of a wireworm, *Agriotes obscurus* L. (Coleoptera, Elateridae) by flooding. Crop. Prot. 25, 1057–1061.

Villani, M., Gould, F., 1985. Screening of crude plant extracts as feeding deterrents of the wireworm, Melanotus communis. Entomol. Exp. Appl. 37, 69–75.

Vos, J., Haverkort, A.J., 2007. Water availability and potato crop performance. In: Vreugdenhil (Ed.), Potato biology and biotechnology, advances and perspectives, Elsevier B.V., Amsterdam, The Netherlands, pp. 333–438.

Wang, X.C., Muhammad, T.N., Hao, M.D., Li, J., 2011a. Sustainable recovery of soil desiccation in semi-humid region on the Loess Plateau. Agr. Water. Manage. 98, 1262–1270.

Wang, F.X., Wu, X.X., Shock, C.C., Chu, L.Y., Gu, X.X., Xue, X., 2011b. Effects of drip irrigation regimes on potato tuber yield and quality under plastic mulch in arid Northwestern China. Field. Crop. Res. 122, 78–84.

Wawrzeńczyk, C., Dams, I., Szumny, A., Szczepanik, M., Nawrot, J., Prądzyńska, A., Gabryś, B., Dancewicz, K., Magnucka, E., Gawdzik, B., Obara, R., Wzorek, A., 2005. Synthesis and evaluation of antifeedant, antifungal and antibacterial activity of isoprenoid lactones. Pol. J. Environ. Stud. 14, 69–84.

Weisz, R., Smilowitz, Z., Christ, B., 1994. Distance, rotation, and border crops affect Colorado potato beetle (Coleoptera, Chrysomelidae) colonization and population density and early blight (*Alternaria solani*) severity in rotated potato fields. J. Econ. Entomol. 87, 723–729.

Weisz, R., Smilowitz, Z., Fleischer, S., 1996. Evaluating risk of Colorado potato beetle (Coleoptera, Chrysomelidae) infestation as a function of migratory distance. J. Econ. Entomol. 89, 435–441.

Wieczorek, P.P., Lipok, J., Jasicka-Misiak, I., 2005. Allelochemicals as potential biopesticides in agroecosystems. Pol. J. Environ. Stud. 14, 89–94.

Wratten, S.D., Gurr, G.M., Tylianakis, J.M., Robinson, K.A., 2007. Cultural control. In: van Emden, H., Harrington, R. (Eds.), Aphids as crop pests, CABI Publishing, Wallingford, UK, pp. 423–445.

Wright, R.J., 1984. Evaluation of crop rotation for control of Colorado potato beetles (Coleoptera, Chrysomelidae) commercial potato fields on Long Island. J. Econ. Entomol. 77, 1254–1259.

Wyss, E., Luka, H., Pfiffner, L., Schlatter, C., Uehlinger, G., Daniel, C., 2005. Approaches to pest management in organic agriculture, a case study in European apple orchards. CAB International, Organic-Research.com, May 2005, 33N–36N. http://orgprints.org/8717/

Zalom, F.G., 1981. Effects of aluminum mulch on fecundity of apterous *Myzus persicae* on head lettuce in afield planting. Entomol. Exp. Appl. 30, 227–230.

Žanić, K., Ban, D., Ban, S.G., Čuljak, T.G., Dumičić, G., 2009. Response of alate aphid species to mulch colour in watermelon. J. Food. Agric. Environ. 7, 496–502.

Zehnder, G.W., Linduska, J.J., 1987. Influence of conservation tillage practices on populations of Colorado potato beetle (Coleoptera, Chrysomelidae) in rotated and nonrotated tomato fields. Environ. Entomol. 16, 135–139.

Zehnder, G.W., Hough-Goldstein, J., 1990. Colorado potato beetle (Coleoptera: Chrysomelidae) population development and effects on yield of potatoes with and without straw mulch. J. Econ. Entomol. 83, 1982–1987.

Zehnder, G.W., Gurr, G.M., Kühne, S., Wade, M.R., Wratten, S.D., Wyss, E., 2007. Arthropod Pest Management in Organic Crops. Annu. Rev. Entomol. 52, 57–80.

Chapter 19

Evolutionary Considerations in Potato Pest Management

Andrei Alyokhin[1], Yolanda H. Chen[2], Maxim Udalov[3], Galina Benkovskaya[3] and Leena Lindström[4]

[1]*School of Biology and Ecology, University of Maine, Orono, ME, USA,* [2]*Plant and Soil Science Department, University of Vermont, Burlington, VT, USA,* [3]*Institute of Biochemistry and Genetics, Russian Academy of Science, Ufa, Russian Federation,* [4]*Department of Biological and Environmental Science, University of Jyväskylä, Finland*

INTRODUCTION

Almost 40 years ago, Theodosius Dobzhansky famously said that nothing in biology makes sense except in the light of evolution (Dobzhansky 1973). While it could be argued *ad nauseam* whether that statement was somewhat of an exaggeration, there is little doubt among the scientific community that the modern theory of evolution provides a useful and convenient framework for understanding patterns and processes observed in the biosphere. Unfortunately, applied biology in general, and agricultural pest management in particular, are still lagging behind the other fields of biology in placing their findings in a broader evolutionary context (Smith and Bernatchez 2008, Hendry *et al.* 2010). Addressing this issue is likely to increase both the efficiency and sustainability of future integrated pest management programs.

FUNDAMENTALS OF EVOLUTION

A classical definition of biological evolution was formulated by Dobzhanski *et al.* (1973) as follows:

Organic evolution is a series of partial or complete and irreversible transformations of the genetic composition of populations, based principally upon altered interactions with their environment. It consists chiefly of adaptive radiations into new environments, adjustments to environmental changes that take place in a particular habitat, and the origin of new ways for exploiting existing habitats. These adaptive changes occasionally give rise to greater complexity of developmental pattern, of physiological reactions, and of interactions between populations and their environment.

Insect Pests of Potato. http://dx.doi.org/10.1016/B978-0-12-386895-4.00019-3

Variability, selection, and adaptation are the three essential components of the evolutionary process. No two organisms living on Earth are exactly the same. Variation in observable traits (phenotypes) among organisms is due to both variations in their genetic make-up (genotype) and non-genetic inheritance (e.g., maternal effects), as well as to different environmental influences experienced during their lifetime. Genetic variation arises from mutations, recombination of genes during sexual reproduction, and migration (gene flow) between populations. Natural selection, which is the major driving force behind evolutionary change, can operate whenever organisms differ in their rates of survival and/or reproduction (also known as fitness) under particular environmental conditions (Futuyma 1986). As a result, genotypes resulting in the expression of most adaptive phenotypic traits increase in frequency over successive generations.

There are three general modes of selection (Futuyma 1986). When an extreme phenotype is characterized by superior fitness, selection is *directional*. It results in a shift in the mean value of a particular trait in the population. A typical example of directional selection is insect adaptation to insecticides. When intermediate phenotypes are the most fit, selection is *stabilizing* or *balancing*. For instance, an extreme deviation from a certain shape of mouthparts would interfere with an insect herbivore's feeding behavior on its host plant; as a result, it would be unlikely to persist in a population. If two or more phenotypes have higher fitness than intermediate phenotypes, selection is *disruptive* or *diversifying*. Adaptation to different host plants by insect herbivores with a subsequent formation of host races and/or speciation is an example of disruptive selection.

Not all shifts in phenotypic traits driven by environmental changes are genetically determined and subject to selection. A range of phenotypes may be expressed by a given genotype (or population or species) across a range of different environmental conditions (Stearns 1989, Gluckman *et al.* 2009). For example, an adult Colorado potato beetle (*Leptinotarsa decemlineata* Say) that is starved as a larva will be much smaller than an adult of the same species that is well fed at the same stage. However, due to current genetic constraints, it will never shrink to the size of a springtail, and neither will it grow as big as a Goliath beetle. If the variation among the individuals in different environments has a genetic component, then these reaction norms can also evolve in response to selection (Stearns and Koella 1986, Olsen *et al.* 2004, Lande 2009, Crispo *et al.* 2010).

Depending on its scale, evolution is broadly divided into *microevolution* and *macroevolution* (Dobzhanskii *et al.* 1973). Microevolution refers to the changes in gene frequencies within a population. Processes of agricultural significance, such as development of insecticide resistance or adaptation to new host plants, normally fall into this category. Macroevolution happens above the species level and includes grand events like the colonization of land by vascular plants or the radiation of the dinosaur lineage, as well as smaller events like the evolution of new genera of leaf beetles.

Evolution of traits towards a better adaptation to changing environments might proceed through existing genetic variation or through new mutations. On a short timescale, the former is probably more important (Aitken *et al.* 2008, Barrett and Schluter 2008, Orr and Unckless 2008). As a result, species that are well adapted to unstable natural environments often make the most formidable pests in unstable artificial environments typical of human agriculture (see below). However, new mutations provide fuel for longer-term evolution, with the greatest contribution expected from large populations with short generation times (Hendry *et al.* 2011). Both mechanisms may result in similar phenotypic outcomes and may be indistinguishable from the economic standpoint. For example, Hartley *et al.* (2006) concluded that malathion resistance in blowflies (*Lucilia cuprina*) evolved based on a pre-existing genetic variation, but resistance to diazinon in the same species evolved through a new mutation.

When thinking about evolution, people often subconsciously invoke a geological timescale, with images of trilobites and dinosaurs coming to mind. While this is certainly appropriate for macroevolutionary developments, it is also important to remember that microevolutionary changes may occur in as little as one generation (although, at the other extreme, they may also take many thousands of years) (Hendry and Kinnison 1999). Actually, meta-analysis of the existing data strongly suggested that contemporary microevolution (defined as taking place on the timescale of less than a few centuries) represents typical rates of microevolution in contemporary populations facing environmental change (Hendry and Kinnison 1999). Such rapid changes should be expected to commonly have profound impacts on human economic activities.

APPLIED EVOLUTION

Mismatches between the current phenotypes of organisms and phenotypes that would be best suited for a given environment are a major issue of interest throughout the diverse fields of applied biology (Hendry *et al.* 2011). Occasionally, scientific effort is directed towards finding ways to minimize such mismatches. For example, diversifying field edges may create a more favorable environment for natural enemies by supplying them with additional resources (see Chapter 9). At other times, the desirable outcome is to maximize the discussed mismatches. For example, destroying crop residues may decrease populations of overwintering pests (see Chapter 18). Theoretical evolutionary principles are essential for meeting applied goals because they help to achieve a better understanding of current mismatches and their potential responses to human manipulations (Hendry *et al.* 2011).

"Density makes the pest" is one of the most important principles of scientifically based integrated pest management (IPM) (Rajotte 1993). The idea that keeping populations of established pests below certain economically damaging levels is a more efficient and sustainable approach than a zero-tolerance policy is widely accepted by the scientific community, and is increasingly gaining traction among commercial growers. Clearly, adaptation to a particular environment is likely to affect population growth rate in that environment. Therefore, understanding population-level evolutionary processes is instrumental for designing scientifically sound IPM plans.

Adaptive evolution influences population dynamics and sometimes allows evolutionary rescue of a severely depressed population in a mismatched environment (Saccheri and Hanski 2006, Kinnison and Hairston 2007, Bell and Gonzalez 2009). In the absence of density-dependent regulation, better-adapted populations grow faster due to higher birth rates and/or lower death rates. When density-dependent regulation is important, better-adapted populations sustain more individuals at a given resource level.

Evolutionary principles have been applied in agriculture for thousands of years (Hendry *et al.* 2011). The best phenotypes – from the human perspective – were generated by simultaneously seeking the best genotypes and maintaining them in the most favorable environments. Superior genotypes were obtained through cultivar and variety development, domestication of new species, and introduction of species into new geographic areas. Superior environments were created through cultural practices (e.g., tillage), fertilization, pest control, etc. Optimal interactions between the two were also sought and found. Despite all errors, side effects, and imperfections, this approach allowed for the maintenance of an ever-increasing level of agricultural output. Employing evolutionary principles to a similar extent in pest management is likely to further improve agricultural production. Perhaps more importantly, it is likely to improve its sustainability.

EVOLUTION IN AGRICULTURAL ECOSYSTEMS

Agricultural fields represent a distinct and usually fairly unstable environment. Obviously, there are considerable differences among the various systems of production. However, as a whole they are characterized by low species diversity (especially in monocultures typical of highly industrialized commodity farming), high levels of disturbance, considerable outflow of harvested organic matter, and the presence of large amounts of xenobiotics in the form of pesticides and inorganic fertilizers. Inherent instability of agricultural ecosystems applies a strong selection pressure towards the traits that help organisms survive potentially catastrophic events such as tillage, crop rotation, and/or application of pesticides. At the same time, survivors are rewarded with an ample supply of food and a relative scarcity of natural enemies. This reduces the amount of energy needed for competition and defense, which can then be channeled to other purposes.

Although sometimes taken to an extreme, challenges facing organisms in agricultural systems are not unique. Many natural ecosystems are also highly disturbed and unstable. Furthermore, co-evolution among organisms, defined as reciprocal changes in their genetic compositions (Janzen 1980), applies considerable selection pressure towards overcoming the physical and chemical barriers to resource utilization. For example, insect herbivores evolve physiological and behavioral mechanisms to overcome defensive compounds produced by their host plants, while the host plants evolve new compounds that will be effective against resistant herbivores. Host-plant co-evolution is often lacking from agricultural systems because higher concentrations of defensive compounds in plants often make them less suitable for human consumption. This allows reallocation of resources by pest organisms towards other purposes, such as reproduction or metabolizing toxic xenobiotics.

As a result of prior adaptation to their natural environments, potential pests are likely to possess a suite of heritable traits that facilitate their adaptation to a crop environment. These often include the ability to detoxify a variety of poisons, high mobility concomitant with the ability to escape unfavorable conditions and to distribute offspring in space, diapause associated with an ability to wait out unfavorable periods and to distribute offspring in time, flexible life histories, behavioral plasticity, and high fecundity.

In fact, pest evolution in anthropogenic systems (including agriculture) can be viewed as a co-evolution between humans and their pests. By taking advantage of their existing genetic variations and new mutations, pests can adapt to a potentially very rewarding and resource-rich environment. In turn, humans can use their cognitive abilities to modify the environment in order to make it unfavorable to existing populations. This applies a new selection pressure on pest populations, with only the genotypes that best match the modified environment surviving to the next generation. These genotypes build up in numbers, eventually leading to the failure of whatever control methods had been applied against them and forcing humans to come up with new techniques. Thus far, it seems that the pests are usually one step ahead of the humans.

The potato production system is a vivid example of challenges and opportunities facing organisms that inhabit agricultural ecosystems. Agronomic practices of growing potatoes involve intensive soil disturbance and a low accumulation of organic matter (see Chapter 10). It is an annual crop, with substantial biomass removed from the system every year. Chemical use is very high (see Chapter 13). Furthermore, potato foliage has naturally high concentrations of glycoalkaloids, which are rather toxic to a variety of herbivores. As a result, the pest complex of potato is characterized by a high degree of plasticity and adaptability, and has shown a remarkable ability to persevere in the face of adversity. Not surprisingly, one of the most important potato defoliators, the Colorado potato beetle (see Chapter 2), has historically been a foster child among applied entomologists because of its ability to adjust to human attempts of its control (Weber and Ferro 1994, Alyokhin 2009).

THE EVOLUTIONARY PROCESS OF BECOMING A PEST

A pest is one of those concepts that everybody understands, yet is difficult to define in a uniform and comprehensive way. From the commonsense point of view, a pest is any organism that we do not like. Economically speaking, it is a group of organisms that it is more profitable to control than to tolerate. Ecologically, it is a population that competes with humans for the same limited resource. While the first two aspects of being a pest are rather anthropocentric and are heavily influenced by processes taking place in human society, competition is a biological phenomenon. Furthermore, the competitive abilities of a species or a population are a direct result of the evolutionary process.

Some species become pests of a particular crop as a result of extending their natural habitats. For example, it is doubtful that a highly polyphagous green peach aphid (*Myzus persicae*) had to undergo profound evolutionary changes to become an important pest of potatoes, even though potato may have been a totally new host. In other cases, however, considerable adaptations are required before a population can match its environment well enough to reach damaging levels. Those often involve developing abilities to feed on a new host or prey, to counteract human management effects (in particular, pesticide applications), and to survive in previously unsuitable environments.

AN OBSCURE LEAF BEETLE TURNS INTO A MAJOR PEST OF POTATOES

The Colorado potato beetle, *Leptinotarsa decemlineata*, is the most important insect defoliator of potatoes, and is extremely difficult to control (Alyokhin 2009). It is thought to originate from the central highlands of Mexico, in what is currently the state of Morelos (Tower 1906, Hsiao 1981, Casagrande 1987). Its ancestral host is buffalo burr, *Solanum rostratum*. It has been suggested that the adhesive burrs of *S. rostratum* clung to horses and cattle and were brought northwards with Spanish settlers into what is now the southern and central plains of the United States (Gauthier *et al.* 1981, Casagrande 1987, Hare 1990, Lu and Logan 1994a). The beetles then followed the host plant after northern populations of *S. rostratum* became established (Casagrande 1987). The Colorado potato beetle was first collected and described by Thomas Say on *S. rostratum* in 1824 near the border of Iowa and Nebraska (Casagrande 1985). The beetle's host expansion onto potatoes was first documented in eastern Nebraska in 1859 (Walsh 1865). After acquiring the ability to feed on potatoes, the beetle spread rapidly, reaching Iowa in 1861 (Walsh 1865) and the eastern seaboard in 1874 (Riley 1875).

Within the genus *Leptinotarsa*, the Colorado potato beetle has the widest host range, feeding on at least 10 species of wild and cultivated solanum (Hsiao 1978, Neck 1983, Jacques 1988). Its diet breadth is matched with an extensive geographic range which far exceeds that of all other species within the genus (Hsiao 1978, Neck 1983, Jacques 1988). Throughout its expanded range in the US, beetle populations have been found feeding on both cultivated (*S. tuberosum, S. melongena,* and *S. lycopersicum*) and wild (*S. saccharoides, S. carolinense, S. rostratum,* and *S. elaeagnifolium*) solanaceous plants (Hsiao 1978, Lu and Logan 1994a).

Tower (1906) recorded considerable variations in behavior and performance associated with different host plants among geographically isolated Colorado potato beetle populations. Despite the presence of potato plants in Mexico since at least the 16th century (Ugent 1968), and significant production of potatoes in the Mexican states of Guanajuato, Sonora, Chihuahua, Sinaloa, and Nuevo Leon since at least the 1940s (SAGARPA 2007), the Colorado potato beetle has never been recorded as a pest of potatoes, or of any other solanaceous crops, in that region (Casagrande 1987, Cappaert 1988). Furthermore, it appears that the beetle's ability to exploit cultivated host plants was acquired after its range expansion. Larvae collected from native plants *S. rostratum, S. angustifolium,* and *S. eleagnifolium* from Morelos in Mexico and from Arizona and Utah in the US showed reduced fecundity and survival on potato compared to pest populations collected from potato plants in the northeastern potato-growing region of the United States (Hsiao 1978, Lu and Logan 1995). Horton and Capinera (1988) found variation among beetle populations associated with wild and cultivated *Solanum* spp. in larval development, survival, and their tendency to diapause. Lu and Logan (1995) also found that the Mexican beetles showed strong ovipositional and feeding preferences for wild *Solanum* species (*S. rostratum* and *S. eleangnifolium*), whereas pest populations (Rhode Island, USA) did not discriminate among host plants. Hsiao (1981) crossed beetles from the population adapted to feeding on *S. eleangnifolium* with beetles adapted to feeding on potato but that performed poorly on *S. eleangnifolium.* The survival rate of progeny from the cross on *S. eleangnifolium* was intermediate between the survival rates of the two parental populations. All the information discussed in this paragraph suggests that there is indeed a genetic basis underlying the differences in host use among populations.

It is unclear how much evolutionary change had to occur in order to allow the Colorado potato beetle to expand onto potato. Beetles that feed on potato and wild *Solanum* spp. are able to mate with each other and produce viable offspring (Hare and Kennedy 1986, Lu and Logan 1994b, 1994c). Therefore, they are considered to be the same biological species. Furthermore, the newly acquired ability to utilize potato as a host plant appears to be a form of host expansion rather than a host shift, because potato-feeding beetles have not lost their ability to feed on *S. rostratum* (Lu and Logan 1994a).

Harrison (1987) observed considerable variability in the acceptance of marginal hosts within beetle populations. Beetles feeding on marginal hosts sampled them to a smaller extent before initiating feeding compared to beetles rejecting such hosts. In other words, they perceived such plants as more acceptable. In areas where plentiful alternative hosts are present, their somewhat low quality may be at least partially compensated by their abundance. Based on those findings, Harrison (1987) hypothesized that relaxation of stabilizing selection in the newly colonized areas resulted in populations of more generalist feeders taking advantage of local *Solanum* spp. Unfortunately, that included the cultivated potato, *S. tuberosum.* This hypothesis, however, does not explain why beetle populations in potato-growing areas of Mexico did not expand to take advantage of the new host. Perhaps the ratio of buffalo burr to potato in that region has not favored such a shift.

Another possible hypothesis is that a certain number of genotypes in ancestral Colorado potato beetle populations were specifically pre-adapted to feeding on potato (Hsiao 1982). The increased ability to feed on potato foliage in
derived populations has been the result of directional selection, with an increase
in the frequency of potato-adapted genotypes at the expense of less adapted
genotypes. Acquiring an improved ability to utilize potato was apparently not
associated with a decreased ability to utilize buffalo burr.

The two hypotheses are not mutually exclusive. On the contrary, relaxation
of stabilizing selection is a logical step before directional selection can shift the
mean value of host-plant acceptability towards potato. Over time, fine-tuning
of beetle behavior and physiology may lead to specificity in favor of newly
adopted hosts (Harrison 1987).

Regardless of its exact mechanism, Colorado potato beetle expansion onto
potato is clearly the result of an evolutionary process (Lu and Logan 1993, 1994b,
1994c, 1994d, 1995). Ancestral beetle populations are generally characterized
by poor performance on potatoes. For example, Lu and Logan (1993, 1994d)
were able to induce beetle larvae from Morelos, Mexico to feed on potato, but
larval survival and performance was quite poor. However, there appeared to
be significant variation in larval acceptance of potato as a food source among
families, indicating that larval feeding acceptance is genetically varied within
the Morelos population (Lu and Logan 1993). Although the Mexican beetles
could be reared to adulthood, they were unable to oviposit on potato plants
unless first exposed to *S. rostratum* (Lu and Logan 1994d). If deprived of their
wild *Solanum* host plants and then re-exposed, female Colorado potato beetles
almost immediately began to oviposit (Lu and Logan 1994d).

Adaptation to new hosts by the Colorado potato beetles can proceed at a very
fast pace. For example, beetles collected from their native host *S. eleangnifolium*
in Arizona (Hsiao 1981) suffered more than 80% mortality on *S. tuberosum* in
the first generation. Most mortality came from non-acceptance of foliage by
young larvae. However, after only five generations of selection on potato, mortality had dropped to less than 20% (Cappaert 1988).

Deciphering the evolutionary process of host-range expansion in the Colorado
potato beetle is a fascinating task that might improve our understanding of biological evolution as a whole. It might also have some applied value in forecasting
future host-range expansion by potential pests and biological control agents introduced for suppression of exotic weeds. However, its main practical significance is
likely to be in delaying the beetle's adaptation to resistant potato varieties.

Improving plant resistance to Colorado potato beetle damage is an underuti-
lized yet potentially valuable tool in the beetle control arsenal (a detailed review
of this method is provided in Chapter 15). Unfortunately, there is a serious
concern that beetles can overcome host-plant resistance just as easily as they
can overcome exposure to insecticides (see below). For example, in the study
by Groden and Casagrande (1986), oviposition and survival rates on resistant
S. berthaultii became comparable to those on susceptible *S. tuberosum* after

only two generations of selection. Pelletier and Smilowitz (1991) and França *et al.* (1994) confirmed the existence of genetic variability in several performance attributes for adaptation to *S. berthaultii*, although França *et al.* (1994) argued that adaptation is not always going to be as rapid as stated by Groden and Casagrande (1986). Similarly, Cantelo *et al.* (1987) observed gradual adaptation to feeding on resistant *S. chacoense* after 12 months of selection. Understanding mechanisms of Colorado potato adaptation to host plants is likely to improve the sustainability of using resistant potato varieties in the future.

INSECTICIDE RESISTANCE

The Insecticide Treadmill

Insecticide resistance is a serious worldwide problem, with at least 489 different insect species having become resistant to about 400 different compounds (Whalon *et al.* 2011). It is a typical example of directional selection: the chemical kills off susceptible genotypes, increasing both the frequency of resistant genes in the population and the mean dose of insecticide required to suppress insect densities below economically damaging levels. In some cases, selection by one chemical leads to resistance to other chemicals through shared physiological or biochemical mechanisms – a phenomenon known as cross-resistance.

Since the mid-twentieth century, commercial agriculture has been firmly stuck on the "insecticide treadmill". An insecticide is introduced by the agrochemical industry, pests develop resistance to it, a replacement chemical becomes available and is used until it also fails, and the cycle goes on and on. Although unsustainable, such an approach has been working reasonably well for some time. Unfortunately, the good old times of abundant and cheap broad-spectrum insecticides are coming to an end. Development and registration of new insecticides is an increasingly complicated and costly process. Furthermore, existing chemicals are being lost to resistance or removed from the market because of environmental concerns. As a result, preservation of existing products has become a progressively more important task for those involved in commercial agriculture (Alyokhin *et al.* 2008).

Unfortunately, potato fields have the outstanding distinction of harboring two of the most resistant pests in the world: the Colorado potato beetle and the green peach aphid (Whalon *et al.* 2011). Because of their serious damaging potentials, failure to control these two pests may have dire consequences for commercial growers.

The Colorado Potato Beetle as a Resistant Superbug

The Colorado potato beetle has been a major target for insecticide applications since 1864, and is credited with being an important driving force behind creating the modern insecticide industry (Gauthier *et al.* 1981). However, over the years this species has proven to be remarkably resilient, developing resistance to all major chemicals ever used against it (Alyokhin *et al.* 2008). According to the currently available international database (Whalon *et al.* 2011), the Colorado

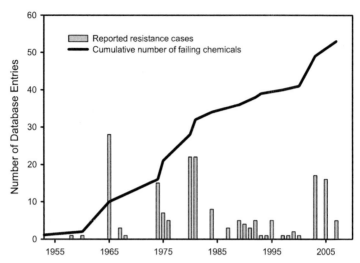

FIGURE 19.1   Progression of insecticide resistance in the Colorado potato beetle (Whalon *et al.* 2011).

potato beetle has developed resistance to 51 different chemicals in 46 world regions (Fig. 19.1). This number is likely an underestimation of the problem because some resistance cases might have been overlooked. For example, none of the cases reported from Russia (Leontieva *et al.* 2006, 2008a, 2008b, Sukhoruchenko and Dolzhenko 2008, Udalov and Benkovskaya 2010) are included.

Insecticides that have so far failed to control the Colorado potato beetle are classified into 10 groups of chemicals and 8 modes of action, including effects on the sodium channel for DDT and pyrethroids, inhibition of acetylcholinesterase for carbamates and organophosphates, blockage of chloride channels for cyclodienes, activation of GABA receptors for avermectins, agonist activity at nicotine acetylcholine receptors for neonicotinoids and spynosins, antagonism for the same receptors for nereistoxin compounds, and binding of receptors in the midgut cells by the endotoxin of *Bacillus thuringiensis* var. *tenebrionis*. Known mechanisms of resistance of the Colorado potato beetle to conventional insecticides include reduced insecticide penetration, target site insensitivity (including knockdown resistance and aceylcholinesterase insensitivity), and enhanced metabolism by esterases, carboxylesterases and monooxigenases (Rose and Brindley 1985, Argentine *et al.* 1989, Ioannidis *et al.* 1991, 1992, Wierenga and Hollingworth 1993, Anspaugh *et al.* 1995, Zhu *et al.* 1996, Lee and Clark 1998, Clark *et al.* 2001). Frequently, multiple mechanisms of resistance have occurred in a single population, and different mechanisms of resistance have occurred in populations from different geographical locations. For instance, acetylcholinesterase from one strain (Michigan) was insensitive to carbamates, and the same enzyme from another strain (Long Island, NY) was insensitive to organophosphates (Wierenga and Hollingworth 1993).

The same Colorado potato beetle populations are often resistant to multiple chemicals (Ioannides *et al.* 1991, Olson *et al.* 2000, Alyokhin *et al.* 2006, 2007). Sometimes this happens due to multiple selection pressures by different chemicals throughout the population's history. For instance, the intense application of organophosphates in the 1980s and pyrethroids in the 1990s in the Rostov region of Russia resulted in the appearance of populations with resistance to both classes of insecticides (Vilkova *et al.* 2005). In other cases, internal mechanisms that confer a resistance to one insecticide may also make the population less susceptible to another insecticide. For example, the rapid development of resistance to pyrethroids in Colorado potato beetle populations has been at least partially attributed to cross-resistance with DDT (Harris and Svec 1981). Similarly, selection with a carbamate insecticide has led to Colorado potato beetle resistance to organophosphate insecticides, and *vice versa* (Boiteau *et al.* 1987, Ioannides *et al.* 1992). Alyokhin *et al.* (2007) found a significant correlation between the toxicities of two neonicotinoid insecticides, imidacloprid and thiamethoxam, even when populations previously exposed to thiamethoxam were excluded from the analysis. There was no statistically detectable difference in the toxicities of those insecticides between populations exposed to both chemicals and populations exposed to imidacloprid alone.

Insecticide resistance in Colorado potato beetle populations had a dubious distinction of being identified as No. 7 in the list of the Top 10 Most Significant Influences on the potato industry over the last 50 years (McCallum 2012). Resistance reached critical levels in the early 1990s, when many potato growers completely ran out of chemical control options. The northeastern United States was the most severely affected area, but insecticide failures have been reported from a wide variety of geographic regions (Alyokhin *et al.* 2008). The situation improved dramatically after neonicotinoid insecticides became commercially available in 1995. Unfortunately, the first cases of beetle resistance to these compounds have been already detected in several places (Mota-Sanchez *et al.* 2006, Alyokhin *et al.* 2007, Udalov *et al.* 2010). While other chemicals still successfully suppress populations of this pest (see Chapter 13), this clearly manifests another turn of the pesticide treadmill.

Being native to North America, the Colorado potato beetle did not undergo the genetic bottleneck typical of introduced pests (Hawthorne 2001, Weber 2003). Therefore, it has been speculated that its populations retained genetic variability necessary to ensure evolutionary plasticity for their adaptations to adverse conditions. However, the relative importance of the last factor is unclear, because resistance is also a problem in Europe, where the Colorado potato beetle is an introduced species (Weber 2003), and where it shows marked reduction in natural genetic variability compared to US populations (Grapputo *et al.* 2005). Furthermore, insecticide applications create genetic bottlenecks of their own by eliminating susceptible genotypes and thus reducing the genetic variability of the surviving population.

Resistant founders may possess genetically determined neutral characteristics. For example, Udalov and Benkovskaya (2011) observed a much higher frequency of phenotypes with certain spot patterns on their heads, pronota, and elytra in the populations of insecticide-resistant beetles compared to the populations of susceptible beetles in the same area of Bashkortostan, Russian Federation. An overall diversity of spot patterns decreased over the 10-year period, presumably due to insecticide selection pressure. A similar process was observed in populations of the Colorado beetle in other areas of the Russian Federation: the Moscow region (Roslavtseva and Eremina 2005), Bryansk region (Oleinikov *et al.* 2006), and Kaliningrad, Rostov, and Vologda regions (Vasil'eva *et al.* 2005). Subsequent laboratory experiments confirmed differential survival following exposure to insecticides for beetles with different spot patterns on their pronota (Udalov and Benkovskaya 2011, Benkovskaya *et al.* 2008a).

Colorado potato mating behavior is strongly directed towards maximizing genetic diversity of its offspring and may at least partially compensate for bottlenecks associated with colonizing new areas, switching to new hosts, and being sprayed with insecticides. Colorado potato beetles are highly promiscuous (Szentesi 1985). Both males and females perform multiple copulations with different partners, with at least three matings required to completely fill the female's spermatheca (Boiteau 1988). For pre-diapause beetles, sperm from different copulations mixes and the female produces offspring that were fathered by different males (Boiteau 1988, Alyokhin and Ferro 1999a, Roderick *et al.* 2003). Males guard females following copulation and display aggressive behavior towards other males (Szentesi 1985). However, the duration of such guarding is usually rather short; therefore, it is unlikely to prevent subsequent mating by other males (Alyokhin, unpubl. data). On the contrary, mated males increase their flight activity, probably to maximize their number of copulations with different mates (Alyokhin and Ferro 1999b). Post-diapause females overwinter some viable sperm from the previous fall; however, mating in the spring significantly increases the number of their offspring (Ferro *et al.* 1991, Baker *et al.* 2005). Sperm from spring mating takes complete precedence over overwintered sperm from the previous year's mating (Baker *et al.* 2005).

The Colorado potato beetle's impressive ability to evolve resistances to insecticides has also often been attributed to high concentrations of toxic glycoalkaloids in the foliage of solanaceous plants. Co-evolution of the beetle and its host plants resulted in the development of the physiological capability to detoxify or tolerate poisons, including human-produced xenobiotics (Ferro 1993, Bishop and Grafius 1996). As a consequence, insecticide-adapted genotypes may already exist in the population when exposed to a newly developed chemical, or they may arise due to a relatively small mutation.

Another contributing factor is generally high Colorado potato beetle fecundity in the environment of a potato field, usually allowing it to reach high population densities. This is a rather prolific species, with a single female laying 300–800 eggs or more (Harcourt 1971). Furthermore, integrating dispersal with diapause, feeding, and reproduction allows the Colorado potato beetle to employ "bet-hedging" reproductive strategies, distributing its offspring in both space (within and between fields) and time (within and between years). This reduces the risk of catastrophic losses of offspring due to insecticides or crop rotation (Voss and

Ferro 1990). The resulting large populations increase the probability of random mutations, while their high growth rates ensure a rapid build-up in numbers of resistant mutants once such a mutation has occurred (Bishop and Grafius 1996).

Although the relative fitness of resistant mutants is often reduced compared to susceptible beetles due to the pleiotropic effects of resistant genes (Alyokhin *et al.* 2008), on rare occasions the selection of superiorly fit genotypes may be also solely responsible for the evolution of resistant strains. In this case, exposure to a toxic compound leads to the selection of the most robust genotypes (Richards *et al.* 2006). For example, Colorado potato beetles selected for feeding on a very unfavorable host *S. berthaultii* produced 1.7 times more eggs compared to the unselected strain when fed on a much more favorable *S. tuberosum*. Fecundity of the selected strain was reduced on *S. berthaultii*, but to levels not statistically different from those recorded for the unselected strain on *S. tuberosum* (Groden and Casagrande 1986).

A completely overlooked factor potentially contributing to the rapid evolution of insecticide resistance in the Colorado potato beetle is hormesis. Hormesis (also known as hormoligosis) is a relatively widespread yet often neglected phenomenon that occurs when a chemical (or some other stressor) that is normally detrimental to an organism at higher doses is stimulatory for some biological parameters at very low doses. The stimulatory effects are believed to be the result of compensatory biochemical processes following a destabilization of normal homeostasis (Calabrese and Baldwin 2001, 2003, Cohen 2006, Dutcher 2007, Calabrese 2009).

Hormesis has been demonstrated in Colorado potato beetles exposed to a number of stressors. Cutler *et al.* (2005) reported increase in the weight of second instars developing from eggs treated with sublethal concentrations of the chitin synthesis inhibitor novaluron compared to the larvae hatching from the control eggs. Sublethal doses of several insecticides enhanced the cold and heat tolerances of adult beetles in experiments performed by Benkovskaya (2009). Also, Alyokhin *et al.* (2009) detected increased oviposition in adult beetles that had been exposed to novaluron soon after eclosing from pupae. However, none of the eggs collected in that experiment hatched; therefore, that particular case could not qualify as true hormesis (Guedes *et al.* 2009). El Tahtaoui (1962) and Bajan and Kmitova (1972) observed increased fecundity in Colorado potato beetles surviving infection by the entomopathogenic fungus *Beauveria bassiana* (Bals.) Vuillemin, but that could have been also explained by selection towards the generally superior genotypes as discussed above.

Field label rates of insecticides applied by farmers to their crops are, by definition, sublethal for resistant beetles. Therefore, it is possible that they have a hormetic effect on resistant organisms, making them phenotypically closer to susceptible organisms despite the commonly present deleterious pleiotropic effects of resistant genes. As a result, the fitness differential between insecticide-exposed resistant insects and unexposed susceptible insects may be smaller than the differential between unexposed resistant and susceptible insects that is typically measured in laboratory resistance studies. Ultimately, this would lead to an increase in the net reproductive rates and intrinsic growth rates of resistant populations.

In other words, hormetic effects may compensate, at least to a certain degree, for the fitness costs of resistant genes. It is also known that although the resistance against antibiotics in bacteria carries costs in the beginning, compensatory mutations will accumulate through repeated directional selection and reduce the costs of resistance, leading to a situation where the resistant bacteria genotypes perform as well as the susceptible genotypes without the antibiotics (e.g., Normark and Normark 2002). Of course, bacteria have much faster growth rates than the beetles, making evolution much faster. Still, it is also possible that this type of selection will take place in beetles, making the resistance problem even bigger.

The selection pressure on Colorado potato beetle populations towards resistance development is usually enormous. Historically, commercial potato growers rely almost exclusively on insecticides for beetle control because, with the exception of crop rotation, other control techniques do not provide a feasible alternative (Harcourt 1971, Casagrande 1987, Bishop and Grafius 1996). Moreover, Colorado potato beetles have a narrow host range, and both larvae and adults feed on the same host plants. This limits the size of an unstructured refuge where susceptible genotypes may escape exposure to chemicals (Bishop and Grafius 1996, Whalon and Ferro 1998) by relocating to untreated volunteer potatoes or closely related solanaceous weeds. Not surprisingly, this is usually insufficient for reducing the frequency of resistant alleles below the economically significant level.

The Green Peach Aphid – Resistance in a Mostly Parthenogenic Organism

The green peach aphid, *Myzus persicae* (Sulzer) (Homoptera: Aphididae), is a highly polyphagous cosmopolitan species that commonly colonizes potato plants. Its populations seldom reach densities sufficient to cause noticeable crop injury by sap feeding. However, the green peach aphid is a very competent vector of the potato leafroll virus and potato virus Y, both of which represent an ominous threat to commercial potato production (see Chapter 3).

Green peach aphids have both winged (alatae) and wingless (aptera) body forms and alternate between several host species. They overwinter as eggs on their primary woody hosts (Rosaceae, especially *Prunus* spp.). After hatching in the spring and reproducing parthenogenetically on the primary host for at least two generations, the aphids start producing winged spring migrants that leave the primary host in search of suitable secondary hosts, which include several hundred species of herbaceous plants in addition to potatoes. Once an acceptable host is found, the spring migrants settle and reproduce (Shands *et al.* 1969, 1972, Shands and Simpson 1970). The majority of the offspring produced by the spring migrants are wingless, but a few winged individuals are produced throughout the summer. The production of winged summer migrants is encouraged by overcrowding and poor quality of host plants (Muller *et al.* 2001). Many overlapping generations are produced during the summer. In the fall, a short day photoperiod induces production of sexual fall migrants that migrate back to the primary winter hosts.

On primary hosts, female fall migrants give birth parthenogenically to wingless females, which then mate with male fall migrants and lay fertilized overwintering eggs. Populations that have both sexual and asexual generations are called holocyclic. In areas with warm climates, some green peach aphid populations do not produce sexual forms and persist year-round as parthenogenic forms on secondary hosts. Such populations are known as anholocyclic.

Similar to Colorado potato beetles, green peach aphids have shown a rather remarkable ability to evolve resistances to a variety of insecticides. According to a currently available data compilation (Whalon *et al.* 2011), they have developed resistances to 69 different chemicals in 43 world regions (Fig. 19.2). As is the case with Colorado potato beetles, this is likely to be an underestimation of the number of resistant populations. Resistances to organophosphates, pyrethroids, and pirimicarb (a dimethyl carbamate) are now widespread (Fenton *et al.* 2010). Resistance to neonicotinoids, particularly to imidacloprid, is also increasingly becoming a problem (Foster *et al.* 2003a, Srigiriraju *et al.* 2010).

Insecticide resistance has arisen independently in different green peach aphid populations on a number of occasions. Its mechanisms include mutations of target proteins – which decreases their affinity for binding insecticide molecules – as well as the enhanced production of metabolic enzymes, which detoxify and/or sequester insecticides (Fenton *et al.* 2010, Bass and Field 2011). For example, target site mutation known as MACE (modified acetylcholinesterase) confers virtual immunity to pirimicarb. Knockdown resistance, which is also known in the Colorado potato beetle (see previous section), is effective against pyrethroid insecticides. Overproduction of metabolic enzymes confers a strong resistance to organophosphates (Fenton *et al.* 2010).

A major evolutionary driver of insecticide resistance in the green peach aphid is gene amplification: the reiteration of a segment of DNA to generate one

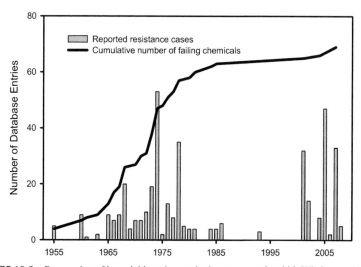

FIGURE 19.2   Progression of insecticide resistance in the green peach aphid (Whalon *et al.* 2011).

or more additional copies in the genome (Bass and Field 2011). This results in the production of metabolic enzymes in amounts sufficient for the detoxification of otherwise lethal concentrations of toxic chemicals. Furthermore, this mechanism may play a role in pesticide resistance conferred by target-site insensitivity. For example, duplication of a γ-aminobutyric acid (GABA) receptor subunit gene has been reported in association with the green peach aphid resistance to the cyclodiene insecticide endosulfan (Anthony *et al.* 1998). Review of recently published studies (Bass and Field 2011) indicates that gene amplification may be a fairly common mechanism of adaptive evolution in arthropods, and that certain genomic loci may be "hot spots" for gene duplication, as evidenced by parallel evolution in several arthropod species. However, the green peach aphid's genome appears to have a special propensity for gene amplification (Bass and Field 2011).

As discussed at the beginning of this chapter, evolution is impossible without initial variation within a given population. In asexual organisms, genotypic variation arises from mutations. In sexual organisms, it is the result of a combination of mutations and recombination during meiosis and fertilization. Green peach aphids are capable of both sexual and asexual reproduction, although some populations have lost their ability for the former. Their sexual stages provide an opportunity to increase the overall genetic diversity of a population, while rapidly reproducing asexual stages provide a means for a quick increase in the frequency of well-adapted genotypes.

Evolutionary successful aphid genotypes are capable of rapidly expanding their ranges and colonizing new geographic areas, both on their own and with human assistance. Winged forms can be carried by wind over considerable distances. Perhaps more importantly, their small body size and somewhat cryptic habits promote the spread of green peach aphids by humans moving around their host plants (Fenton *et al.* 2010). As a result, there are several common and widespread clones, whose successes appear to stem from selection for insecticide resistance in agriculture (Fenton *et al.* 2005, 2010, Zamoum *et al.* 2005, Kasprowicz *et al.* 2008, van Toor *et al.* 2008). The number of such genotypes is still relatively limited, despite the possibility of resistance genes combining into more genotypes in sexual populations each year (Fenton *et al.* 2010). Although gene flow is important in spreading resistance to new areas, similar resistant phenotypes have been also shown to evolve independently in geographically separated green peach aphid populations (Gillespie *et al.* 2008).

Unlike the Colorado potato beetle, the green peach aphid is a highly polyphagous species. Many of its host plants (including potato and tobacco) produce highly toxic phytochemicals in their foliage. Therefore, it is likely that the green peach aphid had naturally evolved physiological mechanisms to deal with a variety of poisons. In theory, polyphagy also somewhat decreases selection pressure on aphid populations by creating refuges on alternative hosts growing outside of the treated crop area. However, extremely high reproductive rates of parthenogenic generations appear to reduce the practical value of such "genetic dilution". Hormetic effects of insecticide applications may further increase the fecundity of exposed resistant aphids (Lowery and Sears 1986).

RESISTANCE TO NON-CHEMICAL CONTROL METHODS

Pesticide resistance is commonly cited as one of the main reasons for switching from a chemical-based pest management system to a system based on using non-chemical alternatives. Although such an argument is definitely valid, it is important to remember that pests can also adapt to non-chemical methods of control. Effective non-chemical techniques apply a considerable selection pressure on pest populations, which in some cases might be even stronger than the pressure applied by synthetic chemicals. Therefore, the evolution of resistances to non-chemical methods should not be a surprise to anyone. Since pest control in commercial agriculture has been dominated by pesticides for many decades, most reported cases of pest adaptation involve resistances to toxic chemicals. As more alternative methods enter (or, in some cases, re-enter) mainstream agriculture, their increasing failures due to resistance development in pest populations should be expected.

Several existing studies confirm that green peach aphids are fully capable of evolving resistances to biological control agents. The parasitoid wasp *Aphidius colemani* (Hymenoptera: Braconidae) is routinely released by greenhouse growers to control this pest, and this method has typically been found to be successful (Gillespie *et al.* 2008). However, in 2002 and 2003, growers in British Columbia experienced severe green peach aphid outbreaks. The clone involved in those outbreaks showed reduced vulnerability to parasitoids and a higher reproductive rate compared to the other two clones tested in the same study (Gillespie *et al.* 2008). Aphids from the resistant clone were less frequently stung by wasps, and a lower proportion of these stings resulted in mummy development. Herzog *et al.* (2007) conducted a laboratory study in which caged aphid populations were maintained for six to eight generations with or without *A. colemani*. Populations confined with *A. colemani* evolved to contain only a single, highly resistant clone. The control treatment consisted of a diverse suite of clones, although their relative frequencies shifted towards predominance of more prolific genotypes.

Except for several cases of adaptation to resistant potato species that have already been discussed above, there are no reports of Colorado potato beetle resistances to non-chemical methods of its control. However, there is certainly a good possibility of such a development. For example, the efficiency of annual crop rotation for reducing field colonization by overwintering adults could be compromised by multi-year diapause. Extended diapause is not unusual in the Colorado potato beetle (Isely 1935, Ushatinskaya 1962, 1966, Biever and Chauvin 1990, Tauber and Tauber 2002). In some populations, as many as 21% of overwintering adults emerge after spending 2 years in the soil (Biever and Chauvin 1990). Selection for multi-year diapause has been responsible for the eventual failure of annual crop rotation to control another leaf beetle, the northern corn rootworm (*Diabrotica barberi*) (Levine *et al.* 1992).

Obviously, potential problems with insect adaptation do not mean that non-chemical techniques should not be used for pest control. However, they should not be treated as an everlasting silver-bullet solution that is sustainable by definition. Resistance management, discussed in the following section, is an important consideration for all methods of pest management.

RESISTANCE MANAGEMENT

Successful pest control depends on increasing mismatches between pestiferous organisms and their environments. Application of toxic chemicals achieves this goal by creating a highly unfavorable environment for target pests. It also applies a very strong selection pressure towards resistance development. After resistant genotypes take over the population, this mismatch effectively disappears. Moreover, conditions may become more favorable for them because of the removal of susceptible competitors and natural enemies.

Resistance management approaches are based either on restoring the mismatch between newly resistant pests and their habitat, or on decreasing selection pressure on pest populations before they become resistant. In practice, the former is usually achieved by rotating insecticides with different modes of action. Although despised by the proponents of sustainable agriculture, the pesticide treadmill has been serving the farming community fairly well. A steady flow of new products to the pesticide market allowed for the quick and relatively painless replacement of failing compounds with new chemistries. However, developing new active ingredients is an increasingly difficult and expensive task (Alyokhin et al. 2008). As a result, it is safe to say that the era of abundant and cheap pesticides is largely over. Therefore, relying on the chemical industry to keep inventing new products is a risky approach.

Selection pressure is usually reduced by leaving untreated crop areas within or near treated fields. Susceptible populations persist within these plots and mate with newly resistant genotypes that arise in adjacent treated areas. However, most commercial growers perceive leaving part of their crop untreated as risky, and are reluctant to do it on their farms. Furthermore, a number of assumptions need to be met for this refuge-based strategy to succeed. First, resistant alleles should be at least partially recessive, which means that progeny (heterozygotes at the resistance locus) of one resistant parent and one susceptible parent are not as resistant as their resistant parent and can be killed by a sufficiently high dose of an insecticide. This is usually true for the Colorado potato beetle (Alyokhin et al. 2008), but less applicable for the green peach aphids, which reproduce parthenogenically for most of their life cycles. Secondly, in the absence of pesticides, resistant alleles should be associated with the decreased fitness of resistant individuals, such as fewer offspring, shorter life spans, longer times of development, lower tolerances of unfavorable conditions, etc. This way, resistant individuals are at a selective disadvantage in the refuge, and their frequency there remains relatively low. Reduced fitness is indeed very common among insecticide-resistant insects, including the Colorado potato beetle and the green peach aphid (Alyokhin *et al.* 2008, Fenton *et al.* 2010). Finally, there should be sufficient movement between the main crop and the refuge with subsequent mating upon arrival, so that resistant individuals do not mate with each other and do not leave highly resistant offspring. Although both the Colorado potato beetle and the green peach aphid can be highly mobile and travel over long distances, there is also evidence that significant segments of their populations do not move far and probably inbreed (Alyokhin *et al.* 2008, Fenton *et al.* 2010).

Practicing integrated pest management by combining multiple techniques and using economic thresholds is by far the best approach to delaying the evolution of resistance. This creates a complex environment with multiple factors that are unfavorable to the target pest. Simultaneous adaptations to multiple unrelated influences, which may be as different as toxic chemicals, mechanical barriers, and predators, are likely to require profound genetic changes. As a result, it is statistically less probable that the adaptation to a single factor stems from a single-gene mutation. Furthermore, resistance to one factor may be offset by increased susceptibility to another factor. For example, esterase-based insecticide resistance was negatively correlated with resistance to parasitoids in several clones of the green peach aphid (Foster *et al.* 2003b, 2007, Gillespie *et al.* 2008). Also, using economic thresholds eliminates unnecessary insecticide applications, thus reducing selection pressure on target pests.

INTERACTIONS WITH THE ABIOTIC ENVIRONMENT

Insect adaptation to abiotic conditions (temperature, water availability, solar irradiation, etc.) is an increasingly important consideration in pest management. First, extensive human traffic and commerce results in the constant introduction of potentially pestiferous species to new areas. Predicting a pest's ability to adapt to conditions typical of a newly colonized location is essential for forecasting future pest outbreaks and taking the actions necessary for their prevention. Secondly, changing climate is likely to apply new selection pressures on pest populations inhabiting a given geographic area, as well as opening previously unsuitable areas for colonization and range expansion.

Range expansion to new environments requires that the species in question adjusts its life cycle to the new environment. The Colorado potato beetle currently continues to expand its range, often with dire economic consequences for newly colonized areas (Alyokhin 2009). Researchers have tried to understand the limits of range expansion of the Colorado potato beetle by testing its performance in various abiotic conditions. Similar to the majority of insects, two important abiotic factors have been suggested for limiting the Colorado potato beetle's range expansion: temperature, and light regime. Temperature is important for ectotherms, as all life stages are affected by its changes. Typically, growth is reduced by temperature decreases from the optimum (Logan *et al.* 1985), which can in turn have various effects on other life-history traits, such as size (Boman *et al.*

2008, Lyytinen *et al.* 2008). Thus, the beetle cannot invade areas where it cannot complete its development from egg to adult, as only the adult stages survive over winter. Temperature can also directly limit range expansion if the individuals have no means of surviving temperature extremes. For instance, the thermal extremes during winters (freezing) have been suggested to limit the range expansion of the Colorado potato beetle in many areas. The supercooling point in Colorado potato beetles ranges between −19°C and −5°C, and is related to the water content of the body (Costanzo *et al.* 1997). However, beetles that dig into the ground to over-winter may not encounter these extreme temperatures because soil temperatures are much warmer than ambient temperatures. Beetles are also rather tolerant to cold exposure during their larval stage, as only 3.1% die when exposed to −4°C (Lyytinen *et al.* 2009). Furthermore, they can adapt, to some extent, to cold temperatures by upregulating heat shock proteins (Lyytinen *et al.* 2012).

Temperature has been also shown to affect Colorado potato beetle interactions with their host plants. De Wilde *et al.* (1969) reported that when given a choice between potato and bittersweet nightshade, *S. dulcamara*, the beetles chose potato more frequently at low temperatures but *S. dulcamara* at high temperatures. Similarly, beetle survival on horse-nettle, *S. carolinense*, increased with increasing temperatures (Hilbeck and Kennedy 1998). The mechanisms behind this phenomenon are unknown, and may involve temperature-mediated changes in both insects and plants. Regardless, the observed temperature mediation may have important implications for range expansion and host adaptation in the Colorado potato beetle (Hilbeck and Kennedy 1998).

A photoperiod gives cues for the timing of the life cycle for many temperate insects. Therefore, it has been identified as an important limiting factor for range expansion. The Colorado potato beetle is a multivoltine species in the areas where conditions are suitable (de Kort 1990), and it has a facultative diapause that has to be initiated at the correct time. Diapause initiation is crucial because the decision cannot be reversed. Since winters in northern latitudes arrive early in relation to the photoperiod, there is selection to enter diapause under relatively long-day conditions. Species invading those areas from the south must therefore be able to adjust their diapause behavior to overwinter successfully (Yamanaka *et al.* 2008). The correct timing of diapause is assured by having a sensitive stage that responds to a shortening of the photoperiod (Lefevere and de Kort 1989, Noronha and Cloutier 1998, 2006). The Colorado potato beetle has a photosensitive phase primarily just after adult emergence from a pupa, but also to some degree at the last larval stage (de Wilde *et al.* 1959). The critical day length is dependent on the population, ranging between 12 hours of light at 32°N latitude (Tauber *et al.* 1988) and 15–16 hours of light at latitudes above 45°N (de Wilde *et al.* 1959, Danilevski 1965, Tauber *et al.* 1988). Furthermore, it has been estimated that some proportion of beetles will enter diapause in photoperiods longer than a 16-hour day length (Danilevski 1965). In Europe, 10–24% of beetles reared in long-day conditions (20L:4D) will enter diapause immediately when allowed, irrespective of their origin, suggesting that not all beetles respond to light conditions similarly (Peferoen *et al.* 1981, Piiroinen 2010). Although these beetles reared in long-day conditions entered diapause, their survival was lower, probably because they had not accumulated enough resources for successful completion of diapause. Nevertheless, polymorphism existing for diapause initiation can be crucial when populations are exposed to new conditions.

The question of range expansion is more complicated than performance at a certain temperature or with a certain light regime. It is also related to the potential to adapt to new environments. Such a potential is determined by genetic variation in traits that are under selection when a population is exposed to new conditions. Although there is extensive experimental evidence for beetle performances under different conditions, the additive genetic variation in many traits has not been adequately assessed. Retrospective analysis of the beetle's geographic distribution (EPPO 2011) suggests that it has had the potential to adapt to various climatic conditions. Although the beetle did lose considerable genetic variation when invading Europe, it managed to expand its range almost as quickly as it did in North America (Grapputo *et al.* 2005). The established populations still have additive genetic variance in development rates, and thus have the potential to respond to changes in summer temperatures (Boman *et al.* 2008, Lyytinen *et al.* 2008). However, successful overwintering also requires adaptive genetic changes in diapause-related behaviors, metabolism, or body mass. Insufficient genetic variation in these traits is likely to limit the Colorado potato beetle's potential to respond to selection due to harsher winters, which in turn could restrict range expansion (Piiroinen *et al.* 2011). Unfortunately, some of these traits are difficult to measure. Furthermore, genetic potential is the property of a population; therefore, results obtained for one population may not be representative of other populations of interest.

HUMANS' TURN TO ADAPT?

Humans are recognized to be the biggest evolutionary force operating at the moment (Palumbi 2001). While this is certainly true, it is important to recognize that we are also at the receiving end of the selection pressures that we generate. Despite all technological advances, pest and disease outbreaks continue to plague humankind, sometimes at an increasing frequency and/or severity. This is in large part due to the plasticity of pestiferous species and their abilities to adapt to whatever poison we are trying to unleash against them. Furthermore, human health and the environment often fall victim to the collateral damage resulting from our mostly xenobiotic-based endeavors. Clearly, a simplistic approach of measuring pest management success as the number of dead "bugs" is extremely near-sighted. Instead, this statistic should be treated as the number of dead susceptible genotypes, and the evolutionary consequences of their removal from a population should be addressed before creating an economically important problem. Incorporating our knowledge of fundamental evolutionary processes into pest control practices will take time and effort, but it is essential for maximizing their efficiency. Essentially, we need to learn to better adapt to the environment of our own creation.

REFERENCES

Aitken, S.N., Yeaman, S., Holliday, J.A., Wang, T.L., Curtis-McLane, S., 2008. Adaptation, migration or extirpation: climate change outcomes for tree populations. Evol. Appl. 1, 95–111.

Alyokhin, A., 2009. Colorado potato beetle management on potatoes: current challenges and future prospects. Fruit Vegetable Cereal Sci. Biotech. 3, 10–19.

Alyokhin, A.V., Ferro, D.N., 1999a. Electrophoretic confirmation of sperm mixing in mated Colorado potato beetles (Coleoptera: Chrysomelidae). Ann. Entomol. Soc. Am. 92, 230–235.

Alyokhin, A.V., Ferro, D.N., 1999b. Reproduction and dispersal of summer-generation Colorado potato beetle (Coleoptera: Chrysomelidae). Environ. Entomol. 28, 425–430.

Alyokhin, A., Dively, G., Patterson, M., Rogers, D., Mahoney, M., Wollam, J., 2006. Susceptibility of imidacloprid-resistant Colorado potato beetles to non-neonicotinoid insecticides in the laboratory and field trials. Am. J. Potato Res. 83, 485–494.

Alyokhin, A., Dively, G., Patterson, M., Castaldo, C., Rogers, D., Mahoney, M., Wollam, J., 2007. Resistance and cross-resistance to imidacloprid and thiamethoxam in the Colorado potato beetle. Pest Manage. Sci. 63, 32–41.

Alyokhin, A., Baker, M., Mota-Sanchez, D., Dively, G., Grafius, E., 2008. Colorado potato beetle resistance to insecticides. Am. J. Potato Res. 85, 395–413.

Anspaugh, D., Kennedy, D.G.G., Roe, R.M., 1995. Purification and characterization of a resistance-associated esterase from the Colorado Potato Beetle, *Leptinotarsa decemlineata* (Say). Pestic. Biochem. Physiol. 53, 84–96.

Anthony, N., Unruh, T., Ganser, D., ffrench-Constant, R., 1998. Duplication of the Rdl GABA receptor subunit gene in an insecticide-resistant aphid, *Myzus persicae*. Mol. Gen. Genet. 260, 165–175.

Argentine, J.A., Clark, J.M., Ferro, D.N., 1989. Genetics and synergism of resistance to azinphosmethyl and permethrin in the Colorado potato beetle (Coleoptera, Chrysomelidae). J. Econ. Entomol. 82, 698–705.

Bajan, C., Kmitova, K., 1972. The effect of entomogenous fungi *Paecilomyces farinosus* (Dicks) Brown and Smith and *Beauveria bassiana* (Bals.) Vuillemin, on the oviposition by *Leptinotarsa decemlineata* (Say) females and the survival of larvae. Pol. J. Ecol. 20, 423–432.

Baker, M.B., Alyokhin, A., Dastur, S.R., Porter, A.H., Ferro, D.N., 2005. Sperm precedence in the overwintered Colorado potato beetles (Coleoptera: Chrysomelidae) and its implications for insecticide resistance management. Ann. Entomol. Soc. Am. 98, 989–995.

Barrett, R.D.H., Schluter, D., 2008. Adaptation from standing genetic variation. Trends Ecol. Evol. 23, 38–44.

Bass, C., Field, L.M., 2011. Gene amplification and insecticide resistance. Pest. Manag. Sci. 67, 886–890.

Bell, G., Gonzalez, A., 2009. Evolutionary rescue can prevent extinction following environmental change. Ecol. Lett. 12, 942–948.

Benkovskaya, G.V., 2009. The stress-reaction as the mechanism of realization of adaptive potential in insect individuals and populations of insects. Doctoral dissertation, Institute of Animal Systematics and Ecology. Russia, Novosibirsk [in Russian].

Benkovskaya, G.V., Leontieva, T.L., Udalov, M.B., 2008a. Colorado beetle resistance to insecticides in South Urals. Agrochemistry 8, 55–59.

Benkovskaya, G.V., Udalov, M.B., Khusnutdinova, E.K., 2008b. The genetic base and phenotypic manifestations of Colorado potato beetle resistance to organophosphorus insecticides. Russian J. Genetics 5, 553–558.

Biever, K.D., Chauvin, R.L., 1990. Prolonged dormancy in a Pacific Northwest population of the Colorado potato beetle, *Leptinotarsa decemlineata* (Say) (Coleoptera: Chrysomelidae). Can. Entomol. 122, 175–177.

Bishop, B.A., Grafius, E.J., 1996. Insecticide resistance in the Colorado potato beetle. In: Jolivet, P., Hsiao, T.H. (Eds.), Chrysomelidae Biology, vol. 1. SBP Academic Publishing, Amsterdam, The Netherlands.

Boiteau, G., Parry, R.H., Harris, C.R., 1987. Insecticide resistance in New Brunswick populations of the Colorado potato beetle (Coleoptera: Chrysomelidae). Can. Entomol. 119, 459–463.

Boman, S., Grapputo, A., Lindström, L., Lyytinen, A., Mappes, J., 2008. Quantitative genetic approach for assessing invasiveness: geographic and genetic variation in life-history traits. Biol. Inv. 10, 1135–1145.

Calabrese, E.J., 2009. Getting the dose-response wrong: why hormesis became marginalized and the threshold model accepted. Arch. Toxicol. 83, 227–247.

Calabrese, E.J., Baldwin, L.A., 2001. The frequency of U-shaped dose responses in the toxicological literature. Toxicol. Sci. 62, 330–338.

Calabrese, E.J., Baldwin, L.A., 2003. Hormesis: the dose-response revolution. Annu. Rev. Pharmacol. Toxicol. 43, 175–197.

Cantelo, W.W., Sanford, L.L., Sinden, S.L., Deahl, K.L., 1987. Research to develop plants resistant to the Colorado potato beetle, *Leptinotarsa decemlineata* (Say). In: Labeyrie, V., Fabres, G., Lachaise, D. (Eds.), Insects-Plants: Proceedings of the 6th International Symposium on Insect-Plant Relationships, Dr. W. Junk Publishing, Dordrecht, p. 380.

Cappaert, D.L., 1988. Ecology of the Colorado potato beetle in Mexico. MS Thesis Michigan State University, East Lansing, MI.

Casagrande, R.A., 1985. The "Iowa" potato beetle, *Leptinotarsa decemlineata*. Bull. Entomol. Soc. Am. 31, 27–29.

Casagrande, R.A., 1987. The Colorado potato beetle 125 years of mismanagement. Bull. Entomol. Soc. Am. 33, 142–150.

Cohen, E., 2006. Pesticide-mediated homeostatic modulation in arthropods. Pestic. Biochem. Physiol. 85, 21–27.

Costanzo, J.P., Moore, J.B., Lee Jr., R.E., Kaufman, P.E., Wyman, J.A., 1997. Influence of soil hydric parameters on the winter cold hardiness of a burrowing beetle, *Leptinotarsa decemlineata* (Say). J. Comp. Physiol. B. 167, 169–176.

Crispo, E., DiBattista, J.D., Correa, C., Thibert-Plante, X., McKellar, A.E., Schwartz, A.K., Berner, D., et al., 2010. The evolution of phenotypic plasticity in response to anthropogenic disturbance. Evol. Ecol. Res. 12, 4766.

Cutler, G.C., Scott-Dupree, C.D., Tolman, J.H., Harris, C.R., 2005. Acute and sublethal toxicity of novaluron, a novel chitin synthesis inhibitor, to *Leptinotarsa decemlineata* (Coleoptera:Chrysomelidae). Pest Manage. Sci. 61, 1060–1068.

Danilevski, A.S., 1965. Photoperiodism and Seasonal Development of Insects. J. Johnston; Oliver & Boyd, Edinburgh.

de Kort, C.A.D., 1990. Thirty-five years of diapause research with the Colorado potato beetle. Entomol. Exp. App. 56, 1–13.

de Wilde, J., Duintjer, C.S., Mook, L., 1959. Physiology of diapause in the adult Colorado beetle (*Leptinotarsa decemlineata* Say) – The photoperiod as a controlling factor. J. Insect Physiol. 3, 75–80.

de Wilde, J., Bongers, W., Schooneveld, H., 1969. Effects of host-plant age on phytophagous insect. Entomol. Exp. Appl. 12, 714–720.

Dobzhansky, T., 1973. Nothing in biology makes sense except in the light of evolution. Am. Biol. Teacher 35, 125–129.

Dobzhansky, T., Ayala, F.J., Stebbins, G.L., Valentine, J.W., 1973. Evolution. W.H. Freeman and Co., San Francisco, CA.

Dutcher, J.D., 2007. A review of resurgence and replacement causing pest outbreaks in IPM. In: Ciancio, A., Mukerji, K.G. (Eds.), General Concepts in Integrated Pest and Disease, Springer, New York, NY, pp. 27–43.

El Tahtaoui, M., 1962. L'influence du champignon *Beauveria bassiana* (Bals.) Vuill. Sur la fecondite et la diapause du doryphore, *Leptinotarsa decemlineata* Say. Entomophaga 2, 549–553.

EPPO, 2011. PQR – EPPO database on quarantine pests (available online). http://www.eppo.int.

Fenton, B., Malloch, G., Woodford, J.A.T., Foster, S.P., Anstead, J., Denholm, I., et al., 2005. The attack of the clones: tracking the movement of insecticide-resistant peach-potato aphids *Myzus persicae* (Hemiptera: Aphididae). Bull. Entomol. Res. 95, 483–494.

Fenton, B., Margaritopoulus, J.T., Malloch, G.L., Foster, S.P., 2010. Micro-evolutionary change in relation to insecticide resistance in the peach-potato aphid, *Myzus persicae*. Ecol. Entomol. 35, 131–146.

Ferro, D.N., 1993. Potential for resistance to *Bacillus thuringiensis*: Colorado potato beetle (Coleoptera: Chrysomelidae) – a model system. Am. Entomol. 39, 38–44.

Ferro, D.N., Tuttle, A.F., Weber, D.C., 1991. Ovipositional and flight behavior of overwintered Colorado potato beetle (Coleoptera: Chrysomelidae). Environ. Entomol. 20, 1309–1314.

Foster, S.P., Denholm, I., Thompson, R., 2003a. Variation in response to neonicotinoid insecticides in peach-potato aphids, *Myzus persicae* (Hemiptera: Aphididae). Pest Manage. Sci. 59, 166–173.

Foster, S.P., Kift, N.B., Baverstock, J., Sime, S., Reynolds, K., Jones, J., et al., 2003b. Association of MACE-based insecticide resistance in *Myzus persicae* with reproductive rate, response to alarm pheromone and vulnerability to attack by *Aphidius colemani*. Pest Manage. Sci. 59, 1169–1178.

Foster, S.P., Tomiczek, M, Thompson, R., Denholm, I., Poppy, G., Kraaijeveld, A.R., Powell, W., 2007. Behavioural side effects of insecticide resistance in aphids increase their vulnerability to parasitoid attack. Anim. Behav. 74, 621–632.

França, F.H., Plaisted, R.L., Roush, R.T., Via, S., Tingey, W.M., 1994. Selection response of the Colorado potato beetle for adaptation to the resistant potato. *Solanum berthaultii*. Entomol. Exp. Appl. 73, 101–109.

Futuyma, D.J., 1986. Evolutionary Biology, second ed. Sinauer, Sunderland, MA.

Gauthier, N.L., Hofmaster, R.N., Semel, M., 1981. History of Colorado potato beetle control. In: Lashomb, J.H., Casagrande, R. (Eds.), Advances in Potato Pest Management, Hutchinson Ross Publishing Co., Stroudsburg, PA, pp. 13–33.

Gillespie, D.R., Quiring, D.J.M., Foottit, R.G., Foster, S.P., Acheampong, S., 2008. Implications of phenotypic variation of *Myzus persicae* (Hemiptera: Aphididae) for biological control on greenhouse pepper plants. J. Appl. Entomol. 133, 505–511.

Gluckman, P.D., Hanson, M.A., Bateson, P., Beedle, A.S., Law, C.M., Bhutta, Z.A., Anokhin, K.V., et al., 2009. Towards a new developmental synthesis: adaptive developmental plasticity and human disease. Lancet 373, 1654–1657.

Grapputo, A., Boman, S., Lindström, L., Lyytinen, A., Mappes, J., 2005. The voyage of an invasive species across continents: genetic diversity of North American and European Colorado potato beetle populations. Mol. Ecol. 14, 4207–4219.

Groden, E., Casagrande, R.A., 1986. Population dynamics of the Colorado potato beetle, *Leptinotarsa decemlineata* (Coleoptera: Chrysomelidae), on *Solanum berthaultii*. J. Econ. Entomol. 79, 91–97.

Guedes, R.N.C., Magalhães, L.C., Cosme, L.V., 2009. Stimulatory sublethal response of a generalist predator to permethrin: hormesis, hormoligosis, or homeostatic regulation? J. Econ. Entomol. 102, 170–176.

Harcourt, D.G., 1971. Population dynamics of *Leptinotarsa decemlineata* (Say) in eastern Ontario. III. Major population processes. Can. Entomol. 103, 1049–1061.

Hare, J.D., 1990. Ecology and management of the Colorado potato beetle. Ann. Rev. Entomol. 35, 81–100.

Hare, J.D., Kennedy, G.G., 1986. Genetic variation in plant–insect associations – survival of *Leptinotarsa decemlineata* populations on *Solanum carolinenese*. Evolution 40, 1031–1043.

Harris, C.R., Svec, H.I., 1981. Colorado potato beetle resistance to carbofuran and several other insecticides in Quebec. J. Econ. Entomol. 74, 421–424.

Harrison, G.D., 1987. Host-plant discrimination and evolution of feeding preference in the Colorardo potato beetle *Leptinotarsa decemlineata*. Physiol. Entomol. 12, 407–415.

Hartley, C.J., Newcomb, R.D., Russell, R.J., Yong, C.G., Stevens, J.R., Yeates, D.K., La Salle, J., et al., 2006. Amplification of DNA from preserved specimens shows blowflies were preadapted for the rapid evolution of insecticide resistance. Proc. Natl. Acad. Sci. USA 103, 8757–8762.

Hawthorne, D.J., 2001. AFLP-based genetic linkage map of the Colorado potato beetle *Leptinotarsa decemlineata*: Sex chromosomes and a pyrethroid-resistance candidate gene. Genetics 158, 695–700.

Hendry, A.P., Kinnison, M.T., 1999. The pace of modern life: measuring rates of contemporary microevolution. Evolution 53, 1637–1653.

Hendry, A.P., Lohmann, L.G., Conti, E., Cracraft, J., Crandall, K.A., Faith, D.P., Hauser, C., et al., 2010. Evolutionary biology in biodiversity science, conservation, and policy: a call to action. Evolution 64, 1517–1528.

Hendry, A.P., Kinnison, M.T., Heino, M., Day, T., Smith, T.B., Fitt, G., Bergstrom, C.T., Oakeshott, J., Jørgensen, P.S., Zalucki, M.P., Gilchrist, G., Southerton, S., Sih, A., Strauss, S., Denison, R.F., Carroll, S.P., 2011. Evolutionary principles and their practical application. Evol. Appl. 4, 159–183.

Herzog, J., Muller, C.B., Vorburger, C., 2007. Strong parasitoid-mediated selection in experimental populations of aphids. Biology Lett. 3, 667–669.

Hilbeck, A., Kennedy, G.G., 1998. Effects of temperature on survival and preimaginal development rates of Colorado potato beetle on potato and horse-nettle: potential role in host range expansion. Entomol. Exp. Appl. 89, 261–269.

Horton, D.R., Capinera, J.L., 1988. Effects of host availability on diapause and voltinism in a non-agricultural popualtion of Colorado potato beetle (Coleoptera: Chrysomelidae). J. Kans. Entomol. Soc. 61, 62–67.

Hsiao, T., 1978. Host plant adaptations among geographic populations of the Colorado potato beetle. Entomol. Exp. App. 24, 437–447.

Hsiao, T., 1982. Inheritance of three autosomal mutations in the Colorado potato beetle, *Leptinotarsa decemlinieata* (Coleoptera: Chrysomeldiae). Genome 24, 681–686.

Hsiao, T.H., 1981. Ecophysiological adaptations among geographic populations of the Colorado potato beetle in North America. In: Lashomb, J., Casagrande, R. (Eds.), Advances in Potato Pest Management, Dowden Hutchinson & Ross, New York, NY, pp. 69–85.

Ioannidis, P.M., Grafius, E., Whalon, M.E., 1991. Patterns of insecticide resistance to azinphosmethyl, carbofuran, and permethrin in the Colorado potato beetle (Coleoptera: Chrysomelidae). J. Econ. Entomol. 84, 1417–1423.

Ioannidis, P.M., Grafius, E., Wierenga, J.M., Whalon, M.E., Hollingworth, R.M., 1992. Selection, inheritance and characterization of carbofuran resistance in Colorado potato beetle (Coleoptera: Chrysomelidae). Pestic. Sci. 35, 215–222.

Isely, D., 1935. Variations in the seasonal history of the Colorado potato beetle. J. Kans. Entomol. Soc. 8, 142–145.

Jacques Jr., R.L., 1988. The Potato beetles: the genus *Leptinotarsa* in North America (Coleoptera: Chrysomelidae). E. J. Brill, New York, NY.

Janzen, D.H., 1980. When is it coevolution? Evolution 34, 611–612.

Kasprowicz, L., Malloch, G., Foster, S., Pickup, J., Zhan, J., Fenton, B., 2008. Clonal turnover of MACE-carrying peach-potato aphids (*Myzus persicae* (Sulzer), Homoptera: Aphididae) colonizing Scotland. Bull. Entomol. Res. 98, 115–124.

Kinnison, M.T., Hairston Jr., N.G., 2007. Eco-evolutionary conservation biology: contemporary evolution and the dynamics of persistence. Funct. Ecol. 21, 444–454.

Lande, R., 2009. Adaptation to an extraordinary environment by evolution of phenotypic plasticity and genetic assimilation. J. Evol. Biol. 22, 1435–1446.

Lee, S., Clark, J.M., 1996. Tissue distribution and biochemical characterization of carboxylesterases associated with permethrin resistance in a near isogenic strain of Colorado potato beetle. Pestic. Biochem. Physiol. 56, 208–219.

Lee, S.H., Clark, J.M., 1998. Purification and characterization of multiple-charged forms of permethrin carboxylesterase(s) from the hemolymph of resistant Colorado potato beetle. Pestic. Biochem. Physiol. 60, 31–47.

Lefevere, K.S., de Kort, C.A.D., 1989. Adult diapause in the Colorado potato beetle, *Leptinotarsa decemlineata*: Effects of external factors on maintenance, termination and post-diapause development. Physiol. Entomol. 14, 299–308.

Leontieva, T.L., Benkovskaya, G.V., Udalov, M.B., Poscryako, A.V., 2006. Insecticide Resistance Level in *Leptinotarsa decemlineata* Say Population in the South Ural. Resist. Pest Manage. Newsletter 2, 25–26.

Levine, E., Oloumi-Sadeghi, H., Fisher, J.R., 1992. Discovery of multiyear diapause in Illinois and South Dakota Northern corn rootworm (Coleoptera: Chrysomelidae) eggs and incidence of the prolonged diapause trait in Illinois. J. Econ. Entomol. 85, 262–267.

Logan, P.A., Casagrande, R.A., Faubert, H.H., Drummond, F.A., 1985. Temperature-dependent development and feeding of immature Colorado potato beetles, *Leptinotarsa decemlineata* (Say) (Coleoptera: Chrysomelidae). Environ. Entomol. 14, 275–283.

Lowery, D.T., Sears, M.K., 1986. Effect of exposure to insecticide azinphosmethyl on reproduction of green peach aphid (Homoptera: Aphididae). J. Econ. Entomol. 79, 1534–1538.

Lu, W.H., Logan, P., 1993. Induction of feeding on potato in Mexican *Leptinotarsa decemlineata* (Say) (Coleoptera: Chrysomelidae). Environ. Entomol. 22, 759–765.

Lu, W.H., Logan, P., 1994a. Geographic variation in larval feeding acceptance and performance of *Leptinotarsa decemlineata* (Coleoptera, Chrysomelidae). Ann. Entomol. Soc. Am. 87, 460–469.

Lu, W.H., Logan, P., 1994b. Genetic variation in oviposition between and within populations of *Leptinotarsa decemlineata* (Coleoptera, Chysomelidae). Ann. Entomol. Soc. Am. 87, 634–640.

Lu, W.H., Logan, P., 1994c. Inheritance of larval body color in *Leptinotarsa decemlineata* (Coleoptera: Chrysomelidae). Ann. Entomol. Soc. Am. 87, 454–459.

Lu, W.H., Logan, P., 1994d. Effects of potato association on oviposition behavior of Mexican *Leptinotarsa decemlineata* (Coleoptera: Chrysomelidae). Environ. Entomol. 23, 85–90.

Lu, W.H., Logan, P., 1995. Inheritance of host-related feeding and ovipositional behaviors in *Leptinotarsa decemlineata* (Coleoptera, Chrysomelidae). Environ. Entomol. 24, 278–287.

Lyytinen, A., Lindstrom, L., Mappes, J., 2008. Genetic variation in growth and development time under two selection regimes in *Leptinotarsa decemlineata*. Entomol. Exp. Appl. 127, 157–167.

Lyytinen, A., Boman, S., Grapputo, A., Lindström, L., Mappes, J., 2009. Cold tolerance during larval development: effects on the thermal distribution limits of *Leptinotarsa decemlineata*. Entomol. Exp. App. 133, 92–99.

Lyytinen, A., Mappes, J., Lindström, L., 2012. Variation in Hsp70 levels after cold shock: Signs of evolutionary responses to thermal selection among *Leptinotarsa decemlineata* populations. PLoS ONE 7 (2), e31446.

McCallum, M., 2012. Top 10 List. Spudman 50, 30–31.

Mota-Sanchez, D., Hollingworth, R.M., Grafius, E.J., Moyer, D.D., 2006. Resistance and cross-resistance to neonicotinoid insecticides and spinosad in the Colorado potato beetle, *Leptinotarsa decemlineata* (Say) (Coleoptera: Chrysomelidae). Pest Manage. Sci. 62, 30–37.

Muller, C.B., Williams, I.S., Hardie, J., 2001. The role of nutrition, crowding and interspecific interactions in the development of winged aphids. Ecol. Entomol. 26, 330–340.

Neck, R.W., 1983. Foodplant ecology and geographical range of the Colorado potato beetle and a related species (*Leptinotarsa* spp.) (Coleoptera: Chrysomelidae). Coleop. Bull. 37, 177–182.

Normark, B.H., Normark, S., 2002. Evolution and spread of antibiotic resistance. J. Int. Med. 252, 91–106.

Noronha, C., Cloutier, C., 1998. Effect of soil conditions and body size on digging by prediapause Colorado potato beetles (Coleoptera: Chrysomelidae). Can. J. Zool./Rev. Can. Zool. 76, 1705–1713.

Noronha, C., Cloutier, C., 2006. Effects of potato foliage age and temperature regime on prediapause Colorado potato beetle *Leptinotarsa decemlineata* (Coleoptera: Chrysomelidae). Environ. Entomol. 35, 590–599.

Oleinikov, A.V., Yakovleva, I.N., Roslavtseva, S.A., 2006. Resistance to insecticides, phenetic structure, and enzyme activity in populations of the Colorado potato beetle *Leptinotarsa decemlineata* (Say) in the Bryansk Oblast. Agrokhimiya 3, 46–51 [in Russian].

Olsen, E.M., Heino, M., Lilly, G.R., Morgan, M.J., Brattey, J., Ernande, B., Dieckmann, U., 2004. Maturation trends indicative of rapid evolution preceded the collapse of northern cod. Nature 428, 932–935.

Olson, E.R., Dively, G.P., Nelson, J.O., 2000. Baseline susceptibility to imidacloprid and cross resistance patterns in Colorado potato beetle (Coleoptera: Chrysomelidae) populations. J. Econ. Entomol. 93, 447–458.

Orr, H.A., Unckless, R.L., 2008. Population extinction and the genetics of adaptation. Am. Nat. 172, 160–169.

Palumbi, S.R., 2001. Humans as the world's greatest evolutionary force. Science 293, 1786–1790.

Peferoen, M., Huybrechts, R., Loof, A., 1981. Longevity and fecundity in the Colorado potato beetle, *Leptinotarsa decemlineata*. Entomol. Exp. Appl. 29, 321–329.

Pelletier, Y., Smilowitz, Z., 1991. Biological and genetic study on the utilization of *Solanum berthaultii* Hawkes by the Colorado potato beetle (*Leptinotarsa decemlineata* (Say). Can. J. Zool. 69, 1280–1288.

Piiroinen, S., 2010. Range expansion to novel environments. PhD Thesis University of Jyväskylä, Jyväskylä, Finland.

Piiroinen, S., Ketola, T., Lyytinen, A., Lindstrom, L., 2011. Energy use, diapause behaviour and northern range expansion potential in the invasive Colorado potato beetle. Funct. Ecol. 25, 527–536.

Rajotte, E., 1993. From profitability to food safety and the environment: Shifting the objectives of IPM. Plant Dis. 77, 296–299.

Richards, C.L., Bossdorf, O., Muth, N.Z., Gurevitch, J., Pigliucci, M., 2006. Jack of all trades, master of some? On the role of phenotypic plasticity in plant invasions. Ecol. Lett. 9, 981–993.

Riley, C.V., 1877. Ninth annual report on the noxious, beneficial, and other insects of the state of Missouri. Regan and Carter, Jefferson City, MO.

Roderick, G.K., de Mendoza, L.G., Dively, G.P., Follett, P.A., 2003. Sperm precedence in Colorado potato beetle, *Leptinotarsa decemlineata* (Coleoptera: Chrysomelidae): Temporal variation assessed by neutral markers. Ann. Entomol. Soc. Am. 96, 631–636.

Rose, R.L., A Brindley, W., 1985. An evaluation of the role of oxidative enzymes in Colorado potato beetle resistance to carbamate insecticides. Pestic. Biochem. Physiol. 23, 74–84.

Roslavtseva, S.A., Eremina,Yu, O., 2005. Resistance to Pyrethroids and Phenotypic Analysis of the Colorado Potato Beetle Populations from Mozhaisk District, Moscow Oblast. In: Proc. Second All-Russian Congress on Plant Protection "Phytosanitary Remediation of Ecosystems", St. Petersburg, Russia. pp. 57–58.

Saccheri, I., Hanski, I., 2006. Natural selection and population dynamics. Trends Ecol. Evol. 21, 341–347.

SAGARPA, 2007. Governing Plan for National Potato Production. Publication of Secretaría de agricultura g, desarollo rural, pesca y alimentación, D.F., Mexico [In Spanish].

Shands, W.A., Simpson, G.W., 1971. Seasonal history of the buckthorn aphid and suitability of alder-leaved buckthorn as a primary host in northeastern Maine. Maine Life Sci. Agric. Exp. Stn. Bull 51.

Shands, W.A., Simpson, G.W., Wave, H.E., 1969. Canada plum, *Prunus nigra* Aiton, as a primary host of the green peach aphid, *Myzus persicae* (Sulzer), in northeastern Maine. Maine Agric. Exp. Stn. Bull 39.

Shands, W.A., Simpson, G.W., Wave, H.E., 1972. Seasonal population trends and productiveness of the potato aphid on swamp rose in northeastern Maine. Maine Life Sci. Agric. Exp. Stn. Bull 52.

Smith, T.B., Bernatchez, L., 2008. Evolutionary change in human altered environments. Mol. Ecol. 17, 1–8.

Srigiriraju, L., Semtnera, P.J., Bloomquist, J.R., 2010. Monitoring for imidacloprid resistance in the tobacco-adapted form of the green peach aphid, *Myzus persicae* (Sulzer) (Hemiptera: Aphididae), in the eastern United States. Pest Manage. Sci. 66, 676–685.

Stearns, S.C., 1989. The evolutionary significance of phenotypic plasticity – phenotypic sources of variation among organisms can be described by developmental switches and reaction norms. Bioscience 39, 436–445.

Stearns, S.C., Koella, J.C., 1986. The evolution of phenotypic plasticity in life-history traits – predictions of reaction norms for age and size at maturity. Evolution 40, 893–913.

Sukhoruchenko, G.I., Dolzhenko, V.I., 2008. Problems of resistance development in arthropod pests of agricultural crops in Russia. EPPO Bull. 1, 119–126.

Szentesi, A., 1985. Behavioral aspects of female guarding and inter-male conflict in the Colorado potato beetle. In: Ferro, D.N., Voss, R.H. (Eds.), Proceedings, Symposium on the Colorado potato beetle. XVIIth International Congress of Entomology. Res. Bull 704 Mass. Agric. Exp. Stn. Circ. 347, pp. 127–137.

Tauber, M.J., Tauber, C.A., 2002. Prolonged dormancy in *Leptinotarsa decemlineata* (Coleoptera: Chrysomelidae): a ten-year field study with implications for crop rotation. Environ. Entomol. 31, 499–504.

Tauber, M.J., Tauber, C.A., Obrycki, J.J., Gollands, B., Wright, R.J., 1988. Voltinism and the induction of aestival diapause in the Colorado potato beetle, *Leptinotarsa decemlineata* (Coleoptera, Chrysomelidae). Ann. Entomol. Soc. Am. 81, 748–754.

Tower, W., 1906. Investigation of evolution in chrysomelid beetles of the genus *Leptinotarsa*. Carnegie Institution of Washington, Washington, DC.

Udalov, M.B., Benkovskaya, G.V., 2010. Polymorphism CoxI gene of Colorado potato beetle in South Ural populations. Resist. Pest Manage. Newsletter 2, 29–32.

Udalov, M.B., Benkovskaya, G.V., 2011. Change in the polymorphism level in populations of the colorado potato beetle. Russian J. Genetics: Appl. Research 5, 390–395.

Udalov, M.B., Benkovskaya, G.V., Khusnutdinova, E.K., 2010. Population structure of the Colorado potato beetle in the Southern Urals. Russian J. Ecology 2, 159–166.

Ugent, D., 1968. The potato in Mexico: geography and primitive culture. Econ. Bot. 22, 108–123.

Ushatinskaya, R.S., 1962. Colorado potato beetle diapause and development of its multi-year infestations. Zashchita Rastenii 6, 53–54 [in Russian].

van Toor, R.F., Foster, S.P., Anstead, J.A., Mitchinson, S., Fenton, B., Kasprowicz, L., 2008. Insecticide resistance and genetic composition of *Myzus persicae* (Hemiptera: Aphididae) on field potatoes in New Zealand. Crop Prot. 27, 236–247.

Vasil'eva, T.I., Ivanova, G.P., Ivanov, S.G., et al., 2005. Changes of the Phenotypic Structure of the Colorado Potato Beetle Population Depending on the Intensity of Using Insecticides. Proc. Second All-Russian Congress on Plant Protection "Phytosanitary Remediation of Ecosystems", Russia, St. Petersburg, pp. 14–15.

Vilkova, N.A., Sukhoruchenko, G.I., Fasulati, S.R., 2005. Strategy of agricultural plants protection from adventive species of phytophagous insects by the example of Colorado potato beetle *Leptinotarsa decemlineata* Say (Coleoptera, Chrysomelidae). Newslett. Plant Prot. 1, 3–15.

Voss, R.H., Ferro, D.N., 1990. Phenology of flight and walking by Colorado potato beetle (Coleoptera: Chrysomelidae) adults in western Massachusetts. Environ. Entomol. 19, 117–122.

Walsh, B.D., 1865. The new potato bug and its natural history. Practical Entomologist 1, 1–4.

Weber, D.C., Ferro, D.N., 1994. Colorado potato beetle: diverse life history poses challenge to management. In: Zender, G.W., Jansson, R.K., Powelson, M.L., Raman, K.V. (Eds.), Advances in Potato Pest Biology and Management, APS Press, St Paul, MN, pp. 54–70.

Weber, D., 2003. Colorado beetle: pest on the move. Pesticide Outlook 14, 256–259.

Whalon, M.E., Ferro, D.N., 1998. Bt-potato resistance management. In: Mellon, M., Rissler, J. (Eds.), Now or never: serious new plans to save a natural pest control, Union of Concerned Scientists (UCS), Cambridge, MA, pp. 107–136.

Whalon, M.E., Mota-Sanchez, D., Hollingworth, R.M., Duynslager, L., 2011. Arthropod Pesticide Resistance Database. www.pesticideresistance.org, accessed on August 25, 2011.

Wierenga, J.M., Hollingworth, R.M., 1992. Inhibition of insect acetylcholinesterase by the potato glycoalkaloid -chaconine. Natur. Tox. 1, 96–99.

Yamanaka, T., Tatsuki, S., Shimada, M., 2008. Adaptation to the new land or effect of global warming? An age-structured model for rapid voltinism change in an alien lepidopteran pest. J. Anim. Ecol. 77, 585–596.

Zamoum, T., Simon, J.C., Crochard, D., Ballanger, Y., Lapchin, L., Vanlerberghe-Masutti, F., et al., 2005. Does insecticide resistance alone account for the low genetic variability of asexually reproducing populations of the peach-potato aphid *Myzus persicae*? Heredity 94, 630–639.

Zhu, K.Y., Lee, S.H., Clark, J.M., 1996. A point mutation of acetylcholinesterase associated with azinphosmethyl resistance and reduced fitness in Colorado potato beetle. Pestic. Biochem. Physiol. 55, 100–108.

Part V

Current Challenges and Future Directions

Chapter 20

Epilogue: The Road to Sustainability

Andrei Alyokhin[1], Charles Vincent[2] and Philippe Giordanengo[3]

[1]School of Biology and Ecology, University of Maine, Orono, USA, [2]Agriculture and Agri-Food Canada, Saint-Jean-sur-Richelieu, Québec, Canada, [3]Université de Picardie Jules Verne, Amiens, France; CNRS, and INRA, Institut Sophia Agrobiotech, Sophia Antipolis, France; Université de Nice Sophia Antipolis, Sophia Antipolis, France

Potato is a very valuable crop that provides high-quality nutrition to billions of people around the world. As reviewed in Part II of this book, potatoes are attacked by a number of potentially devastating insect pests. The extent of damage is difficult to generalize, as it is highly dependent on specific conditions. However, in a world where close to 1 billion people are lacking basic food security, any amount of prevented losses is precious. Moreover, on an anecdotal level all three editors of this book (and probably all of its authors) can attest that, in a bad year, the entire potato crop can be wiped out in the absence of adequate control measures. This can be devastating even for an industrialized farmer in a wealthy nation, with government-subsidized crop insurance and a savings account in a local bank.

Early development of the modern pesticide industry was largely shaped by the demands of potato production, particularly the sudden need to control rapidly spreading Colorado potato beetles (Chapter 2). The quest for insecticides suitable for the control of the Colorado potato beetle has continued since 1864 (Gauthier et al. 1981), and includes such milestones as Paris Green (copper[II]-acetoarsenite), lead arsenate, DDT, cyclodiene organochlorines, organophosphates, carbamates, and neonicotinoids (Alyokhin 2009). Application equipment has also come a long way, with major steps including perforated tin cans, powder guns with hand-cranked gear drives, and wheel-drawn pressurized pump sprayers (Sanderson 1912, Gauthier et al. 1981). The first successful steam-powered sprayer appeared in 1894, quickly followed by a gasoline-powered sprayer in 1900. Finally, the first successful airblast sprayer was manufactured around 1937 (Gauthier et al. 1981).

Currently, chemical control still remains the foundation of the insect pest management on commercial potato farms (Chapter 13). While it is not a politically correct statement in academic circles, insecticides do provide a

Insect Pests of Potato. http://dx.doi.org/10.1016/B978-0-12-386895-4.00020-X

convenient, relatively cheap, and fairly reliable method of pest control. As a result, they are readily embraced by a majority of potato farmers. However, overreliance on chemicals (or, for that matter, on any single management approach) is never going to be sustainable in the long run. Older products fail due to the evolution of resistance by the ever increasing number of pest species and populations (Chapter 19). These products are also being withdrawn from the market due to more stringent safety regulations.

So far, the pesticide industry has been able to come up with newly developed replacements in a timely manner (although it was a close call for the Colorado potato beetle in some areas of the northeastern United States in the mid-1990s). However, the discovery and commercialization of new insecticides is an increasingly complicated and expensive procedure. Entire classes of insecticides, such as organophosphates, are no longer acceptable to large segments of society due to their impact on non-target organisms, including humans. Developing similar broad-spectrum poisons is almost unthinkable in current economic and political environments. Finding biorational substitutes that are highly successful in suppressing target pests but have little or no effect on other species is a noble task. However, it is also difficult and time consuming. As a result, it is naïve to expect that the "pesticide treadmill" will forever continue turning with the same efficiency as it has in the past.

More than 8000 years after potato domestication by humans, insects remain a major challenge to its production. Insects that damage potatoes include some of the most adaptable pests known to man (Chapter 19). Humans are also a very adaptable species – a characteristic that has allowed us to spread from a small valley in Africa to all continents on Earth, and to have an environmental impact that is probably comparable to that of vascular plants colonizing the land. Yet, for some reason, this adaptability seems to fail when we are dealing with insect pests of potatoes and other crops. After the major breakthrough of inventing modern synthetic insecticides and machinery for their application, we seem to have lost momentum and limited further adaptation to an often mindless replacement of one chemical with another. If we are to maintain an edge over our six-legged competitors, we should be able to do better than that.

Relying on a single method for crop protection is essentially no more sustainable than relying on a single crop to feed the impoverished and exploited population of 19th century Ireland (Chapter 1). A change of the entire system is needed to provide a comprehensive and lasting solution to a perennial problem. Moving alternative methods of pest control into the mainstream of modern agriculture and integrating them into a unified, knowledge-based approach is the next important step in the development of farming. Calendar sprays of broad-spectrum pesticides should join slash-and-burn practices as a once essential, but now desperately outdated, technique.

To be changed, a system should be first recognized as a system. As an unfortunate side effect of agricultural industrialization, we started treating agroecosystems as a form of machinery that produces desired outputs based on external inputs. This makes day-to-day farm management much easier. However, it also fails to take advantage of possible ecosystem services (Chapters 9 and 10), and often disrupts the intricate balance among the ecosystem's components. Furthermore, such an attitude often fails to account for the dynamic nature of biological systems. Unlike factory machinery, agroecosystems constantly change, driven both by external influences and by internal evolution. Also, much to our chagrin, they cannot be turned off for maintenance and upgrades.

A better understanding of the structure and function of a potato ecosystem is necessary for developing truly sustainable crop protection solutions. We already have a multitude of tools at our disposal (Part IV). Some of them are relatively new (e.g., genetic engineering, Chapter 15), others have been around for centuries if not millennia (e.g., biological control, Chapter 14, and crop rotation, Chapter 18). Some of them are very strong (e.g., insecticides, Chapter 13), while others are rather weak (e.g., organic soil amendments, Chapters 10 and 18). To further complicate the matter, some of them are not compatible with each other (e.g., many insecticides and natural enemies), or with economic demands (e.g., extended crop rotations and market prices for non-potato crops).

Integrating these techniques into an economically viable and environmentally friendly system that can be constantly adjusted to the changing conditions is a daunting task. It is also an open-ended process, as developing a universal set of approaches that will work forever in every potato field in the world is impossible. However, our predecessors managed to turn a small toxic plant growing at high altitudes in remote mountains into a major staple crop that is effectively grown around the globe. If they could succeed, so can we.

REFERENCES

Alyokhin, A., 2009. Colorado potato beetle management on potatoes: current challenges and future prospects. Fruit Vegetable Cereal Sci. Biotech. 3, pp. 10–19.

Gauthier, N.L., Hofmaster, R.N., Semel, M., 1981. History of Colorado potato beetle control. In: Lashomb, J.H., Casagrande, R. (Eds.), Advances in Potato Pest Management, Hutchinson Ross Publishing Co., Stroudsburg, PA, pp. 13–33.

Sanderson, E.D., 1912. Insect Pests of Farm, Garden, and Orchard. Wiley, New York, NY.

Index

Page numbers followed by *f,* indicate figure and *t,* indicate table.